SHARKS
OF THE
WORLD

A COMPLETE GUIDE

SHARKS
OF THE
WORLD

A COMPLETE GUIDE

David A. Ebert, Marc Dando
and Sarah Fowler

Published by Princeton University Press,
41 William Street, Princeton, New Jersey 08540
In the United Kingdom: Princeton University Press, 6 Oxford Street,
Woodstock, Oxfordshire OX20 1TR

press.princeton.edu

First edition published in 2013

British Library Cataloging-in-Publication Data is available

Library of Congress Control Number 2021933867
ISBN 978-0-691-20599-1
Ebook ISBN 978-0-691-21087-2

Production and design by **WILD**NATUREPRESS Ltd., Plymouth, UK
Printed in Slovakia

10 9 8 7 6 5 4 3 2 1

Half title page: Leopard Catshark, *Poroderma pantherinum*, in reef. Simon's Town, Western Cape, South Africa.

Contents

Acknowledgements

We wish to thank all of those who have been extremely helpful and generous with their time in responding to our numerous questions, providing data and information from their own research (some of it unpublished), and much needed literature. The species accounts were improved immensely from the contribution of colleagues, friends. We extend our particular thanks to the following individuals for general discussions, contributions and information on various aspects of this project, and apologise to those whom we have missed:

Mark Harris (F.F.C., Elasmobranch Studies, Florida, USA), particularly for the tooth counts

John Richardson, for impeccable proofing

Charlie Underwood, for the evolution pages

Dan Abel (Coastal Carolina University, USA)

K.V. Akhilesh and K.K. Bineesh (Central Marine Fisheries Research Institute, India)

Ivy Baremore, Rachel Graham and Zeddy Seymour (MarAlliance, Belize)

Rhett Bennett and Dave Van Beuningen (Wildlife Conservation Society, South Africa)

Dean Grubbs (Florida State University, USA)

David Catania and Jon Fong (California Academy of Sciences, San Francisco, USA)

Demian Chapman and Diego Cardeñosa (Florida International University, Miami, USA)

Patricia Charvet (Brazil)

Gustavo Chiaramonte (Museo Argentino de Ciencias Naturales 'Bernardino Rivadavia', Buenos Aires, Argentina)

Francisco Concha (Universidad de Valparaíso, Chile)

Justin Cordova, Kelley van Hees (Pacific Shark Research Center/Moss Landing Marine Laboratories, California, USA)

Paul Cowley, Angus Paterson, Elaine Heemstra, Roger Bills, Mzwandile Dwani, Nkosinathi Mazungula and Vuyani Hanisi (South African Institute of Aquatic Biodiversity, South Africa)

Clinton Duffy (Department of Conservation, New Zealand)

Nick Dulvy and the IUCN Shark Specialist Group programme officers (Simon Fraser University, Canada)

Edoardo Mostarda (FAO, Rome, Italy)

Nicolás R. Ehemann (Centro Interdisciplinario de Ciencias Marinas, Baja California, Mexico)

Malcolm Francis (NIWA, New Zealand)

Otto Gadig (Instituto de Biociências, UNESP, Brazil)

Hsuan-Ching (Hans) Ho (National Museum of Marine Biology and Aquarium, Taiwan)

Hua-Hsun Hsu (King Fahd University of Petroleum and Minerals, Saudi Arabia)

Hajime Ishihara (W&I Associates Co. Ltd, Japan)

Rima Jabado (Gulf Elasmo Project, Dubai, U.A.E.)

Salvador Jorgensen (Monterey Bay Aquarium, California, USA)

Shoou-Jeng Joung and Chi-Ju (Debbie) Yu (National Taiwan Ocean University, Taiwan)

Robert Kirk and colleagues at Princeton University Press, UK and USA

Peter Kyne (Charles Darwin University, Australia)

Peter Last (CSIRO Marine Laboratories, Hobart, Tasmania, Australia)

Robin Leslie (Rhodes University, South Africa)

Mabel Manjaji Matsumoto (Universiti Malaysia Sabah, Malaysia)

Ryo Misawa (Tohoku National Fisheries Research Institute, Japan)

Eva Meyers (Angel Shark Project, Research Museum Alexander Koenig, Germany)

Kazuhiro Nakaya (Hokkaido University, Hokkaido, Japan)

Gavin Naylor (Florida Program Shark Research, University of Florida, USA)

Caroline Pollack (IUCN Red List Programme)

Keiichi Sato (Okinawa Churaumi Aquarium, Okinawa, Japan)

Michael Scholl, Clarens, Switzerland

Fabrizio Serena (National Research Council, Institute for Marine Biological Resources and Biotechnologies, Italy)

Bernard Séret (Ichtyo-Consult, Paris, France)

Mahmood Shivji (Nova Southeastern University, Florida, USA)

Jeremy Stafford-Deitsch (Minions, UK)

Matthias Stehmann (ICHTHYS, Hamburg, Germany)

Andrew Stewart (Te Papa Museum, New Zealand)

Sho Tanaka (Tokai University, Japan)

Simon Weigmann (Elasmo-Lab, University of Hamburg, Germany)

William White (CSIRO Australian National Fish Collection, Hobart, Tasmania, Australia)

Sabine Wintner (University of KwaZulu-Natal, South Africa)

Atsuko Yamaguchi (University of Nagasaki, Japan)

The book has greatly benefited by the generous donation of photographs by the following photographers and organisations; Sirachai Shin Arunrugstichai, Fundação Projeto Tamar, Hawaii Undersea Research Laboratory Archive, Andrea Marshall, Mark Mohlmann, Monterey Bay Aquarium Research Institute, Naiti Morales, Ocean Exploration Trust, Okinawa Churaumi Aquarium, Al Reeve, Shark Spotters and Peter Verhoog.

David Ebert would like to especially thank Greg Cailliet (Pacific Shark Research Center, Moss Landing Marine Laboratories, Moss Landing, California, USA) for his encouragement and support throughout my career. To Marsha Englebrecht, Earl and Margaret (Peggy) Ebert, Austin Ebert, and Lana Boyovich for their support and encouragement.

Sarah Fowler thanks Matt and Becky Spencer for their tolerance of her preoccupation with sharks; Sonny Gruber and Jack Musick, with fond memories of their introduction to and decades of guidance on shark research; and of course Leonard Compagno, for his friendship and in awe of his remarkable career untangling the taxonomy of sharks, rays and chimaeras.

Marc Dando would especially like to thank Julie for her encouragement, support, tolerance and advice, without it his part in this book would never have been realised. Also a special thank you to Ryan, Megan and Darren and family for their understanding in the all-consuming time spent on this project.

All the authors thank Julie Dando, the mastermind behind the scenes, for her support, patience, diligence and for making this book happen.

Photographic credits

Foreword

I grew up obsessed with sharks. Back then, I had no idea how many species existed around the world, that some could venture into fresh water or even migrate across vast oceans, that their sizes could range from 14cm to over 18m, that some could have specialised cells in their skin making them bioluminescent, or even that some could live for over 300 years! The more I learnt about them, the more I was hooked and wanted to know more. Books like *Sharks of the World* were my window to the world of sharks and I started using such resources to teach myself how to identify them.

I also learnt that sharks were growing in commercial importance in industrial and small-scale artisanal fisheries. Sharks are targeted or caught as bycatch with their meat, fins, cartilage, teeth, jaws, liver, and internal organs being processed and used in various forms. With fisheries continuing to expand around the world, a number of species are now considered to have an elevated risk of extinction with many populations having drastically declined. Given the importance of sharks from a biodiversity, livelihood, and economic perspective, one would think that we would have more information about these animals. But there are still so many unknowns and little data from many parts of the world. In fact, even though species identification is the cornerstone of effective fisheries management, the challenges associated with identifying sharks means that there is still limited reporting in terms of catch and effort, landing volumes, and trade data in many fisheries around the world. Yet, these data are key to accurately provide population estimates and improve our capacity to manage them. And this is only possible through improved identification and reporting at the species-level.

Beyond the importance and role of species identification for shark conservation, who doesn't want to know more about sharks? Throughout history, these animals have fascinated and inspired humans, they have been revered and celebrated in myths and legends, and have shaped stories featuring these 'large dangerous predators'. Increasingly, curiosity and admiration for sharks from afar is turning into an urge to see and encounter sharks.

Sharks of the World delivers all of that for us. As the only book covering all species of sharks known to date, it also allows anyone interested in sharks, be it by fascination or fear, to access the most updated and comprehensive synthesis of the taxonomy and biology of all described shark species. The first incarnation of this book was published in 2005, and since then every update has been critical in consolidating information painstakingly collected by scientists and researchers globally. This is because our understanding of species diversity has been altered over the last few years based on discoveries of new species from expanding research surveys as well as molecular advances. *Sharks of the World* is a significant contribution to knowledge of sharks globally and this breadth of information will help improve the capacity of researchers, fisheries managers, observers, the fishing industry, enforcement bodies, and policy makers to monitor catches and landings of sharks and therefore the accuracy of information required for their management.

In featuring over 500 species of sharks with beautiful and scientifically accurate colour illustrations, this book highlights the diversity of species, describes key morphological features for identification purposes, provides insights into their known biology, habitat, and distribution, as well as updates their current conservation status. Sharks continue to fascinate me, and I have no doubt that anyone that picks up this book will be drawn to these enigmatic fish, their beauty, and their fragility. I hope readers will also be inspired to keep learning about them and appreciate the need for their conservation.

Rima Jabado
Elasmo Project

Hidden in plain sight, a pair of Tassled Wobbegongs, *Eucrossorhinus dasypogon*, cryptically resting on a large circular plate coral on a coral reef in Raja Ampat, West Papua, Indonesia.

INTRODUCTION

Blue Shark, *Prionace glauca*, often migrate vertically after their prey, following squid as they dive into deep water during daylight hours, then ascending back to the surface at night.

The first, single volume, illustrated *Field Guide to the Sharks of the World*, by Leonard Compagno, Sarah Fowler and Marc Dando (published 2005), covered some 400 valid shark species described and named by scientists plus around 50 more that had not yet received formal descriptions and scientific names. Eight years later, in 2013, the first edition of this book, *Sharks of the World: a fully illustrated guide*, presented 501 shark species, including 90 whose names had not appeared in the 2005 field guide. A companion *Pocket Guide to Sharks of the World* added keys to teeth and featured the most common shark fins seen in international trade. The authors, who include David Ebert, one of Leonard's former students, are now delighted to add another 51 species to this second edition, bringing the total to 536 in 2021. Forty-three of these have formally been named since the 2013 edition. The other eight were formerly considered 'junior synonyms', but are now known to be valid species. (When two or more species are found to be the same, the oldest name takes precedence and the more recently published name becomes a junior synonym.) Sixteen species in the previous edition, including some formerly 'well-known' sharks, have now been removed because they are identical to other species described under different names, or because their name had actually been applied earlier to another taxon – in both cases the earlier scientific description takes precedence.

Adding so many new species highlights their astonishing discovery rate: scientists have named nearly 40% of all new species of sharks and their relatives (including skates, rays and chimaeras) in the last 50 years. More than 325 species have been described during the past 20 years alone. In addition to completely new discoveries, we now know that some species earlier described from different regions or oceans under the same name are actually several quite distinct, isolated species with a similar appearance (known as sibling species) or are regional subspecies that are genetically isolated. This book includes over 120 shark species named since 2005. The increasing use of DNA analysis to determine the relatedness of species is also having a significant impact upon our understanding of shark taxonomy and biodiversity (see p. 61), although there is still debate on how to interpret these results. Our rapidly growing knowledge illustrates how difficult it has been to stop adding new species in order to publish this volume – which currently (if briefly) presents the most up-to-date review of some shark taxa.

These updates presented the authors and illustrator with a massive task: reviewing the extensive literature (recently published and still in preparation) and examining huge numbers of photographs and illustrations of sharks (living, dead and preserved museum specimens). The decision to include so much hitherto unpublished information on shark teeth presented a major challenge. We benefited greatly during this process from the generous assistance of Mark Harris.

We hope that this guide will encourage more people to study sharks – whether by observing and photographing sharks in the sea, at fish markets, or in museum collections – and to discover new information that challenges or amplifies the descriptions and illustrations in this book. Indeed, we actively encourage readers to do so; knowledge of these amazing animals is seriously lacking in many parts of the world, and there is still so much to learn about shark biodiversity and biology.

How to use this book

The structure of this book will be familiar to anyone who has used other species identification guides. Illustrations on pp. 13–15 identify and name external features mentioned in the text, and p. 16 explains the sequence of the species pages (by order, then family). We then briefly present some background information on sharks, summarising their evolution, body structure, biology, life history and behaviour, and various aspects of their relationship with humans (e.g. sharks in science, legend, fisheries, sport and tourism, food, and their conservation and management).

We recommend always using the illustrated dichotomous (branching) identification key when trying to identify a particular shark species (even if you already think that you know what it is). Start working through the numbered boxes from p. 89 onwards until you have found the order or family within which the shark occurs. Then turn to the page indicated in the key to study the detailed illustrations, confirm your identification, and read more about the species. There are keys to identify detached fins and teeth on pp. 592 and 595.

Species descriptions occupy most of the book, beginning with the most primitive sharks, the frilled sharks (p. 96), and ending with the most recently evolved, the hammerheads (p. 569). Each major taxonomic section starts with a brief introductory description of the characteristics of each order and/or family, including the most important external diagnostic identification characteristics (e.g. the number and shape of fins, fin spines, gills, position of mouth etc.). We recommend reading these basic descriptions first, to ensure that they match the animal you saw, before trying to identify the species from the colour plates. The plates illustrate the animals as they appear in life, while the text alongside provides a small amount of information on the species illustrated and directs readers to the pages with more detailed material. Each introductory section also reviews briefly the biology and status of the taxonomic group.

One of the challenges faced when preparing this book has been the large number of apparently very rare species that are known from just one or two preserved specimens. Often these examples have been lost from collections; in some cases there are not even illustrations available to work from (and certainly no colour pictures or photographs of freshly killed or live animals). For these rarities, the illustration is based upon very sparse information indeed.

The introduction to each family is followed by a description of every known species, identified by one widely used vernacular name and its scientific name. Species are mostly listed in the alphabetical order of their scientific names within each family, firstly the genus, then the species name (see Naming species, p. 17). Each species description is accompanied by line drawings of the side view of the animal, and (if possible) the underside of the head and the teeth. The tooth count figure used in this book is the total number of tooth rows along both jaw lines (left and right) of the upper and lower jaws, where known. This figure is not the total number of teeth to be found in the jaw of that species, as this would have to include all the series of teeth in the rows (see p. 26). The outermost series, the functional teeth, are the rows counted. (The tooth size and count information are not always readily available or recorded. Any further information would be greatly appreciated by the authors; please send this to us via the publishers.)

All body measurements provided are given as total length in centimetres (cm), measured as the 'point to point' distance (i.e. not over the curve of the body) from the tip of the snout to the tip of the caudal fin (tail). Where available, these include total length at birth, maturity, and maximum size. For egg-laying species, we have tried to indicate the size of eggcase and length of young when they hatch. 'Mature' indicates the size at which the species first reaches sexual maturity (this is often different for males and females; the latter are generally larger). If this information is not available, we may provide 'Immature', 'Adolescent' or 'Adult' lengths instead. The 'Maximum' length is often the largest recorded, not the maximum that a species could reach (which may be much larger). Anglers often like to know the weight rather than, or as well as, the length of the sharks that they catch, but we could not provide this information. Weights are not often recorded in the scientific literature because they are very variable, depending upon the time of year, state of pregnancy, recent meals and so on. Length, on the other hand, is a more constant, reliable and useful measure of the size and age of the animal. Length:weight conversion tables are available for some species, to allow anglers to estimate the weight of their catch without removing it from the water, but there is no space to include them here.

The **identification** text includes each species' characteristic coloration, shape and other distinctive features, some of which are illustrated. It must be read in conjunction with the earlier section that summarises the overall identification characteristics for the order and family. The drawings on pp. 13–15 identify the body parts referred to in the text. If doubt remains (it is very difficult to distinguish some species), you can consult more detailed regional identification field guides for European, Mediterranean and North American waters (also published by Princeton), see p. 600. Other regions may be covered by the most recent FAO regional identification guides and species catalogues (download from www.fao.org/fishery/fishfinder/publications/en).

Distribution lists the oceans and (for sharks with a limited geographic range) the countries where the species has been recorded. The distribution **maps** provided have been updated since 2013, but may still be incomplete because there has been so little shark research in many parts of the world. It is possible that many species have a wider regional distribution than illustrated, but it is unlikely that a Pacific Ocean shark will turn up in, for example, the Mediterranean or Caribbean Seas. On the other hand, it is also possible that a single shark species that is recorded from widely separated locations (particularly if it is small, a poor swimmer, and lives on or near the seabed) may turn out to be several similar species. Solid blue in the maps refers to known distribution, tint of blue is possible distribution.

Habitat describes where the animal most commonly lives (e.g. seabed or pelagic, intertidal or deepsea), including the depth below sea level in metres, where the species has been reported. **Behaviour** may include movements and migrations, and characteristic feeding activities, where known. **Biology** summarises what (if anything) is known about the reproduction of the species (see also pp. 48–51) and what it eats. **Status** includes the December 2020 IUCN Red List Threatened Species Assessments (p. 82, and check www.iucnredlist.org for updates), and brief information about the conservation, management and fisheries status of the animal, including whether it is known to have caused injury to people.

A **glossary** of technical terms is provided on pp. 580–585 and a list of **further reading** on p. 600.

Topography

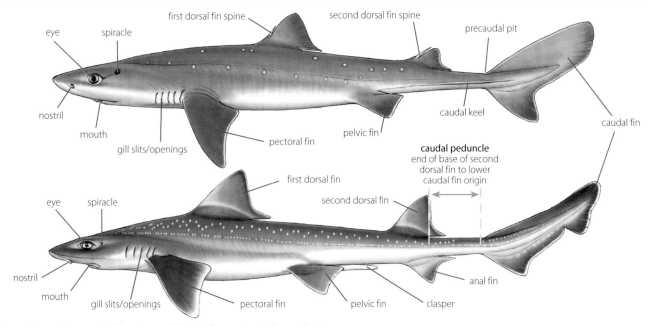

Figure 1: Lateral views – (top) female squalid shark; (bottom) male houndshark.

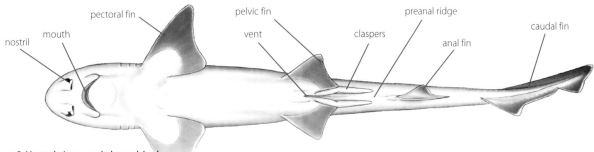

Figure 2: Ventral view – male houndshark.

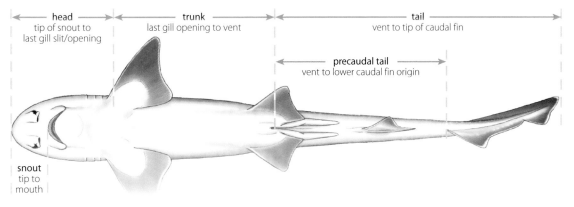

Figure 3: Ventral view showing body regions referred to in text.

Figure 4: Caudal fin topography.

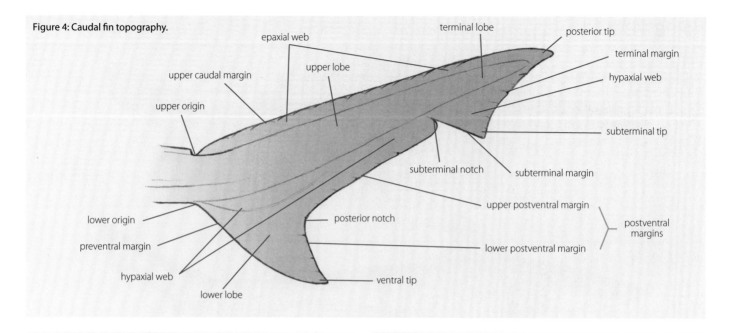

terminal lobe
posterior tip
epaxial web
terminal margin
upper lobe
hypaxial web
upper caudal margin
upper origin
subterminal tip
subterminal notch
subterminal margin
upper postventral margin
lower origin
posterior notch
preventral margin
lower postventral margin
hypaxial web
postventral margins
lower lobe
ventral tip

Figure 5: Dorsal fin topography.

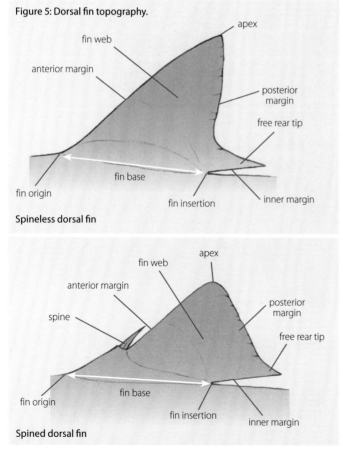

apex
fin web
anterior margin
posterior margin
free rear tip
inner margin
fin origin
fin base
fin insertion

Spineless dorsal fin

apex
fin web
anterior margin
posterior margin
spine
free rear tip
fin origin
fin base
fin insertion
inner margin

Spined dorsal fin

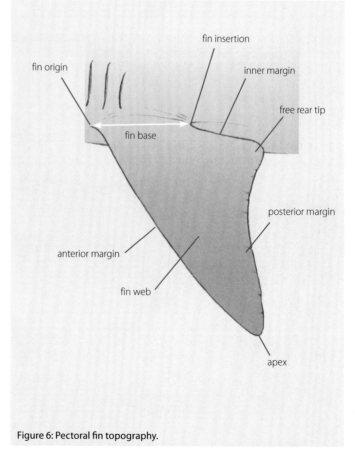

fin insertion
fin origin
inner margin
free rear tip
fin base
posterior margin
anterior margin
fin web
apex

Figure 6: Pectoral fin topography.

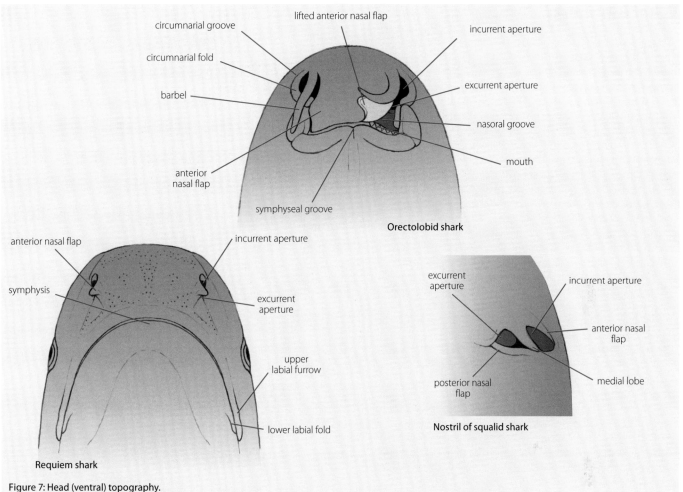

Figure 7: Head (ventral) topography.

Orectolobid shark

- lifted anterior nasal flap
- circumnarial groove
- circumnarial fold
- barbel
- incurrent aperture
- excurrent aperture
- nasoral groove
- mouth
- anterior nasal flap
- symphyseal groove

Requiem shark

- anterior nasal flap
- symphysis
- incurrent aperture
- excurrent aperture
- upper labial furrow
- lower labial fold

Nostril of squalid shark

- excurrent aperture
- incurrent aperture
- anterior nasal flap
- posterior nasal flap
- medial lobe

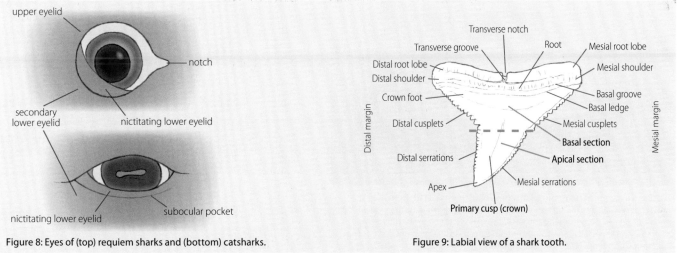

Figure 8: Eyes of (top) requiem sharks and (bottom) catsharks.

- upper eyelid
- notch
- secondary lower eyelid
- nictitating lower eyelid
- nictitating lower eyelid
- subocular pocket

Figure 9: Labial view of a shark tooth.

- Transverse notch
- Transverse groove
- Root
- Mesial root lobe
- Distal root lobe
- Mesial shoulder
- Distal shoulder
- Crown foot
- Basal groove
- Basal ledge
- Distal cusplets
- Mesial cusplets
- **Basal section**
- Distal serrations
- **Apical section**
- Mesial serrations
- Apex
- **Primary cusp (crown)**
- Distal margin
- Mesial margin

What is a shark?

Sharks belong to the taxonomic class Chondrichthyes, or cartilaginous fishes. As the name suggests, these fishes have a simple internal skeleton formed from tough, flexible cartilage; no bone is present in the skeleton, fins or scales. Unlike primitive, jawless, cartilaginous lampreys, sharks have jaws and nostrils on the underside of their head. Mature males always have claspers. Other features unique to chondrichthyans include placoid scales and rows of teeth that are continually replaced.

There are two main groups of chondrichthyans. The largest is subclass Elasmobranchii ('elasmo' means plate and 'branchii' means gills), which includes the sharks and rays. This group is characterised by multiple (usually five) paired gill openings on their heads. Subclass Holocephali ('holo': whole, 'cephali': head) contains the chimaeras, which are a much smaller group of living animals (although very diverse in the fossil record). A soft gill cover with one opening on each side of the head protects the four pairs of gill openings in holocephalans.

The sharks comprise over 536 (and rising – more will have been described before this book is printed) of known elasmobranch species. They are usually cylindrical in shape (sometimes flattened) and easily recognised from the five to seven paired gill openings on the side of their head and pectoral fins that are not attached to the head above the gill openings. The vertebral column extends into the top lobe of the large, usually asymmetrical, caudal fin (tail) and they have one or two dorsal fins (sometimes with spines). A few species are flattened, like most rays, but the location of their gill openings and shape of their pectoral fins are still diagnostic.

Rays, or batoid fishes (not included in this book), are sometimes known affectionately as 'flat sharks' or 'pancake sharks'. There are about 670 described species of ray, characterised by expanded, flattened, wing-like pectoral fins that are completely fused to the sides of the head above the gill openings on a generally short, flat body. Some species of ray, however, look more like flattened sharks than the more familiar disc-shaped stingrays and skates. Examples of these are sawfishes and guitarfishes, which look similar to sawsharks (p. 221) and angelsharks (p. 227).

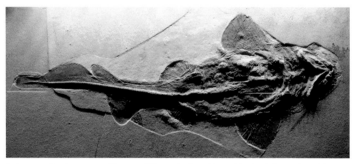

Palaeocarcharias stromeri is a small (max 1m long) benthic shark from the mid-Jurassic period, about 165 million years ago. It may look like a modern carpet shark, but researchers studying the structure of its teeth have identified this small shark as a mackerel shark and therefore the oldest known ancestor of the Megamouth, Shortfin Mako and White Sharks.

CHONDRICHTHYES

Doliodus and other basal forms (Extinct; Silurian to Devonian)

HOLOCEPHALI

CHIMAERIFORMES (Carboniferous to Recent)

Multiple Extinct holocephalan orders

SYMMORIIFORMES (Extinct; Devonian to Permian, possibly Cretaceo

CLADOSELACHIFORMES (Extinct; Devonian)

ELASMOBRANCHII

XENACANTHIFORMES (Extinct; Devonian to Triassic)

CTENACANTHIFORMES (Extinct; Devonian to Triassic)

EUSELACHII

HYBODONTIFORMES (Extinct; Devonian to Cretaceous)

BATOIDEA

MYLIOBATIFORMES – STINGRAYS (Cret

RHINOPRISTIOFORMES – GUITARFISH

TORPEDINIFORMES – ELECTRIC RAYS (

RAJIFORMES – SKATES (Cretaceous to I

NEOSELACHII

SELACHII

LAMNIFORMES – MACKEREL SHARKS (Cre
8 families   10 genera   Pages 301–323

CARCHARHINIFORMES – GROUNDSHARKS (J
10 families   52 genera   Pages 324–579

ORECTOLOBIFORMES – CARPET SHARKS (Jurass
7 families   13 genera   Pages 258–300

HETERODONTIFORMES – BULLHEAD SHARKS (J
1 family   1 genus   Pages 247–257

HEXANCHIFORMES – FRILLED AND COW SHARK
2 families   4 genera   Pages 96–101

SQUALIFORMES – DOGFISH SHARKS (Cretaceou
6 families   22 genera   Pages 105–219

SQUATINIFORMES – ANGELSHARKS (Jurassic to
1 family   1 genus   Pages 227–246

ECHINORHINIFORMES – BRAMBLE SHARKS (Cre
1 family   1 genus   Pages 102–104

PRISTIOPHORIFORMES – SAWSHARKS (Cretaceo
1 family   2 genera   Pages 220–226

SYNECHODONTIFORMES (Extinct; Permian to Paleogene

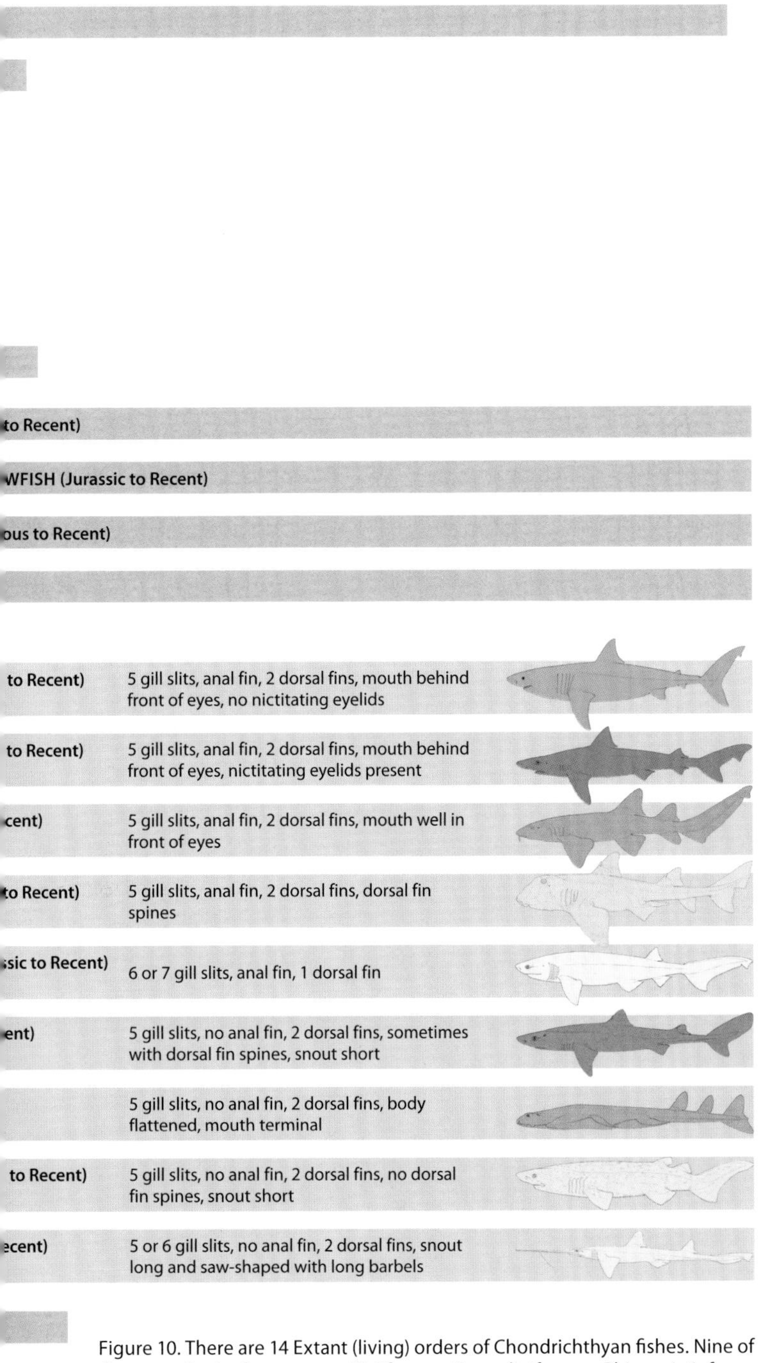

to Recent)
WFISH (Jurassic to Recent)
ous to Recent)
to Recent)
5 gill slits, anal fin, 2 dorsal fins, mouth behind front of eyes, no nictitating eyelids
to Recent)
5 gill slits, anal fin, 2 dorsal fins, mouth behind front of eyes, nictitating eyelids present
cent)
5 gill slits, anal fin, 2 dorsal fins, mouth well in front of eyes
to Recent)
5 gill slits, anal fin, 2 dorsal fins, dorsal fin spines
sic to Recent)
6 or 7 gill slits, anal fin, 1 dorsal fin
ent)
5 gill slits, no anal fin, 2 dorsal fins, sometimes with dorsal fin spines, snout short
5 gill slits, no anal fin, 2 dorsal fins, body flattened, mouth terminal
to Recent)
5 gill slits, no anal fin, 2 dorsal fins, no dorsal fin spines, snout short
ecent)
5 or 6 gill slits, no anal fin, 2 dorsal fins, snout long and saw-shaped with long barbels

Figure 10. There are 14 Extant (living) orders of Chondrichthyan fishes. Nine of them are sharks, four are rays (Rajiformes, Torpediniformes, Rhinopristioformes and Myliobatiformes) and the chimaeras (Chimaeriformes).

Naming species

Every scientifically-described species has a unique two-part (bi-nominal) scientific name, usually based on Latin grammatical rules and written in *italics*. This two-part name is always used to identify that species, everywhere in the world. It is important to use this, rather than the vernacular 'common' name, because there can be many different vernacular names for one species, and several similar species can be called by the same vernacular name. The scientific name can be simple, or refer to the animal's appearance or location, or verge upon nonsensical. About 157 shark species are named after people – including some well-known scientists and conservationists.

To minimise the risk of duplicate scientific descriptions, research must ensure that a species is indeed new, and be published in a peer-reviewed scientific journal. Description includes the location and date of capture, the name of the collector, and must be based upon 'type material'. Unless the species is extremely rare or threatened, the 'type' is normally a preserved specimen stored in a museum or other collection so that it can be examined (and verified) by other scientists.

The first part of the scientific name indicates the species' **genus**, the second part the individual **species** name. For example, order Lamniformes, the mackerel sharks (p. 301), includes eight families. One of these is **family** Lamnidae (p. 319), which includes three **genera** (the plural of genus): *Carcharodon*, *Isurus* and *Lamna*. There are two **species** of *Lamna*: the Salmon Shark *Lamna ditropis* from the North Pacific, and the Porbeagle Shark *Lamna nasus* from the North Atlantic and southern oceans. In this book, all species in a family are mostly listed and presented in alphabetical order, first by genus, then by species. Thus, within family Lamnidae, the White Shark *Carcharodon carcharias* appears first (p. 320), and Porbeagle Shark *Lamna nasus* last (p. 323). Some large genera are further subdivided into clades (subgenera) of species that are very similar in appearance (e.g. *Etmopterus*).

When we use vernacular names for a species, these are capitalised (e.g. Spiny Dogfish), but lower case is used for a group of sharks from the same genus or family (e.g. dogfish sharks).

The Spiny Dogfish was named *Squalus acanthias* by Carl Linnaeus in 1758. This scientific name is still in use today.

THE EVOLUTION OF FISHES
by Sarah Fowler and Charlie Underwood

All vertebrates (mammals, birds, reptiles, amphibians and fishes) are descended from small (~3cm), simple, superficially fish-like creatures that appeared in the world's oceans more than 500 million years ago (mya). Geologists call this period the Cambrian Explosion, because about a dozen major animal groups with internal and external skeletons 'suddenly' (over a period of 10 million years or so) appeared in the fossil record, then diversified amazingly. The proto-fishes, ancestors of all vertebrates, were leaf-shaped (long and flattened from side to side). They had muscle blocks attached to a rod (notochord) running along their body and a simple gut, but no eyes, fins or bones, and were likely poor swimmers. The lancelet, *Branchiostoma lanceolatum*, which burrows in sand in shallow water, is the closest example of this form of animal still in existence.

Over the next few tens of millions of years, these simple animals evolved and began leaving recognisably fish-like fossils. The notochord became a cartilaginous backbone, eyes and unpaired fins appeared, and bony plates protected the head and upper body. Lacking jaws, they probably sucked food off the seabed. Lampreys and hagfish descended from a common jawless ancestor. Biting jaws evolved from the first set of gill arches, and paired fins improved swimming ability. Such improvements through natural selection enabled the jawed fishes to feed upon more diverse prey in a wider range of habitats. In due course they evolved into far more varied forms – including some that could breathe air and crawl out on land.

Scientists now believe that modern sharks are likely descended from some of the earliest jawed vertebrates, the acanthodians or spiny sharks. These appeared 440mya, during the Silurian Period, leaving fossil evidence of their scales and spines. The placoderms (plated fishes) appeared around the same time. The huge diversification and abundance of these groups and the primitive Agnatha, jawless fishes, in oceans and fresh water during the Devonian Period led to it being called the Age of Fishes. Most of these ancestral fishes and their descendants are now extinct and known only from fossils, but all modern jawed fishes descend from groups that evolved during about 50 million years of the Devonian. The earliest known shark fossil dates

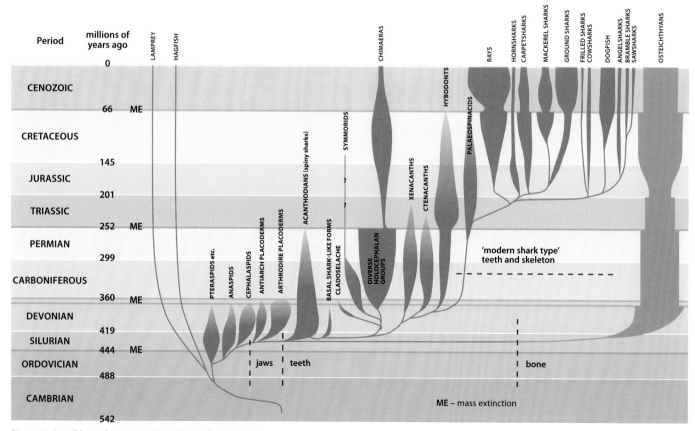

Figure 11: Possible evolutionary relationships of Chondrichthyes.

from the early Devonian, about 409mya (200 million years before dinosaurs walked the earth).

Chondrichthyans (the name is derived from the Greek 'chondros': cartilage and 'ichthos': fish), which kept the ancestral cartilaginous skeleton and lack internal bones and external flattened bony scales, were particularly widespread and biodiverse during the Age of Fishes. These ancestors of modern sharks, rays and chimaeras made up 60% of the fish species present in later Carboniferous limestone fossil deposits from shallow water and are known from over 3000 fossil species (many more, including oceanic and deepsea species, must have become extinct without trace). While some of these fossils are bizarre (e.g. *Helicoprion* and *Stethacanthus*), others, such as *Cladoselache*, appear similar to some modern sharks.

This ancient chondrichthyan diversity ended during the mass extinction at the end of the Permian Period (~252mya), which is thought to have been caused by massive volcanic eruptions, leading to acid rain, catastrophic sediment run-off, global warming and ocean deoxygenation. It killed off more than 90% of all marine species. By the Jurassic, 145–200mya, a resurgence was underway, with modern forms of sharks, rays and chimaeras appearing and diversifying in shallow water habitats.

Only two of the major fish groups from the Age of Fishes remain abundant today: class Chondrichthyes, and class Osteichthyes ('osteos': bone, 'ichthos': fish), the bony fishes or teleosts. The latter have a bony skeleton instead of cartilage and are now by far the most abundant of vertebrates. The third surviving group, Sarcopterygii ('sarcos': flesh, and 'pteryx': fin or wing), which once contained the ancestors of all terrestrial vertebrates, is now represented by a few coelacanths and lungfishes. There are an estimated over 34,000 or so species of teleost living today, but only some 1,265 chondrichthyans, about 292 of which have only been described by scientists in the past 15 years.

SOLVING EVOLUTIONARY PUZZLES

In a new twist to our understanding of shark evolution, researchers recently made a remarkable discovery when studying the fossil of a newly described species of heavily armoured placoderm fish in Mongolia – its brain case was made from bone! This is notable because the placoderm fishes, which became extinct ~300mya, during the Carboniferous period, were also descended from the ancestors of bony and cartilaginous fishes (Figure 11) – one might call them cousins. Until this discovery, palaeontologists had agreed that all ancestors of jawed vertebrates possessed cartilaginous skeletons and, while sharks and their relatives retained this structure, bone first evolved in the osteichthyans (the group that contains bony fishes and limbed vertebrates). The existence of this bony placoderm skull, however, indicates that bone evolved much earlier than had previously been believed. This suggests another possibility: that the ancestors of the cartilaginous fishes had bony skeletons, but that sharks and their relatives subsequently reverted back to cartilage because this lighter more flexible structure was more advantageous for their lifestyle.

On the other hand, evolution does not proceed in a linear fashion. Parallel evolution describes the process by which related taxa evolve independently but in the same general direction as they adapt to similar ways of life – the caudal fins (tails) of Mako Shark and Bluefin Tuna are examples of parallel evolution in fishes that have become fast-swimming pelagic predators. Convergent evolution is the process in which distantly related species develop, completely independently, similar characteristics that did not exist in their common ancestors; they become more similar to each other than they are to their forebears. Examples include the development of complex eyes in octopi and vertebrates, and echolocation in bats and whales. Perhaps this ancient bony skull is evidence of the convergent evolution of bone in some placoderms and the bony fishes.

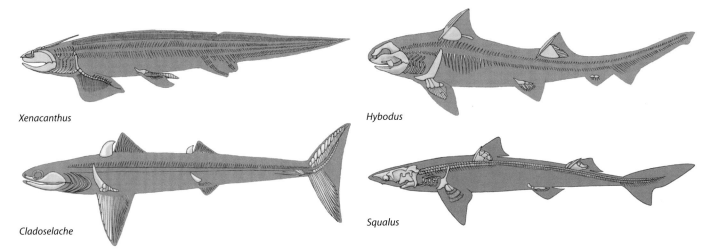

Xenacanthus

Hybodus

Cladoselache

Squalus

Figure 12: Left: Reconstruction of two Palaeozoic skeletons, a xenacanth (Euselachii) and a cladoselachian. Right: Reconstruction of a hybodont and modern shark skeletons (after Schaeffer 1967).

Evolution of sharks

Many shark enthusiasts enjoy reminding dinosaur devotees that their favourite reptiles are no more than 'johnny-come-latelies' in the family tree of charismatic vertebrates. This timeline illustrates just how recently (in geological time) dinosaurs and mammals appeared on earth, compared with the lengthy and sophisticated evolutionary history of sharks and their relatives during the past 440 million years.

Two main techniques are used to study shark evolution: the traditional investigation of fossil evidence, augmented today by sophisticated modern techniques (such as CT scanning) to examine structures hidden inside fossils; and the genetic analysis of living species to back-cast the rate at which they evolved from extinct common ancestors – vital when no fossils have been found or fossils are difficult to interpret. Scientific knowledge is evolving faster than sharks; this page illustrates current knowledge, which is changing rapidly as new finds are made and specimens analysed by new methods.

Our timeline begins with the acanthodians, fishes with scales like those of bony fish but fin spines and often teeth like those of sharks; they are now considered to be the earliest known cartilaginous fishes. It continues through the Devonian–the Age of Fishes, which ended with a paired mass extinction

event that killed at least 75% of all species on Earth and many lineages of fishes, including all armoured forms. This was followed by the Carboniferous Period–the Golden Age of Sharks, when the diversity of cartilaginous fishes expanded and became more abundant than any other types of fishes in both rivers and seas. Another mass extinction event wiped out about 96% of marine life at the end of the Permian Period; only a few shark lineages survived into the Triassic. These included the early 'modern type' sharks with a stiffer backbone allowing them to swim faster and flexible, protruding jaws, enabling them to eat more diverse prey. The first modern forms of chondrichthyans, Hexanchiformes (frilled and cow sharks), appeared during the Early Jurassic 'Age of Reptiles', followed by most other modern shark groups by the end of the Jurassic. Sharks were diverse and common during the Cretaceous Period, which ended with an asteroid strike and the fifth mass extinction. This killed all non-avian dinosaurs and many shallow water species. The smallest, fish-eating, deep water sharks were most likely to survive. Modern sharks arose from these survivors, including Megalodon, the largest shark that ever lived, which became extinct before the early hominins appear in the fossil record.

Figure 13: Timeline of shark evolution (blue), illustrated alongside other major evolutionary events (green).

* The classification 'acanthodians' is used by some paleontologists for a general grab-bag of fish-like forms. Some of these appear to be osteichthyans but true acanthodians are all on the chondrichthyan line.
** Whilst their first appearance in the fossil record is earlier, the well-preserved skeletons of *Protospinax*, *Palaeocarcharias* and Jurassic rays are all from this time.

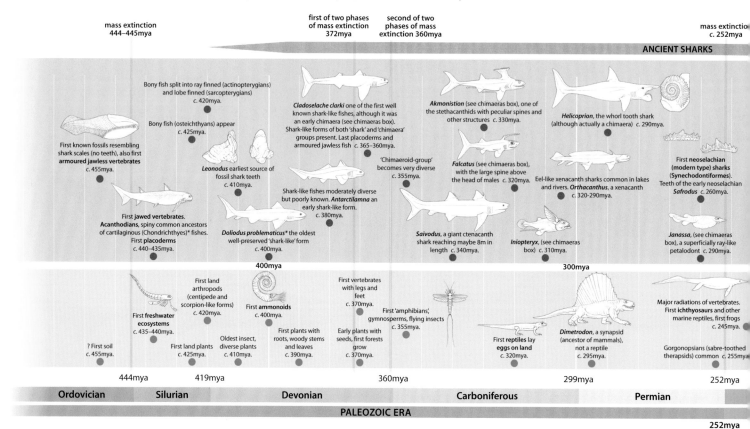

CHIMAERAS

Chimaeras today are a minor group of distinctive and usually deep water species. However, the group to which they belong, the Holocephali, were very diverse in Devonian–Permian seas and rivers and contained some very distinctive forms. It is now known that some of the most bizarre prehistoric 'sharks' evolved within the chimaera lineage. These include the small *Akmonistion*, with spikes on its head and a bizarre anvil-shaped dorsal fin in males; *Helicoprion*, a large chondrichthyan known from spiral tooth whorls in its bottom jaw; and tiny *Falcatus*, whose males had a long spine jutting forwards from their back over the top of the head.

EVOLUTION OF THE RAYS

The oldest fossils of batoid fishes (skates and rays) adapted to life on the seabed are *c.* 180 million years old, but genetic studies suggest a rather earlier origin. Jurassic rays looked like modern guitarfish, but many retained primitive features such as fin spines. Rays diversified within the Cretaceous, into the first skates, stingrays and sawfish-like forms, the last in both the seas and fresh water. After the K–Pg extinction event rays radiated rapidly, with the first true sawfish, electric rays, freshwater stingrays and filter-feeding mobulid rays all appearing near the base of the Eocene, about 55mya.

Mass extinction events

Five events are particularly significant, but there were many more smaller or local extinction events.

Late Ordovician (445–444mya) A very short lived ice age within an otherwise warm world caused two extinction events, one as the oceans cooled and one as they warmed and became anoxic. The cause of the ice age is poorly known.

Late Devonian (372 and 359mya) This was a catastrophic two (or more) stage extinction event, with ocean anoxia impacting many animals, of at least 70% of species. Armoured jawless 'fish' (ostracoderms) were wiped out by the first phase and placoderms in the second phase, clearing the space for sharks to occupy.

Permo–Triassic (252mya) Massive volcanic eruptions caused global warming and disrupted ocean and chemistry; the 'Great Dying' was the largest extinction event, killing 90%–96% of all species (96% of marine species, including trilobites, corals and many other groups and over 70% of land species). Some shark-like forms survived, only to fade out during the Triassic as new forms evolved.

Triassic–Jurassic (228–201mya) This two stage event caused 70% to 75% of all species to become extinct, including most large reptiles and amphibians, allowing terrestrial dinosaurs to become dominant. It is unclear whether the main extinctions on land and in the sea were synchronous.

Cretaceous–Paleogene/Tertiary (66mya, the K–T or K–Pg event) The impact of a massive asteroid caused a global winter. About 75% of all species became extinct, including the dinosaurs and all other land animals larger than 25kg, all marine ammonites, plesiosaurs and mosasaurs. Shark and ray extinctions were mostly of large predators and fresh water and coastal species.

mass extinction 201mya

mass extinction 66mya

MODERN SHARKS

Diverse hybodont sharks. *Hybodus*, one of a number of hybodont sharks with similar body form but very varied teeth *c.* 240–230mya.

First cow shark (Hexanchiformes) and galeomorph (horn/nurse) shark teeth known. *Notidanoides*, tooth of a Jurassic cow shark *c.* 195mya.

Earliest teeth of catsharks. The possible mackerel shark ancestor *Palaeocarcharias* *c.* 168mya.

First mackerel sharks and stingrays *c.* 135–130mya.

Aquilolamna, the earliest plankton-feeding shark *c.* 90mya.

Last hybodont sharks, first eagle rays *c.* 70–66mya.

Ancestors of White Shark evolved from early broad-toothed mako sharks. First tiger, hammerhead and requiem sharks *c.* 66–47.8mya.

Otodus megalodon – Megalodon, the largest carnivorous fish that ever lived (not an ancestor of White Shark), preyed on coastal whales *c.* 20–3.5mya.

Synechodontiform sharks present but usually rare in Triassic. *Synechodus*, an early neoselachian.

Earliest teeth of rays, nurse and horn sharks. *Protospinax* (relative of angelsharks) *c.* 185–180mya.

Diverse mackerel sharks appear. *Scapanorhynchus*, a relative of the modern goblin shark *c.* 100mya.

Diverse ground sharks, diverse rays with all major modern groups present, first plankton-feeding mobulid rays and sharks *c.* 55mya.

Requiem sharks are the dominant large predators in tropical seas *c.* 30mya.

Teeth of diverse sharks of uncertain affinities. Row of teeth of *Pseudodalatias*, superficially similar to those of the modern *Dalatias licha c.* 210mya.

Diverse and well-preserved sharks and rays from Germany and France**. The Jurassic ray *Spathobatis c.* 155mya.

First dogfish (Squaliformes). The sclerorhynchid ray *Libanopristis c.* 125mya.

First lantern sharks. Distinctive crushing teeth of *Ptychodus* common *c.* 90mya.

Otodus obliquus, the first megatoothed shark *c.* 60–50mya.

Heliobatis, one of the first freshwater stingrays *c.* 50mya.

200mya

100mya

present

First dinosaurs, pterosaurs, turtles, teleost (modern bony fishes), crocodiles *c.* 240–230mya.

First true ammonites *c.* 200mya.

Earliest preserved feathers on dinosaurs *c.* 168mya.

First seabirds, most pterosaurs very large, first snakes *c.* 110mya.

Mammals diverse but are all small *c.* 90mya.

Rapid radiation of mammals, birds and fish *c.* 65mya.

First bats *c.* 55mya.

Early hominins *c.* 6–7mya.

First very large (10 tonne +) dinosaurs *c.* 185–180mya.

First flowering plants, *Iguanodon c.* 135–130mya.

First primates, elephants, horses *c.* 55mya.

First grasslands and grazing mammals, first baleen whales *c.* 55mya.

First mammals, first plesiosaurs *c.* 210mya.

Archaeopteryx (link between dinosaurs and birds), many famous and huge dinosaurs *c.* 155mya.

First marsupial and placental mammals, true birds *c.* 125mya.

Spinosaurus (aquatic sail-backed dinosaur), flowering plants dominant on most continents *c.* 100mya.

Tyrannosaurus, *Triceratops c.* 70–66mya.

First marine whales and sea cows *c.* 45mya.

Early humans, ice ages start 3mya.

201mya | 145mya | 66mya | 28mya | 2.6mya

Triassic	Jurassic	Cretaceous	Paleogene	Neogene	Q

MESOZOIC ERA | CENOZOIC ERA

66mya

Body structure

The basic body structure of sharks has remained virtually unchanged for hundreds of millions of years. There are three main sections: the head, trunk, and tail. The head has three regions: the snout (known as the rostrum in sawsharks), the orbital region (including eyes and mouth) and the branchial region (which bears gills on its sides and sometimes spiracles). The trunk, or body, extends from the paired pectoral fins to the paired pelvic fins (which include claspers in males) and vent. It usually bears the first dorsal fin, although in some species (e.g. Bluntnose Sixgill Shark, p. 100) the first dorsal fin is absent. The tail is subdivided into the precaudal tail (which bears the anal fin and one dorsal fin) and the caudal fin, sometimes separated by a precaudal pit, which increases tail flexibility.

While these features are constant, there is considerable variation in shark size and shape, depending upon habitat and feeding strategy. The largest, the plankton-feeding sharks (p. 308), are gigantic, reaching a length of up to 20m in the Whale Shark (p. 300). Major predators, such as mackerel sharks (p. 319), are also large, but not as big as the plankton-feeders. The great majority of sharks are around one metre long, but others (including some lanternsharks, p. 160) are tiny and mature at less than 30cm in length.

Although most sharks are fairly cylindrical in shape, variations are associated with different lifestyles. The benthic species, which spend most of their time resting on the seabed, have a body shape adapted for this lifestyle and are not as 'cigar-shaped' as the fast-swimming pelagic sharks. Indeed, the angelsharks (p. 227) and wobbegongs (p. 267) are very flat. They can still be distinguished from rays because the pectoral fins of sharks are not fused to the side of the head, and their gill slits are on the side of the body (although angelshark gills are hard to see, hidden beneath anterior fin lobes). In contrast, rays' gill slits are located beneath the wings, which are formed by the complete fusion of the pectoral fins to the sides of the head. These diagnostic characters (fins fused to the head and the position of the gill slits) also distinguish the shark-shaped rays, such as guitarfishes and the Shark Ray, *Rhina ancylostoma*, from true sharks.

Returning to the more typical sharks, even their basic cylindrical body shape exhibits considerable variation. The primitive frilled sharks (p. 97) are long and eel-like, with a very short snout and almost terminal mouth. These species, and most other sharks, have a heterocercal caudal fin, the upper lobe being considerably longer than the lower. Strong-swimming sharks are propelled by a fairly large, stiff, lower caudal fin lobe. In the fastest swimmers, like the Shortfin Mako illustrated here, the upper and lower lobes of the caudal fin are almost symmetrical. Furthermore, the precaudal tail is flattened dorso-ventrally, providing a pronounced lateral keel that provides stability at high speed. The very similar tail structure in makos and fast oceanic teleosts, such as the Bluefin Tuna and marlins, is an excellent example of convergent evolution – these species have evolved the same body and tail shape because this is the most efficient structure for their open ocean environment. There is a major difference, though: tuna and billfishes, like all teleosts, have gas-filled swim bladders to provide them with neutral buoyancy, as well as very narrow, scythe-like pectoral fins. Sharks have no swim bladders and therefore require very broad pectoral fins to produce lift. These, along with the pelvic fins, control the shark's angle of pitch (up and down movement) and prevent rolling from side to side.

Caudal fin
In most sharks, the upper caudal lobe is much longer than the lower. In mako sharks, which are very fast swimmers, the tail is nearly symmetrical

Lateral keels
Usually found in fast-swimming sharks, these are formed by a dorso-ventral flattening and a lateral widening of the caudal peduncle just before the caudal fin. They are thought to provide stability when swimming

Second dorsal fin
Absent in frilled and cow sharks

Anal fin
Absent in many shark families

Pelvic fins
Help to provide lift when swimming. Associated with claspers in males

Figure 14: External features of the Shortfin Mako.

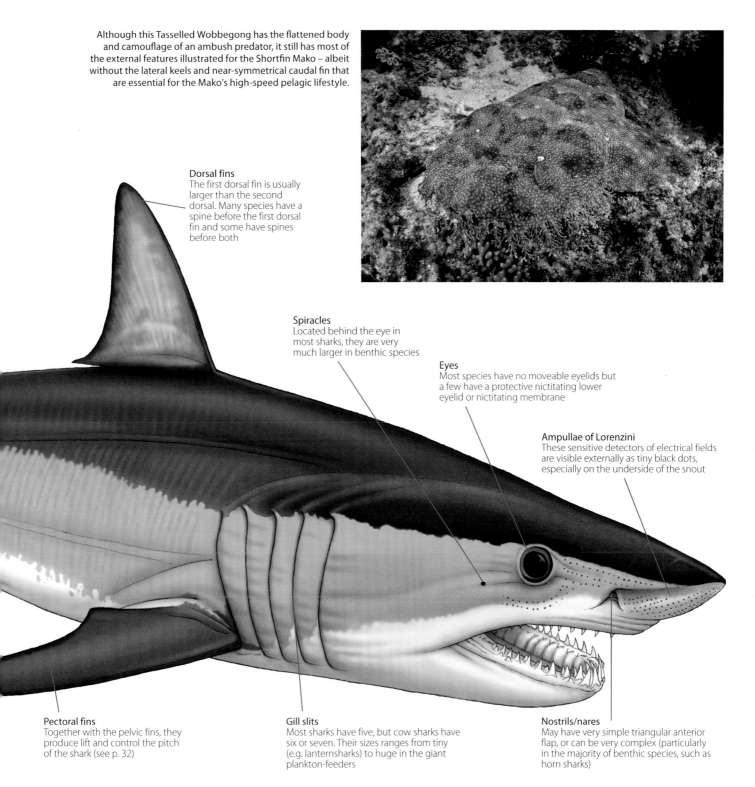

Although this Tasselled Wobbegong has the flattened body and camouflage of an ambush predator, it still has most of the external features illustrated for the Shortfin Mako – albeit without the lateral keels and near-symmetrical caudal fin that are essential for the Mako's high-speed pelagic lifestyle.

Dorsal fins
The first dorsal fin is usually larger than the second dorsal. Many species have a spine before the first dorsal fin and some have spines before both

Spiracles
Located behind the eye in most sharks, they are very much larger in benthic species

Eyes
Most species have no moveable eyelids but a few have a protective nictitating lower eyelid or nictitating membrane

Ampullae of Lorenzini
These sensitive detectors of electrical fields are visible externally as tiny black dots, especially on the underside of the snout

Pectoral fins
Together with the pelvic fins, they produce lift and control the pitch of the shark (see p. 32)

Gill slits
Most sharks have five, but cow sharks have six or seven. Their sizes ranges from tiny (e.g. lanternsharks) to huge in the giant plankton-feeders

Nostrils/nares
May have very simple triangular anterior flap, or can be very complex (particularly in the majority of benthic species, such as horn sharks)

Skeleton

Sharks have a very simple internal skeleton formed from cartilage (the same substance that supports our ears and noses). Unlike bone, cartilage does not contain nerves or blood vessels. This strong material is lighter and more flexible than bone because it contains fewer minerals; it is made up mostly of proteins although the cartilage of older, larger sharks may become partly calcified, harder and more bone-like. The advantages of possessing a light, manoeuvrable skeleton include more efficient swimming (to catch prey or escape from predators) and the ability to exploit different habitats and hiding places. The swim bladders possessed by most teleosts developed to compensate for the weight of their heavy, bony skeletons and provide neutral buoyancy but, as already noted, these are not found in cartilaginous fishes.

Shark skeletons also have remarkably few individual parts. The head contains an unseamed, box-like skull (chondrocranium, or braincase), cartilaginous structures supporting the gills, and the jaws (which are thought to have developed from the first gill arch and are not attached to the skull). A long vertebral column runs the length of the body from the skull into the upper caudal lobe. This takes the form of a string of hourglass-shaped vertebrae (equally as strong and stiff as bone vertebrae) beneath an arch that protects the spinal cord. Finally, cartilaginous structures support the fins and, in males, the claspers. The lack of any connection between most of these skeletal parts makes sharks incredibly flexible; as a result, most species are capable of turning rapidly in a very tight circle.

Figure 15 illustrates the evolution of shark skulls. The braincase of a primitive cladodont (top left), fossilised in Lower Carboniferous limestone, contains just three elements. Later fossils have more, and living species are yet more complex. Over time, the snout (containing complex sensory organs) has moved forward, brain size increased, and the underslung lower jaw become more heavily muscled (see p. 28). However, even the most advanced living sharks have only ten cartilaginous elements in their braincase, compared with over 60 bones in a teleost fish skull, and 22 in the human cranium.

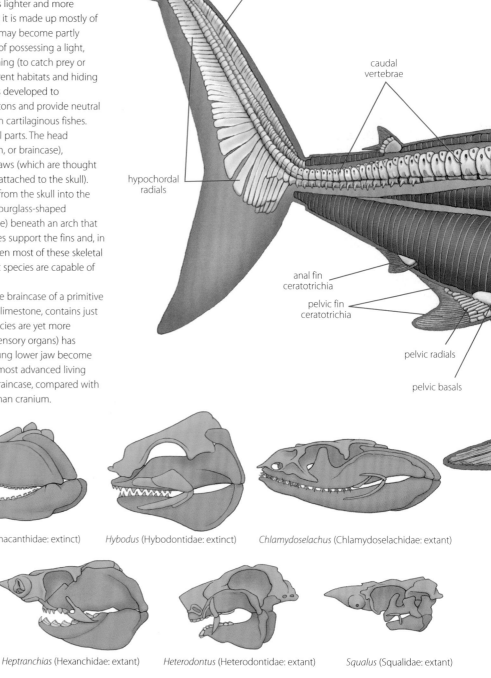

pteryquadrate
labial cartilage
chondrocranium
hyomandibula
mandible
hyoid

Cladodus (Cladodontidae: extinct)   *Xenacanthus* (Xenacanthidae: extinct)   *Hybodus* (Hybodontidae: extinct)   *Chlamydoselachus* (Chlamydoselachidae: extant)

Heptranchias (Hexanchidae: extant)   *Heterodontus* (Heterodontidae: extant)   *Squalus* (Squalidae: extant)

Figure 15: Skulls of elasmobranchs: earliest, top left, to modern, bottom right (after Schaeffer 1967).

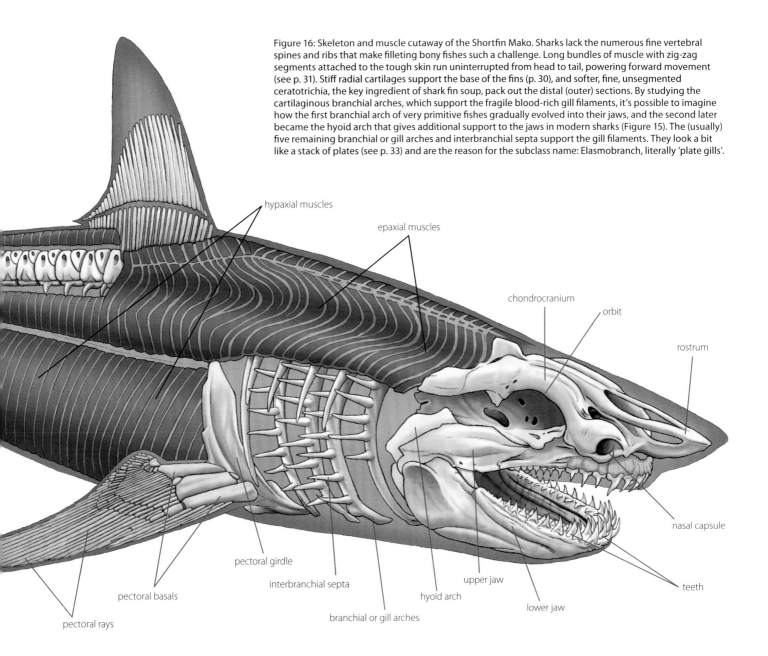

Figure 16: Skeleton and muscle cutaway of the Shortfin Mako. Sharks lack the numerous fine vertebral spines and ribs that make filleting bony fishes such a challenge. Long bundles of muscle with zig-zag segments attached to the tough skin run uninterrupted from head to tail, powering forward movement (see p. 31). Stiff radial cartilages support the base of the fins (p. 30), and softer, fine, unsegmented ceratotrichia, the key ingredient of shark fin soup, pack out the distal (outer) sections. By studying the cartilaginous branchial arches, which support the fragile blood-rich gill filaments, it's possible to imagine how the first branchial arch of very primitive fishes gradually evolved into their jaws, and the second later became the hyoid arch that gives additional support to the jaws in modern sharks (Figure 15). The (usually) five remaining branchial or gill arches and interbranchial septa support the gill filaments. They look a bit like a stack of plates (see p. 33) and are the reason for the subclass name: Elasmobranch, literally 'plate gills'.

hypaxial muscles

epaxial muscles

chondrocranium

orbit

rostrum

nasal capsule

teeth

lower jaw

upper jaw

hyoid arch

pectoral girdle

interbranchial septa

branchial or gill arches

pectoral basals

pectoral rays

CARTILAGE, WHAT IS IT?

Cartilage is the smooth, stiff but flexible material that supports our ears and nose and covers the ends of bones to allow the joints to move freely. It is formed from special cells, called chondrocytes, surrounded by a flexible collagenous matrix. Cartilage does not contain nerves or blood vessels, so nutrients diffuse slowly through it and, if damaged, is slow to repair itself. Bone is much harder and heavier, because it contains minerals. The canals that carry blood vessels and nerves through bone give it a spongy appearance in cross section, compared with the uniform appearance of cartilage. Damaged bone is able to mend more quickly than damaged cartilage (as anyone with a torn knee cartilage can attest).

Teeth and jaws

Most people see many more shark teeth than they ever encounter whole sharks. That is unsurprising, when one considers the uncountable numbers of elasmobranch teeth that have been produced by thousands of species during the past 400 million years, and their extraordinary durability. Oceans and sedimentary rocks are full of them; indeed, most fossil species are known only from these remains. There are two reasons for this abundance. Firstly, shark teeth are incredibly persistent and numerous in the fossil record because their enamel surface is so hard. Secondly, sharks produce and discard thousands of teeth during their life-time – they are not rare. It is possible to find apparently pristine, razor-sharp teeth, hundreds of millions of years old, in sedimentary rocks (ancient or recent), in rivers, and on the seabed. The huge teeth of the gigantic Megalodon, *Otodus megalodon*, can sometimes still appear relatively fresh, even though this species became extinct about four million years ago.

All sharks have multiple (up to ten) longitudinal rows of teeth running along the edges of their upper and lower jaw. New teeth form in a deep groove just inside the mouth, making up a series behind every front (functional) tooth in the row. These series of replacement teeth (there are usually 20–30 in sharks, but Whale Sharks (p. 300) have more than 300) move forward from inside the mouth on a sort of conveyor belt, which is formed by the skin in which they are anchored. Thus, new teeth can continuously replace the oldest front teeth

This fossil Megalodon tooth has been stained by the sediments in which it lay buried for millions of years.

as these become worn and fall out. Estimates of discard and replacement rates vary enormously between species and even within a single animal, depending upon its age, diet, the water temperature or other seasonal changes, and can be different in the upper and lower jaw. Teeth may fall out in just 8–10 days per row (e.g. in the Lemon Shark, p. 561), one month for the Horn Shark (p. 252), or not for several months. Nurse Sharks (p. 298) lose teeth every 9–28 days in summer, but this process takes 51–70 days in winter. It is commonplace to see many recently shed teeth from living sharks lying on the bottom of an aquarium. In most species only a few teeth are shed at a time, but cookiecutter sharks (p. 212) shed their entire rows of huge lower teeth as a complete set. Whatever the timing and sequence, that adds up to a lot of teeth over a lifetime for a shark that may live for 40 years or more, and was shedding teeth before it had even been born (see p. 33).

Shark teeth take many forms and sizes. They range from thin spear-shaped or fang-like teeth used for seizing slippery fishes and squid before swallowing

row 1   row 2   row 3   row 4

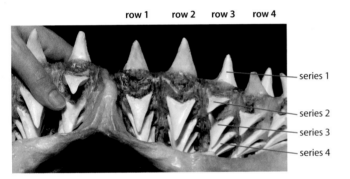

series 1
series 2
series 3
series 4

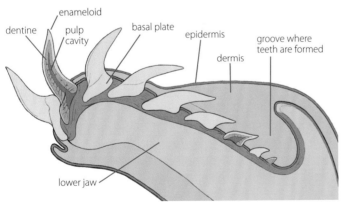

dentine
enameloid
pulp cavity
basal plate
epidermis
groove where teeth are formed
dermis
lower jaw

Figure 17: Series and rows of teeth can be seen very clearly in this huge White Shark jaw (above). The illustration (below) shows a section through the jaw and a single row of teeth – the oldest on the left is about to fall out and be replaced by the tooth shown in cross section, while the newest tooth on the right is only just beginning to form deep inside the mouth.

This close-up shows a few of the hundreds of 'homodont' teeth in a Whale Shark jaw. In this image, the worn and missing teeth on the outside of the jaw are on the right, and new sharp teeth are moving out from the inside (left) to replace them.

them whole, to very large and flat crushing teeth (sometimes arranged like a pavement) specialised for breaking up shellfish, and saw-edged teeth for cutting bites out of large prey. It is fairly easy to guess what a shark eats from a quick look at its jaws, but many teeth are so tiny that a microscope is needed to examine them properly. Only a very few species have 'homodont' teeth (all the same size and shape). The Whale Shark (p. 300) and Basking Shark (p. 311), plankton-feeders that do not need teeth to catch their prey, fall into this group (see photo opposite).

For the majority of species, tooth shape varies considerably, usually depending upon their position in the jaw, but also on the age of the shark. Changes that occur with age are known as 'ontogenetic'; they take place during the animal's lifetime, alongside alterations in diet and often habitat. As with changes to the form of the dermal denticles from which teeth were originally derived, it can be a mistake for scientists to describe a new species solely because it has a different tooth shape from another closely related shark; it might simply be younger or older than the specimen on which the original description was based.

Horn sharks (p. 247) are named 'Heterodonts' ('different teeth') for a very good reason. The teeth at the front of both jaws (upper and lower) are pointed, for grabbing prey, while those at the back are flattened, for crushing invertebrates. The Lamniform shark pups (p. 301), which feed on eggs and other embryos inside the uterus, protect the mother from bites by starting off with blunt teeth, whereas the teeth that develop after birth have sharp points and cutting edges. The young in many shark species have spear-like teeth for feeding on small, soft prey. The teeth of horn sharks then flatten downwards with age, to enable them to crush hard-bodied shellfish. White Shark teeth (p. 320) change from being long and narrow in fish-eating youngsters (less than 1.5m long) to the triangular, flat teeth in the upper jaw that adults use to bite chunks out of marine mammals. Cookiecutter (p. 212) and Kitefin Shark (p. 210) shark jaws are arranged the other way around to the White Shark: they have many pointed teeth for clutching prey in the top jaw, and huge, flat cutting teeth (arranged in a single functional series) in their lower jaw. Cow sharks (p. 98) also have short, sharp teeth in the top jaw and long, serrated teeth in the lower; spinning their bodies to saw bites out of their prey is made easier because their single, small dorsal fin is set a long way back near the tail to minimise resistance against the water. Tiger Shark teeth (p. 568) are designed like a notched paper cutter or scissor on one side, and a saw on the other. Both cutting surfaces are brought into play alternately as the shark shakes its head after fastening onto its prey. This enables Tiger Sharks to bite right through large, hard-shelled sea turtles.

Figure 18: key characteristics of teeth for major taxonomic groups (continues on p. 28).

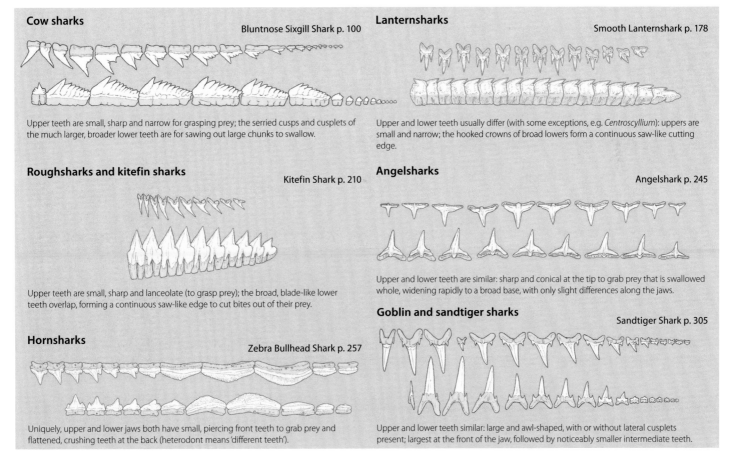

Cow sharks
Bluntnose Sixgill Shark p. 100

Upper teeth are small, sharp and narrow for grasping prey; the serried cusps and cusplets of the much larger, broader lower teeth are for sawing out large chunks to swallow.

Roughsharks and kitefin sharks
Kitefin Shark p. 210

Upper teeth are small, sharp and lanceolate (to grasp prey); the broad, blade-like lower teeth overlap, forming a continuous saw-like edge to cut bites out of their prey.

Hornsharks
Zebra Bullhead Shark p. 257

Uniquely, upper and lower jaws both have small, piercing front teeth to grab prey and flattened, crushing teeth at the back (heterodont means 'different teeth').

Lanternsharks
Smooth Lanternshark p. 178

Upper and lower teeth usually differ (with some exceptions, e.g. *Centroscyllium*): uppers are small and narrow; the hooked crowns of broad lowers form a continuous saw-like cutting edge.

Angelsharks
Angelshark p. 245

Upper and lower teeth are similar: sharp and conical at the tip to grab prey that is swallowed whole, widening rapidly to a broad base, with only slight differences along the jaws.

Goblin and sandtiger sharks
Sandtiger Shark p. 305

Upper and lower teeth similar: large and awl-shaped, with or without lateral cusplets present; largest at the front of the jaw, followed by noticeably smaller intermediate teeth.

Thresher sharks

Bigeye Thresher p. 318

Upper and lower teeth are similar: relatively large to very large, varying from broadly triangular and serrated, to long and slender with smooth-edged cusps.

Catsharks

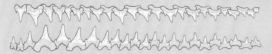

Smallspotted Catshark p. 439

Upper and lower teeth are similar: they are small with acute, narrow cusps, often lateral cusplets; not blade-like, but posterior teeth can be comb-like.

Requiem sharks

Sandbar Shark p. 542

Upper and lower jaw teeth dissimilar: uppers usually broader, with erect to slightly oblique cusps, usually serrated; lowers usually with oblique to erect, narrow, serrated cusp.

Mackerel sharks

White Shark p. 320

Upper and lower teeth similar: relatively large to very large; vary from broadly triangular and serrated (e.g. White Shark), to long and slender with smooth-edged cusps (e.g. Porbeagle); ontogenetic (size and shape may alter with age).

Houndsharks *Mustelus* sp.

Smoothhound p. 485

Upper and lower teeth similar: small and arranged in an interlocking pavement pattern, sometimes with cusps and cusplets.

Tiger shark

Tiger Shark p. 568

Upper and lower jaws similar: characteristic cockscomb shape with heavy, bent, oblique cusps, strong distal cusplets and prominent serrations, but no smooth blades.

The jaws of sharks are most unusual compared with those of mammals and bony fishes, because the upper jaw is not rigidly attached to the rest of the skull. Instead, it can move independently, so that both jaws may be protruded away from the skull during feeding. This enables the teeth to be rotated outwards, to improve the efficiency of bites and the manipulation of prey. It also helps sharks to take deep bites out of much larger prey, for example when scavenging upon whale carcasses. Protrusion of both jaws also makes suction-feeding more efficient in plankton-gulping Whale Sharks (p. 300), and ambush-feeders like the angelsharks (p. 227) and wobbegongs (p. 267). Slowed down film footage of a Goblin Shark (p. 304) lunging at its prey provides the most extreme example yet known of jaw protrusion in sharks, extending by an extraordinary 8.6 to 9.5% of its body length.

Because neither jaw is attached to the skull, obvious clues to their orientation are lacking. As a result, the jaw sets removed from sharks and put on display are often mounted upside down!

Figure 19: Jaw action of a ground shark: it takes about one second for a shark to complete the sequence of movements needed to bite its prey. The shark tilts its snout up slightly, drops its lower jaw, and moves its mobile upper jaw forward, out, and downwards, exposing the upper teeth. Meanwhile, the hyoid arch rotates forward to brace the open jaws. Fractionally before the bite, the nictitating membrane moves across to cover and protect the eye – blinding the shark as it strikes. The bite takes place as the lower jaw snaps shut, and ends when the snout drops back to its original position.

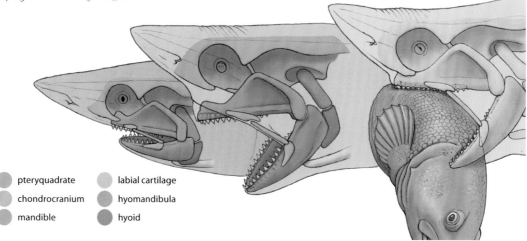

- pteryquadrate
- chondrocranium
- mandible
- labial cartilage
- hyomandibula
- hyoid

TOOTH TYPES AND TOOTH COUNT

Since the earliest documented taxonomic studies of sharks, dental morphology and meristics (tooth shape and tooth count) have been key factors for identifying and distinguishing between shark species. This is why so many technical scientific terms are used to describe the wide variety of teeth shapes and their positions, labelled in the diagram below.

For the most part, the shape and number of a shark's teeth reflect their lifestyle and preferred prey, which vary between species. But replacement teeth can gradually change shape as a shark grows and its predation behaviour changes, as it switches to larger and/or different prey. Teeth may also vary between males and females of the same species, because they are used during mating (when a male bites the female), as well as due to variations in diet. However, the most noticeable differences can be seen within a single set of jaws. Though not obvious in all shark species, this variation (known as heterodonty – literally 'different teeth') can include the development of broader, larger teeth in one jaw and narrower, less robust teeth in the other (Figure 18). In nearly all species, the largest, most useful teeth grow towards the front of the jaw (the anterior section) and smaller, lower-crowned teeth at the back (posterior). In species with tall, spiky teeth (high crown development and longer cusps), the teeth to the rear of the jaw are variably concave to angled distally (towards the jaw articulation) and slightly convex mesially (towards the centre).

Each quadrant of a jaw can be broken down into sections of rows (or files) that include one or more of the following tooth types: symphyseal, anterior, intermediate, anterolateral (where there is no differentiation between anterior and lateral teeth), lateral, anteroposterior (teeth in the symphyseal region where there is no differentiation between anterior and posterior teeth), and posterior teeth.

Tooth counts are sometimes recorded as the total in each quadrant of the upper and lower jaw, but may be further subdivided according to tooth type and placement. The most widely used method, total count, may be broken down by upper left side total, centrally located teeth, and then upper right side total, followed by a similar count of the lower jaw. Each jaw is divided at the centre, called the symphysis, by small groups of symphyseal teeth, or sometimes even smaller teeth, known as alternates and medials. As a simple rule of thumb, teeth at the centre of a jaw that are one half the size of adjacent teeth or smaller are usually shown as a 'symphyseal count' in a meristic formula. For example, the jaw illustrated below would likely be counted as 24-2-24 / 16-3-16 or as a total of 50 upper jaw, 35 lower jaw (as used in this book). It should be noted, though, that in some genera, e.g. *Mustelus*, the tooth count may strongly vary within each species and therefore may not be diagnostic. Regardless, whenever a specimen is documented for scientific purposes and the jaws cannot be kept, it is advisable to carefully record the tooth count.

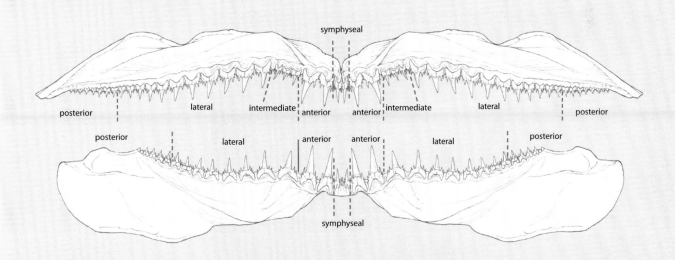

Figure 20: Smalltooth Sandtiger, *Odontaspis ferox*, jaws.

Fins

There are three main categories of shark fin: paired, median, and caudal.

The **paired (or bilateral) pectoral and pelvic fins** are located on each side of the body, roughly equivalent to the position of limbs (legs, arms or wings) in land vertebrates. They control pitch (seesawing up and down) and roll (sideways), provide essential lift when swimming (see p. 32), and can be used by some bottom-dwellers such as carpetsharks to crawl over the seabed). The largest pectoral fins are found in the least active and the fastest shark species: those that spend a lot of time resting on the seabed, and those that need a lot of lift to make sure that they remain level in the water column at high speeds. Every vertebrate has (or had) these paired fins, although one or both pairs may be vestigial or lost in some groups – e.g. whales and snakes.

The **unpaired (or median) dorsal and anal fins** are found along the centre line of the body and provide critical stability. Most sharks, including ancestral species such as *Hybodus* (Figure 12), have two dorsal fins (sometimes with spines) and an anal fin. The spines are important for defence – for example, the very large dorsal spines of horn or bullhead sharks (Heterodontidae) prevent predators from swallowing them whole. However, they create drag and are therefore very small or absent in faster swimmers. Some orders have lost one of their median fins. Several (Squaliformes: dogfish sharks, Echinorhiniformes: bramble sharks, Squatiniformes: angelsharks and Pristiophoriformes: sawsharks) no longer have anal fins, and in many other species (particularly the fastest swimmers) this fin is greatly reduced in size, presumably to reduce drag. Since scientists have not been able to decide what useful function is provided by the anal fin that cannot be derived from the dorsal fin(s), it is presumably no great loss! The function of dorsal fins is well understood: they act as keels, improving manoeuvrability by allowing quick turns and preventing rolling from side to side – just as the keel of a sailing boat reduces the risk of capsize. The dorsal fin is particularly large in fast-moving species, such as mako (Figure 16) but, because only one fin is needed for stabilisation, the second is often greatly

reduced in size to reduce drag. The Hexanchiformes, cow and frilled sharks, do not have a second dorsal fin, and the first is small and set far back on the body; this might be a useful adaptation that helps them intentionally to roll their bodies when using their saw-like teeth to cut bites out of large prey.

The **caudal fin**, the tail, provides the most important source of propulsion in sharks. It comes in many shapes and sizes, depending upon the species' life style (Figure 21, p. 31). Slow, sluggish sharks have a relatively narrow, flexible caudal fin fringing the tail end of the vertebral column, and the lower lobe is small. Active swimmers have a much stiffer caudal fin and more pronounced lower caudal lobe. In the fastest sharks, the makos upper and lower caudal lobes are virtually the same size. Angelsharks are unique in that their lower caudal lobe is larger than the upper lobe. This is an adaptation for their unusual life style: the large lower lobe helps to push up the head when the shark lunges at prey from its ambush position on the seabed.

Ironically, it is the structure of shark skeletons, which for hundreds of millions of years has made them some of the oceans' most advanced and efficient animals, that recently became one of the greatest threats to their future survival. Shark fins are supported at their base by substantial cartilaginous basal and radial rods and by the vertebrae in the upper lobe of the caudal fin (see Figure 16, p. 25). Most of the outer fin web, however, is supported and stiffened by fine, elongated fibres of collagenous tissue known as ceratotrichia, instead of the bony spines found in teleost fins. These long, unsegmented fibres, which are similar to keratin (the protein that makes up hair and nails) occur in all shark fins, but are particularly densely packed into the lower caudal lobe of fast-swimming sharks, which has little radial cartilage. Although flavourless, ceratotrichia are highly valued for their texture and are the most important ingredient of shark fin soup. This makes shark fin one of the most valuable seafood products entering international trade (far more expensive by weight than shark meat) and drives many unsustainable shark fisheries (see pp. 65–73).

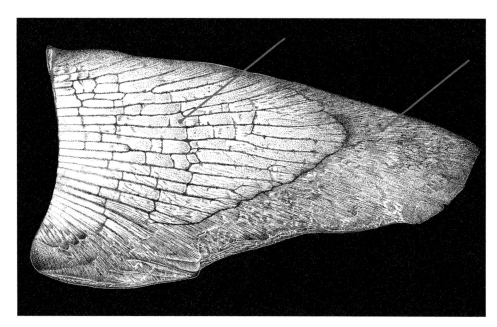

Pectoral fin radials and much finer ceratotrichia along the outer fin web.

Movement

The characteristic zigzag, segmented muscle fibres of sharks can easily be recognised in the skinned carcasses of dogfish seen on the fishmonger's slab, while shark steaks, cut across the body, show the long bundles of muscle fibres that run from head to tail. Forward movement is produced when these muscle fibres, which are anchored onto the skin, contract first on one side of the body, then on the other. This pulls them against the central vertebral column to produce a series of undulations along the body and a sinuous swimming movement powered by the caudal fin. Unusually, angelsharks (p. 227) may also supplement this form of propulsion with undulations of their large pectoral fins, similar to the swimming movements of skates and rays.

These muscle contractions produce forward thrust and acceleration, but the more a shark's body bends, the less efficient is its use of energy. Efficient swimmers, therefore, initially accelerate with a series of waves along the body which they then stiffen, resulting in less pronounced undulations and the expending of less energy for cruising. The most inefficient form of swimming is known as 'anguilliform', referring to the very eel-like movements of long, thin fish that swim by undulating their entire trunk and tail laterally, with more than one lateral wave being present (Figure 22). Examples include frilled sharks (p. 96), catsharks (p. 407) and wobbegongs (p. 267). More efficient swimmers, such as the pelagic carcharhiniforms, only undulate the posterior half of their body and less than one wave is present ('carangiform' propulsion, named after the group of teleosts, the jacks, that employ it). The really high-speed pelagic species, like the makos (p. 321), move only their caudal peduncle and caudal fin. The similarity of body form between mackerel sharks (p. 301) and tunas (teardrop in body shape, with a crescent-shaped tail and keels for stability) has already been mentioned, and this mode of propulsion is named 'thunniform', after the tunas.

Tail shape and presence or absence of a precaudal pit also provide clues to swimming mode and speed. A long upper caudal lobe and small lower lobe, particularly when these are very flexible, probably indicates anguilliform propulsion. Thresher sharks (p. 314) are a different case: their extremely long upper caudal lobe appears to be an adaptation for herding and striking prey at speeds of up to 21.8m/second, almost 50mph, rather than for propulsion, and they have a large, rigid lower caudal lobe.

Shark fins help to stabilise the animal as it moves through the water. Unlike the fins of teleost fishes, they are quite stiff and cannot be folded alongside the body. The large paired pectoral fins work in a similar way to the hydroplanes of a submarine or the elevators of an aircraft, by reducing rolling and providing lift as the shark swims (Figure 23). This lift, as well as lift possibly provided by the angle of the underside of the head, counters the tendency to sink caused by the shark's negative buoyancy, allowing the animal to move more-or-less horizontally through the water column. Adjustments to the angle of the large paired pectoral fins (and to a lesser extent the pelvic fins) also controls pitching (up and down motion) and yawing (lateral side to side motion); the latter is more difficult to control and requires a combination of fine fin and tail adjustments (Figure 24). Even then, many benthic sharks tend to have a slight lateral nod of the head as they swim through the water.

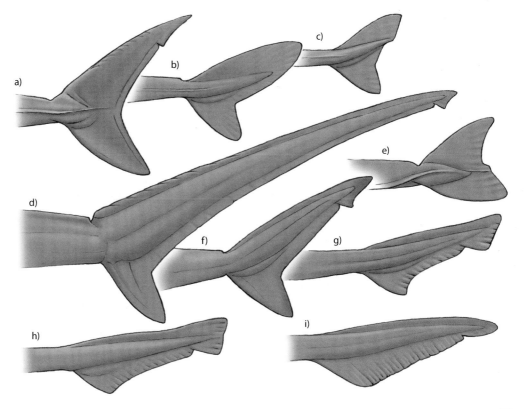

Figure 21: Tail shapes of nine families of shark.

a) Lamnidae (very fast swimmers)

b) Squalidae (moderately active swimmers)

c) Dalatiidae (moderately active swimmers)

d) Alopiidae (very fast swimmers, with highly specialised upper caudal lobe used to stun prey)

e) Squatinidae (benthic ambush predators; poor swimmers)

f) Carcharhinidae (fast swimmers)

g) Triakidae (slow swimmers)

h) Scyliorhinidae (benthic; sluggish swimmers)

i) Chlamydoselachidae (primitive, relatively sluggish swimmers)

Maintaining their position in the water column is a problem for sharks that spend their lives swimming off the seabed, because they lack the gas-filled swim bladders that give bony fishes neutral buoyancy. While their relatively light cartilaginous skeleton is a help, most buoyancy is provided by low density oils stored in the large liver (which may represent up to 25% of the total weight of some deep water sharks, compared with around 5% in mammals). This is still not enough to make them neutrally buoyant. Sandtiger sharks (p. 305), commonly displayed in aquaria, have an interesting trick to enable them to hover virtually motionless in mid water – they gulp air at the surface and store it in their stomachs. Any other species, however, will sink if it stops moving forward and loses the lift provided by its pectoral fins.

Negative buoyancy is not a problem for those species that live on the seabed. In fact, it could be a positive advantage for the surprisingly large number of sharks, including some bullhead sharks (p. 247), which probably spend as much time walking around on the seabed as they do swimming. They have developed well-muscled, mobile paired fins to enable them to do this. Some species live in the intertidal and certain bamboosharks (p. 284) are even reported to scuttle around out of the water from rockpool to rockpool, moving in a very similar manner to a walking salamander (an amphibian).

Finally, time of day has a significant influence upon the movement patterns of sharks that spend much of the time resting on the seabed. Many species, including the Horn Shark (p. 252), swellsharks (p. 415), the Lemon Shark (p. 561) and Smallspotted Catshark (p. 439), are nocturnal (mostly active at night).

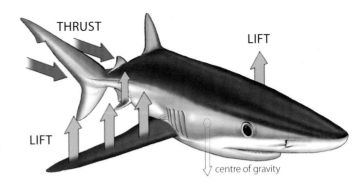

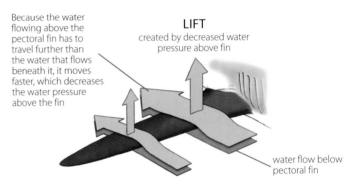

Figure 23: Physical forces acting on a shark during swimming. The shark is usually heavier than the surrounding seawater and has to maintain its mid-water position by continual swimming. The asymmetrical tail produces a forward movement and a lifting force is generated by horizontal pectoral fins, pelvic fins and head region.

anguilliform   carangiform   thunniform

Figure 22: The three major swimming methods of sharks, seen from above, illustrate (from left to right) ever more efficient propulsion methods derived from constraining the waves produced by muscular contractions.

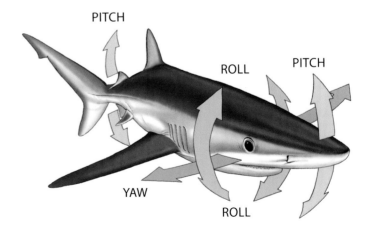

Figure 24: Movement of a shark showing roll, pitch and yaw.

Respiration and circulation

Many sharks spend considerable periods of time resting on the seabed, using their mouth cavities to pump water over their gills so that they can extract oxygen and discard carbon dioxide. This is known as buccal pumping, or branchial ventilation. Although some energy is needed for this process, these benthic sharks have a fairly low metabolic rate. Actively-swimming sharks can ventilate their gills simply by holding their mouth open, but their ability to pump water over their gills, when their movement slows or stops, varies. The most active fast-swimming species, including carcharhinids (p. 502) and hammerheads (p. 569), are completely dependent upon 'ram ventilation': they are unable to pump water over their gills and must swim continuously if they are to stay alive. These more active species have a much higher metabolic rate and larger hearts than benthic species; they must use more energy to remain in constant motion throughout their lives. Indeed, for sharks of similar sizes, there is a close relationship between their normal swimming speed and their metabolic rate (this is the rate at which they use energy, which is closely related to their oxygen consumption).

We (authors and readers) have a slower heart rate when resting, at 60–100 beats per minute (bpm), than when we move around. When we exercise strenuously, our heart beats much faster (150–200bpm) to pump more blood to our lungs where it is oxygenated, then returned to our heart to be sent all around the body, including to the hard-working muscles, finally returning to the heart after most of the oxygen has been removed. This double circulatory system is found in all vertebrates except for fishes, which have one circuit that sends blood from the heart, to the gills, then around the body and back to the heart. More exercise is associated with a faster heartbeat in virtually all vertebrates. For example, juvenile Bluefin Tuna hearts normally beat at 20–50bpm, but increase up to 130bpm when they are being fed. The chondrichthyan fishes are the exception. Sharks and their relatives change the quantity of blood sent around the body during a given period of time not by increasing the stroke rate, but by increasing the volume of blood pumped by each heartbeat – in other words, instead of beating more rapidly during exercise, each heartbeat ejects a larger volume of blood. This is possible because the four muscular chambers of a shark's heart do not only pump blood, they also store it until needed; this is why heart size is significantly larger for more active shark species (although still much smaller than the heart of a similar-sized mammal). Added to this, muscle contractions in the shark's body during faster swimming also help pump de-oxygenated blood back to the heart. The average shark heart rate does not vary much between relatively sluggish benthic species (40bpm), slow swimmers (50bmp), and the very fast makos (60–70bpm), although the relative volume of blood sent around the body is very much larger in the latter.

The process by which the gills extract oxygen from seawater is a simple and effective countercurrent exchange system (Figure 25). Blood flows from the heart through capillaries (tiny, thin-walled blood vessels) in the gill lamellae in

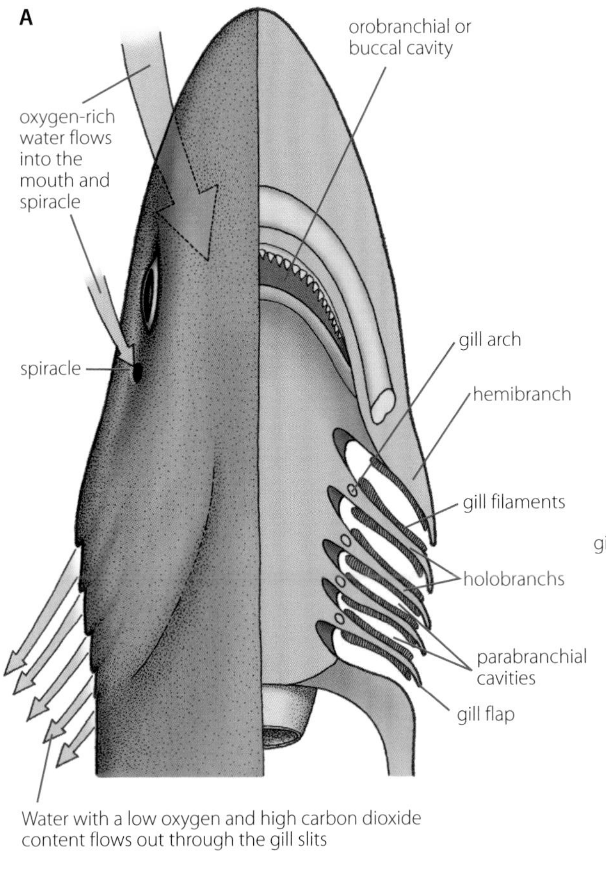

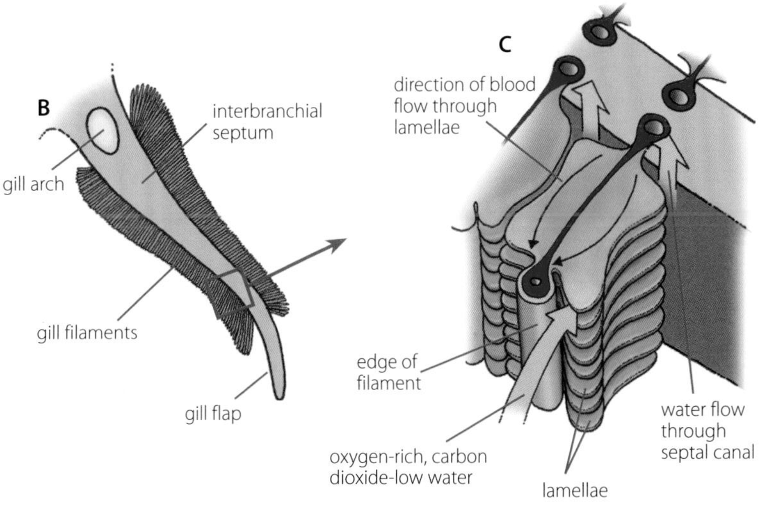

A

oxygen-rich water flows into the mouth and spiracle

spiracle

orobranchial or buccal cavity

gill arch

hemibranch

gill filaments

holobranchs

parabranchial cavities

gill flap

Water with a low oxygen and high carbon dioxide content flows out through the gill slits

B

gill arch

interbranchial septum

gill filaments

gill flap

C

direction of blood flow through lamellae

edge of filament

oxygen-rich, carbon dioxide-low water

lamellae

water flow through septal canal

Figure 25: Respiration in a typical shark. **A**. External and cutaway view of water flow in through the mouth and out through the gill slits. **B**. Enlarged lateral cutaway view of one gill. **C**. Further enlarged gill septum with attached lamellae (the blood flows through the lamellae in the opposite direction to the flow of water over the lamellae). (Based on *Sharks, Skates and Rays* 1999.)

the opposite direction to the incoming flow of oxygenated seawater from the mouth. Meanwhile, the gases dissolved in both liquids, blood and water, diffuse across the very large surface area of the thin-walled lamellae, always in the direction from highest to lowest concentrations. Blood entering the gills from the heart, low in oxygen and high in carbon dioxide, comes into close contact with the seawater that is just about to leave through the gill slits, but which is still relatively high in oxygen and lower in carbon dioxide. The net gaseous exchange across the gills at this point is therefore that of oxygen in and carbon dioxide out of the blood. By the time the blood reaches the other end of the lamellae, it is richer in oxygen and has a lower level of carbon dioxide. However, at this point it is surrounded by water that has only just entered the gill chamber, and which contains even higher levels of oxygen and lower carbon dioxide. As a result, there is still a net flow of oxygen into the blood and carbon dioxide out, until the point where the capillaries converge into the blood vessels that take the blood around the body, before returning it to the heart. This contraflow system, along with the very large surface area of the gill filaments, allows sharks to respire effectively even when there are only very low levels of oxygen in their environment. Land animals and marine mammals have an easier time, by comparison, because there is seven times as much oxygen in air as there is dissolved in well-aerated water.

During aerobic metabolism, oxygen is used to release energy from glucose (the byproducts are water and carbon dioxide), and the amount of work undertaken by red muscle is limited by the oxygen available. Despite their ability to pump larger volumes of blood around the body during exercise, anaerobic metabolism is important in sharks that need to undertake sudden bursts of swimming at high speeds and must use energy at a faster rate than oxygen can be re-supplied to the muscles. Some sharks undertake a significant amount of these 'high burst' events: Blue Sharks (p. 562) can perform short duration bursts at speeds of up to 2m/second while pursuing prey, and Blacktip (p. 536) and Spinner Sharks (p. 528) presumably also require a great deal of anaerobic energy for breaching and spinning. These activities are powered by white muscle, which is present in varying amounts in different species, but makes up the majority of the muscle in the endothermic (warm-blooded) lamnid sharks.

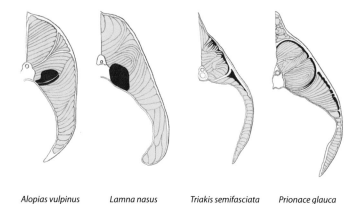

| Alopias vulpinus | Lamna nasus | Triakis semifasciata | Prionace glauca |

Figure 26: The red muscles of warm-blooded, fast swimming sharks (left) are located in the well-insulated body core, for efficient conservation of the heat that they generate, rather than under the skin (right).

Warm-blooded sharks

Water temperature is one of the most important environmental factors influencing the distribution of sharks. Most sharks and bony fishes, and all invertebrates, are 'cold-blooded' or 'ectotherms'; their body temperature is very similar to the outside water temperature, because 20–40% of the heat produced by their muscles is rapidly lost across the gills during respiration. This occurs in the same way that gases diffuse across the narrow-walled blood vessels in the gills (see p. 33): heat moves out of the blood into the colder seawater, and cold, oxygenated blood returning to the body cools the muscles down. Many ectothermic shark species require a relatively warm temperature to function effectively. They are therefore confined to warm-temperate and tropical waters and may migrate seasonally to remain within optimum temperatures. Although large ectothermic sharks may dive regularly into very deep, cold water, they warm up again when they return to the surface. Other ectotherms, including deepsea sharks (more than half of all species), are adapted to cold environments and will die in warm water. Cold water ectothermic sharks are often small and very slow-growing, because a low body temperature slows chemical processes such as digestion and results in slow growth.

In contrast, the large-bodied threshers (p. 314) and the mackerel or lamnid sharks (p. 319) maintain a much warmer core body temperature than the surrounding water: from 8°C in makos, to 14°C in White Sharks, and over 20°C warmer in the case of Salmon Sharks (p. 322). These 'endotherms' are more efficient predators (a 10°C increase in body temperature makes muscles three times more efficient, and a warm brain improves sensory functions), faster growing than similar sized ectothermic sharks, and can occupy a broader range of water temperatures and habitats. On the other hand, warm-blooded animals have a much higher metabolic rate and, despite being more efficient at digesting their prey, endotherms therefore need to eat more food than the equivalent sized ectotherm. Although tropical oceans are high in biodiversity, they do not support a high level of biomass, so warm-blooded sharks are most common in more productive cold-temperate waters, where some of their prey (oily fishes and in some cases marine mammals) also have a higher calorific value.

It is difficult enough for marine mammals to maintain body temperature in the cold ocean, with the help of thick layers of fat under the skin and lungs buried deep inside their warm body. With sharks, the greatest challenge is to stop heat loss through the gills (which continually expose all the blood in the shark's body to seawater temperatures and provide a huge surface area for heat exchange), eyes, gut and skin. The warm-blooded sharks achieve this with the aid of a specialised network of blood vessels not found in ectotherms: paired arteries and veins which branch into structures formed from tiny blood vessels (capillaries) packed tightly together into a contraflow heat exchange system. This is known to scientists as a 'rete mirabile' or 'marvellous net'. Retia (the plural) occur in the muscles, near the eyes and brain, and near the gut (Figure 27). Porbeagle and Salmon Sharks also have a rete for the kidney, which is how they manage to maintain the highest internal body temperatures of all species studied.

Within a rete, capillaries carrying cold blood run alongside, but in the opposite direction to, capillaries carrying warm blood from the muscles inside the body. These two sets of vessels exchange heat so effectively as they pass each other, that the warmth from the body ends up returning to the centre instead of dissipating into the sea. This heat conservation system is further aided, at least in some endotherms, by the heat-generating red muscles lying inside the body where they are well-insulated, next to the spine, instead of near the skin as in ectotherms (Figure 26). Large size is essential for efficient endothermy, because big animals generate more body heat and have a smaller relative surface area than little animals, so lose less heat through their skin.

When an endothermic shark enters warmer water, it does not need to keep so much of the warmth generated by its muscles. Some of the blood flow is therefore directed through other pathways, around instead of through the retia, allowing the excess heat to be lost into the sea.

Like lamnid sharks, a warm-blooded, active lifestyle characterised by high speed swimming and heat conservation is also a feature of some of the tunas, swordfish and marlins, and is an impressive example of convergent evolution.

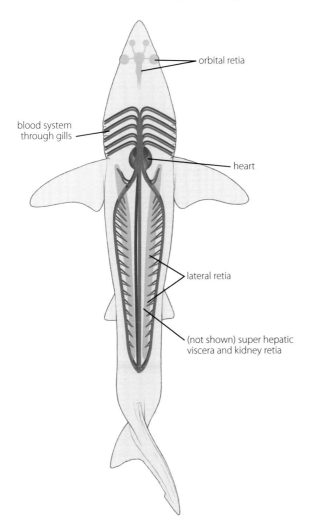

orbital retia

blood system through gills

heart

lateral retia

(not shown) super hepatic viscera and kidney retia

Figure 27: Rete mirabile are the heat-exchanging networks of blood vessels that enable warm-blooded sharks to maintain a high body temperature, instead of losing warmth into the sea.

Osmoregulation

Fishes face the relentless challenge of maintaining a constant concentration of water and salts in their body tissues, because their skin and gill membranes are so porous (leaky). In freshwater, water diffuses across these membranes into the body and salt leaks out the other way. In seawater, which is usually more salty than living tissues, body water is drawn out and salts tend to enter. This process happens because (as for gases, p. 34) water and salts will always diffuse across a permeable membrane in the direction that will eventually result in a uniform concentration on both sides. Osmoregulation is the process by which animals counteract this natural trend and maintain stable salt and water gradients between the external environment and their body fluids. Depending upon the salinity of their environment, osmoregulation can be achieved by excreting excess water or salts, by minimising loss, and/or by compensating for the loss by actively taking one or the other up from the environment.

Saltwater teleosts cope with the loss of water across their gills by taking in lots of seawater (drinking like a fish!) and actively excreting the excess sodium and chloride salts through special cells in their gills or gut. Sharks have a very different strategy. Instead of drinking seawater, they retain high concentrations of metabolic waste chemicals or 'salts' in their body so that it has a similar overall osmolarity, although comprised of 'salts' in different proportions to those found in seawater. This changes the direction in which water diffuses: out of the sea and into their more concentrated body fluids, where any excess is excreted by the kidneys. One of the most important chemicals retained in sharks is urea (which is a very toxic chemical that is rapidly excreted in the urine of mammals and most other animals). Sharks also excrete excess salt across their gill membranes, like bony fishes, and have a special 'rectal gland' that extracts salt from the blood for excretion through the gut.

These adaptations mean that most sharks are confined to fully marine environments, because they are unable to change the regulation of their body fluids to cope with less salty water. A few species, however, are frequently found in estuaries and occasionally even in brackish (very low salinity) water. Exceptionally, so far as we know, only the Bull Shark (p. 535) has the ability to move freely and regularly between the fully saline open sea and freshwater rivers and lakes. These animals have to cope with huge quantities of water flooding into their bodies when they enter fresh water. They do so by completely changing their kidney function (reducing the quantity of urea and other metabolic wastes in their body by excreting large quantities of urea in watery urine) and reversing the direction of movement of salts across the gills (by absorbing salts from the environment instead of excreting them).

The fossil record shows that, hundreds of millions of years ago, numerous shark species occurred in freshwater. The tropical river sharks, *Glyphis* species (p. 548), are the closest to fresh water specialists known today, although they are also recorded from very weakly saline habitats and may move between adjacent rivers during the wet season (when estuaries discharge so much rainwater that nearshore coastal waters also become brackish). Although osmoregulation has not been studied in these very rare species, their occasional presence in estuaries suggests that their osmoregulatory functions are similar to those of the Bull Shark in fresh water. In contrast, the distantly related South American freshwater stingrays, family Potamotrygonidae, have lived for so long in rivers that they have completely lost their ability to produce urea, and quickly die if placed in salt water.

Skin and scales

Sharks are typically protected by a very tough skin, formed from a dense network of collagen (protein fibres). Their muscle blocks are firmly attached to the inside of the skin, rather than to the skeleton. Because male sharks frequently hold onto the female with their teeth during mating and inflict fairly serious bites (see p. 49), the skin of females is often very much thicker than males. This is particularly obvious in the Blue Shark (p. 562).

Sharks occur in many colours, produced by pigment cells in the skin (see opposite). Most are fairly drab shades of grey and brown, but some (including many benthic sharks) are spectacularly patterned. These patterns camouflage the shark and obscure its silhouette, allowing it to hide from predators and/or ambush prey. Long-tailed carpetsharks of the genus *Hemiscyllium* (pp. 288–292), are small, bottom-dwelling sharks with large markings that look like huge eyes. These may confuse and deter predators. Even very bright colour markings, as in the wobbegongs (pp. 274–279), provide incredibly good camouflage when the shark is lying motionless on the seabed waiting for a meal to swim by (p. 44). For pelagic species, camouflage takes the form of counter-shading: the silhouette of a Shortfin Mako (p. 321) cruising overhead is hard to spot against the bright surface of the sea because its ventral surface is white, while the dark coloured back conceals it against the depths of the ocean when it viewed from above. Even the largely unpatterned requiem sharks (pp. 502–567) are counter-shaded. In contrast, this is not a useful feature for deepsea species that live in a uniformly dark environment, such as the dark brown sleeper sharks (plates 15–17) and the bizarre, pink-grey Goblin Shark, p. 304).

Externally, shark skin is usually protected by small, sharp, tooth-like placoid scales covered by a hard layer of enamel, although some species do not have scales on their underside. These scales, also known as dermal denticles, are very similar in structure to shark teeth (see Figure 28) – and indeed, teeth were originally formed from scales that migrated into the mouth of ancestral sharks. The shapes of scales are incredibly varied, including between species and on different parts of the body of a single animal. They include low, flattened plates or shields, complex ridged and pointed crowns, and long, hooked structures. Furthermore, their shape may change as a shark grows and matures, because

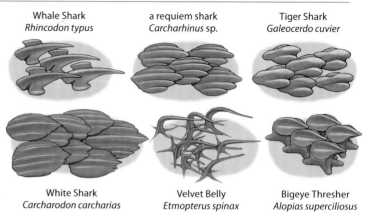

| Whale Shark
Rhincodon typus | a requiem shark
Carcharhinus sp. | Tiger Shark
Galeocerdo cuvier |
| White Shark
Carcharodon carcharias | Velvet Belly
Etmopterus spinax | Bigeye Thresher
Alopias superciliosus |

Figure 28: Not only do dermal denticles vary greatly between species, as shown here, but also between different zones of the body surface within a single animal. Furthermore, the shape and spacing of juvenile denticles may be completely different to those in adults of the same species.

dermal denticles fall out continuously (albeit, not as fast as shark teeth) and are replaced as new ones grow through the skin. Some species described on the basis of their differing scale morphology and named in the first edition of this book have been removed from this edition. That is because these descriptions turned out to be based on specimens of different size and age from the same species. Despite these complications, denticle structure can be used to distinguish between some very similar species. It can also be used to identify unprocessed shark fins, for example to find out whether they come from protected species.

In some species, the dermal denticles are enlarged to form specialised structures such as fin spines, rostral teeth in sawsharks (pp. 222–226), and the enlarged thorns or bucklers of bramble sharks (p. 104); these do not fall out but grow endlessly (annual growth rings in fin spines may sometimes be used to calculate the age of a shark, where they have not been worn away at the tip).

One of the most important functions of the dermal denticles, apart from physical protection, is to provide a surface that minimises surface drag, or friction, and maximises swimming efficiency. The surface patterns of the denticles achieve this by producing a laminar flow along the microscopic gullies and ridges on the surface of the skin, and tiny vortices (eddies) at the tips of the denticles that block the development of large-scale turbulent crossflow that would otherwise increase friction. In the fastest swimmers, like makos, the denticles are thin, light, densely packed in longitudinal patterns, and slightly flexible to enhance 'burst swimming'. This is in contrast to the dermal denticles of slow-moving sharks that are less densely packed and not aligned in rows. Experiments in the laboratory have found that real shark skin was still 12.3% faster than a fabric designed to mimic shark skin but, more recently, an artificial skin produced by a 3D printer, based on scans of Shortfin Mako skin, increased speed through the water by 6.6% over a completely smooth skin. This is such a significant advantage that 'biomimetic shark skin' is now being created to produce more energy-efficient surfaces for swimming costumes, underwater vehicles and even aircraft.

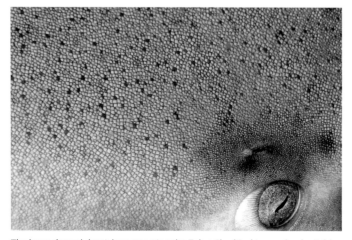

The large dermal denticles protecting the Zebra Shark's skin are clearly visible in this close-up.

In addition to being energy-efficient, materials based on shark skin can also reduce bio-fouling (the growth of seaweed and barnacles on the hulls of ships) by making it more difficult for planktonic larvae or algal spores to settle. Experimental materials have reduced barnacle settlement by 67% and algal growth by 85%. The first stage in the development of bio-fouling is the development of a film of bacteria over the clean, uncolonised surface. Producing a plastic coating with a micro pattern of ridges and ravines has also been found to reduce the growth of bacteria on surfaces, and is being used in hospitals.

Pigmentation, bioluminescence and phosphorescence

Sharks exhibit a wide range of skin colours and patterns, to conceal, deceive, or even to advertise their presence (see opposite page). These colours come from the pigments produced by cells known as chromatophores, located just beneath the skin surface. Each chromatophore contains one or more colours – dark coloured melanin is produced by melanophores, red and yellow pigments in xanthophores, etc. Some fish and invertebrate species can rapidly change colour by blocking or altering the size of these cells, for camouflage or to send social signals, but the pigment of an adult shark changes only very slowly, or not at all. The unique patterns of spots on its flanks or fins can therefore be used to recognise the same shark, year after year. Many researchers use photo-identification to monitor the movements of individual White Sharks, even if the shape and size of some markings changes slightly over time. Catsharks can slowly change the shade of their skin: those moved from a dark to a light-coloured tank will eventually become paler, to more closely match their background, but the underlying pattern of spots will not change.

Researchers studying juvenile Scalloped Hammerhead pups kept in a shallow seawater pond in Hawaii noticed the pups slowly becoming darker. They wondered whether they might be developing a suntan. This was easy to investigate: the scientists fixed opaque filters over sections of the pectoral fins of pale pups newly-caught from deeper murky waters nearby, then released them into the pool. Sure enough, a few weeks later these pups were also getting darker. Their exposure to ultraviolet in sunlight had stimulated the production of melanin in their skin – but colour of the shielded parts of their fins had not changed. The sharks were suntanning, just like holidaymakers lying on nearby beaches. However, unlike some humans and other fish species, no signs of skin damage caused by UV radiation have been recorded in sharks. Leucism and albinism (partial loss and total lack of melanin, respectively) have been recorded in several species, including Tope, Bamboo, Blacktip and Zebra Sharks. This is rarely seen, particularly in adults, because shark pups are unlikely to survive for very long in the wild without their normal camouflage patterning.

Camouflage is so important for the survival of shark pups that the young of some species look completely different from their parents, so that they can hide more effectively on the seabed. This is particularly noticeable in the Zebra Shark (p. 293) – but in this case the babies aren't trying to avoid attention. When the tiny eel-like pups emerge from their eggcases, not only do they have bold black and white stripes, but instead of hiding on the seabed they swim around at the surface, making themselves as conspicuous as possible. Their Zebra phase lasts for less than a year; the stripes start breaking up when the pups are about one month old, turning them increasingly mottled, until only small, scattered spots remain in adults. The strange behaviour of the newly hatched pups is a form of Batesian mimicry: they escape predation during the most vulnerable period of their lives by looking and behaving like the highly poisonous sea snakes that are avoided by virtually all marine predators.

There are other options besides skin pigmentation available to sharks needing to seek or avoid attention, particularly for species that live in the deep ocean where skin colours are invisible. Lanternsharks (p. 165) use bioluminesce: a chemical reaction producing visible light from special organs called photophores. These are made up of clusters of light-generating cells called photocytes, surrounded by a pigmented sheath and topped at the surface by pigment and lens cells. The photophores can be switched on and off as required. When turned off, the photophores are covered with a layer of pigment, but when the pigment cells contract, light is transmitted out through the lenses. The 'on switches' are hormones, one of which acts rapidly, perhaps when light is needed relatively quickly for signalling to other lanternsharks. Another acts more slowly, probably to produce the longer-term counter-illumination of their underside that camouflages lanternsharks from predators looking up towards the bright surface of the sea.

Finally, a few catsharks have recently been discovered to biofluoresce. This is the process by which an organism's pigments absorb dim light at a low wavelength, then emit brighter light at a higher wavelength. Dim blue light is the only natural illumination in the dark ocean, but many small catsharks can transform this into very bright green light, making them glow in the dark. Most fluorescent species studied produce a special protein to generate this light, but the recently discovered tiny molecule used by the Chain Catshark (p. 444) and Swellshark (p. 427) to glow in the dark is completely different. It also has another useful property: it kills harmful bacteria.

This newly-hatched Zebra Shark (top) not only looks similar to several species of highly poisonous sea snakes (below), but also swims like them.

Senses

Sharks are highly evolved predatory animals with a sophisticated central nervous system and large brain (which is comparable in its relative size and complexity to those of birds and mammals). Although the sections of the brain that deal with muscle control, learning, memory and the senses (i.e. sight, pressure and hearing, electro-senses, smell and taste) have been identified, the functions of some other parts are still not understood.

Figure 29 illustrates the great variation in brain size and structure between species. These have developed in order to process sensory information from different sources, or to control a variety of behaviours (i.e. catching prey, avoiding predators and interacting socially). The brains of angelsharks (p. 234 – ambush predators with few natural enemies) and some dogfishes lie at the relatively simple end of the scale. Those of carcharhinids (not illustrated) are significantly larger, while hammerheads (p. 569) possess the largest and most complex brains of all shark species (although mantas and devil rays have the biggest and most sophisticated of all elasmobranch brains). Within the lamnid sharks, the enormous plankton-feeding Basking Shark (p. 311) has the smallest brain relative to its body size (catching plankton is not very challenging!) and the Sandtiger Shark (p. 305) the largest, while that of the White Shark (p. 320) is intermediate.

Olfaction (smelling and tasting) are of huge importance to sharks, particularly for long-distance sensing in migratory pelagic sharks. These are forms of chemoreception – the process of detecting the many chemicals found in the environment. The paired nostrils, or nares, (located beneath the head, in front of the mouth) do not play any role in respiration. They are connected to the mouth (which contains taste buds) in some species, but in others they are completely separate. A complex system of nasal flaps directs incoming water flow over highly sensitive membranes, which can pick up even a tiny trace of amino acids (the chemical building blocks of proteins). Initial processing of these signals takes place in the olfactory bulb, an extension of the brain, before the large olfactory lobe of the forebrain analyses this

information and enables the shark to zero in upon the distant source of a scent by swimming up current. It is possible that sharks typically use a sinuous track, changing direction constantly, to ensure that the signal received by each nostril is as strong as that received by the other. If one nostril is plugged, a shark hunting for a hidden bait will never turn towards its blocked side. Some scientists believe that when White Sharks (p. 320 – 'swimming noses', with the largest olfactory sac known in sharks, at 18% of total brain weight) and Oceanic Whitetips (p. 537) put their snouts out of the water, they are tracking scents dispersed more rapidly by air. Olfaction is not only used to detect

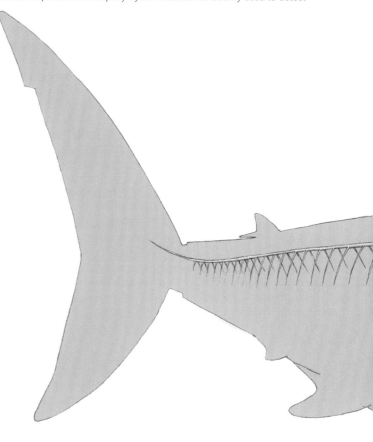

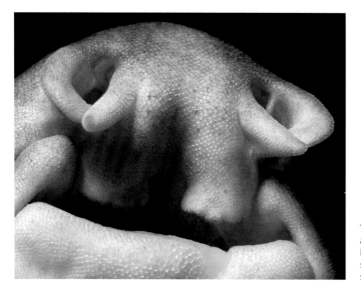

The skin flaps, folds, channels and barbels around the mouth and nostrils of this Bamboo Shark direct incoming water over sensitive receptors that can detect very small traces of organic compounds.

Figure 29: Dorsal views of four shark brains, showing the comparative sizes of the forebrain and the cerebellum (based on Hamlett (ed.) 1999). From the left: the forebrain includes the olfactory organs, cerebrum (cerebral hemispheres controlling decisions, learning and memory), hypothalamus and pituitary gland. The midbrain includes the large optic lobes. In the hind brain, the cerebellum regulates muscle coordination, including the integration of sensory processing with movement (e.g. reflex responses such as 'fight or flight'), while the brain stem receives input from the inner ear, lateral line and electro-sensory systems. The forebrain of the Bonnethead Shark (a hammerhead shark) is the largest because of the need to process so much information from its complex sensory organs.

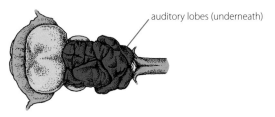

auditory lobes (underneath)

Bonnethead, *Sphyrna tiburo*

Dusky Smoothhound, *Mustelus canis*

Pacific Angelshark, *Squatina californica*

- ● medulla oblongata
- ● auditory lobe
- ● cerebellum
- ● forebrain
- ● optic lobe
- ● olfactory lobe

Piked Dogfish, *Squalus acanthias*

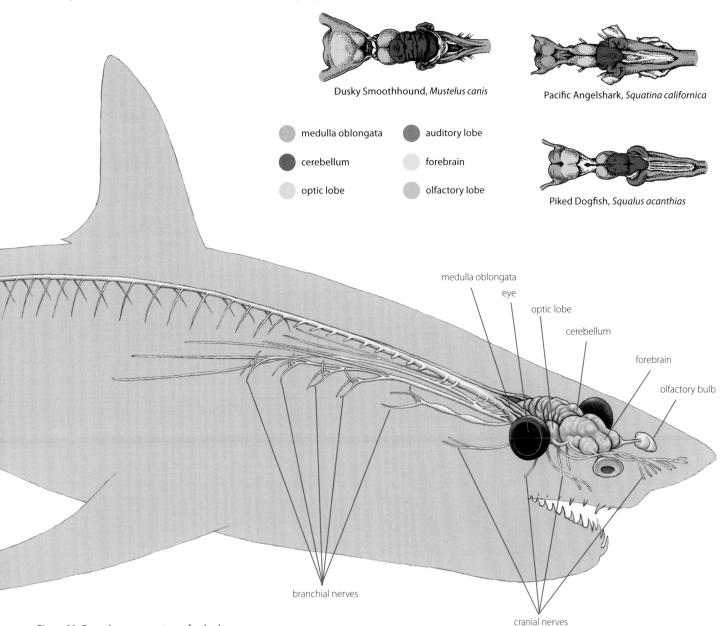

medulla oblongata
eye
optic lobe
cerebellum
forebrain
olfactory bulb

branchial nerves

cranial nerves

Figure 30: Central nervous system of a shark.

potential prey or other food items at long distance, but probably also to find mates (by sensing pheromones). It may even help guide some long, transoceanic migrations between feeding, mating and pupping grounds.

Vision is highly important for sharks living in clear water to locate prey, and takes over from taste and scent at close quarters (from about 15m). Tiny eyes are found in those species (such as the river sharks, p. 548) which spend their lives in such turbid water that vision is virtually useless. The eyes of top visual predators, like the White Shark (p. 320), are usually large, very sophisticated and extremely similar to those of mammals, and have good binocular vision (where the eye's field of view overlaps). Vision in the hammerhead sharks covers 360 degrees – the eyes located at the ends of their wing-like heads can see all around them, while the visual field of benthic angelsharks is directed up and forward. The iris surrounds a pupil (round, crescent-shaped, or a vertical, horizontal or diagonal slit), which can be opened wide to admit light in dim conditions, or (in shallow water species) contracted to a pinhole in bright light. Behind this, a crystalline lens focuses images onto the retina, which contains structures known as 'cones' for good vision (colour vision in many species), and rods for high sensitivity to low levels of light. Like many other animals, a reflective mirror or 'tapetum' behind the retina allows light to bounce around inside the eye in order to improve vision in very dark conditions. This is the reason why cats' eyes (and those of many other active, nocturnal mammals) shine when caught in torch light. Some sharks that live in shallow water have an additional layer of dark pigment cells that can expand to cover this 'mirror' by day, then contract to expose it at night. This adaptation allows these sharks to function over a wide range of light conditions. Deepsea and nocturnal sharks are remarkable for their huge, glowing, green eyes, which are designed to capture as much light as possible in the darkness of the deep ocean and cannot contract their pupils. Most sharks cannot close their eyelids to protect their eyes, but some groups have developed a third eyelid called the nictitating lower eyelid to do this. White Sharks and Whale Sharks (p. 300) lack this structure and instead protect their vulnerable eyes by rotating them by

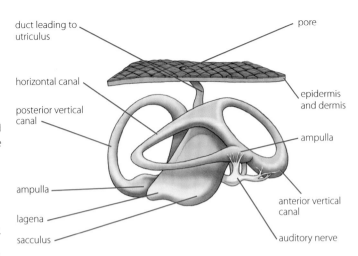

Figure 32: The ear of a shark is comprised of a cartilaginous sac-like structure (the sacculus) attached to three D-shaped canals set at right angles to each other. The canals contain hair cells and can detect acceleration in each plane (up/down, forwards/backwards, and sideways). The sacculus contains ear stones and sensory cells that are sensitive to gravity and vibrations.

180 degrees in the eye sockets to expose the tough sclera, while Hexanchid sharks have muscles that can pull the eye several centimetres back into the orbit of the cranium.

The forebrains of all vertebrates contain a pineal organ, located near the top of the head. The structure of the cranium in sharks transmits light to this organ, which is sensitive to day length and possibly even to the position of the sun; this may be an essential navigation aid.

Pressure detection in mammals is restricted to sound (changes in pressure that cause vibrations in air or water, which are detected by the ears) and touch (direct pressure detected by the skin). Sharks have an additional pressure detection sense, which is far more advanced but so totally unfamiliar to us that it is virtually impossible to find the words to describe it in everyday terms. To start with the familiar:

Hearing is the process by which sound vibrations are detected by the inner ear. The inner ears of sharks (they have no outer ears, only a small opening behind each eye) are very similar in structure to those of mammals: three semi-circular canals set at right angles to each other contain 'hair cells' embedded in gel which detect vibrations and gravity (Figure 32). They are incredibly sensitive to the vibrations caused by sound waves, particularly low-frequency signals. Many shark species can even identify the precise direction from which the sound originated, and are most strongly attracted to irregular pulses of low frequency. Sharks also learn that certain noises are associated with food – they gather at popular shark feeding sites when dive boats are heard approaching, because they expect to be fed.

Mechanosensing is the sensory category that includes touch as well as the remote hydrodynamic process (unknown in mammals) using the lateral line and associated minute sense organs. The lateral line (which is not the same as the lateral line on teleost fishes) is made up from a row of small, open pores along the side of the body leading into a water-filled canal system lined with hair cells, known as neuromasts, beneath the skin. Pressure changes in the water outside are transferred through the pores into the canal, causing the tiny

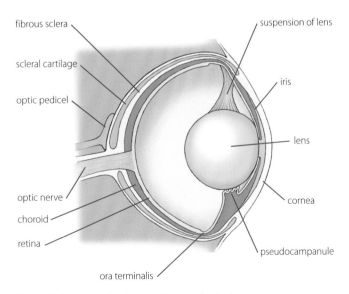

Figure 31: Lateral section through the eye of a shark.

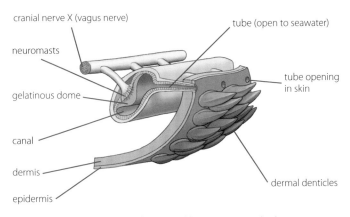

Figure 33: Close-up cutaway of the lateral line system in a shark.

hairs to move and send nerve impulses to the brain. In addition to the posterior lateral lines, which run (one on each side) from the tip of the tail along the side of the body to the top of the shark's head, there are additional series of lateral lines (rows of sense organs) around the head and mouth, and sometimes scattered elsewhere along the body, that are sealed away from the surrounding water. To further complicate matters, these systems include several different types of sensory cells. Pressure changes in the water outside the shark are transferred into this variety of sensory organs where they also cause tiny hairs inside the organs to move, sending nerve impulses to the brain. The brain can interpret these signals as nearby prey, predators or other sharks, whether received in combination with other stimuli such as sight or smell, and even in conditions when no other senses can be used. This system is certainly of far greater use than vision for those sharks with tiny eyes that live in turbid waters, or which need to detect prey moving unseen beneath their snout.

Electrosensing is another sensory category that has no equivalent in mammals. This, the most extraordinary special shark sense, detects the incredibly tiny electric fields given off by living animals and inanimate objects, and even by water moving through the earth's magnetic field.

The electrosensory system of sharks relies upon unique receptors housed within tiny jelly-filled sensory organs lined with a sensory epithelium, which are known as ampullae of Lorenzini (Figure 34). Clusters of these bulb-shaped ampullae are embedded in the surface muscle just beneath the skin, concentrated around the shark's head and mouth. One end of each ampulla joins to a canal about one millimetre wide that extends through the surface of the skin and is open, making contact with the surrounding water. The other end is attached to a nerve that connects clusters of adjacent ampullae to the lateral line nerve. These clusters pick up the electric signals produced by nearby living animals and by inanimate sources. Sharks can use these organs to detect prey at close quarters (up to 50cm or so), even when the prey is hidden. Hammerheads (p. 569), with their broad, wing-like heads, are particularly skilled at this, presumably because they have more ampullae than any other group of sharks. Using the plethora of sense organs in their enlarged heads, they can quickly and accurately locate prey entirely buried in sand, before devouring it. With an ability to detect electric fields of as little as a millionth of a volt, it is unsurprising that sharks have been known to bite seabed cables or other metal objects that produce electric fields. The Kitefin Shark (p. 210) has been identified as the culprit that has chewed upon very deep water transatlantic telephone cables. However, it is not all about finding things to eat: electro-reception is also used by the embryos of egg-laying sharks to detect approaching predators and freeze in response, and to orientate sharks within the earth's magnetic field – a very important ability when undertaking transoceanic migrations. For example, it has been discovered that Scalloped Hammerheads (p. 576) use their built-in compasses to follow 'magnetic highways' along the seabed, which link their nocturnal feeding areas and the aggregation sites where they gather by day. Some scientists are concerned that the electromagnetic fields generated by cables from offshore wind farms (essential to provide carbon-free energy and combat climate change) might disrupt these behaviours. For example, if cables are laid across shark or ray egg-laying grounds, embryos may not be able to detect the electric fields of approaching predators (p. 49), while cables across migration routes might affect seasonal movements; others point out that warming ocean temperatures will have a far more serious impact upon sharks and their prey.

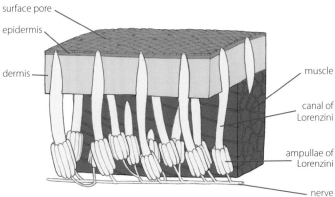

Figure 34: The snout of this Sandtiger Shark (left) is marked by several meandering lines of tiny pores; these mark the open ends of tubes leading into the ampullae of Lorenzini (see cutaway diagram on the right).

Feeding and digestion

Although sharks have a reputation for eating just about anything that fits into their mouths, this is based mainly upon the odd habits of Tiger Sharks (p. 568). These notorious consumers of garbage have been found with an extraordinary range of inedible objects in their stomachs, in addition to more conventional foods. In reality, although some sharks may have a broad diet, the majority are specialised feeders and take a relatively small range of prey animals, although their preferred range of prey species can change as the shark grows. While Bonnethead Sharks (p. 578) hunting for Blue Crab hiding in seagrass meadows swallow a substantial amount of grass and can digest about half of it, this is accidental: no shark is known to be intentionally omnivorous, and certainly none are herbivorous.

Their teeth usually provide important clues to their favoured foods and feeding methods (p. 26). Many sharks have a combination of differently shaped teeth: pointed teeth used to seize and hold their prey, flat teeth to crush invertebrates, or saw-like teeth to remove chunks or even dismember large animals when combined with vigorous shakes of the head. Shark hunting strategies mostly fall into two main groups: active searching for prey (which may include many different techniques and tactics), and a wide range of 'sit and wait', or 'lure and ambush' stratagems.

Many of the kitefin sharks, family Dalatiidae (p. 207), are highly specialist feeders. Most of these small to medium-sized sharks are poorly muscled and have small fins. However, their narrow heads and rounded snouts conceal powerful jaws and remarkable tooth sets: long, narrow and pointed in the top jaw, and large with broad, interlocked blades in the bottom jaw. This characteristic is most pronounced in the cookiecutter sharks (p. 212), which have thick lips, short snouts and large forward-facing eyes. These sharks have adopted an ectoparasitic lifestyle: they feed by cutting large plugs of flesh out of much larger, living animals, including whales, oceanic tunas and billfishes. Although never observed feeding, they are thought to latch onto their prey with their sharp upper and blade-like lower teeth, use their thick lips to seal off the area being bitten, apply suction, and then twist their bodies around while their lower teeth cut out a neat, circular plug of flesh. Even more remarkable, however, is their ability to get close enough to their prey (which include huge ocean racers) because cookiecutters do not appear to be built for speed. These little sharks have light-emitting organs and move from the dark depths to near the surface at night. Scientists speculate that they give off light as a decoy to attract prey into a painful ambush – chasing some of the fastest animals in the sea

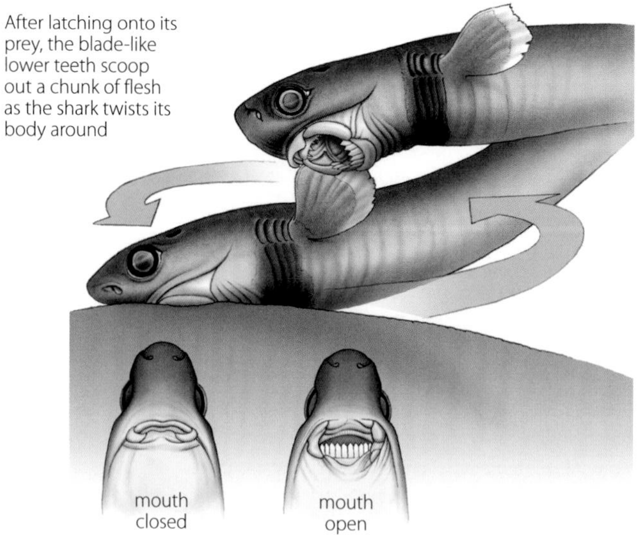

After latching onto its prey, the blade-like lower teeth scoop out a chunk of flesh as the shark twists its body around

mouth closed   mouth open

Figure 35: Cookiecutter Shark feeding.

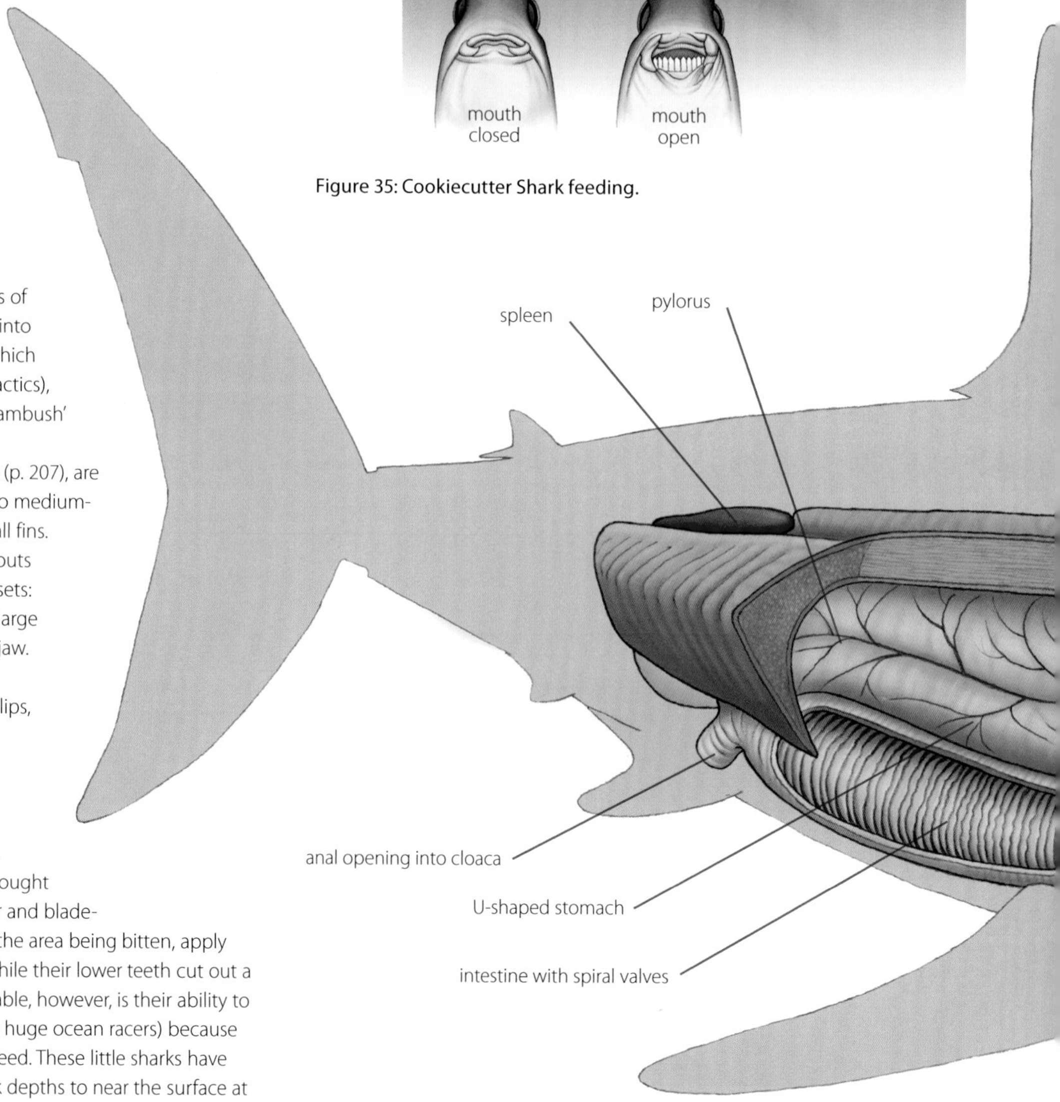

spleen

pylorus

anal opening into cloaca

U-shaped stomach

intestine with spiral valves

certainly does not seem a likely strategy for these species. Large fishes that have been caught in fishing gear must, by comparison, be sitting targets for cookiecutters: the huge tunas and billfishes landed from oceanic fishing vessels often have numerous, fresh, tell-tale circular wounds inflicted by these tiny parasites. Whatever tactics they use to get close enough to feed on living prey should not also work on inanimate objects, but these little sharks are also infamous for biting holes in the rubber hydrophone covers of nuclear submarines.

The enormous plankton-feeding Basking Shark (p. 311) has an extremely specialised diet. It is a ram-feeder: holding its enormous mouth wide open, it swims steadily through the dense blooms of planktonic crustaceans that form in highly productive, cool-temperate waters. The key to this strategy are the gill rakers: long, slender filaments, similar to baleen plates in the great whales, which efficiently sieve the water passing out through this shark's broad gill slits. Planktonic prey is trapped in these filaments and gulped down in huge mouthfuls. The numerous tiny, pointed teeth of these giant sharks are not needed for feeding, but apparently still have a function: mature male Basking Shark's teeth are worn down, presumably from holding onto the rough skin of females during mating.

This view of the open mouth of a filter-feeding Basking Shark shows its pale, branchial arches. Zooplankton are caught as seawater is sieved through the dark gill rakers and flows out through the five enormous gill openings on each side of the head (only four are visible from this angle).

Interestingly, the three plankton-feeding sharks (Whale, Basking and Megamouth Sharks, p. 308) are not closely related; all developed this lifestyle independently (as did the mobulid rays), and they have different strategies for consuming their planktonic prey. Ram-feeding would be unsuccessful in unproductive tropical waters. Instead, Whale Sharks (p. 300) use their powerful throat and gill muscles to suck in the isolated patches of zooplankton and associated small fishes and squids characteristic of tropical oceans, sometimes hanging vertically near the surface and pumping water containing concentrations of prey over their gills. They also travel long distances to locate the brief (but predictable) blooms of food produced during coral, crustacean and fish spawning events.

liver

pharynx with internal gill openings

duodenum

pancreas

oesophagus

heart

Figure 36: Internal organs (excluding the urogenital system – see p. 48) of a Shortfin Mako.

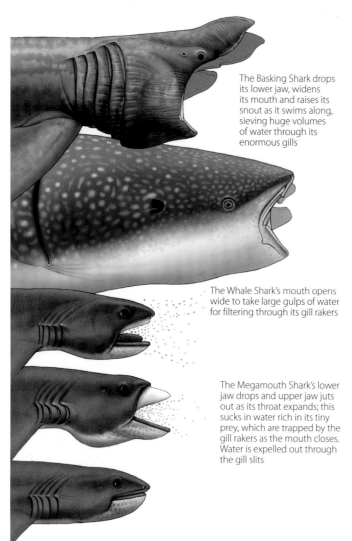

The Basking Shark drops its lower jaw, widens its mouth and raises its snout as it swims along, sieving huge volumes of water through its enormous gills

The Whale Shark's mouth opens wide to take large gulps of water for filtering through its gill rakers

The Megamouth Shark's lower jaw drops and upper jaw juts out as its throat expands; this sucks in water rich in its tiny prey, which are trapped by the gill rakers as the mouth closes. Water is expelled out through the gill slits

Figure 37: Mouth positions of a feeding Basking Shark (top), Whale Shark (centre) and Megamouth Shark (bottom). Profiles of Basking and Whale Sharks with closed mouths appear in the background.

The mysterious deepsea Megamouth Shark (p. 310), the smallest of the three, has not been observed feeding, but one animal tracked using an electronic tag followed the daily vertical migrations of a dense layer of deep water zooplankton. Like the Whale Shark, the Megamouth probably also uses its large mouth for suction feeding, and there is a theory that it has luminescent tissue in its upper jaw which attracts planktonic prey in deep, dark water.

The highly distinctive, prominent white blotches on the fin tips of Oceanic Whitetip Sharks (p. 537) make them easy to identify in photographs and when they appear in shipments of shark fin. In life, these could, however, be mistaken at a distance for bait fish, thus luring in the medium-sized, oceanic predators that are themselves eaten by Oceanic Whitetips. Divers often report

Small schools of Pilot Fish often accompany Oceanic Whitetip Sharks.

shoals of very small fishes surrounding large, slow-moving predatory sharks, including Sand Tigers and some carcharhinids. These bait fish could benefit from the association because they are too small to be of interest to the shark, but the shark's presence protects them from the medium-sized predators that would eat them. The benefit could be mutual, if the shark is able to make use of the camouflage provided by the small fish to ambush their approaching attackers.

Ambushing prey is a more common feeding strategy among shark species that live on the seabed, particularly those which lie motionless in wait for their prey to come to them. The flattened angelsharks (p. 234) are highly effective ambush predators that are perfectly camouflaged for this task, suddenly lunging forward or upwards to suck passing fishes into their cavernous mouth. They will return regularly to locations that have successfully provided them with a meal. The beautifully disguised Tasselled Wobbegong (p. 274) has a similar, but even more sophisticated, approach: it does not simply lie flat on the seabed, waiting in camouflage for passing prey, but instead slowly waves around the tip of its tail to mimic the movement of a small fish. This encourages the approach of fish similarly sized to the lure (seeking safety in numbers), or entices slightly larger predators to investigate. The only difference in outcome for the wobbegong is the size of the dinner it can grab with its powerful jaws and swallows whole, once the curious fish comes within striking range.

Figure 38: Angelshark ambush-feeding.

Other bottom-dwelling species search more actively for their meals. Sawsharks (p. 221) may use their tooth-studded rostra either to stir prey up from the seabed or slash at swimming prey. Bullhead sharks (p. 247) mostly feed on hard-shelled crustaceans, sea urchins and molluscs, which they grab with their pointed front teeth then grind up with the large, flat 'molariform' teeth of the lower jaw. Many carpetsharks (p. 259) have a relatively eel-like existence on the seabed, where they use their large pectoral fins to move around and their small mouth and large buccal cavity to create the forceful suction needed to pull invertebrates out of crevices. There is amazing film footage of packs of Whitetip Reef Sharks (p. 567) searching through complex reef habitat, even moving small boulders to locate and capture prey hiding in holes.

Moving off the bottom, many sharks are specialist feeders on small fishes or squids. They seize these with their long, pointed teeth, which are often directed back towards the throat, and swallow them whole (they do not possess the cutting teeth needed to tear apart larger items). The Viper Dogfish (p. 186) is an extreme example of this feeding strategy: it has extraordinary pointed fangs to strike at its prey as its jaws protrude and is remarkably similar in appearance to the deepsea teleost Viper Fish, whose fangs are so huge that they do not fit inside its mouth – this is an example of convergent evolution.

Cooperative feeding occurs in several species (including the Whitetip Reef Shark, mentioned above), which hunt in packs in order to herd and capture their prey. This strategy not only allows elusive animals to be cornered, but also enables small sharks to dismember prey that are very much larger than themselves. The Broadnose Sevengill Sharks that are such a great attraction to divers by day as they peacefully hang around in South African giant kelp forests are reputed to undergo a complete character change at dusk. They form hunting packs to pursue prey much larger than themselves, including pinnipeds, which they tear apart with their large saw-like teeth. Perhaps the spy-hopping reported for this species is used to check on the activity and location of pinnipeds hauled out on nearby rocks. Thresher sharks (p. 315) use their long tails to stun the shoaling fishes that they hunt, and have been observed feeding in cooperative groups. Packs of small Blacktip Reef Sharks have been filmed herding small bait fishes into very shallow water where they are much easier to catch. Some of these sharks will even chase these fishes right out of the water and follow to grab them on the sand, before wriggling back into the sea. Many of the deepsea squaloid dogfishes appear to be social and highly migratory, travelling huge distances around ocean basins in large schools. These continual migrations may partly be necessary to find new sources of food in their low productivity environment.

We conclude with the iconic species that is usually the first to spring to mind when shark feeding is discussed, and which has been studied by so many scientists: the White Shark (p. 320). This species, and other large, predatory carcharhinid sharks, can swallow small prey whole after grabbing them. It also utilises its saw-like teeth to take huge bites out of larger animals (see Figure 19) by shaking its head vigorously, or, when dealing with large potentially dangerous prey, ambush, bite and spit, leaving its victim to bleed to death so that it can return safely later to consume it. There is not room in this

Whitetip Reef Sharks eating Soldierfish, *Myripristis* sp., trapped on a coral reef at night. Cocos Island, Costa Rica.

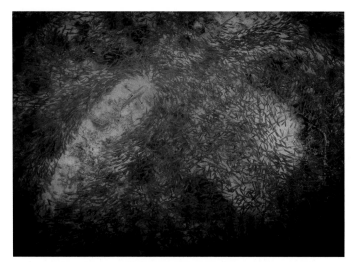

The members of this dense shoal of bait fish keep a cautious synchronised distance away from the Blacktip Reef Sharks moving through them.

book to review the huge range of strategies researchers have seen employed by White Sharks to capture their prey, including lunging from a deep water ambush towards a target swimming on the surface, or launching an attack with the sun behind them. What is clear, however, is that these sharks are able to adapt their behaviour to improve their success rates – and this is not one of the shark species with the largest brain size relative to its body!

Once the prey has been captured and swallowed, digestion can be a long business in sharks, particularly in cold-blooded species that process their food slowly. Starting at the head: food is chewed up in the mouth, if necessary, then quickly moves into the pharynx, is swallowed and enters the oesophagus. The muscular walls of the oesophagus then move it into the stomach (Figure 36). Numerous folds in the stomach wall relax to allow it to expand and accommodate arriving food.

Most shark stomachs are J-shaped. They store food and some initial digestion also occurs there, with the churning achieved by contractions of the stomach muscles helping acidic secretions and enzymes to begin breaking down the food. Inedible items may never get any further than the stomach and are, in due course, 'coughed' back up again.

Many sharks can turn their intestine inside out and evert it out of the mouth to eject any unwanted contents. This is also a fairly common response to being caught by hook and line – it looks horrible and may be mistaken for a fatal injury. However, because sharks regularly do this voluntarily, an everted

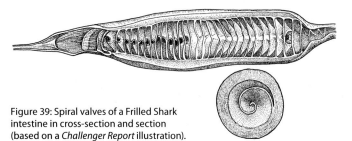

Figure 39: Spiral valves of a Frilled Shark intestine in cross-section and section (based on a *Challenger Report* illustration).

stomach is no reason to destroy a captured shark that would otherwise have been returned to the water alive.

Partly digested food passes from the stomach to the intestine, where most digestion takes place. A large intestinal surface area is essential to secrete digestive enzymes and to absorb the nutrients released by the digested food. Mammals have a very long intestinal tube to provide this large surface area but, in comparison, the intestine is extremely short in sharks. This compact package is possible because of the spiral valve, which fits multiple turns and a remarkably large internal surface area inside a short and stout section of the shark's gut. The spiral valve takes three main forms: a conicospiral valve, angled anteriorly in the intestine and resembling an auger; a ring valve, with numerous short turns resembling a stack of washers; or a scroll valve (found in the requiem and hammerhead sharks), in which the valve has partly uncoiled and resembles a rolled up scroll with one edge attached to the intestinal wall. The length of the spiral valve and the number of turns it contains depend upon the type of food typically eaten and the time needed for meals to be fully digested; some spiral valves contain only a few turns, others many dozen. Once the food has passed around and around inside the valve and has been fully digested, remaining waste products pass out of the spiral valve into the last section of the intestine, where faeces accumulate and dry out before passing into the cloaca and being discharged through the vent. Sharks may also evert their spiral valves through their cloaca, presumably to rinse out indigestible contents.

All animals have internal parasites. However, among the many fascinating discoveries by scientists studying sharks is the remarkable diversity of their intestinal parasites. Every few turns of spiral valve contains a different species of parasitic worm, specialised to feed on that particular stage of the partly digested food.

A Broadnose Sevengill Shark everting and rinsing its conicospiral valve.

TROPHIC ECOLOGY: ARE SHARKS WHAT THEY EAT – AND WHERE DID THEY EAT IT?

While studying their teeth (p. 26) provides some clues, there used to be only one way to find out what sharks eat: examining what is actually in their stomachs. Traditionally, this involved dissecting dead sharks and picking through the half-digested contents. Researchers then discovered that it is possible to pump out the stomach of a live shark; still yucky, potentially more hazardous for the scientist, but survivable for the shark. Either way, the result is just a snapshot of recent meals. With luck, fishes and invertebrates recently swallowed whole will still be recognisable, but many sludgy, lumpy and spiky bits and pieces are very hard to identify. One finding of these research efforts is that a large number of some shark species swim around with empty stomachs. In many cases that could be why they grabbed the baited hook, but the same is sometimes true for sharks caught in trawls and gillnets. Scientists therefore needed a different approach to find out how the marine food web is structured and what the shark's position is within it. That means knowing more than just the content of their last meal, but also the full range of their usual diet and where they found that food. This basic knowledge is very important for the successful management of fisheries, migratory species, ecosystems, marine protected areas, and for species conservation.

The modern way of studying tropic ecology is 'stable isotope analysis'. This only requires a tiny sample of shark tissue – such as a plug of muscle or a clipping from a fin – to be collected before the shark is quickly released back to the wild. The isotopes that can be extracted from this tissue sample are the slightly different forms of very common and abundant chemical elements, such as carbon, nitrogen and sulphur. The number of neutrons in the nucleus, which determines the weight of a molecule of a given element, varies between isotopes, and this can be measured in the laboratory. Some isotopes are stable – they do not lose their neutrons, others are radioactive (unstable) and not useful for this research. Variations in the ratio between different stable isotopes of a single element provide some really important information about the shark's general diet, incorporating the entire range of food eaten over the past few months. The ratio between different stable isotopes of carbon in the tissue provides information on what type of marine plants are at the bottom of the food web within which the individual shark was feeding (e.g. seaweed, seagrass, phytoplankton or a mixture of these). Sulphur isotope ratios differ between pelagic (e.g. open ocean) and benthic (e.g. coastal) food webs. Nitrogen isotope ratios provide information about the shark's relative position in the food web, because this particular ratio changes at every step in the food web, from the primary producers (plants) at the bottom, through herbivores (plant-eaters) at level 2, to low, middle (meso) and eventually top (apex) predators at trophic levels 3 to 5. Thus, if Species A only feeds on Species B, Species A will occupy a higher trophic level than Species B. If A feeds on the same range of species as B, they will be at the same tropic level; they will also be at a similar trophic level if A feeds on a mixture of B and other species that are lower in the food web.

Most sharks, even small ones, are meso-predators, or secondary consumers, located in the middle of the food web at levels 3–4. A very large, toothy shark that eats large marine mammals definitely looks like an apex predator. However, if it is mainly eating herbivores, such as Dugong, a trophic ecologist will tell you that it is not at the apex; it probably occupies a similar, or even lower, level than the zooplankton-feeding Basking Shark (p. 311) and invertebrate-eating Zebra and Horn Sharks (pp. 293 and 252), which are located around trophic level 3.1–3.2. This is because there are only two steps in the food chain from plants (level 1), via Dugong, to big scary-looking shark. This is fewer than the steps from plants, to grazing and filter-feeding invertebrates, to predatory crabs (some of which also scavenge dead vertebrates) to, for example, Starry Smoothhound (p. 478); this small shark is estimated to be at level 4.3. This is similar to the tiny Cookiecutter Shark (p. 212), because the latter's varied diet ranges widely up and down the food web and can include mouthfuls of apex predator, such as Sperm Whale and White Shark (p. 320). Very few sharks are truly apex predators. An adult White Shark meets the grade, at 4.5, and certainly the Broadnose Sevengill (p. 101) at 4.7, but these species would have occupied a lower trophic level as juveniles.

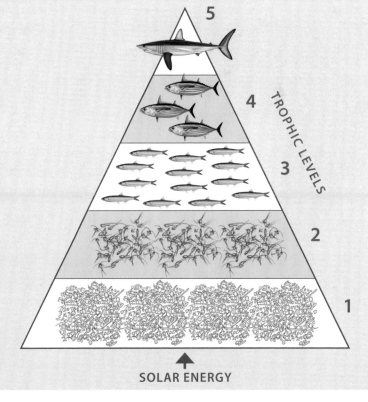

Figure 40: This trophic pyramid illustrates the five levels of a marine food web, from a broad base of primary producers (marine plants absorbing the sun's energy), to apex predators at the top.

Life history

The life history of sharks is remarkably different from that of the majority of the bony fishes. Most bony fishes reproduce by spawning: shoals of mature adults gather to release vast numbers of microscopic eggs and sperm into the sea or onto the seabed, where external fertilisation takes place. The mortality of these eggs and the tiny larvae that hatch from them is very high – only a few may survive from thousands or even millions of eggs laid. However, this still means that large numbers of offspring can, in a good year, be produced by a relatively small number of adults, and the young may also reach maturity and start breeding at a fairly early age. (Exceptions are the small minority of bony fishes that give birth to live young, guard their broods, or even protect them by carrying them around in their mouths or pouches… fascinating animals, but this book is about sharks.)

In contrast, shark life history and reproductive strategies have more in common with those of birds and mammals than with the majority of bony fishes: sharks produce only a small number of relatively large young, which stand a better chance of surviving to adulthood (see p. 52). Unlike birds and mammals, however, sharks produce fully-developed young that do not need any maternal care after birth. Furthermore, shark life history strategies are far more varied and, in many cases, even more advanced than those of other vertebrates. Sharks and their relatives have, after all, spent hundreds of millions of years developing internal fertilisation and live birth, encompassing a remark-ably wide variety of methods to nourish their embryos. Placental reproduction was well established in sharks long before mammals came on the scene.

Mating and fertilisation

Reproduction takes many forms in sharks, but all require internal fertilisation, usually following a period of courtship to obtain the cooperation of the female. Mating has not been observed in many shark species, but in larger sharks the male bites the female to hold her alongside, while using the closest of his paired claspers (these are grooved extensions of the pelvic fins – Figure 42) to insert packages of sperm into the female reproductive tract. Smaller species twine around each other. Female shark reproductive organs

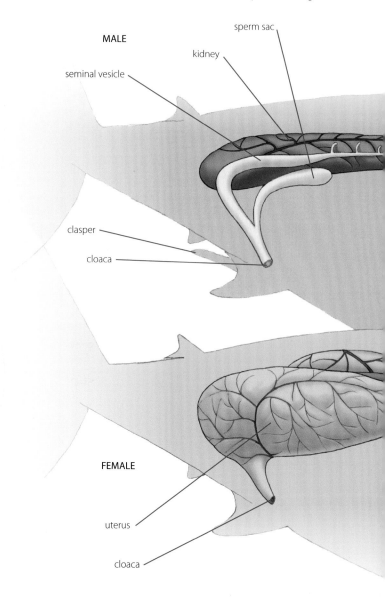

Figure 42: Reproductive organs in male and female Shortfin Makos.

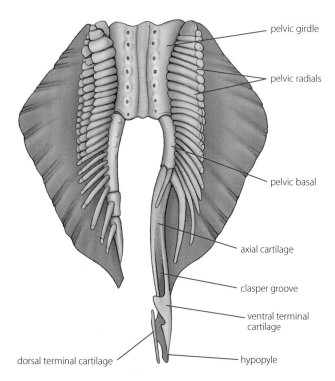

Figure 41: Cartilage skeleton of a male shark's pelvic fin and clasper.

are also paired (Figure 42): there are two ovaries supplying the two uteri (the plural of uterus), which join the single cloaca. After internal fertilisation of the large, yolky eggs has taken place (often by more than one male – see p. 53), a tough protective covering is laid down. Strategies then diverge remarkably in different species, ranging from oviparity, when the eggs are laid and hatch outside the mother, to various forms of viviparity, in which the females give birth to fully developed live young.

Recent studies have concluded that live birth (viviparity) was the ancestral form of reproduction in chondrichthyans, with several taxonomic groups subsequently adopting egg-laying (oviparity), and some of these later reverting back to viviparity. The most successful strategies are those that produce the optimum number of adults in future generations – but this can be achieved in several different ways.

Egg-laying (oviparity)

About 40% of all shark species, including the large families Pentanchidae and Scyliorhinidae, are oviparous, often laying their eggs almost immediately after fertilisation and deposition of the protective eggcase. Eggs are generally produced in pairs (one from each ovary and uterus) at daily or weekly intervals. Spiky horns or curly tendrils on the corners of the eggcase are used to anchor it carefully onto the seabed, or to entangle it around tough seaweeds or invertebrate colonies, such as seafans. Hornshark eggcases (p. 247) are unmistakable because they have a unique screw-shaped design, sometimes also with tendrils or horns. In some species, as soon as these eggs are laid, and before the outer casing hardens, the mother has been observed carefully picking them up with her mouth and wedging them into a rock crevice, pointed end down, so that they are not washed away. This ensures that the egg remains in relative safety in the nursery ground that she selected.

The advantage of this 'single' or 'immediate' oviparity, is that the mother can be small and mature at a young age because each egg takes up very little space inside the uterus, and it does not remain there for long. She may therefore continue to lay pairs of eggs over a long reproductive season, and potentially produce much larger numbers of young than live-bearing sharks. The disadvantage, however, is the length of time that each pup takes to develop in its eggcase, slowly absorbing the yolk-sac, until it is ready to hatch as a miniature version of the adults. This process may take over a year, during which time the eggs (however well camouflaged they may be) are exposed to seabed predators. This becomes particularly risky towards the end of development, when slits appear in the eggcase allowing seawater to enter and the embryo starts using its tail to flush the capsule and pumping water over its gills. The scent released, and these movements of the embryo, are likely to attract predators. The embryo's only defence against detection is to freeze if it perceives the electric fields of a potentially threatening animal approaching.

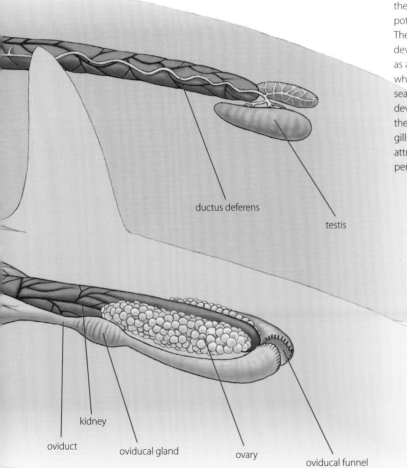

ductus deferens

testis

kidney

oviduct

oviducal gland

ovary

oviducal funnel

Mating scars above the left pectoral fin of a female Tiger Shark.

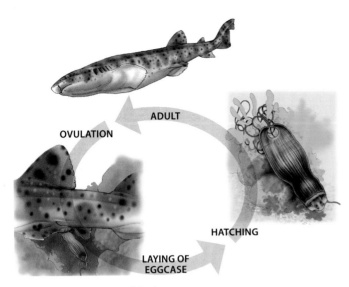

Figure 43: Oviparity in a swellshark.

Many oviparous sharks therefore retain their fertilised eggs for several months, sometimes laying them only a few weeks before hatching, thus reducing the risk of egg predation and increasing survival rates. This strategy is called 'retained' or 'delayed' oviparity, because up to eight eggs (depending upon species) are retained in each uterus. These eggcases are often characterised by thinner, shorter tendrils and they spend less time on the seabed before hatching. Females that retain several eggs (multiple oviparity) require larger bodies to accommodate their eggs than sharks that only retain one (single oviparity); the former usually take longer to mature than the latter, which can start reproducing at a younger age. The Sarawak Swellshark (p. 423) only produces one egg in each uterus, but retains them until both embryos are very large (perhaps as much as 80% of their size at hatching). She becomes very fat; two embryos, each about one quarter of her own body length, occupy most of her abdominal cavity, but she is still able to reproduce at a young age. The unique characteristic of this species' eggcase is that it is completely transparent, showing the embryo inside. However, by the time it is laid, the embryo has developed a striking spotted pattern that is presumably an effective camouflage, even though the pup is fully visible inside its glassy capsule. Regardless of how long the eggs spend in or outside the mother, newly hatched oviparous pups are very small; only a few will escape predation and reach adulthood, unless they mature very quickly.

Figure 44: A hornshark eggcase (left) and catshark eggcases (right).

Live birth (viviparity)

Most shark species (60%) are viviparous. This appears to be the ancestral strategy for sharks and rays (although not all scientists agree). Oviparity subsequently developed because it enabled small-bodied species to produce larger numbers of young more rapidly. Subsequently, some taxa reverted back to viviparity, including a few of the predominantly egg-laying catsharks. For example, most sawtailed sharks, *Galeus* species, are oviparous (lay eggs), but the very closely related African Sawtail Catshark (p. 383) retains its eggs in the uterus until after the pups hatch out. Some researchers classify as oviparous all species whose eggs had a thick protective case at some stage during their development, including those which hatch inside their mother before birth – it is a continuum, but here we use the simplest definition of viviparous species: those in which the mother gives birth to fully developed young.

About 40% of shark species, including the Lollipop Catshark (p. 371), Piked Dogfish (p. 117) and the Whale Shark (p. 300), have 'yolk-sac viviparity'. In these species, the single egg yolk is the only source of food for the embryos. The Lollipop Catshark gives birth to just one pup per uterus – more than twins would be physically impossible in one of the world's smallest sharks. The slightly larger African Sawtail Shark has 5–13 pups; Piked Dogfish, several times larger, can give birth to over 30 (bigger mothers have more pups); and the only pregnant Whale Shark recorded had about 300 small pups and eggs at various stages of development. Other species with yolk-sac viviparity include the frilled and cow sharks, the largely deepsea Squaliformes – dogfish, gulper, lantern and sleeper sharks, sawsharks and angelsharks. Some of the

Courtship in Nurse Sharks: the male (left) uses its mouth to latch on to the female by her left pectoral fin.

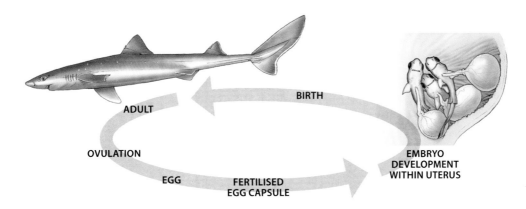

Figure 45: Yolk-sac viviparity in the Piked Dogfish.

Squaliforme sharks have among the longest gestation periods known in vertebrates (over two years) but, after all that time, may only give birth to one or two pups.

A pup that is nourished solely from a yolk-sac is inevitably born some 20–25% lighter than the original weight of the egg. That is because there is never a perfect conversion from food source to weight gained by an animal – some of the nourishment must be converted to energy needed by the growing pup, and to various waste products. Predictably, therefore, a Gulper Shark pup (p. 140) is born at about 80% of the original weight of the egg – not a bad conversion rate for a two-year pregnancy. A neonate Great Lanternshark pup (p. 183), however, has only lost about 8% of the weight of its original egg. This must be one of many taxa that have developed a way to increase the amount of food made available to the young inside the female, so that they are much larger and better developed at birth and more likely to survive the first few dangerous weeks of life in the ocean. There are several strategies for achieving this. It is likely that the Great Lanternshark is one of many species that employ 'histrophy': they secrete nutritious mucous (histotroph) through the wall of the uterus. Although less in the Lanternshark, other taxa produce so much histotroph through outgrowths of the uterus wall that a pup can be born up to 350% heavier than its original egg.

Lamnid sharks (p. 319) and a few carcharhinid and orectolobid sharks produce a great many infertile eggs that are steadily released from the ovaries to feed the growing young inside the uteri – a process known as 'oophagy'. Some species (e.g. the Pelagic Thresher, p. 318) produce only one fertile egg from each ovary; all other eggs are infertile and destined to feed the single pup that develops in each uterus, until the female gives birth to large twins. Other species produce several fertilised eggs as well as the infertile eggs, and give birth to fairly large litters.

The Sandtiger Shark (p. 305) is notorious for 'intrauterine cannibalism' or 'embryophagy'. Its pups do not just eat infertile eggs inside the uteri, they will also eat the other developing pups (their brothers and sisters) until just one pup survives in each uterus. Researchers have discovered that although female Sandtigers will mate with several different males, the last two surviving pups often have the same father. Presumably the eggs fertilised during the first mating hatch first and are therefore older, larger and more dangerous than their younger half-siblings. This also explains why male Sandtigers guard

females from other males after mating. The would-be fathers are (albeit unknowingly) giving their own pups a head-start over those from later encounters, thus increasing the chance that their own offspring will survive.

The most advanced form of reproduction, used by about 18% of living sharks including the Bull Shark (p. 535), Blue Shark (p. 562) and the hammerheads (p. 501), is placental viviparity. This is believed to have evolved independently up to 20 times in sharks and rays and is fairly variable in form. In these species, the yolk-sac provides the initial source of nutrition, until exhausted. It then becomes attached to the wall of the uterus, forming a placenta, while the yolk stalk (the connection between the yolk-sac and the embryo) turns into a placental cord. Unlike mammals, the cord doesn't directly

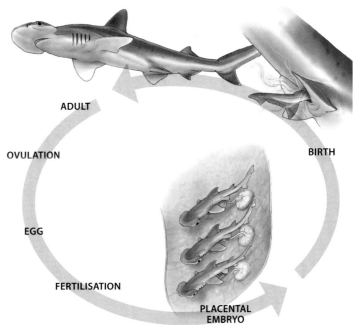

Figure 46: Viviparity in a Bonnethead Shark.

transfer nutrients from the mother to the pup through its blood stream. Instead, the mother feeds her embryos through uterine secretions. Large litters can be nourished in this way (up to 135 pups have been recorded for a Blue Shark). This strategy is very similar to mammalian reproduction, but it developed in sharks long before mammals evolved.

The Tiger Shark (p. 568) has recently been removed from family Carcharhinidae and assigned its own family partly because it does not use placental viviparity; this is the only known shark with 'embryotrophy'. Unborn Tiger Shark pups are enclosed within sacs filled with fluid, or embryotroph, which provides them with so much nutrition that they can increase their body weight by over 1000% during their gestation. Furthermore, Tiger Sharks can give birth to 45–60 pups in a single litter, which is among the largest litter sizes known in sharks.

Some of these viviparous reproductive strategies require such a huge investment on the part of the female that several species are unable to reproduce annually. The mothers need to take one or two 'resting years' to recover and rebuild their liver's energy stores.

Virgin birth (parthenogenesis)

Sharks have one final reproductive strategy: parthenogenesis, or 'virgin birth'. Various theories had been suggested to explain the mysterious births of pups, or fertile eggs produced by female sharks that have been kept for years in aquaria without encountering a male. Genetic research into the Zebra Shark (p. 293), Blacktip Shark (p. 536), Bonnethead Shark (p. 578) and Whitespotted Bambooshark (p. 287), among others, has now confirmed that these really were virgin births, with no contribution of paternal DNA. However, the pups (which are always female) are not identical to their mother; they are 'half clones'. The nucleus of an unfertilised egg cell contains only half of the mother's genetic material, in the form of a set of unduplicated strands of chromosomes, or chromatids. The egg cannot start to develop until a matching set of chromatids has fused with this nuclear DNA to produce duplicated chromosomes. The second set of chromatids normally comes from a sperm cell (the paternal DNA). In parthenogenesis, however, an identical set of chromatids from a 'polar cell' (a normally inactive by-product of the cell division process that produced the egg) fuses with the chromatids in the nucleus. Because the pup's nuclear DNA is formed from two identical sets of chromatids, she only carries half of the genetic variation present in her mother. Any pups that inherit a deleterious mutation in one chromatid will not survive, because the mutation will not be masked by a 'good' copy of the faulty gene in the second chromatid. Parthenogenetic offspring that live, however, are actually purged of any lethal recessive alleles (alternative forms of genes) that may be lurking in the genome (the set of genetic material) of their mother and so may be fitter than her sexual offspring, which are likely to carry these alleles. The type of parthenogenesis observed in sharks in captivity is known to occur in wild snakes, so its discovery in wild sharks may only be a matter of time.

This Tiger Shark pup and its several dozen siblings were nourished before birth by a rich fluid called embryotroph. The mottled camouflage patterns on its skin will change into stripes as it matures.

LOW COST *VS* HIGH COST REPRODUCTION

Most species of bony fish have an 'r-selected' reproductive strategy, which is characterised by the production of numerous, small 'low-cost' young, few of which survive, and a fairly short interval between birth and maturity. This results in a high rebound potential, meaning that populations can recover rapidly from depletion. Their short life cycles also make them relatively adaptable to changing environmental conditions (for example, fishing pressure may result in stocks becoming dominated by smaller individuals that reproduce at an earlier age). In contrast, sharks are 'K-selected': they produce only a small number of relatively large 'high cost' young, which stand a better chance of surviving to adulthood. A K-selected reproductive strategy, which includes the production of large offspring, requires a relatively large body size; even the smallest sharks are very much bigger than the smallest bony fishes. Furthermore, most sharks grow slowly; it may take many years until they are large enough to reproduce (particularly for species from the cold, low-energy deep ocean environment, where food is scarce). Producing large young also takes time: the gestation (length of pregnancy) in some species takes several years, longer than any mammal. Overall, therefore, sharks are long-lived, slow-growing animals that produce only a relatively small number of pups during their lifetime. This is a good strategy for large apex predators, which have very few or no natural enemies and need only produce a small number of young to maintain a stable population. Alas, the recent development of large-scale fisheries now threatens many of these species. Long-lived, slow-growing sharks can't suddenly (within one or two lifetimes) start growing and maturing faster, or producing more and smaller young. This is why populations of some of the largest shark species are in steep decline. In contrast, those smaller sharks that grow and mature relatively quickly are better able to withstand exploitation by fisheries. Intriguingly, every major taxonomic group contains species that lie at different locations along the spectrum of r- to K-selected life history strategies.

A Lemon Shark pup swims away, moments after its birth in a shallow nursery lagoon. The umbilical cord and remains of its yolk-sac placenta trailing behind are about to drop off, then the tiny umbilical scar left between its pectoral fins will heal and vanish.

Multiple paternity

The opposite of 'no dad' (above) might be 'many dads'. The first record of multiple paternity in a single litter of sharks, made possible through advances in genetic analysis, caused considerable excitement in the scientific world. Today, however, there are numerous papers on the subject. It has been recorded in Shortfin Mako, Sandbar Shark, Dusky Shark, Nurse Shark, Leopard Shark, Lemon Shark, two hammerheads, several species of spiny dogfish and smoothhounds, a catshark, and probably many other species. Polyandry (mating with more than one male) has been observed in several shark species. It is also possible for females to store sperm in oviducal glands and use this to produce viable pups for several years, so the widespread incidence of multiple paternity (half siblings) in a single litter should not be surprising. Indeed, one of the most unusual scientific papers on this subject describes NOT finding evidence of multiple paternity in 112 pups from four monogamous, pregnant Tiger Sharks.

Hermaphrodites

To further complicate the picture, a few cases of hermaphroditism in sharks are recorded. These include infertile hermaphrodites of the Brown Lanternshark (p. 180), and fertile hermaphrodites in some deep water catsharks, *Apristurus* spp. The former was attributed to pollution, but the latter… well, these rare species apparently have small population sizes. If *Apristurus* do not often encounter each other in the deep ocean, being hermaphroditic could double their chances of breeding successfully.

Age and longevity

Although we have given examples of a few sharks whose reproductive strategies are known, the taxonomic sections of this guide illustrate just how little we know about most species' reproduction. We also do not know the age at which the majority of sharks mature, their growth rates, or how long they can live. Bony fishes contain otoliths (ear bones) which deposit a new layer each year and can relatively easily be sectioned and counted using a microscope (rather like counting growth rings in tree trunks). Sharks, however,

do not have otoliths. Although their vertebrae and fin spines may contain distinct growth rings, it can be hard to tell whether these were laid down annually, seasonally, or more or less frequently than that. Fin spines may also be abraded and difficult to analyse (particularly for very old animals). Because sharks kept in aquaria are fed freely and not exposed to seasonal changes, growth rates are different from those in the wild. Tag and release programmes, as well as the examination of landed sharks by researchers can provide vital information about natural growth and maturation rates. These are fairly well known in the fastest growing, rapidly maturing and fecund coastal sharks, as well as in other species that are commercially important. It is far more difficult to study rare species or those whose growth slows with age, particularly sharks that live in the deep ocean. Researchers are now finding ways to overcome these challenges with the use of bomb carbon dating of shark vertebrae and eye lenses, confirming that some slow-growing sharks may not reach maturity until they are over 40 years old, and can live for well over 100 years.

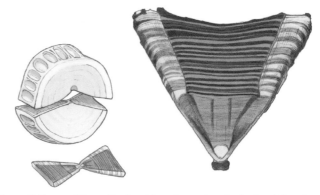

Figure 47: If a very thin section is cut from the centre of a shark vertebra, the growth rings can be counted under a microscope. Unfortunately, these rings are not always laid down annually; scientists need to find out how many rings appear each year before the shark's age can be calculated – harmless chemical markers such as tetracycline can be used for this (see pp. 58 and 62).

Behaviour

A lot of shark behaviour is innate – it is genetically-programmed and can be performed in response to a cue without the shark having to learn the actions. For example, newborn pups (and unborn Sandtiger pups (p. 305)) can sense and capture prey. Those pups that are not eaten on the day that they are born (or even before they hatch – p. 51) have applied certain innate behaviours to avoid predation. Sharks that undertake long-distance migrations for the first time also display innate behaviour. Even more remarkable than this, however, is the capacity for sharks to hone these innate skills, and to learn new behaviours. Cognition is the process of acquiring knowledge and understanding by thinking, experience and through the senses – as well as remembering what has been learnt and using this information for judging and problem-solving. The high cognitive abilities of numerous shark species are reflected in their remarkable range of behaviours and have been the subject of many research studies in recent years.

Many people asked to identify an example of shark behaviour will focus on well-known, even notorious, predatory behaviours, such as the breaching of a White Shark (p. 320), launching itself at surface-swimming prey from deep water, the ability of sharks to follow faint scent trails over long distances, and perhaps the cunning ambushes of angelsharks (p. 227) and wobbegongs (p. 267). At least during their earliest life stages, however, all sharks urgently need to engage in behaviour that helps them to avoid being eaten by larger predators (potentially including their parents), and most spend their entire life doing so. While eating and not being eaten have to be the top priorities for any shark, their ultimate life goal is reproduction. Surviving for long enough to

achieve this is made possible by many fascinating behaviours, including navigation and often complex social interactions.

Aggregating behaviour has been observed at all life stages in various shark species. Newborn and juvenile sharks tend to aggregate in certain sheltered habitats, such as mangroves and shallow lagoons, because these provide a refuge from larger predators (particularly other sharks) and a reliable source of food. Juvenile and subadult sharks often congregate in groups, sometimes segregated by sex, adult males are usually found together, and mature females may divide into schools of pregnant and non-pregnant animals. Blue Sharks (p. 562) and the Piked and North Pacific Spiny Dogfishes (pp. 117 & 118) spend most of their adult lives in single sex groups that are also segregated by age. Once they reach maturity, they may meet with members of the opposite sex just once a year to mate. Large sharks may aggregate seasonally to take advantage of particularly rich sources of food, or while migrating. For example, Bronze Whalers (p. 527) aggregate when they follow and prey upon the sardine run along the coast of South Africa. Basking and Whale Sharks aggregate when filter-feeding upon particularly rich sources of food – copepod zooplankton concentrated along temperate marine frontal systems for the former, and pelagic fish eggs following mass spawning events for the latter. Many of the predictable, seasonal aggregations of Whale Sharks so popular with divers and snorkellers are comprised almost entirely of immature males (p. 300). Biologists immediately question what happens to these individuals when they mature and do not return, where the immature females are, and how do the mature males and females meet.

Tiger Sharks migrate huge distances to arrive at the remote northwest Hawaiian Islands just in time for the first training flights (and crash-landings) of Laysan Albatross fledglings – a valuable seasonal food source for this Tiger Shark population.

Researchers using shark pens to study the behaviour and personality traits of juvenile Lemon Sharks born in a sheltered lagoon at Bimini, Bahamas.

Hundreds of Scalloped Hammerheads, predominantly females, aggregate in schools near seamounts during the day, then disperse to feed at night. White Sharks and Tiger Sharks aggregate at certain times of year to feed upon rich seasonal sources of prey in very specific locations – the former at seal colonies while juveniles are learning to swim, and the latter at albatross nesting sites on islands just as the fledglings are learning to fly. There are clear hierarchies apparent in some of these aggregations, with the largest animals dominant over smaller individuals within a species. Where aggregations contain more than one species, there is also a hierarchy of species dominance.

A relatively new behavioural research theme is the study of social groups in sharks. Social groups are not the same as aggregations. Aggregations are caused by external influences and conditions. Social groups are formed by interactions at individual level. They are difficult to study in large sharks with extensive ranges, so most research has been focused on smaller, less mobile species such as juvenile Lemon Sharks and Smallspotted Catsharks. The results have shown that individuals in both species are more likely to be found in the company of other particular sharks – they 'prefer' each other's company – or at least are more likely to associate with familiar rather than unfamiliar animals. Social links usually occur with sharks of a similar size, and also with genetically related sharks. The largest individual in any social group is more likely to lead it.

Linked with the above work is a new field of investigation: that of shark personalities, particularly their sociability, activity and aggressiveness, and traits of shyness/boldness and exploration/avoidance. In one recent study, for example, some Smallspotted Catsharks were found to be relatively solitary; they spent a lot of time hiding alone, camouflaged on the seabed, while others in the same study were gregarious and literally piled up in large obvious groups. Only half of all Lemon Sharks born in the lagoon at Bimini, Bahamas,

survive their first year of life. Most are very cautious, spending a lot of time hiding in mangrove forests and only venturing out to feed when the tide is low and larger sharks, unable to reach them. These pups share a relatively limited food resource and grow slowly. Other pups are consistently more adventurous, taking excursions into new habitats and deeper water where they may find more prey and grow faster, but the cost is a greater immediate risk of predation. Trade-offs are always involved. However, because not all shark pups behave the same, the population as a whole is more likely to survive environmental changes and alterations in their habitat that affect food sources and risk of predation.

Another important aspect of social behaviour in sharks is their ability to learn from each other. Sharks are, individually, very good at learning how to perform tasks that result in a reward, such as distinguishing between the shapes or movements of targets that will (or will not) produce a food reward, or navigating a maze. Individuals can remember these tasks and repeat their performance many months later. It is unsurprising that wild sharks regularly fed at a particular location quickly learn to associate the noise of a boat engine with their next meal and turn up for the benefit of tourists at a shark feeding site – or disconcertingly, many months later, if a boat full of anglers anchors in the same place. More interestingly, however, a naive (untrained) shark will begin to perform tasks more rapidly in the company of an experienced shark than if it is learning alone. The same is true for groups of naive sharks, which are slower to solve a puzzle than mixed groups of experienced and naive sharks. The exception is when naive sharks have been kept in an adjacent holding area that allows them to see experienced sharks completing a task and winning their reward. In this situation, naive sharks learn more rapidly than if they had not previously seen how the task was done. Sharks learn by watching others.

Migrations

The migratory behaviour of sharks is another fascinating area of research, greatly aided not only by the development of highly sophisticated electronic tagging technology and expanding networks of satellites and marine listening stations to monitor signals from tagged sharks (p. 58), but also molecular genetics (p. 60). These powerful new scientific tools have revolutionised this field of research. As a result, some shark migration patterns are known to be even more complex than those of birds, with different long-distance journeys carried out by mature and immature animals, males and females. Migrations may regularly cross ocean basins and, even more interestingly, are not always annual. Tagging studies have demonstrated that many species of shark with a two year reproductive cycle also exhibit a two year migration pattern, only visiting their pupping grounds every other year in order to give birth, while pregnant females may occupy different areas from those that are not pregnant, but recovering between litters or visiting mating grounds.

As an example of how scientific knowledge progresses, initial genetic studies of White Sharks, twenty years ago, discovered that mature females were philopatric: they tended to give birth near where they had been born. Using genetic analyses (see p. 60), researchers could even determine which region a female shark came from. In contrast, there were no clear regional differences in the paternal contribution to litters. An obvious explanation was that adult males are highly migratory, roving long distances to breed with relatively sedentary females. Satellite tagging soon refuted that hypothesis, when a well-known, mature female White Shark named Nicole was tagged off the coast of South Africa, where she had been recorded several times over the previous four years. Four months later, on schedule, her satellite tag was released and popped up – off the coast of Western Australia! Geolocation data collected by the tag indicated that she swam 11,000km in 99 days, following a relatively straight route across the Indian Ocean. This was the first confirmed transocean movement of a White Shark. However, it was not a one-off. Nicole's distinctive dorsal fin was recognised again, back on the coast of South Africa, six months later, demonstrating that there is, of course, more than one way to move male genetic material around the world. More recent data from White Sharks tagged in New Zealand (Figure 48) have confirmed similar very long distance transocean migrations in the West Pacific and very clear philopatry in mature females.

Every year sees an increase in our awareness of the complexity of the behaviour of these intelligent animals. More and more shark species are discovered to be social, living in large groups and hunting their prey in packs. Even predominantly solitary sharks (which meet only occasionally to breed, pup, or to aggregate on rich hunting grounds) exhibit a complex system of signals and behaviours that minimise potentially dangerous conflict between individuals and structure their social interactions. Genetic analyses combined with the results from electronic tracking projects have helped scientists to assess the extent to which different populations mingle or are isolated from each other, and whether the long-distance movements of, and genetic exchange between, populations are made by males or females, mature or immature sharks. The more scientists find out, the more questions arise.

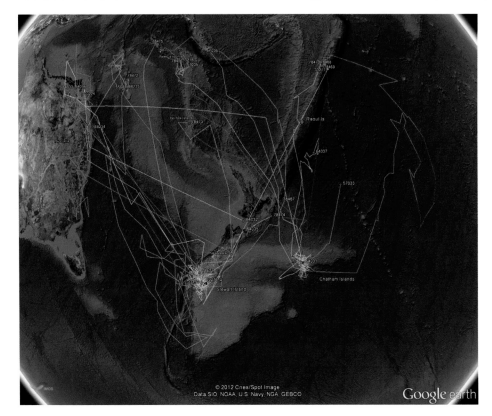

Figure 48: Electronic tagging of White Sharks at two important aggregation sites in New Zealand waters has enabled scientists to discover formerly unimaginable seasonal migrations. These include long-distance journeys to spend the winter in tropical waters up to 3,300km away. New Zealand White Sharks have been tracked to locations off Fiji, the Australian Great Barrier Reef, New Caledonia and Tonga. One female tracked for two years returned each winter to the same area of the Coral Sea and has been sighted four summers running off Stewart Island (south of South Island, New Zealand). Intriguingly, White Sharks tagged off the Chatham Islands have never been tracked to Stewart Island, and vice versa. www.niwa.co.nz/our-science/oceans/research-projects/all/white-sharks (*Malcolm Francis (National Institute of Water and Atmospheric Research) and Clinton Duffy (Department of Conservation)*).

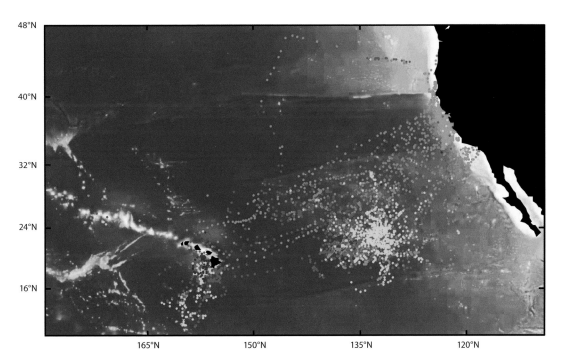

The first edition of this book featured a research project that was still underway: the study that had identified the mid Pacific 'White Shark Café'. Electronic pop-up archival transmitting (PAT) tags, attached to the dorsal fins of White Sharks off the North American coast, had enabled researchers to discover this area and to describe the behaviour of the White Sharks that visit it. What we did not know, almost ten years ago, was why this area was so popular with White Sharks, and what they were doing there. Figure 42, which plots median daily position and behaviour (colour-coded) from 53 White Shark tracks (taken from a 2012 paper), illustrates this story.

The Café is about halfway between the autumn to winter coastal feeding grounds of North America (orange), which the sharks leave in winter, and the warm waters near the island of Hawaii where (unknown to the majority of human residents and visitors) many White Sharks spend part of the summer (May to July) foraging. Intriguingly, it was initially hard for researchers to explain why this area of open ocean appears to be so attractive to White Sharks. There appeared not to be much on the menu at the Café, with satellite imagery of the area indicating that, at least on the surface, it is almost a deep water desert. It therefore seemed unlikely to be important for feeding. Nonetheless, White Sharks chose to swim 2,000km in about 100 days (mostly travelling just below the surface, with only occasional dives – green dots) to reach it, then hang out there for months on end. Shark foraging away from the coast usually involves hunting at a depth of between 350–500m by day, and around 200m at night (this is known as diurnal vertical migration, shown in pink). Once large numbers of sharks have gathered at the Café, however, their behaviour changes completely. Males arrive first, followed by females. The males started to undertake numerous, repetitive and rapid vertical dives (yellow), from the surface to about 500m and back, sometimes up to 150 times in 24 hours. This very unusual behaviour occurs in a relatively small area right in the centre of the Café, marked in Figure 49 by a very dense yellow patch.

In contrast, the smaller number of females (only half the tagged females turn up) tend to move in and out of the edges of the male aggregation and dive less frequently, mainly during daylight hours. After a few months of this characteristic Café behaviour, the males abruptly discontinued this activity and the sharks departed (green and pink tracks), before returning to their mainland feeding grounds at the end of the summer.

The researchers originally suggested that this could be the first known example of 'lekking' in sharks. Lekking (groups of males taking part in competitive displays to attract females to mate) is usually associated with breeding birds, but mammals, insects and some reptiles also lek. It is entirely reasonable to believe that large-brained social animals like the White Shark would lek to attract mates, and that the White Shark Café might have been the hitherto unknown White Shark mating ground. More recent research, however, assisted by the use of high-tech remotely operated vehicles (ROVs) to track tagged sharks and investigate their habitat, discovered that the Café is a very important feeding ground. The reason it is invisible from satellites is because the foraging zone is so far below the surface. The White Sharks visit this area to feed in the highly productive midwater zone, which contains abundant deepsea fishes and squids that live below the depths reached by sunlight (about 400m during the day and 200m at night). It is still unclear, however, why males and females undertake such different diving patterns – and of course there is no reason why this might not also be a White Shark mating area. Some other shark species which aggregate to feed on abundant prey will certainly take the opportunity to mate in the same location. As for the other half of the female White Sharks: well, their gestation (pregnancy) may last up to 18 months; females that mated the previous summer will still be heavily pregnant one year later and have no need to mate again so soon. They may also prefer to avoid such strenuous diving activity and grow their pups in warmer, shallow water elsewhere.

TAGGING AND TRACING SHARKS

Traditional shark tagging programmes utilise huge numbers of inexpensive numbered visual tags, often relying heavily upon the voluntary help of sports anglers and commercial fishermen. Today, electronic tags are providing far more detailed information on the movement and activities of individual sharks. These sophisticated tools have revolutionised our knowledge of shark life history, behaviour and migration.

Marker tags (numbered plastic discs, streamer or 'spaghetti' tags) rely upon the tag number and other details (e.g. species, sex and length of shark) being reported when recovered. Huge numbers have been deployed over the years, contributing significantly to age and growth studies, and information on movement from tagged to recovery location. They may also indicate when a shark has been injected with a chemical marker to study growth (see p. 53), or contains an internal archival tag recording its location, movements and behaviour over time. For example, an internal accelerometer records when prey has been chased, and changes in internal temperature indicate that a meal has been swallowed.

Passive Integrated Transponder (PIT) tags contain a tiny antenna connected to a coded chip, completely encased in a glass capsule. They are placed under the skin with a hypodermic needle or a manual insertion gun. When the antenna detects a radio signal from a portable reader, held a few inches away, the chip returns its identification code – no battery is necessary. They should last for 75 years or more.

Satellite tags may report in real time or collect and archive data for later transmission. **SPOT** (Smart Position Transmitting) tags track horizontal migrations by reporting to satellite whenever the shark surfaces. **PAT** (Pop-up Archival Transmitting) tags are used on sharks that do not spend a lot of time at the surface. They can store depth, temperature and light data several times a minute for up to two years. An electric current is switched on after a predetermined interval to corrode a metal attachment and release the buoyant tag. Downloaded light and sea surface temperature data identify the shark's position over time, while depth and water temperature provide information on its diving behaviour and oceanic habitat.

Acoustic tags also track movements. They emit coded sound signals (to identify individual animals) which are picked up by receivers. During active tracking, the tagged shark is followed using a directional hydrophone so that the researcher can collect environmental information (e.g. depth, water temperature) in real time. Receivers are also deployed in seabed networks (which can provide precise locations by triangulating signals picked up by several receivers), on oceanic buoys and even in self-propelled drone vessels capable of crossing major oceans and shadowing tagged animals. These either transmit the signal back to the researcher, or store it for later download and analysis.

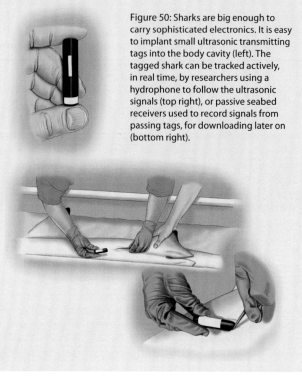

Figure 50: Sharks are big enough to carry sophisticated electronics. It is easy to implant small ultrasonic transmitting tags into the body cavity (left). The tagged shark can be tracked actively, in real time, by researchers using a hydrophone to follow the ultrasonic signals (top right), or passive seabed receivers used to record signals from passing tags, for downloading later on (bottom right).

BRUVS

The use of baited remote underwater video systems (BRUVS) to survey and monitor shark populations has become widespread in recent years. It is a relatively cheap and easy survey technique, requiring a simple weighted frame with an underwater camera pointing at the attached bait container. The frames are left on the seabed or suspended in mid-water for the camera to record species attracted by the bait. BRUVS are fishery-independent, non-destructive, not constrained by scuba diving depths (which limit the use of traditional underwater visual census), and produce a standardised, repeatable overview of predators occurring in otherwise inaccessible survey areas. In addition to identifying and counting sharks and other fishes, they also can be used to study the behaviour and interactions of fishes in the absence of divers, who might otherwise change the natural behaviour of the study animals. Scientists do, however, need to collect a lot of footage from many BRUVS, using the same methodology, before these surveys can provide sufficiently robust data for analysis.

The largest single standardised BRUVS survey of shark populations, Global FinPrint, was launched in 2015 in a collaboration of over 120 researchers. The programme focused on coral reef habitats in the west Atlantic, Indian Ocean, and the west and central Pacific. During 2016–17, more than 400 reefs were surveyed across 58 countries and territories in these four regions. Some 30–100 BRUVS were set at each location, each of them recording a target of 60 minutes of footage. This produced over 21,000 hours of video!

BRUVS do not have to sit on the seabed. A BRUVS project supported by the Save Our Seas Foundation surveyed Galapagos Sharks in the waters of Easter Island, Chile (southeast Pacific).

All sequences were viewed by trained volunteers, who noted when sharks appeared and flagged these records for confirmation by the scientists who analysed the data. The results are being made available to the public through a global open-access database, and ten continuous hours of video footage are available to view online (globalfinprint.org).

A BRUVS system placed on a reef in the Maldives, as a contribution to the Global FinPrint research programme.

Genetics

Genetics is the study of heredity, the biological process by which parents pass their genes on to their offspring. Genes are made up of DNA (deoxyribonucleic acid), a complex molecule containing the biological instructions that make each species unique and enables cells to function. DNA occurs in the nucleus and mitochondria of cells. In sexual reproduction, organisms receive half of their nuclear DNA from each parent, but mitochondrial DNA only from their mother. The tiny mutations that occur over time as DNA molecules are replicated make it possible not only to distinguish between species, populations and individuals, but also to find out how closely related they are.

This field of research, known as molecular genetics, has been used to study sharks since the 1980s, but its application to shark research, conservation and management has increased at a remarkable rate in recent years. It is proving to be a hugely valuable addition to traditional methods (based on visual descriptions) for identifying and classifying species. Rapid and relatively inexpensive modern techniques now allow scientists to sequence the entire genetic code ('genome') of organisms, including the White Shark, which has 24,520 genes (compared to about 30,000 in humans). Genes comprise only a very small proportion of the total genome (~1% in mankind) and only mutate (change) very slowly. It is still unclear what functions are delivered by the other, non-coding, 99%, but sections of non-coding DNA mutate much faster than the genes. Researchers can therefore use small sections of DNA from the cell nucleus (which is inherited from both parents) or from a mitochondrion (inherited only from the mother) to identify and/or distinguish between species and between individuals of a single species. The results of these molecular-level studies can show how closely related different species are, and even determine relatedness between individuals of the same species.

Barcoding is the process of describing and comparing very short sections of DNA from a single location on the same highly-conserved (stable) gene that occurs in all animals or species. It relies upon the 'barcoding gap', in which the genetic variation between species is greater than the variation within species. This technique has now been used to identify hundreds of thousands of species through various 'Barcode of Life' initiatives worldwide; these 'barcode' sequences are now stored online so that any researcher can access them. However, the value of these online databases depends wholly upon the accurate identification of the specimens from which the tissue samples were originally obtained. A quick look through this book will show that it is not that easy to distinguish between some shark species. If the original identification of the whole animal was incorrect, then the barcode sequence identifying that species will also be incorrectly labelled. Barcoding is vital for forensic analyses, e.g. to identify meat collected from an illegal catch.

Metabarcoding is the process of simultaneously identifying many taxa from a single sample. It may be used, for example, to scan large random samples taken from bulk shipments of shark fin to determine the presence or absence of protected species (a positive result stimulates a more detailed investigation), and for studies of environmental DNA (see box opposite).

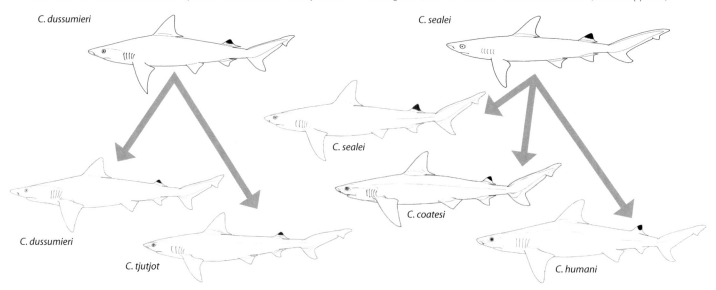

C. dussumieri

C. sealei

C. sealei

C. dussumieri

C. coatesi

C. tjutjot

C. humani

Figure 51: Many closely related shark species are extremely similar in appearance and hard to identify, particularly because some of their distinguishing morphological features change significantly as the shark grows (these are called ontogenetic changes). Since each species has consistently different nuclear and mitochondrial DNA, genetic analyses are a very important complementary tool to traditional identification methods based solely upon morphology. A recent taxonomic revision of some small, wide-ranging, inshore species of whaler sharks (genus *Carcharhinus*) provides a good example. A review of this genus some 30 years ago used morphological characteristics and fin colour to identify two very similar species, Whitecheek Shark (*C. dussumieri*, p. 530) and Blackspot Shark (*C. sealei*, p. 544). Also 'lumped' into this pair were some other species (all with black-tipped second dorsal fins) that had been described under different names by much earlier taxonomists. Genetic analyses of the many tissue samples collected in recent years from the geographic range of these 'two species' enabled the whole *sealei-dussumieri* group to be re-evaluated. The result has been to split *C. sealei* into *C. sealei* and *C. coatesi* (p. 529), while *C. dussumieri* has been split into *C. dussumieri*, *C. tjutjot* (p. 545) and possibly a fifth, undescribed, species (*Carcharhinus* sp.).

Microsatellite sequences provide information on the genetic differences between, and hence unique nature of, individuals within a single species. They are repetitive, non-coding segments of the genome. Because these sequences mutate rapidly, they can be used as markers to determine genetic inheritance, and since DNA can be extracted even from a single cell, molecular analysis needs only a tiny amount of material. For example, blood samples or fin clips can easily be taken from living animals, including newborn pups when scientists find a shark giving birth, before releasing them back into the wild. Collecting DNA samples is even less invasive if parasites containing blood can be removed or mucus wiped off the skin of a shark as it swims past. Dried tissue from ancient samples in museums can also be used to confirm species identity; study species that are now extinct, or extremely rare and difficult to find in the wild; and potentially even estimate the size and diversity of shark populations hundreds of years ago, before modern fisheries depleted stocks.

A long-term research programme focusing on the Lemon Shark (p. 561) began collecting tissue samples in Bimini, Bahamas, long before these molecular genetic techniques were developed. It has now used these samples to prove that mature female sharks return every two years to give birth in the nursery grounds where they were born, after mating with males that were mostly not born there. Fewer than 100 breeding females have been identified returning to Bimini during this programme, but their pups shared over 400 different fathers. Some of the many examples of multiple paternity in sharks,

identified through microsatellite sequencing, are described on p. 53.

Molecular genetics is also vital for forensic analyses, for example confirming the identification of shark fins or fish steaks to species level – and sometimes even to the ocean basin where they were caught – to find out whether products on sale are correctly labelled, or protected species are being traded illegally. These represent extraordinarily powerful tools.

In some cases, the visual identification of whole animals is so difficult that genetic analyses have thrown up huge surprises: the range of the Scalloped Hammerhead (p. 576) in western Atlantic waters is now known to be occupied by a second, genetically distinct, hammerhead species, the Carolina Hammerhead (p. 575) that is visually indistinguishable. Equally puzzling was the finding that several gulper shark species that were originally described on different sides of the world (in the North Atlantic and off South Australia) turned out to be genetically identical; they had only been described as distinct species because their appearance changes so much as they grow (e.g. p. 36). This raises new questions. Can these small, apparently slow-moving deepsea sharks really undertake the long-distance migrations necessary to maintain regular gene flow between such isolated populations? Alternatively, is the lack of genetic variation because the cold, low-energy, sparsely-inhabited deep sea environment not only decelerates the life history of species living there, making the gulper sharks remarkably slow growing and long-lived, but also brakes their evolution to an effectively indiscernible rate?

ENVIRONMENTAL DNA

'Scene of crime' investigations are shown frequently on TV (very occasionally in news programmes, more often in box sets). They feature white-suited investigators, wearing shower caps and plastic foot coverings to avoid contamination, searching for traces of DNA left behind by the perpetrator. We all shed DNA, wherever we go, particularly the tiny scales of skin which make up most household dust. Sharks (and other aquatic species) do the same. If water samples are collected, filtered, and the filters returned to the laboratory for analysis (with cross-contamination carefully avoided – just as in scene of crime protocols), it is possible to extract the environmental DNA (eDNA) and use metabarcoding to identify the shark species present.

A recent study in New Caledonia compared the results of surveys using BRUVS and traditional underwater visual census (UVC, carried out by divers) with eDNA analysis. Figure 52 illustrates the results. Nine shark species were identified by BRUVS and UVC (during a combined total of over 3,000 surveys), but 85% of the UVC and 46% of the BRUVS surveys failed to detect any sharks. Thirteen species were recorded from just 22 eDNA samples, and only 9% (2) of the eDNA samples failed to collect any evidence of sharks. There is no doubt which of these techniques provides the greatest return for the least sampling effort. eDNA analyses have a marked advantage over BRUVS and UVC surveys because they are collecting data from a much larger area, due to water mixing and dispersion, and over a greater length of time than the other visual techniques. All of the sampling methods, however, missed at least one species, and only six species were detected by all three methods.

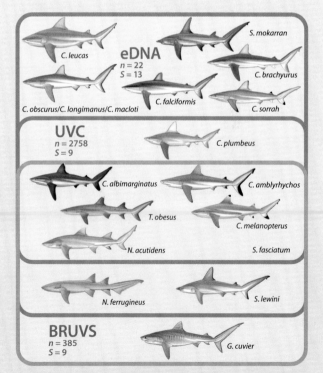

Figure 52: A comparison of shark species identified from eDNA, BRUVs and UVC surveys (see text). From Boussarie *et al.* 2018.

Research priorities – how to help

Many early studies of sharks were driven by the public fear of shark attack and the need to understand more about the animals responsible. Later, shark research was often viewed as the 'Cinderella' of fisheries investigations, largely because important commercial species have a higher funding priority (sharks are generally low value and low volume in fisheries). Interest in shark research has now risen steeply because of increased awareness of their threatened status – and hence the need to be able to identify and deliver appropriate biodiversity, conservation and fisheries management objectives – and their remarkable charisma. New sources of funding are available (including from private foundations and non-governmental organisations) and additional regional scientific organisations have been established. The powerful new research tools described above are also becoming less expensive and more widely available. Exciting discoveries are now regularly published in the scientific literature, popular magazines and online, and disseminated at a growing number of annual and biennial shark research conferences. Despite all this, many countries' researchers and fisheries managers still lack expertise in shark research, and capacity-building is urgently needed.

Rare, vulnerable and threatened species require special attention (but rarely receive it). Fisheries monitoring and survey effort is inadequate in many regions, with the result that shark biodiversity and distribution is very poorly known. The highest shark research priorities are still to improve species-specific identification, which will lead to an improved understanding of their biology and ecology, critical habitats, population structure, age and growth, reproduction and life history, as well as collecting species-specific catch data (particularly catch per unit effort (CPUE)). This information is needed to enable stock assessments to be undertaken, population status monitored, and conservation and management advice provided, including recommendations for sustainable catches and stock recovery. The authors hope that this book may help to redress this situation.

In addition to recording landings of sharks from commercial and sports fisheries (which can provide data on species identification, population size and structure, size at birth and maturity, and litter size), tag and release programmes form an important source of data on distribution, life cycle, migration, age and growth (see pp. 56 & 58). Recreational anglers, commercial fishermen and divers are increasingly becoming involved in such research programmes and, if conscientious with their observations, can be a hugely valuable asset to researchers. Even if not involved in a shark tagging programme, observers should still look out for any tag on a shark and record its details (if possible, by following instructions on the tag) before carefully releasing the animal. At the very least, it is important to describe the tag, record its number, the total length of the shark tagged, and the date and location of capture. If this information is sent to any regional or national fisheries laboratory, it will automatically be forwarded to the correct research group.

Researchers occasionally need to be able to study the whole animal, or at least parts of it, when tagged sharks are recaptured after some time at liberty. If so, these instructions are usually provided on the tag. This is particularly important if the tagged shark also contains implanted tags beneath the skin or in its body cavity (p.58), or if it had been injected with a chemical (tetracycline) which leaves a visible marker in its vertebrae. If the animal is later recaptured and the tag and vertebrae are returned to the research team, the number of rings added after injection can be counted (p.53). This provides information on age and growth rate which may be essential for the development of effective management strategies.

Never remove an electronic tag from a living fish! These tags are expensive and difficult to deploy, and are designed either to transmit information while still attached, or are programmed to be released automatically after a set period. They must never be removed prematurely (unless found on a dead animal), as this will probably result in the loss of valuable information. On the other hand, adding an additional visual tag before releasing a previously tagged animal, and after recording the above information, can be helpful.

Divers and snorkellers can also take part in research projects that use photo-identification to monitor migration patterns and estimate population size. The highly migratory White, Whale and Basking Sharks are most commonly studied in this way and it is possible for anyone to upload their own photographs to online collaborative databases. Various carcharhinids may also bear characteristic scars that enable individuals to be re-identified from photographs.

This Caribbean Reef Shark has been measured, biopsied and tagged by a MarAlliance research team studying methyl mercury pollution.

Shark legends

Sharks feature in the oral histories and legends of many coastal communities, including the Eastern Mediterranean, Polynesia, Caribbean and South America. The oldest known shark tales are possibly a bloodthirsty and unflattering collection of myths from the Bronze Age Mycenaean civilisation in the Mediterranean, 3,000–4,000 years ago, where there was an early cult to the sea god Poseidon and White Sharks (p. 320) must have been feared by coastal fishers. One of Poseidon's daughters, Lamia, had an affair with Zeus which ended badly when Zeus' wife found out, killed Lamia's children, and Lamia went mad. Zeus thoughtfully turned Lamia into a gigantic shark, so that she could take revenge by eating other people's children. Their illegitimate son, Akheilos, was also turned into a (smaller) shark after boasting that he was more attractive than the beautiful Aphrodite, goddess of love, the sea, and seafaring (and occasional lover of Poseidon, among others). Poseidon had his own attack-shark, or whale, named Cetus. When the king and queen of Aethiopia boasted that their daughter Andromeda was more beautiful than Poseidon's offspring, Cetus was dispatched to ravage the coast of Aethiopia. Andromeda's parents decided to stake her out on the shore as a sacrifice to placate Cetus, but Perseus intervened, killed Cetus, and married Andromeda. Incidentally, many scientific names for species are derived from Greek, and these myths strongly influenced 18th century scientific nomenclature for the order Lamniformes and some Lamnid species, including the Basking Shark (*Cetorhinus maximus*, p. 311) and Porbeagle (*Lamna nasus*, p. 323). The White Shark's scientific name (*Carcharodon carcharias*, p. 320) is also derived from Greek, but not from the name of a god or monster. The prefix *carchar- is* derived from 'Karcharos' meaning sharp, and the suffix *-odon* from 'odous' meaning teeth.

Slightly later stories from the Mediterranean, like those in most newspapers today, focused mainly upon shark attacks (although not always fatal ones). Which biblical animal swallowed Jonah: a shark or a whale? There are records of White Sharks discovered with animals completely intact in their stomachs – dolphins, a 50kg sea lion and, apparently, an intact and fully clothed sailor (the latter reported in the 16th century by French naturalist Guillaume Rondelet), and sharks can certainly disgorge their stomach contents. So perhaps a White Shark is a more likely culprit than Jonah's whale described in modern translations of the Bible (although it is an impossibly tall tale, whichever species was involved). The Greek writer Herodotus described shark attacks on shipwrecked Persian sailors in the 5th century BC. Aristotle's reports of his research, in the 3rd century BC, accurately included several aspects of shark biology, but misunderstood their feeding strategies (he suggested that they have to turn upside down to take food – a constraint placed upon them by nature in order to give their prey an opportunity to escape). In his *Natural History* encyclopaedia, the Roman writer, Pliny the Elder recorded that Mediterranean sponge divers were having uncomfortably close encounters with sharks over 2000 years ago. He warned that small sharks might bite the exposed white parts of divers' bodies unless deliberately frightened away (he recognised the timid nature of these animals), anticipating modern records of sharks biting the flashing white hands or feet of swimmers and surfers presumably misidentified as small fishes.

Sharks are of great importance in the oral legends and culture of the people of Palau, which is why they feature so prominently among the decorative symbols on the front of this Abai (a traditional men's meeting house). Palau was the first island nation to declare its entire exclusive economic zone as a shark sanctuary.

The verbal folklore of Polynesians, who were colonising the Pacific around the time of the collapse of the Mycenaean civilisation on the other side of the world, relates a broad spectrum of stories that, in some respects, present a remarkably balanced picture of what might happen to those who swim with sharks. In these legends, sharks may be benign gods, guardians, or friends with supernatural powers, protecting rather than threatening fishermen. They do, however, also relate cautionary tales of encounters that led to bites and sometimes death. The latter are generally portrayed as the consequences of not taking sufficient care when encountering a natural hazard, or as retaliation to deliberate or accidental provocation – not including insults to a lady's beauty.

In contrast to the vengeful gods of the ancient Greeks, the highly revered shark gods of the Hawaiians are benign; they helped to protect people and the islands. In some legends, Hawaii was created by the volcano goddess Pele, who arrived by canoe from Tahiti under the protection and guidance of her brother Kamohoali'i, the king of the shark gods and among the most important spirit guardians (Aumakua) of the Hawaiian Islands. A skilled navigator, he could change from a shark into a human or other marine animals in order to help people, including by leading lost voyagers home. He also taught another sister, Hi'iaka, to surf! Hawaiian devotees of Kamohoali'i would never eat, harm or harass any shark. Ka'ahupahau was born a human, but became a shark goddess dedicated to protecting people from shark attacks, particularly around her underwater cave at Pearl Harbour. The shark god Kane'i'kokala specialised in saving people from shipwrecks, while the implausibly enormous Kuhaimoana ensured that fishermen had good catches. The god Keali'ikau'o Ka'u fell in love with a human woman, who subsequently gave birth to a green shark that also helped people. The trickster shark god, Kane'apua, brother of Pele and Kamohoali'i, was an entertainer who could perform magic.

Dakuwaqa (half shark, half man) was a major god of the Fijian Islands, 5,000km southwest of Hawaii. He would help fishermen avoid danger at sea, provide them with a bountiful catch, and protect people from ferocious sea monsters by guarding the entrance to the lagoons. In return, local fishermen ceremoniously poured a bowl of yaqona (kava drink) into the sea for Dakuwaqa, to ensure their safe return from a fishing trip. In Tonga, Dakuwaqa was known as the warrior god, Takuaka, who protected people from other vicious gods. In the Cook Islands, he was Avatea (shown below).

The story of Ina and the shark that helped her is so well-known across Polynesia that they appear on Cook Islands bank notes.

A basalt carving from Rarotonga illustrates the half man, half shark deity known as Avatea in the Cook Islands.

There are several versions of a widespread Pacific legend describing the fate that befell a foolish girl, Ina, who injured the shark that had kindly offered to ferry her between islands to see her boyfriend, the king of fishes. When she became thirsty, the shark allowed her to use his dorsal fin to open a coconut. This story is commemorated on Cook Islands bank notes. Later on, instead of asking permission, she cracked a second coconut open on his head. She was immediately thrown off by the shark, whose hammer-shaped head is still misshapen from the injury. On some islands the legend concludes with the shark eating her, or her drowning, but the Cook Islands version ends with Tekea the Great, king of all sharks, taking pity on her. He rose from the depths, rescued and carried her safely to her boyfriend's floating island.

Shark worship was once fairly common across the Indo-Pacific, possibly sometimes including human sacrifice. Shark calling is still practiced in several locations there, including by fishers from Papua New Guinea, who sing and shake a coconut rattle against the submerged part of their canoe's hull to attract a shark. Some tribes claim sharks as their ancestors. Sharks are important in the culture and folklore of certain peoples in Melanesia, West Africa, Australasia, the Caribbean and the Amazon basin. A Maori myth tells the story of a princess, Kawariki, who fell in love with a peasant boy Tutira. Her sorcerer father disapproved and turned Tutira into a shark. The lovers continued to meet in secret and swim together. When a tsunami destroyed the village and swept the villagers out to sea, Tutira the shark man rescued them all. As a reward, the king turned Tutira back into a human and allowed him to marry Kawariki.

Sharks also appear in stories about the stars. Torres Strait Islanders use the stars to track the changing seasons and as a reminder of when to plant crops or hunt particular prey. Baidam, the shark constellation, is comprised of stars in the Big Dipper, part of Ursa Major (the Great Bear). When these stars appear to the north of the Straits, over Papua New Guinea, the islanders know that the shark mating season is starting. This also means that it is the right time of year to plant sweet potato, banana and sugar cane. In the west Atlantic, tribal people of Brazil and Guyana call Orion's Belt 'the leg of Nohi Abassi'. Nohi, totally fed up with his mother-in-law, trained a shark to eat her. However, she discovered his plans and disguised her other daughter as the shark. Instead of going for his mother-in-law, Nohi Abassi's sister-in-law attacked him and sawed his leg off. The leg became the constellation (and presumably an important reminder always to be nice to your mother-in-law).

The remains of large sharks have been discovered entombed underneath the ruins of the Aztec Great Temple beneath central Mexico City, together with human and crocodile bones. Perhaps they represented the great sea monster whose body, ripped apart by the gods in Aztec mythology, first formed the earth, accompanied by sacrifices to placate the civilisation's blood-thirsty gods.

It is saddening to discover that the modern myths surrounding sharks and shark attacks, depicted in film and other media, have today reached even the most remote of coastal fishing communities in developing countries. Lack of a reliable electricity supply has not prevented widespread viewing of shark horror films and, despite no known local reports of shark attack, has resulted in a widespread fear of sharks that is not supported by centuries of experience at sea.

Fisheries

Unmanaged or inadequately managed fisheries are, without doubt, the single most serious threat to the world's sharks. The impacts of habitat loss and climate change are (presently) minor in comparison. Sharks are caught in fisheries almost everywhere that they occur. This is virtually throughout the world's oceans (except for polar regions and the very deepest areas) and even in a few large rivers and lakes. These catches can be classified as industrial (commercial, large scale), artisanal (small scale), and subsistence (very small scale, to feed the fishers' families and community). The catches made by these different scales of fishery can be subdivided into target and bycatch.

The greatest volume and variety of shark species are taken in coastal multi-species or 'catch all' fisheries, using a variety of gear types, from handlines, to longlines, set nets and trawlers. These fisheries harvest and utilise a very wide range of marine fishes, but some may discard the least valuable portion of the catch. Discarded bycatch often used to include large shark carcasses, although their fins were usually retained. Where finning prohibitions have been adopted, and are being enforced effectively, such discards should no longer occur – not least because the markets for, and value of, shark meat have risen significantly in recent years. In developing countries, where many coastal fisheries are largely unmonitored and unmanaged, landing sites provide a rich hunting ground for shark enthusiasts seeking new species, or researchers wishing to investigate the biology of poorly known species. For those lucky enough to visit such areas, it is often worth rising early enough in the morning to be able to watch fishing boats unloading their catches.

Targeted shark fisheries are those that set out to capture and land sharks, whether as a primary or a secondary catch. The primary catch is (at least in theory) the main target of the fishery. Examples of targeted shark fisheries have included those for North Atlantic Porbeagle Shark (p. 323), and several fisheries around the world for deepsea sharks, spiny dogfishes (p. 117) and smooth-hound sharks (p. 477); these species were the primary catch of those fisheries. Some of these target fisheries were only short-lived before collapsing due to overexploitation – the sharks were being caught faster than they could reproduce and replace themselves. Others, which are under careful manage-ment and lower fishing pressure (e.g. Rig, p. 482), are sustainable and support viable target fisheries.

In contrast, industrial longline fisheries for oceanic Swordfish usually catch much larger numbers of sharks than they do Swordfish, and these sharks (particularly the valuable Shortfin Mako, p. 321) are often retained and landed alongside the Swordfish. Because there are now restrictive quotas (catch limits) for Swordfish, which in this fishery is nominally the primary catch, a larger, unregulated shark catch may be the only way to make the fishery economically viable; the sharks are, therefore, a vital secondary target. Indeed, several pelagic fisheries might more accurately be described as shark fisheries with a secondary catch of Swordfish or tuna. Specimens of slightly less valuable species, such as the Blue Shark (p. 562), may be discarded from a Swordfish fishery, depending upon the amount of space in the vessel's hold and level of market demand on shore. Although pelagic sharks form a large proportion of the total catch of oceanic fisheries and, because of their very broad distribution, are probably the world's most heavily fished sharks, they are not equally threatened. Blue Sharks, for example, appear sufficiently fecund and resilient to survive very high, mostly unregulated fishing pressure, despite

Almost 80% of shark species have been reported in the bycatch of large-scale, industrial fisheries. Many, like this dogfish shark, are caught in trawl fisheries and discarded, unwanted, back into the sea.

comprising 15% of the world's reported chondrichthyan catch – greater than any other single shark species. Other pelagic sharks that were formerly an unmanaged secondary target (or utilised bycatch) have become seriously threatened with extinction, with some countries and fishery bodies taking steps to protect them. Oceanic Whitetip (p. 537) the three largest species of hammerheads (p. 574) and thresher sharks (p. 315) are now strictly prohibited species within the high seas fisheries managed by some Regional Fisheries Management Organisations (RFMOs), yet the Endangered Shortfin Mako remains largely unmanaged at the time of writing.

Bycatch refers to the accidental catch made during fisheries targeting other species. It is responsible for substantially the largest volume of global shark fisheries mortality. Shark bycatch can be even less sustainable than target fisheries, because the fishers are targeting faster-growing species that can withstand a much higher fishing effort and the viability of the fishery is not detrimentally affected by a decline in the number of sharks caught. Indeed, sharks can be driven to extinction in bycatch fisheries, something less likely to happen in a target shark fishery because fishing becomes uneconomic before all the sharks have been taken. Shark bycatch may be valuable to the fishers, in which case it is kept for eating or to sell. For example, virtually everything caught is landed in some indiscriminate, small-scale artisanal and subsistence fisheries. Alternatively, if the bycatch is unwanted, it is likely to be discarded at sea, alive or dead (unless discarding is prohibited in that fishery or country).

Long-term studies of fish landing sites and markets in tropical countries report a change in the species composition of shark bycatch over time. Initially, shark landings are dominated by large species, such as Scalloped Hammerheads (p. 576). Following decades of intensive fishing effort, the large sharks are no longer seen. Instead, a variety of small catsharks and bamboo sharks are on sale. The latter had always been a bycatch of local fisheries, but had been discarded at sea in earlier years, when larger, more valuable species

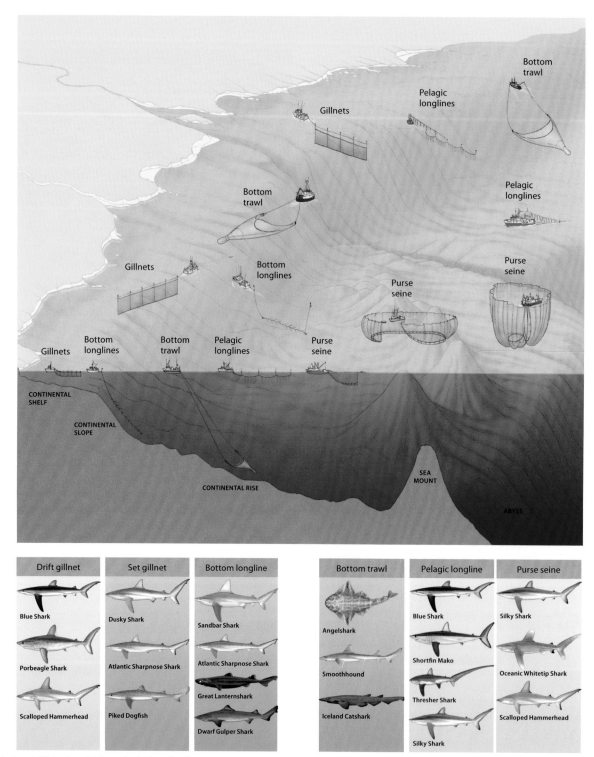

Figure 53. Industrial fisheries which catch sharks. This illustrates the larger coastal and offshore fishing methods with a bycatch and, in some cases, a targeted catch of sharks.

were still being caught. Although larger numbers of sharks, and a far greater range of species, are killed as a bycatch in coastal and deepsea fisheries than on the open ocean, the latter catches are often unrecorded, even when brought back to shore. If they are recorded by fisheries staff, the records are unlikely to be sufficiently detailed to identify the catch by species. There is still an urgent need to identify, quantify and report shark bycatches in most countries; we hope that this book will help.

Gillnets are a cheap and widely available form of fishing gear used extensively in industrial, artisanal and subsistence fisheries – with the length of net and size of mesh often in proportion to the scale of fishery. Modern gillnets are made of clear artificial fibres and very hard to see underwater compared with traditional cotton or flax nets. They are called 'gill' nets because the meshes were originally designed to catch behind the gill covers of fishes that are unable to swim all the way through them because they are too large, thus preventing them from escaping. It can be very hard to untangle and release gillnet bycatch alive. Nets with a wider mesh are used to entangle large-bodied species and are notoriously indiscriminate fishing gears, killing marine mammals and turtles as well as sharks and other fishes. Drift gillnets (drift nets) are so destructive to wildlife and fish stocks that a 1991 United Nations Resolution banned the use of pelagic drift nets longer than 2.5km in international waters, but they can still be used legally elsewhere and illegal use continues.

Bottom-set gillnets entangle animals swimming near the bottom and damage invertebrates attached to the seabed. Some are designed specifically to catch particular shark species, including the escatera (Spain), squaenera (Italy), sklatara (Croatia) and martramaou (France), all of which were designed to catch angelsharks. Large-mesh nets are used in a few target shark fisheries (legal and illegal) but also take a substantial bycatch of sharks in many other fisheries. When used in shallow coastal areas with a greater diversity of threatened species, these nets may be even more damaging to fish stocks and wildlife than pelagic drift nets.

Gillnets that have been lost at sea by fishers are known as 'ghost nets'. The weight of any catch entangled in a ghost net causes it to sink to the seabed, to

Large-mesh shark nets may become so entangled on the seabed that they cannot be retrieved. As well as damaging fragile seabed features, they may continue to 'ghost fish' indefinitely.

decompose and/or be eaten by other species, whereupon the net can float back into the water and begin fishing again. Gillnets are particularly unpopular with many fisheries managers, conservationists, scientists and some fishers, because of the damage they cause, but it is still rare for their use to be prohibited. Fishers often need financial and technical help to give up using their gillnets, so that they can purchase less damaging fishing gears and retrain to use them effectively.

Longlining is a significantly less destructive fishing technique. Pelagic longline fisheries usually target tunas and billfish (e.g. marlin, Swordfish) and, as noted above, can catch more sharks than teleosts, sometimes intentionally when the latter stocks are depleted and sharks are needed to make the fishing trip profitable. Careful choice of bait, hook size and shape, visibility of leaders and the time of day and depth at which the lines are set can greatly reduce unwanted bycatch and make live release much easier. Researchers have also tried to minimise shark bycatch by using chemicals, rare metals or electric fields that only repel sharks, not teleosts, but these measures are often too expensive for large-scale deployment on longlines. Bottom longlines are, or have been, used in some target shark fisheries, including for large coastal sharks such as Sandbar Shark (p. 542), and gulper sharks (p. 140) in very deep water. Longlining is the best fishing technique for shark research programmes that monitor population trends over time and catch sharks for tag and release programmes.

Purse seine nets encircle dense schools of fishes, including anchovies, mackerel and tunas, which are then scooped or pumped into the seiner's hold. When well-managed, these fisheries can be very 'clean'; this means that they mostly catch the target species and have very little bycatch. This is not the case for tuna purse seine fisheries that use artificial fish aggregating devices (FADs) to make it easier to locate and catch schools of tuna. The original FADs were natural: floating logs, tangles of vegetation, Whale Sharks (which are good indicators of the isolated rich patches of prey where tunas feed) and, particularly, schools of dolphins. Dolphins were considered the best, because they swim with the most valuable, fast-moving, large adult tunas; furthermore, only a few other bycatch species could keep up and be caught alongside them. Although purse seine fishers could become skilled at releasing dolphins from the nets before scooping up the tunas, public awareness of this practice soon made it virtually impossible to sell any tuna that were not marketed as 'dolphin-friendly'. Setting nets on dolphins is now widely prohibited and not tolerated by any of the major tuna companies that purchase and process tuna. Other forms of FAD were needed. There are not enough Whale Sharks to support industrial purse seine fisheries, and intentionally setting nets on this shark was subsequently discouraged, then prohibited in most tuna fisheries (perhaps tins of tuna advertising 'Whale Shark-safe fisheries' will soon join those for 'Dolphin-safe fisheries' on supermarket shelves). Purse seine fishers therefore created their own, increasingly complex, artificial FADs. These are rafts, usually equipped with radio transmitters so that they can easily be relocated. Fishers discovered that the more complicated the structure, the more effective it was in attracting tuna; hanging nets beneath the FADs created a whole new floating ecosystem which attracted even more tunas. Unfortunately these 'entangling FADs' also attracted many other species, including marine turtles and oceanic sharks, and sharks became a major purse seine bycatch. It is much harder to release unwanted pelagic sharks unharmed from a purse seine as it was to release dolphins and Whale Sharks.

FINNING

Finning is defined as cutting off and retaining shark fins, and then discarding the carcass at sea. Removing fins on land while processing a carcass for the utilisation of its meat, fins and other products is not finning. Finning is widely condemned for many reasons:

- It is wasteful and threatens food security in developing regions where shark meat is an important source of protein.
- It increases the risk of illegal, unreported and unregulated (IUU) fishing, because fishing effort is not limited by the need to store and process bulky carcasses onboard vessels (fins are small and easy to keep and conceal, dried or frozen). Finning also makes it very difficult to monitor and record the species and numbers of sharks caught (these data are essential for stock assessments and science-based fisheries management).
- It rapidly depletes shark stocks, threatening marine ecosystem stability, sustainable fisheries, and non-consumptive uses such as shark ecotourism operations.
- There are serious welfare concerns when sharks are finned and discarded alive.

Finning is now prohibited in more than 40 countries, including more than half of the world's largest shark fishing nations, and by at least eight of the RFMOs that manage oceanic pelagic fisheries. Different rules, however, have been set to implement and monitor compliance with these finning bans – and a few do not specify any such measures. Most early finning bans allowed fins to be removed from shark carcasses on board vessels, but set a maximum fin:body weight ratio to prevent illegal discard of carcasses at sea. Monitoring compliance requires weighing of fins and carcasses together on land, which is labour-intensive, complicated when ratios vary between fleets, and impossible if fins and carcasses are landed in different ports. Species specific identification of catches is also more difficult. Careful fin cutting practices to minimise fin weight can still allow a certain amount of finning to take place undetected, particularly when high ratios have been set. Requiring fins and carcasses to be landed together, with the fins still naturally attached, avoids these problems, and 'fins naturally attached' is now the preferred option when implementing shark finning bans.

Silky Sharks (p. 531) are most seriously affected by the use of FADs in purse seine fisheries. This is because many juvenile Silky Sharks spend the first few years of their life living around drifting objects. These young sharks (100–120cm in length) stay close to their 'home' FAD by day, move deeper to feed at night, then return to their chosen FAD for the following day. Large numbers of juveniles are therefore bycaught when purse seines are set on drifting FADs. However, this bycatch is not the greatest threat to Silky Sharks. Researchers discovered that the nets hung beneath FADs were catching and killing 5–10 times more juvenile sharks than the observed purse seine bycatch – some 0.48 to 0.96 million sharks per annum in the Indian Ocean alone. Because the sharks caught died and fell out of the nets within two days, this catch was largely unseen. It was estimated, however, that there were so many FADs in use about ten years ago that a Silky Shark had a 29% chance of surviving one year, 9% chance of survival to two years, and only a 3% chance of surviving until it reached three years old. It seemed as though very few juveniles might survive long enough to reach maturity (which is 15 years old, for Indian Ocean females).

Once the destructive nature of FADs became apparent, the tuna Regional Fisheries Management Organisations began to regulate their use, for example by mandating FAD management plans; limiting total numbers deployed; prohibiting entangling FADs; collecting and destroying lost, ghost-fishing FADs; and requiring new FADs to be made from biodegradable materials. Because Silky Sharks may live for over 30 years, however, it may be another decade before the impact of these missing age classes becomes apparent as birth rates crash due to a lack of mature sharks, and longer still before the benefits of these recent management measures can be assessed.

Trawling takes several forms. Midwater trawls may, like purse seines, take a fairly clean catch of shoaling fish species, which are located using sonar, with very little bycatch and usually even fewer sharks. Even a clean catch can be unsustainable, however, if it is targeting teleost spawning aggregations, which concentrate a large proportion of the entire population into a small area. Basking Sharks (p. 311) have unexpectedly been caught in midwater trawls, presumably because they were feeding on the dense, nutritious soup of fish eggs produced by the spawning event.

Bottom trawling, however, is one of the most damaging fishing operations known, particularly when very heavy fishing gear is used. Deepsea trawl fisheries can operate to depths of over 3,000m, beyond the maximum depths at which sharks occur, and over 40% of the world's trawling grounds are now off the edge of the continental shelves. Trawling is so damaging because the heavy gears used damage fragile seabed habitats and are completely indiscriminate, scooping up everything in their path. Trawl bycatch can be many times larger than the target catch. In very deep water, where productivity is low and species are particularly long-lived, deepsea trawling is more akin to a mining operation than a fishery. There are numerous examples around the world of deepsea shark fisheries that have initially produced high yields of valuable shark meat and liver oil before crashing after only a few years and failing to recover. Gulper sharks, for example, are particularly vulnerable to overexploitation in deepsea trawl and longline fisheries.

Trends in fisheries

While small-scale fishers have always caught small quantities of sharks, including the largest filter-feeding sharks, the majority of large-scale shark fisheries are a relatively recent phenomenon. Coastal fishing pressure increased sharply in the 1950s when cheap, artificial fibre, monofilament nets became widely available; widespread depletion and perhaps the unrecorded disappearance of some coastal shark species, including the Lost Shark (p. 539), probably took place around that time. Large-scale target shark fisheries really began to expand in the mid 1980s, as demand increased for their fins, meat and cartilage (e.g. Figure 5, below), then intensified still further when shark fin markets soared (pp. 80). Although reported shark catches more than tripled between the 1950s and the 1990s, they have always contributed less than 1% of the world's wild capture marine fisheries. Global shark catches peaked in 2000, a few years after total marine fisheries, then slowly fell as stocks collapsed and in some cases catch limits (including zero quotas or prohibitions) were introduced to prevent overexploitation or allow populations to recover.

Reported global landings from wild capture fisheries (all fishes) have shown this same increase from the 1950s, when UN Food and Agriculture Organisation (FAO) records began, to the 1980s, reaching a peak in 1996 before stabilising (fluctuations are driven by variations in catches of small schooling fishes). It is noticeable that shark landings peaked slightly later than all other global fish landings and then declined more steeply. What makes this a particularly worrying pattern is that the quality and quantity of shark fisheries data submitted to the FAO in recent decades has significantly improved (Figure 54). Early catch records are believed to have been significant underestimates of

shark mortality, due to incomplete data (a few major fishing nations have never reported their shark catch) and unrecorded discards at sea. It is also difficult to separate sharks from rays to determine precise trends, because they were often recorded as 'sharks, rays, skates etc.' in official data. In the 1950s, over 70% of shark catches were lumped into unidentified groupings such as 'sharks, rays, skates etc. not elsewhere included', but this unidentified fraction of the world shark catch has fallen to about 30% in recent years while the proportion of catches identified to shark species has risen (Figure 54). Today, about 40% of shark catch is reported by species name, and an additional 15% by genus or family. Had shark stocks been broadly stable, we might have expected this better reporting to produce a steadily increasing trend in shark catches, while moderate declines would probably have been masked. Unfortunately, although the catch data are improving, the volume of reported catches are not.

Furthermore, the relatively simple global trend in reported shark landings masks a far more complicated and serious picture: the serial depletion of shark fisheries in one region after another. In other words, for each shark fishery that completed a 'boom and bust' cycle (initial high landings followed by a rapid population crash and collapse), fishing efforts were expanding elsewhere, leading to yet more short-term, local shark landing increases. Thus, although the huge northeast Atlantic shark landings have halved since their peak in the 1970s, the steep downward trend in European catches during the 1980s (particularly of Piked Dogfish, Figure 57) was obscured in global statistics by huge, expanding shark fisheries in the Americas and Asia. This cycle initially enabled the apparent global increase in shark landings to continue – until the

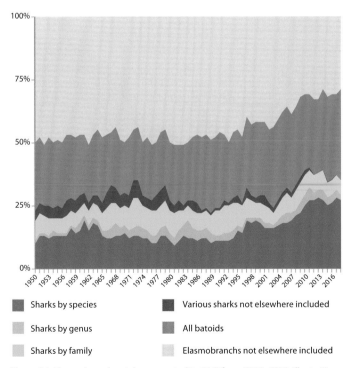

Sharks by species

Sharks by genus

Sharks by family

Various sharks not elsewhere included

All batoids

Elasmobranchs not elsewhere included

Figure 54. Elasmobranch catches reported to FAO from 1950–2018, illustrating improved shark species identification since the mid 1990s.

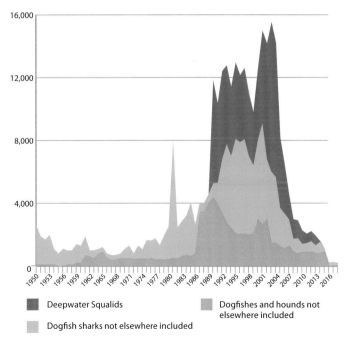

Deepwater Squalids

Dogfish sharks not elsewhere included

Dogfishes and hounds not elsewhere included

Figure 55. Dogfish shark catches (t) in the northeast Atlantic, 1950–2018 (extracted from FAO FishStatJ). Catches rose to supply EU demand for meat, causing stock collapses and resulting in fisheries closures.

Blue Shark is the world's most common pelagic shark species and an abundant (but declining) bycatch of oceanic longline fisheries.

majority of fisheries were in decline and the long-term downward trend could no longer be hidden. Unfortunately, as noted above, even when target shark fisheries cease, the remnants of the population often continue to be taken as bycatch in other fisheries for more abundant and resilient bony fish species; sharks can be driven to extinction by this 'incidental fishing pressure'. It is, however, important to note that catch trends do not necessarily represent trends in the population size. Declines in catches may be caused by declining stocks, but could also reflect falling fishing effort, the introduction of quotas, or even fisheries prohibitions. Scientific stock assessments give a much better picture of stock status, but are often inconclusive for shark species due to a lack of good fisheries and fishery-independent monitoring data and/or incomplete understanding of shark life history and population structure. This is particularly challenging when fisheries are only targeting certain age classes for long-lived species (see Piked Dogfish and Shortfin Mako trends, below).

Target shark fisheries are uncommon by comparison with the major 'catch-all' and oceanic bycatch fisheries. This is partly because shark stocks are fairly small, and these fisheries have tended to be relatively short-lived. Examples of historic target fisheries that have collapsed include those for North Atlantic Porbeagle (p. 323), Tope (p. 471) in the southwest Atlantic, and Piked Dogfish

(p. 117) in the northeast Pacific and European waters. Unfortunately, management was introduced too late to avoid the closure of these fisheries, but there is hope that the stocks will recover under management in the coming decades. More recent target fisheries include those for the fins of a variety of reef shark species in the Red Sea and Indo-Pacific waters, mostly unmanaged, and many of them illegal. These fisheries have wiped out many reef shark populations (see Global FinPrint, p. 59) and it may be many decades before signs of recovery are seen, even in the huge new shark sanctuaries now being established across the vast exclusive economic zones (EEZ) of some small island states.

Well-managed, sustainable shark fisheries include the northwest Atlantic Piked Dogfish fishery, which crashed in the late 1990s but has recovered so well under management that it was certified as sustainable by the Marine Stewardship Council (MSC) in 2012; the Southern Australian Gummy Shark fishery (p. 477); Tope and Piked Dogfish fisheries in New Zealand; and some coastal fisheries for a variety of large sharks in Australia and the USA. The problem for mixed species fisheries is that management measures appropriate for the sustainable exploitation of smaller and more resilient species are often inadequate to protect the larger, more heavily K-selected sharks (p. 52) which occur in the same habitats and are caught by the same fishing gears. On the other hand, reducing fishing pressure sufficiently to protect and rebuild stocks of the most threatened species may effectively close the fishery for other species. These significant management challenges can be resolved through a

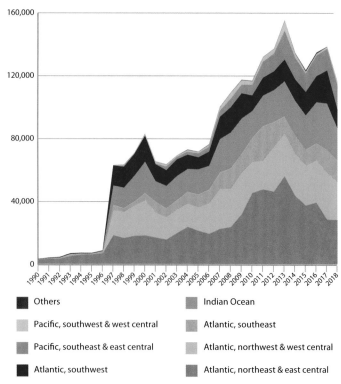

Figure 56. Cumulative Blue Shark landings (t) by ocean region, as reported to FAO, 1990–2018. Catches before 1997 may have been discarded or misreported.

Juvenile Shortfin Mako on sale for their meat in Milan, Italy. This is a high value, utilised bycatch of pelagic swordfish and tuna longline fisheries in the northeast Atlantic.

although mostly discarded at sea and seriously under-recorded until at least the late 1990s. Today, Blue Shark accounts for almost 20% of global reported shark landings, but 34% of all the shark fins in Hong Kong markets. There are two possible reasons for this discrepancy: that many Blue Shark are still being included in non species-specific catch records, and/or that up to one-third of Blue Shark catches are still being finned and discarded at sea – despite widespread regional and national shark finning prohibitions – because their meat is relatively low value compared with the price paid for their large fins. Blue Sharks are certainly biologically more resilient to fishing pressure than other large oceanic sharks – they mature at a younger age and have large litters of pups. It is also possible that they are benefiting, through competitive release, from the depletion of larger predatory sharks and teleosts that eat Blue Shark pups or competed with Blues for food, and survival rates are therefore improving. However, FAO records also show that, after rising steadily to a peak of over 160,000t in 2013, reported global Blue Shark catches have since declined by close to 30% in only five years, to 100,000t in 2018 (Figure 56). Scientists have expressed concern that initial assessments for the South Atlantic suggest that this stock may not be in good shape and central Atlantic catches (which were very high) have halved in five years, as have those in the western Pacific. On the other hand, catches appear to be rising in the eastern

combination of the collection and analysis of accurate scientific data and close collaboration with the fishing industry. For example, the selectivity of fishing gear can often be improved through measures such as setting hooks at times and depths where threatened species are less likely to encounter them, and avoiding the use of wire traces on longlines, to allow sharks to escape more easily. The largest, most vulnerable species, or the largest breeding females, can be avoided in a net fishery by ensuring that the mesh size used is small enough for the largest sharks to bounce off, while smaller, more abundant and fecund species can still be captured. This approach is used in the Southern Shark Fishery in Australia, which used to target large Tope and a bycatch of small Gummy Shark, but now targets Gummy Shark and endeavours to minimise catches of the overfished, Critically Endangered Tope.

Blue Shark is the world's most heavily fished shark, as a target or bycatch of pelagic longline fisheries. It is the pelagic shark most at risk from these fisheries because of the very high global overlap between pelagic longliners and tagged Blue Shark oceanic habitat (greater than any other shark species studied, at 49% globally and 76% in the North Atlantic). This is also one of the very few shark species reported every year since FAO catch records began,

RESILIENCE TO FISHERIES: SHARKS VERSUS BONY FISHES

Fisheries mortality, whether in targeted fisheries or caused by accidental bycatch, is far more likely to result in the depletion of shark populations than when bony fish populations are fished. This is because the biology of sharks, which produce relatively small numbers of large young (p. 52), makes them less resilient to overfishing than bony fishes. The majority of bony fishes produce vast quantities of tiny young that are quick to grow, particularly when competition from larger older members of the population has been diminished by their removal in fisheries. While removing a large proportion of adult bony fishes generally leads to increased survival of young, therefore to higher fisheries yields, lowering the number of adult sharks means that fewer pups are born. Fisheries managers generally aim to fish populations of commercially exploited bony fishes down to 30–70% of their unexploited size (depending upon the species), because this maximises the amount of fish that can be taken by fishers. This is known as the maximum sustainable yield – but intentionally reducing fish stocks to this point is risky, because it allows no margin for miscalculation, management error, or events and activities outside the control of decision-makers. Efforts to apply the same logic and traditional fisheries management tools to sharks has unfortunately often led to failure, because permitted catches were set too high. Careful management of shark fisheries has been successful in several parts of the world, and has also allowed seriously depleted shark stocks to be rebuilt. Unfortunately, examples of good management are, at the time of writing, in the minority; most of the world's shark fisheries remain unmanaged, despite sharks needing more careful management than other fisheries.

Pacific and southwest Atlantic (despite concerns for the stock in the latter) and to be stable in the Indian Ocean. Inadequate data mean that stock assessments for this species, where they exist, are uncertain. At the time of writing, the world's first regional Blue Shark catch limits have just been adopted by the tuna RFMO (International Commission for the Conservation of Atlantic Tunas (ICCAT)) that manages oceanic fisheries in the north and south Atlantic.

Shortfin Mako and Silky Shark (pp. 321 and 531) are the other two remaining, commercially important, globally distributed, pelagic sharks. They are listed as Endangered and Vulnerable, respectively, on the IUCN Red List of Threatened Species. Shortfin Mako is of very high commercial value for its meat, which means it is more likely to be kept than discarded. It is second only to Blue Shark in terms of its global exposure to pelagic fisheries, with a 37% overlap between the tracks of satellite tagged Shortfin Makos and the locations of longline vessels, rising to 62% overlap in the Atlantic. It is initially surprising that Shortfin Mako catch per unit effort (CPUE) has remained fairly stable since longline fisheries expanded throughout global oceans during the 1980s, until this species' life history is taken into account. Oceanic longline fisheries mostly catch juveniles, aged 3–10 years. Shortfin Makos older than 10 years become too large for the hooks and lines that are used in longline fisheries, and virtually vanish from the catch for the remaining 20 or so years of their life. Maturity for 50% of Shortfin Makos occurs at age 21 (a little bit older for the remaining 50%); the females then produce about 12 pups per litter, at 2–3 year intervals, until a maximum age of 30-something years. That is only a few litters in a lifetime; the number of pups born is intrinsically linked to the abundance of mature females in the population. If, as seems likely, very few juveniles survive the gauntlet of eight years of intensive longline fishing effort,

this low number of survivors will produce significantly fewer pups than the unfished adult generation that they are replacing. The 2019 stock assessment model for the Atlantic indicated that an immediate prohibition on the retention of Shortfin Makos would mean that the stock will continue to decline for at least 15 years, before rebuilding became possible, with a 53% probability of rebuilding the stock by 2045 (this scenario did not take into account some continuing, unavoidable bycatch mortality). By 2020, only one country (Canada) had adopted the urgent scientific advice to prohibit the landing of Shortfin Makos.

As noted on p. 67, there may be a similar problem for Silky Sharks in the Indian Ocean (and perhaps elsewhere). In this case rather than juvenile bycatch in longline fisheries, the culprit is the huge pup mortality beneath the entangling FADs used by purse seiners. With so few Silky pups surviving to reach maturity, another demographic timebomb may be ticking for this (formerly?) very common pelagic shark. Although landings and mortality data are very sparse for this species, its abundance in Hong Kong shark fin markets (where it is the second most common species after Blue Shark) is being monitored and this may produce the best early warning of stock collapse.

Increased awareness of the plight of these and other large pelagic oceanic sharks has now led to the protection of most of the species most seriously threatened with extinction. However, RFMOs and the majority of their members are, at the time of writing, still resisting the introduction of catch limits for Shortfin Makos and Blue Sharks – now the two remaining commercially important oceanic shark species. This is disappointing; it is surely preferable to introduce sustainable management sooner, rather than resorting to catch prohibitions later on, once stocks have declined perilously close to the point of no return.

Combined **Piked Dogfish** fisheries produce the world's second highest total shark catch – this is also a very widely distributed species. The largest fishery for Piked Dogfish was in the northeast Atlantic (Figure 57), where the species (known regionally as Spurdog) was targeted for well over 100 years for its meat, highly valued in European markets. Catches peaked at around 50,000t per annum in 1972, then decreased steeply. EU target fisheries closed in 2006 because stock assessments showed that the population had declined by over 90%, to less than 10% of its original size, and was in danger of collapse. The decline stabilised over the next ten years and the stock is now very slowly recovering (but is still below safe limits). Foreign vessels also fished Piked Dogfish off the USA and Canada during the 1960s and '70s, until excluded from these countries' EEZs. Nearly 20 years later, a new northwest Atlantic fishery started exporting to European markets, with the stock subsequently declining by more than 80% during the 1990s (Figure 57). Management of Piked Dogfish fisheries is particularly challenging because this is such a slow-growing, late maturing, long-lived shark species, with low intrinsic rates of population increase, and among the longest gestation period (two years) of any vertebrate. Furthermore, populations are usually strongly segregated by age and sex. The biggest females give birth to the largest number of pups, and fishers preferentially target the largest, usually pregnant, females for their meat. The selective removal of mature females over a long period of time can lead to reproductive failure of the stock. Because juveniles take 15 years to reach maturity, it then takes over a decade for a lost generation of pups to be expressed in the population as missing adult age classes and the stock will continue to decline even after closure of the fishery. Given predictions that

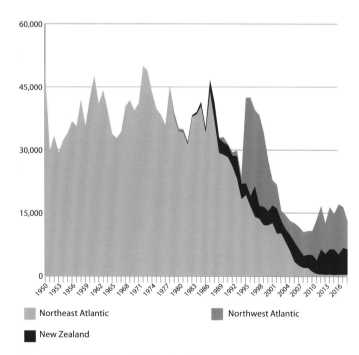

Figure 57. Major Piked Dogfish fisheries (t) 1950–2018.

Northeast Atlantic   Northwest Atlantic

New Zealand

stock rebuilding would take 15–30 years, some observers were surprised that the US Atlantic stock recovered more rapidly than expected following significant quota reductions; this fishery was certified as sustainable by the MSC as early as 2012 (albeit with a very much lower fishing effort and low quotas). In contrast to these dramatic examples of stock crashes, New Zealand introduced quota management for
its Piked Dogfish stocks in the early 2000s, following increased targeting for export. Catch rates and estimates of biomass have remained stable or increasing.

Tope is another globally distributed, long-lived species that has been fished for almost 100 years, frequently unsustainably. Tope has extremely slow, or no recovery following 'boom and bust' fisheries, as even a small amount of bycatch can prevent stock rebuilding. While under-reporting to the FAO means that Figure 58 does not include all landings since 1950, but the collapse of fisheries in the southwest Atlantic are clearly illustrated. The 2020 IUCN Red List assessment also identified similar population declines (to less than 20% of the original stock size) during the 1940s in the northeast Atlantic, Mediterranean, South Africa and Australia. Exceptions are New Zealand, where the stock has declined by less than 50% and quota management is in place, and California. The last target fishery in California collapsed nearly 80 years ago and showed no hints of rebuilding until inshore net fishing was prohibited in 1994, when signs of recovery started to appear (presumably due to increased survival of juveniles). This last example shows that recovery is possible, albeit very slow.

The most recent fisheries frontier has been in the deep sea, off the edge of continental shelves and around oceanic islands. A remarkable diversity of shark

STATUS OF WORLD FISHERIES

The FAO has been monitoring assessed marine finned fish stocks (which include only a very few shark stocks) since 1974, and recording their declines. The proportion of fish stocks that are within biologically sustainable levels fell from 90% in 1974 to 66% in 2017 (the remaining 34% of assessed stocks are overfished). Stocks classified as being at maximum sustainable yield fell until 1989, but have since increased to 60%, partly due to improved management, leaving just 6% of stocks underfished. The FAO estimated that 79% of marine fish landings in 2017 came from biologically sustainable stocks. The Mediterranean and Black Sea have the highest proportion (63%) of unsustainably fished stocks, followed by the southeast Pacific (55%) and southwest Atlantic (53%). Fish stocks are in the best condition in the eastern central, southwest, northeast and western central Pacific, with only 13–22% overfished. It is clear from the FAO's analysis that fishing pressure has decreased and stock biomass has increased to biologically sustainable levels in intensively managed fisheries. Unsurprisingly, stocks are in a poor state where fisheries management is inadequate or absent.

species adapted to relatively narrow depth bands and a very stable, low-energy environment occurs in these deep habitats. **Deepsea sharks** tend to grow and reproduce even more slowly than sharks found in coastal or oceanic waters (some species only have one pup every two years). This makes them the least resilient to fishing pressure of all sharks (and indeed all vertebrates) and far more vulnerable to depletion by fisheries than their shallow water cousins. Truly, deepsea fisheries (particularly those for shark meat and liver oil) have more in common with mining operations than with traditional fisheries, because the stock recovery rates are so extraordinarily slow. These short-lived fisheries have exploited unnamed and unknown shark species never seen by scientists, and could be driving some of these species to extinction before they can even be described. Figure 55 (p. 69) illustrates northeast Atlantic landings of deepsea Squaliform sharks including Portuguese Dogfish (p. 193) and gulper sharks (p. 140), and other unspecified groups of dogfishes which may include deepsea species, targeted collectively as 'Siki' sharks during the 1990s and early 2000s. This fishery appeared stable for almost 20 years, but that was only because it incorporated a large number of much shorter, sequential fisheries, pursued by different fleets using different gear types in different locations. Target deepsea fishing for 'Siki' is now prohibited in the northeast Atlantic and other regions, following shocking stock declines in only a few years. Their recovery is unlikely during the authors' lifetimes. There are also proposals to close larger areas of the deep sea to all fishing activity. However, fishermen whose shallow water catches have declined still want to be able to fish for other valuable deep water species that occur within the range of these deepsea sharks. The problem, if such fisheries are permitted, is how to limit the bycatch mortality that could so easily drive these biologically vulnerable sharks to extinction.

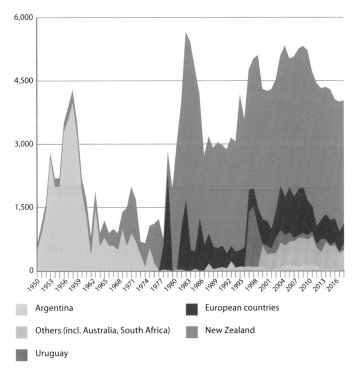

Argentina

Others (incl. Australia, South Africa)

Uruguay

European countries

New Zealand

Figure 58. Tope catches (t) by top fishing nations, 1950–2018.

Recreational angling

Recreational shark fisheries, while originally tiny in scale compared with industrial fisheries, are rapidly increasing in scale, economic importance and environmental impact around the world. During the 1950s, when sharks and rays were a lot more abundant, they comprised less than 1% of the total recreational catch; today it is closer to 6% and rising fast, particularly in Oceania and South America. Anglers value sharks, particularly large sharks, very highly as game fish and many are prepared to travel long distances and spend more on fishing trips if there is a chance that sharks might be caught. As fish stocks decline, recreational fishing is producing greater economic benefits to some local coastal communities than commercial fisheries, due to the high income from boat charters, sales of fishing gear, local accommodation and food, and because there are just so many sea anglers. At 220 million globally, there are more sea anglers than there are commercial fishing vessels! In Europe, anglers now take 27% of the total catch of seabass, which is substantial. Fortunately, catch and release fishing has a smaller impact on the health of fish stocks than traditional commercial fisheries, and sea anglers in many countries now release the sharks they catch.

Several decades ago, the majority of sports anglers brought their catch back to shore, to weigh and photograph it. It was only possible to claim a record weight if the dead fish was provided as proof. Some sports fisheries, particularly the large game fishing tournaments, are reported to have killed higher numbers of sharks than the commercial fishers working in the same waters. Over time, anglers realised that they were having a detrimental impact on shark stocks. Bag limits were introduced in some countries to regulate catches. Rare species were sometimes deleted from sports angling record books, to reduce the incentive to catch them (e.g. Angelshark in Ireland). Scientists developed length:weight conversion tables, allowing anglers to measure their catch and claim records without killing or even removing sharks from the water. Codes of conduct have been developed to avoid damage to fishes and maximise their chances of survival (see Appendix III p. 588). Today, catch and release has become the norm in many countries, where sea anglers are among the most active shark conservationists, and thousands contribute regularly to scientific research, through citizen science projects and the use of sea angling smartphone apps.

A recent UK study, in collaboration with over 1,500 anglers, assessed the economic contribution of sea angling for all fish species. It found that around 800,000 adults went sea fishing at least once a year, fishing for 7 million days. During this time, about 50 million fish were caught (including dogfish and Blue Shark), and 80% of them were released. Sea anglers spent an average of ~£1,000 (US$1,300) annually, and directly or indirectly supported about 15,000 jobs. In comparison, UK commercial fishing and fish processing industries employed 24,000 people. Across the whole of Europe, 8.7 million sea anglers (1.6% of the population) fished for 77.6 million days and spent €5.9 billion (US$7 billion, £5.3 billion) annually. Europe is mid-range, globally; sea anglers make up a larger proportion of the population in Australia, New Zealand and the USA, but smaller in South America and Africa (although rising very fast on the latter continents). Only in the USA do sea anglers spend more on their sport than the Europeans. In Florida, which has the largest recreational shark fishery in the USA, shark angling alone generated US$8 billion (£6 billion/€ 6.8 billion) in sales in 2011.

The world's largest and longest running sea angling citizen science project is the Cooperative Shark Tagging Program (CSTP), which uses tagging, release and recapture to study the life history of sharks in the Atlantic Ocean. This voluntary collaboration between sea anglers, commercial fishers and biologists began in the USA during the early 1960s. Over 50 years later, the CSTP has expanded to cover the whole of the North American and European Atlantic Ocean and Gulf of Mexico coasts. Thousands of participants from nearly 60 countries have contributed by tagging 35 species of shark or returning these tags. Almost 118,000 Blue Sharks have been tagged since the project began (representing 51% of the ~230,000 tags issued) and over 8,200 of them were recaptured. A long-distance record-breaking Blue Shark (p. 562) was tagged off Long Island, New York, and recaptured more than eight years later in the South Atlantic, off Africa, almost 4,000 nautical miles (nm) away – and had likely repeated that journey several times during its regular migrations. This Blue Shark was one of 20 species for which the CSTP has produced new distance records – individual Tiger Shark and Shortfin Mako journeys have exceeded 3,000nm, while Bigeye Thresher, Dusky and Sandbar Sharks have all been recorded travelling over 2,000nm. One Sandbar Shark (p. 542) was recaptured almost 28 years after it was first tagged. The CSTP and other shark tagging projects have not only produced vast quantities of scientific information, but have also changed the nature and impact of sea angling by encouraging the careful live release of catches and rewarding tag returns.

It is very important for anglers to follow good practice when catching and returning sharks to the water; they are more fragile than many fishers realise. Certain species, particularly the large hammerheads, are more likely than others to die after being caught (although this may not happen until after they have been put back into the sea). The gills and other internal organs are very easily damaged once the shark has been removed from the water, because they are not protected by the internal skeleton. Injury to the spinal column and internal organs is particularly likely to occur if the animal is lifted by its head or tail, or forced down onto a hard, dry surface. Gaffing eventually causes the death of many sharks and rays. Deep hooking in the gills or gut causes permanent damage to internal organs. Playing the fish for a long time can also result in sufficient stress and muscle damage to cause post-release mortality. Finally, removing a tag from a shark that can be released alive with the tag still attached is not always the right thing to do. Check for instructions on the tag, certainly record its number and take photos, but remember that a large electronic tag might still be recording and transmitting data to a researcher. Buying, programming and going out to sea to attach one of these tags is very expensive, and having the tag removed prematurely can ruin an important research programme that is helping shark conservation efforts. See Appendix III for good practice in shark angling and how to report tagged sharks.

Shark watching

Those lucky enough to see a large shark swimming freely in its natural environment are unlikely ever to forget their first encounter. This is one of the world's greatest wildlife experiences. What is more, there are so many ways to see sharks. As photos on these pages show: getting into the water is not essential, because good views of large sharks are possible from a boat. With a bit of practice, non-divers can use a snorkel on the surface, inside a cage or just holding onto a rope. Many more species are accessible to snorkellers and scuba divers in hundreds of places around the world. Divers (with deep pockets) can easily experience good views of well over 30 shark species at well-known diving hotspots. These range from Greenland Sharks in the cold waters of the Arctic; to Basking, Sixgill, Blue and White Sharks in temperate waters; and a wide variety of reef, hammerhead, thresher and other large sharks in warm seas. That is without counting the many species of beautifully patterned small sharks that live on or near the seabed in shallow coastal waters around the world. Angelsharks, for example, are increasingly important for attracting divers to the Canary Islands, off the west coast of northern Africa.

Having said that, just a few shark species in warm clear waters dominate the shark tourism industry. A global survey of shark watching sites found that Whale Shark encounters contributed ~30% of locations (many are seasonal aggregations), while reef sharks were present at 33% of sites. A few species can only be seen reliably in a few countries, locations or seasons. For example, White Shark cage diving is most important in South Africa, northwest Mexico, South Australia, USA and New Zealand.

Benefits of shark tourism

As with angling, shark watching can have major socio-economic benefits for the coastal communities that provide the boats, guides, accommodation and other facilities needed by visitors. In 2010, an estimated 590,000 recreational divers were thought to be taking part in shark diving in numerous locations across 45 countries worldwide (with dedicated shark tourism operations in 29 of these), directly supporting about 10,000 local jobs. Other researchers believed that these figures were underestimates. Growth in the industry was projected to reach almost 2.5 million shark divers by 2021.

The economic benefits of shark watching to small Pacific island states, particularly those with few other sources of revenue, are frequently quoted. In 2010, the year after Palau was declared a shark sanctuary, shark diving was estimated to be worth US$18 million, 8% of gross domestic product (GDP), the third highest contributor to Palau's gross tax revenue, and the source of US$1.2 million in local salaries. The 100 resident reef sharks at the main dive sites had an estimated lifetime value of US$1.9 million each (compared with US$108 if killed for their meat and fins). In Fiji, 78% of visiting divers went on shark dives, contributing over US$42 million to the economy, with at least US$4 million directed to local communities and US$17.5 million in tax revenues. Tourism's contribution to these economies has grown since then.

In 1990 the value of reef shark diving in the Maldives, Indian Ocean, was US$2.3 million, far more than the value of commercial shark fishing. Subsequently, all sharks were legally protected: reef sharks to defend the tourism industry, and pelagic sharks to support the tuna pole and line fishery (fishers believe that Silky Sharks aggregate tuna and make them easier to

catch). Whale Shark tourism became important and national revenues from shark diving alone doubled, to over US$43 million in 2016. Additional expenditure by diving visitors increased this income to almost US$155 million. In 2010, 78,000 Whale Shark tourists spent US$9.4 million visiting just one atoll. By 2014, tourism in the Maldives accounted for 27% of GDP, and most tourists dived or snorkelled. Those who visited specifically to see reef, Whale, Tiger and Thresher Sharks, said that they would be less likely to return if shark abundance fell or if they saw illegal shark fishing activity or shark products on sale, potentially halving dive tourism income. Conversely, higher numbers of sharks made return visits more likely and could substantially increase tourism income. Many of these divers visit other countries, where their expenditure is likely to be similarly linked to future trends in shark abundance and management.

Whale Shark provisioning has recently become established in the Philippines, where immature male sharks had started stealing the bait from handlines operated from small boats. The fishermen's solution (throwing shrimp to the sharks to draw them away from their lines) has since developed into a very successful, locally managed tourist attraction, but is controversial due to debate over whether provisioning is harmful to the sharks, to overfished shrimp stocks, and/or to the environment.

Highly inquisitive Oceanic Whitetip Sharks (p. 537) are now Critically Endangered globally and hard to find outside shark sanctuaries, such as The Bahamas, where they can be viewed from boats.

The Bahamas is the world's largest shark diving destination. Over 20 years, after banning longline shark fishing in 1993, the country hosted more than one million shark:diver interactions worth an estimated US$800 million (gross income). This represented over 70% of all Greater Caribbean ecotourism operations. Each reef shark at popular dive sites was worth tens of thousands of dollars in tourist revenue every year. A shark sanctuary was therefore declared in 2011. By 2015, diving contributed over US$110 million annually to the Bahamian economy, 99% of that from sharks and rays, including significant benefits to disadvantaged outer islands. Caribbean Reef Sharks (p. 541), which are now scarce or absent from many other Caribbean countries, contributed nearly 94% of this income, with the balance primarily from Great Hammerhead, Oceanic Whitetip and Tiger Sharks. These figures exclude diving from visiting liveaboard vessels. The latter are mostly from Florida, where divers were estimated to be spending some US$126 million annually on targeted shark dives (excluding the value of dives when encountering sharks was a priority, but not the target).

Balancing fisheries and tourism

These examples illustrate the high value of shark tourism to countries with long-established shark sanctuaries, strong management and healthy shark populations. The challenge for other nations is how to distribute income from shark tourism to those communities that presently only benefit from or depend upon fisheries.

Studies in Asia have recognised potential conflicts between tourism and shark fishing, highlighting the need to ensure that income from shark tourism supports the livelihoods of local communities and encourages their diversification away from shark fishing. However, it can be very difficult to ensure that fishers receive the direct benefits from tourism that would make this possible. Shark diving in Semporna, Malaysia, provided regional revenues of almost US$10 million in 2012, generating >US$2 million in government taxes and US$1.4 million in local salaries. Researchers also found that most

White Shark (p. 320) cage diving at Guadalupe Island, west coast of Mexico, is a huge tourist attraction in the region.

The popularity of Basking Shark (p. 311) watching is rising in the northeast Atlantic, from northwest France, through Ireland and the British Isles, to as far north as Iceland (where this photograph was taken).

visiting divers would be willing to pay additional fees to support marine protected area management and enforcement. In contrast, the economic value of commercial and subsistence shark fisheries in Semporna was less than 5% of the value of shark diving, but still important for the communities who were unable to benefit from tourism revenues.

Indonesia reports the world's largest national shark fishery catch, and overfishing is the primary threat to shark populations. Yet the country attracts almost 190,000 shark tourists annually to 24 shark watching hotspots. In 2017, shark tourism generated income of US$22 million, 1.45 times higher than the value of shark product exports. The loss of sharks from these hotspots might forfeit about 25% of dive tourist expenditure, causing an economic hit to shark and ray tourism of over US$121 million by 2027. To avoid this, and collateral damage to marine species, ecosystems, fisheries and coastal communities, it is vitally important to provide local fishers with strong incentives to conserve sharks.

Acceptable or dangerous?

Unfortunately, there can be unintended consequences to shark watching, and some aspects of this activity are very controversial. These include 'provisioning', meaning the use of chum to attract, or bait to feed sharks, thus conditioning them to tolerate and even approach watchers. This can be associated with changes to shark numbers, movements, stress levels and behaviour. It may also raise safety concerns for human participants, the sharks themselves and potentially even other water users. Provisioning is therefore one of the activities that may be strictly regulated by governments that licence shark ecotourism operations, to ensure that they do not damage the resources upon which they rely. Other concerns include unequal distribution of the costs and benefits of shark tourism within local communities, and unsustainable practices associated with shark watching. The latter range from damage to coral reefs, noise and pollution, to overfishing of the species used in provisioning or to feed the huge numbers of human visitors that flock to some of the

DOS AND DON'TS FOR SHARK WATCHING

There are many sources of guidance for taking part in safe, responsible and respectful shark tourism, whether shark watching, cage diving, or boat trips (the Shark Trust website is very helpful). The following are important considerations.

1. Choose a tourist operator who places top priority on their customers' safety (in and out of the water), stresses education rather than 'thrills', and supports shark research and conservation activities. Make sure that their operating licence and environmental and tourism accreditations are up to date. Check their reviews to assess their reputation and whether staff are well-trained. Whenever possible, support local businesses and communities.

2. Never try to touch a shark, even if the dive guide does so. These are large predators and can be unpredictable. You may not notice the danger signs that precede a bite. Riding, tail-pulling, and even touching a wild animal is harassment, which is often illegal as well as dangerous and disrespectful. Please don't do it!

3. Be aware that many dive operators use bait to attract sharks, to maximise good sightings for their clients. This activity should be licenced and use sustainable sources of bait. In some places, actually feeding the sharks is illegal (the scent trail is sufficient to attract them); elsewhere feeding is allowed. If the latter, please leave this to the experts! Some very experienced dive operators have been bitten while feeding sharks. Indiscriminate chumming around boats is bad practice, particularly when people share the water with sharks; feeding or chumming for sharks is now illegal in a few places.

4. Shark watching does not always involve bait, particularly where sharks are common and/or form natural aggregations. For example, it is possible to watch filter-feeding Basking and Whale Sharks, groups of Sandtiger and Scalloped Hammerhead Sharks, and various species of juvenile sharks in their nursery grounds – but please be careful not to disturb or harass these animals in their natural habitats.

5. Diving in aquariums is guaranteed to produce close-up views – but points 1 and 2 still apply.

world's popular shark watching centres. The box above suggests some pointers for ethical shark watching that help to minimise these concerns.

On the other hand, shark watchers can contribute to research, through citizen science projects such as photo-identification or counting sharks. Shark tourism raises awareness of the importance of healthy marine habitats, ecosystems and shark populations, and encourages their conservation and management by locals. Ideally, it provides a profitable alternative to unsustainable shark fishing and spreads economic benefits across the whole community, including to former shark fishers. By helping to dispel some of the myths and preconceptions that have often hindered efforts to protect sharks, the groundswell of public opinion supporting shark conservation influences decision-makers in host countries and across the world – shark watching is a very powerful argument for protecting marine biodiversity.

Shark encounters

Sharks live in an environment that is alien to humans. Many people perceive the sea as deep and dark, mysterious, dangerous, full of shipwrecks, and (until recently) largely unknown. Countless numbers have drowned since humans first ventured out from shore. Even today, hundreds of thousands drown accidentally every year; only motor vehicles are more dangerous than water. Strangely, though, it is not the ocean (or the journey to the beach) that scares the public, but the incredibly small risk of an encounter with some of the most elusive and rarely seen animals it contains. Many tens of millions of people use the sea for work or recreation, but there are only about 100 shark bite incidents reported globally each year. Furthermore, some 40% of these bites are provoked – defined as a human initiating the interaction with the shark. Provoked, or not, these incidents result in an annual average of four or five deaths. Although the risk of serious injury from sharks is incredibly small, and the fear produced may seem to be completely out of proportion to this risk, the shock and horror engendered by the rare events of serious injury or death are understandable; these are tragedies – not least because the likelihood of them occurring at all is so slim.

Very few of the coastal species most likely to come into contact with people pose a serious danger (exceptions are the Bull, White, and Tiger Sharks). Most shark encounters are sporadic, erratic, and almost accidental, with many having been avoidable. The huge majority of unprovoked shark bites are caused by mistaken identity – a shark identifies a sequence of noise, vibration or movement as a potential prey item and strikes at the pale flashing hand or foot of a swimmer or surfer that it has mistaken for a fish. It then retreats instantly when it realises its mistake – termed a 'hit and run'. Most times, the injury is so minor that no medical attention is needed – indeed, the attacker is usually much smaller than the victim. In the less common, but more dangerous attacks, large sharks 'bump and bite' the victim, or carry out a 'sneak' attack 'out of nowhere' to deliver a single bite – this a common strategy for disabling large marine mammal prey. One bite may be sufficient to immobilise the quarry, even if the shark doesn't return, and it may result in death if the casualty is not immediately helped to shore and given medical attention. Fortunately, people are far more likely than seals to receive the help necessary for them to survive such incidents. Examples of multiple bites and sharks actually feeding on human prey are extremely rare and, above all, rogue sharks that become 'serial man-eaters' seem to be myths generated by the fictional Jaws films.

Shark bites can occur anywhere that sharks and people may meet in the water (although none have been recorded in the cold, overfished waters of the United Kingdom) and in almost any water depth sufficient to allow a shark to swim. Incidents are more likely to occur when larger numbers of people are using the same water as sharks, simply because this increases the mathematical probability of an encounter between them. However, the incidence of shark attacks has not risen as steeply as the numbers of people entering the water for recreational purposes has increased – perhaps because water users are taking more care to avoid risky behaviour. Despite booming beach tourism and a 400% increase in PADI-registered divers from 1990–2008, shark attacks have only increased by some 25% per decade since the 1990s – and the numbers are still very small. In Florida, USA, where most bites occur, the number has remained at around one bite per half a million beach visitors since the early 1990s; twice as many people die in alligator attacks and 30 times as many from domestic dog attacks. Globally, there is one tragic drowning fatality among every 3.5 million beach visits, compared with one (non-fatal) shark bite in 11.5 million visits.

In addition to increased human coastal populations, research into the potential causes of unprovoked shark bites has identified habitat destruction, water quality, climate change and changes in the abundance of normal shark prey as possible contributing factors, but recovering shark numbers in recent decades off the coasts of Australia and USA appear unlikely to have influenced bite rates there.

Most shark bite incidents recorded in shark attack databases fit into one of the following classic patterns. Many bites occur near dawn or dusk, when sharks are hunting actively. Many occur in murky water, for example near river mouths, after high rainfall, or during onshore winds. These poor visibility conditions seem more likely to lead to cases of mistaken identity resulting in bites, although large predatory sharks, such as Bull Sharks, can also approach their prey more closely in these conditions without being observed. Many bites seem to occur because the casualty was swimming erratically (splashing around) or swimming with dolphins, which are often accompanied by sharks. Swimming alone, rather than with a buddy, can also increase chances of injury from a shark, as can urinating in the water; feeding sharks; spearfishing; swimming near boats that are fishing, discarding bait, slurry or dead animals from their cargo; wearing strongly contrasting, coloured clothing and/or jewellery; or simply ignoring the warnings provided to swimmers when large, dangerous sharks are known to be present. Swimming right up to sharks, following them around closely, picking them up or pulling their tails are also practically fail-safe ways to attract a shark bite (who would not expect to be bitten if they tried that with an unknown dog?). Avoiding the above behaviour patterns and treating sharks with respect will greatly decrease your (already tiny) chances of a damaging encounter with a shark.

Apart from changing human behaviour in the water, other ways to reduce the risk of shark bites include culling sharks; using physical barriers to separate sharks and people; surveillance systems to warn swimmers when dangerous large sharks are present; and personal shark deterrents.

Shark culls were historically the favoured way to mitigate the risk of shark bite: by reducing the number of sharks present, the chances of an encounter between sharks and water users is also reduced. Since most large, dangerous sharks are highly migratory, short-term, local culls generally do not lessen the risk of shark bite. Long-term beach meshing programmes covering large areas of coast during the 20th century did reduce shark populations and the incidence of shark bites in Australia and South Africa (commercial shark fisheries also contributed). Unfortunately, however, shark nets kill indiscriminately, including larger numbers of harmless and protected species such as other sharks, sea turtles and cetaceans. The nets do not form a continuous physical barrier to separate sharks and swimmers, and can be replaced by drumlines with large baited hooks to target the largest and likely most dangerous sharks. However, any culling of top predators can have damaging ecological and economic consequences – the latter when the large sharks targeted also support important ecotourism operations. Arguably the strongest lesson learnt from these programmes is that, once begun, it is virtually impossible to give them up due to fear of the public relations fallout associated with a shark bite at an important bathing beach from which nets and drumlines had been withdrawn.

A major research programme into shark hazard mitigation, initiated after a series of White Shark attacks in Western Australia, recommended against using beach nets and drumlines. Net barriers, extending from shore to shore, seabed to surface, were suggested instead. These have been used in a few locations around the world to enclose relatively small areas of water, but are vulnerable to damage in bad weather. Since 2013 a 350m shark exclusion net at Fish Hoek, South Africa, has been deployed successfully by hand almost daily (weather permitting) during the summer, and during weekends and holidays in spring and autumn. Watchers ensure that whales and dolphins are not entangled. This is one of several beaches covered by Cape Town's Shark Spotters Programme, which uses trained observers to look for large sharks before they reach water users and deploy signals alerting beach users when a shark is nearby, or when the weather conditions prevent spotters from scanning for sharks.

Alternative mitigation measures adopted under Western Australia's Sharksmart and Sea Sense initiatives include a smartphone app providing near real-time information on shark activity; beach and aerial surveillance; beach enclosures; warning lights and sirens to alert beach users to approaching sharks; and a $200 rebate to residents who purchase an approved electronic personal shark deterrent device.

The search for an effective personal shark deterrent dates back to the 1940s in the USA and has accelerated in recent years in Australia and South Africa. Hundreds of years ago, fishermen reportedly used to hang a decaying shark from their boats to keep sharks away from their catch. With this in mind, shark scientists initially focused their attention on the chemicals produced during the decomposition of shark flesh (known as necromones). These certainly have a repellent effect on sharks, even when they are actively feeding. A small Red Sea flatfish, the Moses Sole, secrets a particularly noxious toxin from glands at the base of its dorsal and anal fins that defends it from shark predation; this chemical has also been tested successfully as a shark deterrent. However, because these chemicals diffuse away rapidly in sea water, they only work well if sprayed directly at a shark. Since many dangerous sharks make ambush attacks, there may be no opportunity for a diver or snorkeller to 'mace' one as it approaches. Necromones have, however, been incorporated into slow-release shark-repellent chum for anglers fed up with losing their catch to sharks.

Experiments with acoustic or visual deterrents haven't been particularly successful. Mimicking the noise of approaching orcas – major predators of large sharks – might have worked if sharks could clearly hear that sound frequency. Flashing a large black and white orca pattern at approaching sharks can give them a nasty surprise – at least the first time it happens, but the image needs to be large. The drawback to using the banded warning patterns of seasnakes is that these are prey species for some sharks. As an aside, a few shark researchers refer to the 'safety yellow' colour of some buoyancy aids as 'yum-yum yellow' – some sharks reportedly find this shade particularly attractive.

Devices that emit electric currents that over-stimulate a shark's electrosensory system (p. 41) are now being commercially produced for use as personal deterrents by surfers, divers and kayak fishers. Although sharks are attracted to the electric currents that mimic the bioelectric fields produced by their prey, they are repelled by much stronger signals. Though it becomes complicated, as the most effective current strength varies between species. Sharks may become habituated (develop increased tolerance) to these fields with repeated exposures, and the effect is restricted to a radius of less than 2m from the device – so it is important to use it correctly. Research is also underway into the development of much smaller devices that might be used on commercial longlines to prevent shark depredation of hooked tunas or billfishes. The challenge is to find a cost-effective model that could be deployed on thousands of hooks. The research continues.

Many swimmers choose to use this corner of Fish Hoek Beach (near Cape Town, South Africa), protected by a net barrier that excludes sharks and other marine life. To minimise bycatch and damage to the net, it is set and removed every morning and evening in summer and for other major holidays.

Eating sharks

Historically, shark meat has been an important food for coastal communities and transported, dried salted, via the great continental trading routes for thousands of years; it is still widely traded in this traditional form. Other products – teeth, fins, cartilage, skin and oil – travelled even further and, when they reached their final destination, were no longer associated with the sharks that had provided them. 'Shark fin' is 'fish wing' in Chinese, and their teeth were once thought to be dragons' teeth. Shark products are still sold under a wide variety of trade names, for marketing reasons and to disguise their origin. This makes it difficult for the responsible consumer to identify shark products to species level, or to find commodities derived from sustainably managed fisheries.

Shark meat was traditionally unpopular because the high levels of urea in the flesh necessary for osmoregulation (p. 35) are rapidly converted to ammonia by bacterial enzymes after the shark has died. Only a tiny quantity of urea (1.8%) and even less ammonia (0.03%) is enough to taint shark meat and make it unpleasant to eat. These problems are overcome by careful handling and storage; over 80% of shark meat is now traded globally as frozen steaks and fillets. The quantity of shark meat in trade doubled between 1985 and 2001, peaking at about 100,000 tonnes, then declined as catches fell during the past decade. Most of it is destined for Europe (particularly Spain, Italy, Portugal and France) and South America (Brazil, Uruguay and Peru). Public health warnings are regularly issued against eating too much shark meat, because of the toxins they contain (see box opposite).

Shark fin has been one of the world's most valuable seafood products for over two thousand years. It was once so rare, precious and difficult to prepare well that it was originally only eaten at the banquets of Chinese emperors (Ming Dynasty, 14th to 17th century). Until recently, this was still a dish served only at the most important occasions, but its consumption rose greatly with the liberalisation and growth of major Asian economies during the 1980s.

These newborn sharks are being sold for their meat.

Shark fin processing involves bleaching, removing the skin and basal cartilage, then carefully detaching the two layers of valuable fin rays from each side of the radial cartilage (p.25), shown here with some vertebral columns.

Associated with this was an increase in disposable income and huge growth in consumer demand for luxury products, including shark fin soup.

Shark fins consist primarily of cartilage, skin and ceratotrichia (p. 30). They are therefore lightweight and easy to air-dry or freeze and store for long periods onboard ship or on land, until they can be sold. Interestingly the amount, size and consistency of ceratotrichia within shark fins varies between species, which means that fins from some species are more valuable than those of an equal size from others. Because of this, fin traders have learned accurately to identify the dried fins of many species simply by eye. Value is also determined by fin position and size: not only are the pelvic and anal fins less valuable because they are smaller, but pectoral, dorsal and caudal fins contain a larger density of high-value ceratotrichia because they have to be very stiff and solid for the shark to swim efficiently.

Comprehensive fin trade data published during the 1990–2000s by China, the major processing and consuming country, recorded the doubling of reported imports of raw shark fin to Hong Kong and the steepest increase in global consumption of shark fin soup. Since then, the quantity and value of shark fins traded has fallen slightly. Hong Kong Special Autonomous Region (SAR) remains by far the largest importer, followed by Malaysia, mainland China and Singapore. Several conservation organisations operate consumer education campaigns that aim to reduce consumption of shark fin soup within Asian communities. These campaigns highlight the threat that overfishing poses to shark populations, ethical problems associated with shark finning, and the human health risk associated with eating shark products (see box).

Cartilage appears in various medicinal products, primarily as a human and pet health supplement for the treatment of joint problems. Similar products are available that do not contain shark cartilage. Cartilage has also, notoriously, been marketed as a cancer cure, primarily on the basis that sharks do not get cancer (which is untrue, sharks suffer from a great many cancers). Scientific and medical trials of shark cartilage concluded that it is ineffective in treating cancer or improving quality of life for cancer patients, and the companies that had promoted cartilage as a cancer treatment were required to compensate their customers. Furthermore, many potentially detrimental side effects associated with consuming shark cartilage have been identified. Astonishingly, consumers still spend millions of dollars annually on shark cartilage supplements.

Oil, which can be easily extracted from shark livers (traditionally by chopping it up and leaving it to decay in containers in the sun), has been used for hundreds of years to waterproof boats, for lighting, and for a variety of medical and industrial purposes. The high levels of vitamin A in shark liver oil prompted some important shark fisheries in the early 20th century, until synthetic vitamin A became available and the fisheries producing health supplements collapsed. Today, demand for shark liver oil is rising again, for use in cosmetics (creams, lipsticks and gloss), pharmaceuticals and as a machine oil. This is because it contains high levels of squalene, a very light non-greasy oil that moisturises skin and lubricates machinery very effectively. Squalene is certainly a valuable compound, but it does not have to be extracted from shark livers. It can also be synthesised or derived from plants – good news for threatened deepsea sharks whose valuable liver oil was once a major product and driver of target shark fisheries. Consumer awareness campaigns opposing the use of shark oil in cosmetics are now gaining publicity.

Shark skin makes a very durable, thick leather (far stronger than cow hide) when correctly treated and is commonly manufactured into belts and wallets. Dried, untreated skin, with the denticles still embedded, is called 'shagreen' and makes a very effective sandpaper and, traditionally, a non-slip binding for sword hilts. Polished shark leather with denticles is known as 'boroso' leather and used for a wide variety of very expensive products (although ray leather is more commonly seen).

Among the newest products to arrive on the shelves of 'traditional' Chinese medicine stores are sets of the fern-like dried **gill plates** (gill rakers) of mantas and devil rays. It is possible that Whale Shark and Basking Shark gill plates could also be of interest to traders, although apparently not yet reported in this market.

BIOACCUMULATION AND BIOMAGNIFICATION

A variety of environmental toxins and pollutants, including mercury (an extremely poisonous heavy metal) and polychlorinated biphenyls (PCBs), occur in seawater. Once absorbed or ingested, these persistent toxins cannot be excreted because they are not soluble; they therefore **bioaccumulate** in the tissues of marine animals. The quantity of toxins present in an animal depends upon its age, diet, and how polluted its environment is. Marine algae absorb very low levels of toxins, which then enter the diet of herbivorous animals. Herbivores only accumulate small quantities of toxins from algae, although older, larger animals will contain more than younger animals in the same population because they have been bioaccumulating the toxins for longer. Larger fishes, slightly higher up in the food chain, eat the herbivores and are, in their turn, preyed upon by even larger predators. **Biomagnification** is the process by which a predatory animal acquires, in its diet, all of the toxins that have accumulated throughout the lives of all of the plants and animals lower down in the food chain. The larger and older the predator, the greater the concentration of toxins in its body tissues. Thus, the bodies of small, short-lived planktonic fishes contain less than 1/100th of the amount of mercury found in a shark. Humans, at the very top of the food chain, also bioaccumulate mercury and other environmental toxins, with potentially serious health effects. This is why public health and food standards authorities issue warnings on the danger, particularly for pregnant women and children, of eating large predatory fishes such as sharks and marlin too frequently.

Below: Overconsumption of shark cartilage may present more health hazards than benefits (see text).

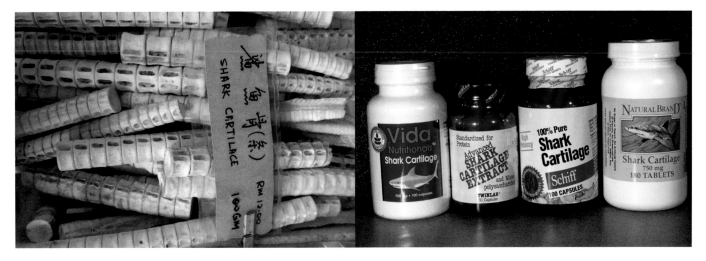

Conservation and management

Species conservation

Before the 1980s, the idea of 'shark conservation' was inconceivable and shark fisheries management barely warranted a second thought. Then, in 1984, the state of New South Wales (Australia) adopted the world's first ever legal protection for a shark: the Sandtiger Shark, (p. 305, locally known as Grey Nurse), possibly the world's most threatened shark population. This iconic and harmless species, which provided one of Australia's most thrilling dive experiences, had been depleted by a combination of accidental bycatch in fisheries and deliberate persecution by anglers and divers, and was subsequently classified as Critically Endangered in the region. The Sandtiger Shark was certainly not the world's only threatened shark, but it remained the world's only protected shark until 1991, when South Africa adopted national protection for the White Shark (p. 320). This was also the year of the world's first International Shark Conference, convened in Australia. The media response to such a novel concept was mixed (one local radio station called for an improved shark eradication programme), but the event marked the beginning of international shark conservation and management efforts. Other countries began to enact national protections for threatened sharks, followed eventually by listings in some United Nations Regional Seas Conventions and Action Plans (RSCAPs – see box opposite) and global biodiversity conventions. Resources devoted to shark conservation even began to exceed those that had, in several countries, formerly been deployed in eradication.

The Sandtiger Shark became the world's first protected shark species in 1984 – but only in Australia, where it is known as Grey Nurse Shark.

Figure 60 (p. 85) illustrates the rate of international shark conservation progress over the last 20 years. Protected status for threatened sharks initially focused on the high-profile 'big three': Whale Shark, Basking Shark and White Shark. These were the first shark species to be listed in the two main international biodiversity agreements: the Convention on Migratory Species (CMS) and the Convention on International Trade in Endangered Species (CITES). Listing sharks in the CITES Appendices (see box – p. 85) was a particularly slow process, due to the difficulty of obtaining agreement to binding trade regulations from two-thirds of the signatory states. Although CITES started discussing unsustainable trade in shark products and adopted the first CITES Shark Resolution in 1994, the first shark species were not listed until 2002, with commercially important species not listed until 2013 (see Table opposite). Progress in listing shark species under CMS has been slightly quicker: in 1999, the Whale Shark was the first listed, and by 2020 the Appendices included 19 shark species. A similar (but not identical) species list is annexed to the voluntary CMS Memorandum of Understanding on Migratory Sharks, adopted in 2010. This is the world's only international agreement dedicated to sharks and their relatives. It encourages its Parties (see box – p. 85) and additional states that are not party to CMS to sign and collaborate in the implementation of an international migratory Shark Action Plan. The regional biodiversity agreements for the North Atlantic (OSPAR Convention), Mediterranean (Barcelona Convention) and Wider Caribbean (Cartagena Convention) have also listed several shark species, some of which are threatened regionally rather than globally, and a few possibly not threatened at all – but still in need of management.

Table 1 (see opposite) illustrates the significant overlap between the species lists for these environmental agreements, which include a combined total of just 30 of almost 140 threatened shark species. Although over 70 shark species have been identified in the fin trade, only 14 are listed in CITES, while CMS includes only 19 of the 70 species considered to be migratory. (Although these two sets of 70 traded/migratory species overlap, they are not identical, and both include some species that are not threatened with extinction.) The reason for such a short list of protected sharks, when so many species are threatened, is partly because the large, charismatic species are the most widely recognised and more likely to be proposed for listing. The majority of threatened sharks are small, do not migrate, and are a bycatch rather than a fisheries target. Furthermore, although some of their products likely end up in international trade, these are not of high value and are hard to identify to species level by eye. Two-thirds of CITES Parties will not agree to add sharks to the Appendices unless the species are clearly threatened and their products are sufficiently easy for a customs officer to recognise, so that the implementation of trade regulations and tracing of shipments through the supply chain is practicable. Ultimately, the responsibility for managing shark species, listed or not, lies with individual countries, whether through fisheries or environmental regulations in their own waters, and/or regionally through RSCAPs or Regional Fishery Bodies (RFBs).

Fisheries management

Fortunately, partly in response to the rising biodiversity conservation profile of sharks, and partly due to their increased economic importance, the management of shark fisheries has also gradually improved over the past 30 years. Fisheries management takes place at national and regional level – the former within the areas of sea controlled by coastal states (territorial waters close to shore and larger, offshore EEZs), and the latter on the high seas under the remit of RFBs. Nevertheless even on the high seas, it is the countries where fishing vessels are flagged that are required to mandate and monitor compliance with fisheries management measures.

Table 1. Conservation and fisheries management measures for shark species.

English name	Scientific name	IUCN	Biodiversity Conventions		Tuna Regional Fisheries Management Organisations (tRFMOs)					Other Regional Fisheries Management Organisations (RFMOs)			Regional Seas Conventions and Action Plans (RSCAPs)			
			CITES	CMS	CCBST	IATTC	ICCAT	IOTC	WCPFC	GFCM	NEAFC	NAFO	BarCon	CCAMLR	OSPAR	WCR
Sharpnose Sevengill Shark	*Heptranchias perlo*	NT											III			
Spurdog	*Squalus acanthias*	VU		II, MOU							o		III		λ	
Gulper Shark	*Centrophorus granulosus*	EN											III		λ	
Leafscale Gulper Shark	*Centrophorus squamosus*	EN													λ	
Portuguese Dogfish	*Centroscymnus coelolepis*	NT													λ	
Greenland Shark	*Somniosus microcephalus*	VU										o				
Sleeper sharks	*S. antarcticus/S. longus*	LC/DD												o		
Angular Roughshark	*Oxynotus centrina*	VU								x			II			
Sawback Angelshark	*Squatina aculeata*	CR								x			II			
Smoothback Angelshark	*Squatina oculata*	CR								x			II			
Angelshark	*Squatina squatina*	CR		I, II, MOU						x			II		λ	
Whale Shark	*Rhincodon typus*	EN	II	I, II, MOU	x	x		x	x							III
Smalltooth Sandtiger	*Odontaspis ferox*	VU								x			II			
Bigeye Thresher	*Alopias superciliosus*	VU	II	II, MOU	x		x	x								
Common Thresher	*Alopias vulpinus*	VU	II	II, MOU	x		x**	x					III			
Pelagic Thresher	*Alopias pelagicus*	EN	II	II, MOU	x			x								
Basking Shark	*Cetorhinus maximus*	EN	II	I, II, MOU						x	o		II			
White Shark	*Carcharodon carcharias*	VU	II	I, II, MOU						x			II		λ	
Shortfin Mako	*Isurus oxyrinchus*	EN	II	II, MOU			x**			x			II			
Longfin Mako	*Isurus paucus*	EN	II	II, MOU												
Porbeagle	*Lamna nasus*	VU	II	II, MOU	x		x			x	o		II			
Tope	*Galeorhinus galeus*	CR		II						x			II		λ	
Starry Smoothhound	*Mustelus asterias*	LC											III			
Smoothhound	*Mustelus mustelus*	VU											III			
Blackspotted Smoothhound	*Mustelus punctulatus*	DD											III			
Silky Shark	*Carcharhinus falciformis*	VU	II	II, MOU	x	o x**	x	o, x	x							III
Oceanic Whitetip Shark	*Carcharhinus longimanus*	CR	II	I, MOU	x	x	x	x	x							III
Dusky Shark	*Carcharhinus obscurus*	EN		II, MOU												
Sandbar Shark	*Carcharhinus plumbeus*	VU											III			
Blue Shark	*Prionace glauca*	NT		II	o		o						III			
Scalloped Hammerhead	*Sphyrna lewini*	CR	II	II, MOU	x		x			x			II			III
Great Hammerhead	*Sphyrna mokarran*	CR	II	II, MOU	x		x			x			II			III
Smooth Hammerhead	*Sphyrna zygaena*	VU	II	II, MOU	x		x			x			II			III
Deepsea sharks											o				o	
All species															o	

Biodiversity Conventions (see p. 85)

CITES: Convention on International Trade in Endangered Species of Wild Fauna & Flora
Appendix I: Species whose international commercial trade is prohibited
Appendix II: Species whose international trade is regulated under permit

CMS: Convention on the Conservation of Migratory Species of Wild Animals
Appendix I: Species are to be strictly protected
Appendix II: CMS Parties cooperate for these species' conservation
MOU: Memorandum of Understanding on the Conservation of Migratory Sharks (Annex 1)

Tuna Regional Fisheries Management Organizsations (tRFMOs)

CCBST: Commission for the Conservation of Southern Bluefin Tuna
Requires Members to follow measures adopted by IOTC, WCPFC and ICCAT

IATTC: Inter-American Tropical Tuna Commission
o management measure applies; ** exemptions apply; x species prohibited
Shark finning regulated through a fin:carcass ratio. Particular attention paid to FAD management

ICCAT: International Commission for the Conservation of Atlantic Tunas
o management measure applies; ** exemptions apply; x species prohibited
Shark finning regulated through a fin:carcass ratio

IOTC: Indian Ocean Tuna Commission
o management measure applies; ** exemptions apply; x species prohibited
Shark finning regulated through 'fins attached' (fresh sharks) or fin:carcass ratio (frozen sharks)
Particular attention paid to management of FADs

WCPFC: Western and Central Pacific Fisheries Commission
o management measure applies; ** exemptions apply; x species prohibited
Shark finning regulated through 'fins naturally attached', or bagged, tied together, labelled/shark

Other Regional Fisheries Management Organisations (RFMOs)

GFCM: General Fisheries Commission for the Mediterranean
* Prohibits retention, transshipment, landing, storage, display and sale of species listed in BarCon SPA/BD. Annex II. Shark finning regulated through 'fins naturally attached'

NAFO: Northwest Atlantic Fisheries Organisation
o target fishing prohibited; Shark finning regulated through 'fins naturally attached'

NEAFC: North East Atlantic Fisheries Commission
o target fishing prohibited; Shark finning regulated through 'fins naturally attached'

Regional Seas Conventions and Action Plans

BarCon: Barcelona Convention for the Protection of the Mediterranean
SPA/BD: BarCon Protocol concerning Specially Protected Areas and Biological Diversity
Annex II: Endangered and threatened species. Annex III: Species whose exploitation is regulated

CCAMLR: Commission for the Conservation of Antarctic Marine Living Resources
o prohibits targeting and deep-sea gillnetting; sleeper shark *Somniosus* spp. bycatch limits

OSPAR Convention for the Protection of the Marine Environment of the North-East Atlantic
λ included in OSPAR list of threatened and/or declining species

WCR: Convention for the Protection and Development of the Marine Environment in the Wider Caribbean Region (Cartagena Convention)
SPAW: Protocol Concerning Specially Protected Areas and Wildlife
Annex III: species whose exploitation is regulated

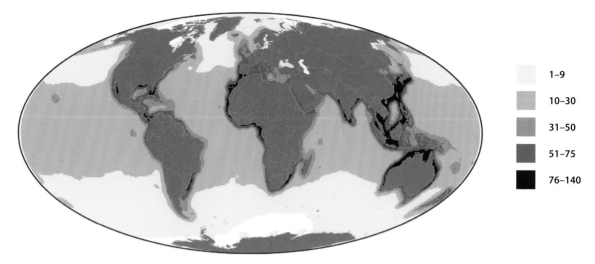

Figure 59. Global distribution of chondrichthyan fish diversity (shading indicates numbers of species present).

	1–9
	10–30
	31–50
	51–75
	76–140

Australia, the USA, and Canada were the first countries to adopt shark fishery management plans, in the late 1980s and early 1990s. At a global level, the FAO International Plan of Action for the Conservation and Management of Sharks (IPOA–Sharks, which also covers skates, rays and chimaeras) was formally adopted in 1999. This optimistically urged all shark fishing states to produce Shark Assessment reports and adopt a National Shark Management Plan (NPOA) by 2001 – but it is voluntary, and progress was disappointing. By 2012, only one-third of the 143 countries reporting shark and ray catches to the FAO during the previous decade had adopted an NPOA. However, these encompassed nearly two-thirds of the 26 fishing nations that each reported at least 1% of the global total, including some European Union Member States operating under the EU Community Plan of Action for sharks. Together, they accounted for 84% of the world's catch. By 2020, the ranking of countries by their shark catch had changed slightly; the seven largest fishers were the same, but they were now reporting almost 60% of the world's entire shark and ray catch between them, an increase from about 48% a decade earlier. Other countries now report significantly fewer shark landings due to declining stocks, restrictive management or both. Concerningly, a few of the world's major fishing countries are still not reporting any shark catch,while roughly a quarter of the world's largest shark reporting countries have apparently still not adopted formal shark management plans. At the same time, some of the largest shark fisheries still do not have any effective management – several are in the regions with the highest biodiversity of sharks, rays and chimaeras (Figure 59). The Shark Plans that do exist are a mixed bunch. Some are well integrated into effective fisheries management frameworks; others were apparently paper exercises and have had little or no impact on fisheries or even catch reporting in the countries that published them. Clearly, lack of management capacity (e.g. trained staff) and fisheries data (often due to species identification problems) are common challenges for many countries that are struggling to implement the IPOA–Sharks – sustainable shark fisheries management is difficult to achieve, even in developed countries. The IPOA-Sharks did, however, highlight the problem of shark finning, a relatively easy management issue to address, helping to promote national and regional finning bans that prevented one unsustainable form of exploitation (see Box, p. 68).

National Shark Plans mostly govern fisheries on the continental shelf, inside states' EEZs, although the EEZ of island states (that lack a shelf) include large areas of open ocean. Instead of managing shark fisheries within their EEZ, some countries (most of them small island states) decided simply to prohibit all commercial shark fishing within their waters. Others went further, also prohibiting the retention of shark bycatch and the sale, possession and trade of any sharks or shark products. Between 2009 and 2017, at least 13 nations and territories established shark sanctuaries (other fish species are not protected), some of them huge and when combined, cover millions of square miles of open ocean. The benefits of protecting large predators include healthier marine ecosystems and, in some cases, extremely valuable shark ecotourism operations (see p. 75). Since these areas are still open to fisheries for other pelagic species, including by foreign fleets, enforcing these shark fishing bans may be challenging for small island states with limited fisheries management resources. However, some high seas tuna fleets, recognising the marketing value of shark-friendly fisheries, have independently adopted a voluntary policy of not allowing any bycaught sharks (whole or parts) to be kept on board.

Another relatively simple, way to reduce shark mortality is by restricting use of the fishing gears that catch the most sharks (or the most threatened shark species). As described on p. 67, gillnets are among the most destructive of fishing gears because of their indiscriminate bycatch. It is surprising, therefore, that bans on the possession and use of gillnets in national fisheries, as adopted in Belize in 2020, are not more widespread – see Box p. 86. Similarly, prohibitions on bottom trawling, similar to Sri Lanka's in 2017, would benefit many sharks and other species, and their habitats.

High seas fisheries fall outside coastal states' jurisdiction (other than through national regulations that affect all flagged vessels). Over 50 RFBs have been established by international agreement to support the management of fish stocks fished by several states, most of them by providing technical non-binding advice. These RFMOs can, however, adopt mandatory conservation and fisheries management measures for certain fish stocks within specified ocean

BIODIVERSITY CONVENTIONS

Convention on International Trade in Endangered Species of Wild Fauna and Flora (CITES)

CITES was established to protect wild species from overexploitation through international trade, by promoting cooperation between the 183 CITES Parties (the 95% of the world's countries that have signed the Convention). Commercial trade is prohibited for species threatened with extinction, which are listed in Appendix I, and strictly regulated for species listed in Appendix II, that may become threatened unless trade in their products is legal, sustainable and traceable. Adding species to these two Appendices must be agreed by at least two-thirds of Parties voting at a meeting of the Conference which is held every two to three years. Parties that do not wish to be bound by listings can take out a reservation (but still have to abide by the provisions of CITES if they trade with Parties that have not taken out a reservation). Appendix III listings are requests from individual Parties for international assistance to support their national legislation (e.g. implementing trade bans for protected species). Resolutions and Decisions adopted by the Conference also contribute to shark conservation and management activities. See www.cites.org

By 2020, 14 shark species had been listed in CITES Appendix II (none in Appendix I). Their products can be traded if the exporting Party confirms that the species were obtained legally, and certifies that the trade is not detrimental to their wild populations (i.e. sustainable). Export, import and re-export permits ensure that the trade is traceable. Live animals must be transported according to welfare guidelines. Exporting and importing Parties need to be able to identify the products being traded, which for sharks are often their fins (meat is also traded). A quick guide to the visual identification of fins of CITES-listed species is provided on p. 592. CITES Parties may use genetic analyses to confirm the identification of shark products, when illegal trade is suspected.

Convention on the Conservation of Migratory Species (CMS)

CMS aims to conserve migratory terrestrial, aquatic and avian species throughout their range. In January 2021, it had 132 Parties, but non-Parties can also take part in CMS initiatives. Migratory species threatened with extinction are listed in Appendix I; Parties have obligations to strictly protect these species, their habitats and migratory routes. Appendix II lists species that need or would benefit from international co-operation. Unlike CITES, a species may be listed in both CMS Appendices. In 2020, 19 shark species were listed: four in both Appendix I and II, 14 only in Appendix II, and one shark species only in Appendix I. IN 2010, CMS also adopted the world's first international instrument for the conservation of sharks and their relatives, the Memorandum of Understanding on Migratory Sharks. This voluntary MOU aims to achieve and maintain a favourable conservation status for the migratory sharks in its Annex, which in 2020 included all but two of the species listed in the main CMS Appendices. See www.cms.int and www.sharkmou.org

Regional Seas Conventions and Action Plans (RSCAPs)

The UN Regional Seas Programme is implemented through 18 RSCAPs, with over 140 participating countries. In 2020, only three of the RSCAPs listed shark species that require regional conservation and management action: the OSPAR Convention for the Protection of the Marine Environment of the North-East Atlantic, the Barcelona Convention for the Protection of the Marine Environment and Coastal Region of the Mediterranean, and the Cartagena Convention for the Protection and Development of the Marine Environment in the Wider Caribbean Region. Other RSCAPs, in the South Pacific, East Africa and the Red Sea and Gulf of Aden, also have shark conservation programmes. The General Fisheries Council for the Mediterranean has adopted the list of Endangered and Threatened shark species from Annex II of the Barcelona Convention's Specially Protected Areas and Biological Diversity (SPA/BD) Protocol as a GFCM prohibited species list.

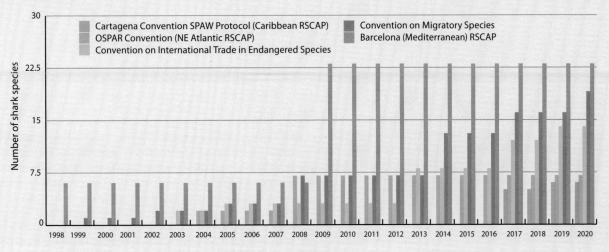

Figure 60. Numbers of shark species listed in biodiversity conventions.

MAKING A NET DIFFERENCE IN BELIZE

Traditional fishers were the first to see the losses in Belize's fish populations and coastal wildlife that followed the introduction of unselective monofilament gillnets last century. 'I haven't seen a sawfish since the 90s, nets killed them off' shared Dan Castellanos, traditional fisher and former shark and net fisherman turned guide in Belize. Nets have been responsible for significant reductions in the abundance and diversity of sharks (of great importance for ecotourism), continued threats to protected species such as turtles and manatees, and the loss of fishes of significant non-consumptive economic value to recreational sea anglers. For 20 years, over 2,300 of the country's 2,500 registered fishers had unsuccessfully supported a range of local efforts to prohibit the use of gillnets in Belize. The first national ban on the use of these nets was eventually gazetted on 6th November 2020, after all key sectors had joined together to call for government action. The possession and use of gillnets is now no longer permitted throughout the waters of Belize.

The benefits of banning fishing with gillnets will extend far beyond a reduction in fishing effort and conservation of sharks and other threatened species, and also include:

- Restoration of depleted fisheries
- Promotion of fair and equitable fishing
- Support for alternative livelihoods
- Boosting tourism, through the survival of species with great tourism value (e.g. sharks)
- Increasing effectiveness of protected areas
- Improved safeguarding of protected species that were being captured in nets
- Protection of fragile habitats
- Reduction in plastic pollution and ghost fishing by discarded nets
- Simplification of enforcement, and ability to include the public in vigilance
- Improved fisheries management
- Increased food security

The recipe for Belize's net ban is highly replicable to other countries. It relied upon the formation of a cross-sector Coalition for Sustainable Fisheries that included fisher representation (Belize Fishermen Federation and the Belize Game Fish Association), the scientific and conservation sector (MarAlliance), the tourism sector (Belize Tourism Industry Association, Turneffe Atoll Trust, Yellow Dog Fishing and Conservation), and the advocacy NGO Oceana Belize, and was underpinned by extensive consultations with net fishers and tourism guides. The Coalition also undertook major fundraising to support fishers with their transition to economic alternatives and net buy-back. These efforts helped to generate support at ministerial level within the government. Setting a precedent is often the hardest part of creating a movement that includes letting go of unsustainable practices. The Belize net ban represents a much-needed precedent to help other nations shift fisheries away from the use of nets, and enable the restoration of their fish populations, fisheries and food security.

regions, including on the high seas. Five RFMOs manage oceanic fisheries for tuna and tuna-like species (e.g. swordfishes and marlins), while also catching huge numbers of the threatened large pelagic sharks that are increasingly being listed in biodiversity conservation agreements. The other RFMOs are responsible for a wider range of highly migratory and straddling (shared) fish stocks. Until recently, very few RFMOs were even monitoring shark catches, let alone managing shark fisheries. Now, however, many have adopted shark finning bans and some of the tuna RFMOs have prohibited the retention of a few of the most threatened oceanic sharks (e.g. the Oceanic Whitetip, Silky Shark and some thresher and hammerhead sharks), although the measures adopted and species protected vary between different RFMOs. In 2019, the first international oceanic shark catch limits were established by a tuna RFMO: for Blue Shark in the North and South Atlantic.

Conservation status of sharks

Support for proposals to conserve and manage sharks has strengthened during the past few decades. This is the result of steady progress in the huge task of assessing and regularly reassessing the conservation status of all sharks and other chondrichthyan fishes (over 1,100 species). This work has been undertaken by the International Union for the Conservation of Nature (IUCN) Shark Specialist Group volunteer network, under the aegis of the IUCN Red List of Threatened Species (www.iucnredlist.org). This massive project took nearly 20 years to produce the first-ever systematic evaluation of the relative risk of extinction for an entire class of fishes, the Chondrichthyes, using the IUCN Red List categories and criteria. Under the Global Shark Trends project, which monitors trends in shark populations and global extinction risk, it is now nearing the end of the first complete reassessment. The conservation status of the class has changed over time in response to fisheries exploitation, the implementation of conservation and fisheries management measures, and as a

IUCN Red List Category	Number of shark species	Percentage of shark species	
		Including Data Deficient	Data sufficient
Extinct	0	0%	0%
Extinct in the Wild	0	0%	0%
Critically Endangered	31	5.8%	7.2%
Endangered	43	8%	10%
Vulnerable	64	11.9%	14.9%
Near Threatened	49	9.1%	11.4%
Least Concern	242	45.1%	56.4%
Data Deficient	96	17.9%	
Not Evaluated	11	2.1%	
Total	**536**		

Table 2. Summary of Global Shark Red List Assessment Results.

result of improved scientific knowledge. The 2020 Red List Category is provided for every shark species in this book, but a regular programme of updates means that changes are made several times each year, so please always consult the Red List website for up-to-date information.

Table 2 and Figure 61 summarise the Red List status of the 536 shark species described in this book, as published at the end of 2020. Eleven species (2%) are not yet evaluated in the Red List. 138 species (26%) are classified as threatened: 31 (6%) as Critically Endangered (one of these possibly Extinct), 43 (8%) as Endangered and 64 (12%) as Vulnerable. Another 49 sharks (9%) are Near Threatened, which means that they are close to qualifying, or likely to qualify for, a threatened category in the near future. 96 species (18%) are Data Deficient, meaning that there is not enough information to accurately assess their status; this is a great improvement on the 45% of species assigned to this category in 2012. Large numbers of poorly known deepsea sharks (particularly lanternsharks, catsharks and kitefin sharks) are now assessed as Least Concern

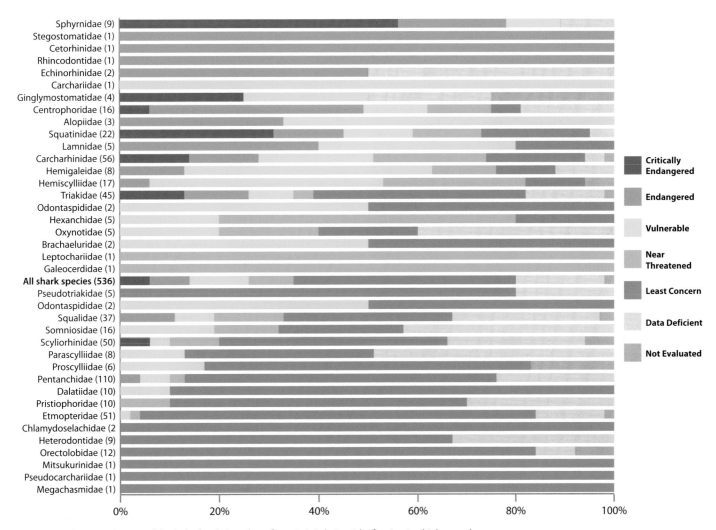

Figure 61. Threatened status of sharks by family (number of species). Relative risk of extinction highest at the top.

DRIVERS OF BIODIVERSITY LOSS: OCEANS AND FISHING

Many readers will have heard of the IPCC (Intergovernmental Panel for Climate Change), which provides the world's governments with regular scientific reviews that set out the status of human-induced climate change, its natural, political, and economic impacts and risks, and possible responses. However, you may not have heard of the IPBES (Intergovernmental Science-Policy Platform on Biodiversity and Ecosystem Services), sometimes referred to as 'the IPCC for wildlife'. The 7th IPBES reports, assessing the drivers of biodiversity loss, were published in 2019, following a three-year investigation. These reports cover four regions of the world – the entire planet except for the poles and the open oceans. Their results for 'Oceans and Fishing' are grim reading:

- 66% of the marine environment is severely altered by human actions
- 55% of the ocean area is covered by industrial fishing
- 33% of the global fish catch is illegal or unregulated
- 33% of fish stocks are being harvested unsustainably
- 60% of stocks are at maximum sustainable yield
- 100–300 million people are at increased risk of floods and hurricanes because of loss of coastal habitats and protection
- Seagrass beds, home to diverse communities of species (and fish nursery grounds), have decreased by 10% per decade
- Live coral reefs have halved in the past 150 years, with climate change causing a dramatic decline in recent decades
- There are over 400 ocean dead zones caused by fertiliser pollution (a combined area greater than the United Kingdom)

instead of Data Deficient because of their less accessible deepsea habitat and low fisheries value. Overall, 45% of species (242 sharks) are now assessed as Least Concern.

The two most threatened orders, based on their IUCN Red List Index, are the angelsharks and mackerel sharks. At family level, those at greatest risk are the hammerheads, gulper sharks and threshers, followed by angelsharks, requiem sharks, weasel sharks and mackerel sharks. These are mostly coastal and oceanic, pelagic species that are exposed to intensive target and bycatch fisheries (demand for shark fin is a significant driver for shark mortality, and in some species also the meat), while the bigger sharks also tend to be K-selected and not resilient to high levels of exploitation. Other threats include coastal, estuarine and riverine habitat damage, and persecution. Only 18 shark species, mostly shallow water and coral reef species in the Indo-Pacific, are currently considered threatened by climate change, but this number may rise with future updates. In contrast, most deepsea shark families, even the large-bodied sleeper sharks, are generally not seriously threatened. The gulper sharks, assessed as at very high risk (see above), are a clear exception because they are targeted for their meat and liver oil, and are remarkably slow growing, late to mature and produce so few young.

The portrayal of sharks in the classic 1975 horror film Jaws, and its sequels and imitations, had a huge influence on perceptions of the relationship between people and sharks. It encouraged us to think of sharks as a serious

and ever-present threat to anyone who dared to venture into their environment – just at a time when larger numbers of people were spending more leisure time at the coast and in the sea. While it is certainly true that a few shark species are potentially dangerous to people, and the relatively small number of shark bites every year can have truly tragic consequences, the broader picture is very different. It is sharks that are under threat worldwide due to human impacts.

While far more shark species have become extinct than are living today, those extinction events occurred in the very distant past and were driven by major environmental changes in world oceans or global-scale, natural catastrophic events (see p. 19). Industrial fisheries have completely changed the picture. We are now heading for a new extinction event in the oceans, this time driven by our unsustainable exploitation of what we arrogantly term 'marine resources' and by human-induced damage to the marine ecosystem. What is most extraordinary is the very recent nature of this impact. While most extinctions of large land animals took place relatively slowly as humans spread across the world over millennia, our impact upon most large marine animals is extremely recent. This latest extinction crisis dawned less than a century ago, but the major impact on marine animals has only occurred during the past 30 years. To put this into context: not only is this less than the life span of many shark species, but some sharks born around 30 years ago may, if they have survived, only just be mature and starting to reproduce.

Fortunately, there is still time to reverse this trend. Today, we are entering a new phase in the relationship between sharks and people – one of improved understanding and awareness, and striving for conservation rather than eradication. Some of the strongest support for shark conservation comes from those who have encountered sharks face to face and probably understand them best: divers, anglers, surfers and scientists, supported by younger generations whose perceptions were formed not from Jaws, but from exposure to sharks in public aquaria, and documentary films and their education messages. We now understand that it is important to ensure that sharks remain in the world's oceans, not just because they are amazing, beautiful creatures, but also because we need them. Well-managed shark populations have an important direct value to poor coastal communities, both for food security and for income (including ecotourism). Sharks also play a vital role in maintaining the biodiversity, structure, function and stability of the marine ecosystem. Removing large predators can have counterintuitive and unpredictable results – not necessarily causing an increase in numbers of the fishes that they were eating and more food for us. Because sharks also reduce the numbers of other predators, their loss may result in reduced populations of important commercial species lower down the food chain.

The authors hope that readers of this book will realise that, despite media hyperbole, the true story is not one of man-eating sharks, but of shark-eating man. Sharks, not humans, are the species in danger and some species are being driven rapidly towards extinction. We all need sharks in our oceans. We need them for food, particularly in countries where protein is scarce. We need them for the recreational benefits they bring, for the maintenance of the entire marine ecosystem, and for the sheer joy of knowing that they are still somewhere out there – some of the most beautiful, highly evolved, streamlined and quite simply coolest animals in the world.

Please help us to help them: support shark conservation organisations and campaigns, and act responsibly to minimise your impact upon their populations and habitats – whether you are lucky enough to see them or not.

KEY TO ORDERS AND FAMILIES OF LIVING SHARKS

1a Anal fin absent, *drawing 1*. → **2**

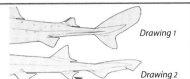

Drawing 1

1b Anal fin present, *drawing 2*. → **10**

Drawing 2

2a [1a] Mouth at end of head; body flat and ray-like; very large pectoral fins with triangular anterior lobes that overlap gill slits; caudal fin with base slanted ventrally (hypocercal) see *drawing 3*.

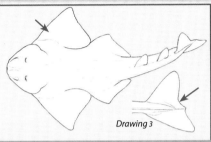

Drawing 3

▷ Order Squatiniformes, family Squatinidae. Angelsharks: pages 227–246

2b [1a] Mouth beneath head; body cylindrical, compressed or slightly flattened, not ray-like; pectoral fins smaller, without anterior lobes; caudal fin with base slanted dorsally (heterocercal) see *drawing 4* or horizontal (diphycercal). → **3**

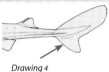

Drawing 4

3a [2b] Very long, flattened, saw-like snout, with rows of large and small sharp-pointed tooth-like denticles on sides and underside; a pair of long, tape-like barbels on lower surface of snout in front of nostrils.

▷ Order Pristiophoriformes, family Pristiophoridae. Sawsharks: pages 221–226

3b [2b] Snout normal, not saw-like. Order Squaliformes. Dogfish sharks. → **4**

4a [3b] Spiracles small and well behind eyes; fifth gill slits much longer than first four; body covered with moderately large, close set, thorn-like denticles or sparse, large, plate-like denticles; pelvic fins much larger than second dorsal fin; first dorsal fin origin over or behind pelvic fin origins.

▷ Order Echinorhiniformes. Bramble sharks: pages 102–104

4b [3b] Spiracles larger and close behind eyes; fifth gill slits not significantly larger than first to fourth; denticles variable in shape but small to moderately large; pelvic fins usually as large as second dorsal fin or smaller; first dorsal fin origin well in front of pelvic origins. → **5**

5a [4b] Body very high and compressed, triangular in cross-section with heavy lateral keels between pectoral and pelvic bases; dorsal fins extremely high.

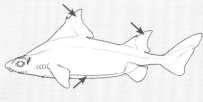

▷ Family Oxynotidae. Roughsharks: pages 204–206

5b [4b] Body low and more cylindrical in cross-section, with low lateral keels between pectoral and pelvic bases or no keels; dorsal fins low. → **6**

6a [5b] Teeth similar and blade-like in both jaws, with a deflected horizontal cusp, a low blade and no cusplets see *drawing 5*; caudal peduncle usually with an upper precaudal pit (weak or absent in *Cirrhigaleus*); strong lateral keels present on caudal peduncle see *drawing 6*; dorsal fin spines without grooves; subterminal notch absent from caudal fin.

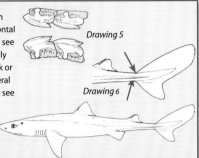

Drawing 5

Drawing 6

▷ Family Squalidae. Dogfish sharks: pages 114–134

6b [5b] Teeth dissimilar in upper and lower jaws, see *drawing 7*; caudal peduncle without precaudal pits; lateral keels usually absent on caudal peduncle, see *drawing 8* (weak ones present in some members of Family Dalatiidae); dorsal fin spines, when present, with lateral grooves; subterminal notch usually present and well developed on caudal fin. → **7**

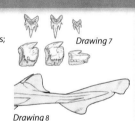

Drawing 7

Drawing 8

7a [6b] Teeth either hook-like or with cusps and cusplets in both jaws, or upper teeth with cusps and cusplets and lower teeth compressed, blade-like and more or less overlapping; underside of body, flanks, and tail usually with more or less conspicuous, dense, black markings (photomarks) with light organs (photophores).

▷ Family Etmopteridae. Lanternsharks: pages 160–186

7b [6b] Upper teeth with strong cusps but without cusplets; lower teeth laterally expanded, compressed, blade-like and overlapping, much larger than uppers see *drawing 9*; underside of body, flanks and tail without conspicuous, dense, black markings that have light organs, though light producing organs may be present elsewhere. → **8**

Drawing 9

8a [7b] Upper teeth relatively broad and blade-like; lower teeth low, wide and blade-like, see *drawing 10* below.

Drawing 10

▶ Family Centrophoridae. Gulper sharks: pages 135–147

8b [7b] Upper teeth relatively narrow and not blade-like; lower teeth high, wide and blade-like. → **9**

Drawing 11

9a [8b] Head moderately broad and somewhat flattened or conical; snout flat and narrowly rounded to elongate-rounded in dorsoventral view; abdomen usually with lateral keels; both dorsal fins either with or without (*Somniosus*, *Scymnodalatias*) fin spines.

▶ Family Somniosidae. Sleeper sharks: pages 192–201

9b [8b] Head narrow and rounded-conical; snout conical and narrowly rounded to elongate-rounded in dorsoventral view; abdomen without lateral ridges; most genera lacking dorsal fin spines (*Squaliolus* with a small first dorsal spine only).

▶ Family Dalatiidae. Kitefin sharks: pages 201–215

10a [1b] 6 or 7 gill slits on each side of head; one dorsal fin present, set far back. Order Hexanchiformes; cow and frilled sharks. → **11**

10b [1b] Two dorsal fins (except the scyliorhinid *Pentanchus profundicolus* with one dorsal fin); five gill slits on each side of head. → **12**

11a [10a] Mouth terminal on head; teeth tricuspidate, see *drawing 12*, and similar in both jaws; six pairs of gill slits, first pair connected across the underside of the throat; body elongated and eel-like.

Drawing 12

▶ Family Chlamydoselachidae. Frilled sharks: pages 96–97

11b [10a] Mouth subterminal on head; front teeth unicuspidate in upper jaw and comb-shaped and blade-like in lower jaw, see *drawing 13*; six or seven pairs of gill slits, first not connected across underside of throat; body fairly stocky, not eel-like.

Drawing 13

▶ Family Hexanchidae. Cow sharks: pages 98–101

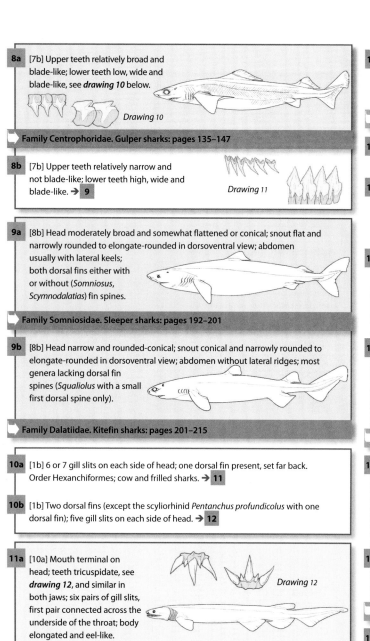
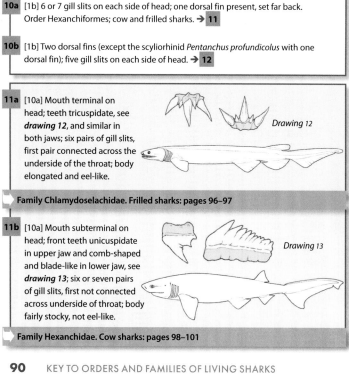

12a [10b] A strong spine on each dorsal fin.

▶ Order Heterodontiformes, family Heterodontidae. Bullhead sharks: pages 247–257

12b [10b] Dorsal fins without spines. → **13**

13a [12b] Eyes behind mouth; deep nasoral grooves connecting nostrils and mouth; a pair of barbels just medial to incurrent apertures of nostrils, see *drawing 14* (rudimentary in Family Rhincodontidae). Order Orectolobiformes. Carpet sharks. → **14**

Drawing 14

13b [12b] Eyes partly or entirely over mouth; nasoral grooves usually absent, when present (a few members of the Family Scyliorhinidae) broad and shallow; barbels, when present, developed from anterior nasal flaps of nostrils, not separate from them, see *drawing 15*. → **20**

Drawing 15

14a [13a] Mouth huge and nearly at end of head; external gill slits very large; caudal peduncle with strong lateral keels; caudal fin with a strong lower lobe, but without a subterminal notch.

▶ Family Rhincodontidae. Whale Shark: page 300

14b [13a] Mouth smaller and subterminal; external gill slits small; caudal peduncle without strong lateral keels; caudal fin with a weak lower lobe or none, but with a strong terminal lobe and subterminal notch, see *drawing 16*. → **15**

Drawing 16

15a [14b] Caudal fin about as long as rest of shark.

▶ Family Stegostomatidae. Zebra Shark: page 293

15b [14b] Caudal fin much shorter than rest of shark. → **16**

16a [15b] Head and body greatly flattened, head with skin flaps on sides; two rows of large, fang-like teeth at symphysis of upper jaw and three in lower jaw.

▶ Family Orectolobidae. Wobbegongs: pages 267–279

16b [15b] Head and body cylindrical or moderately flattened, head without skin flaps; teeth small. → **17**

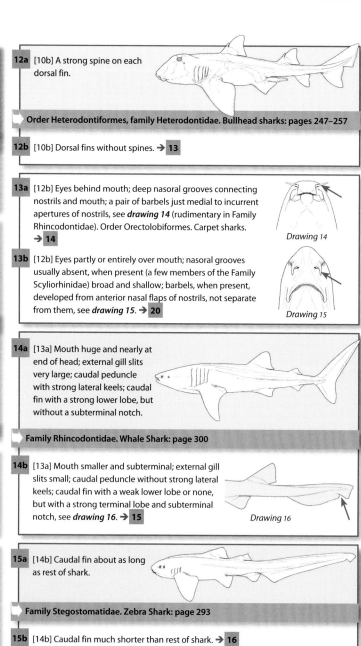
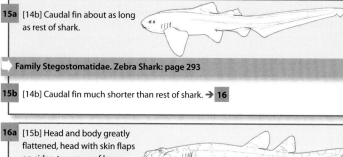
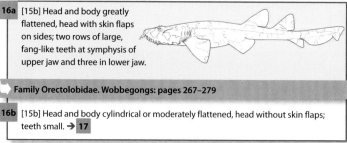

17a [16b] No circumnarial lobe and groove around outer edges of nostrils, see *drawing 17*.

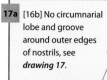

Drawing 17

▷ **Family Ginglymostomatidae. Nurse sharks: pages 296–299**

17b [16b] A circumnarial lobe and groove around outer edges of nostrils, see drawing 18. ➜ **18**

Drawing 18

18a [17b] Spiracles minute; origin of anal fin well in front of second dorsal origin, separated from lower caudal origin by space equal or greater than its base length.

▷ **Family Parascylliidae. Collared carpetsharks: pages 259–265**

18b [17b] Spiracles large; origin of anal fin well behind second dorsal origin, separated from lower caudal origin by space less than its base length. ➜ **19**

19a [18b] Nasal barbels very long; distance from vent to lower caudal origin shorter than distance from snout to vent; anal fin high and angular.

▷ **Family Brachaeluridae. Blind sharks: pages 265–266**

19b Nasal barbels short; distance from vent to lower caudal origin longer than distance from snout to vent; anal fin low, rounded and keel-like.

▷ **Family Hemiscylliidae. Longtailed carpetsharks: pages 284–292**

20a [13b] No nictitating eyelids; largest teeth in mouth usually are two or three rows of anteriors on either side of upper and lower jaw symphyses; upper anterior teeth separated from large lateral teeth at sides of jaw by a gap that may have one or more rows of small intermediate teeth, *drawing 19*; plankton-feeding species (Families Megachasmidae and Cetorhinidae) have reduced teeth with anterior, intermediate and lateral teeth poorly differentiated; intestine with ring valve. Order Lamniformes. Mackerel sharks. ➜ **21**

Drawing 19

20b [13b] Nictitating eyelids present; largest teeth in mouth are well lateral on dental band, not on either side of symphysis; no gap or intermediate teeth separating large anterior teeth from still larger teeth in upper jaw, see *drawing 20*; intestine usually with spiral or scroll valve. Order Carcharhiniformes. Ground sharks. ➜ **27**

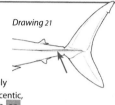

Drawing 20

21a [20a] A strong keel present on each side of caudal peduncle; caudal fin crescentic and nearly symmetrical, with a long lower lobe, see *drawing 21*. ➜ **22**

Drawing 21

21b [20a] No keels on caudal peduncle, or weak ones (Family Pseudocarchariidae); caudal fin asymmetrical, not crescentic, with lower lobe relatively short but strong, or absent. ➜ **23**

22a [21a] Teeth large and few, sharp-edged; gill openings large but not extending onto upper surface of head; no gill rakers on internal gill arches.

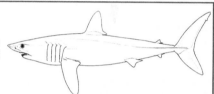

▷ **Family Lamnidae. Mackerel sharks: pages 319–323**

22b [21a] Teeth minute and very numerous, not sharp-edged; gill openings huge, extending onto upper surface of head; gill rakers present on internal gill arches, sometimes absent after shedding.

▷ **Family Cetorhinidae. Basking Shark: page 311**

23a [21b] Snout elongated and blade-like; anal fin larger than dorsal fins; no precaudal pits; caudal fin without lower lobe.

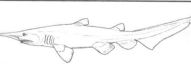

▷ **Family Mitsukurinidae. Goblin Shark: page 304**

23b [21b] Snout conical or flattened, short and not blade-like; anal fin subequal to dorsal fins in size or smaller than them; upper and sometimes lower precaudal pits present; caudal fin with strong lower lobe. ➜ **24**

24a [23b] Caudal fin about as long as rest of shark.

▶ Family Alopiidae. Thresher sharks: pages 314–318

24b [23b] Caudal fin less than half the length of rest of shark. ➔ **25**

25a [24b] Mouth huge, terminal on head, level with snout; teeth small, very numerous, and hook-shaped; internal gill openings screened by numerous long, papillose gill rakers.

▶ Family Megachasmidae. Megamouth Shark: page 310

25b [24b] Mouth smaller, subterminal on head, behind snout tip; teeth blade-like with large anterior teeth, intermediate teeth, and lateral teeth in upper jaw; internal gill openings without gill rakers. ➔ **26**

26a [25b] Eyes very large; gill slits extending onto upper surface of head; both upper and lower precaudal pits present; a low keel on each side of caudal peduncle.

▶ Family Pseudocarchariidae. Crocodile Shark: page 307

26b [25b] Eyes smaller; gill slits not extending onto upper surface of head; lower precaudal pit absent; no keels on caudal peduncle.

▶ Families Carchariidae and Odontaspididae. Sandtiger sharks: pages 305–307

27a Head with lateral expansions or blades, like a double-edged axe.

▶ Family Sphyrnidae. Hammerhead sharks: pages 569–579

27b [20b] Head normal, not expanded laterally. ➔ **28**

28a [27b] Origin of first dorsal fin over or behind pelvic in bases. ➔ **29**

28b [27b] Origin of first dorsal fin well ahead of pelvic fin bases. ➔ **30**

29a [28a] Supraorbital crest absent on cranium above eyes.

▶ Family Pentanchidae. Deepsea catsharks: pages 325–406

29a [28a] Supraorbital crest present on cranium above eyes. (Crest can be felt by running your fingers over the eye orbits)

▶ Family Scyliorhinidae. Catsharks: pages 407–446

30a [28b] No precaudal pits, dorsal caudal fin margin smooth. ➔ **31**

30b [28b] Precaudal pits and rippled dorsal caudal margin present, see *drawing 22*, (ripples sometimes irregular in *Scoliodon* and *Triaenodon* of family Carcharhinidae). ➔ **34**

Drawing 22

31a [30a] Labial furrows very short or absent, when present confined to mouth corners; posterior teeth on dental bands comb-like, see *drawing 23*. ➔ **32**

Drawing 23

31b [30a] Labial furrows relatively long with uppers extending partway or all the way anterior to level of symphysis, see *drawing 24*; posterior teeth on dental bands not comb-like. ➔ **33**

Drawing 24

32a [31a] Snout bell-shaped in dorsoventral profile, with a deep groove in front of eye, see *drawing 25*; internarial space over 1.5 times nostril width; inside of mouth and edges of gill arches without papillae; first dorsal fin more or less elongated, base closer to pectoral fins than pelvic fins.

Drawing 25

▶ Family Pseudotriakidae. False catsharks: pages 453–456

32b [31a] Snout rounded-parabolic or subangular in dorsoventral profile, see *drawing 26*, without a deep groove in front of eye; internarial space less than 1.3 times nostril width; inside of mouth and edges of gill arches with papillae; first dorsal fin short, base closer to pelvic fins than pectoral fins.

Drawing 26

▶ Family Proscylliidae. Finback catsharks: pages 447–452

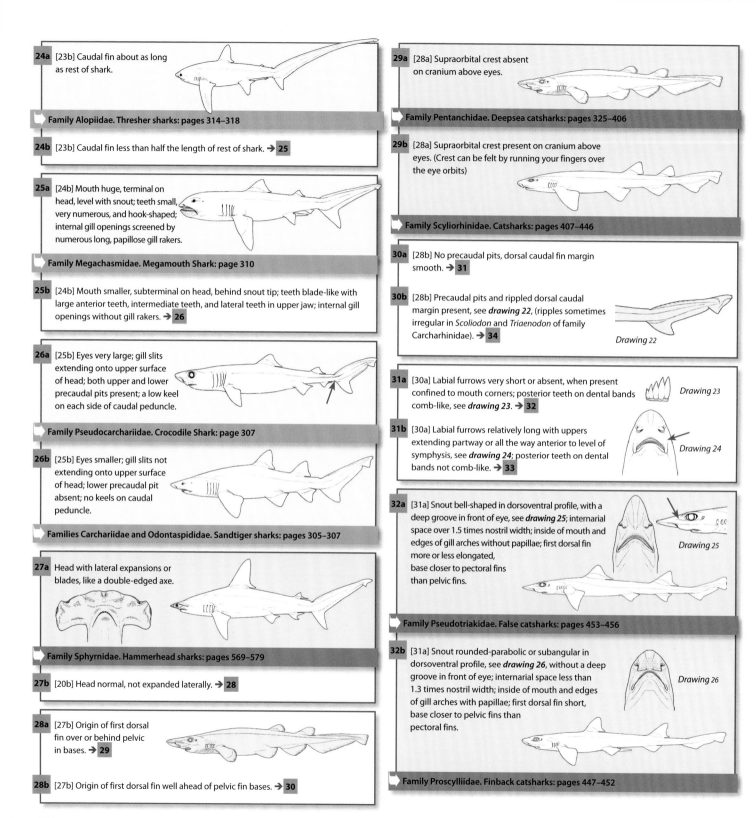

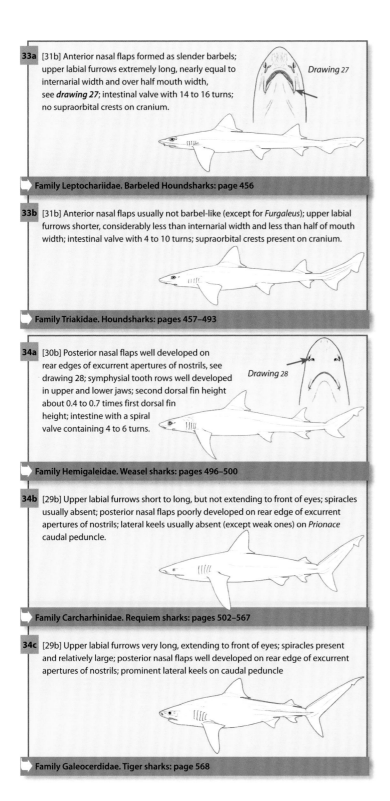

33a [31b] Anterior nasal flaps formed as slender barbels; upper labial furrows extremely long, nearly equal to internarial width and over half mouth width, see *drawing 27*; intestinal valve with 14 to 16 turns; no supraorbital crests on cranium.

Drawing 27

➥ **Family Leptochariidae. Barbeled Houndsharks: page 456**

33b [31b] Anterior nasal flaps usually not barbel-like (except for *Furgaleus*); upper labial furrows shorter, considerably less than internarial width and less than half of mouth width; intestinal valve with 4 to 10 turns; supraorbital crests present on cranium.

➥ **Family Triakidae. Houndsharks: pages 457–493**

34a [30b] Posterior nasal flaps well developed on rear edges of excurrent apertures of nostrils, see drawing 28; symphysial tooth rows well developed in upper and lower jaws; second dorsal fin height about 0.4 to 0.7 times first dorsal fin height; intestine with a spiral valve containing 4 to 6 turns.

Drawing 28

➥ **Family Hemigaleidae. Weasel sharks: pages 496–500**

34b [29b] Upper labial furrows short to long, but not extending to front of eyes; spiracles usually absent; posterior nasal flaps poorly developed on rear edge of excurrent apertures of nostrils; lateral keels usually absent (except weak ones) on *Prionace* caudal peduncle.

➥ **Family Carcharhinidae. Requiem sharks: pages 502–567**

34c [29b] Upper labial furrows very long, extending to front of eyes; spiracles present and relatively large; posterior nasal flaps well developed on rear edge of excurrent apertures of nostrils; prominent lateral keels on caudal peduncle

➥ **Family Galeocerdidae. Tiger sharks: page 568**

Plate 1 HEXANCHIFORMES

One spineless dorsal fin, anal fin present; six or seven pairs of gill slits; large mouth; tiny spiracles.

○ **Southern African Frilled Shark** *Chlamydoselachus africana*　　　　　　page 97

Southeast Atlantic, possibly southwest Indian Ocean; 300–1400m. Benthic, epibenthic and pelagic. Similar to and difficult to distinguish from Frilled Shark (below), but usually has a longer head and shorter body.

○ **Frilled Shark** *Chlamydoselachus anguineus*　　　　　　page 97

Worldwide patchy distribution; 17–1520m. Benthic, epibenthic and pelagic. Large terminal mouth with widely spaced slender three-cusped needle-sharp teeth; six pairs of gills; low dorsal fin smaller than anal fin, pectoral fins smaller than pelvic fins.

○ **Sharpnose Sevengill Shark** *Heptranchias perlo*　　　　　　page 98

Worldwide except northeast Pacific; patchy in tropical and temperate waters; 0–1000m. Mainly deepsea. Slender body; acutely pointed head with narrow mouth and large eyes; juveniles with black dorsal fin apex which fades with growth.

○ **Bluntnose Sixgill Shark** *Hexanchus griseus*　　　　　　page 100

Worldwide except in polar regions; patchy distribution; 0–2500m. Juveniles found inshore in cool waters; adults mostly deepsea, occasionally in shallow waters near head of submarine canyons. Large heavy body; broad head with wide mouth and small white-ringed eyes; soft supple fins; light coloured lateral line and posterior fin edges.

○ **Bigeye Sixgill Shark** *Hexanchus nakamurai*　　　　　　page 99

Indian Ocean and west Pacific; patchy in warm-temperate and tropical waters; 0–700m. On or near seabed, occasionally near surface or inshore. Slender body; narrow head with narrow mouth and large eyes; fins with white posterior edges and tips, caudal fin with deep subterminal notch.

○ **Atlantic Bigeye Sixgill Shark** *Hexanchus vitulus*　　　　　　page 99

North Atlantic, rarely Mediterranean; patchy in warm-temperate and tropical waters; poorly known, 0–700m. Slender body; fairly narrow head with bluntly pointed snout and broadly acute mouth; fins with white posterior edges and tips.

○ **Broadnose Sevengill Shark** *Notorynchus cepedianus*　　　　　　page 101

Worldwide except North Atlantic; patchy distribution in cool-temperate waters; surfline to at least 570m, but mostly inshore less than 100m. Bluntly pointed broad head with wide mouth and small eyes; numerous small black spots (some are plain or white-spotted), black dorsal fin apex in newborns fades with growth.

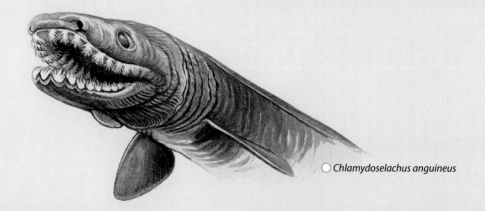

○ *Chlamydoselachus anguineus*

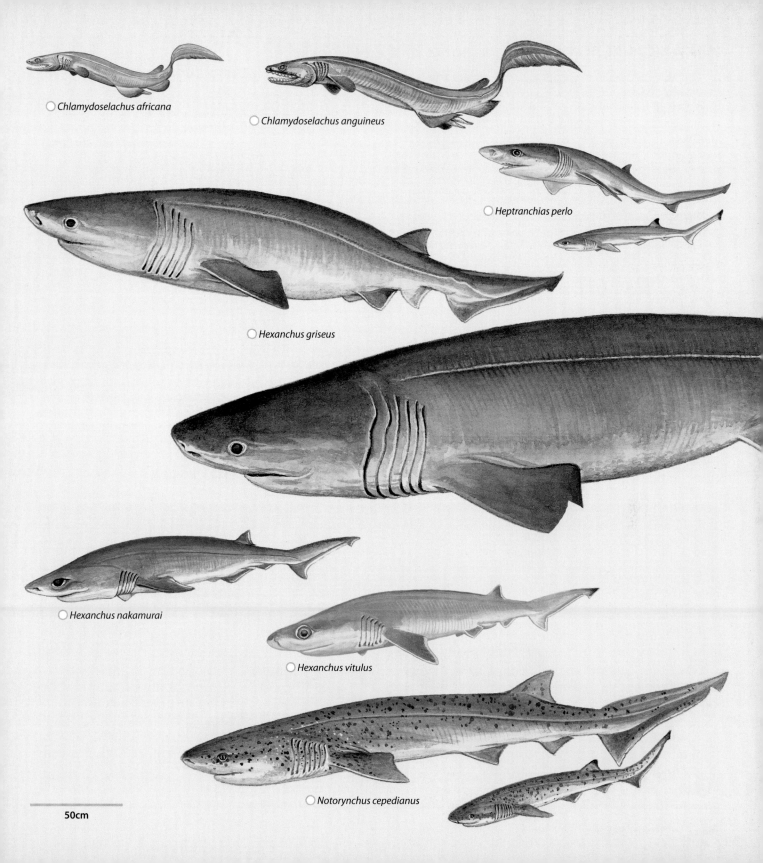

Chlamydoselachus africana

Chlamydoselachus anguineus

Heptranchias perlo

Hexanchus griseus

Hexanchus nakamurai

Hexanchus vitulus

Notorynchus cepedianus

50cm

HEXANCHIFORMES: frilled sharks and cow sharks

This order contains one of the oldest lineages of living sharks. Two extremely distinctive families: Chlamydoselachidae, now known to contain at least two very similar species, and Hexanchidae, with five species.

Identification Six or seven pairs of gill slits in front of the pectoral fins; one spineless dorsal fin over or behind the pelvic fins; anal fin present. The vertebral column extends into the caudal fin's long upper lobe; the lower lobe is short or absent. Large mouth. Eyes on the side of the head; spiracles very small, located well behind and above eyes. Medium to large sized.

Biology Most species are widespread worldwide, from tropical to temperate and boreal waters. Generally found in deeper, cooler waters in the tropics, but also inshore in temperate seas; some species migrate diurnally.

Status Taken as bycatch and in targeted trawl, net, line and sports fisheries. Some species are popular with sports divers. Most species are assessed as Least Concern or Near Threatened in the IUCN Red List of Threatened Species.

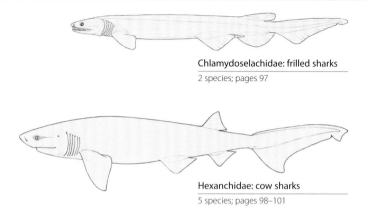

Chlamydoselachidae: frilled sharks

2 species; pages 97

Hexanchidae: cow sharks

5 species; pages 98–101

CHLAMYDOSELACHIDAE: frilled sharks

Two living species, very similar externally (with some local variability), until recently thought to be one wide-ranging species. Frilled sharks off Mozambique, on seamounts south of Madagascar, and in the Maldives require species confirmation.

Identification Long, brownish, eel-like body; flattened, short-snouted, snake-like head; large terminal mouth with widely spaced, slender, needle-sharp three-cusped teeth; large green eyes. Frilly gills protrude from six pairs of gill slits; first gill slit is continuous across throat. Dorsal fin low and much smaller than anal fin; pectoral fins smaller than pelvic fins. Internal structures of chondrocranium and vertebrae, vertebral counts and calcification patterns, pectoral fin skeletal morphology and radial counts, and intestinal valve counts distinguish the southern African species from frilled sharks off Japan and Taiwan.

Biology Viviparous; usually small litters and possibly very long gestation periods (up to 3.5 years).

Status Occasional bycatch in deepsea fisheries. Both species are assessed as Least Concern in the IUCN Red List of Threatened Species.

Frilled Shark, *Chlamydoselachus anguineus*, photographed by the E/V *Nautilus* Exploration Program, Ocean Exploration Trust, at Gorringe Bank off Portugal (p. 97).

SOUTHERN AFRICAN FRILLED SHARK *Chlamydoselachus africana* FAO code: **HWR** Plate page 94

~5mm

Teeth
Upper 24–30
Lower 23–27

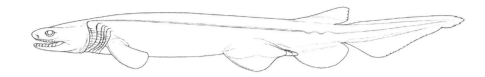

Measurements Born: ~50cm. **Mature:** males ~92cm, females at least 117cm.
Max: females at least 117cm.

Identification See opposite. Head length more than 17% of total length, but otherwise difficult to distinguish from *Chlamydoselachus anguineus* based on external features.
Distribution Southern Africa (southern Angola, Namibia and South Africa); records from the southwest Indian Ocean require confirmation.
Habitat Benthic, epibenthic and pelagic. Upper continental slopes from 300–1400m.
Behaviour Feeds on actively swimming deepsea catsharks, dogfish of the genera *Apristurus*, *Galeus* and *Squalus*, and bony fishes.
Biology Viviparous, at least 3 pups per litter, but nothing else known.
Status IUCN Red List: Least Concern. An occasional unutilised bycatch in hake fisheries.

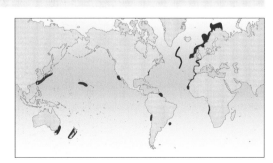

FRILLED SHARK *Chlamydoselachus anguineus* FAO code: **HXC** Plate page 94

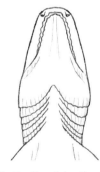

5mm

Teeth
Upper 19–28
Lower 21–29

Measurements Born: ~39–60cm. **Mature:** males ~118cm, females 126–150cm.
Max: males 165cm, females 196cm.

Identification See opposite. Head length less than 17% of total length.
Distribution Widely but patchily distributed worldwide; rare to uncommon, but locally common in some areas.
Habitat Benthic, epibenthic and pelagic. Offshore shelves and upper continental and island slopes, usually in deep water (17–1520m). Frequently migrates to the surface at night.
Behaviour Feeds on actively swimming deepsea squid and fishes, including smaller sharks. Habit (in captivity) of swimming with mouth open may lure prey to conspicuous white teeth. These sharks move widely within the water column: one individual was caught at less than 20m over a water depth of 1500m.
Biology Viviparous; 2–15 pups per litter feed on huge uterine eggs (11–12cm diameter). Pregnant females have very large abdomens. May reproduce all year in deep water, but mates in spring off Japan. Gestation period probably very long (1–3 years). The smallest free-swimming individual was 54cm long. Diet includes small dogfish and catshark species, bony fishes and midwater crustaceans.
Status IUCN Red List: Least Concern. Bycatch of deep bottom trawls and gillnets. Utilised for fishmeal and meat. Occasionally kept in aquaria. Harmless to humans.

HEXANCHIDAE: cow sharks

Three genera and five species: *Hexanchus* (three species), *Heptranchias* (one species) and *Notorynchus* (one species). Cow sharks are mostly found in cold water: deep water in warm-temperate and tropical regions, but may enter shallow water in cool-temperate areas. Only the Broadnose Sevengill Shark permanently inhabits shallow coastal areas.

Identification Moderately slender to stocky cylindrical sharks with six or seven pairs of gill slits (first pair not connected across the throat) in front of pectoral fins. Ventral mouth. Large compressed comb-like teeth in the lower jaw, smaller cuspidate teeth in the upper jaw. Single spineless dorsal fin, relatively high, angular and short. Pectoral fins angular, larger than pelvic fins. Anal fin smaller than dorsal fin. Caudal fin with marked sub-terminal notch.

Biology Viviparous. Some are migratory, moving inshore seasonally to feed or pup.

Status Taken as a bycatch and in some commercial and target sports fisheries. Important for dive tourism in a few shallow water locations. Most species are assessed as Near Threatened or Vulnerable in the IUCN Red List of Threatened Species.

Broadnose Sevengill Shark, *Notorynchus cepedianus* (p. 101).

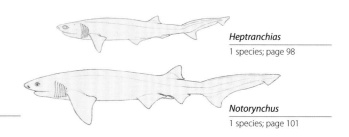

Heptranchias
1 species; page 98

Notorynchus
1 species; page 101

Hexanchus
3 species; pages 99–100

SHARPNOSE SEVENGILL SHARK *Heptranchias perlo* FAO code: **HXT** Plate page 94

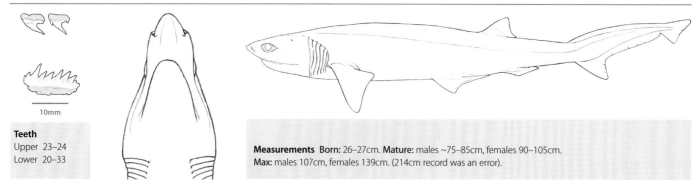

Teeth
Upper 23–24
Lower 20–33

10mm

Measurements Born: 26–27cm. **Mature:** males ~75–85cm, females 90–105cm.
Max: males 107cm, females 139cm. (214cm record was an error).

Identification Acutely pointed head. Seven pairs of gill slits. Narrow mouth, five rows of comb-shaped teeth in lower jaw. Large eyes. Black blotch on tip of dorsal fin and upper caudal lobe prominent in young, but faded or absent in adults.

Distribution Wide-ranging but patchily distributed. Tropical and temperate seas, not in northeast Pacific.

Habitat Mainly deep water (0–1000m), continental and island shelves and upper slopes, occasionally shallower water close inshore. Benthic and epibenthic; may also swim well off the bottom.

Behaviour Poorly known. Probably a strong, active swimmer. Feeds mostly on small to moderately large demersal and pelagic fishes, cephalopods and occasionally on crustaceans. Snaps vigorously when captured.

Biology Viviparous, 6–20 pups per litter. Apparently reproduces year round.

Status IUCN Red List: Near Threatened. Relatively uncommon. Sometimes a utilised bycatch in bottom trawl and longline fisheries. Occasionally kept in aquaria.

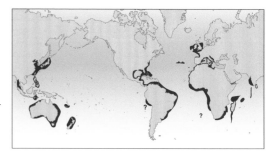

BIGEYE SIXGILL SHARK   *Hexanchus nakamurai*   FAO code: **HXN**   Plate page 94

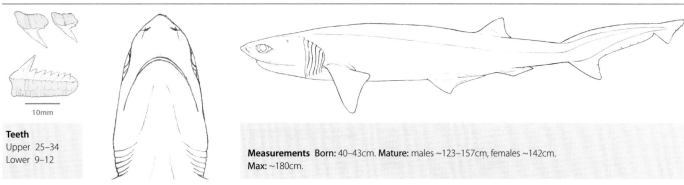

Teeth
Upper 25–34
Lower 9–12

Measurements Born: 40–43cm. **Mature:** males ~123–157cm, females ~142cm.
Max: ~180cm.

Identification Slender shark; body and fins quite firm. Narrow head; narrow ventral mouth (width ~1.5 times length); five rows of large comb-shaped teeth in lower jaw on each side. Large eyes. Upper caudal lobe deeply notched; short lower caudal lobe (strong in adults, weak in young). Colour sharply divided between dark above and light below. Fins usually with white trailing edges and tips, sometimes dusky.
Distribution Widely but patchily distributed in most warm-temperate and tropical seas, excluding the Atlantic and eastern Pacific. Often confused with the Bluntnose Sixgill Shark.
Habitat Continental and island shelves and slopes; on or near bottom, 0–700m, occasionally near surface or inshore.   **Behaviour** Little-known, primarily deepsea shark. Approaches submersibles cautiously.
Biology Viviparous, 13–26 pups per litter. Feeds mostly on small to medium-sized bony fishes and occasionally on crustaceans.
Status IUCN Red List: Near Threatened. Uncommon to rare, but possibly misreported as other Hexanchid species. Taken as bycatch in some of the many fisheries that operate across its range, not commercially important.

ATLANTIC BIGEYE SIXGILL SHARK   *Hexanchus vitulus*   FAO code: **HXW**   Plate page 94

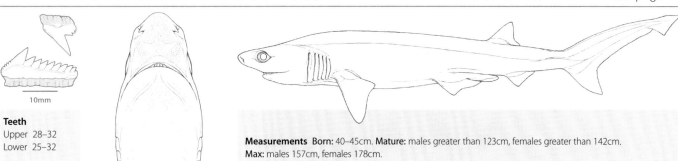

Teeth
Upper 28–32
Lower 25–32

Measurements Born: 40–45cm. **Mature:** males greater than 123cm, females greater than 142cm.
Max: males 157cm, females 178cm.

Identification Slender, medium-sized shark. Fairly narrow head; bluntly pointed snout; broadly acute mouth. Six paired gill slits. Single small dorsal fin with the origin from over the posterior half of the base to just behind the insertion point of the pelvic fins. Anal fin smaller than dorsal fin. Uniform dark to light brownish grey above, becoming lighter to white below. Upper caudal lobe with black tip in young, fading in adults; trailing fin edges with white margins.
Distribution North Atlantic: Bay of Biscay to Mediterranean (rarely) in east; Bahamas, Gulf of Mexico and Caribbean to Venezuela and Guyanas in west.
Habitat Poorly known, primarily along continental and insular slopes, 0–700m.
Behaviour This shark has been observed approaching bait stations set by submersibles, but appears to be more cautious than the larger Bluntnose Sixgill Shark *Hexanchus griseus*.
Biology Viviparous, 13–26 pups per litter, but reproductive cycle unknown. Feeds mostly on bony fishes and cephalopods, sometimes on crustaceans.
Status IUCN Red List: Least Concern. Occasionally taken as incidental bycatch, but few deepwater fisheries operate in its range. The species name was recently resurrected based on molecular data, but its separation from the Indo-Pacific Bigeye Sixgill Shark *H. nakamurai* is unclear based on their morphology. Presently, the only way to separate these species is by location: those in the North Atlantic are referred to *H. vitulus* and those in the Indo-Pacific are assigned to *H. nakamurai*.

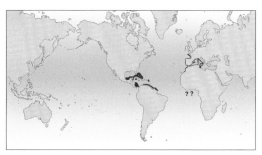

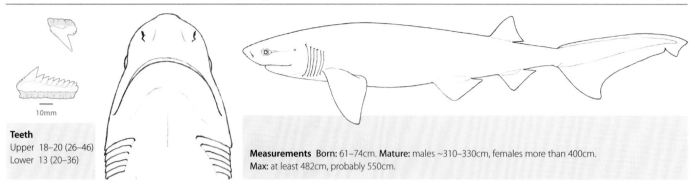

Teeth
Upper 18–20 (26–46)
Lower 13 (20–36)

Measurements Born: 61–74cm. **Mature:** males ~310–330cm, females more than 400cm.
Max: at least 482cm, probably 550cm.

Identification Large, heavy, powerful body; soft, supple fins. Broad head; ventral mouth (width more than two times length); six rows of large comb-shaped lower teeth on each side (tooth count in parentheses above includes smaller posterior teeth). Small eyes; dark pupil ringed with white, fluorescent blue-green in life. Skin grey or tan to blackish, sometimes darker spots on sides. Light coloured lateral line and trailing fin edges. Underside lighter than upper in newborns.

Distribution Patchy distribution worldwide; possibly absent from Arctic and Antarctic.

Habitat Shelves and slopes of continents, islands, seamounts and mid-ocean ridges, usually 200–1100m, to at least 2500m. These sharks are often associated with areas of upwelling and high biological productivity. Hydrographic data from areas where Bluntnose Sixgill Sharks occur reveal a bottom temperature of 6–10°C in waters with high nutrient levels. Young may occur close inshore in cool water, and adults are more likely to be found in shallow water close to submarine canyons. Adult females move inshore seasonally to give birth, with the newborns occurring in very shallow water nursery areas off the west coast of North America in Monterey Bay, Puget Sound, and adjacent waters of British Columbia. There is also a nursery off the coast of southern Namibia.

Behaviour Observed by divers in shallow water, filmed from submersibles, and tracked for short distances. Occurs alone and in groups. A slow but strong swimmer. Adult sharks are more sensitive to light than juveniles and less likely to be seen in clear shallow water, but may approach the surface at night or during

dense plankton blooms. At least one confirmed non-fatal attack by this species on a hookah diver gathering clams in Puget Sound, Washington, USA. A possible second non-fatal attack by a large 'cowshark' at the Farallon Islands, off San Francisco, may have been this species or the large, and potentially aggressive, *Notorynchus cepedianus*, (unfortunately the tooth fragments, clearly of a hexanchid, removed from the victim were discarded).

Biology Viviparous, very large litters of 47–108 pups. Young and adults may be segregated, with the young using inshore nursery grounds. Possibly long-lived. Feeds on squid, benthic and pelagic bony fishes, small sharks and rays. Large sharks (at least 2m) take cetaceans and seals.

Status IUCN Red List: Near Threatened. Can be locally common in bycatch and target fisheries for food, fishmeal and oil, and in sports fisheries, but vulnerable to overfishing and requires careful management. A deepsea fishery collapsed in the Maldives. Declining Bluntnose Sixgill Shark numbers near Seattle, USA, and in adjacent British Colombia, Canada, led dive operators and dive clubs to campaign for its protection; as a result, targeting and possession of this species is now legally prohibited there. The Bluntnose Sixgill Shark is now the focus of an important seasonal dive ecotourism industry in British Columbia, Canada and Washington State, USA.

A young Bluntnose Sixgill Shark, *Hexanchus griseus*.

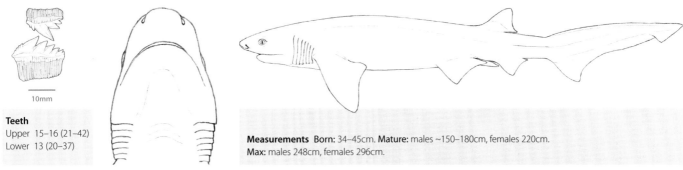

10mm

Teeth
Upper 15–16 (21–42)
Lower 13 (20–37)

Measurements Born: 34–45cm. **Mature:** males ~150–180cm, females 220cm.
Max: males 248cm, females 296cm.

Identification Broad, bluntly pointed head; wide mouth; six rows of large, comb-shaped lower teeth on each side (tooth count in parentheses above includes smaller posterior teeth). Small eyes. Grey to brown body, usually with many small black spots, occasionally plain or with white spots. Black tips on newborns' dorsal fin and upper caudal lobe fade with age.

Distribution Wide-ranging, found in most temperate inshore continental waters, but absent in the North Atlantic. Some populations may be isolated.

Habitat Coastal, common in shallow bays and close to shore. From the surf line (less than 1m) to at least 570m, but mostly less than 100m. Large sharks in bays and estuaries are seasonally common in deeper channels, but will search for food over mud flats at high tide.

Behaviour A strong swimmer, often cruises steadily and slowly near the bottom, sometimes at the surface. Attacks prey at high speed. Most active at night, during overcast conditions, and in turbid water. Often moves into and out of bays with the rising and falling tide, and out of low salinity conditions, and is migratory in some temperate areas. For example, in Australia, males move from Hobart (Tasmania) north into warmer southern New South Wales waters in late autumn, then return to Hobart in early summer. Mating and birthing occurs in some bays and estuaries in temperate regions. Adjacent populations may use different breeding grounds and be isolated from each other. The Broadnose Sevengill Shark is apparently social: it is often found in aggregations and may hunt large prey, including pinnipeds, cooperatively. Some have been observed 'spy-hopping'.

Biology Viviparous. Mating may occur in spring/summer. There is probably a one year gestation followed by one year recovery. Half the adult female population gives birth each spring to 67–104 pups in shallow nursery bays. Males mature at 4–5 years and females 11–21 years; longevity is 30–50 years. Newborn length doubles in 6 months; mature adults grow only 0–9cm a year. This is a powerful top predator on marine

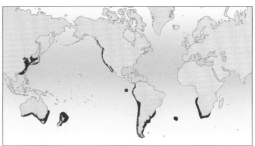

vertebrates (chondrichthyans, bony fishes and marine mammals, such as seals and small cetaceans), but also eats carrion. Orca have been observed killing these sharks and consuming their livers in False Bay, South Africa, whereupon the surviving sharks abandoned their daylight kelp forest resting area.

Status IUCN Red List: **Vulnerable**. Important in target and bycatch commercial fisheries, with intensive fishing pressure occurring across much of its distribution, and for sports diving and dive tourism in southern California and South Africa. Utilised for meat, hide, liver oil and display in aquaria. Rarely harmful to people (captive sharks have occasionally bitten divers and anglers), but there are verified reports of unprovoked biting of people in the wild and attacks on domestic dogs wading and swimming in shallow water.

Broadnose Sevengill Shark, *Notorynchus cepedianus*.

Echinorhinidae: bramble sharks

Two species of large, sluggish, deepsea sharks. A third undescribed species may occur in the northern Indian Ocean.

Identification Skin denticles very large and thorn-like. Stout, cylindrical body. Five gill openings in front of pectoral fin, fifth larger than others. Broad, flat head and snout, very small spiracles far behind the eyes. Two similar-sized, unspined, small dorsal fins, set close together and close to the caudal fin; origin of first dorsal slightly behind pelvic fin origin. No anal fin. Lower caudal lobe poorly developed in adults, absent in young, subterminal caudal notch lacking or not obvious.

Biology Widely distributed on soft-bottom habitats over temperate and tropical continental and insular shelves and upper slopes, sometimes entering shallow cold water. Large mouth and pharynx may be used to suck in prey (bony fishes, small chondrichthyans and invertebrates) as it comes within range.

Status Uncommon to rare and poorly known. Utilised for their liver oil (reportedly higher value than oil from other deepwater sharks) and susceptible to overexploitation and population depletion.

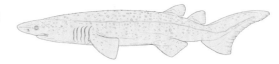

Echinorhinidae: bramble sharks
2 species; page 104

Prickly Shark, *Echinorhinus cookei* (p. 104).

Plate 2 ECHINORHINIFORMES

Broad, flat head with tiny spiracles; two similar-sized dorsal fins set close together and well back, no anal fin; large thorn-like skin denticles.

○ **Bramble Shark** *Echinorhinus brucus* — page 104

Atlantic, Mediterranean, Indian Ocean, west Pacific; 10–1200m. On or near seabed, occasionally inshore. Irregular scattered whitish conspicuous denticles, which can fuse into plates; uniformly grey or brownish to black on back and sides, fin edges blackish, usually light below.

○ **Prickly Shark** *Echinorhinus cookei* — page 104

Pacific; 4–1100m. Close to bottom. Light coloured inconspicuous denticles numerous and regularly spaced, but few below snout; white around mouth and snout, uniformly brown to salty grey or black, lighter colouring around mouth and ventral surface, posterior fin edges blackish.

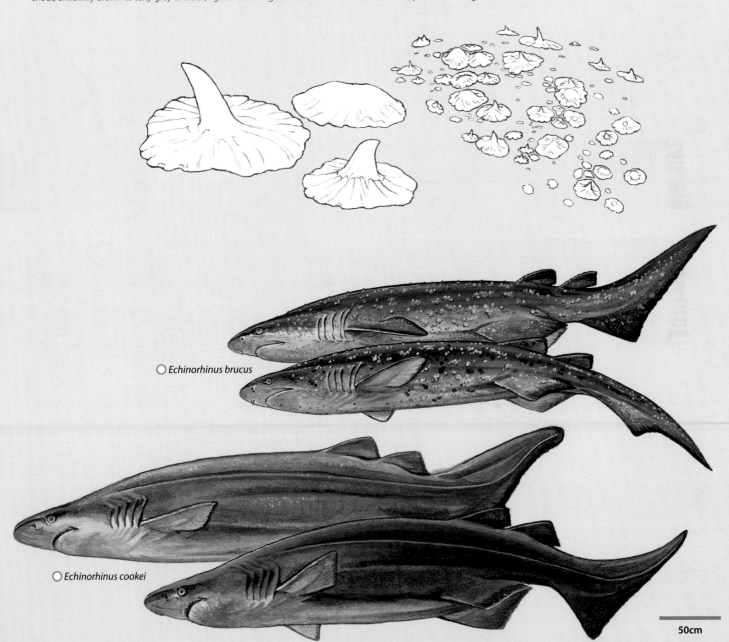

○ *Echinorhinus brucus*

○ *Echinorhinus cookei*

50cm

BRAMBLE SHARK *Echinorhinus brucus*

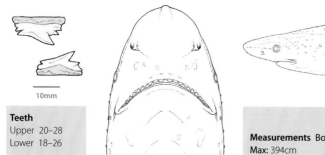

Teeth
Upper 20–28
Lower 18–26

Measurements Born: 40–55cm. **Mature:** males ~150–190cm, females 190–230cm. **Max:** 394cm.

Identification Adults have large (greater than 1cm) sparsely and irregularly scattered, whitish, thorn-like denticles with smooth margins, some fused into multi-cusped plates. Young (less than 90cm) have small, close-set denticles below snout and around mouth, becoming large, scattered and conspicuous in larger sharks. Grey, brownish or blackish above, often lighter below; may have red or black spots or blotches on back and sides. Fin edges blackish.
Distribution Atlantic, Mediterranean, west Pacific and Indian Ocean.
Habitat Deep water, continental and insular shelves and slopes, on or near bottom (10–1214m). In shallow cold water upwelling areas this species seasonally comes close inshore, to just beyond the surfline.
Behaviour Poorly known. Sluggish.
Biology Viviparous, 10–52 pups per litter. Eats mostly bony fishes and crustaceans, but also small sharks.
Status IUCN Red List: Endangered. This species has been a target and retained bycatch of many deepsea fisheries for its valuable liver oil. Populations been depleted and possibly extirpated in parts of its range.

PRICKLY SHARK *Echinorhinus cookei*

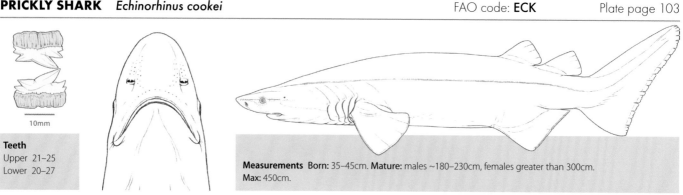

Teeth
Upper 21–25
Lower 20–27

Measurements Born: 35–45cm. **Mature:** males ~180–230cm, females greater than 300cm. **Max:** 450cm.

Identification Numerous, regularly spaced (not fused), light coloured, inconspicuous denticles with single cusps and scalloped bases (less than 5mm); very few below snout and around mouth. Colour light to mid grey, grey-brown, or blackish, often lighter below. White around mouth and beneath snout; trailing fin edges blackish.
Distribution Pacific; distribution patchy.
Habitat Continental and insular shelves and upper slopes, seamounts. Close to bottom, preferring soft sediments, 4–1100m.
Behaviour Sluggish and docile; seen stationary, swimming slowly alone and in groups of 30 or more. Can enter the pelagic zone during vertical diel migrations; exhibits high site fidelity. May investigate divers and submersibles.
Biology Viviparous, one litter of 114 pups recorded. Eats bony and cartilaginous fishes, octopus and squid.
Status IUCN Red List: Data Deficient. Occasional bycatch in deepsea fisheries. Rare, but likely often misidentified as *Echinorhinus brucus*, thus population trends are unknown.

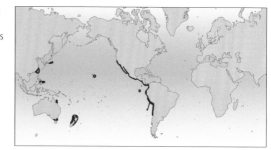

SQUALIFORMES: dogfish sharks

A large and varied order with about 135 species in six families: Squalidae, dogfish sharks (37 described species); Centrophoridae, gulper sharks (16 species); Etmopteridae, lanternsharks (51 described species, others undescribed); Somniosidae, sleeper sharks (16 species); Oxynotidae, roughsharks (five species); and Dalatiidae, kitefin sharks (ten species).

Identification Two dorsal fins (with or without spines); no anal fin. Caudal fin with vertebral column elevated into a moderately long upper lobe; lower lobe absent to strong. Five gill slits, all in front of pectoral fin origins. Nostrils not connected to the mouth by grooves. Spiracles behind and about opposite or above level of eyes. Eyes on side of head; no nictitating lower eyelids. Size ranges from dwarf to huge.

Biology All species for which reproductive mode is known are viviparous (with yolk-sac); litter size from one or two to over 50 pups. Some dogfish sharks have the oldest known age at maturity, lowest fecundity (one foetus per litter in a few *Centrophorus*) and longest known gestation period (18 to 24 months for Spurdog, *Squalus acanthias* and *S. suckleyi*). Some species are solitary, others form huge nomadic schools that range long distances on annual migrations. Some hunt and feed cooperatively. A few (cookiecutter sharks) are parasitic.

Status Dogfish sharks occur in a wide range of marine and estuarine habitats and depths in all oceans worldwide and include the only sharks found at high latitudes close to the poles. Their greatest diversity occurs in deep water (many species occur nowhere else). Many species are fished commercially for meat or liver oil. Those with slow growth and low reproductive capacity are highly vulnerable to depletion by overfishing. The IUCN Red List classifies 17% of this order as threatened, and 51% as Least Concern, while 22% of species are Data Deficient.

Spined Pygmy Shark *Squaliolus laticaudus*: young, close-up of head, light-emitting organs visible on underside (p. 215).

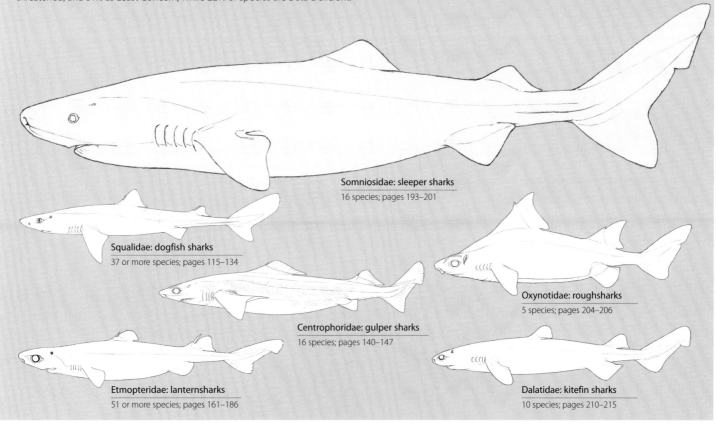

Somniosidae: sleeper sharks
16 species; pages 193–201

Squalidae: dogfish sharks
37 or more species; pages 115–134

Centrophoridae: gulper sharks
16 species; pages 140–147

Oxynotidae: roughsharks
5 species; pages 204–206

Etmopteridae: lanternsharks
51 or more species; pages 161–186

Dalatiidae: kitefin sharks
10 species; pages 210–215

Plate 3 SQUALIDAE I

Head with moderately long snout and large spiracle close behind eye; cylindrical body; two dorsal fins, each preceded by a strong ungrooved spine, no anal fin; many *Squalus* species are difficult to distinguish.

○ **Roughskin Spurdog** *Cirrhigaleus asper*   page 115

West Indian Ocean, west Atlantic, and central Pacific (latter is likely a different species); warm temperate to tropical; 73–600m. On or near the bottom, sometimes off bays and river mouths. Stocky rough-skinned body; broad flat head with short broadly rounded snout, large stubby barbels do not reach mouth; two high stout dorsal spines; conspicuous white-edged fin margins.

○ **Southern Mandarin Dogfish** *Cirrhigaleus australis*   page 115

Southwest Pacific: east coast of Australia from New South Wales to Tasmania and New Zealand (possibly also west Australia, Vanuatu, Fiji); 90–1100m. On or near the bottom. Similar to *C. asper*, but barbels on anterior nasal flaps are less than 2.5 times prenasal length and reach to the mouth.

○ **Mandarin Dogfish** *Cirrhigaleus barbifer*   page 116

West Pacific: southern Japan, possibly to New Zealand; patchy; 146–640m. On or near the bottom. Similar to *C. australis*, but distinguished by long moustache-like barbels greater than 2.5 times prenasal length, which reach mouth; conspicuous white-edged fin margins.

○ **Eastern Highfin Spurdog** *Squalus albifrons*   page 123

West Pacific: east Australia; 131–450m. Snout short, small nostrils, first dorsal fin erect, originating anterior to free rear pectoral fin tips, dorsal spines prominent and robust, second dorsal spine broad-based, caudal fin short; dusky dorsal fin apices and edges, caudal fin with light posterior margins and lower lobe, no dark marks on caudal fin.

○ **Western Highfin Spurdog** *Squalus altipinnis*   page 124

Central Indo-Pacific: northwest Australia; 298–305m. Similar to *S. albifrons*, but distinguished by slightly longer snout; slender second dorsal spine slightly lower than fin apex; greyish dorsal fins with pale apices, caudal fin dusky with narrow pale posterior margins.

○ **Fatspine Spurdog** *Squalus crassispinus*   page 127

Central Indo-Pacific: northwest Australia; 187–262m. Slender body; broad head with broad short snout and small medial lobes; first dorsal fin moderately high, dorsal spines stout, first dorsal spine lower than fin apex, second dorsal spine same height as fin apex; pale fins, dusky dorsal fin apices, caudal fin with dusky dorsal margin and almost white lower lobe, posterior margins and posterior tip.

○ **Eastern Longnose Spurdog** *Squalus grahami*   page 129

Central Indo-Pacific: east Australia; 148–504m. Slender body, snout elongated; dorsal spines pointed and slender, first dorsal spine less than fin height, second dorsal spine about equal to fin height; caudal fin relatively short, less than one-fifth total length. Uniformly greyish above and on flanks, paler below and along pectoral fin margins, dorsal fin apices dark, and caudal fin with dark bar along postventral margin.

○ **Western Longnose Spurdog** *Squalus nasutus*   page 133

Central Indo-Pacific: west Australia to Philippines and South China Sea; 300–850m. Slender body; narrow head with long narrow snout and small medial lobes; first dorsal fin low, dorsal spines slender, first dorsal spine much shorter than second; dorsal fins with dusky apices and posterior margins free tips white, caudal fin with dark dorsal margin and dark blotch on upper ventral margin (fades in adult).

○ **Bartail Spurdog** *Squalus notocaudatus*   page 133

Central Indo-Pacific: northeast Australia; offshore, 225–454m. Fairly slender body; broad head, short broad snout with large medial lobes; high short first dorsal fin with spine lower than apex, second dorsal fin smaller with spine higher than apex, pectoral fins' posterior margin deeply concave with narrow tip; dorsal fin apices dusky, caudal fin with dark dorsal margin and white posterior margins, obvious dark bar along caudal base.

○ *Cirrhigaleus australis*

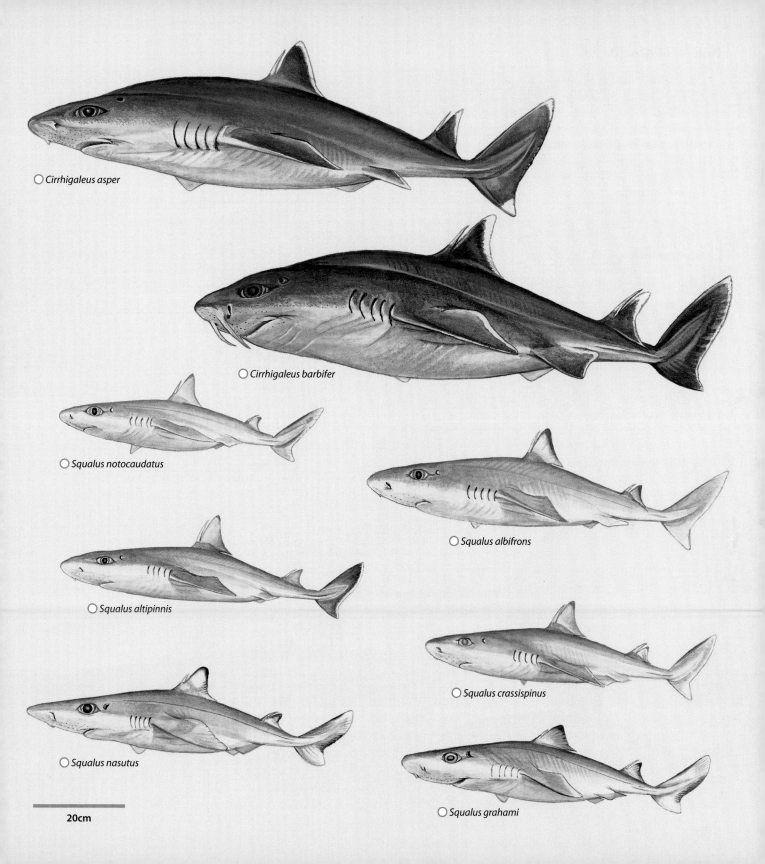

○ *Cirrhigaleus asper*

○ *Cirrhigaleus barbifer*

○ *Squalus notocaudatus*

○ *Squalus albifrons*

○ *Squalus altipinnis*

○ *Squalus crassispinus*

○ *Squalus nasutus*

○ *Squalus grahami*

20cm

Plate 4 SQUALIDAE II

○ **Piked Dogfish** *Squalus acanthias*   page 117

Worldwide except North Pacific, west Indian Ocean, tropical waters and near poles; 0–1978m possibly deeper, 0–200m in epipelagic cold water. Slender body; narrow head, long pointed snout; first dorsal fin low, first dorsal spine slender and very short; often with white spots on back and sides, black dorsal fin tips in young, caudal fin with no blackish marks.

○ **Longnose Spurdog** *Squalus blainville*   page 125

East Atlantic: temperate to tropical; 15–1500m or deeper. On or near the muddy bottom. Heavy body; broad head, relatively short broad snout; erect high first dorsal fin and robust tall first dorsal spine; dorsal fins white-edged.

○ **Cuban Dogfish** *Squalus cubensis*   page 121

West Atlantic: warm temperate to tropical; 10–731m. On or near the bottom. Slender body; broad head, short rounded snout and small medial lobes; first dorsal moderately high, dorsal spines slender and high, pectoral fins deeply concave; pectoral fins with white tips and posterior margin, dorsal fins with black patch at tips, pectoral and caudal fins with white posterior margins.

○ **Japanese Spurdog** *Squalus japonicus*   page 131

Northwest Pacific; 52–400m. On or near the bottom. Fairly slender body; long narrow tapering snout, small medial lobes; first dorsal fin moderately high, long slender dorsal spine; dorsal fin apices dusky, dorsal caudal margin dusky, pectoral and caudal fins with white posterior margins.

○ **Shortnose Spurdog** *Squalus megalops*   page 122

Central Indo-Pacific: Australia (in the northeast Atlantic and Mediterranean this species is referred to as *S. megalops*, but its taxonomic validity is under investigation); 0–732m. On or near the bottom. Small, slender body; broad head, short broad snout, small medial lobes; first dorsal fin moderately high, short slender dorsal spine; dorsal fin apices dark, dorsal caudal margin dark, pectoral and caudal fins with white posterior margins.

○ **Blacktailed Spurdog** *Squalus melanurus*   page 131

Southwest-Pacific: New Caledonia; 34–790m. Slender body; broad head, very long broad snout, small medial lobes; first dorsal fin high, long slender spine; dorsal fin apices black, dorsal caudal margin partially black and lower lobe with black patch.

○ **Shortspine Spurdog** *Squalus mitsukurii*   page 132

Northwest Pacific; 4–954m. On or near the bottom. Fairly slender body; broad head, relatively long broad snout, large medial lobes; first dorsal spine stout and short; dorsal fin apices dusky, pectoral, pelvic and caudal fins with white posterior margins, posterior caudal notch dusky.

○ **Cyrano Spurdog** *Squalus rancureli*   page 134

Central Indo-Pacific: New Hebrides; 210–500m. Slender body; broad head, very long broad snout, small medial lobes; first dorsal fin high, long slender dorsal spine; dorsal fin webs dusky, apices blackish, pectoral and pelvic fins with white posterior margins, caudal fin web dusky, upper postventral margin and ventral tip white.

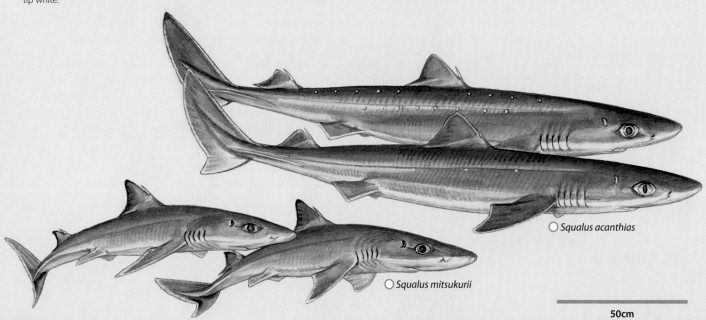

○ *Squalus acanthias*

○ *Squalus mitsukurii*

50cm

○ *Squalus blainville*

○ *Squalus cubensis*

○ *Squalus japonicus*

○ *Squalus megalops*

○ *Squalus melanurus*

○ *Squalus rancureli*

20cm

Plate 5 SQUALIDAE III

○ **Japanese Shortnose Spurdog** *Squalus brevirostris*

Northwest Pacific, possibly more widespread but needs confirmation; 40–163m and deeper. Small stout body; head short; dorsal fins unequal in size; dorsal spines slender, tapering towards tip; no white spots or other obvious dark fin markings.

○ **Bighead Spurdog** *Squalus bucephalus*

Southwest Pacific: northern Tasman Sea; 448–880m. Large stout body; broad head, short snout; dorsal spines slender, tapering towards tip; first dorsal fin apex darker along margins, most prominent at fin tip; second dorsal fin lighter coloured, free rear tip greyish with a narrow white margin.

○ **Greeneye Spurdog** *Squalus chloroculus*

Southeast Indian Ocean to southwest Pacific: southern Australia; 216–1360m. Moderately elongated snout, bluntly rounded at tip; dorsal spines prominent, slender, tapering to tip; dorsal fins mostly grey except for blackish margins extending from above fin spine and along the outer margin to the notch in posterior margin, posterior edge of caudal fin at notch and centre of lobe usually dark.

○ **Edmund's Spurdog** *Squalus edmundsi*

Central Indo-Pacific: Western Australia and eastern Indonesia; 204–850m. Long, narrow snout; prominent first dorsal spine; second dorsal spine slightly higher than first. Grey to greyish brown above, lighter grey below; first dorsal fin dusky except for black-tipped margin at fin apex.

○ **Taiwan Spurdog** *Squalus formosus*

Northwest Pacific: Taiwan and southern Japan; shallower than 300m. Medium-sized, slender-bodied, with a short narrow snout; prominent, robust, erect, dorsal spines; colour strongly demarcated on head through gill openings, dorsal fins blackish on posterior margin and caudal fin with posterior white margin.

○ **New Zealand Dogfish** *Squalus griffini*

Southwest Pacific: around New Zealand; 15–700m. Large slender body; long narrowly rounded snout; first dorsal fin nearly triangular, dorsal spine about half height of dorsal fin, second dorsal spine height about equal to second dorsal fin height; dorsal and caudal fins grey with first dorsal fin margin base and free rear tip lighter than rest of fin, second dorsal fin base uniformly coloured except for narrow black posterior margin.

○ **Indonesian Shortsnout Spurdog** *Squalus hemipinnis*

Central Indo-Pacific: endemic to eastern Indonesia; deeper than 100m. Moderate-sized, slender body; narrow, short bluntly pointed snout; dorsal fins unequal in size, first much larger than second; slate grey above sharply demarcated with light and dark areas on head and extending through top of gills; first dorsal fin grey except at darker apex; caudal fin mostly grey except for broad white posterior margin.

○ **Seychelles Spurdog** *Squalus lalannei*

West Indian Ocean: Seychelles; 1000m. Moderate-sized, slender body; short head with rounded snout tip; dorsal fins rounded at apices, dorsal spines tapering toward apices, first dorsal spine height less than fin height, second dorsal spine height about equal to fin height; uniformly grey body with blackish dorsal fins.

○ **Philippines Spurdog** *Squalus montalbani*

Western Pacific, Taiwan to Australia and east Indian Ocean; 154–1370m. Body moderately elongated; snout narrow and pointed at tip; dorsal spines prominent, slender, first dorsal spine height three-quarters height of dorsal fin, second dorsal spine about equal height to dorsal fin; caudal fin notch and upper lobe usually darker colour.

○ **Kermadec Spiny Dogfish** *Squalus raoulensis*

Southwest Pacific; 250–500m. Small-bodied; short narrow head, snout bluntly pointed at tip; dorsal spines slender, moderately robust at base tapering towards apices, first dorsal spine height shorter than fin apex, second dorsal spine height slightly less than fin height; colour strongly demarcated from snout tip through gill openings, first dorsal fin with narrow black margin on apex, second dorsal fin with darker edge on fin apex and with posterior margin edge white, caudal fin with well-defined white border extending along posterior fin margin.

○ **North Pacific Spiny Dogfish** *Squalus suckleyi*

North Pacific; 0–1236m. Similar to *Squalus acanthias* (p. 117), except distance from pelvic fin midpoint to first dorsal fin insertion usually more than 13% TL for *S. suckleyi* versus 10% or less for *S. acanthias*; flanks usually with conspicuous white spots, which may be absent in larger individuals.

○ *Squalus suckleyi*

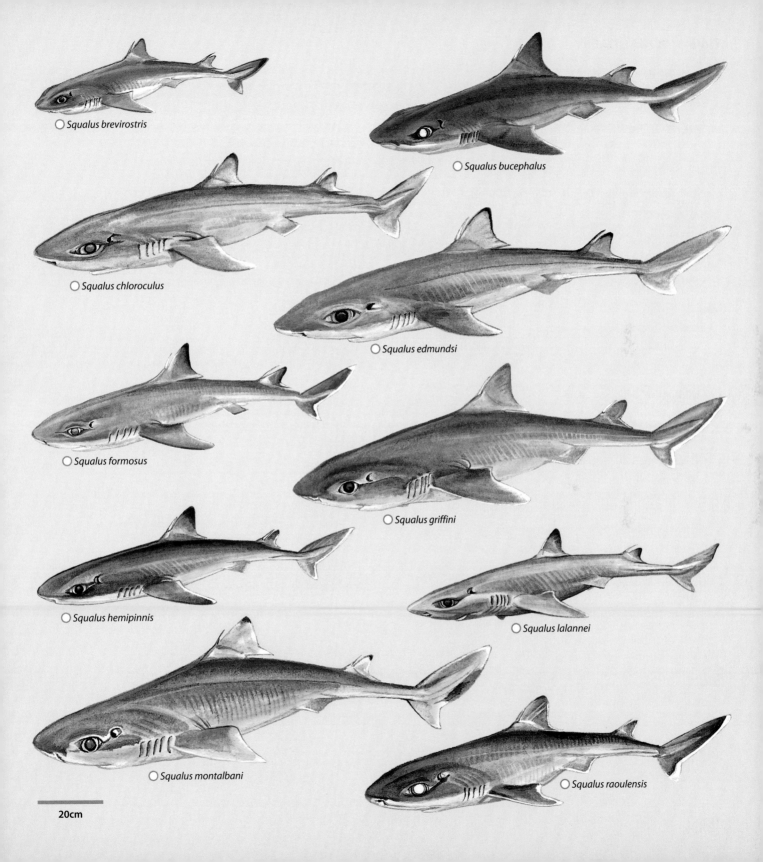

○ *Squalus brevirostris*

○ *Squalus bucephalus*

○ *Squalus chloroculus*

○ *Squalus edmundsi*

○ *Squalus formosus*

○ *Squalus griffini*

○ *Squalus hemipinnis*

○ *Squalus lalannei*

○ *Squalus montalbani*

○ *Squalus raoulensis*

20cm

Plate 6 SQUALIDAE IV

○ **Bluntnose Spiny Dogfish** *Squalus acutipinnis*

page 119

Southwest Atlantic to southeast Indian Ocean: southern Angola and South Africa; inshore to 450m on upper shelf in large dense schools. Newborns pelagic, adults benthic. Small-bodied; bronzy-grey above, lighter below; dorsal fins black-tipped, white-edged, fade with age.

○ **Brazilian Whitetail Dogfish** *Squalus albicaudus*

page 119

Southwest Atlantic; poorly known endemic, often misidentified; continental slope, 195–421m. On or near the bottom. Small stout body; short pointed snout; first dorsal fin larger and darker brown at tip with spine about half fin height, second spine equal or less than fin height; brownish grey above, white below, no spots; white posterior fin margins, very distinct on caudal fin.

○ **Northeastern Brazilian Dogfish** *Squalus bahiensis*

page 124

Southwest Atlantic; poorly known endemic, unknown habitat, on coast of Salvador Bahia. Slender dogfish with elongated blunt snout; first dorsal fin larger with spine about half fin height, second dorsal fin smaller, with fin spine height less than fin height; grey above, white below, no spots; white posterior fin margins, caudal fin with whitish centre and margins.

○ **Long-snouted African Spurdog** *Squalus bassi*

page 125

Southeast Atlantic to southwest Indian Ocean: southern Africa; Namibia to eastern South Africa, possibly southern Mozambique; large schools aggregate on shelf and upper slopes, 159–591m. Large stout body; long angular snout; first dorsal spine shorter than fin base, lower than fin tip, originating above pectoral inner margin; pearl grey to brownish above, white below, no light spots on flanks; white-edged fins.

○ **Emperor Dogfish** *Squalus boretzi*

page 126

Central North Pacific; 100–525m on slopes of Emperor Seamount. Small slender body with short rounded snout; brownish grey above, lighter on sides and below; coloration sharply demarcated below eyes and mid-gills, and level with pectoral and pelvic fin bases; dorsal fins dark-edged; white edges to caudal fin and posterior pectoral fin edges.

○ **Gulf Dogfish** *Squalus clarkae*

page 126

West Atlantic continental slopes: North Carolina to Venezuela, and Caribbean; 137–750m. Large stout body; short snout; grey above fading to white below; white/light pectoral fin edges (more distinct in juveniles); black-tipped first dorsal, more than one and a half height of second; light caudal fin margins, dark bar extends across fork, mostly white lower lobe.

○ **Hawaiian Spurdog** *Squalus hawaiiensis*

page 130

North Pacific: Hawaiian Islands' slopes and seamounts; 100–500m. Large stout body; short snout, slight hump at nape; first dorsal height more than one and a half of second, short stout spines; dark grey to brown body, light grey to white below; greyish pectoral and pelvic fins with white margins; dusky caudal fin with broken white edges, triangular dark bar across fork.

○ **Atlantic Lobed-fin Dogfish** *Squalus lobularis*

page 122

South Atlantic; southern Brazil to Patagonia. Slender body, elongated snout with rounded tip; first dorsal fin higher than second with spine about half fin height, second dorsal spine less than fin height; darkish grey, pale below, no spots; most fins grey with white posterior margins.

○ **Humpback Western Dogfish** *Squalus quasimodo*

page 134

Southwest Atlantic; a poorly known southern Brazilian endemic. Stout, hump-backed, slenderer towards caudal fin; elongated pointed snout; first dorsal with short spine higher than second dorsal fin, with spine of equal height; dark brown body and fins, paler below; blackish dorsal fin apices, whitish caudal margin may be preceded by faint black stripe.

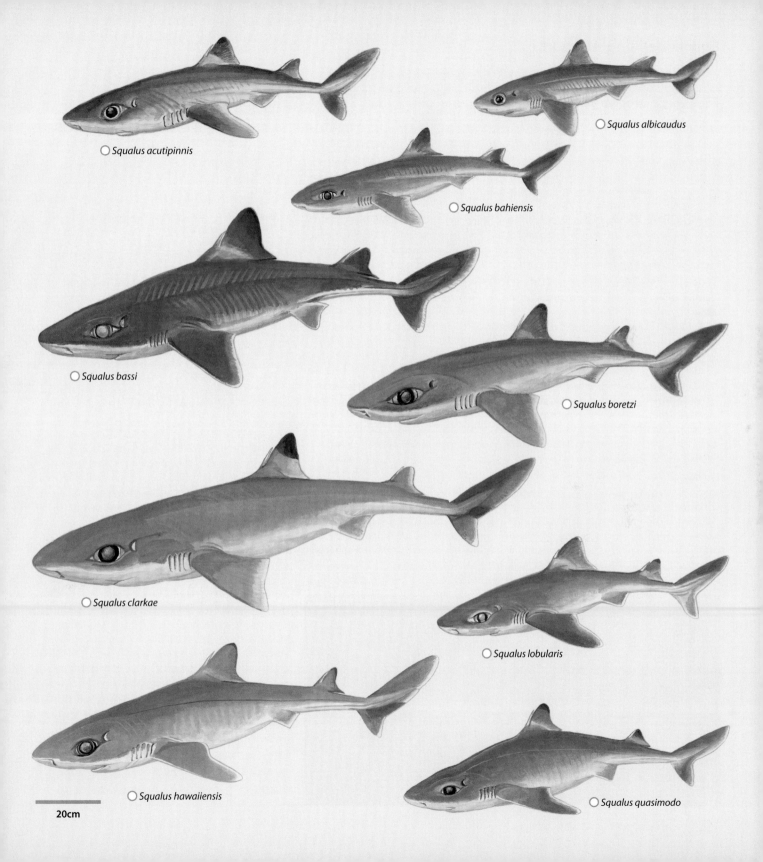

○ *Squalus acutipinnis*

○ *Squalus albicaudus*

○ *Squalus bahiensis*

○ *Squalus bassi*

○ *Squalus boretzi*

○ *Squalus clarkae*

○ *Squalus lobularis*

20cm

○ *Squalus hawaiiensis*

○ *Squalus quasimodo*

Squalidae: dogfish sharks

Two genera, *Cirrhigaleus* (three species) and *Squalus* (34 or more species). *Squalus* is subdivided into three main subgroups known as clades: the *'acanthias'* subgroup (two species), *'megalops'* subgroup (nine species) and *'mitsukurii'* subgroup (at least 23 species, several recently discovered and not yet described). Dogfishes in the *S. acanthias*-clade have a fairly long snout and usually white spots or a pattern of spots and dashes on the sides of their body. The other two clades can be separated by examining the approximate distance from the snout tip to the inner nostril margin, and the distance from the inner nostril edge to front of labial furrow. Species in the shortnose dogfish *S. megalops*-clade have a short, broad snout, with the distance from the snout tip to the inner margin of the nostril being shorter than the distance from the inner edge of nostril to the front of upper labial furrow. Those in the *S. mitsukurii*-clade have a short, broad to greatly elongated and narrow snout, with the distance from the snout tip to inner edge of the nostril equal to or longer than the distance to the front of the upper labial furrow, and lack white spots or dashes on their sides. Dogfish sharks are recorded almost worldwide in boreal, temperate and tropical seas, except in the tropical eastern Pacific. Tropical records may be incomplete because cooler offshore and deepsea habitats are poorly surveyed.

Identification Two dorsal fins with strong ungrooved spines. No anal fin. Snout short to moderately long, mouth short and transverse, low blade-like cutting teeth in both jaws. Spiracles large and close to eyes. Body cylindrical in cross section. First dorsal origin opposite or slightly behind the pectoral fins, second dorsal fin strongly falcate. Pelvic fins smaller than pectoral fins. Caudal peduncle with strong lateral keels, caudal fin without subterminal notch. Colour light grey to medium brown, not black, no luminous organs. *Cirrhigaleus* is stockier with rough skin, enlarged nasal barbels, precaudal pits vestigial or absent, a short caudal peduncle, equal-sized first dorsal fins, and a weak lower caudal lobe. *Squalus* is usually slenderer, smooth-skinned, lacks barbels but has medial lobes on the anterior nasal flaps, strong precaudal pits, an elongated caudal peduncle, the second dorsal fin often much smaller than the first, and a strong lower caudal lobe. Many species of *Squalus* are very hard to distinguish without precise measurements of features and vertebral counts. Their location can, therefore, be the best way to identify *Squalus* to species.

Biology All are viviparous (with yolk-sac), 1–32 pups per litter. Several species are highly social and may segregate by age and sex. Some *Squalus* form huge highly nomadic schools that undertake local and long-distance annual migrations and feed communally, sometimes cleaning out or driving away local populations of prey species. Other species seem to be solitary or only occur in small groups. Prey includes mainly bony fishes, cephalopods and crustaceans, also other cartilaginous fishes and invertebrates. All species have powerful jaws and can dismember prey larger than themselves. Some species are wide-ranging in both northern and southern hemispheres, but absent from intervening tropical seas (e.g. *S. acanthias*) or equatorial

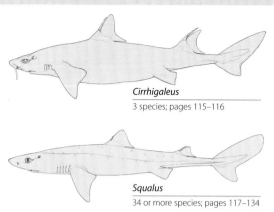

Cirrhigaleus
3 species; pages 115–116

Squalus
34 or more species; pages 117–134

zone. Others are localised endemics. Most are benthic (although some have pelagic young). They occur from the intertidal (mainly in cool-temperate waters) down to depths of 600m, with some to over 1000m, but they are usually replaced by other deepsea shark families below 700–1000m).

Status Despite the considerable (original) abundance and wide range of some species, they are highly vulnerable to overfishing because of their late maturity, longevity, low fecundity and long intervals between litters; *S. acanthias* and *S. suckleyi* have the longest known gestation period of any vertebrate: 18–24 months. Several stock declines have occurred, and recovery of depleted populations can be very slow. Rare species of little or no commercial value are taken as bycatch in fisheries for more common and abundant species. They may become threatened if restricted to small endemic or localised populations or isolated habitats, such as seamounts, targeted by fisheries. Their abundance and ease of capture make some species of *Squalus*, particularly *S. acanthias*, among the most important targets of commercial shark fisheries. Dogfish sharks are landed by up to 50 countries, mostly in bottom trawl fisheries, but also by hook and line, gill nets, seines, pelagic trawls and fish traps. Their meat, liver oil, fins and occasionally leather are of high value. Sports anglers target some species, but they are not important game fish. A few species are regularly displayed in aquaria. Some species use their mildly toxic fin spines and teeth for defence, a hazard to human handlers. Some are considered 'trash fish' because they can cause damage to fishing gear and prey on, or drive away, more valuable fisheries species. The IUCN Red List assesses 35% of species as Least Concern, 30% as Data Deficient, and the rest as Endangered, Vulnerable or Near Threatened.

Piked Dogfish, *Squalus acanthias*, British Columbia, Canada (p. 117).

ROUGHSKIN SPURDOG *Cirrhigaleus asper*

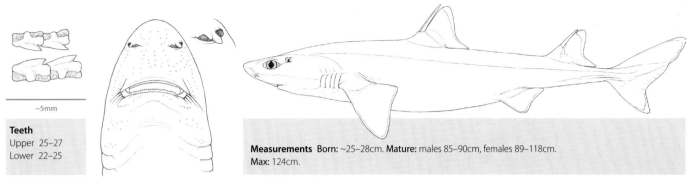

Teeth
Upper 25–27
Lower 22–25

~5mm

Measurements Born: ~25–28cm. **Mature:** males 85–90cm, females 89–118cm.
Max: 124cm.

Identification Stocky, rough-skinned body. Broad, flat head. Short, broadly rounded snout. Anterior nasal flaps with short, stubby medial barbel. Dorsal fins equal-sized with very high, stout spines; first dorsal origin behind pectoral bases. Upper precaudal pit weak or absent. No white spots on sides of body. Fins conspicuously white-edged, without dark markings.
Distribution West Indian Ocean, west Atlantic, and central Pacific in warm-temperate to tropical waters (west Atlantic and Pacific Ocean forms are likely different species).
Habitat On or near bottom; upper and outer continental and insular shelves and slopes (73–600m), sometimes off bays and river mouths.
Behaviour Unknown. Heavy body may indicate a moderately inactive benthic life.
Biology Viviparous, 18–22 pups per litter. Eats bony fishes and squid.
Status IUCN Red List: Data Deficient. Not commercially fished, probably taken as bycatch.

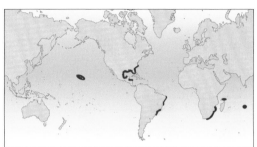

SOUTHERN MANDARIN DOGFISH *Cirrhigaleus australis*

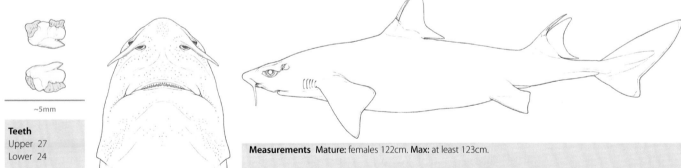

Teeth
Upper 27
Lower 24

~5mm

Measurements Mature: females 122cm. **Max:** at least 123cm.

Identification As for *Cirrhigaleus barbifer*, with long moustache-like anterior nasal barbels reaching the mouth, but shorter prenarial snout length, less than 2.5 times horizontal snout length. Uniform grey-brown above, lighter grey below.
Distribution Southwest Pacific: east Australia, from New South Wales to Tasmania, and New Zealand. Records from west Australia, Vanuatu and Fiji may also be this species.
Habitat Continental upper slopes; 360–640m off Australia, and 90–1100m off New Zealand, with one exceptionally shallow record from 18m. Usually on or near bottom.
Behaviour Unknown. Long barbels may contain chemosensory receptors for prey location.
Biology Viviparous, with litters of at least 10 pups. Feeds on cephalopods and bony fishes.
Status IUCN Red List: Data Deficient. Uncommon to rare, taken as bycatch.

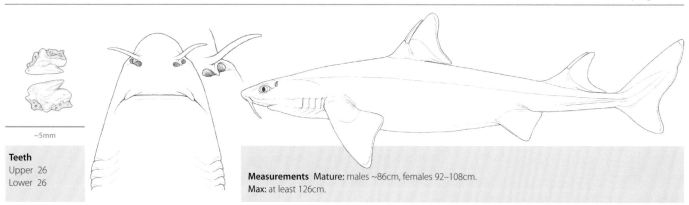

~5mm

Teeth
Upper 26
Lower 26

Measurements Mature: males ~86cm, females 92–108cm.
Max: at least 126cm.

Identification As for *Cirrhigaleus asper*, but barbels on anterior nasal flaps very elongated and moustache-like, reaching the mouth; similar to *C. australis*, but barbels more than 2.5 times length.
Distribution West Pacific: southern Japan, Taiwan and Indonesia; records from outside this area require confirmation.
Habitat Continental and insular upper slopes and outer shelves, 146–640m, on or near bottom.
Behaviour Unknown. Long barbels may contain chemosensory receptors for prey location.
Biology Viviparous. One reported litter of 10 pups (5 per uterus). Prey unknown.
Status IUCN Red List: **Least Concern**. Uncommon to rare, taken by bycatch.

Mandarin Dogfish, *Cirrhigaleus barbifer*, Okinawa Churaumi Aquarium.

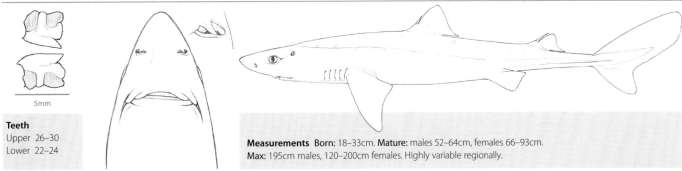

Teeth
Upper 26–30
Lower 22–24

5mm

Measurements Born: 18–33cm. **Mature:** males 52–64cm, females 66–93cm.
Max: 195cm males, 120–200cm females. Highly variable regionally.

Identification (S. acanthias clade) See family. Slender dogfish. Narrow head; relatively long, pointed snout; no medial barbel on anterior nasal flaps. First dorsal fin low, origin usually behind or sometimes over pectoral free rear tips; first dorsal spine slender and very short with origin well behind pectoral free rear tips. Pectoral fins with shallowly concave posterior margins and narrowly rounded free rear tips. Body grey to bluish grey above, often with white spots on sides; lighter to white below. Dorsal fin tips and edges dusky or plain in adults, with black apices and white posterior margins and free rear tips in young; pectoral fins with light posterior margins in adults; no conspicuous black blotches on fins.

Distribution Almost worldwide, except absent from North Pacific, west Indian Ocean, tropical waters and near poles. Little or no mixing between northern and southern hemisphere populations and only limited genetic mixing between some stocks with overlapping range and feeding grounds but different migration patterns.

Habitat Boreal to warm-temperate continental and insular shelves, occasionally slopes (0–600, possibly 1978m or deeper), from surface to bottom. Epipelagic (0–200m) in cold water. Usually near bottom on continental shelf, near surface in oceanic waters. Often on soft sediments in enclosed and open bays and estuaries, where most nursery grounds occur.

Behaviour Mainly demersal, apparently also epipelagic. Sometimes solitary or schooling with other small sharks, often forming immense dense feeding aggregations on rich feeding grounds. Segregates by size and sex into packs or schools of small juveniles (mixed sexes), mature males, larger immature females, or large mature females (often pregnant). Mixed adult schools occasionally reported. Pregnant females usually pup in shallow inshore nursery grounds; some populations pup in deep water on outer shelves and upper slopes. Slow swimmer but undertakes long-distance seasonal migrations north to south or deep to shallow as water temperature changes (prefers 7–8°C to 12–15°C). Other populations are resident year-round. Long-distance movements reported from tag returns include a 1600km journey in the northwest Atlantic.

Biology Extremely long-lived, slow-growing and late-maturing. Viviparous with yolk-sac. Litter size (1–32) varies regionally and larger females have more and larger pups. Gestation period varies regionally, from 12–24 months, to only 12 months in the Black Sea (where largest females, pups and litters occur). Mating may occur in winter, with pups born in winter, spring or summer. Age of younger fish can be measured from annual growth rings on fin spines (spines wear down in old fish) or seasonal calcium peaks in vertebrae. Two northeast Atlantic males tagged and measured when mature and recaptured over 30 years later grew only 0.27–0.34cm annually. Age at maturity: 10–20 years, with longevity varying between populations; maximum age over 40 years. Feeds mainly on bony fishes and invertebrates, occasionally other chondrichthyans. Can capture and dismember prey larger than itself. Predators include large sharks, teleosts and marine mammals.

Status IUCN Red List: **Vulnerable** globally (**Endangered** in northeast Atlantic and Mediterranean, **Vulnerable** in the northwest Atlantic and South America, **Least Concern** in Australasia and southern Africa). Very well-studied. Possibly once the world's most abundant and important commercial shark species, utilised for meat (high value in Europe), liver oil and fins, and supporting large target trawl and line fisheries (comparable scale to teleost fisheries). Also of commercial fisheries significance because large numbers may damage fishing gear and affect catches of other species. Targeted by sports anglers, displayed in public aquaria, and important for scientific research and teaching. Mildly toxic dorsal spines can cause painful lacerations. Its very low reproductive capacity and the targeting of aggregations of large pregnant females makes it highly vulnerable to overfishing; North Atlantic commercial fisheries collapsed. Exploitation continues under management in the northwest, but fisheries are closed in the northeast Atlantic. Northern hemisphere Piked Dogfish stocks are listed by several environmental conventions: CMS (Appendix II and MOU), OSPAR and Barcelona (Appendix III).

Piked Dogfish, *Squalus acanthias*, newborn young with yolk-sac still attached.

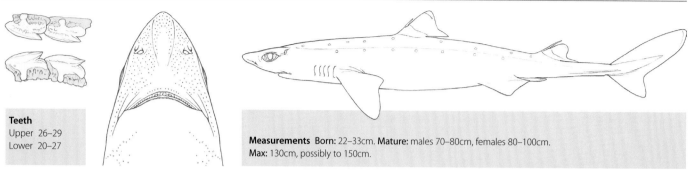

Teeth
Upper 26–29
Lower 20–27

Measurements Born: 22–33cm. **Mature:** males 70–80cm, females 80–100cm. **Max:** 130cm, possibly to 150cm.

Identification (S. acanthias clade) Same as *Squalus acanthias* except distance from pelvic fin midpoint to first dorsal fin insertion usually 13–15% of total body length for *S. suckleyi* versus 8.8–9.8% for *S. acanthias*. White spots usually on flanks with greyish background colour, lighter ventrally.

Distribution Endemic to North Pacific: from Baja California to the Bering Sea and to China, including Taiwan.

Habitat Boreal to warm-temperate continental and insular shelves, occasionally slopes (0–1236m), from surface to bottom. Epipelagic (0–200m) in cold water. Usually near bottom on continental shelf, near surface in oceanic waters. Often on soft sediments in enclosed and open bays and estuaries, where most nursery grounds occur.

Behaviour Mainly demersal, also epipelagic; usually schooling, often forming immense, dense feeding aggregations over rich feeding grounds. Segregates by size and sex into packs or schools of small juveniles (both sexes), mature males, larger immature females, or large mature females (often pregnant). Pregnant females usually pup in deep water on outer shelves and upper slopes. Slow swimmer but undertakes long-distance seasonal migrations north to south or deep to shallow as water temperature changes (prefers 7–8 to 12–15°C). Other populations are resident year-round. Long-distance movements reported from tag returns include a 6500km Pacific Ocean crossing.

Biology Extremely long-lived, slow-growing and late-maturing. Viviparous, 2–12 pups per litter, increasing with female size. Gestation period from 18–24 months. Mating may occur in winter, with pups born in winter, spring or summer. Age is measured from annual fin spine growth rings (in young fish) or seasonal calcium peaks in vertebrae. Age at maturity is 14–35 years, longevity 70 to more than 100 years, varying between populations. Feeds mainly on bony fishes and invertebrates, occasionally other chondrichthyans. Can capture and dismember prey larger than itself. Predators include large sharks, teleosts and marine mammals.

Status IUCN Red List: **Least Concern**. Synonymised with *S. acanthias* until 2010 and listed as that species in Appendix II of the Convention for the Conservation of Migratory Species (CMS) and Annex I to the CMS Sharks MOU. The North Pacific Spiny Dogfish was targeted from the 19th century by fisheries in Japan, and during the 20th century in the northeast Pacific (Canada and USA). Its very low reproductive capacity and the targeting of aggregations of large pregnant females resulted in severe depletion of some stocks, but recovery is taking place in the northeast Pacific.

North Pacific Spiny Dogfish, *Squalus suckleyi*, Sea of Japan, Russia.

BLUNTNOSE SPINY DOGFISH *Squalus acutipinnis*   FAO code: **DGZ**   Plate page 112

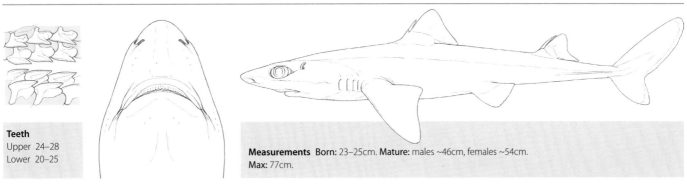

Teeth
Upper 24–28
Lower 20–25

Measurements Born: 23–25cm. **Mature:** males ~46cm, females ~54cm.
Max: 77cm.

Identification (*S. megalops* **clade**) Small-bodied. Short, angular snout; distance from snout tip to inner nasal margin less than from inner edge of nostril to front of upper labial furrow. Relatively wide mouth, almost as long as snout. First dorsal fin height two-thirds or less of length, origin about over pectoral fins; short spine, less than half of fin base length, height below fin tip. Unspotted, bronzy-grey back and sides, lighter below. Dorsal fin black tips and white edges fade and are inconspicuous in adults.
Distribution Southwest Atlantic to southeast Indian Ocean: southern Angola to east coast of South Africa. Records of this species from southern Mozambique, Madagascar and Mauritius require confirmation.
Habitat Inshore to upper continental slopes to 450m depth. Newborn young are mostly pelagic and occur at about 150m depth over deeper water while larger individuals, mostly adults, live close to the bottom.
Behaviour Very social sharks, often forming large dense schools. Strongly segregated by size and sex: adult males mostly occupy the west coast, females mostly the south coast.
Biology Viviparous, 2–4 pups per litter. Continuous reproductive cycle (no gap between pregnancies) in mature females. Gestation period uncertain, may be 12–24 months. Feeds mainly on small bony fishes, cephalopods and crustaceans.
Status IUCN Red List: Near Threatened. Very common off South Africa, often a substantial bycatch, but usually discarded. Has gone by several different scientific names, but based on morphological and molecular data, all are synonymous with *Squalus acutipinnis*.

BRAZILIAN WHITETAIL DOGFISH *Squalus albicaudus*   FAO code: **DGZ**   Plate page 112

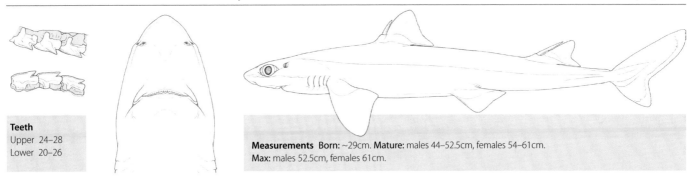

Teeth
Upper 24–28
Lower 20–26

Measurements Born: ~29cm. **Mature:** males 44–52.5cm, females 54–61cm.
Max: males 52.5cm, females 61cm.

Identification (*S. megalops* **clade**) Small, stout dogfish. Short, pointed snout; distance from snout tip to inner nasal margin short, much less than distance from inner edge of nostril to front of upper labial furrow. Nearly transverse mouth, its width more than one-half snout length. First dorsal fin height greater than second; spine height about one-half first dorsal fin height; second dorsal fin spine height about equal or slightly less than fin height. Body colour brownish grey above. First dorsal fin darker brown at apex, but no other distinctive markings; white posterior fin margins and distinctive white caudal fin margin. White below.
Distribution Southwest Atlantic: known only from northeast and southeast Brazil.
Habitat Continental slopes, 195–421m.
Behaviour Unknown.
Biology Viviparous, but nothing known due to misidentification with other regional dogfish species. Its diet likely includes bony fishes, cephalopods and crustaceans.
Status IUCN Red List: Data Deficient. Caught in bottom trawls and longlines. No population trend data available.

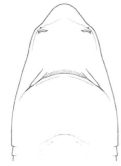

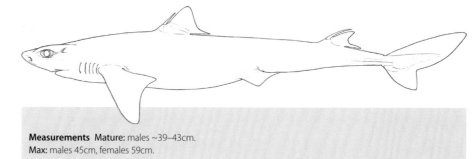

Teeth
Upper 24–25
Lower 20

Measurements Mature: males ~39–43cm.
Max: males 45cm, females 59cm.

Identification (*S. megalops* clade) A small, stout-bodied *Squalus* with a short head. Two unequal-sized dorsal fins; first dorsal fin relatively low, origin over inner margins of pectoral fins. Reddish above. No white spots or obvious dark dorsal or caudal fin markings. Lighter below.
Distribution Northwest Pacific: southern Japan to Taiwan, possibly more widespread but needs confirmation.
Habitat Outer continental shelves and upper slopes, from 40–163m and deeper.
Behaviour Nothing known.
Biology Viviparous, but nothing else known. Possibly eats bony fishes, crustaceans and octopuses.
Status IUCN Red List: Data Deficient. This species has been misidentified with other *Squalus* species and, as such, there is very little information available on the population of this species to assess it beyond Data Deficient.

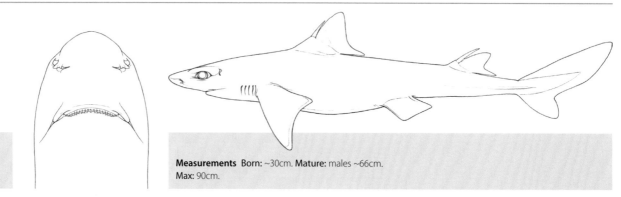

Teeth
Upper 26–27
Lower 22–24

Measurements Born: ~30cm. **Mature:** males ~66cm.
Max: 90cm.

Identification (*S. megalops* clade) A large, stout-bodied dogfish with a prominent hump at its nape. Broad-headed with a short snout. First dorsal fin origin over pectoral fin free rear tip. Colour is a uniform dark brown above, becoming lighter below. First dorsal fin apex darker along outer edges, most prominent at fin tip; second dorsal fin lighter, free rear tip greyish with a narrow white margin.
Distribution Southwest Pacific: known only from the northern Tasman Sea on the Norfolk Ridge, off New Caledonia.
Habitat   Upper continental slopes, from 405–880m.
Behaviour Nothing known.
Biology Viviparous, nothing else known. Possibly eats small fishes and invertebrates.
Status IUCN Red List: Data Deficient. Only known from a very few specimens.

CUBAN DOGFISH *Squalus cubensis* FAO code: **QUC** Plate page 108

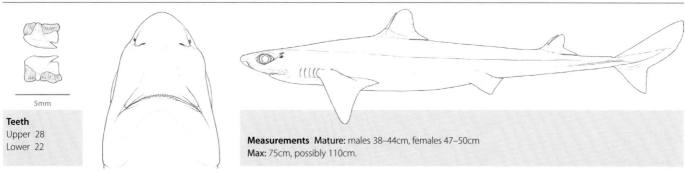

Teeth
Upper 28
Lower 22

Measurements Mature: males 38–44cm, females 47–50cm
Max: 75cm, possibly 110cm.

Identification (*S. megalops* clade) Slender-bodied. Broad head; short, rounded snout; small medial lobe on anterior nasal flaps. First dorsal fin moderately high, spine slender and high; fin and spine origin over pectoral inner margins. Deeply concave posterior pectoral fin margins and angular free rear tips. Body grey, unspotted. Large black or dusky patches on dorsal fin tips; conspicuous white posterior margins on pectoral and caudal fins.
Distribution West Atlantic: warm-temperate and tropical waters.
Habitat Offshore, outer continental shelf and upper slopes, on or near bottom, 10–731m. Young are found shallower than adults.
Behaviour Occurs in large, dense schools.
Biology Viviparous, about 10 pups per litter.
Status IUCN Red List: Least Concern. Apparently common. Most of its geographic and depth range is unfished. Where it is taken as a bycatch, this is usually discarded. Distribution records outside southern Caribbean waters may be of other species.

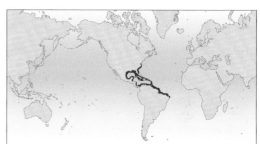

SEYCHELLES SPURDOG *Squalus lalannei* FAO code: **DGZ** Plate page 110

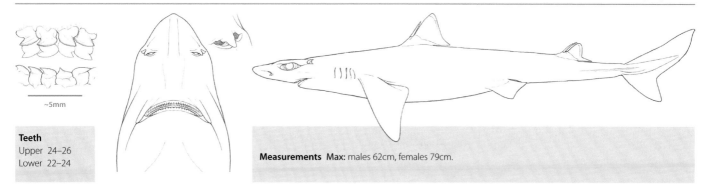

Teeth
Upper 24–26
Lower 22–24

Measurements Max: males 62cm, females 79cm.

Identification (*S. megalops* clade) A poorly known, moderate-sized, slender-bodied dogfish. Short snout, rounded at the tip. First dorsal fin origin over free rear tip of pectoral fin. Uniformly grey with blackish dorsal fins.
Distribution West Indian Ocean: known only from Seychelles.
Habitat Virtually unknown, only two specimens from 1000m.
Behaviour Unknown.
Biology Unknown.
Status IUCN Red List: Least Concern. Additional specimens of this species should be retained for detailed examination.

ATLANTIC LOBED-FIN DOGFISH *Squalus lobularis*   FAO code: **DGZ**   Plate page 112

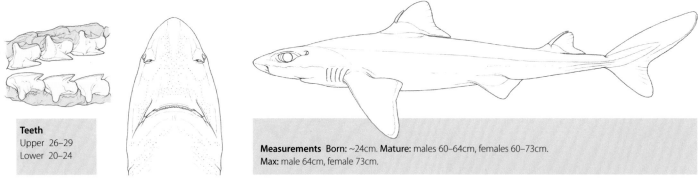

Teeth
Upper 26–29
Lower 20–24

Measurements Born: ~24cm. **Mature:** males 60–64cm, females 60–73cm.
Max: male 64cm, female 73cm.

Identification (*S. megalops* clade) Slender body. Snout elongate, bluntly rounded at tip; distance from snout tip to inner nasal margin greater than distance from inner edge of nostril to front of upper labial furrows. Mouth slightly arched and broad, its width about three-quarters snout length. First dorsal fin height greater than second, spine height about one-half first dorsal fin height; second dorsal spine height less than fin height. Coloration darkish grey above, pale below. Most fins grey with white posterior margins; caudal fin also greyish with partial whitish margins along edges, whitish on central fin; no spots or other distinctive markings.
Distribution Southwest Atlantic: from southern Brazil to Patagonia, Argentina.
Habitat Unknown. **Behaviour** Unknown.
Biology Viviparous, nothing else known. Diet unknown, but likely includes crustaceans, cephalopods and bony fishes
Status IUCN Red List: Data Deficient. Caught in bottom trawls and longlines in Brazilian waters. Too recently classified for an assessment of population status and trends to be possible.

SHORTNOSE SPURDOG *Squalus megalops*   FAO code: **DOP**   Plate page 108

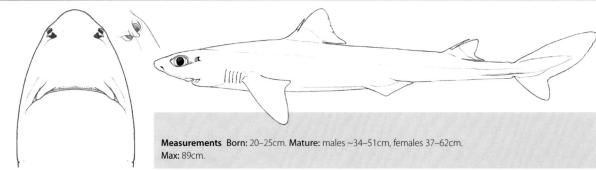

Teeth
Upper 25–27
Lower 22–24

~5mm

Measurements Born: 20–25cm. **Mature:** males ~34–51cm, females 37–62cm.
Max: 89cm.

Identification (*S. megalops* clade) Small, slender dogfish. Broad head; short, broad snout; small medial lobes. First dorsal fin moderately high with short, slender spine; origin over pectoral fin inner margins. Fairly concave posterior pectoral fin margins, angular free rear tips. Coloration grey-brown to dark brown without white spots. Dorsal fin apices, dorsal caudal margin and area of posterior caudal notch dusky or blackish; pectoral and caudal fin lobes with white posterior margins; no conspicuous large black patches on fins.
Distribution Central Indo-Pacific: endemic to Australia. In the northeast Atlantic and Mediterranean this species is referred to as *Squalus megalops*, but its taxonomic validity is under investigation.
Habitat Continental shelves and upper slopes, on or near seabed, 0–732m. Nursery grounds on outer continental shelves.
Behaviour Social, often in large, dense schools; partly sexually segregated.
Biology Viviparous with yolk-sac, 1–6 pups per litter (generally 2–3) after about two-year gestation. Feeds on variety of bony fishes, invertebrates, sometimes batoids.
Status IUCN Red List: Least Concern. A common to abundant species, important as a target and retained bycatch in some fisheries. Monitoring shows increasing fisheries catch rates and no significant population decline.

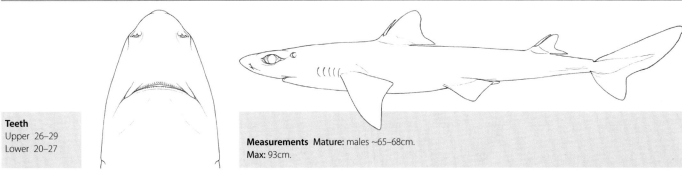

Teeth
Upper 26–29
Lower 20–27

Measurements Mature: males ~65–68cm.
Max: 93cm.

Identification (*S. megalops* clade) A small, narrow-headed dogfish. Short snout. Dorsal fins unequal in size, each preceded by a slender spine. First dorsal fin originates over inner margin of pectoral fins. Uniform reddish brown above, paler below, with variable light and dark areas. Dorsal fins with black margin on apices.
Distribution Southwest Pacific: only known from off northern New Zealand, along insular slopes off Napier Island and Raoul Island on the Kermadec Ridge.
Habitat Deep water, 250–500m, most abundant over 300m on insular slope.
Behaviour Unknown.
Biology Viviparous, nothing else known.
Status IUCN Red List: **Least Concern**. Occurs in a deepsea habitat where fishing has been banned; appears to be relatively abundant where it occurs.

EASTERN HIGHFIN SPURDOG *Squalus albifrons*   FAO code: **DGZ**   Plate page 106

Teeth
Upper 27
Lower 22–23

Measurements Mature: males 61cm, females 74cm.
Max: 86cm (female).

Identification (*S. mitsukurii* clade) Similar to *Squalus altipinnis*. Distinguished by small medial lobes, short first dorsal and caudal fins, and robust second dorsal spine. Broad, triangular, slightly concave pectorals. Caudal fin upper dorsal margin greater than 21% total body length. Dorsal fins with dusky tips and edges; caudal fin with no dark marks; rear caudal margin and lower lobe lighter.
Distribution West Pacific: east Australia from Queensland Plateau (off Cairns) to Montague Island (New South Wales).
Habitat Outer continental shelf and upper continental slope, 131–450m.
Behaviour Unknown.
Biology Viviparous with yolk-sac, nothing else known.
Status IUCN Red List: **Least Concern**. Rare to uncommon, endemic to small area.

WESTERN HIGHFIN SPURDOG   *Squalus altipinnis*   FAO code: **DGZ**   Plate page 106

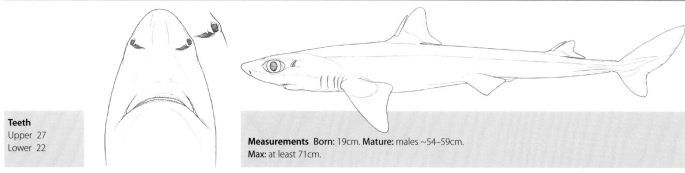

Teeth
Upper 27
Lower 22

Measurements Born: 19cm. **Mature:** males ~54–59cm.
Max: at least 71cm.

Identification (*S. mitsukurii* clade) Similar to *Squalus albifrons* apart from slightly longer snout, slender second dorsal spine slightly lower than fin apex, greyish dorsal fins with pale tips, and caudal fin with narrow pale border (sometimes an inconspicuous dark bar on base).
Distribution Central Indo-Pacific: northwest Australia from off Rowley Shoals.
Habitat Upper continental slope, 130–305m.
Behaviour Virtually unknown. Known from only two adult males.
Biology Viviparous, nothing else known.
Status IUCN Red List: Data Deficient. Endemic, possibly discarded bycatch of deepsea trawl fisheries.

NORTHEASTERN BRAZILIAN DOGFISH   *Squalus bahiensis*   FAO code: **DGZ**   Plate page 112

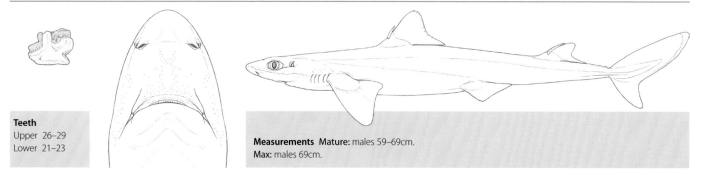

Teeth
Upper 26–29
Lower 21–23

Measurements Mature: males 59–69cm.
Max: males 69cm.

Identification (*S. mitsukurii* clade) Slender-bodied. Elongate, blunt snout, broadly rounded at tip; distance from snout tip to inner nasal margin greater than distance from inner edge of nostril to front of upper labial furrows. Mouth slightly arched and broad, its width more than two-thirds snout length. First dorsal fin height greater than second; spine height less than one-half first dorsal fin height; second dorsal spine height less than fin height. Coloration grey above. Posterior fin margins white; caudal fin whitish along margins and central fin; no spots or other distinctive markings. White below.
Distribution Southwest Atlantic: endemic to the coast of Salvador Bahia, Brazil.
Habitat Unknown.
Behaviour Unknown.
Biology Known from only 3 adult male specimens, females unknown for the species. Presumed viviparous.
Status IUCN Red List: Data Deficient. Caught in intensive bottom trawl and longline fisheries in Brazilian waters, but too recently described and poorly known for population trend data to be available.

LONG-SNOUTED AFRICAN SPURDOG *Squalus bassi*   FAO code: **DGZ**   Plate page 112

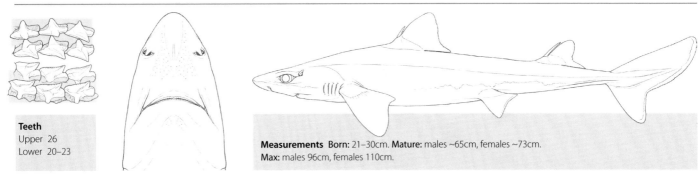

Teeth
Upper 26
Lower 20–23

Measurements Born: 21–30cm. **Mature:** males ~65cm, females ~73cm.
Max: males 96cm, females 110cm.

Identification (*S. mitsukurii* clade) A large, stout-bodied dogfish. Snout long and angular; distance from snout tip to inner nasal margin longer than distance from inner edge of nostril to front of upper labial furrows. Mouth relatively small, width slightly more than half the snout length. First dorsal fin height about two-thirds or less of fin length; spine height shorter than fin base, not reaching fin tip, originating above pectoral inner margin. Pearl grey to brownish above, no light-coloured spots on flanks. White-edged fins. White below.
Distribution Southeast Atlantic to southwest Indian Ocean: southern Africa from northwest of the Orange River, Namibia to the east coast of South Africa and possibly to southern Mozambique.
Habitat Outer continental shelf and upper slopes from 159–591m, with an average depth of about 300m.
Behaviour Usually aggregates in large groups and usually shows strong segregation by size and sex.
Biology Viviparous, 4–9 pups per litter, nothing else known about its reproductive cycle. Feeds mainly on bony fishes, cephalopods and crustaceans.
Status IUCN Red List: Least Concern. A frequent bycatch of bottom trawl and longline fisheries, but discarded. The South African population is rebuilding following a decrease in fishing effort, although fishing pressure may still be high in southern Mozambique. This species in southern Africa waters had formerly been referred to as *Squalus mitsukurii*, but the latter species appears to be restricted to the northwest Pacific.

LONGNOSE SPURDOG *Squalus blainville*   FAO code: **QUB**   Plate page 108

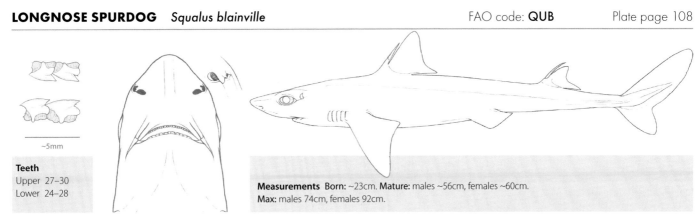

~5mm

Teeth
Upper 27–30
Lower 24–28

Measurements Born: ~23cm. **Mature:** males ~56cm, females ~60cm.
Max: males 74cm, females 92cm.

Identification (*S. mitsukurii* clade) Heavy-bodied. Broad head; relatively short, broad snout. High, erect first dorsal fin, origin over or just behind pectoral bases; origin of long, very heavy spine is over pectoral inner margins. Posterior pectoral fin margins nearly straight to shallowly concave, narrowly rounded free rear tips. Coloration greyish brown, no white spots. White-edged dorsal fins; no obvious dark dorsal or caudal fin markings.
Distribution East Atlantic: temperate to tropical waters including Mediterranean (west Indian and Pacific Ocean forms are a different species).
Habitat On or near muddy bottom, continental shelves and upper slopes, 15–1500m or more.
Behaviour Forms large schools.
Biology Viviparous, 3–4 pups per litter. Eats bony fishes, crustaceans and octopuses.
Status IUCN Red List: Data Deficient. Reportedly common, but possibly confused with *Squalus megalops* and *S. mitsukurii*. Fished with other *Squalus* species.

EMPEROR DOGFISH *Squalus boretzi*

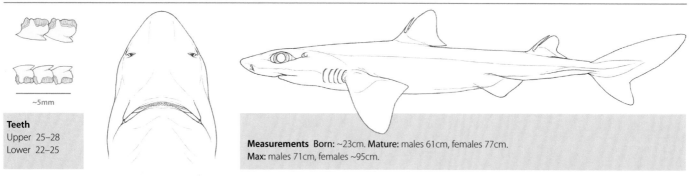

Teeth
Upper 25–28
Lower 22–25

Measurements Born: ~23cm. **Mature:** males 61cm, females 77cm. **Max:** males 71cm, females ~95cm.

Identification (*S. mitsukurii* clade) Small, slender-bodied dogfish. Short, rounded snout; distance from snout tip to inner nasal margin shorter than distance from inner edge of nostril to front of upper labial furrows. Relatively wide mouth, nearly as long as snout. First dorsal fin height about equal to fin length; first dorsal spine origin over pectoral fins; spine height short, not reaching fin tip and less than one-half fin base length. Second dorsal fin less than one-half height first dorsal fin. Coloration is a brownish grey above becoming lighter along sides and below, sharply demarcated below eye and mid-gills and about level with pectoral and pelvic fin bases. Dorsal fins with dark to blackish edges along anterior and posterior margins, including apices; pectoral fin posterior edges white; caudal fin white along lower and upper lobes; no spots or other distinct markings.
Distribution Central North Pacific: Emperor Seamount.
Habitat Insular slopes from 100–525m.
Behaviour Little known, although it appears to occur within water temperature range of 48–61°F (9–16°C).
Biology Viviparous, nothing else known. Diet for this species is unknown, but it likely feeds on bony fishes, cephalopods and crustaceans.

Status IUCN Red List: Not Evaluated.

GULF DOGFISH *Squalus clarkae*

Teeth
Upper 27–29
Lower 23–24

Measurements Born: ~20–30cm. **Mature:** males more than 50cm, females more than 70cm. **Max:** ~125cm.

Identification (*S. mitsukurii* clade) Large, stout-bodied dogfish. Short snout; distance from snout tip to inner nasal margin shorter than distance from inner edge of nostril to front of upper labial furrows. Nearly transverse mouth, width slightly more than half the snout length. Dorsal fins about equal in length, height of first more than one and a half that of second; first dorsal spine height not reaching fin tip, about equal or slightly less than height of second. Coloration grey above becoming pale to white below. Pectoral fin posterior edges lighter to white-edged (more distinct in juveniles). First dorsal fin tip black, grey below; black tip absent on second dorsal fin. Caudal fin margins light-edged but broken by a dark bar extending across the fin fork and a lower lobe that is mostly white.
Distribution West Atlantic: from North Carolina to the northern Gulf of Mexico, the Yucatan Peninsula (Mexico), Central America to Venezuela and throughout the Caribbean.
Habitat Continental slopes, 137–750m.
Behaviour Nothing known.
Biology Viviparous, with 2–15 pups per litter. Feeds mainly on bony fishes, cephalopods and crustaceans.
Status IUCN Red List: Least Concern. An incidental bycatch of trawl, longline and gillnet fisheries, but most of its depth and geographic range is unfished. Previously misidentified as *Squalus mitsukurii*, which only occurs in the northwest Pacific.

GREENEYE SPURDOG *Squalus chloroculus*

FAO code: **DGZ**   Plate page 110

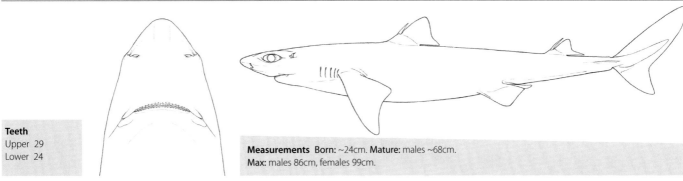

Teeth
Upper 29
Lower 24

Measurements Born: ~24cm. **Mature:** males ~68cm.
Max: males 86cm, females 99cm.

Identification (S. *mitsukurii* clade) Moderately long, blunt snout. First dorsal fin originates over free rear tip of pectoral fins. Prominent dorsal spines. Uniform grey above, except for the upper half of dorsal fins and along caudal fin notch which are darker. Paler below.
Distribution Southeast Indian Ocean to southwest Pacific: southern Australia from Jervis Bay, New South Wales to Eucia, west Australia and Tasmania.
Habitat Continental upper slopes from 216–1360m.
Behaviour Nothing known.
Biology Viviparous, with 4–15 pups per litter. Age at maturity 9–12 years for males and 16 years for females, with maximum age 24 and 26 years, respectively. Diet unknown.
Status IUCN Red List: Endangered. This species has been subject to intense fisheries where it occurs, and population declines of 30–50% have been estimated.

FATSPINE SPURDOG *Squalus crassispinus*

FAO code: **DGZ**   Plate page 106

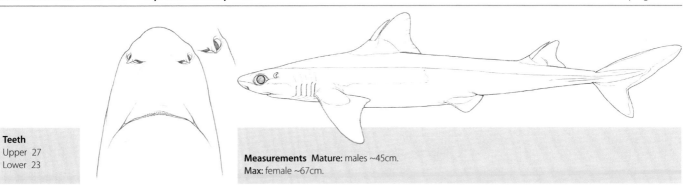

Teeth
Upper 27
Lower 23

Measurements Mature: males ~45cm.
Max: female ~67cm.

Identification (S. *mitsukurii* clade) Small, slender dogfish. Head broad; snout short and broad; small medial lobe on anterior nasal flaps. First dorsal fin moderately high, with origin and spine origin over pectoral fin inner margins; dorsal spines very stout (first is shorter than fin base). Pectoral fins with shallowly concave posterior margins and narrowly rounded free rear tips. Coloration light grey above, paler below; no white spots. Pale fins with dusky dorsal fin tips and front edge of caudal fin; lower caudal lobe, upper caudal tip and trailing edge pale.
Distribution Central Indo-Pacific: northwest Australia, between northwest Cape and north of Rowley Shoals, and Papua New Guinea.
Habitat Outer continental shelf and uppermost slope, 187–262m.
Behaviour Unknown.
Biology Viviparous, nothing else known.
Status IUCN Red List: Least Concern. Endemic to a lightly fished area.

EDMUND'S SPURDOG *Squalus edmundsi*

FAO code: **DGZ**   Plate page 110

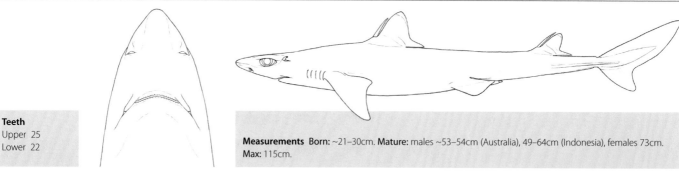

Teeth
Upper 25
Lower 22

Measurements Born: ~21–30cm. **Mature:** males ~53–54cm (Australia), 49–64cm (Indonesia), females 73cm. **Max:** 115cm.

Identification (*S. mitsukurii* clade) Moderate-sized dogfish. Long, narrow preoral snout. First dorsal spine prominent, originating above pectoral inner margins. Second dorsal spine slightly higher than first. Grey to greyish brown above, fading to a lighter grey below. First dorsal fin dusky except for black-tipped margin at fin apex; caudal fin mostly dusky except for narrow black upper caudal fringe.
Distribution Central Indo-Pacific: eastern Indonesia and Western Australia, possibly Papua New Guinea.
Habitat Upper continental slopes from 204–850m, mostly 300–500m.
Behaviour Nothing known.
Biology Viviparous, 3–6 pups per litter. Diet includes small bony fishes, cephalopods and crustaceans.
Status IUCN Red List: Near Threatened. Although there is little information on this species, it occurs throughout areas that are heavily fished and, as such, may be subject to intense exploitation. It is assessed as Near Threatened due to suspected declines in its population.

TAIWAN SPURDOG *Squalus formosus*

FAO code: **DGZ**   Plate page 110

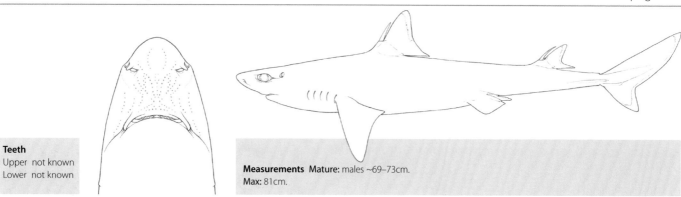

Teeth
Upper not known
Lower not known

Measurements Mature: males ~69–73cm.
Max: 81cm.

Identification (*S. mitsukurii* clade) Medium-sized, slender dogfish. Short, narrow snout. First dorsal fin erect and tall, upper posterior margin nearly straight; first dorsal spine origin just behind pectoral fin insertions. Dorsal spines prominent. Uniform greyish brown above, white below; colour strongly demarcated on head from below the eye to the pectoral fin bases, becoming less distinct posteriorly. Dorsal fin blackish on posterior margin; caudal fin posterior margin whitish.
Distribution Northwest Pacific: known only from Taiwan and Japan.
Habitat Continental shelves and upper slopes, less than 300m.
Behaviour Nothing known.
Biology Viviparous, but nothing else known. Diet presumably includes small fishes, cephalopods and crustaceans.
Status IUCN Red List: Endangered. A retained bycatch of trawl and longline fisheries that have driven major declines in regional elasmobranch stocks. Although fishing pressure is declining around Japan, it remains high across most of this species' range, excluding areas closed to trawling around Taiwan.

EASTERN LONGNOSE SPURDOG   *Squalus grahami*

FAO code: **DGZ**   Plate page 106

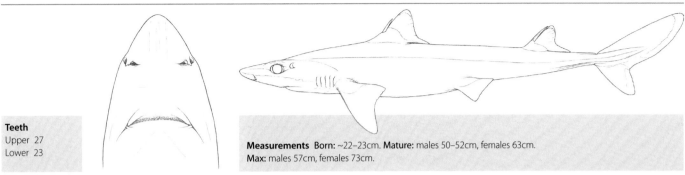

Teeth
Upper 27
Lower 23

Measurements Born: ~22–23cm. **Mature:** males 50–52cm, females 63cm.
Max: males 57cm, females 73cm.

Identification (*S. mitsukurii* clade) Small, slender body. Snout elongated, pointed. Dorsal spines slender; first dorsal spine less than fin height; second dorsal spine about equal to fin height. Caudal fin relatively short, less than one-fifth total body length. Uniformly greyish above and on flanks, paler below and along pectoral fin margins. Dorsal fin apices dark; caudal fin with dark bar along postventral margin.
Distribution Central Indo-Pacific: endemic to east coast Australia, from Cape York, Queensland to Bermagui, New South Wales.
Habitat Upper continental slope, 148–504m, mostly 220–450m.
Behaviour Virtually unknown.
Biology Litters of 1–5 pups, nothing else known.
Status IUCN Red List: Near Threatened. Endemic, seriously depleted by a deepsea trawl fishery in part of range, but most of range unfished.

NEW ZEALAND DOGFISH   *Squalus griffini*

FAO code: **DGZ**   Plate page 110

Teeth
Upper 26–27
Lower 21–24

Measurements Born: ~22–27cm. **Mature:** males 69–76cm; females 86–90cm.
Max: males 90cm, females 110cm.

Identification (*S. mitsukurii* clade) Large, slender-bodied dogfish. Long, narrowly rounded snout. First dorsal spine about half height of dorsal fin. Second dorsal spine height about equal to fin height. Uniform grey-brown above, white below; strongly demarcated on head from snout to below eyes and through gills. Caudal fin with broad, pale posterior margin and on lower lobe.
Distribution Southwest Pacific: known only from around New Zealand and to the north on the Wanganella Bank, Norfolk and Louisville Ridges, Chatham Rise and southern Kermadec Ridge to at least Raoul Island.
Habitat Outer continental shelf to upper continental slope, 15–700m; records from over 700m need confirmation as may be misidentified with larger, undescribed species; most common between 50–300m.
Behaviour Virtually unknown.
Biology Viviparous, with 6–11 pups per litter, average 7–8. Diet includes bottom-dwelling fishes, cephalopods and crustaceans.
Status IUCN Red List: Least Concern. Recent taxonomic re-evaluation of this species indicates that it occurs in an area that is not subjected to intense fishery pressures.

HAWAIIAN SPURDOG *Squalus hawaiiensis*   FAO code: **DGZ**   Plate page 112

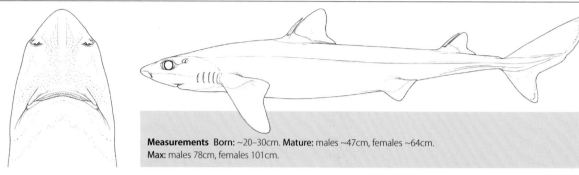

Teeth
Upper 26–28
Lower 23

Measurements Born: ~20–30cm. **Mature:** males ~47cm, females ~64cm.
Max: males 78cm, females 101cm.

Identification (*S. mitsukurii* clade) Large, stout-bodied dogfish with a slight hump at the nape. Snout short; distance from snout tip to inner nasal margin shorter than distance from inner edge of nostril to front of upper labial furrows. Mouth nearly transverse, width slightly more than half the snout length. Dorsal fins nearly equal in length; height of first more than one and a half that of second. Fin spines stout; first less than fin tip and less than one-half height of second fin spine. Uniform dark grey to brown above, becoming light grey to white below. Dorsal fins similar except for black tips that narrow with growth and free rear tips slightly paler. Pectoral and pelvic fins greyish dorsally, slightly darker in the middle with well-defined white posterior margins. Caudal fin dusky with non-continuous white edges at the upper, mid-upper lobe and lower lobe tips; a dark triangular bar extends across the caudal fork to the edge of the lower lobe.
Distribution North Pacific: Hawaiian Islands.
Habitat Insular slopes and seamounts, 100–500m.
Behaviour Unknown.
Biology Viviparous, 3–10 pups per litter. Reproductive cycle is continuous with more than half the adult female population pregnant at any time. Age at maturity is 15 years for males and 8.5 years for females, maximum age is 26 and 23 years, respectively. Feeds mainly on bony fishes, cephalopods and shrimps.

Status IUCN Red List: **Least Concern**. Abundant around the Hawaiian islands, this is a discarded bycatch in some bottom longline fisheries, but unfished in large marine protected areas.

INDONESIAN SHORTSNOUT SPURDOG *Squalus hemipinnis*   FAO code: **DGZ**   Plate page 110

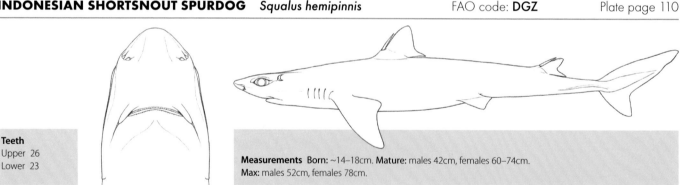

Teeth
Upper 26
Lower 23

Measurements Born: ~14–18cm. **Mature:** males 42cm, females 60–74cm.
Max: males 52cm, females 78cm.

Identification (*S. mitsukurii* clade) Moderate-sized, slender-bodied dogfish. Short, broadly pointed snout. Origin of first dorsal fin slightly posterior to pectoral fin insertion. First dorsal spine broad-based, tapering towards spine tip. Slate grey above, with light and dark areas strongly demarcated on head and through gill openings, indistinct on trunk. Dorsal fin apex slightly darker; caudal fin mostly greyish except for broad white posterior margin; no other obvious dark blotches or markings. White below.
Distribution Central Indo-Pacific: endemic to eastern Indonesia, known from only Cilacap, central Java and Tanjung Luar, eastern Lombok.
Habitat Virtually unknown, but likely at depths greater than 100m.
Behaviour Unknown.
Biology Viviparous, with 3–10 pups per litter, average 6–7. Diet unknown, but likely includes bottom-dwelling fishes, cephalopods and crustaceans.

Status IUCN Red List: **Vulnerable**. Declines are reported from parts of this species' range, all of which is heavily fished. It is landed from target and bycatch fisheries, but there is no management for *Squalus* species.

JAPANESE SPURDOG *Squalus japonicus*

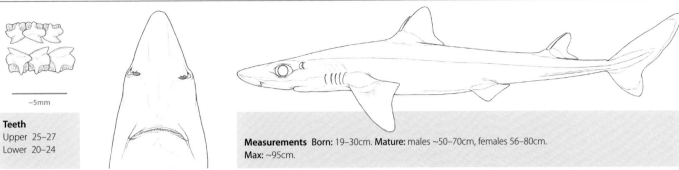

Teeth
Upper 25–27
Lower 20–24

~5mm

Measurements Born: 19–30cm. **Mature:** males ~50–70cm, females 56–80cm.
Max: ~95cm.

Identification (*S. mitsukurii* clade) Fairly slender-bodied. Narrow head; very long pointed snout; small medial lobes. First dorsal fin moderately high; long slender spine, origins over pectoral fin inner margins. Shallowly concave posterior pectoral fin margins and narrowly rounded free rear tips. Grey, reddish grey or bluish brown body without white spots. Dorsal fin apices, dorsal caudal margin and subcaudal notch dusky but no conspicuous black patches; pectorals and caudal fin lobes with white posterior margins.
Distribution Northwest Pacific: Japan, Korea and Taiwan.
Habitat On or near bottom, temperate to tropical outer shelves and upper slopes, 52–400m.
Behaviour Unknown.
Biology Viviparous, 2–8 pups per litter (increasing with female size) each year.
Status IUCN Red List: Endangered. Historic and current fishing pressure is high across this species' range, where it is a retained bycatch for meat and fishmeal and has declined steeply. Decreasing fishing pressure off Japan and inshore gillnet and trawl prohibitions in China and Taiwan, respectively, may allow populations to stabilise. Kept in aquaria.

BLACKTAILED SPURDOG *Squalus melanurus*

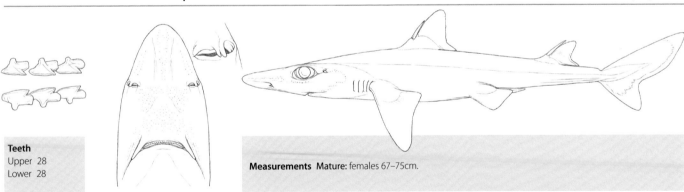

Teeth
Upper 28
Lower 28

Measurements Mature: females 67–75cm.

Identification (*S. mitsukurii* clade) Slender body. Head broad; snout very long and broad; small medial lobes. First dorsal fin high, origin over pectoral insertions; long, slender first dorsal spine, origin over pectoral inner margins. Pectoral fins with straight posterior margins and narrowly rounded free rear tips. Dark grey-brown body without white spots. Black-tipped dorsal fin apices; partial black edge near end of dorsal caudal margin; prominent black patch on lower caudal lobe; pectorals, pelvics and upper caudal lobe with white posterior margins.
Distribution Southwest Pacific: New Caledonia.
Habitat Upper insular slopes, 34–790m, most common from 200–650m
Behaviour Vigorously whips body and long second dorsal spine when captured.
Biology Viviparous, 1 pup in a litter found in a single adult female. Feeds on bony fishes.
Status IUCN Red List: Data Deficient. Very little deepsea fishing occurs within small range.

SHORTSPINE SPURDOG *Squalus mitsukurii*

FAO code: **QUK**   Plate page 108

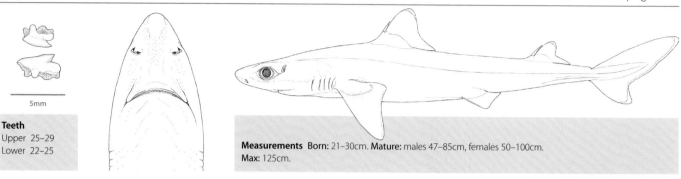

Teeth
Upper 25–29
Lower 22–25

5mm

Measurements Born: 21–30cm. **Mature:** males 47–85cm, females 50–100cm. **Max:** 125cm.

Identification (*S. mitsukurii* clade) Fairly slender dogfish. Head broad; snout relatively long and broad; large medial lobe on anterior nasal flaps. First dorsal fin moderately high; origin over or just behind pectoral bases. First dorsal spine fairly stout and short (much less than fin base); origin over pectoral fin inner margins. Pectoral fins with shallowly concave posterior margins and narrowly rounded free rear tips. Grey or grey-brown body with no white spots, pale below. Dorsal fins with dusky tips and area of posterior caudal notch dusky; these areas blackish in young. Pectoral, pelvic and caudal fins with white posterior margins.
Distribution Northwest Pacific: Japan; records from outside this region are a different species.
Habitat On or near bottom over continental and insular shelves and upper slopes, submarine ridges and seamounts, 4–954m, mostly 100–500m.
Behaviour Often in large aggregations or small schools. Some sexual segregation by depth or latitude.
Biology Viviparous, 2–15 pups per litter (increasing with size of female) after a two year gestation. Age at maturity 4–11 years for males, 15–20 years for females, varying between populations. Maximum age varies from 14–27 years. Powerful predator of bony fishes and invertebrates.
Status IUCN Red List: Endangered. Extremely vulnerable to rapid depletion by fisheries, heavily fished with catches likely utilised, and unmanaged across most of its range. Despite likely widespread depletion, the stock in Japanese waters may now be increasing due to reduced fishing effort.

PHILIPPINES SPURDOG *Squalus montalbani*

FAO code: **DGZ**   Plate page 110

Teeth
Upper not known
Lower not known

Measurements Born: ~20–24cm. **Mature:** males 62–70cm, females ~80–85cm. **Max:** males 93cm, females 116cm.

Identification (*S. mitsukurii* clade) Medium-sized dogfish. Moderately elongated, narrowly pointed snout. First dorsal fin much larger than second; first dorsal fin origin over free rear tips of pectoral fins. Fin spines prominent. Greyish above, becoming paler below. Dorsal fins darker on upper half; caudal fin centre and posterior margin usually darker.
Distribution West Pacific and east Indian Ocean: Taiwan to Australia (east and west coasts), including Philippines and Indonesia.
Habitat Outer continental shelf and upper continental slopes from 154–1370m.
Behaviour Unknown.
Biology Viviparous, with 4–16 pups per litter. Age at maturity ~22 years for males, ~25 years for females, and a maximum age of at least 28 years. Diet includes bottom-dwelling fishes, cephalopods and crustaceans.
Status IUCN Red List: Vulnerable. This species may have experienced declines of 30% or more in some areas. As deepsea fisheries have expanded, there is concern about its population status.

WESTERN LONGNOSE SPURDOG *Squalus nasutus*

FAO code: **DGZ**   Plate page 106

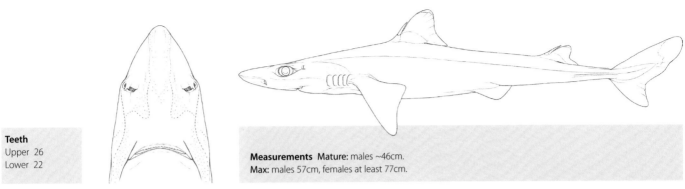

Teeth
Upper 26
Lower 22

Measurements Mature: males ~46cm.
Max: males 57cm, females at least 77cm.

Identification (*S. mitsukurii* clade) Slender body. Head narrow, snout very long and narrow, small medial lobes. Pectoral fins with shallowly concave posterior margins and rounded rear tips. First dorsal fin low, origin opposite pectoral inner margins and spine origin just behind them; first dorsal spine slender and very short. Pale grey above, white below, no white spots. Dark tips and posterior margins to dorsal fins, dark blotch and margin on short caudal fin (less obvious in adults), pale posterior margins on pectorals.
Distribution Central Indo-Pacific: from South China Sea and the Philippines to west Australia.
Habitat Upper continental slope, 300–850m.
Behaviour Virtually unknown. Apparently solitary.
Biology Viviparous, nothing else known. Eats small fishes, cephalopods and crustaceans.
Status IUCN Red List: Near Threatened. Possibly a discarded bycatch of deepsea trawls.

BARTAIL SPURDOG *Squalus notocaudatus*

FAO code: **DGZ**   Plate page 106

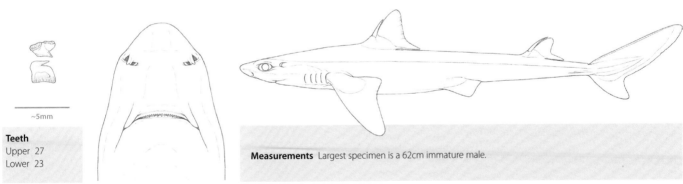

~5mm

Teeth
Upper 27
Lower 23

Measurements Largest specimen is a 62cm immature male.

Identification (*S. mitsukurii* clade) Fairly slender dogfish. Head broad; snout relatively short and broad; large medial lobe on anterior nasal flaps. First dorsal fin high and short; fin and spine origin over pectoral inner margins; first dorsal spine fairly stout and high (same length or just less than fin base). Pectoral fins with deeply concave posterior margins and narrowly rounded free rear tips. Greyish brown above, no white spots. Dorsal fins with dusky tips; dorsal caudal margin dark; posterior caudal fin margin white; obvious dark bar along caudal base, sometimes dark spot above tip of caudal base. White below.
Distribution Central Indo-Pacific: off Queensland, Australia.
Habitat Offshore, upper continental slope, 225–454m.
Behaviour Unknown.
Biology Viviparous, nothing else known. Only known specimens were immature.
Status IUCN Red List: least Concern. Rare or uncommon endemic (known from only 4 immature specimens).

HUMPBACK WESTERN DOGFISH *Squalus quasimodo*

FAO code: **DGZ**   Plate page 112

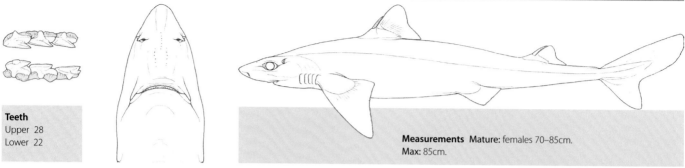

Teeth
Upper 28
Lower 22

Measurements Mature: females 70–85cm.
Max: 85cm.

Identification (*S. mitsukurii* clade) Body stout, 'humped' dorsally, more slender from pelvic fins to caudal fin origin. Snout elongate, narrowly pointed at tip; distance from snout tip to inner nasal margin greater than distance from inner edge of nostril to front of upper labial furrows. Mouth slightly arched to nearly transverse, width about three-quarters snout length. First dorsal fin height greater than second; spine height about half first dorsal fin height; second dorsal spine height about equal to fin height. Dark brown above, pale below. Most fins brown except for blackish dorsal fin apices and caudal fin with whitish posterior margin preceded by a faint black stripe, which may be an artefact of preservation.
Distribution Southwest Atlantic: known only from southern Brazil, and possibly off Margarita Island, Venezuela.
Habitat Unknown.
Behaviour Unknown.
Biology Viviparous, nothing else known. Diet unknown, but likely includes crustaceans, cephalopods and bony fishes.
Status IUCN Red List: Data Deficient. Known from only 4 specimens, all females, 3 of which were mature. Suspected to be caught in bottom trawls and longlines, but too recently described for population status to be known.

CYRANO SPURDOG *Squalus rancureli*

FAO code: **QUR**   Plate page 108

Teeth
Upper 27
Lower 23

Measurements Born: ~24cm. **Mature:** males ~65cm.
Max: 93cm.

Identification (*S. mitsukurii* clade) Slender dogfish. Head broad; snout extremely long and broad; small medial lobe on anterior nasal flaps. First dorsal fin high; first dorsal spine long; origins of both over pectoral fin inner margins. Pectoral fins with broadly concave posterior margins and very narrowly rounded free rear tips. Dark grey-brown above, abruptly light grey below, no white spots. Dorsal fin webs dusky and apices blackish; caudal fin web dusky; pectorals, pelvics and upper caudal lobe with white posterior margins, lower caudal lobe with white tip.
Distribution Central Indo-Pacific: only known from insular slopes of New Hebrides.
Habitat Deep water, 210–500m, but most abundant over 300m on insular slope.
Behaviour Unknown.
Biology Viviparous, 3 pups per litter.
Status IUCN Red List: Near Threatened. Extremely small endemic range.

Centrophoridae: gulper sharks

About 16 mainly deep water, bottom-dwelling species in two genera: *Centrophorus* and *Deania*, recorded almost worldwide from cold-temperate to tropical seas, except in northeast Pacific and very high latitudes. Most diverse in warm waters and in the Indo-West Pacific, where some localised endemic species occur (although many species may prove to be more widely distributed than currently recorded). Main depth range about 200–1500m, but a few records from as shallow as 50m and one *Centrophorus* photographed below 4000m.

Identification Short- to long-nosed cylindrical sharks with huge green or yellowish eyes; blade-like teeth in both jaws, uppers without cusplets, lowers much larger. Two dorsal fins with grooved spines; first dorsal slightly smaller to much larger than second and usually originating in front of pelvic fin origins; second dorsal fin with a straight to weakly concave posterior margin. No anal fin. Very slimy bodies when fresh. *Centrophorus* species have a moderate-sized snout, angular to greatly elongated pectoral free rear tips, and smoother skin with leaf-shaped or block-like denticles. *Deania* species have a very long snout and rough skin with tall slender dorsal dermal denticles topped by pitchfork-shaped crowns.

Biology Little known. Several species are social, forming small schools or huge aggregations (aggregating species are or were among the commonest deepsea sharks). Reproduction is viviparous with yolk-sac (1–17 pups per litter). They feed mainly on bony fishes and cephalopods, also crustacea, small sharks and rays and tunicates.

Status Poorly sampled and identified and imperfectly known, despite their importance in commercial deepsea target and bycatch longline, trawl and gill net fisheries. Their meat is valuable (gulper sharks are among the species known as 'siki shark', highly valued in European markets), as are their large, oily, squalene-rich livers, which are used in cosmetics, health supplements and for machine oil. The gulper sharks are among the most seriously threatened shark families: 63% are assessed as threatened, mostly Endangered, and only one species is Least Concern in the IUCN Red List. Largely unmanaged and unmonitored deepsea fisheries have caused extremely rapid population depletion, particularly among *Centrophorus* species which may collapse after two or three years of target fishing. Gulper sharks have a very limited reproductive capacity, with small litters, long gestation periods, slow growth and late maturity; recovery from overexploitation may therefore be extremely slow.

Centrophorus
12 species; pages 140–145

Deania
4 species; pages 146–147

The gulper sharks are among the more problematic groups for taxonomists, because the precise number of species is uncertain. The lack of preserved type specimens (these are the animals upon which the original descriptions were based) and/or the poor quality of the original written descriptions have caused major identification problems. The dermal denticles were formerly considered a key taxonomic characteristic for separating the gulper sharks. However, very recent examination of a range of species, combined with molecular research, has revealed that at least four previously recognised large gulper sharks, *C. acus*, *C. granulosus*, *C. lusitanicus* and *C. niaukang*, are actually all one and the same species. These four, formerly well-known 'species' were originally described from a wide geographic area and from specimens of different sizes, each of which had very distinct dermal denticles that were used to distinguish between them. These features, however, are now known to be developmental characteristics (or ontogenetic changes) related to fish size, not species. All four species are now classified as *C. granulosus*, since this is the oldest scientific name published, and therefore takes precedence over the three more recent names.

Birdbeak Dogfish, *Deania calcea*, photographed by the E/V *Nautilus* Exploration Program, Ocean Exploration Trust, at Gorringe Bank off Portugal (p. 146).

Plate 7 CENTROPHORIDAE I

Cylindrical; huge green or yellowish eyes; two dorsal fins with grooved spines, no anal fin.

○ **Gulper Shark** *Centrophorus granulosus*

page 140

Atlantic, Indian, west and southwest Pacific Oceans; 50–1500m. On or the near bottom. Largest gulper. Smooth skin; short thick snout; second dorsal fin shorter but nearly as high as long first dorsal, pectoral fin free rear tip long and acute; fin webs dusky, dark tips only in juveniles.

○ **Dumb Gulper Shark** *Centrophorus harrissoni*

page 141

Southwest Pacific: eastern Australia and New Zealand; 220–1050m. Smooth skin; long flat narrow snout; second dorsal fin shorter and lower than first dorsal, pectoral fin free rear tip very long and narrowly angular, notch in terminal margin; dorsal fins with dark oblique bar.

○ **African Gulper Shark** *Centrophorus lesliei*

page 142

East Atlantic to southwest Indian Ocean; 340–610m. Slender body, long head, short snout; both dorsal fins high, first with very long base, second much shorter, large pectoral fins with elongate free rear tip in large adults; no distinct markings on fins.

○ **Longfin Gulper Shark** *Centrophorus longipinnis*

page 142

West Pacific; 330–460m. Relatively slender body and long head, relatively short snout; dorsal fins similar height, first with extremely long base, second much shorter; large pectoral fins with elongate free rear tip in large adults; no distinct markings on adult fins; near-term embryo had blackish dorsal and caudal fins and anterior fin margins, and narrow white posterior margins on dorsal and paired fins.

○ **Smallfin Gulper Shark** *Centrophorus moluccensis*

page 143

Scattered Indian Ocean to west Pacific; 125–823m. Smooth skin; short thick snout; second dorsal fin less than half height of first dorsal, pectoral fin free rear tips long and narrowly angular, caudal fin deep lower lobe; first dorsal fin with dark blotch near apex in juveniles, caudal fin and sometimes pectoral and pelvic fins with narrow pale posterior margins.

○ **Leafscale Gulper Shark** *Centrophorus squamosus*

page 144

Atlantic, Indian, west and southwest Pacific Oceans; 0–3366m. Rough skin; short thick slightly flattened snout; first dorsal fin long and low, second dorsal shorter and higher, pectoral fin free rear tip short, caudal fin posterior margin weakly concave; dusky fins.

○ **Mosaic Gulper Shark** *Centrophorus tessellatus*

page 144

West and central Pacific, possibly Atlantic and Indian Oceans; 260–732m. On or near the bottom. Smooth skin; fairly long thick snout; second dorsal fin about the same as first dorsal, pectoral fin free rear tips long and narrowly angular; light margins to fins.

○ **Little Gulper Shark** *Centrophorus uyato*

page 145

East Atlantic to Indian Ocean; 115–745m, possibly to 1400m. Slender body, moderately long head, relatively short snout; first dorsal fin originates behind pectoral fin insertion and is slightly larger (higher and longer) than the second dorsal fin; juveniles have dark markings on dorsal and caudal fins.

○ *Centrophorus lesliei*

○ *Centrophorus longipinnis*

○ *Centrophorus tessellatus*

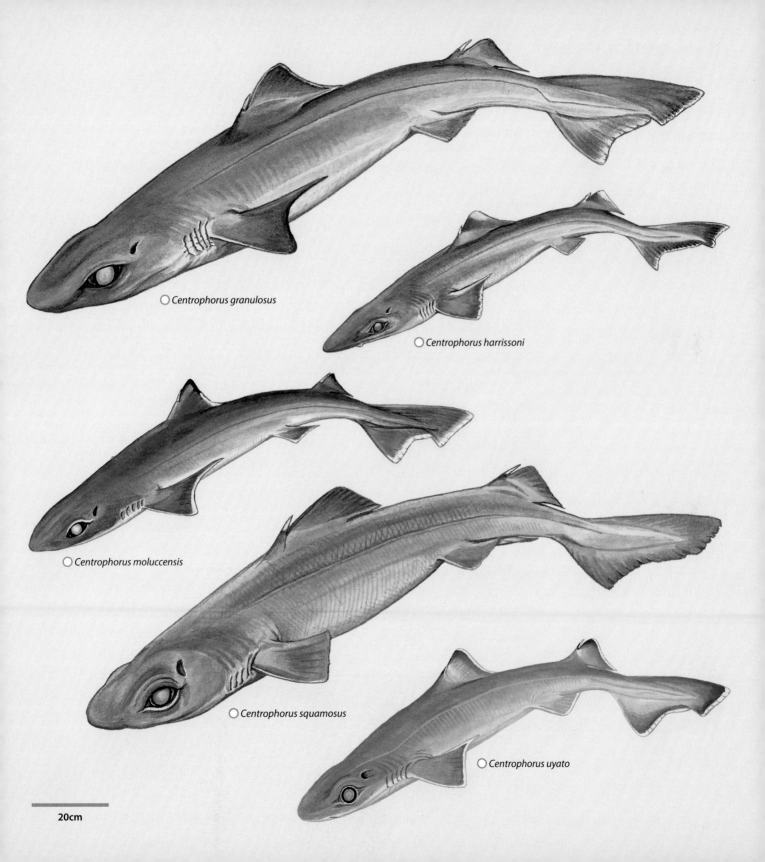

○ Centrophorus granulosus

○ Centrophorus harrissoni

○ Centrophorus moluccensis

○ Centrophorus squamosus

○ Centrophorus uyato

20cm

Plate 8 CENTROPHORIDAE II

○ **Dwarf Gulper Shark** *Centrophorus atromarginatus*

North Indian Ocean and west Pacific; 150–450m. Smooth grey or grey-brown skin; fairly long thick snout; first dorsal fin short and higher than second, pectoral fin free rear tips long and narrowly angular; prominent dusky fin webs, dark fin tips only in juveniles.

○ **Blackfin Gulper Shark** *Centrophorus isodon*

West Pacific and possibly Indian Ocean; 435–770m. Smooth skin; long flat snout; first dorsal fin short, second dorsal lower but as long, pectoral fin free rear tips long and narrowly angular; blackish fin webs especially second dorsal fin and caudal fin.

○ **Seychelles Gulper Shark** *Centrophorus seychellorum*

Indian Ocean: Seychelles; 490–1000m. Relatively long snout; high first dorsal fin, second dorsal base long, pectoral fin angular and rounded at apex, with free rear tip reaching midbase of first dorsal; uniform grey, with blackish margins on dorsal fin apices.

○ **Western Gulper Shark** *Centrophorus westraliensis*

Southeast Indian Ocean: west Australia; 600–750m. Skin smooth; snout elongate, bluntly pointed at tip; pectoral fins extended at tips into angular lobe; first dorsal fin originates well behind pectoral fin axil, second dorsal fin smaller than first; grey above with dorsal fins having a narrow blotch on upper anterior margin.

○ **Birdbeak Dogfish** *Deania calcea*

East Atlantic, Indian, west and southeast Pacific Oceans; 60–1504m. Rough skin; extremely long flat snout; no subcaudal keel; first dorsal fin low and long, second dorsal shorter and higher with longer spine; juveniles with blackish markings on fins, dusky on head.

○ **Rough Longnose Dogfish** *Deania hystricosa*

West Pacific, records from the east Atlantic may be *Deania calcea*; 470–1300m. Very rough skin; extremely long flat snout; no subcaudal keel; first dorsal fin low, second dorsal shorter and higher; no obvious markings.

○ **Arrowhead Dogfish** *Deania profundorum*

West Pacific, Atlantic and Indian Oceans: patchy distribution; 205–1800m. On or near bottom. Smooth skin; extremely long flat snout; subcaudal keel; first dorsal fin relatively long and low, second much taller with much higher fin spine.

○ **Longsnout Dogfish** *Deania quadrispinosa*

Southeast Atlantic, Indian and west Pacific Oceans; 150–1360m. Rough skin; extremely long flat snout; no subcaudal keel; first dorsal fin relatively high, angular and short, second dorsal higher with longer spine; sometimes white-edged fins.

○ *Centrophorus isodon*　　　　○ *Centrophorus atromarginatus*

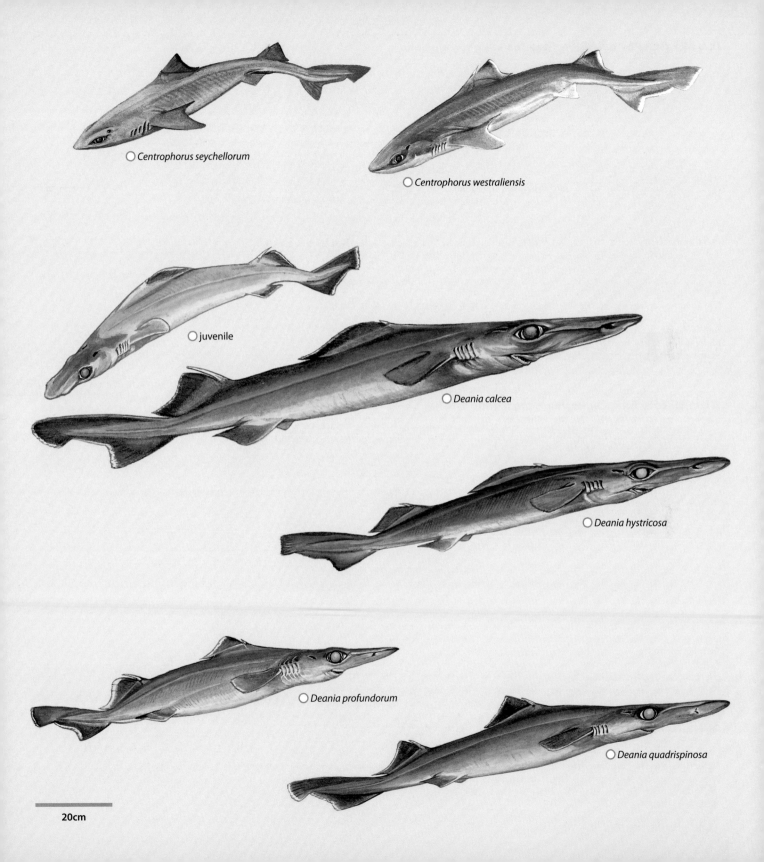

○ *Centrophorus seychellorum*

○ *Centrophorus westraliensis*

○ juvenile

○ *Deania calcea*

○ *Deania hystricosa*

○ *Deania profundorum*

○ *Deania quadrispinosa*

20cm

DWARF GULPER SHARK   *Centrophorus atromarginatus*

FAO code: **GVA**   Plate page 138

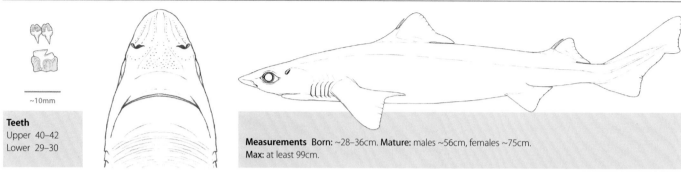

~10mm

Teeth
Upper 40–42
Lower 29–30

Measurements Born: ~28–36cm. **Mature:** males ~56cm, females ~75cm.
Max: at least 99cm.

Identification Fairly long, thick snout. Two dorsal fins with large, grooved spines; first dorsal fin short and higher than second; spine base of second over pelvic fin inner margins or free rear tips. Free rear tips of pectoral fins narrowly angular and greatly elongated. Skin smooth (denticles block-shaped and widely spaced, not overlapping). Grey or grey-brown above, lighter below. Prominent blackish markings on all or most fins (not always on pelvics).
Distribution Northwest Indian Ocean: Sri Lanka, India, Oman. West Pacific: Japan, Taiwan, north Papua New Guinea. (Range is probably wider.)
Habitat Outer continental and insular shelves, upper slopes, 150–450m.   **Behaviour** Unknown.
Biology Viviparous with yolk-sac, 1–2 pups per litter. Diet includes small bony fishes and shrimps.
Status IUCN Red List: **Critically Endangered**. Formerly common, now extirpated from parts of its range. Liver oil utilised.

GULPER SHARK   *Centrophorus granulosus*

FAO code: **GUP**   Plate page 136

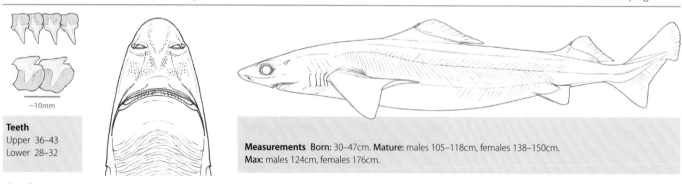

~10mm

Teeth
Upper 36–43
Lower 28–32

Measurements Born: 30–47cm. **Mature:** males 105–118cm, females 138–150cm.
Max: males 124cm, females 176cm.

Identification A large gulper shark with a heavy, robust body. Moderately long head. Long, low first dorsal fin; second dorsal fin similar in height, but shorter in length; each dorsal fin preceded by a fin spine. Teeth dissimilar; uppers with erect to slightly angled cusps; lowers strongly angled and blade-like. Coloration is a uniform brown, greyish or brownish grey; may be slightly paler below.
Distribution Widespread: Atlantic, Indian and southwest Pacific Ocean, and possibly central and southeast Pacific. Absent from the Mediterranean Sea, where previous records were based on misidentification with the Little Gulper Shark, *Centrophorus uyato*.
Habitat Continental shelves and slopes, on or near bottom, 50–1500m, mostly deeper than 600m. The depth of occurrence changes with growth, with smaller individuals occurring in shallower water.
Behaviour Unknown.
Biology Viviparous, 1–13 pups per litter. Age at maturity has been estimated at 8.5 years for males and 16.5 years for females; maximum age estimates are 25 and 39 years, respectively. Feeds mainly on bony fishes, also squid and crustaceans.
Status IUCN Red List: **Endangered** globally, **Critically Endangered** in the northeast Atlantic. Target and bycatch deepsea fisheries for liver oil and meat have caused population declines, to less than 1% of original population size in European waters before a zero quota was set. The Gulper Shark has undergone major taxonomic revisions: there were, until recently, thought to be 4 distinct species (*Centrophorus acus, C. granulosus, C. lusitanticus* and *C. niaukang*), but all are *C. granulosus*, which is the largest species of the genus and family. See p.135.

DUMB GULPER SHARK *Centrophorus harrissoni*

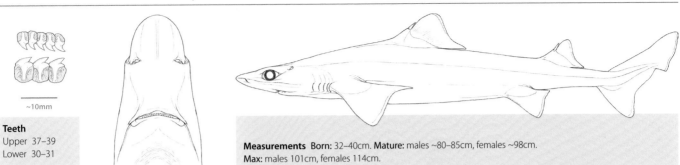

Teeth
Upper 37–39
Lower 30–31

~10mm

Measurements Born: 32–40cm. **Mature:** males ~80–85cm, females ~98cm.
Max: males 101cm, females 114cm.

Identification Snout long, narrow and flat. First dorsal fin short and high. Second dorsal fin lower with spine base over pelvic fin inner margins or free rear tips. Pectoral fin free rear tips narrowly angular and greatly elongated. Shallow notch in postventral caudal fin margin of adults; lower lobe moderately long. Smooth skin (block-shaped denticles widely spaced, not overlapping). Light greyish, paler below. Prominent dark oblique blotch or bar extending back from leading edge of dorsal fins; more diffuse mark on caudal fin.
Distribution Southwest Pacific: eastern Australia, from Tasmania to Queensland, seamounts north of New Zealand and possibly New Caledonia.
Habitat Continental slope, 220–1050m.
Behaviour Unknown.
Biology Viviparous with yolk-sac, 1–2 pups per litter born every 1–2 years. Late-maturing: females at 23–26 years and males at 15–34 years.
Status IUCN Red List: Endangered. Severe population decline in the centre of Australian range since mid 1970s caused by demersal trawl fishery. Fisheries restrictions to protect deepwater sharks in several areas and prohibited status for *Centrophorus harrissoni* are now in force.

BLACKFIN GULPER SHARK *Centrophorus isodon*

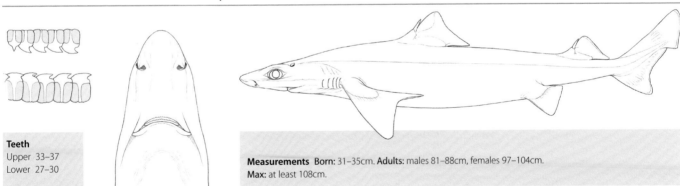

Teeth
Upper 33–37
Lower 27–30

Measurements Born: 31–35cm. **Adults:** males 81–88cm, females 97–104cm.
Max: at least 108cm.

Identification Long, flat snout. First dorsal fin short and high. Second dorsal fin lower with spine base over pelvic fin inner margins. Free rear tips of pectoral fins narrowly angular and greatly elongated. Shallow notch in postventral caudal fin margin of adults; lower lobe moderately long. Smooth skin (block-shaped denticles widely spaced, not overlapping). Blackish grey above, lighter below. Blackish fin webs.
Distribution West Pacific and possibly Indian Ocean. This rare species may be more wide-ranging than reported, with possible records from the North Atlantic.
Habitat Upper continental slopes, 435–770m.
Behaviour Unknown.
Biology Viviparous, 2 pups per litter. Diet includes bony fishes and cephalopods.
Status IUCN Red List: Endangered. Rare and/or often misidentified as other species of *Centrophorus*, but landed from target and bycatch fisheries for meat, oil, fins and fishmeal, causing serious gulper shark population declines.

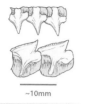

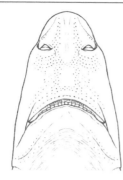

Teeth
Upper 33–42
Lower 29–31

Measurements Born: ~40cm. **Mature:** males 72–77cm, females 86–99cm.
Max: ~100cm.

Identification A medium-sized, slender-bodied gulper shark. Moderately long head; relatively short snout. First dorsal fin high and with a very long fin base; second dorsal fin height about the same as first, but with a much shorter fin base; each dorsal fin proceeded by fin spine. Pectoral fins large; free rear tips elongate in larger individuals. Medium brown to greyish brown above and on sides, slightly paler below. Fins without distinct markings.
Distribution East Atlantic: Canary Islands and Morocco to Angola. Southwest Indian Ocean: Mozambique and Madagascar.
Habitat Continental slopes, 340–610m.
Behaviour Unknown.
Biology Viviparous, 1–2 pups per litter. Feeds mainly on bony fishes, cephalopods and crustaceans.
Status IUCN Red List: Endangered. A virtually unknown species, because previously misidentified with other regional gulper sharks. Declines inferred from the complete overlap of this species' range with intensive fisheries that land shark bycatch.

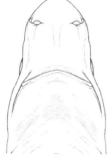

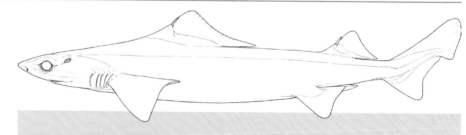

Teeth
Upper 38–43
Lower 29–31

Measurements Born: ~40cm. **Mature:** males 68–78cm, females 87–89cm.
Max: 91cm.

Identification A medium-sized gulper shark with a relatively slender body. Head moderately long; relatively short snout. First dorsal fin base extremely long, second dorsal fin base much shorter; height of dorsal fins similar; each dorsal fin proceeded by fin spine. Pectoral fins large; free rear tips elongate in larger individuals. Brownish to reddish hue, sometimes grey above and on flanks, paler below. Fins without markings in larger specimens; near-term embryo with blackish dorsal and caudal fins, black anterior fin margins and narrow white posterior margin on dorsal and paired fins.
Distribution West Pacific: off Taiwan, Indonesia, Papua New Guinea, possibly Philippines.
Habitat Continental slopes, 330–460m.
Behaviour Unknown.
Biology Viviparous, 1–2 pups per litter. Diet unknown, but likely feeds mainly on bony fishes, cephalopods, and crustaceans.
Status IUCN Red List: Endangered. A target and utilised bycatch of deepsea fisheries, one of which has recorded a 98% reduction in gulper shark catch per unit effort in just two years. It is unfished in another part of its range. This species had previously been misidentified with other regional gulper sharks that are now known to be referable to *Centrophorus granulosus*.

SMALLFIN GULPER SHARK *Centrophorus moluccensis*

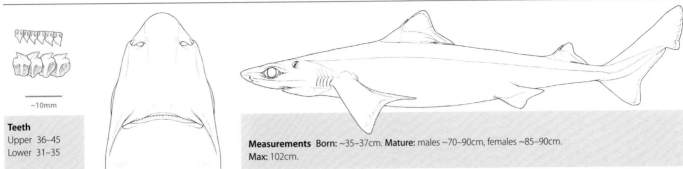

Teeth
Upper 36–45
Lower 31–35

~10mm

Measurements Born: ~35–37cm. **Mature:** males ~70–90cm, females ~85–90cm. **Max:** 102cm.

Identification Short, thick snout. Short first dorsal fin; second less than half height of first, with spine base usually well behind pelvic fins. Pectoral fin free rear tips narrowly angular and greatly elongated. Deeply notched postventral caudal fin margin and deep lower lobe in adults. Smooth skin (block-shaped denticles widely spaced, not overlapping). Greyish brown, much paler below. Dusky fin webs; dark blotch below first dorsal fin tip in juveniles; narrow pale border to caudal fin and sometimes to pectoral and pelvic fins.
Distribution Scattered, Indian Ocean to west Pacific: records from the northern Indian Ocean, particularly off India and Sri Lanka, require confirmation. Distribution and taxonomy are uncertain, this may prove to be a group of visually very similar species.
Habitat Outer continental and insular shelves, upper slopes, 125–823m.
Behaviour Demersal.
Biology Viviparous, 2 pups per litter every 2 years. Feeds mainly on bony fishes and cephalopods, also elasmobranchs, crustaceans.
Status IUCN Red List: **Vulnerable**. Part of this species' range is intensively fished, but it is unfished around some west Pacific islands. Former stock depletion around the Australian coast has ceased since fishing pressure fell. The species is assessed as **Near Threatened** off the east coast of Australia, where New South Wales trawl fisheries had historically depleted that part of the population to less than 5% of its original level, and **Least Concern** on the Australian west coast.

SEYCHELLES GULPER SHARK *Centrophorus seychellorum*

female

male

~10mm

Teeth
Upper 32–33
Lower 29–30

Measurements **Mature:** males ~80cm; largest female still immature at 65cm. **Max:** 80cm.

Identification Snout relatively long. First dorsal fin high. Second dorsal fin base long. Pectoral fins angular, rounded at apex and with free rear tip reaching midbase of first dorsal fin. Lateral trunk denticles thorn-like or rhomboid. Body uniformly grey. Dorsal fin apices and posterior margins blackish.
Distribution Known only from the Seychelles in the west Indian Ocean.
Habitat Insular slopes, 490–1000m.
Behaviour Unknown.
Biology Unknown.
Status IUCN Red List: **Least Concern**. No deepsea fisheries operate within this endemic species' distribution.

LEAFSCALE GULPER SHARK   *Centrophorus squamosus*

FAO code: **GUQ**   Plate page 136

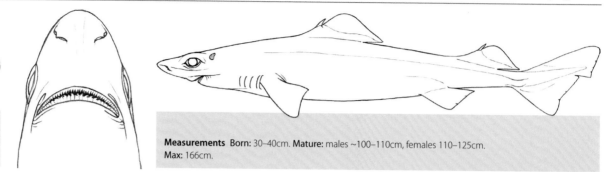

Teeth
Upper 30–38
Lower 24–32

~10mm

Measurements Born: 30–40cm. **Mature:** males ~100–110cm, females 110–125cm. **Max:** 166cm.

Identification Short, thick or slightly flattened snout. First dorsal very long and low. Second dorsal shorter, higher, more triangular; spine base usually opposite pelvic fin inner margins or free rear tips. Pectoral fin free edges short, not particularly angular and elongated. Posterior margin of caudal fin almost straight or slightly concave in adults; lower lobe weakly developed. Rough skin (dermal denticle crowns leaf-shaped in adults, bristle-like in juveniles). Uniform grey, grey-brown or reddish brown. Dusky fins; no prominent markings.
Distribution Atlantic, Indian, west and southeast Pacific Oceans.
Habitat Continental slopes, 0–3366m (rare shallower than 1000m in northeast Atlantic).
Behaviour Apparently demersal and pelagic, this species has been collected from the open ocean, between 0–1250m in water about 4000m deep.
Biology Viviparous, 5–8 pups per litter. Feeds on bony fishes, cephalopods and crustaceans.
Status IUCN Red List: Endangered. An important catch in several deepsea fisheries, which utilise liver oil and meat. Serious declines reported in most of this species' range, but some fisheries are now closed or managed.

MOSAIC GULPER SHARK   *Centrophorus tessellatus*

FAO code: **CEE**   Plate page 136

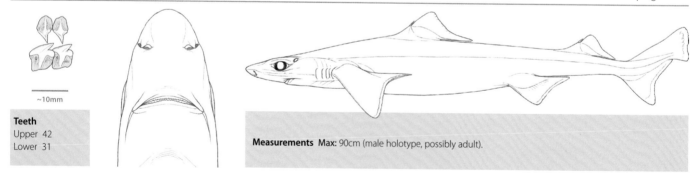

Teeth
Upper 42
Lower 31

~10mm

Measurements Max: 90cm (male holotype, possibly adult).

Identification Fairly long, thick snout. First dorsal fin fairly high and short. Second dorsal fin almost as large; spine origin over free rear tips of pelvic fins. Pectoral fin free rear tips narrowly angular and greatly elongated. Caudal fin with shallowly notched rear margin. Smooth skin (dermal denticles block-like and widely spaced, not overlapping). Light brownish above. Light margins to fins. White below.
Distribution West and central Pacific, possibly the northwest Atlantic and Indian Oceans (provisional records, may actually represent different species).
Habitat Insular slopes, on or near bottom, 260–732m.
Behaviour Unknown.
Biology Viviparous, but otherwise unknown.
Status IUCN Red List: Data Deficient. Rare, may be misidentified in other areas. May be of limited interest to fisheries.

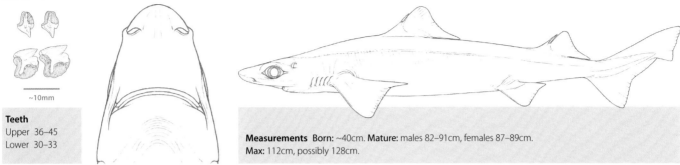

Teeth
Upper 36–45
Lower 30–33

Measurements Born: ~40cm. **Mature:** males 82–91cm, females 87–89cm.
Max: 112cm, possibly 128cm.

Identification A medium-sized, slender-bodied gulper shark. Moderately long head; relatively short snout.
First dorsal fin originates behind pectoral fin insertion and is slightly larger (higher and with a longer fin
base) than the second dorsal fin; a fin spine proceeds each dorsal fin. Uniform greyish brown above,
becoming lighter below. Juveniles have dark markings on dorsal and caudal fins.
Distribution East Atlantic, including Mediterranean, north Indian Ocean and southern Australia; also
possibly west Atlantic and west Indian Oceans, but requires confirmation there due to misidentification with
other regional gulper sharks.
Habitat Continental slopes, 115–745m, possibly to 1400m.
Behaviour Unknown.
Biology Viviparous, 1–2 pups per litter. Diet unknown, but likely feeds mainly on bony fishes, cephalopods
and crustaceans.
Status IUCN Red List: Endangered. Most of this species' range is overlapped by intensive fisheries, which
are known or suspected to have caused stock declines. The species was previously referred to as the
Southern Gulper Shark *Centrophorus zeehaani*, an endemic to southern Australia, but recent molecular and
morphological data have shown these two species to be the same. The distribution, habitat, and biology of
this species are also uncertain due to misidentification with *C. granulosus*.

WESTERN GULPER SHARK *Centrophorus westraliensis*   FAO code: **CWO**   Plate page 138

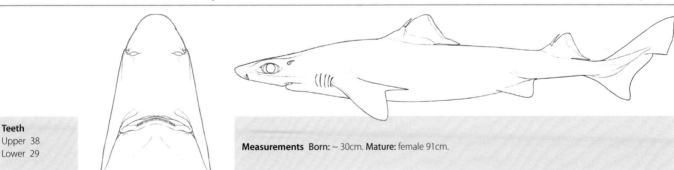

Teeth
Upper 38
Lower 29

Measurements Born: ~ 30cm. **Mature:** female 91cm.

Identification Snout elongate, bluntly pointed at tip. First dorsal fin originating well posterior to pectoral
axil. Second dorsal fin slightly smaller than first. Pectoral fins extended at tips into angular lobe. Caudal fin
relatively long; postventral margin moderately concave. Skin smooth (dermal denticles small, block-like, not
overlapping). Light grey above, pale below. Dorsal fin with a narrow blotch on upper anterior margin.
Distribution Southeast Indian Ocean: endemic to west Australia.
Habitat Upper slopes on or near bottom, 600–750m.
Behaviour Unknown.
Biology Viviparous, but otherwise unknown.
Status IUCN Red List: Data Deficient. The distribution of this poorly known species is covered by the
Western Australia trawl fishery. Fishing effort is low, but population trend is unknown.

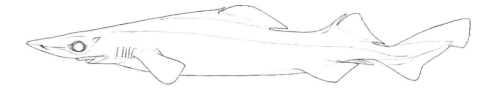

Teeth
Upper 25–35
Lower 27–33

Measurements Born: at least 28–34cm. **Mature:** males ~73–94cm, females 94–106cm. **Max:** 162cm.

Identification Extremely long, flattened snout. Extremely long, low first dorsal fin. Second dorsal fin much shorter, taller with longer fin spine. No subcaudal keel beneath caudal peduncle. Rough skin (lateral trunk denticles pitchfork-shaped, about 0.5mm long). Grey to dark brown with darker fins. Juveniles with dark posterior dorsal fin margins, dark patches above eyes and gill regions and on caudal lobes.

Distribution East Atlantic: Iceland to South Africa. Indian Ocean: South Africa, southern Madagascar, southern Australia. Pacific Ocean: Indonesia, Japan, Taiwan, southeast Australia, New Zealand, and Peru to Chile.

Habitat Continental and insular shelves and slopes, 60–1504m (usually 400–900m).

Behaviour Sometimes schools in large groups.

Biology Viviparous, 1–17 pups per litter. Age estimates for size at maturity are 17 and 25 years for males and females, respectively, with a maximum age of 35 years. Feeds on bony fishes and shrimps.

Status IUCN Red List: Near Threatened. There is a high overlap of intensive fishing pressure across this species' range (except in very deep water) and some well-documented population declines. Management has been introduced and signs of stock recovery reported in some areas. **Endangered** regionally in northeast Atlantic. Formerly abundant in deepsea fisheries, but became seriously depleted before EU catch limits were imposed. Recovery now underway but projected to be extremely slow.

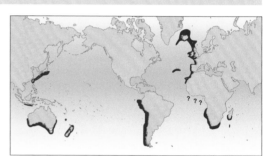

ROUGH LONGNOSE DOGFISH *Deania hystricosa*   FAO code: **SDH**   Plate page 138

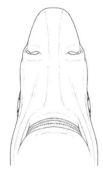

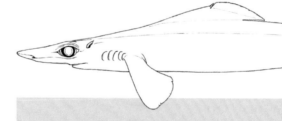

Teeth
Upper 33
Lower 30

Measurements Mature: males ~81–84cm, females 92–106cm. **Max:** 120cm.

Identification Very long, flattened snout. Extremely long, low first dorsal fin. Second dorsal fin much shorter, taller. No subcaudal keel beneath caudal peduncle. Skin very rough due to large pitchfork-shaped lateral trunk denticles (about 1mm long). Blackish brown to grey-brown.

Distribution West Pacific: Japan and New Zealand. Records from the east Atlantic may be *Deania calcea*.

Habitat Insular slopes, 470–1300m; more abundant below 1000m off South Africa.

Behaviour Probably benthic and epibenthic.

Biology Viviparous, one female had 12 large eggs.

Status IUCN Red List: Data Deficient. Low fisheries interest, possible bycatch in deepsea trawl fisheries. The status of this species is in question, with some authors considering it to be identical to *D. calcea*. The morphology of these species is very similar and molecular data have been inconclusive between different studies. We leave it here, but with caution. The issue is currently under investigation.

ARROWHEAD DOGFISH   *Deania profundorum*   FAO code: **SDU**   Plate page 138

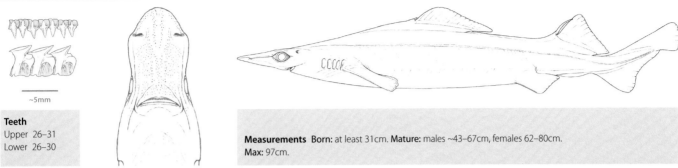

Teeth
Upper 26–31
Lower 26–30

~5mm

Measurements Born: at least 31cm. **Mature:** males ~43–67cm, females 62–80cm.
Max: 97cm.

Identification Smaller than other *Deani* species. Extremely long, flattened snout. First dorsal fin relatively long and low. Second dorsal fin much taller with much higher fin spine. Only *Deania* species with a subcaudal keel beneath the caudal peduncle. Dermal denticles small (0.25mm long). Dark grey or brown.
Distribution East Atlantic: Azores, western Sahara to South Africa. West Atlantic: Gulf of Mexico and Caribbean. West Indian Ocean, including the Gulf of Aden. West Pacific, including the Philippines, possibly Taiwan. Similar to the other members of this genus, its precise distribution is uncertain due to misidentification with other regional *Deania* species.
Habitat On or near bottom, upper continental and insular slopes, 205–1800m.
Behaviour Sometimes occurs in huge aggregations.
Biology Viviparous, 5–7 pups per litter. Feeds on small bottom and midwater bony fishes, squid and crustaceans.
Status IUCN Red List: **Near Threatened**. There is a high level of overlap between intensive fishing pressure and this species' distribution. Not commercially important, but an unmanaged fisheries bycatch.

LONGSNOUT DOGFISH   *Deania quadrispinosa*   FAO code: **SDQ**   Plate page 138

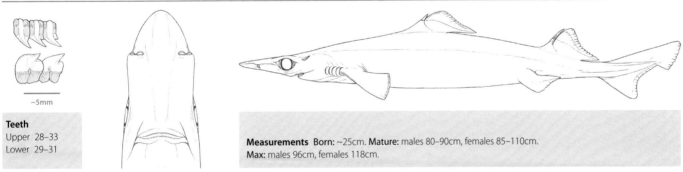

Teeth
Upper 28–33
Lower 29–31

~5mm

Measurements Born: ~25cm. **Mature:** males 80–90cm, females 85–110cm.
Max: males 96cm, females 118cm.

Identification Extremely long, flattened snout. First dorsal fin relatively high, angular and short. Second dorsal fin slightly taller and with much higher fin spine. No subcaudal keel beneath caudal peduncle. Dermal denticles pitchfork-shaped (usually four-spined) and quite large (0.75mm long). Grey or grey-brown to blackish, sometimes with white-edged fins. Juveniles with dark blotch near anterior margin of dorsal fins.
Distribution Southeast Atlantic, Indian and west Pacific Oceans.
Habitat Outer continental shelf and slope, 150–1360m (usually below 400m).
Behaviour Unknown.
Biology Viviparous, with 5–18 pups per litter, average 10. Eats bony fishes.
Status IUCN Red List: **Vulnerable**. A target and bycatch of intensive fisheries that operate across most of this species' range, with the possible exception of very deep water refuges. There are reports of population declines in parts of its distribution, but stocks appear to be stable in others. Some management of deepsea shark fisheries has been introduced.

Plate 9 ETMOPTERIDAE I – *Aculeola, Centroscyllium* and *Trigonognathus*

Dwarf to medium-sized with photophores either inconspicuous or forming distinct black marks; two dorsal fins with strong grooved spines, second dorsal fin and spine much larger than first dorsal, no anal fin and no keel; many species difficult to distinguish.

○ **Hooktooth Dogfish** *Aculeola nigra* .... page 161

Southeast Pacific: South America; 110–735m. Benthic and epibenthic. Stocky; broad blunt snout with broad long arched mouth; gills quite large; short dorsal spines much less than low dorsal fins, pectoral fins with rounded tips.

○ **Highfin Dogfish** *Centroscyllium excelsum* .... page 161

Northwest Pacific: 800–1000m. Deep seamounts. First dorsal fin high and rounded with short spine, second dorsal spine very long reaching higher than fin apex, short caudal peduncle; intense black marks around mouth and beneath pectoral fins; dorsal denticles sparse and irregular, ventral none.

○ **Black Dogfish** *Centroscyllium fabricii* .... page 162

Widespread, temperate Atlantic; 130–2250m. Fairly stout, compressed long abdomen; arched mouth; first dorsal fin low, short caudal peduncle; numerous close-set denticles.

○ **Granular Dogfish** *Centroscyllium granulatum* .... page 162

Southeast Pacific: South America; 100–610m. Small, slender and cylindrical; long abdomen; mouth narrowly arched; first dorsal fin small, much smaller than second dorsal; second dorsal spine very large, higher than fin apex; long caudal peduncle; closely covered by sharp denticles.

○ **Bareskin Dogfish** *Centroscyllium kamoharai* .... page 163

West Pacific and southeast Indian Oceans; 500–1225m. Benthic. Stout and compressed; very short broadly arched mouth; first dorsal fin very low and rounded, second dorsal fin slightly higher; second dorsal fin spine about the same height as fin apex; short caudal peduncle; skin smooth, almost naked.

○ **Combtooth Dogfish** *Centroscyllium nigrum* .... page 163

Central and east Pacific; 32–1212m. On or near bottom. Fairly stout; short broadly arched mouth; dorsal fins about the same size, first dorsal spine short, second dorsal about the same height as the fin apex; prominent white fin tips.

○ **Ornate Dogfish** *Centroscyllium ornatum* .... page 164

North Indian Ocean; 521–1262m. Near the bottom. Narrowly arched mouth; first dorsal fin low and rounded, very long spine nearly as high as second dorsal spine; denticles close-set and numerous.

○ **Whitefin Dogfish** *Centroscyllium ritteri* .... page 164

Northwest Pacific; 150–1100m. Only *Centroscyllium* with obvious black markings on underside of head, abdomen, pectoral fins and stripe below caudal peduncle; mouth broadly arched; first dorsal fin low and rounded with very short spine and about the same height as second dorsal fin, second dorsal spine higher than fin apex; numerous close-set denticles; white posterior fin margins.

○ **Viper Dogfish** *Trigonognathus kabeyai* .... page 186

Northwest and central Pacific; 150–1000m. Mostly on bottom. Very long snake-like mouth with huge curved fang-like teeth, deep pockets on head in front of very large elongated spiracles; grooved spines on both dorsal fins; black underside, black caudal fin markings.

○ *Trigonognathus kabeyai*

20cm

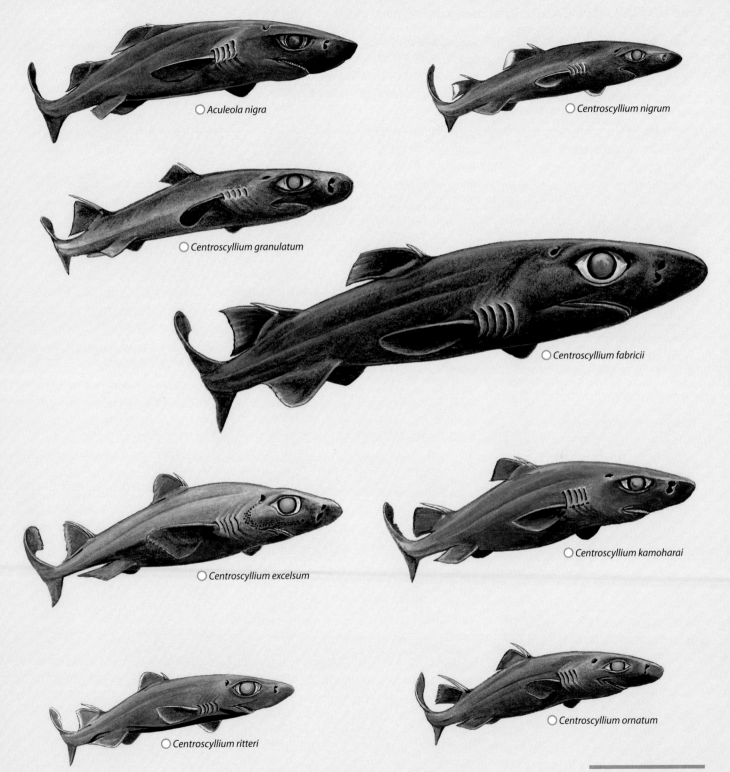

○ *Aculeola nigra*

○ *Centroscyllium nigrum*

○ *Centroscyllium granulatum*

○ *Centroscyllium fabricii*

○ *Centroscyllium excelsum*

○ *Centroscyllium kamoharai*

○ *Centroscyllium ritteri*

○ *Centroscyllium ornatum*

20cm

Plate 10 ETMOPTERIDAE II – *Etmopterus* *gracilispinus* clade

Species in the *Etmopterus gracilispinus* clade have flank markings with a long, thick, curved anterior branch and a short thick posterior branch.

○ **Broadbanded Lanternshark** *Etmopterus gracilispinis* — page 165

West Atlantic; 25–1200m. On or near bottom. Stout; gills very short; slender short caudal fin; second dorsal fin about twice area of first dorsal; no regular rows of denticles; underside grading to black, inconspicuous black ventral and caudal fin markings.

○ **Dwarf Lanternshark** *Etmopterus perryi* — page 165

West central Atlantic; 230–530m. One of smallest living sharks; long broad flat head; second dorsal fin height higher than first dorsal; underside black, conspicuous black ventral and caudal fin markings.

○ **African Lanternshark** *Etmopterus polli* — page 166

East central Atlantic; 300–1000m. On or near bottom. Fairly stout; gills short; fairly long caudal fin; second dorsal fin same height or higher than first dorsal; denticles widely spaced, no regular rows, largely cover snout; blackish underside, black ventral and caudal fin markings.

○ **West Indian Lanternshark** *Etmopterus robinsi* — page 166

West central Atlantic; 412–787m. Moderately stout; gills very short; second dorsal fin much higher than first dorsal but less than twice area of first dorsal; denticles widely spaced, no rows, largely cover snout; underside abruptly black, black ventral and caudal fin markings.

○ **Fringefin Lanternshark** *Etmopterus schultzi* — page 167

West central Atlantic; 220–915m. On or near bottom. Slender; gills very short; moderately long caudal fin; second dorsal fin about twice area first dorsal; denticles widely spaced, no rows, largely cover snout; fin margins naked; dark ventral and caudal fin markings.

○ **Green Lanternshark** *Etmopterus virens* — page 167

Northwest Atlantic; 196–915m. Moderately slender; gills very short; long narrow caudal fin; second dorsal fin area greater than twice area of first dorsal; denticles widely spaced, no rows, largely cover snout; underside black, black ventral and caudal fin markings.

Uncertain clade

○ **Rasptooth Dogfish** *Etmopterus sheikoi* — page 185

Northwest Pacific; 340–370m, possibly to ~1000m. Unmistakable long flat snout with short mouth; grooved dorsal spines, second dorsal spine higher than first dorsal fin; black underside, black caudal fin markings.

○ *Etmopterus gracilispinis*

○ *Etmopterus perryi*

○ *Etmopterus polli*

○ *Etmopterus robinsi*

○ *Etmopterus schultzi*

○ *Etmopterus virens*

○ *Etmopterus sheikoi*

20cm

Plate 11 ETMOPTERIDAE III – *Etmopterus* lucifer clade I

Species in the *Etmopterus lucifer* clade have long thin anterior and posterior branches; the branch lengths (e.g. the anterior length relative to the posterior – longer, shorter, or nearly equal) can help distinguish between members of this clade.

○ **Whitecheek Lanternshark** *Etmopterus alphus*

Southwest Indian Ocean; continental slopes, 472–792m. Small, slender; conspicuous longitudinal row of denticles on sides and back; purplish black above and to sides, black below, sharply demarcated by silvery to white lateral stripe below flank marking; prominent white cheek spot.

○ **Shorttail Lanternshark** *Etmopterus brachyurus*

West Pacific and Central Indo-Pacific; 100–696m. Near bottom. Heavy-bodied; broad head with short thick flat snout; short caudal fin; second dorsal fin higher than first dorsal, second dorsal spine strongly curved; conspicuous line of denticles; black ventral and caudal fin markings.

○ **Lined Lanternshark** *Etmopterus bullisi*

Northwest and central Atlantic; 275–824m. On or near bottom. Slender; gills very short; long caudal fin; conspicuous longitudinal row of denticles on sides and back; underside black, light band midline eye to first dorsal fin, black ventral and caudal fin markings.

○ **Broadsnout Lanternshark** *Etmopterus burgessi*

Northwest Pacific: Taiwan; 300–600m. Moderate-sized; broad snout; uniformly grey above, darker grey to black below; flank markings elongated, posterior branch marking about as long as anterior or slightly shorter.

○ **Combtooth Lanternshark** *Etmopterus decacuspidatus*

Northwest Pacific; 512–692m. On or near bottom. Moderately slender; gills very short; fairly long broad caudal fin; second dorsal fin about twice area first dorsal; no regular rows of denticles; underside black, black ventral and caudal fin markings.

○ **Broken Lined Lanternshark** *Etmopterus dislineatus*

Southwest Pacific: northeast Australia; 590–802m. On or near bottom. Attractive elongate lanternshark; first dorsal fin small and low, about half size second dorsal fin; bristle-like denticles not in rows; dark broken lines on flanks with black caudal fin markings.

○ **Blackmouth Lanternshark** *Etmopterus evansi*

Central Indo-Pacific; 430–550m. Weakly defined rows of denticles on dorsal midline and caudal peduncle but not head; dark borders around mouth, above eyes and sometimes gills, also black caudal fin markings.

○ **Laila's Lanternshark** *Etmopterus lailae*

Central Pacific; insular slopes of seamounts, 314–384m. Moderately large; conspicuous longitudinal row of denticles and 1–3 rows of prominent dark photophores from head to upper caudal fin origin; preserved specimen light to medium-brown on sides and back, darker brown below.

○ **Blackbelly Lanternshark** *Etmopterus lucifer*

West Pacific; 158–1357m. On or near bottom. Stocky; gills moderately long; moderately long caudal fin; second dorsal fin very large; longitudinal rows of denticles snout to tail; underside black, black ventral and caudal fin markings.

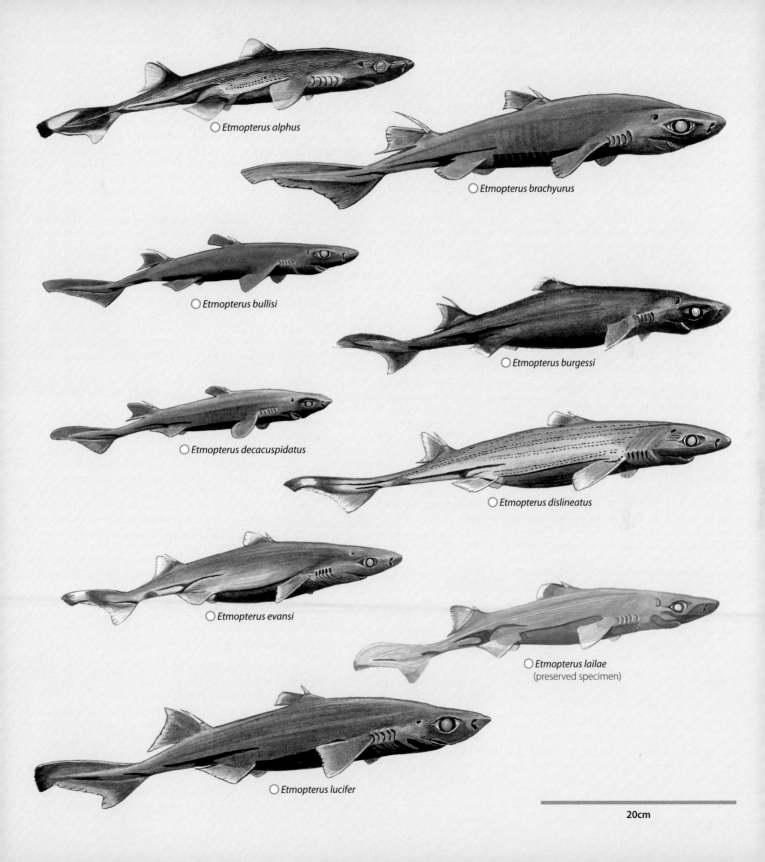

○ *Etmopterus alphus*

○ *Etmopterus brachyurus*

○ *Etmopterus bullisi*

○ *Etmopterus burgessi*

○ *Etmopterus decacuspidatus*

○ *Etmopterus dislineatus*

○ *Etmopterus evansi*

○ *Etmopterus lailae*
(preserved specimen)

○ *Etmopterus lucifer*

20cm

Plate 12 ETMOPTERIDAE IV – *Etmopterus* *lucifer* II and *pusillus* I clades

Etmopterus *lucifer* clade II

Species in the *Etmopterus lucifer* clade have long thin anterior and posterior branches; the branch lengths (e.g. the anterior length relative to the posterior – longer, shorter, or nearly equal) can help distinguish between members of this clade.

○ **Marsha's Lanternshark** *Etmopterus marshae*
page 172

West Pacific; continental slopes, 322–337m. Small, slender; dark purple-black on sides and back, sharply demarcated to black below; conspicuous rows of denticles and 1–3 rows of dark dashes from head to caudal fin origin; paired row of dashes between pectoral and pelvic fins; caudal fin with distinct black bar at upper caudal origin, fading posteriorly to white, with a prominent black caudal fin tip.

○ **Slendertail Lanternshark** *Etmopterus molleri*
page 173

Southwest Pacific; 238–655m. On or near bottom. Slender; second dorsal fin much higher than first dorsal; regular longitudinal rows of denticles snout to caudal fin, except above pectoral fins; underside abruptly black with black stripes on sides.

○ **Densescale Lanternshark** *Etmopterus pycnolepis*
page 173

Southeast Pacific; 330–763m. Slender; narrow head; gills long; moderately long caudal fin; first dorsal fin origin ahead of pectoral fin rear tip, second dorsal fin higher than first dorsal; very small denticles in dense rows head to caudal fin; black body and caudal fin markings.

○ **Papuan Lanternshark** *Etmopterus samadiae*
page 174

Southwest Pacific; continental slopes, 340–785m. Small, slender; greyish to silvery-black on sides and back, sharply demarcated from black underside by lighter to white lateral stripe below black flank marking, which is also sharply demarcated by surrounding lighter lateral markings, and a prominent white cheek spot; conspicuous longitudinal row of denticles.

○ **Sculpted Lanternshark** *Etmopterus sculptus*
page 174

Southeast Atlantic and southwest Indian Ocean; 240–1023m. Moderately large, stout; dark grey-brown above, black below with well-defined narrow, elongated flank markings extending anterior and posterior of pelvic fins; denticles give body a sculpted textured appearance.

Etmopterus *pusillus* clade I

Species in the *Etmopterus pusillus* clade have faint flank markings with a short thick anterior branch and a very short or no posterior branch, or no distinctive marks.

○ **Blurred Smooth Lanternshark** *Etmopterus bigelowi*
page 175

Atlantic, Pacific and Indian Oceans; 0–1000m+. Slender; broad head with long thick flat snout; long caudal fin; first dorsal fin smaller than second dorsal; smooth skin; underside darker, white spot on head, light edges to fins, no conspicuous markings.

○ **Cylindrical Lanternshark** *Etmopterus carteri*
page 175

West central Atlantic; 283–356m. Head semi-cylindrical, about as deep as wide at eyes, snout very short and rounded; gills broad; uniformly dark without concentrations of photophores, fins with pale webs.

○ **Tailspot Lanternshark** *Etmopterus caudistigmus*
page 176

Southwest Pacific: New Caledonia; 638–793m. Slender; narrow head with long thick narrow snout; long caudal fin; second dorsal fin higher than first dorsal; longitudinal row of small close-set denticles on body and caudal fin; obvious photophores on caudal fin.

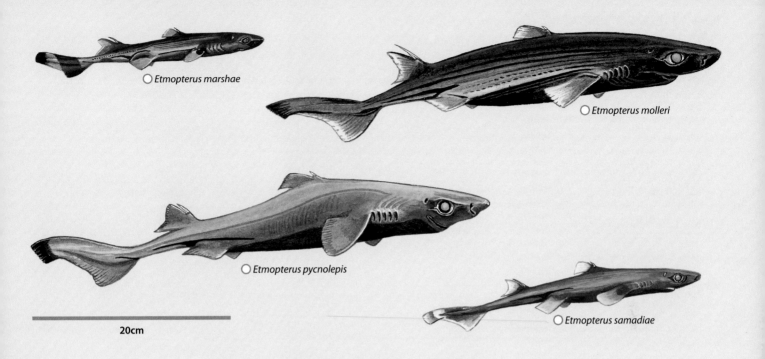

○ Etmopterus marshae

○ Etmopterus molleri

○ Etmopterus pycnolepis

○ Etmopterus samadiae

20cm

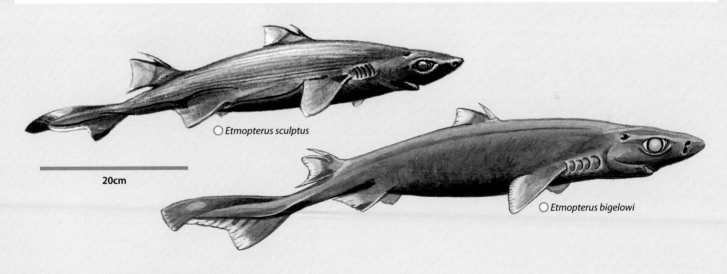

○ Etmopterus sculptus

○ Etmopterus bigelowi

20cm

○ Etmopterus carteri

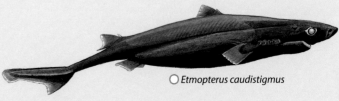

○ Etmopterus caudistigmus

20cm

Plate 13 ETMOPTERIDAE V – *Etmopterus* pusillus II and spinax I clades

Etmopterus pusillus clade II

Species in the *Etmopterus pusillus* clade have faint flank markings with a short thick anterior branch and a very short or no posterior branch, or no distinctive marks.

○ **Pygmy Lanternshark** *Etmopterus fusus*   page 176

Central Indo-Pacific; 430–550m. Cylindrical with long caudal peduncle; second dorsal fin more than twice height first dorsal fin; regular rows of denticles on flanks and caudal peduncle but not head; faint dark marks on flanks and caudal fin, fins pale with dark markings.

○ **Shortfin Smooth Lanternshark** *Etmopterus joungi*   page 177

Northwest Pacific: Taiwan; ~300m or deeper. Moderately elongated body; subconical snout; moderately long caudal fin; dark grey above, darker below with inconspicuous anterior flank marking and no posterior flank marking.

○ **False Lanternshark** *Etmopterus pseudosqualiolus*   page 177

Southwest Pacific; 668–1170m. Fusiform, similar shape to *E. carteri*; very short deep snout with short round eyes; dark brown to black, paler caudal fin with inconspicuous dark marks, distal fin webs pale, terminal caudal lobe dark.

○ **Smooth Lanternshark** *Etmopterus pusillus*   page 178

Widespread, Atlantic, Indian and Pacific Oceans; 0–1120m. On or near bottom. Fairly slender; gills rather long; fairly short broad caudal fin; second dorsal fin less than twice area first dorsal; denticles widely spaced, not in rows, cover snout; obscure black ventral markings.

○ **Thorny Lanternshark** *Etmopterus sentosus*   page 178

West Indian Ocean; 200–500m? Near bottom. Slender; gills quite long; moderately long broad caudal fin; second dorsal fin area greater than twice the first dorsal; two rows denticles on flanks; fin margins largely naked; underside inconspicuous black, black ventral and caudal fin markings.

○ **Splendid Lanternshark** *Etmopterus splendidus*   page 179

Northwest Pacific; 200–300m. Spindle-shaped, similar to *E. fusus*; back purplish black, underside bluish black, precaudal fins pale, red-brown webs and lighter patch on caudal fin.

○ **Hawaiian Lanternshark** *Etmopterus villosus*   page 179

Central Pacific: Hawaiian Islands; 406–911m. On or near bottom. Stout; gills moderately long; short broad caudal fin; second dorsal fin much higher than first dorsal but less than twice area of first dorsal; denticles widely spaced, rows on trunk and caudal fin, cover snout; underside slightly darker, indistinct black ventral markings.

Etmopterus spinax clade I

The *Etmopterus spinax* clade have flank mark shapes (if present) with a long thin anterior branch and a very short or no posterior branch.

○ **Ninja Lanternshark** *Etmopterus benchleyi*   page 180

East Pacific; continental slopes, 836–1443m. Fairly large, slender to stout; uniformly black without distinct markings; short slender dermal denticles, arranged in irregular patches below dorsal fins, and aligned in rows extending onto the fins.

○ **Compagno's Lanternshark** *Etmopterus compagnoi*   page 180

Southeast Atlantic and southwest Indian Oceans; 383–1300m. Moderately stout body; short caudal fin; brown above, becoming black below, with inconspicuous elongated black flank markings above and behind pelvic fins.

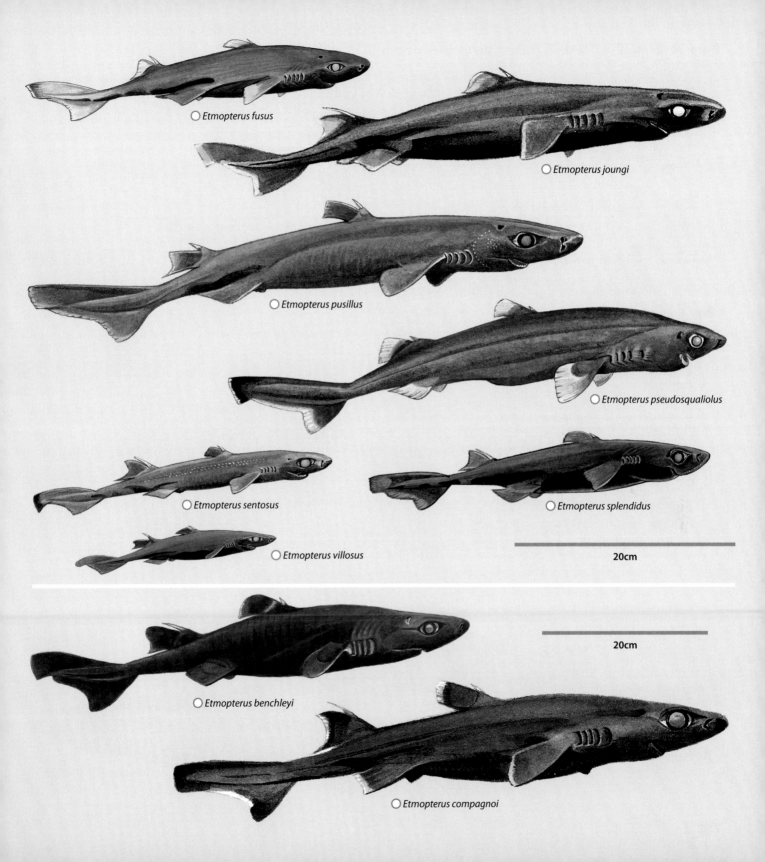

○ *Etmopterus fusus*

○ *Etmopterus joungi*

○ *Etmopterus pusillus*

○ *Etmopterus pseudosqualiolus*

○ *Etmopterus sentosus*

○ *Etmopterus splendidus*

○ *Etmopterus villosus*

20cm

○ *Etmopterus benchleyi*

20cm

○ *Etmopterus compagnoi*

Plate 14 ETMOPTERIDAE VI – *Etmopterus* spinax clade II

The *Etmopterus spinax* clade have flank mark shapes (if present) with a long thin anterior branch and a very short or no posterior branch.

○ **Pink Lanternshark** *Etmopterus dianthus* — page 181

Southwest Pacific; 200–880m. Stout; first dorsal fin small and low with short spine, second dorsal fin less than twice the size of first dorsal with second dorsal spine about the same height as fin; bristle-like denticles not in rows; pinkish fresh (brownish-grey preserved) above, dusky to black below.

○ **Southern Lanternshark** *Etmopterus granulosus* — page 181

Southern oceans; 220–1500m. Heavy-bodied; big head with very short gills; caudal fin short and broad; second dorsal fin much higher than first dorsal; conspicuous lines of large rough denticles on body not head; underside abruptly black, black marks above pelvic fins and caudal fin base.

○ **Caribbean Lanternshark** *Etmopterus hillianus* — page 182

Northwest Atlantic; 180–717m. On or near bottom. Moderately stout; gills very short; moderately long caudal fin; second dorsal fin much higher than first dorsal but less than twice area of first dorsal; no regular rows of denticles but largely cover snout; underside black, black ventral and caudal fin markings.

○ **Smalleye Lanternshark** *Etmopterus litvinovi* — page 182

Southeast Pacific; 630–1100m. Stout; large flat head, gills long; moderately long caudal fin; first dorsal fin origin behind pectoral fin rear tip, second dorsal fin slightly higher than first dorsal fin, interdorsal space short; no rows of denticles; no distinct markings.

○ **Parin's Lanternshark** *Etmopterus parini* — page 183

Northwest Pacific; 40–140m over depths up to 6000m. Pelagic. Relatively small; dark brown back and sides with faint flank markings, black below, fins transparent to white.

○ **Great Lanternshark** *Etmopterus princeps* — page 183

North Atlantic; 350–4500m. On or near bottom. Stout; gills very long; caudal fin moderately long and broad; second dorsal fin much higher than first dorsal but less than twice area of first dorsal; denticles largely cover snout; no conspicuous markings.

○ **Velvet Belly Lanternshark** *Etmopterus spinax* — page 184

East Atlantic and Mediterranean; 70–2490m. Near or well above bottom. Long, fairly stout; gills very short; long caudal fin; second dorsal fin about twice area first dorsal; denticles largely cover snout; underside abruptly black, black marks pelvic fin area to caudal fin.

○ **Brown Lanternshark** *Etmopterus unicolor* — page 184

Northwest Pacific; 120–1500m. Robust; fairly large gills; moderately long caudal fin; long low first dorsal fin with very short spine, second dorsal fin about twice height first dorsal with strong spine; denticles not in rows but cover snout; dark brown to brownish black with darker underside.

○ **Traveller Lanternshark** *Etmopterus viator* — page 185

Southern oceans; 830–1610m; Body stout; snout short; lateral denticles partially form short broken linear rows giving it a rough texture; dark brown to blackish above, darker ventrally, base of black flank markings indistinct in adults, very distinct in subadults; anterior flank marking elongated, posterior branch much shorter.

○ *Etmopterus granulosus*

○ *Etmopterus princeps*

20cm

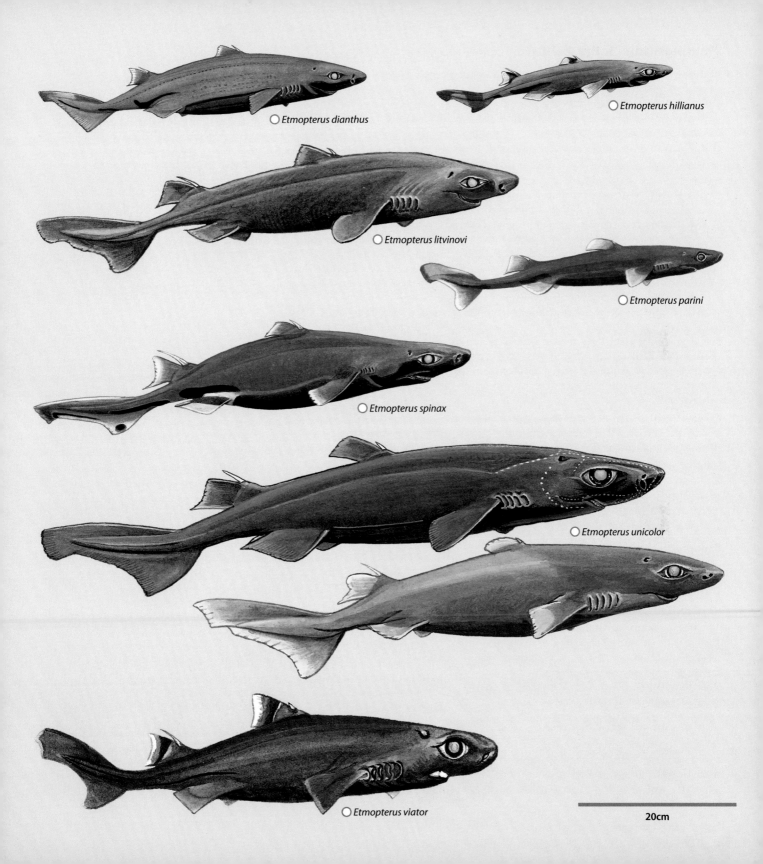

○ *Etmopterus dianthus*

○ *Etmopterus hillianus*

○ *Etmopterus litvinovi*

○ *Etmopterus parini*

○ *Etmopterus spinax*

○ *Etmopterus unicolor*

○ *Etmopterus viator*

20cm

Etmopteridae: lanternsharks

The largest family of squaloid sharks, with over 51 species in four genera (*Aculeola*, *Centroscyllium*, *Etmopterus*, *Trigonognathus*) occurring almost worldwide in deep water, some wide-ranging, many endemic. New species are continually being discovered. The family includes what may be the smallest known species of sharks (*Etmopterus carteri* and *E. perryi*) that mature and give birth to litters of live young when only 10–20cm long.

Identification Dwarf to moderate-sized sharks (adults 10–107cm long, possibly to 120cm) with photophores (light organs), which may be inconspicuous or form distinct black marks on abdomen, flanks or caudal fin (confined to or denser on ventral surface). Two dorsal fins with strong grooved spines (second fin and spine usually larger). No anal fin. No precaudal pits or lateral keels on caudal peduncle. *Centroscyllium* species (combtooth dogfishes) have short to moderately long snouts, comb-like teeth with cusps and cusplets in both jaws, and strong, grooved dorsal spines (second strikingly larger than first). *Aculeola* (hooktooth dogfishes) are very similar except for small hook-like teeth in both jaws and very small equal-sized dorsal spines. Lanternsharks, *Etmopterus* species, often have dark markings (light organs, or photophores) on their flanks and underside. Their upper teeth have a cusp and one or more pairs of cusplets, very different in appearance to the blade-like, cutting lower teeth. Second dorsal fin and spine are much larger than the first. The majority of *Etmopterus* species are assigned to one of the four subgroups of very similar species, known as clades – see below. *Trigonognathus* (viper dogfishes) have very distinctive teeth. These deepsea sharks tend to be in poor condition when collected due to damage by fishing gear. It is also hard to confirm identifications without examining details of denticle and tooth structure, coloration and photophore patterns, and taking precise body measurements; it may not, therefore, be possible to identify some lanternsharks to species using this guide alone.

Biology Most species are bottom-dwelling in deep water, 200–1500m, (range 0–4500m); some semi-oceanic. Several etmopterid dogfishes are social; they form small to huge schools or aggregations. Reproduction, where known, is viviparous with a yolk-sac, with 3–40 pups per litter.

Status Most species are common, but poorly known. Some are known from very few specimens, due to very low or no fishing activity within their range and deepwater habitat. Few are large enough to be of any commercial value and most are discarded if caught as bycatch. For these reasons, the IUCN Red List has assessed 80% of lanternsharks as Least Concern and assigned only one species to a threatened category.

Etmopterus clades

The majority of lanternsharks, genus *Etmopterus*, are assigned to one of the four subgroups of very similar species, known as clades; the species accounts in the following pages are also identified by their clade, as are the plate descriptions. Species in the *Etmopterus gracilispinus* clade have flank markings with a long, thick, curved anterior branch and a short, thick posterior branch. Those in the *E. lucifer* clade have long, thin anterior and posterior branches; the branch lengths (e.g. the anterior length relative to the posterior – longer, shorter, or nearly equal) can help distinguish between members of this clade. Species in the *E. pusillus* clade have faint flank markings with a short, thick anterior branch and a very short or no posterior branch, or no distinctive marks. The *E. spinax* clade have flank mark shapes (if present) with a long, thin anterior branch and a very short or no posterior branch.

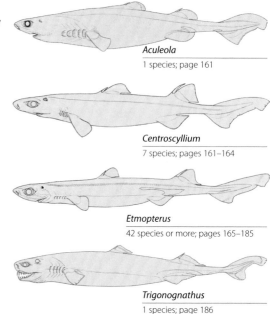

Aculeola
1 species; page 161

Centroscyllium
7 species; pages 161–164

Etmopterus
42 species or more; pages 165–185

Trigonognathus
1 species; page 186

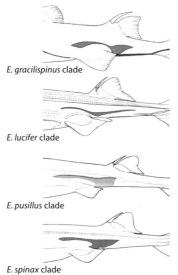

E. gracilispinus clade

E. lucifer clade

E. pusillus clade

E. spinax clade

BIOLUMINESCENCE AND INVISIBILITY CLOAKS

At least 10% of all known shark species have luminescent organs – the lanternsharks and several of the kitefin sharks (Dalatiidae, p. 207), which may have evolved luminous glands completely independently. Lanternshark photophores are small pigment-lined structures, located on the belly, flanks and fins. Each photophore contains up to 12 photocytes and one or two lens cells to focus the light emitted. There can be as many as 500,000 photophores on a single shark, which is more than any other known luminescent animal. Bony fishes and many other luminescent species control their light organs with nerves, but lanternsharks use hormones to switch them on and off. The hormones stimulate pigment cells (melanophores) to cover or uncover the photocytes. The main purpose of these organs is probably to hide these little sharks when viewed by predators from below – a form of invisibility cloak, which makes them vanish from sight against the slightly lighter surface waters; or in scientific terminology 'counter-illumination'. However, there are possibly other reasons for the photophores on the flanks, caudal fin and pelvic region (the latter could be 'Look at me!', or sexual signalling). Photophores may also aid synchronised swimming: perhaps lanternsharks hunt cooperatively. How else would these tiny animals have managed to eat some of the very large animals whose remains have been found in their stomachs?

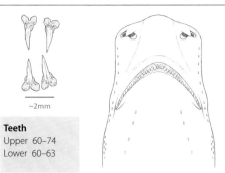

Teeth
Upper 60–74
Lower 60–63

Measurements Born: 13–17cm. **Mature:** males 38–42cm, females 39–52cm.
Max: males 54cm, females ~67cm.

Identification A stocky dogfish. Broad, blunt snout. Broad, long, arched mouth with thin lips; small, hook-like teeth in both jaws. Gill openings quite large. First dorsal fin origin over pectoral fin inner margins; second dorsal fin slightly larger, origin opposite or slightly behind pelvic origins; very short, grooved dorsal spines, much lower than low dorsal fins. Pectoral fin free rear tips rounded. Long upper caudal lobe; lower not differentiated. Uniform blackish brown.
Distribution East Pacific: Peru to central Chile.
Habitat Benthic and epibenthic, continental shelf and upper slope, 110–735m, most common between 200–500m.
Behaviour A demersal species, but it will migrate into the midwater far off the bottom to feed on mesopelagic and pelagic shrimps and bony fishes.
Biology Poorly known. Viviparous, 3–19 pups per litter. Feeds mostly on deepsea shrimps and other crustaceans, and bony fishes including hake.
Status IUCN Red List: Near Threatened. Relatively common within its limited geographical range. The most common (by weight) chondrichthyan bycatch of deepsea crustacean trawls, but also occurs in deeper, untrawled areas.

HIGHFIN DOGFISH *Centroscyllium excelsum*   FAO code: **YCX**   Plate page 148

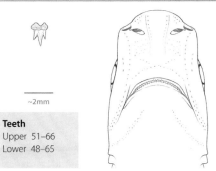

Teeth
Upper 51–66
Lower 48–65

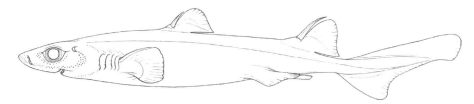

Measurements Born: at least 8–9cm. **Mature:** males 52–62cm, females 53–64cm.
Max: 64cm

Identification Comb-like teeth in both jaws. Very high, rounded first dorsal fin with short spine. Second dorsal much larger than first with very long spine reaching above and behind fin apex. Short caudal peduncle. Dorsal denticles sparse and irregular above, none below. Light brown above, darker below. Lighter fin margins; intense black markings around mouth and on lower surfaces of pectoral fins.
Distribution Northwest Pacific: only known from Emperor Seamount chain east of Japan.
Habitat Deep seamounts, 800–1000m.
Behaviour Unknown.
Biology Viviparous, 10 pups per litter. Feeds on bony fishes.
Status IUCN Red List: Data Deficient. Only 21 specimens known.

BLACK DOGFISH *Centroscyllium fabricii*

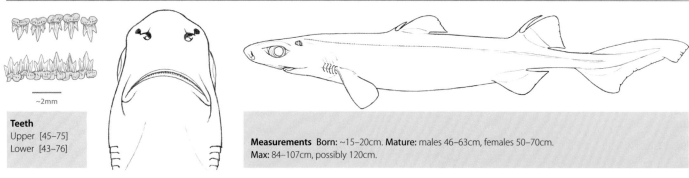

~2mm

Teeth
Upper [45–75]
Lower [43–76]

Measurements Born: ~15–20cm. **Mature:** males 46–63cm, females 50–70cm.
Max: 84–107cm, possibly 120cm.

Identification Arched mouth; comb-like teeth in both jaws. First dorsal fin low, spine short. Second fin and spine larger. Fairly long stout, compressed abdomen. Short caudal peduncle. Numerous close-set denticles. Uniformly blackish brown. No black or white marks.
Distribution Widespread in temperate Atlantic (tropical records uncertain).
Habitat Outer continental shelves and slopes, 180–2250m (mostly deeper than 275m). Found closer to surface at high latitudes and in winter. Bottom temperatures where this shark has been caught are 3.5–4.5°C.
Behaviour Schooling, segregating by sex and size. Larger schools occur in shallower water during winter and spring.
Biology Viviparous, 4–40 pups per litter, average 16. Luminescent organs scattered (irregularly) in skin. Eats crustaceans, cephalopods and small bony fishes.
Status IUCN Red List: Least Concern. Often abundant, it is widespread in its distribution and has a wide depth distribution, which provides refuge from most fisheries. A commonly discarded fisheries bycatch.

GRANULAR DOGFISH *Centroscyllium granulatum*

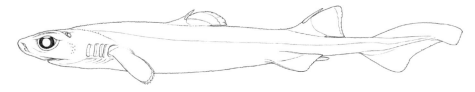

~2mm

Teeth
Upper [45–75]
Lower [43–76]

Measurements Born: ~13–16cm. **Mature:** males 31cm.
Max: at least 62cm.

Identification Very small, slender, cylindrical dogfish. Mouth narrowly arched; comb-like teeth in both jaws. First dorsal fin low, much smaller than second. First dorsal spine short; second very large, extending over and behind fin apex. Long abdomen. Elongated caudal peduncle. Body closely covered by denticles with sharp, hooked cusps. Uniformly brownish black body without conspicuous markings.
Distribution Southeast Pacific: South America, north-central Chile to Magellan Straits.
Habitat Deep water, upper continental slope, 100–610m.
Behaviour Unknown
Biology Viviparous, with litters of at least 16. Feeds on crustaceans and small bony fishes.
Status IUCN Red List: Vulnerable. Bycaught sporadically throughout its narrow depth range in Chilean deepsea shrimp trawl and longline fisheries.

BARESKIN DOGFISH *Centroscyllium kamoharai*

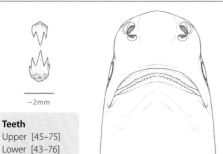

~2mm

Teeth
Upper [45–75]
Lower [43–76]

Measurements Born: 16–31cm. **Mature:** males 40–45cm, females ~55cm.
Max: 63cm.

Identification Small, stout dogfish. Mouth extremely short, very broadly arched; comb-like teeth in both jaws. First dorsal fin very low and rounded with short spine. Second dorsal fin slightly larger than first, spine about as high as fin apex. Caudal peduncle short. Skin smooth, almost naked, fragile and easily damaged (few, wide-spaced denticles). Colour uniformly blackish (except for white patches where very delicate skin has been chafed during capture). Fins with lighter rear margins.
Distribution West Pacific: from south Japan to Okinawa, Taiwan, Philippines, east Australia, New Zealand. Southeast Indian Ocean: west Australia.
Habitat Bottom, deep continental shelf, 500–1225m, mostly deeper than 900m.
Behaviour The sexes may segregate by depth. Very little known.
Biology Poorly known. Viviparous, 3–22 pups per litter, averaging 12. Diet consists of bathypelagic cephalopods, crustaceans and, to a lesser extent, bony fishes.
Status IUCN Red List: Least Concern. Infrequent bycatch of benthic trawls and deepwater longlines in the upper part of its range, but widespread and primarily occurs in deeper water where fisheries do not operate.

COMBTOOTH DOGFISH *Centroscyllium nigrum*

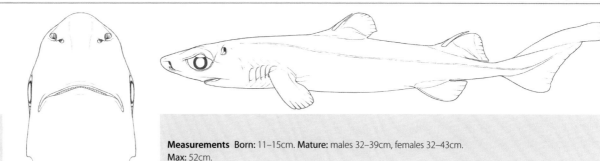

~2mm

Teeth
Upper 40–66
Lower 40–60

Measurements Born: 11–15cm. **Mature:** males 32–39cm, females 32–43cm.
Max: 52cm.

Identification Small dogfish with fairly stout body. Mouth short, broadly arched; comb-like teeth in both jaws. Dorsal fins similar in size; first dorsal spine short, height equal to or shorter than fin apex; second dorsal spine height equal to or higher than fin apex, with origin over or slightly in front of pelvic fin insertions. Colour uniformly blackish brown. Fins with prominent white tips and margins.
Distribution Central and east Pacific: Hawaiian Islands, American continental shelf, Cocos and Galapagos Islands.
Habitat On or near bottom on continental and insular slopes, 32–1212m, mostly below 600m, but one individual was photographed by a SCUBA diver off southern California at about 32m depth.
Behaviour Appears to migrate off the bottom into the midwater based on prey items found in stomach.
Biology Viviparous, 4–15 pups per litter. Feeds mostly on deepsea shrimps, cephalopods and midwater bony fishes.
Status IUCN Red List: Least Concern. Regularly recorded as bycatch of deepsea fisheries, with some local population declines recorded, this species has a very broad geographic and depth range and also inhabits areas where no deepsea fishing takes place.

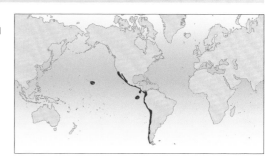

Teeth
Upper [45–75]
Lower [43–76]

Measurements Immatures to at least 30cm.
Max: 51cm.

Identification Mouth narrowly arched; comb-like teeth in both jaws. Low rounded first dorsal fin; very elongated spine, nearly as high as second dorsal spine, reaching above fin apex. Second dorsal fin larger, very long spine extending above fin apex. Denticles close-set and numerous. Blackish body and fins without conspicuous markings.
Distribution North Indian Ocean: Arabian Sea and Bay of Bengal.
Habitat Upper continental slope, near seabed, 521–1262m.
Behaviour Unknown.
Biology Unknown. Presumably viviparous.
Status IUCN Red List: Least Concern. This species' deepsea distribution extends well below the reach of fisheries.

~2mm

Teeth
Upper [45–75]
Lower [43–76]

Measurements Mature: females 42–49cm.

Identification Mouth very broadly arched; comb-like teeth in both jaws. First dorsal fin low and rounded; spine very short. Second dorsal fin about same size, less rounded; spine moderately long, reaching above fin apex. Numerous close-set denticles. Grey-brown above. Fins with white margins. The only *Centroscyllium* with striking black markings (concentrations of photophores) on underside of head, abdomen and pectoral fins; black stripe beneath caudal peduncle extends over pelvic bases.
Distribution Northwest Pacific.
Habitat Deepwater, continental slopes and seamounts, 320–1100m.
Behaviour Unknown.
Biology Unknown. Presumably viviparous.
Status IUCN Red List: Data Deficient. A bycatch of deepwater trawl fisheries, probably discarded due to its small size. This species' limited geographic range and possibly low bycatch survival rates might pose a threat.

BROADBANDED LANTERNSHARK *Etmopterus gracilispinis*

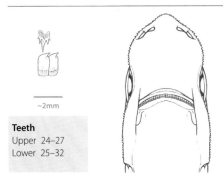

~2mm

Teeth
Upper 24–27
Lower 25–32

Measurements Born: ~13cm or less. **Mature:** males ~26cm, females 33cm. **Max:** ~33cm.

Identification (*E. gracilispinus* clade) Small, stout lanternshark with short slender caudal fin. Gill openings very short. First dorsal fin well behind pectoral fin free rear tips. Second dorsal about twice area of first. No regular rows of lateral trunk denticles. Blackish brown body, grading to black below. Inconspicuous, broad, elongated black mark running above and behind pelvic fins; other elongated black marks at caudal fin base and along axis.
Distribution West Atlantic: USA to Argentina. South African records were misidentification of a different species.
Habitat Outer continental shelves and upper-middle slopes, on or near bottom, 25–1200m; epipelagic and mesopelagic at 70–480m over water 2240m deep off Argentina.
Behaviour Unknown.
Biology Unknown. Presumably viviparous.
Status IUCN Red List: Least Concern. This species is only rarely captured by longlines and midwater trawls, and fishing effort is very low within its range.

DWARF LANTERNSHARK *Etmopterus perryi*

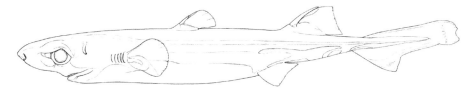

Teeth
Upper 25–30
Lower 32–34

Measurements Born: ~6cm. **Mature:** males ~16–17cm, females 19cm. **Max:** ~21cm.

Identification (*E. gracilispinus* clade) One of the smallest living sharks. Long, broad, flat head. Second dorsal fin larger than first. Brownish above. Conspicuous dark band at tip of terminal lobe of caudal fins; dark blotch on lower caudal lobe. Black below.
Distribution West central Atlantic: Caribbean coast of Colombia.
Habitat Upper continental slope, 230–530m.
Behaviour Unknown.
Biology Viviparous.
Status IUCN Red List: Least Concern. Deepsea trawling occasionally take Dwarf Lanternshark as a bycatch, but fishing pressure is very low within this species' range.

AFRICAN LANTERNSHARK *Etmopterus polli* FAO code: **ETT** Plate page 150

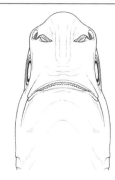

~2mm

Teeth
Upper 27–34
Lower 28–33

Measurements Mature: males 19–23cm, females 24cm.
Max: ~24cm.

Identification (*E. gracilispinus* **clade**) A fairly stout dwarf lanternshark with a fairly long caudal fin. Gill openings short. Denticles widely spaced, not in rows on sides; largely cover snout. Second dorsal fin as large or slightly larger than first. Dark grey body. Underside of snout and abdomen blackish; elongated broad, black mark running above, ahead and behind pelvic fins; other marks at base and along axis of caudal fin.
Distribution East central Atlantic from Mauritania to Angola.
Habitat Upper continental slopes, on or near seabed, 300–1000m, most common from 300–500m.
Behaviour Unknown.
Biology Viviparous. Feeds on cephalopods, crustaceans and small bony fishes.
Status IUCN Red List: Data Deficient. Minor fisheries interest. Bycatch in offshore fisheries sometimes used for food or fishmeal.

WEST INDIAN LANTERNSHARK *Etmopterus robinsi* FAO code: **SHL** Plate page 150

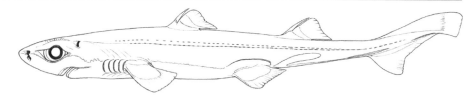

Teeth
Upper not known
Lower not known

Measurements Mature: males 26cm.
Max: males 31cm, females 34cm.

Identification (*E. gracilispinus* **clade**) Moderately stout body, moderately long caudal fin. Upper teeth generally with less than three pairs of cusplets. Gill openings about one third eye-length. Second dorsal fin much larger but less than twice area of first. Lateral denticles with slender, hooked conical crowns; wide-spaced, not in regular rows; denticles largely cover snout. Body grey or dark brown. Underside of snout and abdomen conspicuously dark; elongated broad black mark runs above and behind pelvic fins, others at base and along axis of caudal fin.
Distribution West central Atlantic: Caribbean Sea.
Habitat Continental and insular slopes, 412–787m, mostly deeper than 549m.
Behaviour Unknown.
Biology Viviparous, nothing else known.
Status IUCN Red List: Least Concern. No human impact.

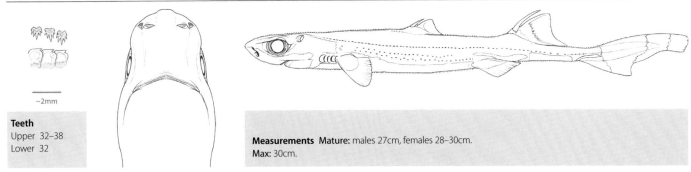

Teeth
Upper 32–38
Lower 32

~2mm

Measurements Mature: males 27cm, females 28–30cm.
Max: 30cm.

Identification (*E. gracilispinus* clade) Slender, moderately long caudal fin (dorsal caudal margin about equal to head length). Gill openings very short. First dorsal far behind pectorals, about equidistant between pectoral and pelvic bases; fin margins naked, prominently fringed margin of naked ceratotrichia. Second dorsal fin about twice area of first. Very slender-crowned, hooked, lateral trunk denticles; wide-spaced; no rows of enlarged thorns on flanks; denticles largely cover snout. Light brown above. Elongated, narrow dusky mark above and behind pelvic fins; elongated black marks at caudal fin base and along axis. Dusky grey below.
Distribution West central Atlantic: northern Gulf of Mexico to off northern Brazil.
Habitat Upper continental slopes, on or near bottom, 220–915m, mostly deeper than 350m.
Behaviour Unknown.
Biology Viviparous, nothing else known.
Status IUCN Red List: Least Concern. No interest to fisheries.

~2mm

Teeth
Upper 29–34
Lower 24–32

Measurements Mature: males 18cm, females 22cm.
Max: 26cm.

Identification (*E. gracilispinus* clade) Slender dwarf lanternshark with long, narrow caudal fin (dorsal caudal margin about the same as head length). Upper teeth generally with less than 3 pairs of cusplets. Gill openings very short, less than one-third eye-length. Second dorsal fin over twice area of first. Lateral trunk denticles very short, stout, hooked, conical crowns; wide-spaced, not in rows; denticles largely cover snout. Colour dark brown or grey-black. Elongated, broad, black mark above and behind pelvic fins; others at caudal fin base and along axis. Underside black.
Distribution Northwest Atlantic: northern Gulf of Mexico and Caribbean, possibly Brazil.
Habitat Upper continental slopes, 196–915m, most deeper than 350m.
Behaviour May school and communally hunt very large prey (squids).
Biology Viviparous. Feeds on cephalopods.
Status IUCN Red List: Least Concern. Relatively common, possibly discarded from deepsea bottom fisheries.

WHITECHEEK LANTERNSHARK *Etmopterus alphus*

FAO code: **EZU**   Plate page 152

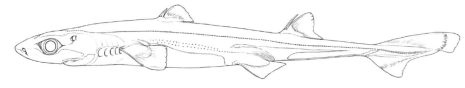

Teeth
Upper 26–30
Lower 31–34

Measurements Born: less than 14cm. **Mature:** males 29–31cm, females 33cm. **Max:** males 34cm, females 39cm.

Identification (*E. lucifer* clade) A relatively small, slender, linear-denticled lanternshark. Teeth dissimilar in upper and lower jaws: uppers multicuspid with strong central cusp flanked by up to three lateral cusplets on each side, decreasing in size distally; lower jaw teeth blade-like, oblique and fused into a single row. Two dorsal fins with fin spines, first smaller than the second. Dermal denticles are hook-like, posteriorly directed in distinct rows. Above and flanks dark purplish black, becoming black below; transition between lateral and ventral surfaces sharply demarcated by prominent silvery to white lateral stripe below the flank marking. Short anterior flank marking branch relative to its posterior branch; black flank marking sharply demarcated by surrounding lighter to whitish lateral markings; prominent white cheek spot.
Distribution Southwest Indian Ocean: off central Mozambique, northwest Madagascar, and southern Madagascar Ridge at Walters Shoal.
Habitat Continental slopes, 472–792m. **Behaviour** Unknown.
Biology Viviparous, at least 5 pups per litter, little else known. Feeds on small bony fishes and crustaceans.
Status IUCN Red List: **Least Concern**. Low numbers bycaught in demersal deepsea trawls, which are not active in the deepest part of this species range.

SHORTTAIL LANTERNSHARK *Etmopterus brachyurus*

FAO code: **ETH**   Plate page 152

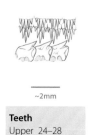

~2mm

Teeth
Upper 24–28
Lower 36–40

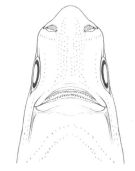

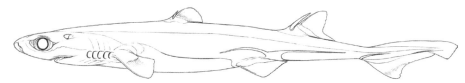

Measurements Born: 11–15cm. **Mature:** males ~24–33cm, females 32–40cm. **Max:** males 44cm, females 50cm.

Identification (*E. lucifer* clade) Small, slender-bodied, broad-headed, short-tailed lanternshark. Short, thick, flat snout. Pointed upper teeth; knife-like lower teeth. Two dorsal fins with spines, first smaller than second; long second spine curved weakly to rear in adults. Moderately rough skin with conspicuous lines of hooked denticles that run from snout tip to caudal fin. Dark brown to grey-black, with a purplish hue on upper surface, conspicuously darker below. Prominent lateral flank and caudal fin markings.
Distribution West Pacific and Central Indo-Pacific: Japan, Taiwan, Philippines, northeast (off Cairns) and Western Australia. Records of this species from elsewhere are likely to be misidentification with other species.
Habitat Near bottom, 100–696m.
Behaviour Unknown.
Biology Viviparous, 4–13 pups per litter, average 8. Feeds on small bony fishes, crustaceans and especially deepsea shrimps.
Status IUCN Red List: **Data Deficient**. No interest to fisheries. The Sculpted Lanternshark *Etmopterus sculptus* off southern Africa has been referred to as *E. brachyurus* or *E.* cf. *brachyurus* but is darker, stouter, larger (adult males to 48cm, adult females to 59cm) and rougher skinned.

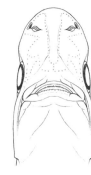

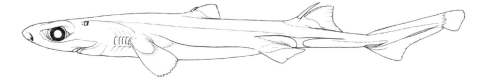

~2mm

Teeth
Upper 18–20
Lower 27–31

Measurements Born: less than 15–16cm. Immature males 26cm, immature female 26cm.
Max: Largest known 30cm individual was immature.

Identification (E. lucifer clade) Slender body, long caudal fin. Gill openings very short. First dorsal origin over inner pectoral fin margins; base closer to pectoral bases than pelvics; very close to larger second dorsal. Conspicuous longitudinal rows of denticles on head, sides and back, extending to caudal fin base. Body dark sooty grey. Lighter band on midline from eyes to first dorsal fin; elongated, narrow black mark running above and behind pelvic fins, others at caudal fin base and along axis. Black below.
Distribution Northwest and central Atlantic: southern USA, Caribbean, Colombia.
Habitat Continental slopes, on or near bottom, 275–824m, mostly below 350m.
Behaviour Poorly known.
Biology Poorly known. Viviparous. Diet includes small crustaceans and squids.
Status IUCN Red List: Least Concern. Probably discarded deepsea fisheries bycatch.

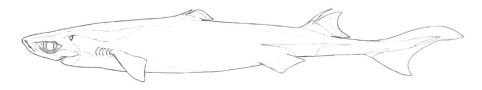

Teeth
Upper 24–26
Lower 32–40

Measurements Born: less than 16cm. **Mature:** males 34cm, females 40.6cm.
Max: 41cm.

Identification (E. lucifer clade) A moderate-sized lanternshark with a broad snout. First dorsal fin spine origin well behind pectoral fin free rear tips. Uniformly dark grey above, becoming darker grey to black below. Flank marking consisting of anterior and posterior branches: anterior branch extending past pelvic fin; posterior branch about equal to or slightly less than anterior branch length and terminating well past second base of dorsal fin insertion.
Distribution Northwest Pacific: known only from off Taiwan.
Habitat Upper continental slopes, 300–600m.
Behaviour Unknown.
Biology Viviparous, little else known. Feeds on small bony fishes.
Status IUCN Red List: Least Concern. Taken as bycatch in demersal trawl fisheries and retained for fish meal, but may also occur in deeper water and other untrawled areas. Catches are monitored and there is no evidence of population decline.

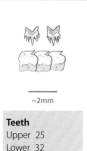

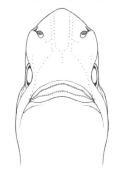

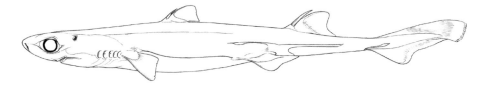

~2mm

Teeth
Upper 25
Lower 32

Measurements Known from one adult male of 29cm.

Identification (*E. lucifer* clade) Moderately slender body. Upper teeth with 4–5 pairs of cusplets on each side. Gill openings very short. First dorsal fin origin slightly behind pectoral fin free rear tips; dorsal base much closer to pectoral bases than to pelvics. Second dorsal about twice area of first. Dorsal caudal fin margin about equal to head length. No regular rows of denticles on sides. Brown above. Elongated, narrow, black mark running above, in front and behind pelvic fins; other elongated black marks at caudal fin base and along axis. Black below.
Distribution Northwest Pacific: Hainan Island, China.
Habitat Collected on or near bottom, 512–692m.
Behaviour Unknown.
Biology Unknown. Presumably viviparous.
Status IUCN Red List: Least Concern. Known from only one specimen caught on a research survey. Unlikely to be taken in fisheries operating in the area, because it is too small to be captured on commercial longlines and trawlers operate in shallower water in the region.

BROKEN LINED LANTERNSHARK *Etmopterus dislineatus*   FAO code: **SHL**   Plate page 152

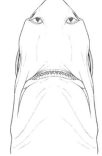

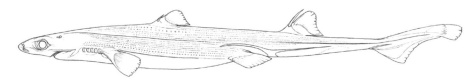

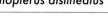

Teeth
Upper not known
Lower not known

Measurements Mature: males ~36cm.
Max: at least 45cm.

Identification (*E. lucifer* clade) Very elongated, attractive lanternshark. First dorsal fin very low and small, about half size of second. Hooked, bristle-like denticles not arranged in regular rows. Light, silvery-brown above, much darker below, with a pattern of dark, broken lines of dots and dashes along the upper flanks. Distinct black markings near pelvic fin and on the caudal peduncle, midcaudal fin and upper fin tip.
Habitat Upper continental slope, on or near bottom, 590–802m.
Behaviour Unknown.
Biology Unknown. Presumably viviparous.
Status IUCN Red List: Least Concern. The small known range is unfished.

BLACKMOUTH LANTERNSHARK *Etmopterus evansi*

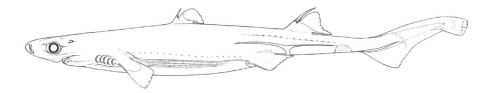

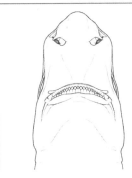

Teeth
Upper not known
Lower not known

Measurements Born: less than 17cm. **Mature:** males 26cm.
Max: at least 34cm.

Identification (*E. lucifer* clade) Small-bodied lanternshark. Hook-like denticles arranged in distinct but weakly defined rows along dorsal midline and on caudal peduncle, not on head. Light brown above, darker below. Dark borders around mouth, above eyes, sometimes around gill slits. Distinct black markings behind the pelvic fin, on caudal peduncle (a dark ventral blotch) and upper caudal fin (dark bands through middle and tip of upper lobe).
Distribution Central Indo-Pacific: northwest Australia, Arafura Sea (Indonesia) and Papua New Guinea.
Habitat Shoals and reefs on continental shelf, 430–550m.
Behaviour Unknown.
Biology Unknown. Presumably viviparous.
Status IUCN Red List: Least Concern. Not currently fished, but could become a bycatch of expanding deepsea fisheries.

LAILA'S LANTERNSHARK *Etmopterus lailae*

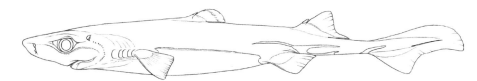

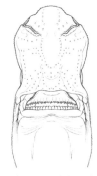

Teeth
Upper 22–24
Lower 26–28

Measurements Born: less than 27cm. **Mature:** males at least 37cm.
Max: at least 37cm.

Identification (*E. lucifer* clade) Moderately large lanternshark. Teeth dissimilar: upper teeth with strong central cusp flanked on each side by two smaller lateral cusplets less than half height of median cusp and decreasing in size distally; lower teeth unicuspid, blade-like, oblique, fused into a single row. Two dorsal fins with fin spines, first smaller than the second. Dermal denticles erect, thorn-like, curved rearwards, in distinct longitudinal rows extending from dorsal head surface to caudal fin. Above and flanks light to medium brown, becoming a darker brown below. Longer anterior flank marking branch relative to its posterior branch; prominent black flank marking, sharply demarcated, but without lighter coloured lateral flanks (described from preserved specimen). Body with 1–3 rows of prominent dark photophores extending from the head posteriorly along flanks to upper caudal fin origin.
Distribution Central Pacific: Koko and South Kanmu Seamounts, northwestern Hawaiian Islands.
Habitat Insular slopes from around seamounts, 314–384m.
Behaviour Unknown.
Biology Known from only 3 immature males, 27–37cm total length. Presumably viviparous. Diet is unknown, but includes small bony fishes, cephalopods and crustaceans.
Status IUCN Red List: Data Deficient. This little known lanternshark, collected during research surveys in the northwestern Hawaiian Islands, was previously misidentified as Blackbelly Lanternshark *Etmopterus lucifer*. It may be a bycatch of commercial fisheries activities on seamounts

BLACKBELLY LANTERNSHARK *Etmopterus lucifer*

FAO code: **ETF** Plate page 152

~2mm

Teeth
Upper 21–26
Lower 29–39

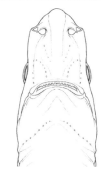

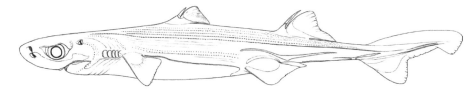

Measurements Mature: males 29–42cm, females 34cm or larger.
Max: ~47cm.

Identification (*E. lucifer* clade) Stocky body, moderately long caudal fin. Teeth blade-like in lower jaw, with cusps and cusplets (generally less than 3 pairs) in upper. Gill openings moderately long. Short interdorsal space. Very large second dorsal fin. Rows of denticles from snout tip to caudal fin. Brown above. Narrow, elongated black mark running above, ahead and behind pelvics; others at caudal fin base and along axis. Black below.
Distribution West Pacific: confirmed from Japan, Taiwan, Philippines, possibly from South China Sea, Australia, New Caledonia and New Zealand. However, specimens from these latter locations should be examined more closely to confirm their identification.
Habitat Outer continental and insular shelves, upper slopes, on or near bottom, 158–1357m.
Behaviour Essentially unknown.
Biology Viviparous, with litters of 7. Age at maturity has been estimated at 10 years for males and 13 years for females. Eats mostly squid and small bony fishes, also shrimp.
Status IUCN Red List: Least Concern. This species has a broad geographic and depth distribution, including areas deeper than current fisheries activity, and likely a low catchability in commercial fisheries because of its small size.

MARSHA'S LANTERNSHARK *Etmopterus marshae*

FAO code: **SHL** Plate page 154

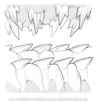

Teeth
Upper 30–36
Lower 30–38

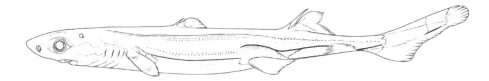

Measurements Born: less than 10cm. **Mature:** males 23cm, females 19cm.or larger.
Max: at least 23cm.

Identification (*E. lucifer* clade) Small, slender lanternshark. Teeth dissimilar: upper teeth with strong central cusp flanked on each side by 1–2 smaller lateral cusplets, each decreasing in size distally; lower teeth unicuspid, blade-like, oblique, fused into a single row. Two dorsal fins with fin spines, first smaller than the second. Dermal denticles hook-like, rearward directed, and organised in distinct longitudinal rows extending from dorsal head surface to caudal fin. Coloration is dark purplish black above and on trunk, becoming dark black below; transition between colours sharply demarcated. Anterior and posterior lateral flank marking branches nearly equal in length. Body trunk with 1–3 rows of dark dashes extending from head to caudal fin origin and a separate paired row of dashes between pectoral and pelvic fins. Caudal fin with distinct black bar at upper caudal origin, fading posteriorly to white, with a prominent black caudal fin tip.
Distribution West Pacific: Philippines between Luzon and Mindoro islands.
Habitat Continental slopes, 322–337m on soft, sandy bottom.
Behaviour Forms small schools, since all 11 known specimens were captured in a single survey trawl.
Biology Viviparous, nothing else known. Feeds mainly on bony fishes, cephalopods and crustaceans.
Status IUCN Red List: Least Concern. This species is likely too small to be caught by most types of fishing gear, although deepwater fisheries are poorly documented within its endemic range.

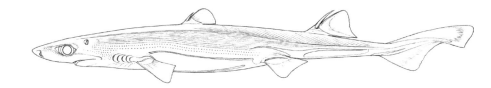

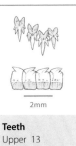

2mm

Teeth
Upper 13
Lower 18

Measurements Born: ~15cm. Mature: males ~33cm. Max: ~46cm.

Identification (*E. lucifer* clade) Similar to *Etmopterus brachyurus* and *E. lucifer*: differs from former in having a largely naked second dorsal fin and the latter in having a taller second dorsal fin and a longer caudal peduncle. Slender-bodied. Second dorsal fin much larger than first. Regular longitudinal rows of denticles on head, flanks, caudal peduncle and caudal base, not on second dorsal fin or above pectoral fins. Light brown above. Dark brown flanks with black photophores; posterior flank marking branch is longer than anterior branch length. Abruptly black below.
Distribution Southwest Pacific: east Australia, New Zealand. Specimens reported from Taiwan and Japan are of a different species.
Habitat Outer continental and insular shelves, upper slopes, on or near bottom, 238–655m.
Behaviour Essentially unknown.
Biology Viviparous, but little else known.
Status IUCN Red List: Data Deficient.

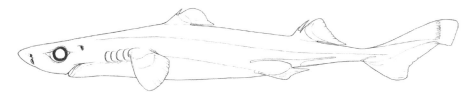

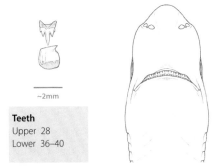

~2mm

Teeth
Upper 28
Lower 36–40

Measurements Max: 41–46cm.

Identification (*E. lucifer* clade) A moderately small lanternshark with a narrow head, slender body and moderately long caudal fin. Upper teeth generally with 1–3 pairs of cusplets. Gill openings long. First dorsal origin ahead of pectoral rear tips. Interdorsal space long. Second dorsal fin larger than first. Denticles hooked, conical, very small and arranged in dense rows on head, trunk and caudal fin. Plain-coloured. Conspicuous black photophores on flanks, caudal fin and underside of body; black-tipped caudal fin. Similar to *Etmopterus brachyurus*, *E. molleri* and *E. lucifer*, but with smaller, more close-set denticles and flank marks with front and rear extensions equal sized.
Distribution Southeast Pacific: Nazca and Sala y Gomez Ridges off Peru and Chile.
Habitat Upper slopes, on or near bottom, 330–763m.
Behaviour Unknown.
Biology Unknown.
Status IUCN Red List: **Least Concern**. Possibly a fisheries bycatch but, if so, interactions are likely limited.

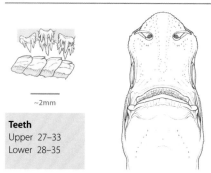

~2mm

Teeth
Upper 27–33
Lower 28–35

Measurements Born: less than 15cm. **Mature:** males 23–27cm, females 28cm.
Max: 28cm.

Identification (E. lucifer clade) A small, linear-denticled lanternshark. Teeth dissimilar: upper teeth with strong central cusp flanked on each side by 2–3 smaller lateral cusplets on each side of median cusp and decreasing in size distally; lower teeth unicuspid, blade-like, oblique, fused into a single row. Two dorsal fins with fin spines, first smaller than the second. Dermal denticles erect, curved rearwards, thorn-like, in distinct longitudinal rows extending along body surface to caudal fin origin. Greyish to silvery black above and on flanks; sharply demarcated below by a paler stripe below the flank markings in most specimens. Anterior flank marking branch slightly shorter relative to its posterior branch; irregular and variable black horizontal dash-like marks on flanks; long caudal base marking. Dorsal, pectoral and pelvic fins dark at bases and along anterior edges, becoming translucent to white on remainder of fins; caudal fin with distinct, large, dark blotch centrally located.

Distribution Southwest Pacific: west of Kairiru Island, Papua New Guinea.
Habitat Continental slopes, 340–785m.
Behaviour Unknown.
Biology Viviparous, little else known. Diet unknown, but likely includes small bony fishes and crustaceans.
Status IUCN Red List: Least Concern. The only known specimens were collected during research surveys. It is likely that the species is too small to be caught by most types of fishing gear and there are no known deepwater fisheries in the region

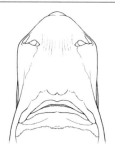

~2mm

Teeth
Upper 23–25
Lower 36–43

Measurements Born: 15–17cm. **Mature:** males 41–48cm, females 45–53cm.
Max: ~59cm.

Identification (E. lucifer clade) A moderately large lanternshark. Dark grey-brown above, black below, with a well-defined narrow, elongated black flank marking extending anteriorly and posteriorly in front of and behind pelvic fins. Dermal denticles uniformly covering the body give it a sculpted textured appearance.

Distribution Southeast Atlantic and southwest Indian Ocean: Namibia to southern Mozambique, and seamounts south of Madagascar.
Habitat Near bottom, 240–1023m, mostly below 450m.
Behaviour Unknown.
Biology Viviparous, but little else known. Feeds on small bony fishes, cephalopods, and crustaceans.
Status IUCN Red List: Least Concern. This species mostly occurs beyond depths where fishing activity currently occurs.

BLURRED SMOOTH LANTERNSHARK *Etmopterus bigelowi*

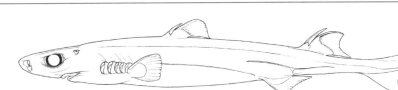

Teeth
Upper 19–24
Lower 25–39

~2mm

Measurements Born: less than 16cm. **Mature:** males ~40–67cm, females 50–65cm.
Max: at least 73cm.

Identification (*E. pusillus* clade) Fairly large, slender, long caudal finned lanternshark. Broad head; long, thick, flat snout. Multicuspid, pointed upper teeth; knife-like lower teeth. Two dorsal fins with fin spines, first smaller than second. This and *Etmopterus pusillus* are the only lanternsharks with flat, block-like dermal denticles and smooth skin texture. Dark brown or blackish (slightly darker below). Small white blotch on top of head; light edges to fins. Photophores present but not as conspicuous dark bands on caudal fin and flanks.
Distribution Atlantic: Gulf of Mexico to Argentina, equatorial west Africa to South Africa. Indian Ocean: off South Africa, open ocean south of Madagascar and west Australia. Pacific: Japan, around the Hawaiian Islands, seamounts and ocean ridges Peru and east Australia.
Habitat Continental and insular shelves and slopes, submarine ridges, from near surface to deeper than 1000m. Partly epipelagic (occurring near surface in open ocean 0–1000m or deeper).
Behaviour Unknown.
Biology Viviparous, poorly known. Feeds on small bony fishes and cephalopods.
Status IUCN Red List: Least Concern. No importance to fisheries, probably discarded bycatch.

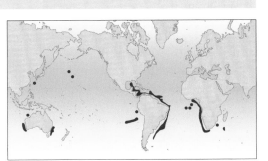

CYLINDRICAL LANTERNSHARK *Etmopterus carteri*

Teeth
Upper 30
Lower 29

Measurements Mature: males and females 18cm.
Max: 21cm.

Identification (*E. pusillus* clade) A tiny lanternshark with semicylindrical head, nearly as deep as wide at eyes. Snout very short and bluntly rounded. Gills broad. Uniformly dark body without concentrations of photophores. Flank markings not visible in life, but become noticeable after preservation; fins with pale webs.
Distribution Central Atlantic: Caribbean coast of Colombia.
Habitat Upper continental slopes, 283–356m, mesopelagic, possibly also epipelagic.
Behaviour Unknown.
Biology Unknown. Presumably viviparous.
Status IUCN Red List: Least Concern. Deepsea fishing has reportedly not yet developed within this species' known range.

TAILSPOT LANTERNSHARK   *Etmopterus caudistigmus*

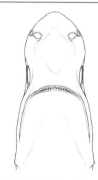

Teeth
Upper 31–35
Lower 37–39

Measurements **Mature:** males 31cm.
Max: females 34cm.

Identification (*E. pusillus* clade) Small, slender-bodied lanternshark. Narrow head; long, thick, narrow snout. Two dorsal fins with short fin spines, first smaller than second; second fin spine weakly curved, pointing posterodorsally in adults. Long caudal fin with obvious photophores. Small, close-set lateral trunk denticles form regular longitudinal rows on body and caudal fin, not head. Dark in colour with black underside.
Distribution Southwest Pacific: New Caledonia.
Habitat Insular slope, 638–793m.
Behaviour Unknown.
Biology Presumably viviparous.
Status IUCN Red List: Least Concern. Known only from three specimens caught on longline in a largely unfished area.

PYGMY LANTERNSHARK   *Etmopterus fusus*

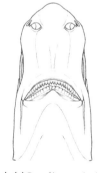

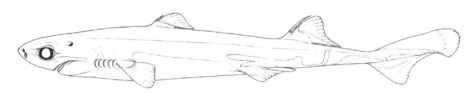

Teeth
Upper not known
Lower not known

Measurements **Mature:** males 25–26cm.
Max: at least 30cm.

Identification (*E. pusillus* clade) Dwarf lanternshark with firm, cylindrical body. Second dorsal fin more than twice height of first. Elongated caudal peduncle. Denticles in regular rows on flanks and caudal peduncle, but not on head. Dark greyish or black. Faint dark markings on flank above and behind pelvic fins and on caudal fin. Fins pale, with dark margins. Similar to *Etmopterus splendidus*.
Distribution Central Indo-Pacific: Western Australia, possibly Java, Indonesia and Papua New Guinea.
Habitat Continental slope, 430–550m (Australia), possibly 120–200m (Indonesia).
Behaviour Unknown.
Biology Unknown. Presumably viviparous.
Status IUCN Red List: Least Concern. Only minor fishing activity in Australian range.

SHORTFIN SMOOTH LANTERNSHARK   *Etmopterus joungi*   FAO code: **SHL**   Plate page 156

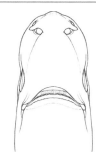

~2mm

Teeth
Upper 27 (25–30)
Lower 34 (33–36)

Measurements Mature: males ~40cm.
Max: at least 47cm for an immature female.

Identification (*E. pusillus* clade) A moderate-sized lanternshark with an elongated body, subconical snout, moderately long caudal fin. Dermal denticles block-like, crowns without any spines. Dark grey above, darker below, with an inconspicuous flank marking that lacks a posterior branch.
Distribution Northwest Pacific: Taiwan.
Habitat Upper continental and insular slopes, ~300m or deeper.
Behaviour Unknown.
Biology Viviparous.
Status IUCN Red List: Least Concern. An uncommon retained bycatch (for fishmeal) in demersal trawls, but its abundance at landing sites has not changed since the species was described 10 years ago.

FALSE LANTERNSHARK   *Etmopterus pseudosqualiolus*   FAO code: **SHL**   Plate page 156

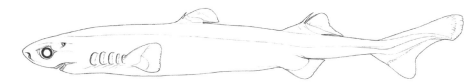

Teeth
Upper 29–34
Lower 31–34

Measurements Mature: males 40–45cm.
Max: 45cm.

Identification (*E. pusillus* clade) Dark, fusiform body. Similar in body shape to *Etmopterus carteri*. Very short, deep snout. Teeth with 3-5 pairs of cusplets in males. Short, rounded eyes. Small pectoral fins. Greatly elongated but inconspicuous caudal base photophore on caudal fin. Dark brown to black above, paler caudal fin. Posterior fin margins pale; conspicuous dark bands absent from middle of caudal fin; terminal caudal lobe dark. Dark below but not abruptly black.
Distribution Southwest Pacific: Norfolk Ridge and Lord Howe Rise, near New Caledonia.
Habitat Oceanic ridges, 668–1170m.
Behaviour Possibly semioceanic.
Biology Viviparous.
Status IUCN Red List: Least Concern. Deepsea fisheries very uncommon within limited range.

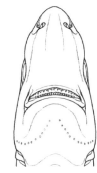

~2mm

Teeth
Upper 23–30
Lower 35–44

Measurements Born: 15–16cm. **Mature:** males 31–39cm, females 38–47cm.
Max: 50cm.

Identification (*E. pusillus* **clade**) Small, fairly slender, smooth-skinned lanternshark. Upper teeth generally with less than three pairs of cusplets. Gill openings rather long. Second dorsal fin less than twice area of first. Fairly short, broad caudal fin. Wide-spaced, low-crowned cuspless denticles (as in *Etmopterus bigelowi*), not arranged in rows, cover snout. Blackish brown body. Obscure broad, black markings above, in front and behind pelvic fins.
Distribution Widespread, Atlantic, Indian, and west Pacific Oceans, also Hawaii (central Pacific).
Habitat Continental slopes, on or near bottom, 274–1200m (possibly to 1998m). Oceanic in south Atlantic, 0–708m over deep water.
Behaviour Unknown.
Biology Viviparous, 1–6 young per litter, average 3.5. Age at maturity 5–9 years males, 8–11 years females, maximum age 13 and 17 years, respectively. Eats fish eggs, lanternfish, hake, squid and other small sharks.
Status IUCN Red List: Least Concern. Utilised bycatch in east Atlantic bottom fisheries.

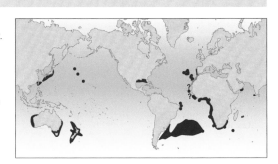

~2mm

Teeth
Upper 24 (holo.)
Lower 37 (holo.)

Measurements Born: 6cm. **Mature:** males 22–26cm, females 25–26cm.
Max: ~27cm.

Identification (*E. pusillus* **clade**) A slender-bodied dwarf lanternshark. 3-4 pairs of cusplets on upper teeth. Gill openings quite long. Distal margins of fins largely naked and more or less fringed with ceratotrichia. Second dorsal fin over twice area of first. Moderately long, broad caudal fin (dorsal caudal margin about equal to head length). 2–3 rows of unique, enlarged, hook-like denticles on flanks, absent from other lanternsharks. Greyish black body, inconspicuously black on its underside. Elongated, broad, black mark above, in front and behind pelvic fins; others at caudal fin base and along axis.
Distribution West Indian Ocean: South Africa to Tanzania, and Madagascar.
Habitat Near bottom, perhaps 200–500m.
Behaviour Unknown.
Biology Viviparous, nothing else known.
Status IUCN Red List: Least Concern. Relatively common off Mozambique.

SPLENDID LANTERNSHARK *Etmopterus splendidus*

FAO code: **ETK**   Plate page 156

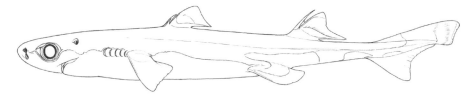

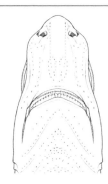

Teeth
Upper 34 (holo.)
Lower 36 (holo.)

Measurements Immature female 31cm.

Identification (*E. pusillus* clade) A spindle-shaped dwarf lanternshark, similar to *Etmopterus fusus*. Above and flanks dark: above purplish black, bluish black below in life (brownish black in preservative). Precaudal fins with pale reddish brown webs; no conspicuous dark bands on caudal fin, but a lighter patch between dark base and terminal lobe of caudal fin.
Distribution Northwest Pacific: Japan and Taiwan.
Habitat Continental slope, 200–300m.
Behaviour Unknown.
Biology Viviparous, nothing else known. Feeds on small fishes, cephalopods and crustaceans.
Status IUCN Red List: Least Concern. Taken in demersal trawl fisheries, retained for fishmeal and possibly liver oil. Rare in Japan, where trawl effort has been declining for 30 years; abundant in Taiwan, where most deepwater trawling takes place below this species' reported depth range.

HAWAIIAN LANTERNSHARK *Etmopterus villosus*

FAO code: **ETV**   Plate page 156

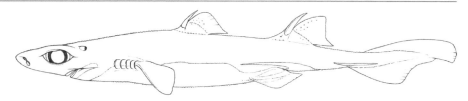

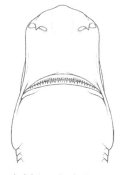

Teeth
Upper 27
Lower 29

Measurements Max: Only known specimen 17cm.

Identification (*E. pusillus* clade) Stout body. Upper teeth with less than 3 pairs of cusplets. Gill openings moderately long, about quarter eye-length. Distal margins of fins largely covered with skin. Second dorsal fin much larger but less than twice area of first, with tall, slightly curved spine. Lateral trunk denticles with slender, hooked, conical crowns, wide-spaced, in regular rows on rear of trunk and caudal fin; no rows of greatly enlarged denticles on flanks above pectoral fins; snout covered with denticles. Dark brown or blackish body, slightly darker below. Indistinct black mark above pelvic fins.
Distribution Central Pacific: Hawaiian Islands.
Habitat Insular slopes, on or near bottom, 406–911m.
Behaviour Unknown.
Biology Viviparous, nothing else known.
Status IUCN Red List: Least Concern. This species occurs below the maximum depth of fisheries in the region.

NINJA LANTERNSHARK *Etmopterus benchleyi*   FAO code: **SHL**   Plate page 156

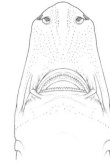

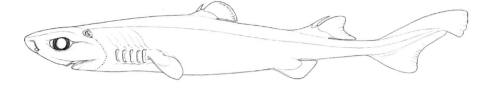

Teeth
Upper 25–30
Lower 30–36

Measurements Born: less than 18cm. **Mature:** males unknown largest was immature at 33cm, females 46cm. **Max:** female 52cm.

Identification (*E. spinax* clade) A moderately large, slender to stout-bodied lanternshark. Teeth dissimilar: upper teeth with single central cusp, flanked by 1–2 shorter lateral cusplets on each side; lower teeth blade-like, angled posteriorly and fused into a single row. Two dorsal fins with fin spines, first smaller than the second. Dermal denticles are short, slender, with slightly hook-like, conical crowns; denticles below dorsal fins arranged in irregular patches and aligned into rows extending onto the fins. Colour is a uniform black above and below, with no distinct markings.
Distribution East Pacific: from Nicaragua south to Panama.
Habitat Continental slopes, 836–1443m.
Behaviour Poorly known, although larger (adult) individuals appear to occur at greater depths than smaller specimens.
Biology Viviparous, 5 pups per litter. Feeds mainly on bony fishes, cephalopods and crustaceans.
Status IUCN Red List: **Least Concern**. Possibly subject to incidental bycatch, but presently there are no deepsea fisheries operating within the depth range of this species. All known specimens were collected during research surveys.

COMPAGNO'S LANTERNSHARK *Etmopterus compagnoi*   FAO code: **ETE**   Plate page 156

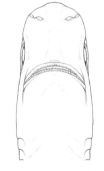

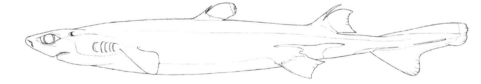

Teeth
Upper 28
Lower 34–38

Measurements Born: ~22cm or less. **Mature:** males 55.5cm, females 62cm. **Max:** males and females 67cm.

Identification (*E. spinax* clade) Moderately stout body. Caudal fin relatively short. Brown above, grading to black below. Inconspicuous elongated, broad, black flank markings above and behind pelvic fins.
Distribution Southeast Atlantic and west Indian Oceans: southern Namibia to South Africa, and on the northern Madagascar Ridge. Records from southern Mozambique require confirmation.
Habitat Collected on or near bottom, 383–1300m, mostly below 600m.
Behaviour Unknown.
Biology Viviparous. Feeds on small bony fishes and cephalopods.
Status IUCN Red List: **Least Concern**. Taken as bycatch on occasion.

PINK LANTERNSHARK   *Etmopterus dianthus*

FAO code: **SHL**   Plate page 158

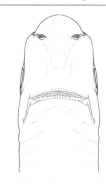

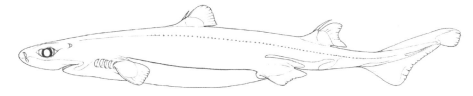

Teeth
Upper not known
Lower not known

Measurements Born: 9–10cm. **Mature:** males at least 34cm.
Max: at least 41cm.

Identification (*E. spinax* clade) Stout lanternshark. First dorsal fin small and low, short spine. Second dorsal fin less than twice size of first, spine nearly as long as fin tip. Fine, bristle-like denticles not arranged in rows. Pinkish above when fresh, brownish grey after preservation. Distinctive black markings behind pelvic fin, on caudal peduncle and upper caudal fin (upper fin tip dark). Dusky to black below.
Distribution Southwest Pacific: northeast Australia and New Caledonia.
Habitat Continental slope from 200–880m.
Behaviour Unknown.
Biology Unknown. Presumably viviparous.
Status IUCN Red List: Least Concern. Range unfished.

SOUTHERN LANTERNSHARK   *Etmopterus granulosus*

FAO code: **ETM**   Plate page 158

~2mm

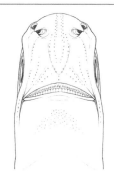

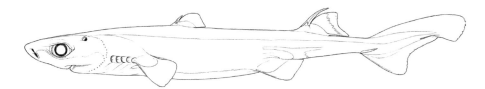

Teeth
Upper 27–35
Lower 28–52

Measurements Born: 17–20cm. **Mature:** males 52–60cm, females 60–75cm.
Max: males 92cm, females 102cm.

Identification (*E. spinax* clade) Heavy-bodied, big-headed lanternshark with conspicuous lines of large, rough, conical denticles on body, not on head. Gill openings very short. Second dorsal fin much larger than first. Short, broad caudal fin. Denticles wide-spaced; randomly on head but in conspicuous lines on sides; snout mostly bare, except for lateral patches of denticles. Colour grey-brown. Short, broad, black mark running above and slightly behind the pelvic inner margins; other elongated black marks at the caudal fin base. Abruptly black below.
Distribution Widespread circumglobal in southern oceans: south Chile to south Argentina, Namibia to eastern Cape, South Africa, south Madagascar Ridge, Southern Ocean islands, south Australia and New Zealand.
Habitat Outermost continental shelves and upper slopes, 220–1500m.
Behaviour Unknown.
Biology Viviparous, 2–16 pups per litter, averaging 8 (South Africa) to 10 (Australia) depending on the region. Age at maturity has been estimated at 20 years for males, and 30 years for females, with a maximum age of 57 years. Mostly feeds on small bony fishes, cephalopods and crustaceans.
Status IUCN Red List: Least Concern. Formerly referred to as *Etmopterus baxteri* in some regions, but that species is now known to be a junior synonym of *E. granulosus*.

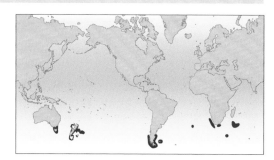

~2mm

Teeth
Upper 24–26
Lower 36–38

Measurements Born: ~9cm. **Mature:** males ~20cm.
Max: males 26cm, females 30cm.

Identification (*E. spinax* clade) A dwarf lanternshark with a stout body, moderately long caudal fin. Upper teeth generally with less than three pairs of cusplets. Gill openings very short. Interdorsal space short. Second dorsal fin much larger, but less than twice area of first. Denticles largely cover snout; not arranged in rows on trunk. Grey or dark brown above. Elongated broad, black mark above and behind pelvic fins; others at caudal fin base and along axis. Black below.
Distribution Northwest Atlantic: Virginia to Florida (USA), Bahamas, Cuba, Bermuda, Hispanola, Lesser Antilles. Not known from western or southern Caribbean.
Habitat Upper continental and insular slopes, on or near bottom, 180–717m.
Behaviour Unknown.
Biology Viviparous, 4–5 pups per litter.
Status IUCN Red List: Least Concern. Low fisheries interest.

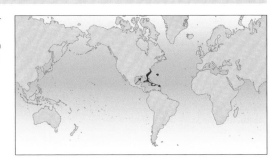

SMALLEYE LANTERNSHARK *Etmopterus litvinovi*   FAO code: **SHL**   Plate page 158

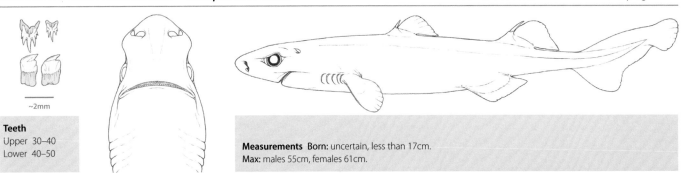

~2mm

Teeth
Upper 30–40
Lower 40–50

Measurements Born: uncertain, less than 17cm.
Max: males 55cm, females 61cm.

Identification (*E. spinax* clade) A moderately large, stout-bodied lanternshark. Large, flat head. Upper teeth generally with 1–3 pairs of cusplets. Gill openings long. First dorsal origin behind pectoral free rear tips. Interdorsal space short. Second dorsal fin slightly larger than first. Moderately long caudal fin. Denticles stout, hooked, conical and not arranged in rows on trunk. Plain-coloured; no obvious markings.
Distribution Southeast Pacific: Nazca and Sala y Gomez Ridges off Peru and Chile.
Habitat Upper slopes, on or near bottom, 630–1100m.
Behaviour Unknown.
Biology Presumably viviparous.
Status IUCN Red List: Least Concern. Apparently very common where it has been caught. Possibly a bycatch of high seas fisheries, but former deepsea fisheries within its range (e.g. for Orange Roughy and Alfonsino) are now closed.

PARIN'S LANTERNSHARK *Etmopterus parini* FAO code: **SHL** Plate page 158

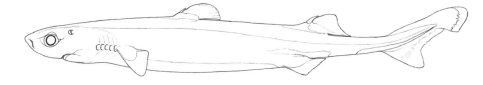

Teeth
Upper 25–29
Lower 30–33

Measurements Mature: females 34–38cm.
Max: 38cm

Identification (*E. spinax* clade) A relatively small-bodied lanternshark. Teeth dissimilar: upper teeth with single central cusp, flanked by one shorter lateral cusplet on each side; lower teeth blade-like, angled posteriorly and fused into a single row. Two dorsal fins with fin spines, first smaller than the second. Rather sparsely arranged, short, hook-like, conical dermal denticles curved rearwards, not in distinct rows. Colour is a dark brown above and to sides, becoming black below. Faint flank markings; anterior flank marking branch is short, while the posterior branch is absent. Fins becoming transparent to white, but no other distinctive markings.
Distribution Northwest Pacific Ocean.
Habitat Pelagic, only 2 known specimens were caught between 40–140m depth over very deep water, up to 6000m depth.
Behaviour Unknown.
Biology Viviparous, possibly with up to 7 pups per litter. Feeds on pelagic octopuses.
Status IUCN Red List: **Not Evaluated.** The only known specimens were caught during a research survey.

GREAT LANTERNSHARK *Etmopterus princeps* FAO code: **ETR** Plate page 158

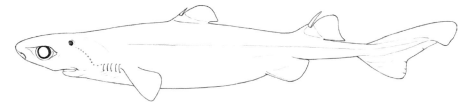

~2mm

Teeth
Upper 29–32
Lower 40–50

Measurements Born: 12–17cm **Mature:** males ~57cm, females 62cm.
Max: 94cm.

Identification (*E. spinax* clade) A large, stout lanternshark. Gill openings very long (half eye-length). Second dorsal fin much larger but less than twice area of first. Moderately long, broad caudal fin. Denticles largely cover snout; widely spaced (not in rows) on sides. Body blackish, no conspicuous dark markings.
Distribution Northwest Atlantic: Nova Scotia to New Jersey. Northeast Atlantic: between Greenland and Iceland to northwest Africa and Azores.
Habitat Continental slopes, on or near bottom, 350–2213m. 3750–4500m on North Atlantic lower rise.
Behaviour Unknown.
Biology Viviparous, 7–18 pups per litter, average 10. These sharks appear to have 2 seasonal reproductive peaks, in June–July and again in October. Feeds on small bony fishes, cephalopods and crustaceans.
Status IUCN Red List: **Least Concern.** Probably a bycatch in east Atlantic slope fisheries.

VELVET BELLY LANTERNSHARK   *Etmopterus spinax*   FAO code: **ETX**   Plate page 158

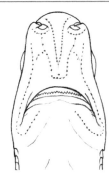
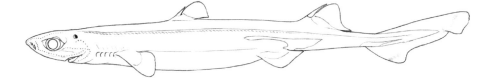

Teeth
Upper 22–32
Lower 26–40

Measurements Born: 8–14cm. **Mature:** males 24–35cm, females 30–41cm.
Max: males 41cm, females 55cm; records of larger sizes may be other species.

Identification (*E. spinax* clade) Long, fairly stout body with long caudal fin. Gill openings very short. Second dorsal fin about twice area of first. No lines of lateral trunk denticles; snout largely covered with denticles. Brown above. Elongated narrow black mark above and behind pelvic fins, others at sides of caudal fin. Abruptly black below.
Distribution East Atlantic: Iceland to Gabon. Mediterranean.
Habitat Outer continental shelves and upper slopes, near or well above bottom, 70–2490m, mostly 200–500m.
Behaviour Unknown.
Biology Viviparous, 1–21 pups per litter, number of young increases with size of the female. Age at maturity 4 years for males and females, maximum age 7–8 years for males, 9–11 years for females. Feeds on small fishes, squid and crustaceans. Smaller sharks mostly feed on invertebrates, but larger sharks have a more diversified diet that includes small fishes.
Status IUCN Red List: **Least Concern** globally. **Near Threatened** in northeast Atlantic where intensive fisheries have caused a population decline. **Least Concern** in Mediterranean. Common. Caught offshore by bottom and pelagic trawls, discarded or occasionally utilised for fishmeal and food.

BROWN LANTERNSHARK   *Etmopterus unicolor*   FAO code: **ETJ**   Plate page 158

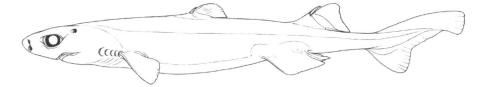

Teeth
Upper 28
Lower 34–38

Measurements Born: ~17cm. **Mature:** males ~48–68cm, females ~53–79cm.
Max: females 79cm.

Identification (*E. spinax* clade) Large, robust lanternshark. Upper teeth generally have less than three pairs of cusplets. Gill openings fairly large, half eye-length. Moderately long caudal fin (length of dorsal margin about equal to head length) and short caudal peduncle. Long, low first dorsal fin with very short spine. Second dorsal about twice height of first with strong spine; fin area about half as large again as that of first. Bristle-like lateral trunk denticles with slender, hooked, conical crowns distributed irregularly, not in rows; denticles largely cover snout. Grey-brown, dark brown or brownish black; dark below. Distinct broad, elongated caudal markings on caudal fin; indistinct pelvic flank markings.
Distribution Northwest Pacific: Japan. Possibly southern Australia and New Zealand. South African records are of a different species, *E. compagnoi*.
Habitat Continental slope and seamounts, 120–1500m, with records of catches in the epipelagic zone up to 120m from the surface, over very deep water.
Behaviour May sometimes swim in midwater.
Biology Viviparous, 2–21 pups per litter, average 12. Feeds mainly on cephalopods, shrimps and small bony fishes.
Status IUCN Red List: **Data Deficient**. Probably a bycatch of trawl and other fisheries. Utilisation unknown.

TRAVELLER LANTERNSHARK *Etmopterus viator*

FAO code: **EZT**   Plate page 158

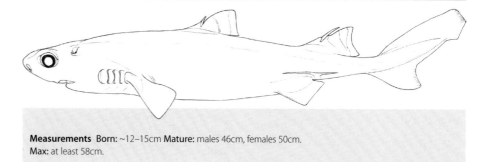

Teeth
Upper 26
Lower 37

Measurements Born: ~12–15cm **Mature:** males 46cm, females 50cm.
Max: at least 58cm.

Identification (*E. spinax* clade) Stout body with short snout. Lateral trunk denticles partially form short, broken, linear rows giving it a rough texture. Dark brown to blackish above, darker below. Base of black flank markings (indistinct in adults, distinct in subadults) posterior to pelvic fins and below second dorsal fin: anterior branch elongated, extending past second dorsal spine; posterior branch much shorter.
Distribution Southern oceans: widespread off New Zealand, and around seamounts south of South Africa.
Habitat Little known lanternshark found from 830–1610m.
Behaviour Unknown.
Biology Viviparous, 2–10 pups per litter. Feeds mostly on bony fishes, but also on crustaceans (euphausiids) and cephalopods (squids).
Status IUCN Red List: Least Concern. This species has a wide geographic and depth range and appears to occur in remote areas with low levels of fishing activity.

RASPTOOTH DOGFISH *Etmopterus sheikoi*

FAO code: **SHL**   Plate page 150

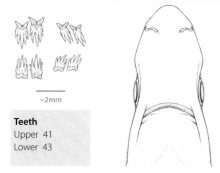

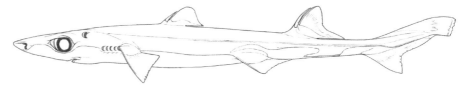

Teeth
Upper 41
Lower 43

Measurements Mature: males 30–40cm.
Max: at least 43cm.

Identification (clade uncertain) Unmistakable long, flat snout. Short mouth with small, compressed, comb-like, teeth with cusps and cusplets in both jaws. Grooved dorsal fin spines on both dorsal fins; second dorsal spine much larger than the first. Dark brown above with black photophores on caudal peduncle and caudal fin. Black below.
Distribution Northwest Pacific: near Japan and off Taiwan.
Habitat Holotype caught on upper slope of submarine ridge, 340–370m, but suspected to range much deeper, perhaps to ~1,000m.
Behaviour Unknown.
Biology Probably viviparous. Feeds on small fishes and invertebrates.
Status IUCN Red List: Least Concern. A bycatch of demersal trawls and retained for fish meal. Never common at landing sites; abundance has not changed since first described over 30 years ago. Very little fishing activity occurs within this species known range. Known only from a small number of specimens.

Teeth
Upper 15–20
Lower 15–20

Measurements Born: ~17cm **Mature:** males 42–47cm, females at least 52cm.
Max: at least 54cm.

Identification Very long, narrow, terminal, snake-like mouth with huge, curved fang-like teeth in front of both highly protrusible jaws. Deep pocket around front of upper jaws. Very large, diagonally elongated spiracles. Grooved spines on both dorsal fins. Body dark brown above, with black photophores on caudal peduncle and caudal fin. Black below.
Distribution Northwest and central Pacific: Japan, Taiwan and Hawaiian Islands.
Habitat Upper continental slopes, on bottom, 250–1000m. Uppermost slope of seamount at 270m. Possibly also oceanic as some specimens have been caught at 150m over water 1500m deep.
Behaviour Long, narrow jaws can probably be protruded to impale relatively large prey (bony fishes and crustaceans) on fang-like teeth, which are then swallowed whole. Apparently luminescent. This species appears to migrate into the water column at night in search of food.
Biology Viviparous, with litters possibly of 25–26 pups. Feeds on bony fishes.
Status IUCN Red List: Least Concern. Apparently rare and localised. Its range extends much deeper than regional fisheries.

Plate 15 SOMNIOSIDAE I

Small to gigantic (40cm to 6m or more); fairly broad head and flat snout, spiracles large and close behind eyes; lateral ridges on abdomen; pectoral fins low with short rounded free rear tips, two small dorsal fins with spines which may be covered with skin, no anal fin, caudal fin heterocercal with strong posterior notch.

○ **Portuguese Dogfish** *Centroscymnus coelolepis*   page 193

Atlantic, Indian and Pacific Oceans; 128–3675m; on or near bottom. Stocky; short snout, short labial furrows; dorsal spines usually just visible, small dorsal fins of equal size, second dorsal fin close to asymmetrical caudal fin; uniform golden-brown to blackish.

○ **Roughskin Dogfish** *Centroscymnus owstonii*   page 193

Atlantic, Indian and Pacific Oceans; 150–1459m; on or near bottom. Similar to *C. coelolepis* but longer snout, longer lower first dorsal fin, taller triangular second dorsal fin with dorsal spines barely exposed.

○ **Longnose Velvet Dogfish** *Centroselachus crepidater*   page 194

East Atlantic, Indian and Pacific Oceans, except for northeast Pacific; 200–2080m; on or near bottom. Slender; very long snout, small mouth with very long encircling upper labial furrows; dorsal spines very small, dorsal fins about the same size, first dorsal fin extends forward in prominent ridge with origin over pectoral bases, second dorsal free rear tip nearly reaches upper caudal origin.

○ **Largespine Velvet Dogfish** *Scymnodon macracanthus*   page 197

Southern hemisphere: Indian and Pacific Oceans; 219–1550m; near bottom. Stocky, tapering strongly behind pectoral fins; short snout; prominent stout dorsal spines, first dorsal fin lower than second and extends forward in prominent ridge; distance from second dorsal fin to upper caudal fin origin about length of dorsal base, free rear tip well in front of upper caudal fin; uniform dark brown to blackish.

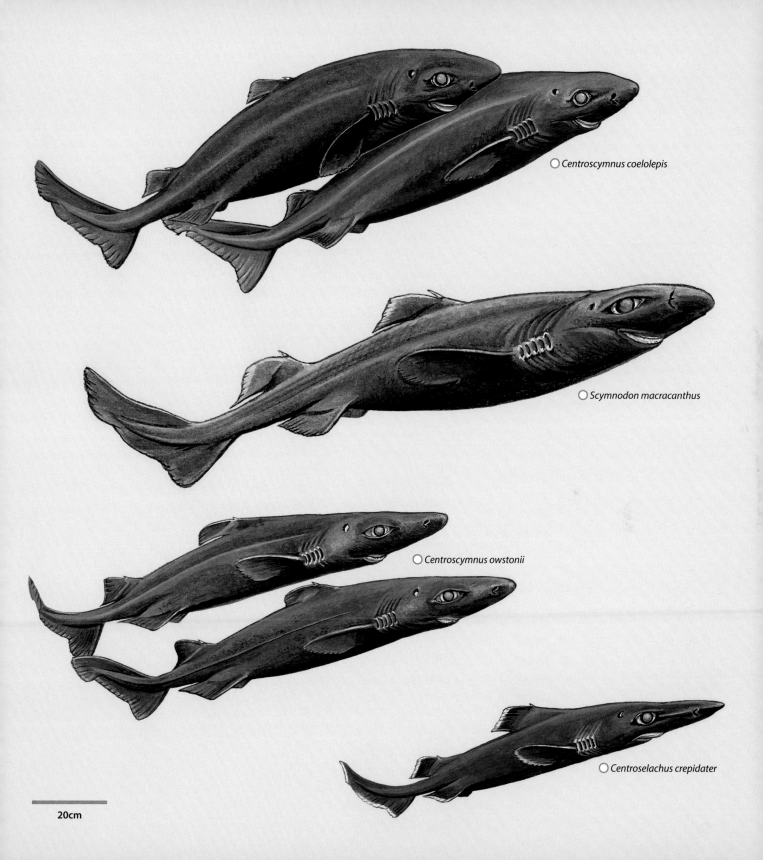

○ *Centroscymnus coelolepis*

○ *Scymnodon macracanthus*

○ Centroscymnus owstonii

○ *Centroselachus crepidater*

20cm

Plate 16 SOMNIOSIDAE II

○ **Whitetail Dogfish** *Scymnodalatias albicauda*   page 194

Southern oceans; 0–572m and deeper. Short broad round snout, long broad arched mouth, horizontally elongate eyes; pectoral fins elongate, no dorsal spines, second dorsal fin slightly larger than first and very close to caudal fin; dark brown or mottled grey, whitish grey fin margins, caudal fin with white blotches on dark terminal lobe.

○ **Azores Dogfish** *Scymnodalatias garricki*   page 195

Central North Atlantic Ridge; oceanic; 300–580m over much deeper water. Small; long broad rounded snout, long broad arched mouth, horizontally elongate eyes; no dorsal spines, first dorsal fin midback; uniform dark brown.

○ **Sparsetooth Dogfish** *Scymnodalatias oligodon*   page 195

Southeast Pacific; oceanic; 0–200m over very deep water. Small; long broad pointed snout, long broad arched mouth, horizontally elongate eyes; no dorsal spines, first dorsal fin midback, caudal fin lower lobe weak; uniform dark brown.

○ **Sherwood Dogfish** *Scymnodalatias sherwoodi*   page 196

Southwest Pacific; 170–500m. Moderately long flat pointed snout, long broad arched mouth, horizontally elongate eyes; pectoral fins leaf-shaped, no dorsal spines, first dorsal fin midback, second dorsal fin origin above rear third of pelvic fin bases, caudal fin lower lobe short and strong; dark brown above, lighter below, with light margins on gills and pectoral fins.

○ **Japanese Velvet Dogfish** *Scymnodon ichiharai*   page 196

Northwest Pacific, possibly Indian Ocean; 450–830m; on or near bottom. Low flat head, moderately long snout, short narrow mouth, short gills less or equal to eye length; caudal peduncle long; pectoral fins narrow leaf-shaped, dorsal spines small, pelvic fins small about the same as second dorsal fin, caudal fin with strong subterminal notch and short lower lobe; uniform black.

○ **Knifetooth Dogfish** *Scymnodon ringens*   page 197

East Atlantic and southwest Pacific; 200–1600m; on or near bottom. Thick high head, short broad snout, very large broad arched mouth, long gills greater than or equal to eye length; dorsal spines small, second dorsal fin slightly larger than first dorsal fin, caudal fin asymmetric with weak subterminal notch and no lower lobe; uniform black.

○ **Velvet Dogfish** *Zameus squamulosus*   page 201

Patchy worldwide; oceanic; 0–1511m; on or near bottom, epipelagic over very deep water. Slender; low flat head, fairly long narrow snout, short narrow mouth, nasoral grooves much longer than upper labial furrows; long caudal fin; dorsal spines small, second dorsal fin larger than first dorsal fin and about the same size as pelvic fins, caudal fin with strong terminal notch and short lower lobe; uniform black.

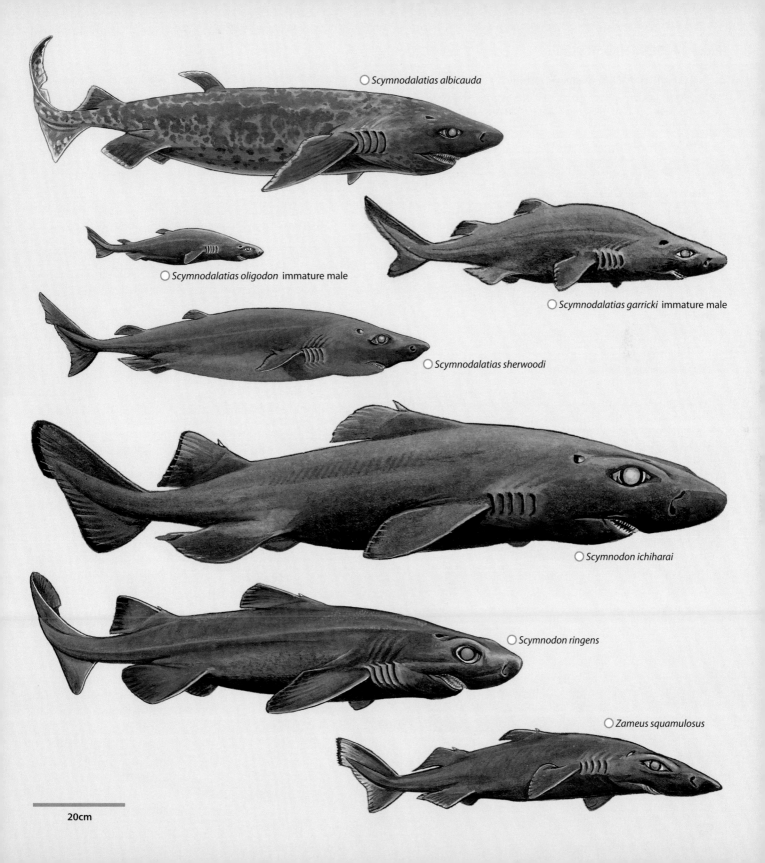

○ *Scymnodalatias albicauda*

○ *Scymnodalatias oligodon* immature male

○ *Scymnodalatias garricki* immature male

○ *Scymnodalatias sherwoodi*

○ *Scymnodon ichiharai*

○ *Scymnodon ringens*

○ *Zameus squamulosus*

20cm

Plate 17 SOMNIOSIDAE III

○ **Frog Shark** *Somniosus longus*   page 198

West Pacific, possibly southeast Pacific and southwest Indian Ocean; 200–1160m; on or near bottom. Slender cylindrical body; short head with short rounded snout; short caudal peduncle, keels on base of caudal fin; no dorsal spines, first dorsal fin about as long as second dorsal fin and closer to pectoral fins than pelvic fins, caudal fin with long lower lobe, relatively short upper lobe; smooth skin; uniformly blackish.

○ **Little Sleeper Shark** *Somniosus rostratus*   page 198

Northeast and northwest Atlantic, west Mediterranean; 180–2734m; on or near bottom. Similar to *S. longus* but smaller eyes, shorter second dorsal fin and shorter gills, about eye length (*S. longus* gills are more than twice eye length).

○ **Southern Sleeper Shark** *Somniosus antarcticus*   page 200

Southern oceans; 0–1440m and deeper. Gigantic, heavy cylindrical body; short rounded snout; short caudal peduncle, keels on base of caudal fin; very small precaudal fins, no dorsal spines, very low dorsal fins equal sized, first dorsal fin closer to pelvic fins than pectoral fins, distance between dorsal fin bases about 80% snout to first gill slit, caudal fin with long lower lobe, short upper lobe; skin rough; uniform grey to blackish.

○ **Greenland Shark** *Somniosus microcephalus*   page 199

North Atlantic and Arctic Oceans; 0–2647m and deeper, inshore in winter. Gigantic, heavy body; short rounded snout; short caudal peduncle, keels on base of caudal fin; small precaudal fins, no dorsal spines, low dorsal fins equal sized, first dorsal fin closer to pectoral fins than pelvic fins, distance between dorsal fin bases about the same as snout to first gill slit, caudal fin with long lower lobe, short upper lobe; skin rough; mid-grey or brown, occasionally transverse bands, small spots/blotches or light spots.

○ **Pacific Sleeper Shark** *Somniosus pacificus*   page 200

North Pacific; 0–2008m and deeper, shallower at high latitudes. Gigantic, heavy (when mature) cylindrical body; short rounded snout; short caudal peduncle, keels absent or present on base of caudal fin; small precaudal fins, no dorsal spines, low dorsal fins equal sized, first dorsal fin closer to pelvic fins than pectoral fins, distance between dorsal fin bases about 70% snout to first gill slit, caudal fin with long lower lobe, short upper lobe; skin rough and bristly; uniformly greyish.

○ *Somniosus longus*

○ *Somniosus rostratus*

20cm

○ Immature *Somniosus pacificus*

○ Immature *Somniosus antarcticus*

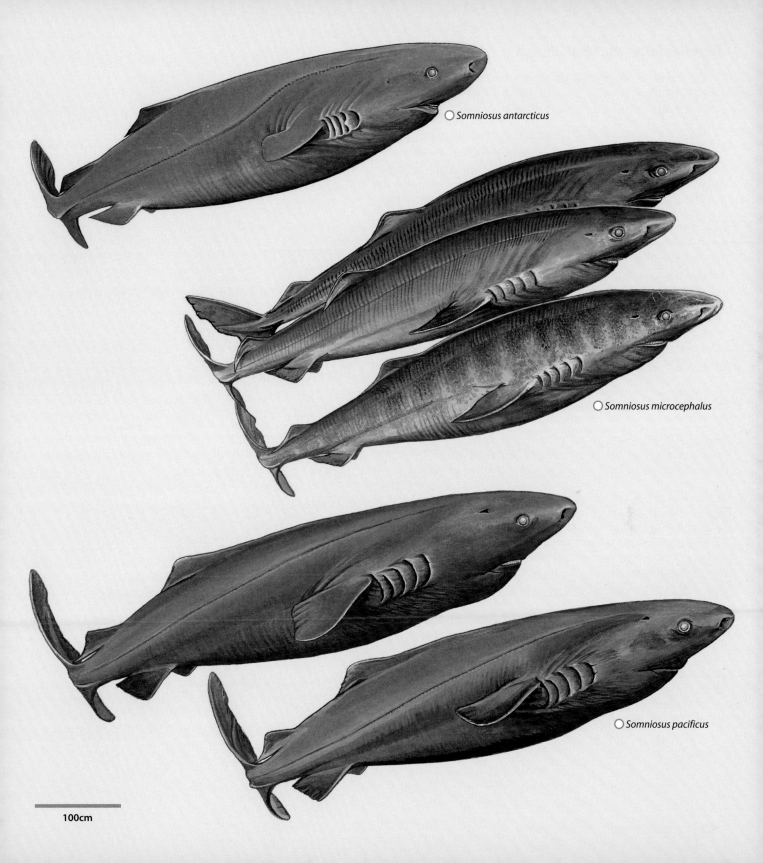

○ *Somniosus antarcticus*

○ *Somniosus microcephalus*

○ *Somniosus pacificus*

100cm

Somniosidae: sleeper sharks

Sixteen deepsea benthic and oceanic species in six genera (*Centroscymnus*, *Centroselachus*, *Scymnodalatias*, *Scymnodon*, *Somniosus* and *Zameus*). Occur circumglobally in most seas, ranging from the tropics to the Arctic and Southern Oceans. Size ranges from small (40–69cm) to gigantic (6m or more).

Identification Fairly broad head and flat snout. Short thin-lipped almost transverse mouth. Small needle-like upper teeth, large compressed blade-like lower teeth. Spiracles large, close behind eyes. Lateral ridges on abdomen, not usually on caudal peduncle (except in most *Somniosus*). Pectoral fins low, angular or rounded, not falcate, rear tips rounded and short. Pelvic fins subequal or larger than first dorsal and pectoral fins, and subequal to or smaller than second dorsal fin. Two small broad dorsal fins, origin of first in front of pelvic fin origins, space between is greater than fin base length, second usually smaller or same size as first. Spines on both fins (may be covered by skin), or on neither (*Scymnodalatias* and *Somniosus*). No anal fin. Caudal fin heterocercal with strong subterminal notch. No photophores.

Biology Poorly known. Viviparous with yolk-sac (four to at least 59 pups per litter). Some researchers estimate female age at maturity in the largest species might be up to 100 years old, and longevity some 400 years (others dispute these figures). Mostly occur near the seabed on continental and insular slopes from 200m to at least 3675m; a few species are oceanic or semi-oceanic. In high northern latitudes, *Somniosus microcephalus* and *S. pacificus* occur on continental shelves, penetrate the intertidal, and occur at the surface in cold water. Sleeper sharks feed on bony fishes, other sharks, rays and chimaeras, cephalopods and other molluscs, crustaceans, seals, whale meat carrion, marine birds, echinoderms and jellyfish. At least one species (*Centroscymnus coelolepis*) uses a similar feeding strategy to the cookiecutter sharks, biting chunks out of live whales, seals and bony fishes. Sleeper sharks do not seem to form large aggregations.

Status Moderately common and an important component of commercial targeted and bycatch deepsea shark fisheries. Caught by line gear, demersal trawls, gillnets, traps, and even by spear or gaff. Their flesh is used for food or fishmeal, and the large very oily livers processed for their high squalene content. In Australia, many species have high mercury levels in their flesh and are discarded. This family's conservation status is poorly known, but of concern because expanding deepsea fisheries are taking large numbers of sharks, including somniosids. Although their life history is sketchily known, it is suspected that reproduction is limited and growth extremely slow. If so, they may be highly vulnerable to overfishing – about 20% are assessed as threatened in the IUCN Red List. Several (e.g. *Scymnodalatias* species) are extremely poorly known, with only a few records and may be distinguished mainly by tooth counts and precise morphological measurements.

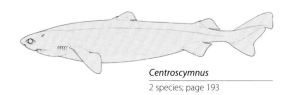

Centroscymnus
2 species; page 193

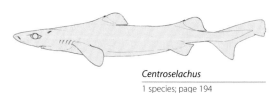

Centroselachus
1 species; page 194

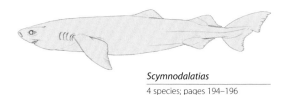

Scymnodalatias
4 species; pages 194–196

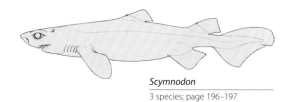

Scymnodon
3 species; page 196–197

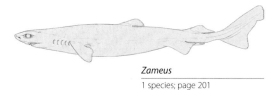

Zameus
1 species; page 201

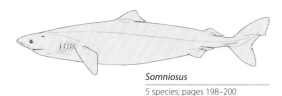

Somniosus
5 species; pages 198–200

PORTUGUESE DOGFISH *Centroscymnus coelolepis*   FAO code: **CYO**   Plate page 186

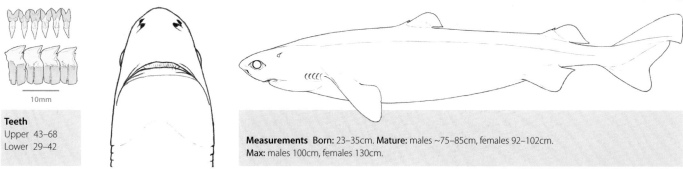

Teeth
Upper 43–68
Lower 29–42

Measurements Born: 23–35cm. **Mature:** males ~75–85cm, females 92–102cm. **Max:** males 100cm, females 130cm.

Identification Stocky, short-snouted dogfish. Large mouth; slender upper teeth; short, bent cusps on broader lower teeth; short labial furrows. Very small dorsal spines (may be covered by skin) on small, equal-sized dorsal fins; second close to asymmetrical caudal fin. Large, round, flat, overlapping denticles in adults. Uniformly blackish to golden-brown.
Distribution Atlantic, Indian and west Pacific Oceans.
Habitat On or near bottom, continental slopes, upper and middle abyssal plain rises, 128–3675m (mostly deeper than 400m).
Behaviour Can take bites out of live prey, including cetaceans, like cookiecutter sharks (*Isistius* species).
Biology Viviparous, 1–29 (mostly 12–14) pups per litter. Eats bony fishes, other sharks, benthic invertebrates and cetacean and seal meat.
Status IUCN Red List: **Near Threatened** globally (out of date). **Endangered** in European Atlantic waters, due to a long history of bycatch and target deepsea fisheries (now prohibited in EU waters), but **Least Concern** in Mediterranean, where it occurs outside the range of fisheries. Utilised for fishmeal and squalene from liver oil.

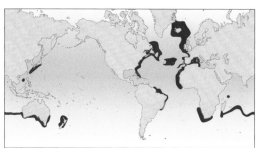

ROUGHSKIN DOGFISH *Centroscymnus owstonii*   FAO code: **CYW**   Plate page 186

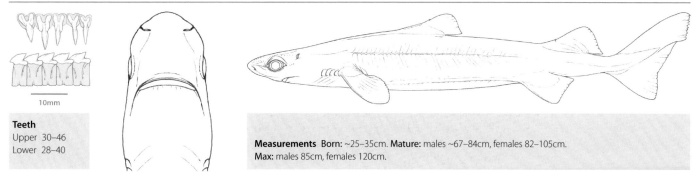

Teeth
Upper 30–46
Lower 28–40

Measurements Born: ~25–35cm. **Mature:** males ~67–84cm, females 82–105cm. **Max:** males 85cm, females 120cm.

Identification Similar to *Centroscymnus coelolepis* but longer snout; smaller denticles with small cusps in adults; longer, lower first dorsal fin; slightly taller, more triangular second dorsal fin. Dorsal spine tips often barely exposed.
Distribution West Atlantic: Gulf of Mexico, Brazil, Uruguay. East Atlantic: Canary Islands and Madeira to South Africa. Indian Ocean. Pacific Ocean: Japan, Australasia and southeast Pacific.
Habitat On or near bottom, upper continental slopes and submarine ridges, 150–1459m, mostly deeper than 600m.
Behaviour Sometimes in schools segregated by sex (females recorded deeper than males off Japan).
Biology Viviparous, up to 34 eggs per female, but litter size unknown. Eats bony fishes and cephalopods.
Status IUCN Red List: **Vulnerable**. Moderately common. Localised fisheries interest (mainly bycatch) and some population declines; sometimes processed for fishmeal and squalene. Formerly recognised as *C. cryptacanthus* in east Atlantic.

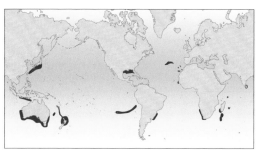

LONGNOSE VELVET DOGFISH *Centroselachus crepidater* FAO code: **CYP** Plate page 186

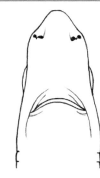

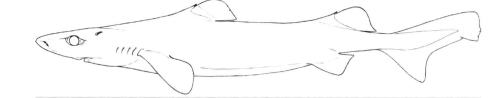

Teeth
Upper 36–51
Lower 30–36

Measurements Born: 28–35cm. **Mature:** males ~60–68cm, females 77–88cm.
Max: males 95cm, females 105cm.

Identification Slender-bodied. Very long snout. Very long upper labial furrows almost encircling small mouth; upper teeth with single cusp; lower teeth with short, oblique cusps. Roughly equal-sized dorsal fins, very small spine tips protruding; first dorsal base expanded forward as prominent ridge, origin over pectoral bases; second dorsal free rear tip nearly reaches upper caudal origin. Large dermal denticles on trunk mostly smooth, circular, overlapping, with or without cusps. Black to dark brown with narrow, light posterior fin margins.
Distribution East Atlantic, Indian and Pacific Oceans, except for northeast Pacific.
Habitat On or near bottom, upper continental and insular slopes, 200–2080m, mostly deeper than 500m.
Behaviour Unknown.
Biology Viviparous, 1–9 pups per litter, average 6, reproduces year-round. Age at maturity estimated at 9 years for males and 20 years for females, with maximum ages of 34 and 54 years, respectively. Eats fishes and cephalopods.
Status IUCN Red List: Near Threatened. Common and wide-ranging, but its distribution is intensively fished (except in very deep water) and it is a utilised bycatch for fishmeal and liver oil. Population declines are reported in some areas, but increasing trends in others.

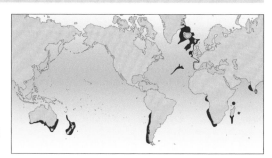

WHITETAIL DOGFISH *Scymnodalatias albicauda* FAO code: **YSA** Plate page 188

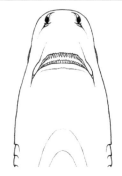

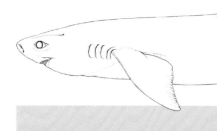

Teeth
Upper 57–62
Lower 35

Measurements Born: longer than 20cm. **Mature:** females 74–111cm.
Max: 111cm.

Identification Short, broad, rounded snout. Long, broadly arched mouth; small upper teeth with very narrow, acute, erect cusps; larger, blade-like, unserrated, interlocked lower teeth with high, erect cusps. Horizontally elongated eyes. Second dorsal fin slightly larger than first and very close to caudal fin; no dorsal spines. Elongated pectoral fins. Lower caudal margin half as long as upper margin. Dark brown or mottled greyish above. Whitish-grey fin margins; obvious white blotches and dark terminal lobe on caudal fin. Lighter below.
Distribution Known from scattered records in the southern oceans and one record from the Coral Sea off northeast Australia.
Habitat Oceanic in epipelagic zone, 0–200m and deeper over 1400–4000m depth; near bottom on submarine ridge, 572m.
Behaviour Possibly mesopelagic or bathypelagic, may rise to near surface at night.
Biology Viviparous, very large litters: at least 59 pups.
Status IUCN Red List: Data Deficient. Very rarely caught by deepsea trawls and tuna longlines.

AZORES DOGFISH *Scymnodalatias garricki*

FAO code: **SHX**   Plate page 188

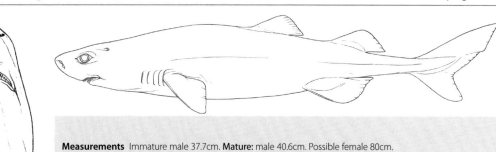

Teeth
Upper 43
Lower 32

Measurements Immature male 37.7cm. **Mature:** male 40.6cm. Possible female 80cm.

Identification Very small dogfish with long head. Snout long, broad and rounded. Mouth long and broadly arched; very small, narrow upper teeth; lower teeth larger, blade-like and interlocked. Eyes horizontally elongated. First dorsal fin located halfway down body, free rear tip just in front of pelvic origins; no dorsal fin spines. Strong lower caudal lobe. Uniform dark brown.
Distribution North Atlantic Ridge.
Habitat Apparently oceanic or deep-benthic. Caught at 300–580m in open ocean over seamounts, where bottom depths are 2000m or more.
Behaviour Unknown.
Biology Unknown. Viviparous.
Status IUCN Red List: Data Deficient. Known from two males (immature and mature), and possibly a third female specimen.

SPARSETOOTH DOGFISH *Scymnodalatias oligodon*

FAO code: **SHX**   Plate page 188

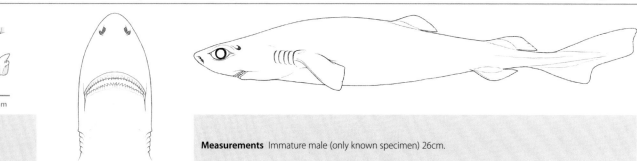

Teeth
Upper 33
Lower 42

Measurements Immature male (only known specimen) 26cm.

Identification Long, broad, pointed snout. Mouth long and broadly arched. Larger lower teeth blade-like, unserrated and interlocked with strongly oblique cusps; fewer, smaller, upper teeth. Eyes horizontally elongated. No dorsal fin spines. Lower caudal lobe weakly developed; lower caudal margin about half as long as upper margin. Uniformly dark brown.
Distribution Southeast Pacific: open ocean.
Habitat Apparently oceanic. Caught near surface, 0–200m in water 2000–4000m deep.
Behaviour Unknown.
Biology Unknown, presumably viviparous.
Status IUCN Red List: Least Concern. Known from one specimen. Not believed to be caught by the longline fisheries operating in this remote area.

SHERWOOD DOGFISH *Scymnodalatias sherwoodi*
FAO code: **YSS** Plate page 188

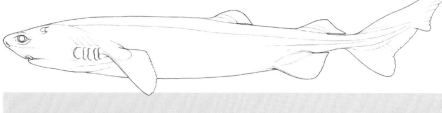

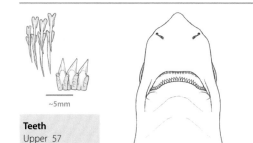

Teeth
Upper 57
Lower 34

Measurements Mature: male 80cm.
Max: 85cm.

Identification Moderately long, flattened, pointed snout. Mouth long and broadly arched; upper teeth small with very narrow, acute, erect cusps; lower teeth larger, blade-like, unserrated and interlocked with high, erect cusps. Eyes horizontally elongated. First dorsal in middle of back, far behind pectoral fins; free rear tip in front of pelvic fins. Second dorsal origin above rear third of pelvic base. No dorsal fin spines. Pectoral fins relatively short, elongated, leaf-shaped. Asymmetric caudal fin: short, strong lower lobe about half as long as upper lobe. Dark brown above, lighter below. Light margins on gill slits and pectoral fins.
Distribution Southwest Pacific and southeast Indian Ocean: off Tasmania and southwest Australia between 40–50°S and New Zealand.
Habitat Deepsea, 170–500m.
Behaviour Unknown.
Biology Unknown, presumably viviparous.
Status IUCN Red List: **Data Deficient.** Infrequent deepsea trawl bycatch. Known only from about five specimens.

JAPANESE VELVET DOGFISH *Scymnodon ichiharai*
FAO code: **QUX** Plate page 188

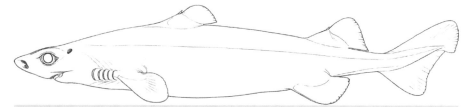

Teeth
Upper 42–48
Lower 28–30

Measurements Mature: males 89–101cm, females 126cm.
Max: 151cm.

Identification Rather low, flat head; moderately long snout. Mouth fairly narrow, short and nearly transverse; small spear-like upper teeth; lowers large, high, knife-cusped cutting teeth. Gill slits rather short (less than half eye-length). Two dorsal fins with small spines. Pectoral fins narrow and leaf-shaped; apices well in front of first dorsal spine. Pelvic fins small, about equal to second dorsal. Caudal peduncle long. Caudal fin with strong subterminal notch and short lower lobe. Colour uniform black.
Distribution Northwest Pacific: Japan and Taiwan. Indian Ocean: Andaman Sea.
Habitat Slope, on or near bottom, 450–830m.
Behaviour Unknown.
Biology Viviparous, litter size unknown, but females recorded with 29–56 ovarian eggs, suggesting these sharks may have large litters. Diet presumed to include fishes and invertebrates.
Status IUCN Red List: **Data Deficient.** Caught in deepset line fisheries.

LARGESPINE VELVET DOGFISH · Scymnodon macracanthus

FAO code: **YSM**   Plate page 186

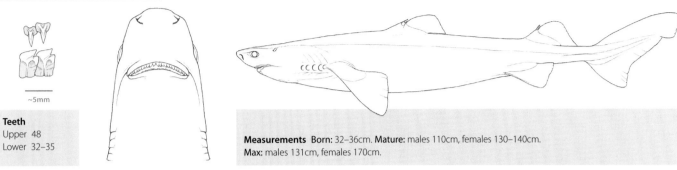

Teeth
Upper 48
Lower 32–35

~5mm

Measurements Born: 32–36cm. **Mature:** males 110cm, females 130–140cm.
Max: males 131cm, females 170cm.

Identification Stocky body, tapering behind the pectoral fins. Very short snout. Thick, fleshy lips; short upper labial furrows. Two dorsal fins with prominent, stout spines; second dorsal fin higher than first; first dorsal fin extends forwards in prominent ridge. Space between second dorsal and upper caudal fin about length of dorsal base; free rear tip well in front of upper caudal origin. Dark brown or blackish in colour.
Distribution Southern hemisphere in the Indian and Pacific Oceans at high latitudes.
Habitat Near bottom, continental and insular slopes, 180–1550m (most common at more than 600m).
Behaviour Forms large schools segregated by size and sex.
Biology Viviparous, up to 36 pups per litter. Males mature at 18 years, females at 29 years, and may live up to 39 years. Feeds on cephalopods and bony fishes.
Status IUCN Red List: Data Deficient. Relatively common at depths where it occurs. Deepsea fisheries bycatch, utilised in New Zealand for fish meal and squalene. Plunket's Shark (formerly *Proscymnodon plunketi*) has been found to be the same as this species. Since the above scientific name is older, it takes precedence.

KNIFETOOTH DOGFISH · Scymnodon ringens

FAO code: **SYR**   Plate page 188

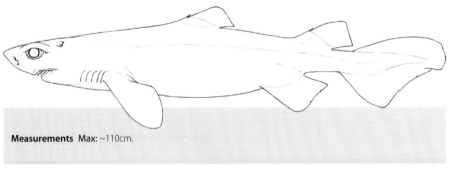

Teeth
Upper 50
Lower 29

5mm

Measurements Max: ~110cm.

Identification Rather thick, high head. Broad, short snout. Very large, broadly arched mouth. Small, lanceolate upper teeth; huge, triangular, cutting lower teeth. Rather long gill slits (more than half eye length). Small dorsal fin spines; second dorsal fin slightly larger than first. Caudal fin asymmetric, not paddle-shaped, with weak subterminal notch and no lower lobe. Uniformly black, no obvious fin markings.
Distribution East Atlantic: Scotland to Senegal. Southwest Pacific: New Zealand.
Habitat Continental slope, on or near bottom, 200–1600m.
Behaviour Immense, triangular, razor-edged lower teeth suggest a formidable predator, capable of attacking and dismembering large prey.
Biology Probably viviparous. Diet includes bony fishes, crustaceans, cephalopods and invertebrates.
Status IUCN Red List: Vulnerable. An incidental catch of intensive trawl and longline fisheries operating throughout its range. Bycatch may be utilised for liver oil and discard survival is probably low. Declines are reported in parts of its range, but other populations appear stable; it may have a refuge from fisheries in very deep water.

FROG SHARK *Somniosus longus*

FAO code: **SHX**   Plate page 190

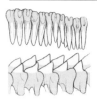

Teeth
Upper 56–57
Lower 31–32

Measurements Born: 21–28cm. **Mature:** males~71cm, females 82–134cm.
Max: 143cm, possibly more than 150cm.

Identification Similar to *Somniosus rostratus*. Small, slender, cylindrical body. Short head; short, rounded snout. Spear-like upper teeth; slicing lower teeth with high, semierect cusps and low roots. Spineless dorsal fins; first dorsal about as long as second and closer to pectorals than pelvic fins. Short caudal peduncle. Caudal fin asymmetrical, paddle-shaped: upper lobe slightly longer than lower lobe; lateral keels on base of caudal fin. Skin smooth: denticles with flat cusps. Uniformly dark body and fins.
Distribution West Pacific: Japan and New Zealand; possibly southeast Pacific and southwest Indian Ocean.
Habitat Outer continental shelves and upper slopes, on or near bottom, 250–1160m.
Behaviour Unknown.
Biology Viviparous. Probably eats deepsea bottom fishes and invertebrates.
Status IUCN Red List: Data Deficient. Rare.

LITTLE SLEEPER SHARK *Somniosus rostratus*

FAO code: **SOR**   Plate page 190

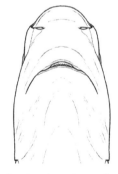

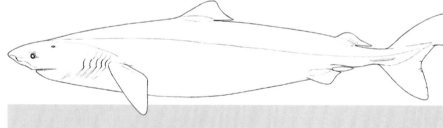

10mm

Teeth
Upper 53–63
Lower 31–38

Measurements Born: 21–28cm. **Mature:** males ~70cm, females 80cm.
Max: 143cm.

Identification Similar to *Somniosus longus* but smaller eyes and shorter second dorsal fin. Slender, cylindrical body. Short head; short, rounded snout. Spear-like upper teeth; slicing lowers with high semierect cusps and low roots. Spineless dorsal fins; first dorsal higher, closer to pectorals than pelvic fins. Short caudal peduncle. Caudal fin asymmetrical, upper lobe slightly longer than lower lobe; lateral keels on caudal fin base. Skin smooth: flat-cusped denticles. Blackish in colour.
Distribution Northeast Atlantic, west Mediterranean, northwest Atlantic (Cuba).
Habitat Outer continental shelves, upper and lower slopes, on or near bottom, 180–2734m.
Behaviour Unknown.
Biology Viviparous, with 6–17 pups per litter. Diet includes cephalopods, bottom-dwelling invertebrates and fishes.
Status IUCN Red List: Least Concern. A bycatch of longlines and bottom trawls in the shallower part of its range, a refuge from fisheries in very deep water, used for fishmeal and possibly food.

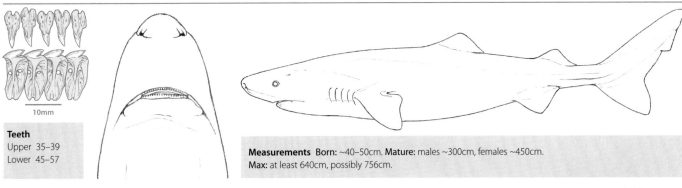

Teeth
Upper 35–39
Lower 45–57

10mm

Measurements Born: ~40–50cm. **Mature:** males ~300cm, females ~450cm. **Max:** at least 640cm, possibly 756cm.

Identification A gigantic, heavy-bodied, cylindrical sleeper shark. Short, rounded snout. Spear-like upper teeth; slicing lower teeth with low, bent cusps and high roots. Small precaudal fins. Spineless, equal-sized, low dorsal fins; first slightly closer to pectoral fins than to pelvics. Distance between dorsal fin bases about equal to snout to first gill slits. Short caudal peduncle. Upper lobe slightly longer than lower lobe; lateral keels present on base of caudal fin. Skin rough: denticles with strong, hook-like, erect cusps. Mid-grey or brown in colour. Sometimes with transverse dark bands, small dark spots and blotches, and small light spots.
Distribution North Atlantic and Arctic, occasionally to Portugal, Gulf of Mexico and off the Caribbean coast of Colombia.
Habitat Continental and insular shelves and upper slopes to at least 2647m, water temperature 0.6–12°C. Inshore in northern winter (from intertidal and surface in shallow bays and river mouths), retreats to 180–1440m when temperature rises. May move into shallower water in spring and summer in north Atlantic.
Behaviour Sluggish, offers little resistance to capture, easily fished through iceholes, but able to capture large active prey including fishes, invertebrates, seabirds and seals. There are reports of Greenland Sharks stalking a wildlife officer walking on pack-ice and a group of divers in the St Lawrence Estuary, suggesting this is an experienced predator on live seals. They also feed on dead cetaceans and drowned horses and reindeer. Greenland Sharks are a major fisheries pest, taking valuable Halibut off hooks and destroying longline gear, and are often bycaught while doing so. Experiments to reduce bycatch by using electropositive metals on longline gear to repel sharks were unsuccessful. A highly conspicuous white copepod parasite often attached to the eye surface is speculated to lure prey species to the shark under a

mutualistic and beneficial relationship – but also causes serious damage to the cornea.
Biology Viviparous, 7–10 pups per litter. The Greenland Shark appears to be a very slow-growing shark and may take more than 100 years to reach maturity. A study suggesting that these sharks live to over 400 years is possibly an over-estimate, but they are no doubt long-lived and among the world's longest-lived vertebrate species. Diet includes bony and cartilaginous fishes, cephalopods, crustaceans, and other invertebrates, and marine mammals including seals and small cetaceans.
Status IUCN Red List: **Vulnerable**. For centuries, Greenland Sharks were traditionally fished by Inuit, using gaffs or hook and line, and utilised for their livers (for lighting oil), skin for boots, and lower dental bands as knives. Artisanal fisheries continue in Greenland and Iceland. They are also caught with longlines, trawls, nets and in cod traps. The Greenland liver oil fishery peaked last century, when up to 58,000 Greenland Sharks were caught annually in Norway, but target fishing for oil ceased in the 1940s. Damage caused to other commercial fisheries by shark predation resulted in eradication programmes; bounties were offered for Greenland Shark in Norway until catches fell due to further population depletion. A bounty is still paid for shark hearts in Greenland and there has been a proposal to burn the carcasses for biofuel. Loss of sea ice due to climate change, increasing access for commercial fishing vessels, poses a new threat. The meat, which contains high levels of urea and the anti-freeze chemical trimethylamine oxide (TMAO), is highly poisonous unless washed, dried, or buried until semi-putrid, but a delicacy in Iceland. Despite being predators of large mammals, there are no confirmed incidents of Greenland Sharks attacking people, not even the divers who approach and tag them underwater, and they are considered harmless to humans.

Greenland Shark, *Somniosus microcephalus*, Baffin Island, Canada.

SOUTHERN SLEEPER SHARK *Somniosus antarcticus* FAO code: **SHX** Plate page 190

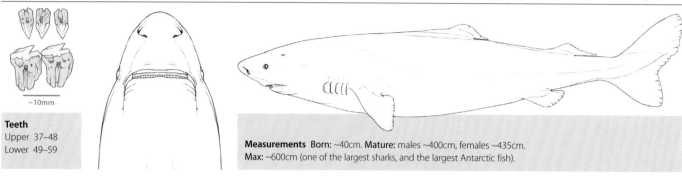

Teeth
Upper 37–48
Lower 49–59

Measurements Born: ~40cm. **Mature:** males ~400cm, females ~435cm.
Max: ~600cm (one of the largest sharks, and the largest Antarctic fish).

Identification Gigantic, heavy, cylindrical body. Short, rounded snout. Spear-like upper teeth; slicing lower teeth with low, bent cusps and high roots. Very small precaudal fins. Spineless, equal-sized dorsal fins; first dorsal slightly closer to pelvics than to pectoral fins. Distance between dorsal fin bases (about 80% of snout to first gill slits) and very low first dorsal fin distinguishes this species from *Somniosus microcephalus* and *S. pacificus*. Caudal fin asymmetrical, upper lobe slightly longer than lower lobe; lateral keels on base of caudal fin. Skin rough: denticles with strong hook-like, erect cusps. Uniform grey to pinkish body and fins.
Distribution Southern oceans and Antarctic.
Habitat Continental and insular shelves, upper slopes down to at least 245–1540m. Water temperature 0.6–12°C.
Behaviour Sluggish, slow-moving epibenthic shark, able to capture large and active prey (fishes and invertebrates). Also feeds on carrion.
Biology Viviparous; 10 pups found in uterus of 5m female. Diet consists of marine mammals, large fishes, and cephalopods.
Status IUCN Red List: **Least Concern**. Discarded bycatch from the toothfish (*Dissostichus* spp.) fishery.

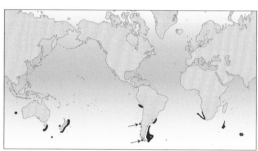

PACIFIC SLEEPER SHARK *Somniosus pacificus* FAO code: **SON** Plate page 190

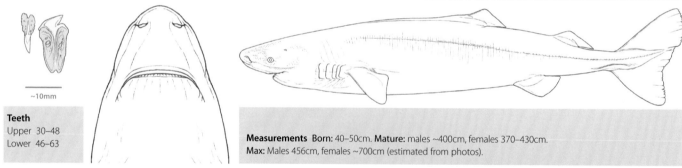

Teeth
Upper 30–48
Lower 46–63

Measurements Born: 40–50cm. **Mature:** males ~400cm, females 370–430cm.
Max: Males 456cm, females ~700cm (estimated from photos).

Identification Giant sleeper shark with heavy, cylindrical body. Short, rounded snout. Spear-like upper teeth; slicing lower teeth with low, bent cusps and high roots. Small precaudal fins. Spineless, equal-sized, low dorsal fins; first dorsal slightly closer to pelvics than to pectoral fins. Distance between dorsal fin bases about 70% of snout to first gill slits. Short caudal peduncle. Caudal fin asymmetrical, upper lobe slightly longer than lower lobe; lateral keels variably present or absent on base of caudal fin. Skin rough and bristly: denticles with strong hook-like, erect cusps. Uniform greyish body and fins.
Distribution North Pacific: Taiwan to Mexico, including Hawaiian Islands.
Habitat Continental shelves and slopes, to over 2008m. Ranges into shallower water at higher latitudes (once found trapped in a tide pool), but occurs in very deep water at lower latitudes.
Behaviour Lumbering sluggish sharks. Small mouth and large oral cavity suggest suction feeding. Takes a wide variety of surface and bottom animals. Seal remains in stomach may be scavenged carrion or taken alive. Possible sexual segregation (pregnant females have not been recorded).
Biology Probably viviparous; up to 300 large eggs per female. Feeds on a variety of bottom dwelling fishes and invertebrates, but also on fast-swimming active fishes such as tunas, and marine mammals including seals, sea lions, and dolphins. Although they are known to scavenge, they may also capture these active prey species by ambushing them.
Status IUCN Red List: **Data Deficient**. Relatively common.

VELVET DOGFISH *Zameus squamulosus*

FAO code: **SSQ**   Plate page 188

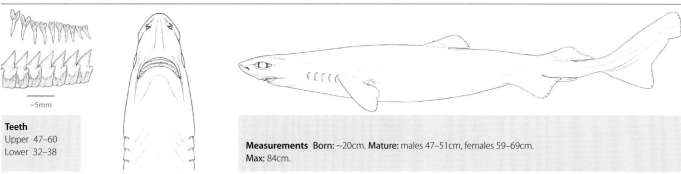

Teeth
Upper 47–60
Lower 32–38

~5mm

Measurements Born: ~20cm. **Mature:** males 47–51cm, females 59–69cm. **Max:** 84cm.

Identification Small, slender dogfish. Low, flat head. Fairly long, narrow snout. Short, narrow mouth; small, spear-like upper teeth; high-cusped, knife-like lower teeth; postoral grooves much longer than short upper labial furrows. Two dorsal fins with small spines; second larger, about the same size as small pelvic fins. Long caudal fin; short lower caudal lobe; strong subterminal caudal fin notch. Denticles tricuspidate with transverse ridges. Uniform black in colour.
Distribution Patchy worldwide.
Habitat Continental and insular slopes, on or near bottom, 550–1511m. Epipelagic and oceanic off Brazil (0–580m in water 2000m deep) and Hawaiian Islands (27–35m in water 4000–6000m deep).
Behaviour Unknown.
Biology Viviparous, 3–10 pups per litter. Diet presumed to include small bony fishes and invertebrates.
Status IUCN Red List: **Least Concern**. A bycatch of surface longline and demersal longline and trawl fisheries, but only in the shallower parts of its large depth range.

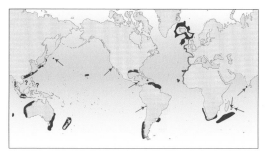

Pacific Sleeper Shark, *Somniosus pacificus*, Central Pacific Basin (p. 200).

Plate 18 OXYNOTIDAE

Unmistakable small sharks, compressed, triangular in cross-section, lateral ridges on abdomen; flat blunt snout, small thick-lipped mouth, large close-set nostrils; two high spined dorsal fins; very rough skin.

○ **Prickly Dogfish** *Oxynotus bruniensis*

Temperate southwest Pacific and southeast Indian Ocean; 45–1120m. Small circular spiracles; dorsal fins with triangular tips, trailing edges straight to slightly concave, first dorsal spine leans forward; uniformly light grey-brown.

○ **Caribbean Roughshark** *Oxynotus caribbaeus*
page 205

West central Atlantic: Caribbean; 218–579m; bottom. Small circular spiracles; dorsal fins with narrow triangular tips, trailing edges concave, first dorsal fin spine leans forward; grey or brownish with dark bands, blotches and small spots separated by prominent light areas over pectoral and pelvic fins.

○ **Angular Roughshark** *Oxynotus centrina*
page 205

East Atlantic, including Mediterranean, southwest Indian Ocean; 35–805m; bottom. Large expanded ridges over eyes covered with large denticles, very large vertical elongate spiracles; first dorsal spine leans forward; grey or grey-brown with dark blotches on head and sides (less prominent in adults) and light horizontal line over cheeks below eye.

○ **Japanese Roughshark** *Oxynotus japonicus*
page 206

Northwest Pacific; 150–350m. Large vertical oval spiracles; dorsal fins with narrow triangular tips, trailing edges shallowly concave, first dorsal spine leans back; uniformly dark brown except white lips, nasal flap margins, fin axils and inner clasper margins.

○ **Sailfin Roughshark** *Oxynotus paradoxus*
page 206

East Atlantic; 265–720m. Small almost circular spiracles; dorsal fins very tall with narrow pointed tips, trailing edges very concave, first dorsal spine leans back; uniformly dark brown or blackish.

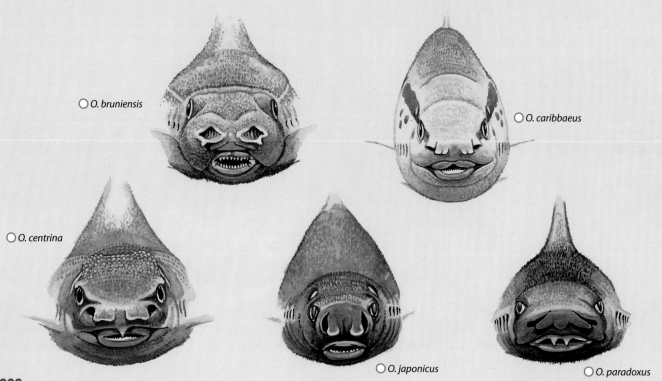

○ *O. bruniensis*

○ *O. caribbaeus*

○ *O. centrina*

○ *O. japonicus*

○ *O. paradoxus*

○ *Oxynotus bruniensis*

○ *Oxynotus caribbaeus*

○ *Oxynotus centrina*

○ *Oxynotus japonicus*

○ *Oxynotus paradoxus*

20cm

Oxynotidae: roughsharks

Identification Five unmistakable species of small sharks. Compressed body, triangular in cross-section with lateral ridges on abdomen and rough skin from large, prickly and close-set denticles. Two high sail-like spined dorsal fins. Rather broad flattened head with flat blunt snout, small thick-lipped mouth encircled by elongated labial furrows, large close-set nostrils, and large to enormous spiracles close behind eyes. Small, spear-like upper teeth form a triangular pad; lower teeth highly compressed to form a saw-like cutting edge, in only 9 to 18 rows. No anal fin.

Biology Poorly known deepsea sharks. Scattered distribution, mainly on temperate to tropical continental and insular shelves. May be weak swimmers, relying on large oily livers for buoyancy. Diet mainly small bottom-dwelling invertebrates (worms, crustaceans and molluscs) and fishes. Females bear litters of 7–23 pups.

Status Uncommon bycatch in deepsea bottom fisheries. May be processed for fish meal, liver oil or occasionally human food. Roughsharks are assessed as Data Deficient, Vulnerable, Near Threatened and Least Concern in the IUCN Red List.

Angular Roughshark, *Oxynotus centrina* (p. 205).

Oxynotus
5 species; pages 204–206

PRICKLY DOGFISH *Oxynotus bruniensis* FAO code: **OXB** Plate page 202

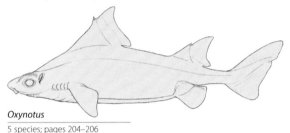

~5mm

Teeth
Upper 14–18
Lower 11–13

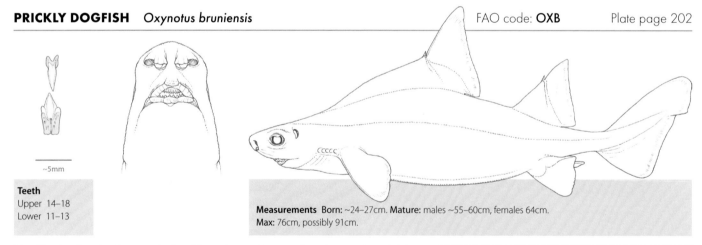

Measurements Born: ~24–27cm. **Mature:** males ~55–60cm, females 64cm. **Max:** 76cm, possibly 91cm.

Identification Spiracles almost circular. Dorsal fins with triangular tips; trailing edges straight or slightly concave. First dorsal spine leans slightly forward. Uniform light grey-brown in colour, no prominent markings.

Distribution Temperate southwest Pacific and southeast Indian Ocean. Locally common off New Zealand and adjacent submarine ridges and seamounts, occasionally caught off southern Australia.

Habitat Deep water on outer continental and insular shelves and upper slopes, 45–1120m, most common at 350–650m.

Behaviour Unknown.

Biology Viviparous, litters of 7–8 pups reported. Feeds exclusively on chimaera eggcases.

Status IUCN Red List: **Near Threatened**. Taken as bycatch in bottom trawls, probably discarded.

CARIBBEAN ROUGHSHARK   *Oxynotus caribbaeus*

FAO code: **OXC**   Plate page 202

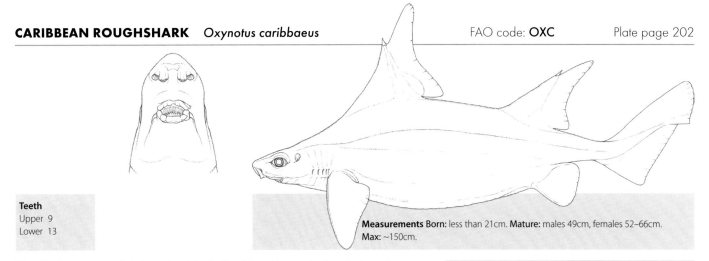

Teeth
Upper 9
Lower 13

Measurements Born: less than 21cm. **Mature:** males 49cm, females 52–66cm.
Max: ~150cm.

Identification Spiracles relatively small and circular. Dorsal fins with narrowly triangular tips and concave trailing edges. First dorsal spine leans forward. Grey or brownish with pattern of dark bands on a light background. Dark blotches and small spots on head, body, caudal fin and fins, separated by prominent light areas over pectoral and pelvic fins.
Distribution West central Atlantic: Gulf of Mexico, Caribbean coast of Central America and Venezuela.
Habitat Bottom on upper continental slopes, 218–579m, water temperature 9.4–16.4°C. Observations by deepsea submersibles reveal this species occurs over soft and hard bottoms, often hovering over rough outcrops of rubble, rock and boulders.
Behaviour Unknown.
Biology Unknown.
Status IUCN Red List: Least Concern. Not important in fisheries.

ANGULAR ROUGHSHARK   *Oxynotus centrina*

FAO code: **OXY**   Plate page 202

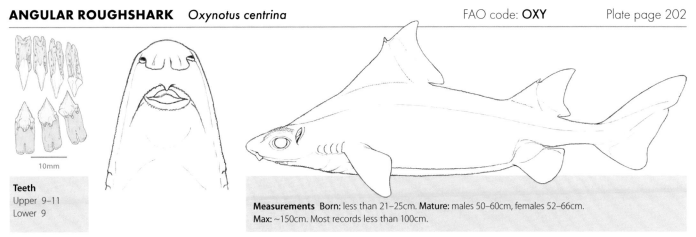

10mm

Teeth
Upper 9–11
Lower 9

Measurements Born: less than 21–25cm. **Mature:** males 50–60cm, females 52–66cm.
Max: ~150cm. Most records less than 100cm.

Identification Ridges over eyes expanded into huge, rounded knobs covered with enlarged denticles (absent in other roughsharks). Spiracles very large and vertically elongated, almost as high as eye-length. First dorsal spine leans slightly forward. Grey or grey-brown. Darker blotches on head and sides (not always clear in adults); light horizontal line crosses cheek below eye.
Distribution East Atlantic: Norway to South Africa and Mediterranean (not Black Sea). Southwest Indian Ocean: off Mozambique and Madagascar.
Habitat Coralline algal and muddy bottom on continental shelves and upper slopes, 35–805m, mostly below 100m.   **Behaviour** Unknown.
Biology Viviparous, 7–8 (Angola) to 23 (Mediterranean) pups per litter, with a 3–12 month gestation. Eats worms, crustaceans and molluscs, and eggcases of catsharks and skates.
Status IUCN Red List: **Vulnerable** globally and in the northeast Atlantic. **Critically Endangered** in the Mediterranean. Rare to uncommon. Minor offshore trawl bycatch.

JAPANESE ROUGHSHARK *Oxynotus japonicus*
FAO code: **OXZ**   Plate page 202

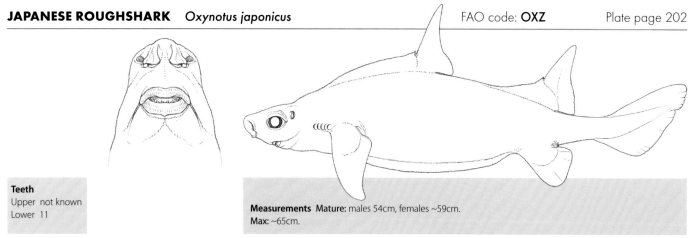

Teeth
Upper not known
Lower 11

Measurements Mature: males 54cm, females ~59cm.
Max: ~65cm.

Identification Spiracles vertically oval, less than half eye-length. Dorsal fins with narrow, triangular tips and shallowly concave trailing edges. First dorsal spine leans slightly back. Uniform dark brown body except for white lips, nasal flap margins, fin axils and inner clasper margins.
Distribution Northwest Pacific: southern Japan (Suruga Bay, Honshu), Ryuku Islands (Okinawa) and northeast Taiwan.
Habitat Upper continental slope, 150–350m. Appears to prefer muddy or sandy bottoms.
Behaviour Unknown.
Biology Viviparous, but nothing else known.
Status IUCN Red List: Data Deficient. Known from a few individuals taken in a bottom trawl fishery that has depleted other shark populations. This roughshark may also prove to be threatened, since its entire known range is fished.

SAILFIN ROUGHSHARK *Oxynotus paradoxus*
FAO code: **OXN**   Plate page 202

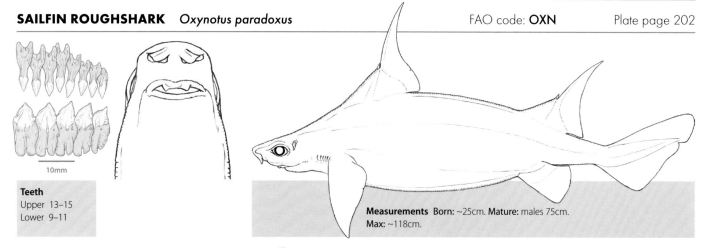

Teeth
Upper 13–15
Lower 9–11

10mm

Measurements Born: ~25cm. **Mature:** males 75cm.
Max: ~118cm.

Identification Spiracles relatively small and almost circular. Dorsal fins tall, narrow and pointed with very concave trailing edges. First dorsal spine leans back. Blackish or dark brown without prominent markings.
Distribution Eastern Atlantic: Scotland to Senegal, including Azores and Canaries and possibly Gulf of Guinea. Absent from the Mediterranean.
Habitat Continental slope, 265–720m.
Behaviour Unknown.
Biology Viviparous, but nothing else known. Likely feeds on bottom invertebrates.
Status IUCN Red List: Data Deficient. Uncommon bycatch of bottom trawls (possibly also longliners) targeting deep-benthic squaloid dogfishes, used for fish meal.

Dalatiidae: kitefin sharks

Ten species in seven genera of dwarf to medium-sized deepsea sharks (*Dalatias*, *Euprotomicroides*, *Euprotomicrus*, *Heteroscymnoides*, *Isistius*, *Mollisquama*, *Squaliolus*), distributed almost worldwide in open ocean or on bottom in mostly temperate to tropical seas. Some species are wide-ranging, others restricted to single ocean basins or ridges, (but might prove to be more widespread with additional sampling).

Identification Head narrow and conical, snout short. Strong jaws with small spear-like upper teeth, large blade-like interlocked lower teeth with smooth or serrated (*Dalatias* only) edges. Pectoral fins with short, broadly rounded free rear tips. Two dorsal fins without spines or with first dorsal spine only (*Squaliolus* species). Second dorsal varies from slightly smaller to much larger than first. No anal fin. Caudal fin with long upper lobe, long to very short or absent lower lobe, and well-developed subterminal notch.

Biology Poorly known deepsea and deep midwater species. The sharks of this family are all viviparous (with yolk-sac); the larger species may produce litters containing as many as three to 16 pups. Some species are powerful predators, others are ectoparasites; some are solitary and others occur in aggregations for at least part of their lifecycle. Some cookiecutter sharks are luminescent, like the lanternsharks (p. 160). However, taxonomists believe that these two families developed their ability to emit bioluminescence independently, at different periods during their evolutionary history; the structure and organisation of the photophores, and the role of their luminescence, is not the same. Like lanternsharks, cookiecutter sharks have photophores on their ventral surfaces, producing counterillumination that will hide them from predators looking up through the water. However, there is a significant difference: the 'collars' below the jaw of cookiecutters lacks these structures. This gap leaves a small dark area which, from below, could look like a small fish. Perhaps cookiecutters uses this as a lure to attract large predators close enough to ambush.

Status One threatened species, *Dalatias licha*, is important in target and bycatch fisheries for meat (which is used for human consumption and/or fishmeal) and squalene oil from its large liver. Others are too small to be of value and are assessed as Least Concern.

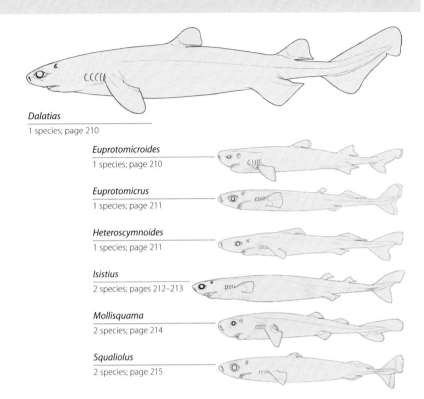

Dalatias
1 species; page 210

Euprotomicroides
1 species; page 210

Euprotomicrus
1 species; page 211

Heteroscymnoides
1 species; page 211

Isistius
2 species; pages 212–213

Mollisquama
2 species; page 214

Squaliolus
2 species; page 215

Kitefin Shark, *Dalatias licha*, photographed by the E/V *Nautilus* Exploration Program, Ocean Exploration Trust, at Gorringe Bank off Portugal (p. 210).

Plate 19 DALATIIDAE

Narrow conical head with short snout; pectoral fins with short broad round free rear tips, two dorsal fins without spines, no anal fin, caudal fin with long upper lobe and well-developed terminal notch.

○ **Kitefin Shark** *Dalatias licha*　　　　　　　　　　　　　　　　　　　　　　　　　　　　　　　　　　　　page 210

Atlantic, Indian and Pacific Oceans; 37–1800m. Medium-sized cylindrical; short blunt snout, thick fringed lips; dorsal fins equal sized, first dorsal fin base closer to pectoral fins than pelvic fins, weak caudal fin lower lobe; uniform brown to blackish.

○ **Taillight Shark** *Euprotomicroides zantedeschia*　　　　　　　　　　　　　　　　　　　　　　　　　page 210

South Atlantic and southeast Pacific; 75–641m. Tiny cylindrical; short blunt snout, thick fringed lips, very long fifth gill slit; luminous gland; first dorsal fin base closer to pelvic than pectoral fins, second dorsal fin origin in front of pelvic fins, caudal fin paddle-shaped; all fins with prominent white margins.

○ **Pygmy Shark** *Euprotomicrus bispinatus*　　　　　　　　　　　　　　　　　　　　　　　　　　　　　page 211

South Atlantic, Indian and Pacific Oceans; from surface to deeper than 1800m. Tiny cylindrical; bulbous snout, large eyes, gills tiny; luminous gland; low lateral keels on caudal peduncle; first dorsal fin tiny flag-like, second dorsal fin origin well behind pelvic fin, caudal fin paddle-shaped nearly symmetrical; black.

○ **Longnose Pygmy Shark** *Heteroscymnoides marleyi*　　　　　　　　　　　　　　　　　　　　　　　page 211

Southern oceans; patchy; possibly surface to 502m. Dwarf cylindrical; very long bulbous snout, thin unpleated lips, small gills; first dorsal fin origin over pectoral fins, second dorsal fin slightly larger than first dorsal, caudal fin paddle-shaped nearly symmetrical with a strong lower lobe; uniformly dark brown.

○ **Cookiecutter Shark** *Isistius brasiliensis*　　　　　　　　　　　　　　　　　　　　　　　　　　　　page 212

Atlantic, Indian and Pacific Oceans; 0–3700m. Small, mid grey to grey-brown, cigar-shaped; very short bulbous snout; ventral luminous organs except on prominent dark collar and light-edged fins; first dorsal fin base over pelvic fin origins, pectoral fins larger than dorsal fins, nearly symmetrical paddle-shaped caudal fin.

○ **Largetooth Cookiecutter Shark** *Isistius plutodus*　　　　　　　　　　　　　　　　　　　　　　　page 213

West and northeast Atlantic and west Pacific Oceans; 30–2060m. Similar to *I. brasiliensis* but larger jaws and teeth, caudal fin smaller and no/faint collar markings.

○ **American Pocket Shark** *Mollisquama mississippiensis*　　　　　　　　　　　　　　　　　　　　page 214

West Atlantic: Gulf of Mexico; 5–580m over depth of 3038m. Small cylindrical; short snout; no dorsal spines; distinctive 'pocket' gland behind pectoral fin; slender caudal fin. Greyish brown body with darker lateral line and lighter fin edges.

○ **Pocket Shark** *Mollisquama parini*　　　　　　　　　　　　　　　　　　　　　　　　　　　　　　　page 214

Southeast Pacific; 330m. Small; short, blunt conical snout with thick, fringed lips; distinctive 'pocket' gland behind pectoral fin; dorsal fins spineless nearly equal-sized, first set just anterior to pelvic fins; caudal fin asymmetrical, not paddle-shaped; dark brown with small light spots on body and white fin margins.

○ **Smalleye Pygmy Shark** *Squaliolus aliae*　　　　　　　　　　　　　　　　　　　　　　　　　　　　page 215

West Pacific and Central Indo-Pacific; 0–2000m. Dwarf spindle-shaped; long bulbous conical snout, lateral papillae on upper lips, eyes smaller than *S. laticaudus*; photophores on upper surface; only first dorsal fin with spine, second dorsal fin base more than twice first, caudal fin paddle-shaped; prominent light fin margins.

○ **Spined Pygmy Shark** *Squaliolus laticaudus*　　　　　　　　　　　　　　　　　　　　　　　　　　page 215

Atlantic, Indian and Pacific Oceans; epipelagic, 10–1800m. Very similar to *S. aliae* but has larger eye with broad arched upper eyelid, lateral papillae absent or weak.

○ *Dalatias licha*

20cm

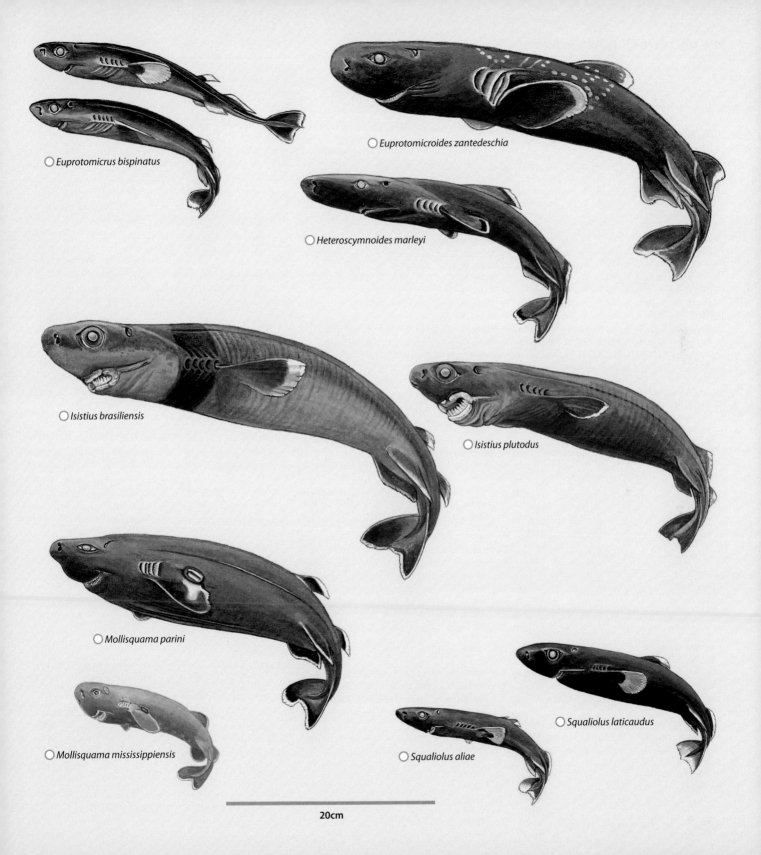

○ *Euprotomicrus bispinatus*

○ *Euprotomicroides zantedeschia*

○ *Heteroscymnoides marleyi*

○ *Isistius brasiliensis*

○ *Isistius plutodus*

○ *Mollisquama parini*

○ *Mollisquama mississippiensis*

○ *Squaliolus aliae*

○ *Squaliolus laticaudus*

20cm

KITEFIN SHARK *Dalatias licha*

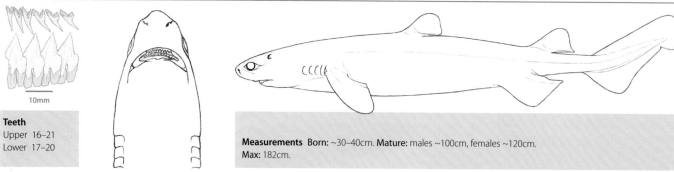

Teeth
Upper 16–21
Lower 17–20

Measurements **Born:** ~30–40cm. **Mature:** males ~100cm, females ~120cm. **Max:** 182cm.

Identification A medium-sized, cylindrical shark. Short, blunt snout. Thick, fringed lips; serrated lower teeth. Spineless dorsal fins; first originates behind pectoral fin free rear tips with base closer to pectoral than pelvic fin bases; second dorsal fin larger. Weak lower caudal lobe. Brown to blackish in colour. Posterior margins of most fins are translucent.
Distribution Atlantic, Indian and Pacific Oceans.
Habitat Deep water (37–1800m, mainly deeper than 200m), warm-temperate and tropical outer continental and insular shelves and slopes, usually on or near bottom.
Behaviour Hovers above bottom (large oil-filled liver provides neutral buoyancy) and swims well off the bottom. Hunters mainly of deepsea fishes, including other sharks, may take bites out of large live prey.
Biology Poorly known. Viviparous, 3–16, pups per litter, averaging 6–8, with reproductive activity occurring throughout the year. A voracious feeder on deepsea fishes, including other sharks and rays.
Status IUCN Red List: **Vulnerable**. Target and utilised bycatch fisheries for meat and squalene have seriously depleted distinct regional populations in southeast Australia, the Mediterranean and Northeast Atlantic. European fisheries were closed due to evidence of collapse and the regional Red List Assessment is Endangered. The Kitefin Shark is also susceptible to fisheries in most other parts of its range.

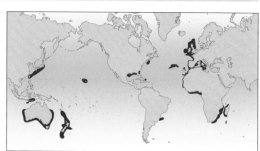

TAILLIGHT SHARK *Euprotomicroides zantedeschia*

Teeth
Upper 29
Lower 29–34

Measurements **Born:** less than 17.6cm. **Mature:** males 41.6–45.5cm, largest female 53cm. **Max:** 53cm.

Identification Compressed body. Short snout with thick, fringed lips. Very wide fifth gill slit (less than twice length of first). First dorsal fin insertion about equidistant between pectoral and pelvic fin bases; slightly smaller than second dorsal; second dorsal origin in front of pelvic origins (behind it in other genera). Large, lobate pectoral fins. A midventral keel on caudal peduncle. Cloaca greatly expanded as a luminous gland containing yellow papillae. Paddle-shaped caudal fin. Colour blackish brown. All fins with prominent white margins.
Distribution South Atlantic, and southeast Pacific off Juan Fernandez Islands, Chile.
Habitat Epipelagic, three of four known specimens were caught in midwater trawls at 75m over very deep water of at least 2000m, while the fourth was caught in a bottom trawl on continental shelf between 458–641m.
Behaviour Unknown.
Biology Unknown, probably viviparous, few young.
Status IUCN Red List: **Least Concern**. Because of low fishing pressure. Four specimens recorded.

PYGMY SHARK *Euprotomicrus bispinatus*

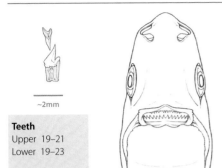

Teeth
Upper 19–21
Lower 19–23

~2mm

Measurements Born: 6–10cm. **Mature:** males 17–19cm, females 22–23cm.
Max: 27cm.

Identification Tiny, cylindrical shark. Bulbous snout. Large eyes. Gill slits tiny. Tiny, flag-like first dorsal fin, quarter length of second; set well behind pectoral fins. Low lateral keels on caudal peduncle. Caudal fin nearly symmetrical, paddle-shaped. Black body. Light-edged fins and luminous organs on underside of body.
Distribution South Atlantic, Indian and Pacific Oceans. Oceanic and amphitemperate.
Habitat Epipelagic, mesopelagic, perhaps bathypelagic in mid-ocean, 0–1829m, and possibly to 9938m. Migrates from surface at night to deeper than 1500m (at least to midwater, perhaps bottom) by day.
Behaviour Eats deepsea squids, bony fishes, some crustaceans, not large prey.
Biology Viviparous, 8 pups per litter.
Status IUCN Red List: Least Concern. This species' small size and pelagic habitat mean that it is unlikely to be captured in most fisheries operating within its broad distribution.

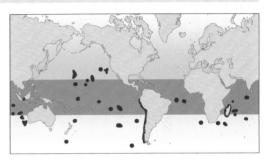

LONGNOSE PYGMY SHARK *Heteroscymnoides marleyi*

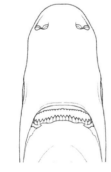

Teeth
Upper 22
Lower 23

Measurements Born: ~12.8cm. **Mature:** males 36cm, females 33.3cm.
Max: males ~37cm.

Identification Dwarf shark. Very long, bulbous snout. Thin, unpleated lips. Small gill slits. First dorsal fin far forward, originating over pectoral bases. Second dorsal only slightly larger than first. Caudal fin almost symmetrical, paddle-shaped with a strong lower lobe. Dark brown. Dark fins with light margins.
Distribution Patchy distribution in southern oceans. May be more wide-ranging.
Habitat Oceanic. Epipelagic in cold, subantarctic, oceanic waters and cold current systems (Benguela and Humboldt). Recorded between surface and 502m in areas 830–4000m and deeper.
Behaviour Unknown.
Biology Presumably viviparous, likely small litters. Prey unknown, presumably small pelagic fishes and invertebrates.
Status IUCN Red List: Least Concern. Only six specimens recorded, from three widely separated locations. Likely more widely distributed but has almost no interactions with fisheries and there are no other known threats.

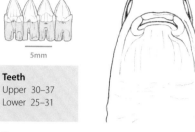

Teeth
Upper 30–37
Lower 25–31

Measurements Born: 14–15cm. **Mature:** males ~31–37cm, females 38–44cm.
Max: males at least 42cm, females at least 56cm.

Alternative name Cigar Shark

Identification Small, cigar-shaped shark. Very short, bulbous snout. Suctorial lips; large, triangular lower teeth in 25–31 rows. Dorsal fins set far back; first dorsal base above pelvic origins. Pelvic fins larger than dorsal fins. Large, nearly symmetrical, paddle-shaped caudal fin with long lower lobe. Luminous organs cover the whole of the lower surface, except for fins and collar, and can glow bright green. Mid-grey or grey-brown in colour. Light-edged fins and prominent dark collar mark around the throat. Cookiecutter Sharks have the largest teeth proportional to their body size of any known shark species, including the White Shark *Carcharodon carcharias* p. 320.

Distribution Scattered throughout the Atlantic, southern Indian and Pacific Oceans.

Habitat A wide-ranging tropical oceanic shark, epipelagic to bathypelagic. Only caught at night, sometimes at surface but usually over very deep water from 0–3700m, frequently near islands.

Behaviour Poor swimmers, they probably migrate vertically from deep water (2000–3000m) to midwater or the surface at night. These relatively small, cigar-shaped sharks are considered ectoparasitic; they are well-known for attacking much larger prey by sneaking up and ambushing them, or perhaps their bioluminescent light organs may attract their intended victims. Whatever the strategy, these are fearless, extremely voracious predators whose large victims include cetaceans, pinnipeds, large fishes, other sharks and (on at least one occasion) a deepsea stingray *Plesiobatis daviesi*. Once close enough, the Cookiecutter Shark's thick lips and modified pharynx are used to attach itself to the prey, then the razor-sharp lower teeth cut into the skin and, with a twisting movement of the body, a plug of flesh is excised. The shark then pulls free, holding the plug by its hook-like upper teeth and leaving behind a crater wound. A dwarf shark that was likely this species nipped an open-ocean swimmer off Hawaii. This species is also reported to have attacked rubber sonar domes on nuclear submarines. Can be active and bite when caught.

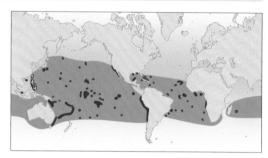

Biology Viviparous with yolk-sac, 6–9 pups per litter. The large oil-filled liver and body cavity and small fins suggest neutral buoyancy (or may compensate for the weight of the highly calcified skeleton). Lower teeth are swallowed as they are replaced in rows (perhaps to recycle calcium). In addition to biting chunks out of large prey, as described above, the Cookiecutter Shark also eats deepsea fishes, squids and crustaceans.

Status IUCN Red List: Least Concern. Too small to be taken by most fisheries, but occasional bycatch in pelagic fish and squid fisheries.

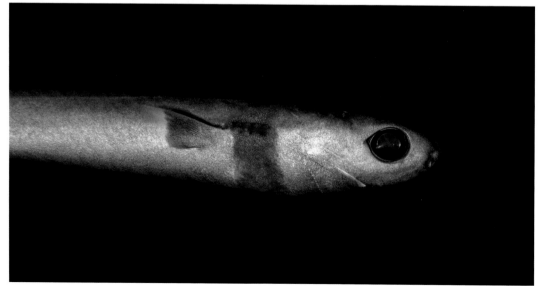

Cookiecutter Shark, *Isistius brasiliensis*, Big Island, Hawaii.

Teeth
Upper 21–29
Lower 17–19

5mm

Measurements Mature: males ~34cm, females 42cm.
Max: females at least 42cm.

Identification Larger jaws and bigger mouth than *Isistius brasiliensis* and only 17–19 rows of enormous lower teeth. Eyes set well forward on short-snouted head provide binocular vision. Small asymmetric caudal fin with short lower lobe. No collar marking or a weak collar marking on throat.
Distribution Possibly circumglobal. Scattered records around the Atlantic and west Pacific.
Habitat Epipelagic, possibly bathypelagic, found from 30–2060m over water up to 6440m deep. Known from a few scattered localities.
Behaviour Smaller fins suggest this is a less active swimmer than *I. brasiliensis*. Feeds in a similar manner, can take even larger and more elongated (twice as long as diameter of mouth) bites out of bony fishes and may do likewise to other prey.
Biology Unknown.
Status IUCN Red List: Least Concern. Although rarely recorded, this species probably has a circumglobal distribution but is infrequently caught in commercial fisheries.

Humpback Whale calf, *Megaptera novaeangliae*, with round scars from Cookiecutter Shark, *Isistius brasiliensis* (p. 212), Tahiti, Society Islands, French Polynesia.

AMERICAN POCKET SHARK   *Mollisquama mississippiensis*   FAO code: **SHX**   Plate page 208

 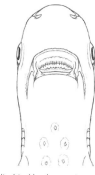 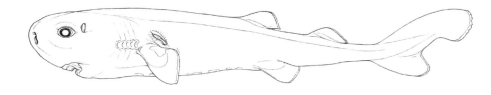

Teeth
Upper 21
Lower 31

~10mm

Measurements Max: at least 14cm.

Identification A small, cylindrical body tapering towards a slender caudal peduncle. A distinctive pocket-like gland in the pectoral fin area. Fin spines absent, and no caudal keels or precaudal pits. Caudal fin asymmetrical. Colour a greyish brown with a dark lateral line against a lighter body background. Lighter fin edges.
Distribution West Atlantic: central Gulf of Mexico.
Habitat Midwater from 5–580m over 3038m depth.
Behaviour Unknown.
Biology Only known from a single immature male, nothing else known.
Status IUCN Red List: Least Concern. The only known specimen was caught in a midwater survey research trawl net over very deep water. None of the commercial fisheries operating in the area are likely to catch this very small shark.

POCKET SHARK   *Mollisquama parini*   FAO code: **SHX**   Plate page 208

 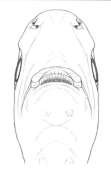 

Teeth
Upper 23
Lower 31

~10mm

Measurements Max: at least 40cm (adolescent female).

Identification Small shark. Short, blunt, conical snout. Thick, fringed lips. Medium-sized gill slits, the fifth almost twice length of first. First dorsal fin set well behind pectoral fins, base just in front of pelvic bases; second dorsal about as large as first. A large pocket-like gland with a conspicuous slit-like opening just above each pectoral base (possibly secreting a pheromone or luminous fluid). Caudal fin asymmetrical, not paddle-shaped, with a weak lower lobe. Dark brown body. Dark fins with light margins.
Distribution Southeast Pacific.
Habitat Only known specimen was collected by a research vessel on the Nazca submarine ridge (about 1200km east of Chile) at 330m depth over very deep water.
Behaviour Unknown.
Biology Unknown.
Status IUCN Red List: Least Concern. This species' small size and midwater habitat make it almost uncatchable in commercial fisheries.

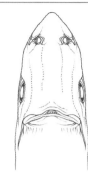

Teeth
Upper 23
Lower 21

Measurements Born: less than 10cm. **Mature:** males ~15cm, females 22cm.
Max: males ~24cm. One of the smallest living sharks.

Identification Spindle-shaped dwarf shark. Long, bulbous, conical snout. Pair of lateral papillae on upper lip partially cover teeth. Eyes smaller than in *Squaliolus laticaudus*; upper eyelid angular. Fin spine (sometimes covered by skin) only on first dorsal fin; origin opposite inner margins or free rear tips of pectoral fins; second dorsal base over twice length of first. Paddle-shaped caudal fin. Colour blackish with prominent light fin margins. Ventral surface covered with photophores.
Distribution West Pacific and Central Indo-Pacific.
Habitat Epipelagic or mesopelagic near land, depth range 0–2000m.
Behaviour May migrate daily from shallow depths at night to deeper water by day.
Biology Virtually unknown. Viviparous, litter size unknown. Feeds on small midwater fishes, cephalopods, and small crustaceans.
Status IUCN Red List: Least Concern. Wide-ranging, considered by some authorities too small to be threatened by fisheries, but a common bycatch of bottom trawl fisheries in the west Pacific.

SPINED PYGMY SHARK *Squaliolus laticaudus* FAO code: **QUL** Plate page 208

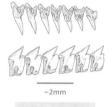

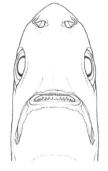

Teeth
Upper 22–23
Lower 16–21

Measurements Born: 8–10cm. **Mature:** males ~15cm, females 17–20cm.
Max: males ~22cm, females ~28cm.

Identification Very similar to *Squaliolus aliae* but has larger eyes with a broadly arched upper eyelid and lateral papillae absent or weakly developed on upper lip. Ventral surface covered with photophores.
Distribution Atlantic, Indian and Pacific Oceans. Oceanic, nearly circumtropical.
Habitat Wide-ranging, tropical, epipelagic species. Highly productive areas near continental and island land masses, sometimes over shelves, usually over slopes, not in mid-ocean. The bottom temperatures where this shark has been caught range from 11.2–26.3°C.
Behaviour Migrates vertically from 200m (night) to 500m (day); caught on the bottom at 700–750m and between 10–80m over water deeper than 1200m, but not caught at surface. Photophores may eliminate shadow caused by light from above.
Biology Viviparous with yolk-sac, 3–5 pups per litter. Feeds on vertically migrating deepsea squids and fishes.
Status IUCN Red List: Least Concern. Despite being very widely distributed, this tiny shark is only occasionally caught by commercial fisheries because of its small size and largely unfished mesopelagic habitat. It also has a refuge from fisheries in very deep water.

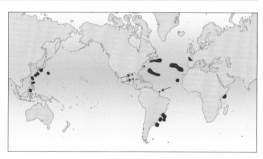

Plate 20 PRISTIOPHORIDAE I

Slender distinctive sharks; flat heads, long flat saw-like snout with long ventral barbels and close-set rows of lateral and ventral sawteeth, large spiracles; thick lateral ridges on caudal peduncle; long caudal fin upper lobe.

○ **Anna's Sixgill Sawshark** *Pliotrema annae* page 222

Western Indian Ocean; shallow shelf; 20–35m. On or near bottom. Slender body; short snout (head <35% TL); barbels midway between snout tip and mouth; lateral teeth 10–11 in front of barbels, 6–7 behind; 6 gills; unstriped medium to dark brown body, white belly with indistinct dark blotches; white-edged posterior fin margins.

○ **Kaja's Sixgill Sawshark** *Pliotrema kajae* page 222

Western Indian Ocean; upper slopes; 214–320m. Slender body; long snout (head >38% TL); barbels midway between snout tip and mouth; lateral teeth 12–14 in front of barbels, 8–17 behind; 6 gills; pale to light brown body with 2 yellowish stripes; white below; narrow white posterior fin margins.

○ **Warren's Sixgill Sawshark** *Pliotrema warreni* page 223

Southeast Atlantic and southwest Indian Ocean; 10–915m. On or near bottom. Slender body; long snout; barbels closer to mouth than other sixgill sawsharks; barbs on posterior edges of larger rostral teeth; lateral teeth 14–18 in front of barbels, 6–19 behind; dark brown body with yellowish longitudinal stripe, uniform white below; translucent dusky fins with narrow white edges.

○ **Longnose Sawshark** *Pristiophorus cirratus* page 223

South Australia; 0–630m. On or near bottom. Stocky sawshark; lateral teeth 9–15 in front of barbels, 9–10 behind; barbels much closer to rostral tip than mouth; pale yellow to greyish brown with sometimes faint blotches spots and bars, dark brown rostral midline stripes and edges, tooth margins blackish.

○ **Japanese Sawshark** *Pristiophorus japonicus* page 224

Northwest Pacific; 50–1240m. On or near bottom. Stocky sawshark; lateral teeth 25–32 in front of barbels, 8–16 behind; barbels closer to mouth than rostral tip; brown or reddish brown, dark brown rostral midline stripes and edges.

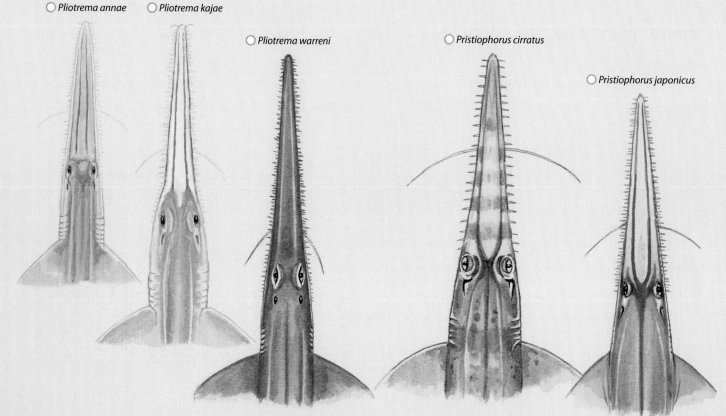

○ *Pliotrema annae*   ○ *Pliotrema kajae*   ○ *Pliotrema warreni*   ○ *Pristiophorus cirratus*   ○ *Pristiophorus japonicus*

Pliotrema annae

Pliotrema kajae

Pliotrema warreni

Pristiophorus cirratus

Pristiophorus japonicus

20cm

Plate 21 PRISTIOPHORIDAE II

○ **Tropical Sawshark** *Pristiophorus delicatus*

East Australia; 239–511m. Slender sawshark; lateral teeth 12–15 in front of barbels, 6–9 behind, juveniles usually with 2–3 smaller teeth between larger teeth; barbels slightly closer to mouth than rostral tip or equidistant; uniformly pale yellow-brown above, no spots or bars.

○ **Lana's Sawshark** *Pristiophorus lanae*

Philippines; 229–593m. On or near bottom. Slender sawshark; lateral teeth 17–26 in front of barbels, 6–17 behind; juveniles have 2–3 smaller teeth between larger teeth; barbels slightly closer to mouth than rostral tip; uniformly dark brown above, no spot or bars.

○ **African Dwarf Sawshark** *Pristiophorus nancyae*

West Indian Ocean; 286–570m. Small sawshark; 2 rows of 4–5 pits beneath rostrum, prominent ridges on large teeth bases; lateral teeth 14–22 in front of barbels, 6–10 behind; barbels much closer to mouth than rostral tip. Broad triangular first dorsal fin; brown above with pale rostrum, dark brown rostral midline and edges, dark anterior edges to pectoral and dorsal fins (conspicuous in juveniles).

○ **Shortnose Sawshark** *Pristiophorus nudipinnis*

South Australia; 0–165m. On or near bottom. Stocky sawshark; lateral teeth 12–13 in front of barbels, 6 behind; barbels closer to mouth than rostral tip; uniformly slate grey above, indistinct dusky rostral midline stripes and edges.

○ **Bahamas Sawshark** *Pristiophorus schroederi*

Northwest Atlantic; 438–952m. On or near bottom. Slender sawshark; 13 lateral teeth in front of barbels, 10 behind, juveniles with 1 smaller tooth between larger teeth; barbels midway between mouth and rostral tip; uniformly light grey above, dark brown rostral midline stripes and edges, juveniles with dark anterior edges to dorsal fins.

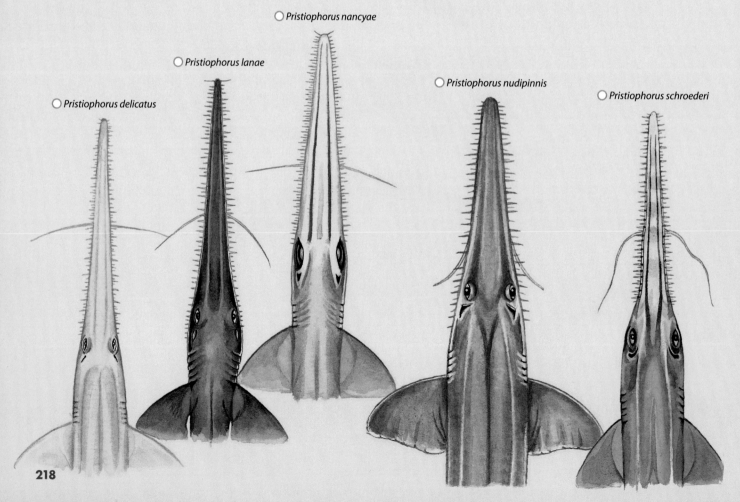

○ *Pristiophorus nancyae*

○ *Pristiophorus lanae*

○ *Pristiophorus delicatus*

○ *Pristiophorus nudipinnis*

○ *Pristiophorus schroederi*

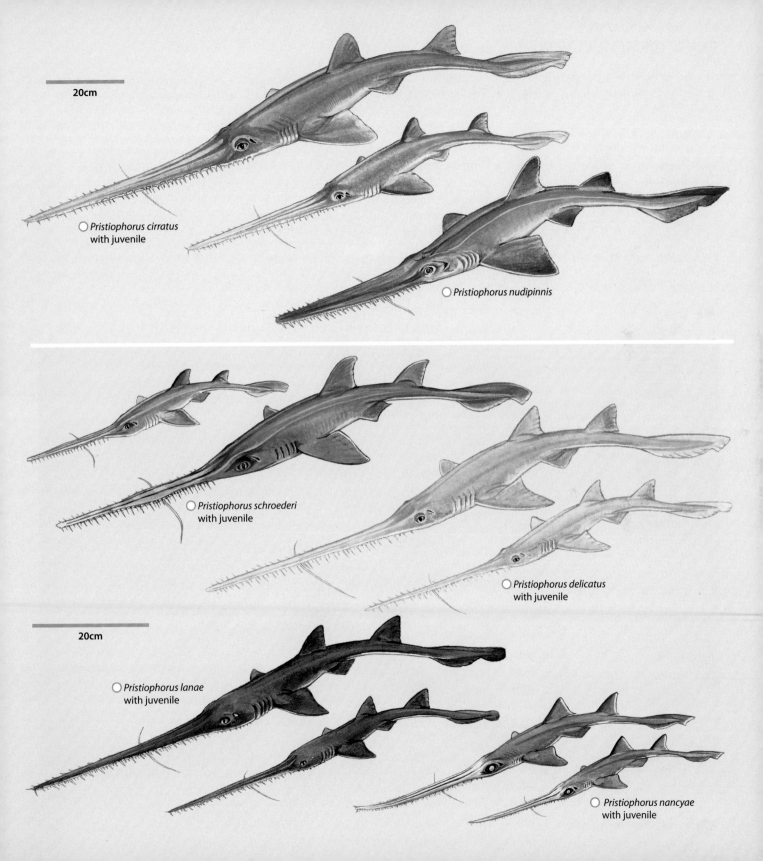

20cm

○ *Pristiophorus cirratus*
with juvenile

○ *Pristiophorus nudipinnis*

○ *Pristiophorus schroederi*
with juvenile

○ *Pristiophorus delicatus*
with juvenile

20cm

○ *Pristiophorus lanae*
with juvenile

○ *Pristiophorus nancyae*
with juvenile

PRISTIOPHORIFORMES   sawsharks

This order contains one family of little-known small sharks, distributed worldwide in the fossil record, but now found only on the continental and insular shelves and upper slopes of the northwest and southeast Atlantic, west Indian and west Pacific Oceans. They are found in shallow water in temperate regions, deeper in the tropics, and sometimes occur in large schools or feeding aggregations. At least one species segregates by depth (adults in deeper water than young). Some have a very restricted distribution. Ten species.

Identification Small, slender sharks (most less than 150cm total length, maximum about 153cm, possibly to 170cm) with cylindrical bodies, flattened heads and a long, flat, saw-like snout with a pair of long, string-like ventral barbels in front of the nostrils and close-set rows of lateral and ventral sawteeth. Eyes on the side of the head, large spiracles. Two spineless dorsal fins, no anal fin. Thick lateral ridges on the caudal peduncle, a long upper caudal lobe and no lower lobe. May be confused with the sawfishes, which are batoids (rays) with flattened bodies, pectoral fins fused to the head, and gill slits underneath the head.

Biology Viviparous; foetuses gain all food from their yolk sac, which is reabsorbed just before the litters of 5–17 pups are born. The large lateral rostral teeth erupt before birth but lie flat against the rostrum until after birth. The tooth-studded rostrum has sensors to detect vibrations and electrical fields and is probably used to capture and kill prey. It may also be used for defence, or when competing or courting with other sawsharks (parallel cuts and scratches occasionally seen on adults are presumably from interactions with other sawsharks). The long rostral barbels may have taste, touch or other sensors and may be trailed along the bottom to locate prey, including small fishes, crustaceans and squids.

Status Harmless, despite the very sharp (non-toxic) teeth; handle with care. Some species are common where they occur and 60% are assessed as Least Concern. They are vulnerable to bycatch where a restricted range overlaps with fisheries, because their saws may easily become entangled in fishing gear including nets and three Data Deficient species may prove to be threatened. Taken as a bycatch in demersal gillnet and bottom trawl fisheries and marketed for food or discarded. Saws may be sold as curios.

Longnose Sawshark, *Pristiophorus cirratus*, southeast Australia (p. 223).

PRISTIOPHORIDAE: sawsharks

Ten species in two genera. The two genera can easily be split based on the number of lateral pairs of gill slits: the *Pliotrema* have six and the *Pristiophorus* have five, but these unusual small sharks are otherwise rather similar in appearance (see above). The sixgill sawshark genus *Pliotrema* was long thought to have just one species, but it is now known to have three, each with relatively restricted distributions in the western Indian Ocean. Unlike the morphologically similar batoid sawfishes, which are all assessed as Endangered or Critically Endangered, most sawsharks are Least Concern. The difference is that the mostly deep water sawsharks appear not to have been seriously affected by target or bycatch fisheries, while most of the sawfishes' shallow coastal range overlaps with intensive human activity and their saws make them extremely vulnerable to net bycatch.

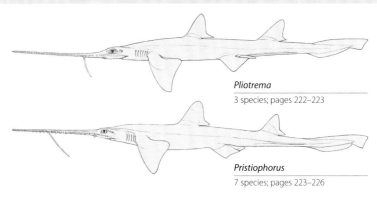

Pliotrema
3 species; pages 222–223

Pristiophorus
7 species; pages 223–226

Sawfish or sawshark?

The family Pristiophoridae, sawsharks, and family Pristidae, sawfish, are named for the toothed rostrum that these species bear, derived from the ancient Greek 'prístēs' for saw, or sawyer. Their similar appearance, however, does not mean that the sawsharks and sawfishes are closely related; not at all. Sawfish are batoids, more closely related to skates and rays (therefore not described in this book, although, confusingly, sawfish are sometimes called carpenter sharks!). Scientists believe that rostral saws evolved completely independently in three families, one of which did not survive the mass extinction about 150mya, at the end of the Cretaceous Period (p. 18). These examples of convergent evolution took place because saw-toothed rostra are useful adaptations for life on the seabed. They can be used to kill larger and more mobile prey than their unsnouted ancestors, and the dangerous armed snouts are also very useful for defence.

Most sawsharks live in deep water and are not kept in aquaria, so there are very few observations of their behaviour. Sawfishes, however, have lived for many years in captivity and their hunting behaviour is well known. Studies of the wear and tear on sawfish and sawshark teeth suggests that their behaviour is very similar. The long rostrum helps them to locate prey hidden on the seabed, because it contains the same sensory organs found along the lateral line and on the snout that can detect pressure changes and electric fields (p. 41). Sawsharks also have the advantage of long, sensitive, electro-sensory and chemo-sensory barbels about halfway along the rostrum, absent in sawfish. Large prey detected on the seabed are pinned down with the rostrum while the animal uses its fins to get into position for the kill. Fishes in the water column can be attacked with swipes and slashes of the rostrum, stunning and knocking them to the seabed or impaling them on the sharp teeth.

Those are the similarities; the obvious differences between sawsharks and sawfishes, apart from the sawsharks' barbels, are the shape and size of the snout and teeth. Sawsharks have a narrow triangular rostrum; sawfish snouts have parallel sides. Sawshark teeth are narrow and conical, rather than wide and blade-like and often alternate in size along the rostrum. The rostral teeth in both families grow continuously from the base, but if a sawshark loses a rostral tooth, it can grow a new one; sawfish cannot do that.

Differences between sawshark and sawfish teeth and rostra.

	Snout shape	Tooth size and shape	Barbels
Sawshark	Elongated isosceles triangle, much broader at base than tip	Alternate small and large conical teeth arranged along the rostrum in most species	Present
Sawfish	Sides almost parallel	Wide blade-like teeth not alternating in size	Absent

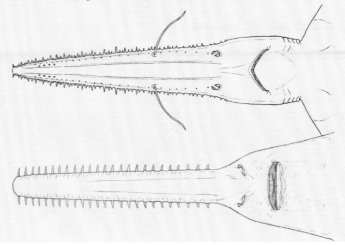

ANNA'S SIXGILL SAWSHARK *Pliotrema annae* FAO code: **PWS** Plate page 216

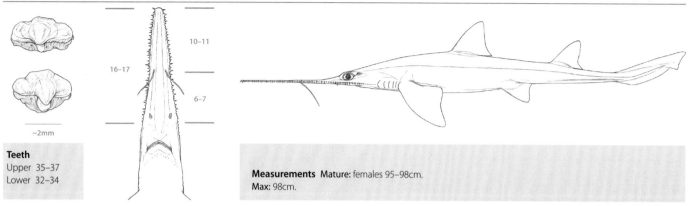

Teeth
Upper 35–37
Lower 32–34

Measurements Mature: females 95–98cm.
Max: 98cm.

Identification Head length less than 35% TL; snout short. Length from rostral tip to barbel origin <13% TL, to eye <22% TL, and to mouth <26% TL. Barbels located halfway from snout tip to mouth. 16–17 large lateral sawteeth: 10–11 in front of barbels, 6–7 behind. Colour dorsally and laterally a uniform medium to dark brown, without longitudinal stripes. Posterior fin margins conspicuously white-edged. White below with a few indistinct dark blotches on belly.
Distribution Known only from Zanzibar (Tanzania), but may occur off Kenya and Somalia.
Habitat Continental shelf at depths of 20–35m.
Behaviour Appears to move into relatively shallow water at night, and into deeper water below 35m during the day. It is frequently landed in fisheries along with deepsea sixgill sharks and spurdogs, coastal species like Tiger Sharks, smoothhounds, and other reefs sharks, and oceanic species such as mako and Silky Sharks.
Biology Viviparous, with 6 eggs found in one adult female. Diet unknown.
Status IUCN Red List: Data Deficient. Rarely recorded. Exposed to intense artisanal coastal fishing effort.

KAJA'S SIXGILL SAWSHARK *Pliotrema kajae* FAO code: **PWS** Plate page 216

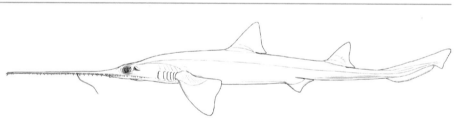

Teeth
Upper 38–43
Lower 35–37

Measurements Born: ~35cm. **Mature:** males 97–102cm, females 117cm.
Max: 143cm.

Identification Head length more than 38% TL; snout long. Length from rostral tip to barbel origin >14% TL, to eye <25% TL, and to mouth <28% TL. Barbels located halfway from snout tip to mouth. 21–31 large lateral sawteeth: 12–14 in front of barbels, 8–17 behind. Colour on dorsal and lateral surface a pale to light brown with two yellowish stripes. Posterior fin margins with narrow white edges. Ventral surface a uniform white.
Distribution Madagascar and Mascarene Ridge.
Habitat Upper slopes from 214–320m.
Behaviour Unknown, the species is known from only about 6 specimens.
Biology Viviparous, litters of up to at least 6. Diet unknown.
Status IUCN Red List: Data Deficient. Most specimens were collected by research surveys. Regional deepwater fishing effort is increasing, but the susceptibility of this sawshark to fisheries bycatch is unknown.

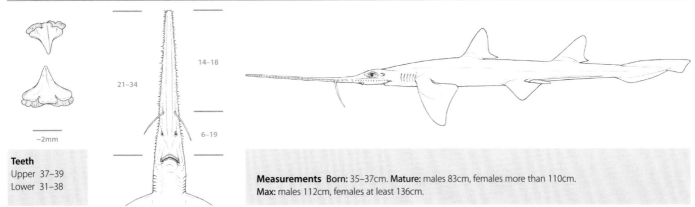

Teeth
Upper 37–39
Lower 31–38

Measurements Born: 35–37cm. **Mature:** males 83cm, females more than 110cm.
Max: males 112cm, females at least 136cm.

Identification Barbels closer to mouth than in other species, about two-thirds way from snout tip to mouth. 21–34 large lateral sawteeth: 14–18 in front of barbels, 6–19 behind. Barbs on posterior edges of larger rostral sawteeth (absent in other sawsharks). Six pairs of gill slits. Dorsal colour dark brown with yellowish longitudinal stripe. Fins a translucent dusky with narrow white edges. Ventral surface uniform white.
Distribution Southwest Indian Ocean: South Africa (most common east of False Bay) to southern Mozambique. Southeast Atlantic records (central Namibia to west coast of South Africa) appear to be vagrants.
Habitat On or near bottom, offshore continental shelf and upper slope from 10–430m, and exceptionally to 915m. Adults usually occur deeper than pups or juveniles. **Behaviour** May use inshore pupping grounds.
Biology Viviparous, 7–17 pups per litter. Feeds on small fishes, crustaceans and squids. Predators include Tiger Sharks. **Status** IUCN Red List: Least Concern. Much of this sawshark's limited southern African range is fished, and it is a bycatch of demersal trawls and possibly longlines. Scientific survey data from South Africa, where trawl fishing effort has declined over the past decade, indicate that the population is increasing.

LONGNOSE SAWSHARK   *Pristiophorus cirratus*   FAO code: **PPC**   Plate page 216

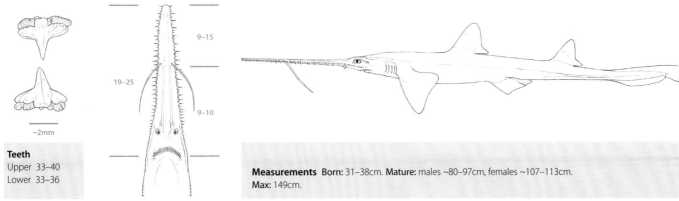

Teeth
Upper 33–40
Lower 33–36

Measurements Born: 31–38cm. **Mature:** males ~80–97cm, females ~107–113cm.
Max: 149cm.

Identification Large, stocky sawshark. Long, narrow rostrum (26–30% TL). Barbels much closer to rostral tip than nostrils. 19–25 large lateral sawteeth: 9–15 in front of barbels, 9–10 behind. Juveniles have 2–3 smaller teeth between large sawteeth. Pale yellow/greyish brown above with variegated (sometimes faint) dark blotches, spots and bars. Darker brown rostral midline and edges, tooth margins blackish. Fin bases blotched. White below.
Distribution Southern Australia. **Habitat** Continental shelf and upper slope, on or near sandy or gravelsand bottom. Occasionally close inshore (bays and estuaries), 40–630m.
Behaviour Occurs in schools or feeding aggregations.
Biology Viviparous with litters of 6–19 pups every other winter. Feeds on small fishes and crustaceans.
Status IUCN Red List: Least Concern. Moderately abundant. Fisheries bycatch levels monitored and stable, used for meat.

TROPICAL SAWSHARK *Pristiophorus delicatus*

FAO code: **PWS**   Plate page 218

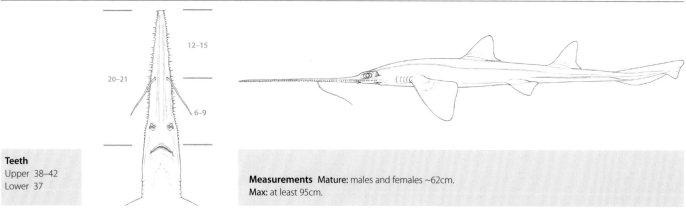

Teeth
Upper 38–42
Lower 37

Measurements Mature: males and females ~62cm.
Max: at least 95cm.

Identification Small, slender sawshark. Rostrum long, narrow, tapering and straight-sided (preoral length 29–31%, length and width at nostrils 5.2–6.0 times preorbital length). Rostral barbels slightly closer to rostral tip than mouth or equidistant between. 20–21 large lateral sawteeth: 12–15 in front of barbels, 6–9 behind. Juveniles usually have 2–3 smaller teeth between large lateral rostral teeth. Uniform pale yellow-brown above, no spots or bars. White below.
Distribution East coast Australia (Queensland).
Habitat Continental slope, 176–511m.
Behaviour Unknown.
Biology Viviparous, otherwise unknown.
Status IUCN Red List: Least Concern. Its range is largely unfished.

JAPANESE SAWSHARK *Pristiophorus japonicus*

FAO code: **PPJ**   Plate page 216

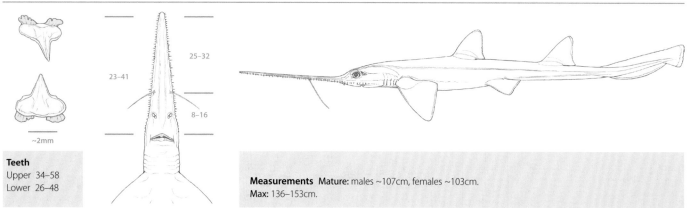

~2mm

Teeth
Upper 34–58
Lower 26–48

Measurements Mature: males ~107cm, females ~103cm.
Max: 136–153cm.

Identification Large and stocky. Moderately long, narrow rostrum (preoral length 26–30%). Barbels closer to mouth than rostral tip. 23–41 large lateral rostral sawteeth: 25–32 before barbels, 8–16 behind. Juveniles with 1–2 smaller teeth between large lateral sawteeth. Plain brown or reddish brown above. Darker brown rostral midline and edges. Juveniles with light trailing edges to pectoral and dorsal fins. White below.
Distribution Northwest Pacific: Japan, Korea, northern China and Taiwan.
Habitat Temperate continental shelves and upper slopes, on or near sand or mud bottom, 50–1240m.
Behaviour Feeds on small bottom fishes and crustaceans; pokes bottom with snout and barbels.
Biology Viviparous, usually 12 pups per litter.
Status IUCN Red List: Data Deficient. A utilised bycatch of many fisheries; particularly vulnerable to entanglement in gillnets. Meat consumed in Japan. Rare throughout its range; possibly already depleted. Kept in aquaria for short periods.

LANA'S SAWSHARK *Pristiophorus lanae* FAO code: **PWS** Plate page 218

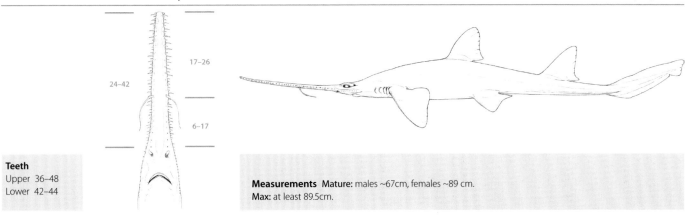

Teeth
Upper 36–48
Lower 42–44

17–26

24–42

6–17

Measurements Mature: males ~67cm, females ~89 cm.
Max: at least 89.5cm.

Identification Small, slender sawshark. Long, very narrow rostrum (preoral length 29–31%, width at nostrils 5.2–6.0 times preorbital length); slightly concave between barbels and nostrils. Barbels slightly closer to mouth than to rostral tip. 24–42 large lateral sawteeth, 17–26 in front of barbels, 6–17 behind. Juveniles with 2–3 smaller teeth between about 21 large lateral teeth. Uniform dark brown above, white below, no spots or bars.
Distribution Philippines: off Apo Island and southern Luzon.
Habitat Upper continental slopes, on bottom, 229–593m.
Behaviour Unknown.
Biology Unknown.
Status IUCN Red List: Near Threatened. May be discarded from the poorly monitored deepsea fisheries operating in its range, but possibly has a refuge in deeper water.

AFRICAN DWARF SAWSHARK *Pristiophorus nancyae* FAO code: **PWS** Plate page 218

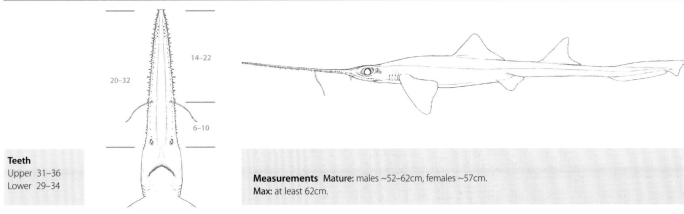

Teeth
Upper 31–36
Lower 29–34

14–22

20–32

6–10

Measurements Mature: males ~52–62cm, females ~57cm.
Max: at least 62cm.

Identification Very small. Two rows of 4–5 enlarged pits on underside (pre-barbel) of rostrum. Barbels much closer to mouth than rostral tip. Prominent ridges on base of large lateral rostral teeth. 20–32 large lateral rostral sawteeth, 14–22 before barbels, 6–10 behind. Broad, triangular first dorsal fin; rear tip extending behind pelvic fin midbases. Uniform brown above, white below. Pale rostrum with dark brown stripes on middle and edges. Pectoral and dorsal fins with dark anterior margins (obvious in juveniles) and prominent light trailing edges.
Distribution West Indian Ocean: southern Mozambique and Madagascar to Somalia and Socotra Islands, possibly to the Arabian Sea off Pakistan.
Habitat Upper continental slope, 286–570m. **Behaviour** Unknown.
Biology Viviparous, but virtually nothing else known. Feeds on small crustaceans.
Status IUCN Red List: Least Concern. Rarely caught in fisheries; further research is needed on its life history and distribution.

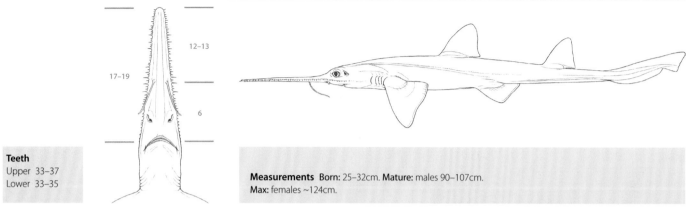

Teeth
Upper 33–37
Lower 33–35

Measurements Born: 25–32cm. **Mature:** males 90–107cm.
Max: females ~124cm.

Identification Large stocky sawshark. Short, broad, tapering rostrum (preoral length 22–24%). Barbels much closer to mouth than rostral tip. Nostrils diagonally oval and elongated. 17–19 large lateral rostral sawteeth, 12–13 before barbels, 6 behind. Juveniles usually with 1 smaller tooth between large sawteeth. Uniform slate grey above, white below. Indistinct dusky stripes along rostrum midline and edges.
Distribution Southern Australia.
Habitat Temperate-subtropical continental shelf, on or near bottom, close inshore 0–165m.
Behaviour Hidden prey may be located with barbels and uprooted by snout.
Biology Viviparous, 7–14 (average 11) pups per litter, produced every other year. Maximum age is ~9 years.
Status IUCN Red List: **Least Concern**. Moderately common within range. Bottom trawl and gillnet bycatch, used for food. The species is actively managed with an annual total allowable catch; catch rates over time have been stable.

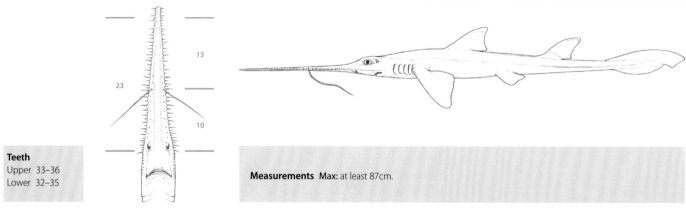

Teeth
Upper 33–36
Lower 32–35

Measurements Max: at least 87cm.

Identification Slender body. Very long, narrow, tapering rostrum (preoral length 31–32%) with a concave pre-barbel edge. Barbels halfway between mouth and rostral tip. 23 large lateral sawteeth: 13 before and 10 behind barbels. Juveniles usually with 1 smaller tooth between large lateral teeth. Uniform unpatterned light grey above, whitish below. Darker brownish stripes along rostrum midline and edges. Light-edged pectoral fins; dark anterior edge to dorsals in juveniles.
Distribution Northwest Atlantic: between Cuba, Florida and Bahamas.
Habitat On or near bottom, continental and insular slopes, 438–952m.
Behaviour Unknown.
Biology Unknown.
Status IUCN Red List: **Least Concern**. There are no deepsea commercial fisheries in the Bahamas.

SQUATINIFORMES angelsharks

Writing in 1555, Guillame Rondelet referred to these sharks as Monkfish for their similarity to a cowled monk. They are found mainly on mud and sand on cool-temperate continental shelves, from intertidal to continental slopes, deeper in tropical water, but are absent from most of the northern Indian Ocean and central Pacific.

Identification Bizarrely shaped medium- to large-sized sharks (up to 244cm, but most smaller than 160cm). May be heavily patterned on dorsal surface, most are uniformly pale below. Similar to rays, with a broad flattened body, short snout and large fins, but with gill openings on the sides of the head, not beneath, and very large pectoral fins not attached to the head opposite the gills (the hindmost gill opening is in front of pectoral fin origins, but covered by triangular anterior fin lobes). Eyes on top of head, close to and level with large spiracles. Large mouth and nostrils (barbels on anterior nasal flaps) at front of snout are separate, mouth extends at sides to opposite or slightly behind the eyes. Very large labial furrows. Two spineless dorsal fins set back on precaudal tail, first over or behind the free rear tips of the very large pelvic fins. No anal fin. Short very thick keels at the base of the caudal peduncle, caudal fin with long dorsal lobe and even longer expanded ventral lobe. Some species are very hard to differentiate.

Biology Poorly known outside Europe and the northeast Pacific. Reproduction is viviparous: litters of 1–25 pups obtain all nourishment from a yolk-sac before birth. Some species may have up to a three-year reproductive cycle. Often lie buried by day in mud and sand (except for eyes and spiracles) and are ambush feeders, using their unusually flexible 'necks' to raise their heads and protruding trap-like jaws to snap up prey at high speed. Food includes small bony fishes, crustaceans, squids, gastropods and clams. At least some species are nocturnally active and swim off the seabed. Most only make localised movements and do not travel far; populations may easily be isolated by deep water or areas of unsuitable habitat.

Status Harmless, unless disturbed or provoked. Many species are intensively fished for food (they have valuable flesh), also oil, fishmeal and leather. Very vulnerable as target and bycatch species in bottom trawl, line gear and fixed bottom nets. Significant population reductions have been reported for many heavily fished species; no other taxonomic order of shark has such a high percentage of Critically Endangered species. Recolonisation of depleted populations from adjacent areas may be very slow. Some tropical species are very poorly known. Divers photographed an apparently undescribed species (not included in this book) on a coral reef in the Andaman Sea.

Australian Angelshark, *Squatina australis*, ambush feeding on schooling fishes (p. 236).

Squatinidae: angelsharks

The order Squatiniformes contains just one family (Squatinidae), which in turn contains only one genus, *Squatina*. Although 22 species are featured in this book, others have not yet been described and named by scientists. Angelsharks pose an interesting conundrum: it is hard to understand why several different species that are so extremely similar in appearance and ecology can occur within a limited geographic area. Why is just one species not sufficient to fill their ecological role? However, multiple angelshark species do occur and appear to overlap in habitat as well as geographic distribution in several regions, including the east Pacific, Taiwan, Australia, west and southeast Atlantic. It is, therefore, worth taking a very careful look at all angelsharks to confirm their identification. For example, there may prove to be up to three species off the Pacific coast of South America. Furthermore, variations in the life history characteristics of the Pacific Angelshark are reported between the Gulf of California and from the Baja Pacific coast, on the other side of the Peninsula. These might prove to be not just variations between populations, but the biological characteristics of distinctly different species. In that case, some of the several discrete populations of Pacific Angelshark along the North American coast might also turn out to be different species Similarly, angelsharks from the western Gulf of Mexico and tropical Caribbean need to be critically examined and their characteristics closely compared with those of the Sand Devil and South American Atlantic coast species. Two new species, *S. david* and *S. varii* included below have been recently described from the Caribbean and southwestern Atlantic. Two other species described from the western Gulf of Mexico, *S. heteroptera* and *S. mexicana*, lack distinguishing characteristics differentiating them from each other and other known regional angelsharks. They are not, therefore, considered valid species, which is why we do not include them in this book.

   The angelsharks are unfortunately facing a higher relative threat of extinction than most other familes of sharks. Over 70% of are assessed as Threatened or Near Threatened with extinction in the IUCN Red List (only the hammerhead sharks have a worse status). However, these declines are increasingly being addressed by fisheries management and/or biodiversity conservation measures. Only five species are listed as Least Concern – most are Australian endemics whose geographic range is either unfished or only very lightly exploited.

Squatina
22 species; pages 234–246

Plate 22 SQUATINIDAE I

Distinct broad flattened body; short snout with large mouth and nostrils, eyes on top of head close to and level with large spiracles, gills on side of head; short thick keel at base of caudal peduncle; very large pectoral fins, two spineless dorsal fins set back on caudal peduncle, first dorsal fin over or behind the very large pelvic fins free rear tip, no anal fin, caudal fin with long upper lobe and larger lower lobe; dorsal surface usually patterned, pale below. Some species are hard to differentiate.

○ **Sawback Angelshark** *Squatina aculeata*   page 234

East Atlantic; mud, 30–500m. Concave between eyes; heavily fringed nasal barbels and anterior nasal flaps; large thorns on head and one row along back; dull grey or light brown, sparsely scattered with small irregular white spots and regular small dark brownish spots, large dark blotches on dorsal surface and tail, no ocelli.

○ **African Angelshark** *Squatina africana*   page 234

Southwest Indian Ocean; sand, mud and surf, 0–600m. Concave between eyes; simple flat nasal barbels with tapering or spatulate tips, anterior nasal flaps smooth or weakly fringed; large thorns on head not back; grey- or red-brown, many light and dark spots with larger symmetrical dark bands or saddles, blotches on pectoral fins, dark tail base with white margins, juveniles often with large ocelli.

○ **Argentine Angelshark** *Squatina argentina*   page 235

Southwest Atlantic; 20–437m. Concave between eyes; simple spatulate nasal barbels, weakly fringed or smooth anterior nasal flaps; large thorns on snout not back; convex anterior margin to pectoral fins forming distinct 'shoulder'; purplish brown with many scattered dark brown spots mostly in circular groups around darker spot, paler dorsal fins, no ocelli.

○ **David's Angelshark** *Squatina david*   page 237

West Atlantic and Caribbean; continental shelf and upper slope, 100–326m. Paired rod-like barbels on nasal flap, inner barbel with no or reduced fringe, outer bifurcate with small branch protruding ventrally near tip; dorsal midback lacks thorns or enlarged dermal denticles; angular dorsal fins, first slightly larger, origin just behind or over free rear tip of pelvic fins; greyish to brownish yellow, numerous dark and whitish spots.

○ **Sand Devil** *Squatina dumeril*   page 239

West Atlantic; inshore to 1290m. On or near bottom. Strongly concave between eyes; simple tapering nasal barbels, weakly fringed or smooth anterior nasal flaps; discrete thorns on snout and between eyes and spiracles in young, more numerous and in patches in adults; fairly broad posteriorly angular pectoral fins; uniform bluish to ash grey, irregular dusky or blackish spots may be present, young with white spots, dorsal and caudal fins darker with light bases, dorsal fins with light tips, underside white with red spots and reddish fin margins.

○ **Angelshark** *Squatina squatina*   page 245

Northeast Atlantic and Mediterranean; mud or sand, <1–150m or more. Large, stocky; lateral head folds with single triangular lobe each side, simple nasal barbels with straight or spatulate tips, smooth or weakly fringed anterior nasal flaps; small thorns on midback in young, adults just very rough, patches small thorns on snout and between eyes; very high broad pectoral fins; grey to red- or green-brown with scattered small white spots and blackish dots and spots, nuchal spot may be present, no ocelli, young often have white reticulations and large dark blotches.

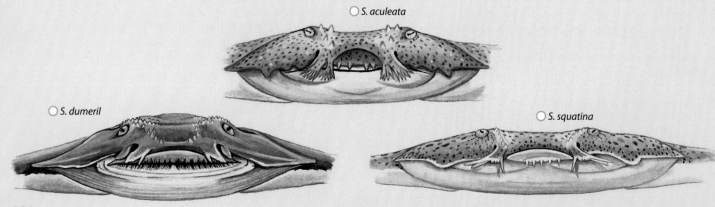

○ *S. aculeata*

○ *S. dumeril*

○ *S. squatina*

○ *Squatina africana*

○ *Squatina aculeata*

○ *Squatina argentina*

○ *Squatina david*

young

○ *Squatina dumeril*

○ *Squatina squatina*

adult

20cm

Plate 23 SQUATINIDAE II

○ **Chilean Angelshark** *Squatina armata*

East to southeast Pacific; 30–75m. Narrow head; large thorns on snout, between eyes and large spiracles, double row on midline of back, between and behind dorsal fins, also row on anterior margin of pectoral fins; grey to reddish brown above, no described markings.

page 236

○ **Australian Angelshark** *Squatina australis*

South Australia; sand, mud, seagrass, surf, 0–130m. Flat or convex between eyes; heavily fringed nasal barbels and anterior nasal flaps, small spiracles; no large thorns in adults, young enlarged denticles on snout, head and predorsal rows on back; dull grey to brown with dense white spots and smaller dark brown spots, no large ocelli.

page 236

○ **Philippines Angelshark** *Squatina caillieti*

Central Indo-Pacific; 363–385m. Upper lip arch semi-oval in shape, height less than one-half arch width; slightly concave between eyes with smooth patch; simple unfringed nasal barbels; no thorns in young, unknown in adults; pelvic fin tips reach first dorsal fin origins; inter-dorsal space greater than dorsal-caudal distance. Greenish brown with numerous darker brown spots outlined in white margins.

page 237

○ **Pacific Angelshark** *Squatina californica*

Northeast (possibly also southeast) Pacific; sand, rocks and near kelp, 1–205m. Concave between large eyes; simple conical nasal barbels with spatulate tips, anterior nasal flaps weakly fringed; thorns in young, small or absent in adults; fairly broad long high pectoral fins; red-brown to blackish, scattered light spots around dark blotches in adults, white-edged pectoral and pelvic fins, pale dorsal fins with dark blotches on base, dark spot at base of pale caudal fin, ocelli in young.

page 238

○ **Taiwan Angelshark** *Squatina formosa*

Northwest Pacific; 100–400m. Concave between large eyes; simple very flat nasal barbels with round tips, anterior nasal flaps smooth or weakly fringed; patches of large denticles on snout and between eyes (rows on back in young); yellow-grey or brown, numerous small dark brown spots and large irregular blotches, small light spots between head and first dorsal fin, saddle/band alongside dorsal fins, small paired dark ocelli.

page 240

○ **Indonesian Angelshark** *Squatina legnota*

Southern Indonesia; habitat and depth unknown. Snout very short; anterior nasal barbel flap with unfringed barbells; no median row of denticles on back; first dorsal fin base longer than second dorsal fin base; interdorsal space weakly concave; dorsal surface uniformly greyish brown with two dark saddles below dorsal fins; anterior ventral surface of pectoral fins blackish.

page 240

○ **Angular Angelshark** *Squatina guggenheim*

Southwest Atlantic; 7–150m. Broadly concave between eyes; lateral head folds without triangular lobes, expanded slightly spatulate unfringed nasal barbels, anterior nasal flaps weakly fringed; short stout symmetrical thorns on snout and between eyes and a pair between spiracles, also row on midline of back; pectoral fins relatively small, high and angular with nearly straight anterior margin; dark tan above with regular pattern of several small to large blackish spots, small irregular dark spots sometimes present, no ocelli.

page 241

○ **Japanese Angelshark** *Squatina japonica*

Northwest Pacific; 10–352m. Concave between large eyes; cylindrical nasal barbels with slightly expanded tips, anterior nasal flaps smooth or weakly fringed; small thorns on snout, between eyes and spiracles, one row on midline of back, skin rough; fairly broad high rounded pectoral fins; rusty or blackish brown with irregular small dark dense white spots, large paired red-brown blotches from base of head to dorsal fins, no ocelli.

page 242

○ *S. caillieti*

○ *S. australis*

○ *S. californica*

○ *Squatina armata*

○ *Squatina australis*

○ *Squatina californica*

○ *Squatina caillieti*
immature (not to scale)

○ *Squatina formosa*

○ *Squatina guggenheim*

○ *Squatina japonica*

○ *Squatina legnota*

20cm

Plate 24 SQUATINIDAE III

○ **Eastern Angelshark** *Squatina albipunctata*   page 235

Southwest Pacific; 35–415m. Low lateral head folds, very short snout, concave between eyes; nasal barbels with expanded tips and lobate fringes; strong thorns above eyes, no thorns on back; yellow-brown to chocolate brown with dense pattern of symmetrical small white dark-edged spots, many large brownish blotches and white nuchal spot, light unspotted dorsal and caudal fins.

○ **Clouded Angelshark** *Squatina nebulosa*   page 242

Northwest Pacific; surf–330m. Lateral head folds with two triangular lobes each side; simple tapering nasal barbels, anterior nasal flaps smooth or weakly fringed; no thorns; very broad low obtuse, rounded pectoral fins; brown to blue-brown with scattered light spots and numerous small black dots, dark spot at base of pectoral fins, dark blotches below dorsal fins, ocelli absent or small.

○ **Hidden Angelshark** *Squatina occulta*   page 243

Southwest Atlantic; 10–350m. Concave between eyes; cylindrical bases to nasal barbels with expanded unfringed tips, anterior nasal flaps very weakly fringed; short stout symmetrical grouped thorns on snout and between eyes, pair between spiracles; pectoral fins quite large high and angular; dark tan with numerous yellow spots and larger blackish marks, a few ocelli on pectoral fins.

○ **Smoothback Angelshark** *Squatina oculata*   page 243

East Atlantic and Mediterranean; 10–500m. Strongly concave between eyes; weakly bifurcate or lobed nasal barbels, anterior nasal flaps weakly fringed; large thorns on snout and above eyes; grey-brown with small white and blackish spots, large dark blotches on base and rear tips of pectoral fins, tail base and under dorsal fins, dorsal and caudal fin margins white, pectoral and pelvic fin margins dusky, white nuchal spot, sometimes symmetrical dark ocelli.

○ **Western Angelshark** *Squatina pseudocellata*   page 244

West Australia; 150–312m. Very short snout, concave between eyes; nasal barbels with expanded tips and lobate fringes; strong thorns above eyes, single row of thorns on back; medium to pale brownish or greyish with widely spaced blue spots and brown blotches, light unspotted dorsal and caudal fins, single white nuchal spot, no ocelli.

○ **Ornate Angelshark** *Squatina tergocellata*   page 244

Central Indo-Pacific; 130–400m. On or near bottom. Strongly concave between large eyes; strongly fringed nasal barbels and anterior nasal flaps; no dorsal thorns; pale yellow-brown with grey-blue or white spots, three pairs of large ocelli with dark rings surrounding thread-like patterned centres.

○ **Ocellated Angelshark** *Squatina tergocellatoides*   page 246

Northwest and Central Indo-Pacific; 100–300m. Concave between eyes; strongly but finely fringed nasal barbels, anterior nasal flaps strongly fringed; no large predorsal thorns on back; light yellowish brown with dense scattering of small round white spots, dorsal fins with black base and anterior margins, six pairs of large ocelli of dark rings around light centres on pectoral and pelvic fins and tail base.

○ **Vari's Angelshark** *Squatina varii*   page 246

Southwest Atlantic; continental slopes, 195–666m. Nasal flaps with slightly fringed median rectangular to rounded barbel between two lateral elongate barbels; no dorsal thorns; angular dorsal fins similar size, first originating just behind pelvic fin free rear tips. Brown above, may have scattered white spots, occasionally irregular dark spots, three pairs of dark blotches on sides of tail; creamy white below with slightly darker fin edges; pectoral fins may have paired white spots.

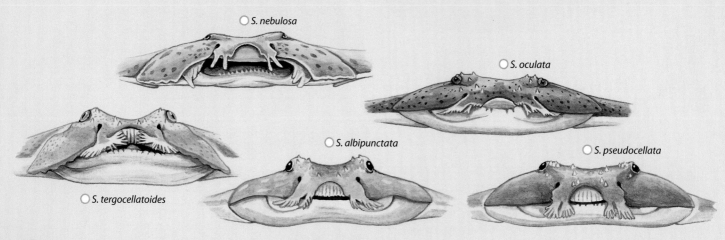

○ *S. nebulosa*

○ *S. oculata*

○ *S. tergocellatoides*

○ *S. albipunctata*

○ *S. pseudocellata*

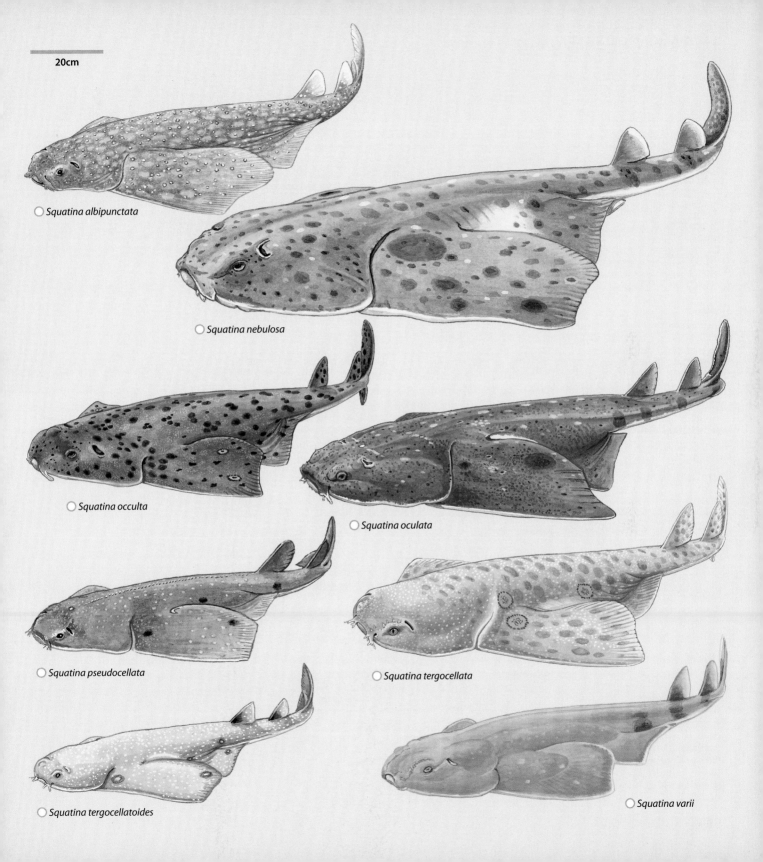

20cm

○ *Squatina albipunctata*

○ *Squatina nebulosa*

○ *Squatina occulta*

○ *Squatina oculata*

○ *Squatina pseudocellata*

○ *Squatina tergocellata*

○ *Squatina tergocellatoides*

○ *Squatina varii*

SAWBACK ANGELSHARK *Squatina aculeata* FAO code: **SUA** Plate page 228

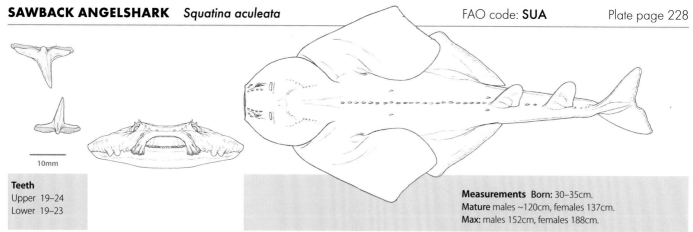

10mm

Teeth
Upper 19–24
Lower 19–23

Measurements Born: 30–35cm.
Mature males ~120cm, females 137cm.
Max: males 152cm, females 188cm.

Identification Concave between eyes; eye–spiracle distance less than 1.5 times eye-length. Heavily fringed nasal barbels and anterior nasal flaps. Large thorns on head and in a row along back. Dull grey or light brown back sparsely scattered with small, irregular, white spots and regular, small, dark brownish spots. No ocelli. Large, dark blotches on head, back, fin bases and tail.
Distribution East Atlantic: Mediterranean, Senegal, The Gambia and Sierra Leone.
Habitat Offshore, outer continental shelf and upper slope, usually on soft muddy bottoms, 30–500m, warm-temperate to tropical waters.
Behaviour Unknown. **Biology** Viviparous, 8–22 pups per litter; gestation is 12 months followed by 12 month resting period. Feeds on small sharks, bony fishes, cuttlefish and crustaceans.
Status IUCN Red List: **Critically Endangered**. Locally extirpated from portions of its range, including the Mediterranean where it is now protected. Caught in target fisheries and as bycatch. See p. 241.

AFRICAN ANGELSHARK *Squatina africana* FAO code: **SUF** Plate page 228

~5mm

Teeth
Upper 20–22
Lower 18–20

Measurements Born: 28–30cm.
Mature: males 77–95cm, females 82–107cm.
Max: ~122cm.

Identification Concave between eyes. Simple flat nasal barbels, tips tapering or spatulate. Anterior nasal flaps smooth or slightly fringed. No angular lobes on lateral dermal flaps. Enlarged thorns on head, not back. Greyish or reddish-brown. Many light and dark spots, often large granular-centred ocelli in young. Larger symmetrical dark bands or saddles, blotches on broad, angular, high pectoral fins. Dark tail base, white margins.
Distribution East and southern Africa: South Africa to Mozambique, Tanzania and Madagascar, possibly Somalia. Nominal west African records were based on other species.
Habitat Continental shelf and upper slope, sand and mud, surf line to 494m (mainly 60–300m).
Behaviour Lies buried to ambush prey.
Biology Viviparous, 1–12 pups per litter (6 average); mating takes place in summer with birth occurring after 12 month gestation period. Eats small bony fishes, cephalopods and shrimp.
Status IUCN Red List: **Near Threatened**. Common only on east coast of South Africa (KwaZulu-Natal). Trawl fishery bycatch.

EASTERN ANGELSHARK *Squatina albipunctata*

FAO code: **ASK**   Plate page 232

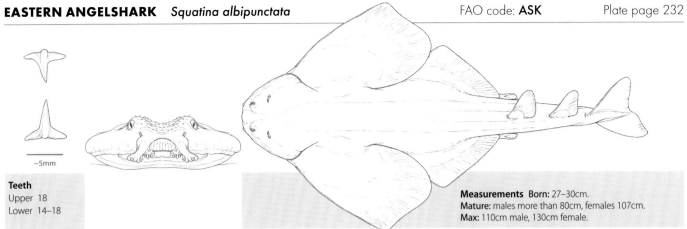

Teeth
Upper 18
Lower 14–18

~5mm

Measurements Born: 27–30cm.
Mature: males more than 80cm, females 107cm.
Max: 110cm male, 130cm female.

Identification Very short snout, concave interorbital space and heavy orbital thorns distinguish this from *Squatina australis*. Nasal barbels with expanded tips and lobate fringes. Low lateral head folds. Spiracles close to eyes, wider than eye-length. Strong orbital thorns; no medial row of predorsal thorns. Yellow-brown to chocolate brown. Dense pattern of symmetrical, small, white, dark-edged spots; many large, brownish blotches; white nuchal spot (no ocelli). Light unspotted dorsal and caudal fins.
Distribution East coast of Australia from Queensland to Victoria.
Habitat Outer continental shelf and upper slope, 35–415m.
Behaviour Unknown. **Biology** Viviparous, litters of up to 20 following a 10 month gestation period. Diet likely includes fishes, crustaceans and cephalopods.
Status IUCN Red List: Vulnerable. Depleted in the heavily fished southern portion of its range, unfished in north.

ARGENTINE ANGELSHARK *Squatina argentina*

FAO code: **SUG**   Plate page 228

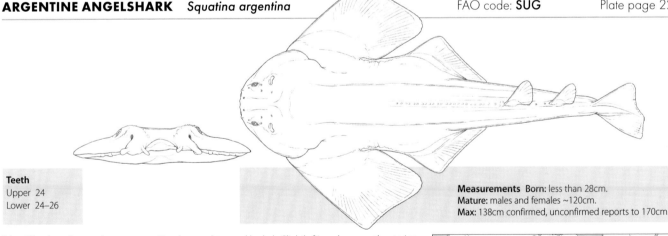

Teeth
Upper 24
Lower 24–26

Measurements Born: less than 28cm.
Mature: males and females ~120cm.
Max: 138cm confirmed, unconfirmed reports to 170cm.

Identification Concave between eyes. Simple, spatulate nasal barbels. Slightly fringed or smooth anterior nasal flaps. No triangular lobes on lateral head folds. Enlarged thorns on snout, not back. Large, broad, obtusely angular pectoral fins; convex leading edge forming a distinct 'shoulder'. Purplish brown with many scattered dark brown spots (no white), mostly in circular groups around a central darker spot; no ocelli. Paler dorsal fins.
Distribution Southern Brazil, Uruguay and Argentina.
Habitat Continental shelf and upper slope, 20–437m (mostly 120–320m).
Behaviour Unknown.
Biology Viviparous, 7–11 pups per litter (average 9–10). Feeds on demersal fishes, shrimp and squids.
Status IUCN Red List: Critically Endangered. Formerly common. Target and bycatch species, with *Squatina guggenheim* and *S. occulta*. Distinguishing between these regional angelsharks is notoriously difficult.

CHILEAN ANGELSHARK   *Squatina armata*

FAO code: **ASK**   Plate page 230

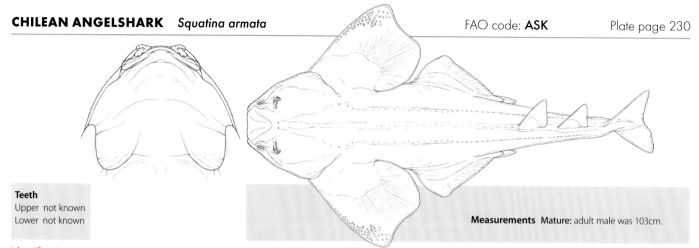

Teeth
Upper not known
Lower not known

Measurements Mature: adult male was 103cm.

Identification Narrow head with very heavy thorns on snout and between eyes and large spiracles; double row of large, hooked thorns on midline of back and between and behind dorsal fins; enlarged thorns on leading edge of pectoral fins. First dorsal origin probably behind free rear tips of pelvic fins. Reddish brown to grey above, paler below.
Distribution Southeast and east Pacific: Chile, Peru, Ecuador, Colombia and Costa Rica.
Habitat Continental shelf from 30–75m.
Behaviour Unknown.
Biology Unknown.
Status IUCN Red List: **Critically Endangered.** This is a bycatch of trawl and gillnet fisheries throughout its range, and retained for its meat. Despite increasing fishing effort, catches have declined steeply and the Chilean Angelshark is no longer landed in some areas.

AUSTRALIAN ANGELSHARK   *Squatina australis*

FAO code: **SUU**   Plate page 230

Teeth
Upper 20
Lower 18

Measurements Mature: males more than 80cm, females 97cm.
Max: males 122cm, females 152cm.

Identification Head flat or convex between eyes. Heavily fringed nasal barbels and anterior nasal flaps. No triangular lobes on lateral head folds. Small spiracles. No large thorns in adults; enlarged denticles on snout, head and multiple predorsal rows in young. Dull greyish brown with dense white spots and small darker brown spots. No large ocelli. White-edged fins; spots on leading edge of pale dorsal fins and lower tail lobe.
Distribution Southern Australia.
Habitat Sand and mud, often in seagrass or near rocky reefs, 0–130m.
Behaviour Lies buried by day, active at night.
Biology Viviparous with up to 20 pups per litter in autumn. Feeds on fishes and crustaceans.
Status IUCN Red List: **Least Concern.** Commercially valuable, trawled, but unfished and common in much of its range.

PHILIPPINES ANGELSHARK   *Squatina caillieti*

FAO code: **ASK**   Plate page 230

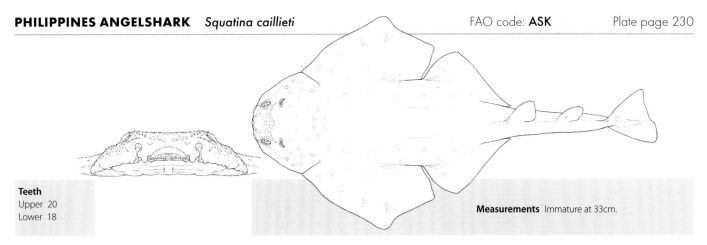

Teeth
Upper 20
Lower 18

Measurements Immature at 33cm.

Identification Head rounded, with moderate tubercles interspersed laterally between mouth and eye crests, with smooth oval patch in between. Simple, unfringed nasal barbels, with rod-like tips. Upper lip arch semi-oval in shape; upper lip arch height less than ½ lip arch width. Eyes almond-shaped, close-set, with relatively short eye–spiracle space. Interdorsal distance greater than dorsal caudal distance (this latter characteristic useful for separating this species from *Squatina formosa* and *S. nebulosa*). Pelvic fin tips reaching first dorsal fin origin. Greenish brown with numerous darker brown spots outlined in white margins. Black subdorsal saddles present along dorsal fins.
Distribution Central Indo-Pacific: Luzon, Philippines.
Habitat Outer continental shelf and upper slope, 363–385m.   **Behaviour** Unknown.
Biology Poorly known. Described from a single immature female.
Status IUCN Red List: **Data Deficient**. The impact of fisheries on this rare species is unknown.

DAVID'S ANGELSHARK   *Squatina david*

FAO code: **ASK**   Plate page 228

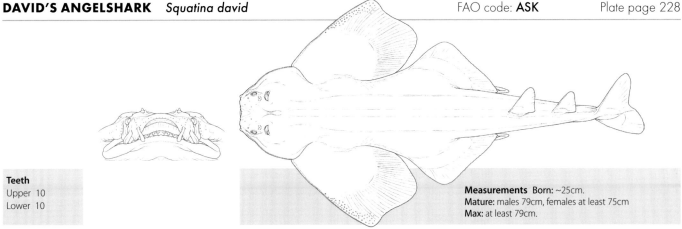

Teeth
Upper 10
Lower 10

Measurements Born: ~25cm.
Mature: males 79cm, females at least 75cm
Max: at least 79cm.

Identification Nasal flap with paired barbels; inner nasal barbel rod-like, with reduced or no fringe; outer nasal barbel rod-like, bifurcate with small branch protruding ventrally near tip. Dorsal midback lacks thorns or enlarged dermal denticles. Dorsal fins angular, first slightly larger than second dorsal fin; first dorsal fin origin just behind or over free rear tip of pelvic fins. Colour dorsally greyish to brownish yellow with numerous dark and whitish spots.
Distribution West Atlantic and Caribbean: Panama, Colombia to Suriname along the northern coast of South America.
Habitat Continental shelf and possibly upper slopes, 100–326m
Behaviour Unknown, species known from only 6 specimens.
Biology Viviparous, nothing else known. Diet unknown.
Status IUCN Red List: **Near Threatened**. This species is caught in gillnets, bottom trawls and longline fisheries throughout most of its range.

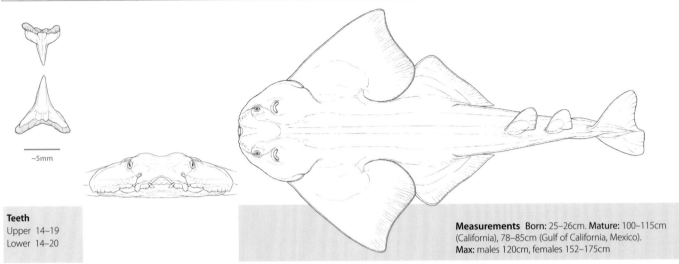

~5mm

Teeth
Upper 14–19
Lower 14–20

Measurements Born: 25–26cm. **Mature:** 100–115cm (California), 78–85cm (Gulf of California, Mexico). **Max:** males 120cm, females 152–175cm

Identification Concave between large eyes; eye–spiracle space less than 1.5 times eye-length. Simple, conical nasal barbels with spatulate tips. Weakly fringed anterior nasal flaps. No triangular lobes on lateral head folds. Thorns prominent in young, small or absent in adults. Fairly broad, long, high pectoral fins. Reddish brown to dark brown or blackish. Scattered light spots (set around dark blotches in adults). Large paired, dark blotches on back and tail form large ocelli in young. White-edged pectoral and pelvic fins. Pale dorsal fins with dark blotches at base. Dark spot at base of pale, spotted tail.

Distribution Northeast Pacific: southeast Alaska to the Gulf of California; possibly also in the southeast Pacific, but this needs confirmation. This species may overlap with several other unnamed angelshark species in the southeast Pacific and, with further investigation, might even turn out to be comprised of more than one species in the northeast Pacific (see below).

Habitat Continental shelf, 1–205m, most common 3–100m. Often around rocks, sometimes near kelp.

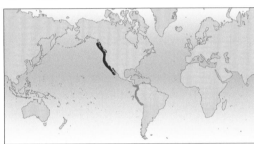

Behaviour Lies buried in flat sand or mud by day, launching from hiding to ambush its fish prey. Most active at night. Believed not to swim long distances, however, on occasion, these sharks are known to move between islands and the mainland off southern California.

Biology Viviparous, 1–13 pups per litter (average 6), after a 9–10 month gestation. About 20% of the young survive to maturity at 9–10 years old. The maximum estimated age is 35 years. The Gulf of California population appears to mature at a smaller size than the Baja Pacific coast population. This variation may indicate that there are two distinctly different Pacific angelshark species on either side of the peninsula, and perhaps other undescribed cryptic species within this species' range.

Status IUCN Red List: **Near Threatened**. This species was abundant in California until a fishery for its meat caused population collapse in the early 1990s. A gill net ban ended this fishery; the population has now recovered and is the subject of dive tourism in California. The species is also taken as bycatch elsewhere and utilised for its meat.

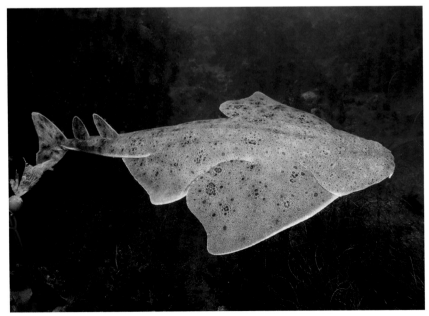

Pacific Angelshark, *Squatina californica*, California.

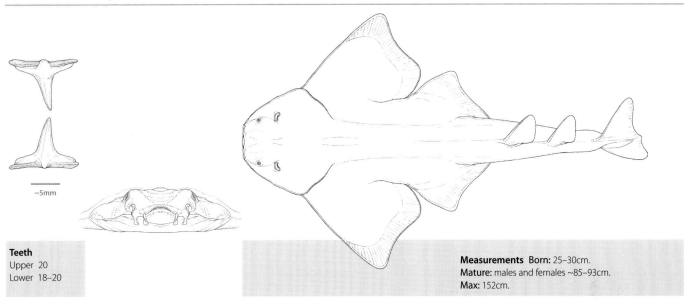

Teeth
Upper 20
Lower 18–20

~5mm

Measurements Born: 25–30cm.
Mature: males and females ~85–93cm.
Max: 152cm.

Identification Almost plain bluish to ashy-grey, no elaborate markings or ocelli, dusky or blackish spots irregularly present or absent. Small white spots often present in young. The underside is white with red spots and reddish fin margins. Dorsal and caudal fins are darker, with light bases. Dorsal fin rear tips are light. Simple, tapering nasal barbels, weakly fringed or smooth anterior nasal flaps, lateral head folds low, no triangular lobes. Strongly concave between the large eyes, eye–spiracle space less than 1.5 times eye length. Fairly broad, posteriorly angular pectoral fins. A few small, discrete thorns on snout, between eyes and spiracles in young, becoming more numerous and forming patches in adults. The thorns along the back of young are reduced, and remain inconspicuous in adults.

Distribution West Atlantic: New England to Gulf of Mexico.

Habitat Continental shelf and slope, on or near bottom, close inshore to 1290m, mostly 40–250m.

Behaviour Appears in inshore shallow water in spring and summer off USA, disappears (presumably into deeper water?) in winter. It is aggressive when captured, hence the English name.

Biology Viviparous, up to 25 pups per litter are born from late winter to early summer. Similarly to the Pacific Angelshark (see previous page), Sand Devil populations appear to have slightly different biological characteristics on the Atlantic coast and in the Gulf of Mexico. Litters off the North American Atlantic coast have 4–25 young per birth, while those in the Gulf of Mexico have only 4–10 young, with an average litter size of 7. The reproductive cycle for this species appears to be at least 2 years, but it is speculated that it may be up to 3 years long. It has a diet of small bottom fishes, crustaceans and bivalves.

Status IUCN Red List: Least Concern. There is no directed fishery for this species, but it is taken as a bycatch in other fisheries in the Gulf of Mexico. It is listed as a prohibited species in US waters (meaning that it must not be targeted in fisheries) due to its life history characteristics, which may make it vulnerable to heavy fishing exploitation.

Angelsharks from the western Gulf of Mexico and tropical Caribbean need to be critically examined and their characteristics closely compared with those of the Sand Devil and South American Atlantic coast species. Two new species, *Squatina david* (p. 237) and *S. varii* (p. 246), recently described from the Caribbean and southwest Atlantic, are included in this book. Two other species described from the western Gulf of Mexico, *S. heteroptera* and *S. mexicana*, lack distinguishing characteristics to differentiate between each other and other known regional angelsharks. They are not, therefore, considered valid species, which is why we do not include them in this book.

TAIWAN ANGELSHARK *Squatina formosa*

FAO code: **SUO**   Plate page 230

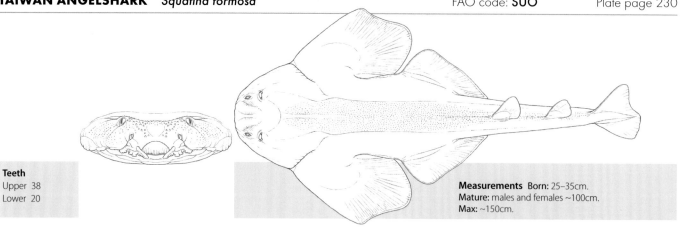

Teeth
Upper 38
Lower 20

Measurements Born: 25–35cm.
Mature: males and females ~100cm.
Max: ~150cm.

Identification Head concave between large eyes; eye–spiracle space less than eye-length. Simple, very flat nasal barbels, tapering round tips. Weakly fringed or smooth anterior nasal flaps. No triangular lobes on lateral head folds. Patches of enlarged denticles on snout, between eyes and in rows on back (in young). Very broad, low, rounded pectoral fins. Yellow-grey or brown with small paired dark ocelli. Small light spots between head and first dorsal fin; many small dark brown spots, larger irregular blotches, and a saddle or band alongside dorsal fins.
Distribution Northwest Pacific: Taiwan and possibly Japan.
Habitat Outer continental shelf and upper slope, 100–400m.
Behaviour Unknown.
Biology Viviparous with litters of up to 16. Feeds primarily on benthic fishes.
Status IUCN Red List: Endangered. Under intensive fishing pressure (as a bycatch) throughout its range.

INDONESIAN ANGELSHARK *Squatina legnota*

FAO code: **ASK**   Plate page 230

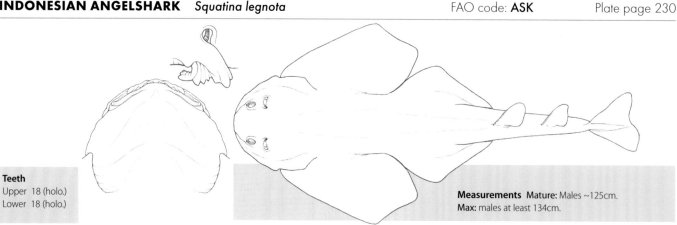

Teeth
Upper 18 (holo.)
Lower 18 (holo.)

Measurements Mature: Males ~125cm.
Max: males at least 134cm.

Identification Head weakly concave between large eyes; snout very short. Anterior nasal barbel flap with unfringed barbels. No median row of denticles on back. Dorsal fins not lobe-like; first dorsal fin base longer than second dorsal fin base. Dorsal surface uniformly greyish brown with two dark subdorsal saddles. Anterior ventral surface of pectoral fin blackish.
Distribution Central Indo-Pacific: southern Indonesia only.
Habitat Unknown. **Behaviour** Unknown.
Biology Unknown.
Status IUCN Red List: Critically Endangered. This species has been recorded at four landing sites in Indonesia, in a region where there is high and intensifying fishing pressure and many other demersal fish resources have been depleted. Fisheries are believed to operate over the entire range of this species, which is one of the most biologically-vulnerable to over-exploitation.

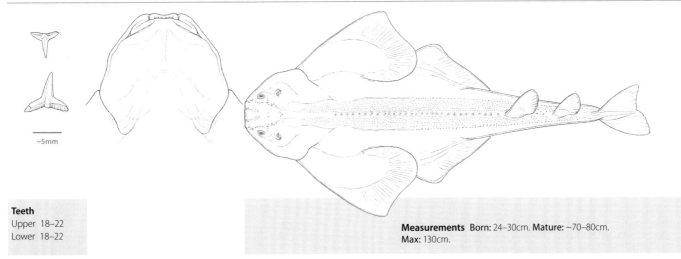

~5mm

Teeth
Upper 18–22
Lower 18–22

Measurements Born: 24–30cm. **Mature:** ~70–80cm.
Max: 130cm.

Identification Head broadly concave between eyes; eye–spiracle space less than 1.5 times eye-length. Nasal barbels with expanded, slightly spatulate, unfringed tips. Anterior nasal flaps weakly fringed. No triangular lobes on lateral head folds. Short, stout thorns in symmetrical groups on snout, interorbital space and a pair between spiracles; a median dorsal row of spines present. Pectoral fins relatively small, high and angular; nearly straight anterior margin. Uniform dark tan above, pale below. Dorsal surface with small, irregular dark spots present or absent in fresh specimens (turning white with fixation); regular pattern of several small to largish blackish spots; no ocelli.

Distribution Southwest Atlantic: Rio de Janeiro state, southern Brazil to Uruguay and Argentina.

Habitat Continental shelf, 7–150m, mostly from 10–80m. Mostly found in water temperatures of 10–22°C.

Behaviour Females migrate in spring, every three years, to shallow coastal waters to give birth; small juveniles remain inshore all year.

Biology Viviparous, 2–10 (usually 3–9) pups per litter develop in utero for 4 months, then move into the greatly enlarged cloaca for the remainder of 11 month gestation, which is followed by a 2 year resting period. Both sexes are estimated to mature at 4 years with a maximum age of 12 years. Eats demersal fishes and shrimps.

Status IUCN Red List: Endangered. Taken in target fisheries and as bycatch with *Squatina argentina* and *S. occulta*; extremely slow to recover from overfishing because of its triennial reproductive cycle. Serious population decline in southern Brazil. Formerly confused with *S. occulta*, but a separate species. The nominal species name *Squatina punctata*, sometimes seen in the literature, is actually a junior synonym of *S. guggenheim*.

EAST ATLANTIC AND MEDITERRANEAN ANGELSHARK PROJECTS

This initiative addresses the very high extinction risk to three angelshark species, *Squatina aculeata* (p. 234), *S. oculata* (p. 243) and *S. squatina* (p. 245), through a regional Conservation Strategy, subregional Action Plans and an Angel Shark Network. It also provides a useful model for the conservation and management of threatened shark species in other regions.

The Angel Shark Conservation Strategy envisages restoring robust populations of these Critically Endangered species and safeguarding them throughout their natural range. It aims to do this by improving their overall profile, increasing reported sightings, generating a better understanding of their distribution and conservation status, and identifying new opportunities for collaboration and conservation action. Separate sub-regional Action Plans for the Canary Islands and Mediterranean set out how to implement the Strategy in these two important but contrasting areas. They encourage governments, organisations and individuals to become involved; citizen science is a very important component of these projects. The Canary Islands are unique because they are the world's last stronghold for the formerly Common Angelshark, *Squatina squatina*, and this Action Plan focuses on promoting coordinated action to ensure that this species remains abundant and protected here. All three regional species occur in the Mediterranean and are, in theory, legally protected here under regional biodiversity and fisheries regulations. However, they are seriously depleted, extirpated from large areas of their former range, threatened where they do still occur by bycatch in fisheries and habitat degradation, and very poorly known. The Mediterranean Action Plan therefore focuses on addressing these threats by promoting regional collaboration and conservation efforts.

The Angel Shark Network angelsharknetwork.com links the above initiatives. It coordinates this work and that of the Welsh Angel Shark Project and collects angelshark records through an online sightings map. It has also initiated a global #AngelSharkDay (26th June) for all 22 species worldwide.

JAPANESE ANGELSHARK *Squatina japonica*

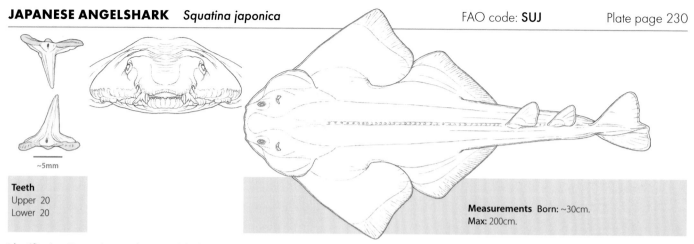

~5mm

Teeth
Upper 20
Lower 20

Measurements Born: ~30cm.
Max: 200cm.

Identification Concave between large eyes. Cylindrical nasal barbels with slightly expanded tips. Weakly fringed or smooth nasal flaps. Lateral head folds without triangular lobes. Small thorns on snout, between eyes and spiracles, and in a row along back; denticles make surface rough. Fairly broad, high, rounded pectoral fins. Rusty or blackish brown; dense dark and small irregular white spots on dorsal surface. Large, paired dark red-brown blotches from base of head to pelvic fins. No ocelli on back. White below with darker margins on fins and tail.
Distribution Northwest Pacific: western Sea of Japan to Yellow Sea, East and South China Seas including Japan, Koreas, northern China and Taiwan.
Habitat Continental shelves and upper slopes from 10–352m.
Behaviour Poorly known.
Biology Viviparous, with litters of up to 10. Poorly known.
Status IUCN Red List: Vulnerable. Taken in trawl fisheries.

CLOUDED ANGELSHARK *Squatina nebulosa*

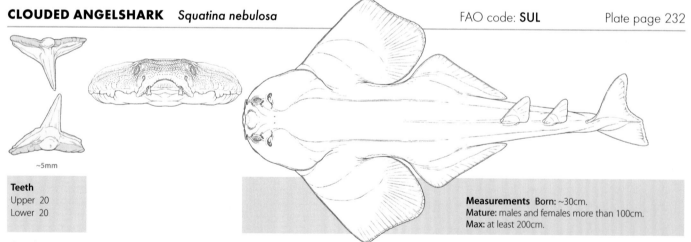

~5mm

Teeth
Upper 20
Lower 20

Measurements Born: ~30cm.
Mature: males and females more than 100cm.
Max: at least 200cm.

Identification Simple, tapering nasal barbels. Weakly fringed or smooth anterior nasal flaps. Lateral head folds with two triangular lobes each side. Large eyes; eye–spiracle space less than 1.5 times eye-length. No enlarged thorns. Very broad, low, obtuse, posteriorly rounded pectoral fins. Brownish to bluish brown with scattered light spots and many small black dots. Large, rounded dark spot at base of pectoral fin; dark blotches below dorsal fins. Ocelli absent or small, scattered and obscure (light rings, dark centres). Pale underside with darker pectoral fin margins. Light dorsal fin margins.
Distribution Northwest Pacific: Japan, Russia, Koreas, China and Taiwan.
Habitat Continental shelves and upper slopes, inshore to 330m.
Behaviour Poorly known.
Biology Poorly known. Viviparous, 12–20 pups per litter. Feeds on benthic teleosts and cephalopods.
Status IUCN Red List: Vulnerable. Taken in demersal fisheries.

HIDDEN ANGELSHARK *Squatina occulta*

FAO code: **ASK** Plate page 232

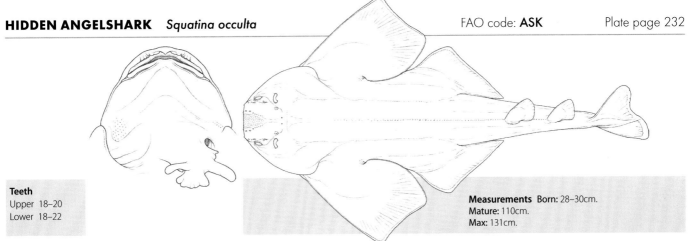

Teeth
Upper 18–20
Lower 18–22

Measurements Born: 28–30cm.
Mature: 110cm.
Max: 131cm.

Identification Head concave between eyes; eye–spiracle space less than 1.5 times eye-length. Nasal barbels with cylindrical bases, expanded slightly spatulate unfringed tips. Anterior nasal flaps barely fringed. Short stout thorns in symmetrical groups on snout and between eyes, a pair between spiracles; no medial dorsal row. Pectoral fins quite large, high and angular; leading edge nearly straight; posterior edge concave. Uniform dark tan with numerous small yellowish spots (no dark edges) and larger blackish marks. A few ocelli on pectoral fins.
Distribution Southwest Atlantic: southern Brazil, Uruguay and Argentina.
Habitat Continental shelf, 10–350m (mostly 50–100m). Mostly found in temperatures of 13–19°C.
Behaviour Unknown. **Biology** Viviparous, 4–10 (average 7) pups per litter in spring after 11 month gestation. Maturity occurs at about 10 years and maximum age is ~21 years. Eats fishes and shrimps.
Status IUCN Red List: **Critically Endangered**. Seriously depleted by fisheries and population slow to rebuild because of its triennial reproductive cycle. Formerly termed *Squatina guggenheim*, but the correct name is *S. occulta* ; the species name *S. punctata* which often appears in the literature is actually true *S. guggenheim*.

SMOOTHBACK ANGELSHARK *Squatina oculata*

FAO code: **SUT** Plate page 232

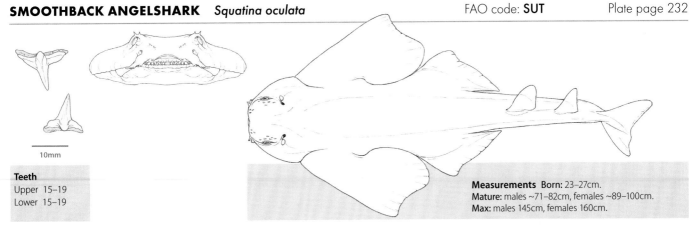

10mm

Teeth
Upper 15–19
Lower 15–19

Measurements Born: 23–27cm.
Mature: males ~71–82cm, females ~89–100cm.
Max: males 145cm, females 160cm.

Identification Strongly concave between eyes; eye–spiracle space less than 1.5 times eye-length. Weakly bifurcated or lobed nasal barbels. Weakly fringed anterior nasal flaps. Large thorns on snout and above eyes. Grey-brown with small, round, white and blackish spots. White nuchal spot. Symmetrical large dark blotches or spots on base and rear tip of pectoral fins, tail base and under dorsal fins. Sometimes symmetrical white-edged dark ocelli. Dorsal and caudal fin margins white; pectoral and pelvic fin margins dusky.
Distribution East Atlantic: Mediterranean, Senegal, The Gambia, Guinea, Sierra Leone and Ghana.
Habitat Continental shelves and upper slopes, 20–500m (mainly 50–100m, deeper in tropics), warm-temperate to tropical waters.
Behaviour Unknown. **Biology** Viviparous, 3–8 (average 6) pups per litter, and a gestation of 12 months; birth occurs in the spring. Eats small fishes, squid, octopus, shrimp and crabs. Misidentification with other angel shark species precludes better information on its biology and distribution.
Status IUCN Red List: **Critically Endangered**. Locally extirpated from portions of its range. Fished off the African coast, bycatch elsewhere. Depleted and protected in Mediterranean. See p. 241.

WESTERN ANGELSHARK   *Squatina pseudocellata*   FAO code: **ASK**   Plate page 232

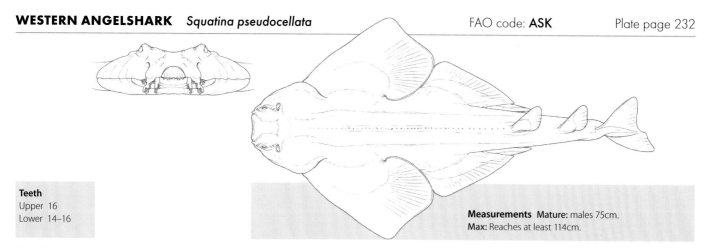

Teeth
Upper 16
Lower 14–16

Measurements Mature: males 75cm.
Max: Reaches at least 114cm.

Identification Differs from *Squatina tergocellata* in colour, pattern and dorsal thorns. Head concave between eyes; very short snout. Nasal barbels with expanded tips and lobate fringes. Strong orbital thorns; a medial row of predorsal thorns. Dorsal surface medium to pale brownish or greyish with a pattern of widely spaced blue spots and brown blotches. No symmetrical small white spots or ocelli, but a single small white nuchal spot. Light unpaired fins without dark spots.
Distribution Western Australia: Cape Leveque to Shark Bay.
Habitat Tropical outer continental shelf and uppermost slope, 150–312m.
Behaviour Unknown.
Biology Unknown.
Status IUCN Red List: Least Concern. Taken as bycatch, but very low fishing effort in range.

ORNATE ANGELSHARK   *Squatina tergocellata*   FAO code: **SUE**   Plate page 232

~10mm

Teeth
Upper 18
Lower 20

Measurements Born: 33–42cm. **Mature:** males ~81–91cm, with adults to 103cm; females ~115–125cm, with adults to 140cm. **Max:** 140cm.

Identification Strongly concave between large eyes; eye–spiracle space less than 1.5 times eye-length. Strongly fringed nasal barbels and anterior nasal flaps. No dorsal thorns. Pale yellow-brown with grey-blue or white spots. Three pairs of large ocelli, dark rings around centres with amitotic pattern. Fins pale with spots or blotches.
Distribution Southwest Australia.
Habitat Continental shelf and upper slope, on or near bottom, 128–400m (most about 300m, young shallower).
Behaviour Partial sexual segregation. May swallow mud to buffer prey toxins.
Biology Viviparous, 2–9 pups per litter (average 4–5), probably every 2 years, 6–12 month gestation. Feeds on fishes and squids.
Status IUCN Red List: Least Concern. Stable bycatch in small fishery, much of range untrawled.

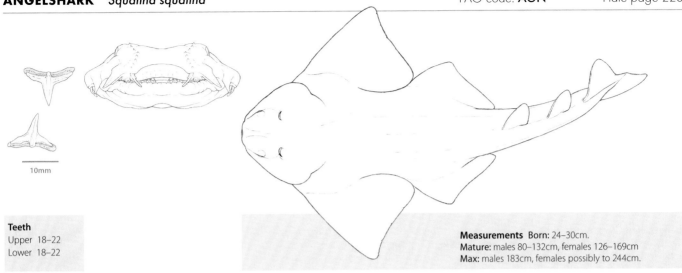

Teeth
Upper 18–22
Lower 18–22

10mm

Measurements Born: 24–30cm.
Mature: males 80–132cm, females 126–169cm
Max: males 183cm, females possibly to 244cm.

Identification Very large and stocky. Nasal barbels simple, with straight or spatulate tip. Smooth or weakly fringed anterior nasal flaps. Lateral head folds with a single triangular lobe each side. Small thorns on middle of back in young disappear with growth, but skin of back very rough; patches of small thorns on snout and between eyes. Very high, broad pectoral fins. Grey to reddish or greenish brown back with scattered small white spots and blackish dots and spots. White nuchal spot may be present. No ocelli. Young often have white reticulations and large dark blotches, adults plainer. Dorsal fins with dark leading edge, pale trailing edge.

Distribution Northeast Atlantic (historically from Norway to Mauritania), Canary Islands, Mediterranean and Black Seas. Relict populations persist off west Ireland and Wales, but it is now only reliably encountered in the Canary Islands (up to 100 animals have been seen in bays off Gran Canaria).

Habitat Mud or sand, inshore (<1m) on coasts and estuaries to deeper than 150m on continental shelf.

Behaviour Uses its large pectoral fins to fan out a depression in the seabed, then partly covers itself with sediment in ambush, leaving its eyes protruding. Very torpid and reluctant to move by day, even when uncovered by divers, but an Angelshark was reported to rise off the bottom and circle a diver with its mouth open when disturbed. Swims strongly off the bottom at night. The Angelshark is (or was) seasonally

migratory, moving north or into deeper water in summer. A tag and release programme off the Irish west coast between 1970 and 2006 received tag returns from northern and south-eastern Ireland, southwest and southeast England, France and northern Spain.

Biology Viviparous, 7–25 pups per litter (number increases with the size of the female). Gestation period is 8–10 months. Pups were born during December to February in the Mediterranean, and July in England. Feeds mainly on flatfishes, skates, crustaceans and molluscs. One record of a cormorant swallowed!

Status IUCN Red List: **Critically Endangered**. This formerly common angelshark is now rare or extirpated throughout most of its former range, due to over-fishing. It may still survive in the southern Mediterranean and northwest Africa, but sightings of all angelshark species have declined steeply in these heavily fished coastal regions. The Canary Islands appears to be the last major refuge for this shark. It is now strictly protected (hopefully not too late) throughout much of its range by fisheries and biodiversity regulations, including in EU, UK and Mediterranean waters. It is listed in Appendix I of the Convention on the Conservation of Migratory Species (CMS), which requires all signatory countries to protect the species and its habitat, and in regional biodiversity conventions (HELCOM, OSPAR Berne and Barcelona). If caught, this shark must be released extremely carefully in the best possible condition. Please report all sightings to the Angel Shark Project (see p. 241).

Angelshark, *Squatina squatina*, Canary Islands.

OCELLATED ANGELSHARK *Squatina tergocellatoides* FAO code: **SUN** Plate page 232

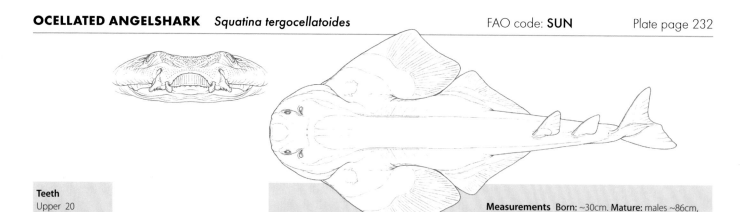

Teeth
Upper 20
Lower 20

Measurements Born: ~30cm. **Mature:** males ~86cm, females ~100cm.
Max: at least 100cm.

Identification Concave space between eyes. Strongly but finely fringed nasal barbels. Strongly fringed anterior nasal flaps. Lateral head folds have two low, rounded lobes each side. No enlarged predorsal thorns on back. Dorsal colour light yellowish brown with a dense scattering of small, round white spots. Six pairs of large ocelli (dark rings around light centres) on pectoral and pelvic fins and on tail base. Dorsal fins with black base and leading edge.
Distribution Northwest and Central Indo-Pacific. South China Sea: strait between China and Taiwan. Southeast Asia and northeast Malaysia.
Habitat Continental shelf, 100–300m.
Behaviour Unknown. **Biology** Viviparous.
Status IUCN Red List: Endangered. A bycatch of intensive trawl and possibly gillnet and longline fisheries operating throughout its range. It is retained for meat, fins, and fish meal.

VARI'S ANGELSHARK *Squatina varii* FAO code: **ASK** Plate page 232

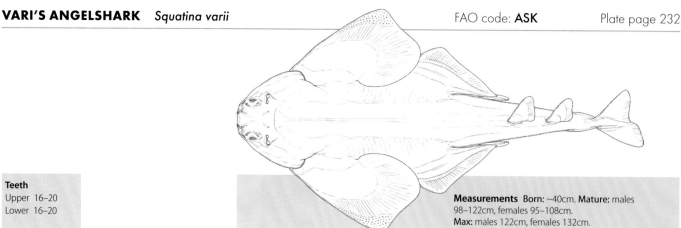

Teeth
Upper 16–20
Lower 16–20

Measurements Born: ~40cm. **Mature:** males 98–122cm, females 95–108cm.
Max: males 122cm, females 132cm.

Identification Nasal flaps with two lateral, elongate barbels flanking a slightly fringed median rectangular to rounded barbel. Dorsal midback lacks thorns or enlarged dermal denticles. Dorsal fins angular, similar in size; first dorsal fin originating just behind pelvic fin free rear tips. Dark to light brown above with scattered white spots variably present or absent on body trunk and pectoral fins; white spots if present on pectoral fins arranged in pairs from near pectoral origin, fin midbase, and posteriorly on pectoral fin. Dark blotches in three pairs located on lateral surface of tail; occasionally dorsal surface has irregular dark spots. Mostly creamy-white below with fin edges slightly darker.
Distribution Southwest Atlantic: Brazil, from Sergipe state to northern Rio de Janeiro state.
Habitat Continental slopes, 195–666m.
Behaviour Unknown.
Biology Viviparous, nothing else known.
Status IUCN Red List: Least Concern. There is no fishing activity within this species' deepwater range.

HETERODONTIFORMES   bullhead sharks

An ancient order with a long fossil record (almost to the beginning of the Mezozoic Era) and many fossil species, now represented by one living family. The taxonomic genus name *Heterodontus*, 'different-teeth', refers to the remarkably dissimilar small, pointed, holding front teeth, compared with the large, blunt rear teeth that are used for crushing invertebrates.

Identification Bullhead sharks are small to medium-sized (maximum total length 165cm, mostly smaller than 100cm) stout-bodied sharks, whose bodies taper posteriorly from the large head down to their tails. They have two dorsal fins (each preceded by a spine) and an anal fin. The caudal fin has a conspicuous sub-terminal lobe. All species have blunt, 'pig-like' snouts, small mouths in front of the eyes, enlarged first gill slits, prominent eye ridges, rough skin and paddle-like paired fins. The teeth are sharp and pointed in the front, but flattened and crushing (molariform) at the back of the jaw.

Biology These are sluggish benthic sharks, which swim slowly or crawl over rocky, kelp-covered and sandy bottom, hunting for prey. They are more active by night than during the day, when some species rest in rocky crevices and caves or with their heads tucked under a ledge. They may return every dawn to the same daily resting place. Bullhead sharks are oviparous, laying unique large screw- or auger-shaped leathery eggcases. Large young (bigger than 4cm) hatch more than five months later (incubation may take up to 12 months in some species, possibly being influenced by water temperature). Where known, the appearance of the eggcases is quite characteristic for each species, although it has not been described for all of them. Most bullhead sharks exhibit strong segregation by sex, size, and life history stage. At least two species lay eggs in particular 'nesting' sites. At least one species is migratory when adult, returning each year after long migrations to its breeding sites, but most have a restricted distribution. They mainly feed on benthic invertebrates (sea urchins, crabs, shrimp, marine gastropods, oysters and worms), rarely on small fishes.

Status Rare to uncommon, not important in commercial fisheries but often taken as a bycatch. They are also caught by sports fishers and divers, but most species are assessed as Least Concern, none are known to be threatened. Some thrive in aquaria and breed successfully under good conditions.

Juvenile Horn Shark, *Heterodontus francisci*, California (p. 252).

Heterodontidae: bullhead or horn sharks

Order Heterodontiformes contains just one family, Heterodontidae: a small taxonomic group with a single genus, *Heterodontus*, containing nine similar-looking species. Both of the widely used common English names for this family are derived from the broad crest over the eyes. These are mostly warm-temperate to tropical sharks, typically nocturnal, usually found where the water temperature is above 21°C in the east and west Pacific and the Indian Ocean. They are absent from the Atlantic Ocean. These are the only living sharks that have a fin spine in front of each dorsal fin and an anal fin. They are unique among sharks in that they may pick up their newly laid eggs and wedge them (while the casing is still soft) into a suitable location for them to develop. Otherwise, not much is known about the life histories of these rather curious-looking species, whose morphological characteristics are rather similar to those of the prehistoric hybodont sharks.

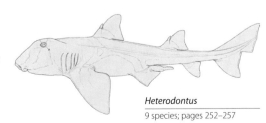

Heterodontus
9 species; pages 252–257

Plate 25 HETERODONTIDAE I

An ancient order with long fossil record, now only one living family of nine species. Stout-bodied; blunt pig-like snout, small mouth, enlarged first gill slit, prominent eye ridges; paddle-like paired fins, two spined dorsal fins, no anal fin; rough skin.

Horn Shark *Heterodontus francisci*   page 252

East Pacific; rock, kelp and sand, intertidal to 152m. First dorsal fin origin over pectoral fin bases; dark to light brown usually with small (less than 1/3 eye diameter) dark spots, dusky patch under eyes, no light bar between eye ridges, no harness pattern; young more vividly coloured, with obvious dark saddles.

Crested Bullhead Shark *Heterodontus galeatus*   page 253

Southwest Pacific: east Australia; rock, seaweed and seagrass, intertidal to 93m. Extremely high eye ridges; light brown or yellow-brown, no spots, broad dark blotch under eyes, dark bar between eye ridges, 5 dark broad bands or saddles; young with more obvious dark bands or saddles.

Japanese Bullhead Shark *Heterodontus japonicus*   page 253

Northwest Pacific; rock and kelp, intertidal to 100m. First dorsal fin origin over pectoral fin bases; tan to brown, no spots, dark blotch under eyes less distinct in adults, about 12 irregular dark saddles and bands; young juveniles more vividly coloured.

Port Jackson Shark *Heterodontus portusjacksoni*   page 255

Australia except north coast; temperate, intertidal to 275m. Grey to light brown or whitish, no spots, dark band under eyes, dark bar between eye ridges, unique dark striped harness pattern.

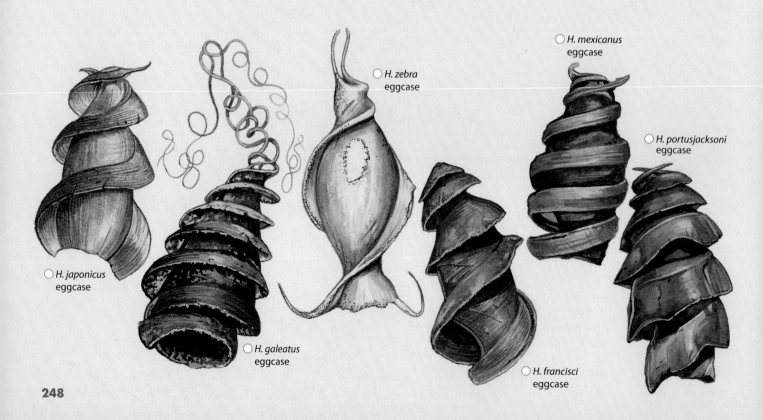

H. mexicanus eggcase

H. zebra eggcase

H. portusjacksoni eggcase

H. japonicus eggcase

H. galeatus eggcase

H. francisci eggcase

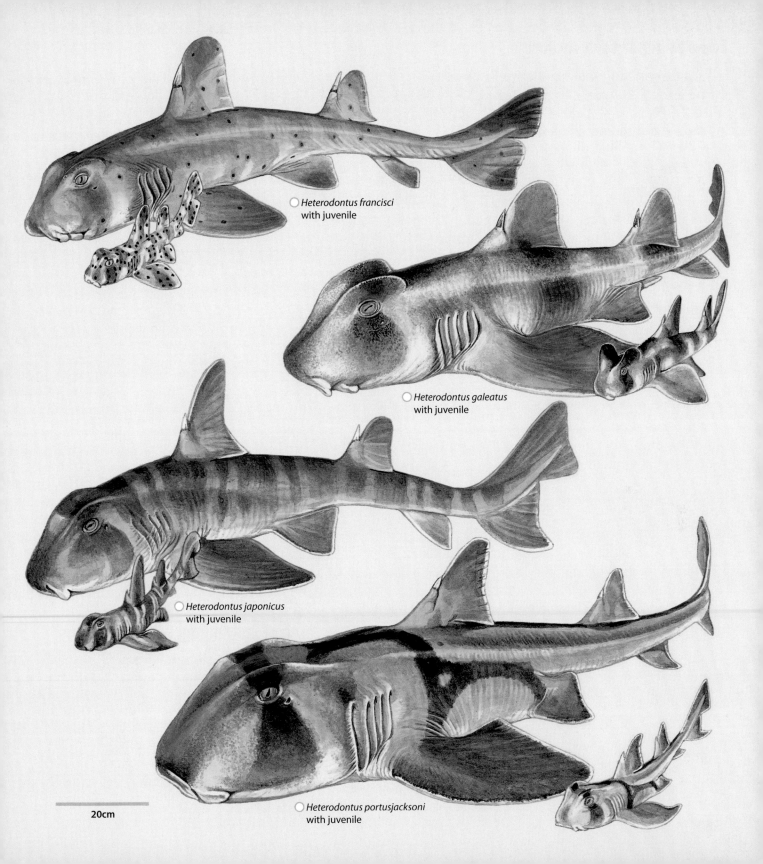

○ *Heterodontus francisci*
with juvenile

○ *Heterodontus galeatus*
with juvenile

○ *Heterodontus japonicus*
with juvenile

○ *Heterodontus portusjacksoni*
with juvenile

20cm

Plate 26 HETERODONTIDAE II

○ **Mexican Hornshark** *Heterodontus mexicanus* page 254

East Pacific; rock, coral reefs and sand, intertidal to 50m. First dorsal fin origin over pectoral fin bases; light grey-brown with large (same size or larger than eye diameter) black spots, 1–2 indistinct blotches under eye, light bar between eye ridges, no harness pattern; young with more obvious dark blotches.

○ **Oman Bullhead Shark** *Heterodontus omanensis* page 254

North Indian Ocean; 72–80m. First dorsal fin origin over pectoral fin inner margins; tan to brown, no spots, dark blotch under eye, dark bar between eye ridges, no harness pattern, 4–5 broad dark brown saddles, dark-tipped fins with white spot on dorsal fin tips, no information on hatchlings.

○ **Galapagos Bullhead Shark** *Heterodontus quoyi* page 256

East Pacific; rock and coral reefs, 3–40m. First dorsal fin origin over pectoral fin inner margins; light grey or brown usually with large (less than ½ eye diameter) black spots, mottled dark spots or blotches under eye, no light bar between eye ridges, no harness pattern; young with less distinct markings and small spots.

○ **Whitespotted Bullhead Shark** *Heterodontus ramalheira* page 257

North and west Indian Ocean; 40–305m. Dark red-brown, usually with small white spots, no dusky patch under eye in adults, no light bar between eye ridges, no harness pattern; hatchlings with unique striking whorl pattern that fades and vanishes with age, parallel dark lines between and under eyes change to dusky patch in large juveniles and are lost in adults.

○ **Zebra Bullhead Shark** *Heterodontus zebra* page 257

West and Central Indo-Pacific; rock, kelp and sand, intertidal to 200m. White or cream, no spots, no light bar between eye ridges, no harness pattern but striking black to dark brown zebra-like narrow vertical saddles and bands; young similar but bands are red-brown.

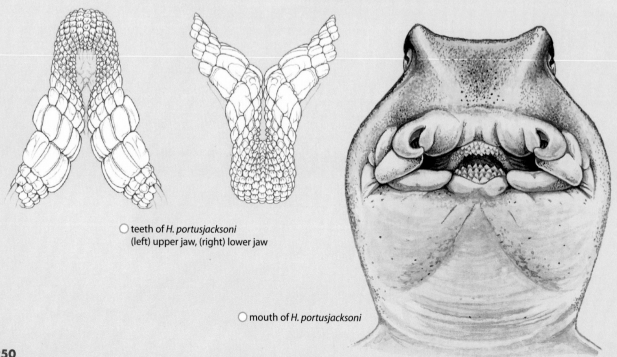

○ teeth of *H. portusjacksoni*
(left) upper jaw, (right) lower jaw

○ mouth of *H. portusjacksoni*

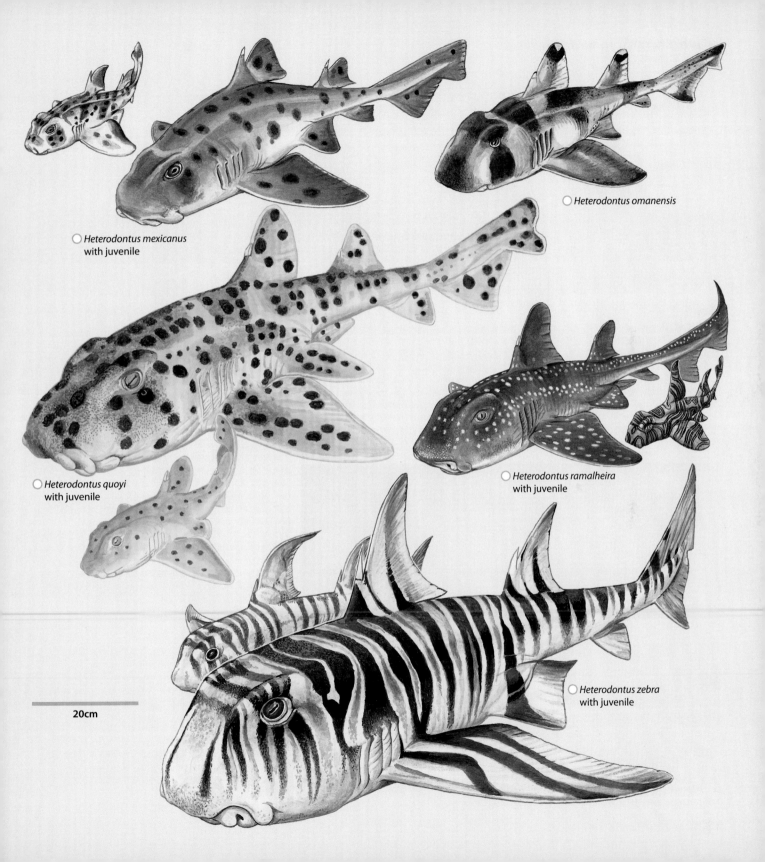

○ *Heterodontus mexicanus*
with juvenile

○ *Heterodontus omanensis*

○ *Heterodontus quoyi*
with juvenile

○ *Heterodontus ramalheira*
with juvenile

20cm

○ *Heterodontus zebra*
with juvenile

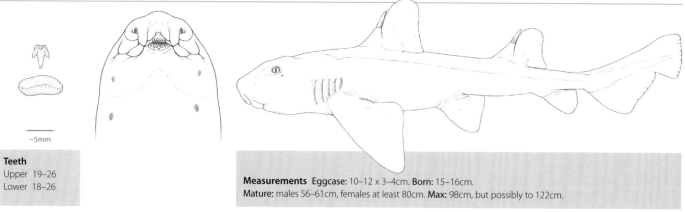

Teeth
Upper 19–26
Lower 18–26

~5mm

Measurements Eggcase: 10–12 x 3–4cm. **Born:** 15–16cm.
Mature: males 56–61cm, females at least 80cm. **Max:** 98cm, but possibly to 122cm.

Identification Dark to light grey or brown, usually with small (less than one-third eye diameter) dark spots. Small dark spots on dusky patch below eye. No light bar between the eye ridges. No harness pattern. First dorsal fin origin over pectoral fin bases, no white fin tips. Young are more brightly coloured, with obvious dark saddles.

Distribution East Pacific: USA (mainly southern California) to Mexico (Baja California, Gulf of California), Colombia, Ecuador and Peru.

Habitat Intertidal to at least 152m (mainly 2–11m). Rocky habitats (often in deep crevices and caves), kelp, sandy gullies and flats. Horn Shark populations are highly segregated depending upon their life stage. Juveniles appear to prefer relatively shallow, sandy habitats, whereas adults tend to occupy either shallow rocky reefs or dense algae. Intermediate-sized adolescent Horn Sharks (35–50cm total length) tend to stay in deeper water, usually 40–150m. Horn Sharks occupy the same habitat as the Swell Shark, *Cephaloscyllium ventriosum*, but their relative abundance shifts depending on the water temperature; Horn Sharks are more abundant in warmer water, usually over 21°C, while Swell Sharks are more abundant when the water temperatures are cooler.

Behaviour These are poor, sluggish swimmers with a small home range, possibly undertaking a limited winter migration into deep water in the north. They use their mobile, muscular, paired fins to crawl over the seabed while hunting for prey. Horn Sharks are mainly solitary (although small aggregations have been reported). The separation of their habitat by size and life history stage, described above, reduces competition for habitat between older, more experienced sharks and juveniles. Furthermore, adult Horn Sharks are nocturnal, active at night and seldom moving from their preferred resting place during the day, while juvenile Horn Sharks tend to be more active by day. Juveniles have an interesting relationship with Bat Rays (*Myliobatis californicus*): the depressions and pits in the sand left by feeding rays provide an important protective habitat for young Horn Sharks on an otherwise featureless bottom and a source of food, as they expose prey items for the young sharks to feed on.

Biology Oviparous. Mating takes place during December–January. Eggs are deposited February–April under rocks or in crevices, and hatch 7–9 months later. In captivity, 2 eggs are laid every 11–14 days for 4 months. Newly hatched Horn Sharks have an internal yolk-sac that supplies nutrition for the first month of life; they only start feeding one month after hatching. Horn Sharks feed on benthic invertebrates and, rarely, small fishes.

Status IUCN Red List: Data Deficient. The Horn Shark has no commercial fisheries interest. It is caught by divers for sport (the fin spines are used for jewellery) and may bite when provoked. Horn Sharks are hardy and easy to maintain and breed in captivity.

Horn Shark, *Heterodontus francisci*, San Benito Island, Baja California.

CRESTED BULLHEAD SHARK *Heterodontus galeatus*

FAO code: **HEG**   Plate page 248

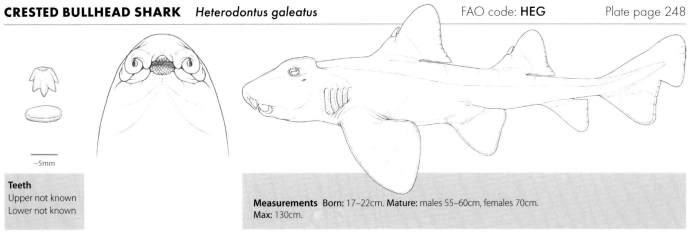

~5mm

Teeth
Upper not known
Lower not known

Measurements Born: 17–22cm. **Mature:** males 55–60cm, females 70cm.
Max: 130cm.

Identification Extremely high, short eye-ridges; depth between ridges about equal to eye-length. Light brown or yellowish brown with 5 dark, broad bands or saddles (more obvious in young). No light or dark spots. Dark bar between eyes; broad dark blotch under eye.
Distribution Southwest Pacific: east Australia; southern Queensland and New South Wales.
Habitat Intertidal to 93m (more common in deeper water), rocky reefs, seagrass beds and in seaweed.
Behaviour Forces its way between rocks to find prey.
Biology Oviparous, lays 10–16 corkscrew-shaped eggcases per year during winter, with long tendrils at ends for anchoring in seaweeds or sponges; these hatch 5–8 months later. Prey are mainly sea urchins, also crustaceans, molluscs and small fishes.
Status IUCN Red List: Least Concern. Relatively uncommon. Occurs only rarely in bycatch, usually released alive. Bred in captivity.

JAPANESE BULLHEAD SHARK *Heterodontus japonicus*

FAO code: **HEJ**   Plate page 248

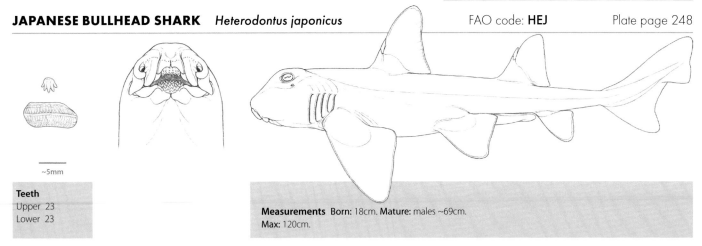

~5mm

Teeth
Upper 23
Lower 23

Measurements Born: 18cm. **Mature:** males ~69cm.
Max: 120cm.

Identification First dorsal fin origin over pectoral fin bases. Tan to brown, with ~12 irregular dark saddles and bands. No spots. Light bar between eyes, dark blotch under eye (indistinct in large adults). Hatchlings brighter coloured.
Distribution Northwest Pacific: Japan, Koreas, north China and Taiwan.
Habitat Bottom, 6–100m. Prefers rocky areas and kelp.
Behaviour Moves by slow, sluggish swimming and 'walking' with mobile paired fins. Protrudes jaws to capture prey. Several females share one 'nest' but do not guard the eggs.
Biology Oviparous, eggs are laid in pairs among rocks or in kelp at depth of 8–9m, during 6–12 spawnings over March–September (mainly March–April in Japan). Eggs hatch in about a year. Feeds on invertebrates and small fishes.
Status IUCN Red List: Least Concern. Low fisheries importance. Kept in aquaria.

MEXICAN HORNSHARK *Heterodontus mexicanus*

FAO code: **HEM**   Plate page 250

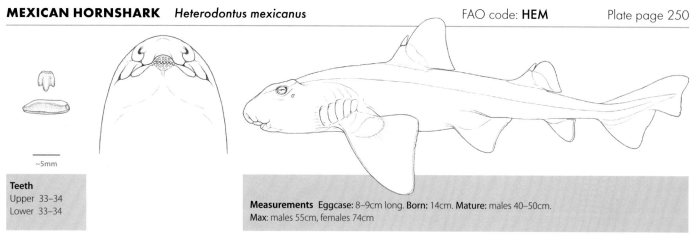

~5mm

Teeth
Upper 33–34
Lower 33–34

Measurements Eggcase: 8–9cm long. **Born:** 14cm. **Mature:** males 40–50cm.
Max: males 55cm, females 74cm

Identification First dorsal fin origin over pectoral fin bases. Light grey-brown to dark grey; large black spots (more than half eye diameter). Light bar between eye ridges; 1–2 indistinct blotches under eye. Eggcases with long tendrils and rigid T-shaped spiral flanges.
Distribution East Pacific: Mexico to Peru.
Habitat Found on rock and coral reefs, on sandy bottoms from close inshore, 0–50m, and around shallow seamounts.
Behaviour Tendrils may be used to anchor eggcases.
Biology Oviparous; corkscrew-shaped eggcases, about 8cm long, with 4 or more continuous flanges and thin tendrils for anchoring at one end. Feeds on crabs and demersal fishes.
Status IUCN Red List: **Least Concern**. Minimal interest to fisheries.

OMAN BULLHEAD SHARK *Heterodontus omanensis*

FAO code: **HDQ**   Plate page 250

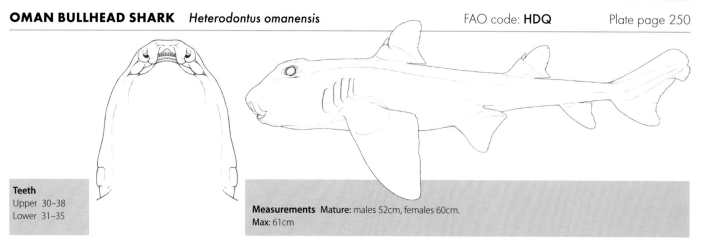

Teeth
Upper 30–38
Lower 31–35

Measurements Mature: males 52cm, females 60cm.
Max: 61cm

Identification First dorsal fin origin over pectoral fin inner margins. Distinguishable from *Heterodontus japonicus* by colour pattern and very low dorsal fins (possibly also a smaller size). Tan to brown with 4–5 broad, dark brown saddles. Dark bar between eyes and blotch under eyes. No spots on body. Dark-tipped fins with a white spot on dorsal fin tips. Hatchling colour pattern unknown.
Distribution North Indian Ocean: Oman and Pakistan.
Habitat Continental shelf, 72–80m, likely occurs mostly on rocky reefs, but also occasionally on soft bottoms.
Behaviour Unknown.
Biology Oviparous; eggcases about 10cm long, corkscrew-shaped, oval with two thin ridges making two turns; one end with two flanges, continuous ridges, and ridges at other end with thin extended tendrils. Diet unknown, but likely includes benthic invertebrates.
Status IUCN Red List: **Data Deficient**. Population size and impact of fisheries are unknown.

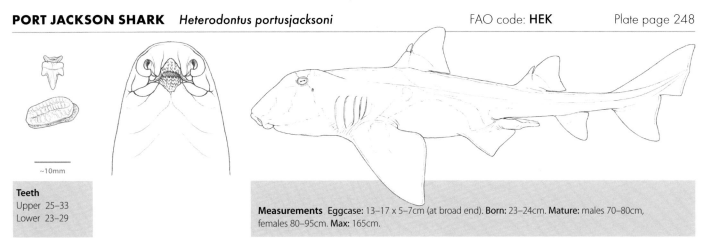

~10mm

Teeth
Upper 25–33
Lower 23–29

Measurements Eggcase: 13–17 x 5–7cm (at broad end). **Born:** 23–24cm. **Mature:** males 70–80cm, females 80–95cm. **Max:** 165cm.

Identification Grey to light brown or whitish with very distinctive black striped 'harness' markings on the body and a band across the eyes. No spots. The eggcases are spiral flange-shaped, about 15cm long and 8cm wide.

Distribution Australia. First described from Port Jackson (the inlet containing Sydney Harbour). Some of the adult females found off the continent in winter will migrate up to 800km south to Tasmania in summer, then return to warmer continental waters in autumn. A New Zealand record may have been a vagrant that overshot Tasmania by more than 2000km, or an aquarium release.

Habitat Temperate waters, intertidal to at least 275m. Often found on sandy bottoms next to rocky reefs, or in caves and gullies floored with sand. Eggs are usually laid on shallow rocky reefs in <5m (occasionally to 20–30m).

Behaviour Port Jackson Sharks are mostly nocturnal. They rest by day, propped up on their large pectoral fins, and are often seen in groups in or near a few favoured caves and gullies within their small home range. If caught and moved several kilometres away, they are able to find their way back to the same location. Similarly to other bullhead sharks, adults forage quite actively at night for food, preferring benthic invertebrates and small fishes. Like other members of this genus, they exhibit strong segregation by sex, size, and life history stage. Mature females use traditional collective egg-laying sites – shallow rocky reefs

with crevices and caves – where they wedge their auger-shaped eggs point downwards. Hatchlings then move to nearby nursery grounds until adolescent, when they move well offshore and segregate by sex, joining the adult population a few years later. Adult males and females also occur in different groups, and undertake complex seasonal breeding migrations. Some of the males and all females in the population move to inshore reefs in July to mate and lay their eggs, then return offshore. Some adults remain in cooler offshore waters all summer, others migrate south, up to 850km from their breeding grounds, and return in the autumn.

Biology As described above, mating and egg-laying occurs in late winter and spring. Oviparous, 10–16 eggs are laid in pairs every 8–17 days, mainly August–September, and their shape ensures that they remain where the female wedges them, in rock crevices on sheltered rocky reefs. The young hatch after about 12 months of development within the eggcase, and will mature in about 8–10 years for males and 11–14 years for females. The size at maturity varies between populations, for example the New South Wales population matures at a larger size and grows to a larger maximum length than the Victoria population, and there are some genetic differences. Port Jackson Sharks reach a maximum age of about 28 years. They feed mainly on sea urchins, other benthic invertebrates, also small fishes.

Status IUCN Red List: Least Concern. Port Jackson Sharks are abundant. Although taken as a bycatch in fisheries, they are of no value for food and are mostly returned alive. They are kept and bred successfully in captivity.

Port Jackson Sharks, *Heterodontus portusjacksoni*, New South Wales, Australia.

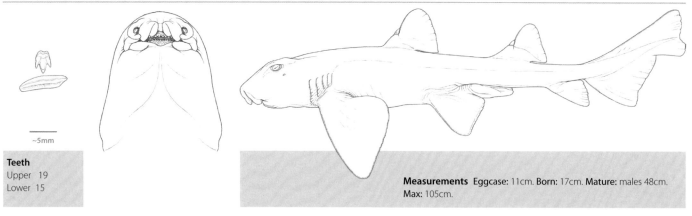

~5mm

Teeth
Upper 19
Lower 15

Measurements Eggcase: 11cm. **Born:** 17cm. **Mature:** males 48cm. **Max:** 105cm.

Identification First dorsal fin origin over pectoral fin inner margins. Light grey or brown with large black spots (greater than ½ eye diameter), smaller and less distinct in young. Mottled dark spots or blotches under eyes; no light bar between eyes.

Distribution East Pacific: Peru (coast and islands) and Galapagos Islands (Ecuador).

Habitat Rocky and coral reefs, rests on ledges of vertical rock surfaces, 3–40m.

Behaviour Nocturnal, very poorly known.

Biology Oviparous. Feeds on crabs.

Status IUCN Red List: Least Concern. This species' inshore rock and reef habitat is mostly unsuitable for commercial fishing operations. Specimens taken in artisanal line and trap fisheries are likely discarded alive, and bullhead sharks have high post-release survival rates.

Galapagos Bullhead Shark, *Heterodontus quoyi*, Punta Vicente Roca, Isla Isabela, Galapagos.

WHITESPOTTED BULLHEAD SHARK   *Heterodontus ramalheira*   FAO code: **HEA**   Plate page 250

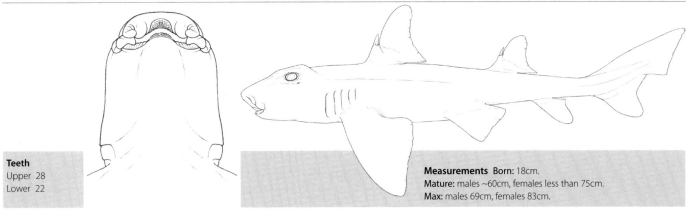

Teeth
Upper 28
Lower 22

Measurements Born: 18cm.
Mature: males ~60cm, females less than 75cm.
Max: males 69cm, females 83cm.

Identification Dark reddish brown with white spots and darker, indistinct saddles in adults. Hatchlings with striking unique whorled pattern of thin, curved parallel dark lines, lost with growth. Parallel dark lines between and under eyes in hatchlings change to a dusky patch in larger juveniles, lost in adults.
Distribution North and west Indian Ocean: southern Mozambique to Somalia, and eastern Arabian Peninsula, Yemen and Oman.
Habitat Outer continental shelf and uppermost slope in deepish water (40–305m, mostly deeper than 100m).
Behaviour Unknown.
Biology Oviparous. Feeds on bottom-dwelling invertebrates, such as crabs.
Status IUCN Red List: Data Deficient. Apparently uncommon. Taken as bycatch.

ZEBRA BULLHEAD SHARK   *Heterodontus zebra*   FAO code: **HEZ**   Plate page 250

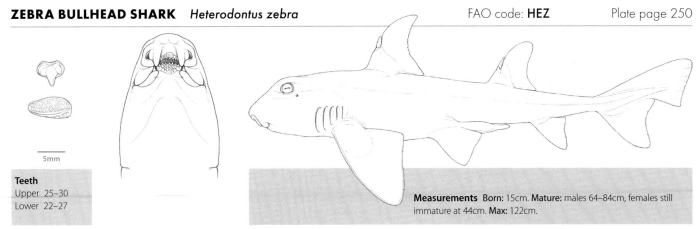

5mm

Teeth
Upper 25–30
Lower 22–27

Measurements Born: 15cm. **Mature:** males 64–84cm, females still immature at 44cm. **Max:** 122cm.

Identification White or cream with striking zebra-striped colour pattern of numerous black or dark brown (red-brown in juveniles) vertical saddles and bands over body, tail and head. No spots.
Distribution West and Central Indo-Pacific: Japan, Koreas, China, Taiwan, Vietnam, Indonesia, northwest Australia.
Habitat Continental and insular shelves of the west Pacific from inshore down to at least 50m in the South China Sea, deeper (150–200m) off western Australia; one specimen was reported from 398m, which represents the deepest known record for a bullhead shark species.
Behaviour Possibly mainly active at night.
Biology Oviparous, poorly known. Feeds on bottom-dwelling invertebrates and small fishes.
Status IUCN Red List: Least Concern. Apparently common. Taken in bycatch. A popular and attractive aquarium fish, both for hobbyists and public display.

ORECTOLOBIFORMES carpetsharks

This order contains about 45 species in seven families: Parascylliidae (collared carpetsharks), Brachaeluridae (blind sharks), Orectolobidae (wobbegongs), Hemiscylliidae (longtailed carpetsharks), Ginglymostomatidae (nurse sharks), Stegostomatidae (Zebra Shark) and Rhincodontidae (Whale Shark). Recent genetic investigation suggests that the Zebra Shark, Whale Shark and the Shorttail Nurse Shark are more closely related to each other than they are to other species of nurse sharks.

Identification Two spineless dorsal fins and an anal fin. Nostrils have barbels (rudimentary in Whale Shark) and are connected via nasoral grooves to short mouths that end in front of the eyes (no nictitating eyelids).

Distribution Worldwide, warm-temperate and tropical seas, from intertidal to deep water, with the greatest diversity and endemism in the tropical Central Indo-Pacific. Smallest are mainly sluggish bottom-dwelling species. Larger species tend to be more active and wide-ranging (e.g. the circumglobal pelagic Whale Shark).

Biology Varied reproductive strategies, including oviparity (egg-laying) and yolk-sac viviparity (foetuses retained inside the female and nourished by their yolk-sac until birth).

Status Some are hardy in captivity. May be taken in fisheries bycatch, retained where traditional target species are depleted, and might become threatened in heavily fished areas.

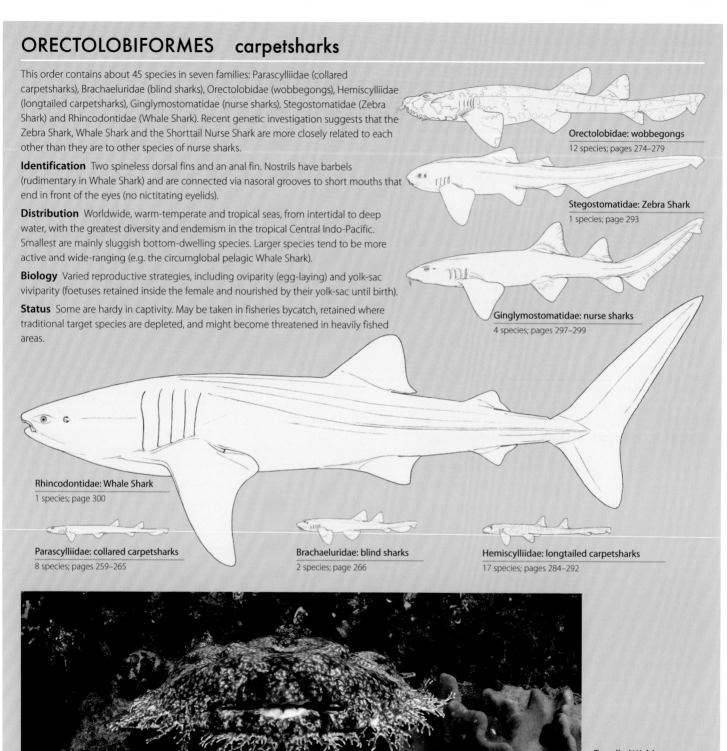

Orectolobidae: wobbegongs
12 species; pages 274–279

Stegostomatidae: Zebra Shark
1 species; page 293

Ginglymostomatidae: nurse sharks
4 species; pages 297–299

Rhincodontidae: Whale Shark
1 species; page 300

Parascylliidae: collared carpetsharks
8 species; pages 259–265

Brachaeluridae: blind sharks
2 species; page 266

Hemiscylliidae: longtailed carpetsharks
17 species; pages 284–292

Tasselled Wobbegong, *Eucrossorhinus dasypogon*. Raja Ampat, Indonesia (p. 274).

Parascylliidae: collared carpetsharks

Two genera in west Pacific, from inshore to deepish continental shelf. All five *Parascyllium* species are Australian endemics. Genus *Cirrhoscyllium* (three species) occurs from Vietnam to Taiwan and Japan.

Identification Small, slender sharks (less than 1m). First dorsal fin originates behind pelvic fin bases, second well behind anal fin origin. Mouth entirely in front of eyes; tiny spiracles. *Cirrhoscyllium* have unique cartilage-cored paired barbels on throat, dark saddles and no spots or collar markings. *Parascyllium* have no barbels on throat and a pattern of saddles and spots.

Biology Little-known benthic species, which can apparently change colour to match the seabed. Some (possibly all) are oviparous, laying elongated, flattened eggcases.

Status Some are hardy in captivity. May be taken in catch, retained where traditional target species are depleted, and may be threatened in heavily fished areas. This is a poorly known family. At the time of writing, half of species were assessed as Data Deficient, 38% as Least Concern, and only one species was assigned to a threatened category in the IUCN Red List.

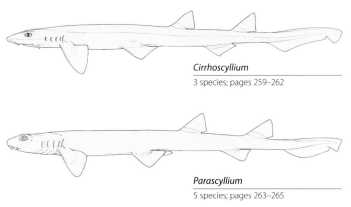

Cirrhoscyllium
3 species; pages 259–262

Parascyllium
5 species; pages 263–265

BARBELTHROAT CARPETSHARK   *Cirrhoscyllium expolitum*   FAO code: **OPC**   Plate page 260

~1mm

Teeth
Upper 28
Lower 25

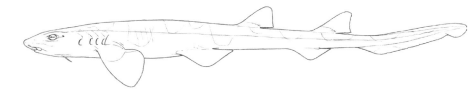

Measurements Known from two females, 30.6cm and (adult) 33.5cm.

Identification Head length 3 times first dorsal fin base. Cartilage-cored paired barbels on throat; nasoral grooves. Mouth in front of eyes. 6–10 diffuse saddlemarks on back, not C-shaped. An elongated, rounded saddle on each side of back and tail between pectoral and pelvic fin bases extends over pelvic fin bases.
Distribution Northwest Pacific: South China Sea off China to Luzon, Philippines. Gulf of Tonkin, Vietnam and off the Tanegashima Islands, Japan.
Habitat Bottom, outer continental shelf, 183–190m.
Behaviour Unknown.
Biology Presumed oviparous, but eggcases have not been found.
Status IUCN Red List: Data Deficient. Known from two specimens, one of which was captured in a research survey. This carpetshark is presumably rare or uncommon, and taken as a bycatch of demersal trawls.

Plate 27 PARASCYLLIIDAE and BRACHAELURIDAE

Parascylliidae: slender; mouth in front of eyes, nostrils with barbels and nasoral grooves, tiny spiracles; first dorsal fin originates behind pelvic fin bases, second dorsal fin well behind anal fin origin. *Cirrhoscyllium*: unique cartilage-cored pair of barbels on throat, saddles, no spots; *Parascyllium*: no barbels on throat, saddles and spots.

○ **Collared Carpetshark** *Parascyllium collare* page 263

Southwest Pacific: east coast Australia; rocky 20–230m. Circumnarial grooves; light yellow- to red-brown with dark unspotted sharp edged collar, 5 dusky saddles, large dark spots except on pectoral fins, fewer than 6 spots on sides of tail between dorsal fins.

○ **Elongate Carpetshark** *Parascyllium elongatum* page 263

Southeast Indian Ocean: west Australia; bottom to at least 50m. Circumnarial grooves; pale brown with white spots in vertical rows separated by darker brown bands over abdomen; no dark collar or black spots.

○ **Barbelthroat Carpetshark** *Cirrhoscyllium expolitum* page 259

Northwest Pacific; bottom 183–190m. Head is 3 times length of first dorsal fin base; 6–10 diffuse saddles not C-shaped, one long and round saddle between pectoral and pelvic fins extends over the pelvic fin bases.

○ **Rusty Carpetshark** *Parascyllium ferrugineum* page 264

Southeast Indian Ocean: south Australia; on or near bottom 5–150m. Circumnarial grooves; grey-brown with indistinct collar, 6–7 dusky saddles, dark spots, more than 6 spots on sides of tail between dorsal fins, more heavily spotted specimens in Tasmania.

○ **Taiwan Saddled Carpetshark** *Cirrhoscyllium formosanum* page 262

Northwest Pacific: southern Taiwan; 110–320m. Head 2.3–2.6 times length of first dorsal fin base; 6 diffuse saddles not C-shaped, 1 long and round saddle between pectoral and pelvic fins extends over the pelvic fin bases.

○ **Saddled Carpetshark** *Cirrhoscyllium japonicum* page 262

Northwest Pacific: southwest Japan; 250–290m. Long snout; 9 bold saddles, 1 C-shaped between pectoral and pelvic fin bases.

○ **Sparsely Spotted Carpetshark** *Parascyllium sparsimaculatum* page 264

Southeast Indian Ocean: west Australia; 205–245m. Circumnarial grooves; pale brown or grey with indistinct unspotted collar, 5 indistinct dark saddles, sparse large spots and blotches, fewer than 6 spots on sides of tail between dorsal fins.

○ **Necklace Carpetshark** *Parascyllium variolatum* page 265

Southeast Indian Ocean: south Australia; on seabed from shallow water to 180m. Dark grey to chocolate brown with dark white spotted collar, unmistakable highly variable pattern of dark blotches and dense white spots, obvious black spots on fins.

Brachaeluridae: stout; mouth in front of eyes, long barbels and nasoral and circumnarial grooves, large spiracles; second dorsal fin well in front of anal fin.

○ **Bluegrey Carpetshark** *Brachaelurus colcloughi* page 266

Southwest Pacific: east Australia; shallow inshore to 217m. Barbels with posterior hooked flap; first dorsal fin larger than second; grey with no spots, young with distinct black markings that fade with age.

○ **Blind Shark** *Brachaelurus waddi* page 266

Southwest Pacific: east Australia; rock and seagrass to 140m. First dorsal fin about the same size as second, anal and caudal fins almost touch; brown with small sparsely scattered white spots, dark saddles in young that become obsolete or indistinct in adults.

20cm

○ *Cirrhoscyllium expolitum*

○ *Cirrhoscyllium formosanum*

○ *Cirrhoscyllium japonicum*

○ Parascyllium collare

○ Parascyllium elongatum

○ Parascyllium ferrugineum

○ Parascyllium sparsimaculatum

○ Parascyllium variolatum

20cm

○ Brachaelurus waddi

○ Brachaelurus colcloughi
with juvenile

TAIWAN SADDLED CARPETSHARK *Cirrhoscyllium formosanum*   FAO code: **OPF**   Plate page 260

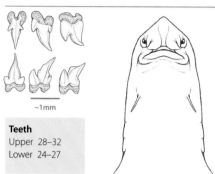

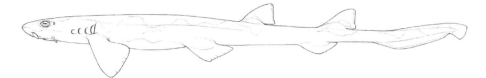

~1mm

Teeth
Upper 28–32
Lower 24–27

Measurements Born: unknown, smallest free-swimming individuals were 19cm. **Mature:** males 32cm, females 38cm. **Max:** known 40cm.

Identification Head length 2.3–2.6 times first dorsal fin base. Cartilage-cored paired barbels on throat; nasoral grooves. Mouth in front of eyes. Six diffuse saddlemarks on back, not C-shaped. An elongated, rounded saddle on each side of back between pectoral and pelvic fin bases extends over pelvic fin bases.
Distribution Northwest Pacific: known only off southern Taiwan.
Habitat Outer shelf, 110–320m.
Behaviour Little-known (collected by bottom trawlers and longlines).
Biology Oviparous, a 38cm female was pregnant, little else known.
Status IUCN Red List: Vulnerable. Virtually the entire restricted depth and geographic range of this endemic species is intensively trawled with small-mesh nets, with the exception of inshore areas where trawls are banned. Other shark populations in the area have declined significantly, but this species is often discarded and other carpetsharks have relatively high discard survival rates. Juveniles are still reported seasonally.

SADDLED CARPETSHARK *Cirrhoscyllium japonicum*   FAO code: **OPJ**   Plate page 260

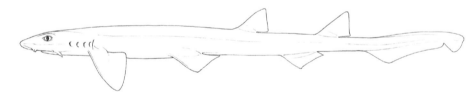

Teeth
Upper 23–32
Lower 22–27

Measurements Born: unknown. **Mature:** males ~ 37cm, females 44cm.
Max: 49cm.

Identification Long snout. Cartilage-cored paired barbels on throat; nasoral grooves. Mouth in front of eyes. Nine boldly marked saddle marks on sides of body, one of them C-shaped between pectoral and pelvic-fin bases.
Distribution Northwest Pacific: southwest Japan (Shikoku and Kyushu to Yakushima Island, possibly Ryukyu islands), and possibly off Guangdong and Hainan, China.
Habitat Uppermost slope, 250–290m.
Behaviour Unknown.
Biology Poorly known. Apparently oviparous (cased eggs discovered in a 45cm female).
Status IUCN Red List: Data Deficient. This species requires reassessment, once its distribution and taxonomy is better understood; it has previously been confused with *C. expolitum*.

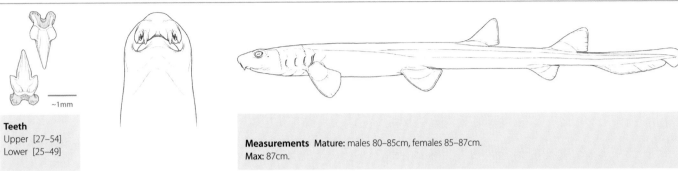

Teeth
Upper [27–54]
Lower [25–49]

~1mm

Measurements Mature: males 80–85cm, females 85–87cm.
Max: 87cm.

Identification Nasal barbels; nasoral and circumnarial grooves. Mouth in front of eyes. First dorsal fin origin behind pelvic fin bases. Anal fin origin well in front of second dorsal origin. Light yellowish to reddish brown. Dark, unspotted, sharp-edged collar over gills. Five dusky saddles on back. Large dark spots on body, tail and fins (except pectorals); less than 6 spots on sides of tail between dorsal fins.
Distribution Southwest Pacific: east coast Australia.
Habitat Rocky reefs and hard-bottomed trawl grounds on continental shelf, 20–230m.
Behaviour Unknown.
Biology Lays flattened, elongate eggcases.
Status IUCN Red List: **Least Concern**. Discarded bycatch of trawlers, high survival rates.

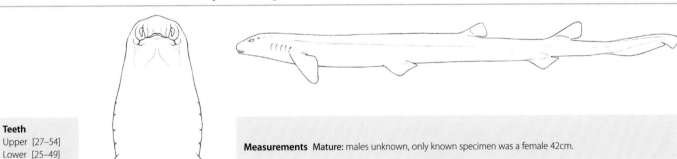

Teeth
Upper [27–54]
Lower [25–49]

Measurements Mature: males unknown, only known specimen was a female 42cm.

Identification Head relatively short, but body extremely elongate. Nasal barbels short; nasoral and circumnarial grooves present. Mouth anterior to very small eyes. First dorsal fin origin posterior to pelvic fin bases. Second dorsal fin origin slightly behind midpoint of anal fin. Pectoral fins rounded. Pale brown with white spots in vertical rows separated by darker brown bands over abdomen. No dark collar or black spots.
Distribution Southeast Indian Ocean: western Australia (south of Cape Leeuwin).
Habitat On or near bottom at 50m; nothing else known.
Behaviour Unknown.
Biology Unknown.
Status IUCN Red List: **Data Deficient**. Rare, known only from a single female specimen taken from the stomach of a Tope *Galeorhinus galeus*.

Teeth
Upper [27–54]
Lower [25–49]

Measurements Born: ~17cm. **Mature:** males ~60cm, females 75cm.
Max: 82cm.

Identification Nasal barbels; nasoral and circumnarial grooves. Mouth in front of eyes. First dorsal fin origin behind pelvic fin bases. Anal fin origin in front of second dorsal origin. Grey-brown. Indistinct dark collar around gills. 6–7 dusky saddles. Dark spots on body, tail and fins; more than 6 on sides of tail between dorsal fins (Tasmanian specimens more heavily spotted).
Distribution Southeast Indian Ocean: southern Australia.
Habitat On or near bottom near rocks, river mouths, in algae on reefs or in seagrass, 5–150m.
Behaviour Nocturnal. Hides in rocky caves and ledges by day.
Biology Oviparous, yellow eggcases with long tendrils laid in summer. Feeds on bottom-dwelling crustaceans and molluscs.
Status IUCN Red List: Least Concern. Not targeted by fisheries, rare in bycatch.

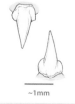

~1mm

Teeth
Upper 43–49
Lower 42

Measurements Max: at least 79cm.

Identification First dorsal fin origin behind pelvic fins. Anal fin origin well in front of second dorsal origin. Pale brownish or greyish above, lighter below. Inconspicuous, unspotted dusky half-collar around gills. Five indistinct, dark saddles on back and tail. Sparse, large, dark spots and blotches on body and fins (less than 6 on the sides of the tail between dorsal fins).
Distribution Southeast Indian Ocean: western Australia. Known from only a very small area.
Habitat Deepsea, upper continental slope, 205–245m.
Behaviour Unknown.
Biology Unknown.
Status IUCN Red List: Data Deficient (only three specimens known).

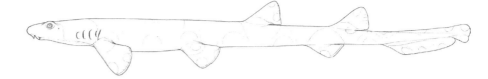

~1mm

Teeth
Upper [27–54]
Lower [25–49]

Measurements Max: ~91cm.

Identification Small dark greyish to chocolate brown shark with beautiful, unmistakable, highly variable pattern: broad, dark, white-spotted collar over gills; obvious black spots on all fins; and dark blotches and dense white spots on body (another white-spotted form from west Australia may be an undescribed species).

Distribution Southeast Indian Ocean: southern Australia. Eastern and western forms may be more than one species.

Habitat A variety of habitats on continental shelf to 180m, including sand, rocky reefs, kelp and seagrass beds.

Behaviour Nocturnal. Juveniles hide under rocks and bottom debris in shallow water.

Biology Virtually unknown, probably oviparous.

Status IUCN Red List: **Least Concern**. Not targeted by fisheries, rare in bycatch.

Brachaeluridae: blind sharks

Named because they close their eyelids when taken out of water. Endemic to east Australian coast, from a few centimetres deep to 217m, on rocky or coral reefs or seaweed. Two species in one genus, *Brachaelurus*.

Identification Small stout sharks (largest 120cm) with two spineless dorsal fins set far back. Anal fin origin well behind origin of second dorsal fin, gap between anal and lower caudal fins shorter than length of anal fin. Large spiracles below and behind eyes, nasoral and circumnarial grooves, long barbels, small transverse mouth in front of eyes, no lateral skin flaps on head.

Biology Viviparous (litters of 6–8 finish absorbing their large yolk-sacs just before birth). Feed on small fishes, crustaceans, squid and sea anemones. At least one species can survive long periods out of water.

Status Very restricted geographic range, one species is very rare. Survive well in captivity.

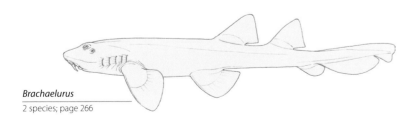

Brachaelurus

2 species; page 266

Blind Shark, *Brachaelurus waddi* (p. 266).

BLUEGREY CARPETSHARK *Brachaelurus colcloughi*

FAO code: **OBH**   Plate page 260

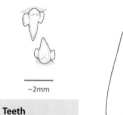

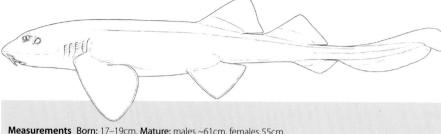

~2mm

Teeth
Upper [32]
Lower [21]

Measurements Born: 17–19cm. **Mature:** males ~61cm, females 55cm. **Max:** at least 85cm.

Identification Small, stout shark. Pair of long barbels with posterior hooked flap. Short mouth ahead of eyes. Large spiracles well behind eyes. Two spineless dorsal fins, first larger and originating over pelvic fin bases. Short precaudal tail and caudal fin. Greyish above, white below. No light spots. Conspicuous black markings on back, dorsal fins and caudal fin of young fade in adults.
Distribution Southwest Pacific: eastern Australia, known from a narrow area <2° latitude.
Habitat Bottom in very shallow (less than 4m) inshore areas to 217m; mostly less than 100m deep. Prefers coarse, silty sands, coral and rocky reefs with caves and crevices.
Behaviour Unknown.
Biology Viviparous with yolk-sac, 2–7 pups per litter. Reproductive cycle may be 2–3 years. Diet is mostly small bony fishes.
Status IUCN Red List: Vulnerable. Known from about 50 specimens in a small, well-surveyed area heavily used by fisheries and for recreation.

BLIND SHARK *Brachaelurus waddi*

FAO code: **OBW**   Plate page 260

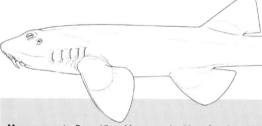

~2mm

Teeth
Upper [32]
Lower [21]

Measurements Born: 17cm. **Mature:** males 58cm, females ~50cm. **Max:** 120cm.

Identification Small, stout shark. Long tapering barbels; nasoral and circumnarial grooves. Small mouth in front of eyes. Large spiracles. Similar-sized dorsal fins set far back, origin of first over pelvic fin bases, second well in front of anal fin origin. Anal and lower caudal fins almost touch. Usually brown above with white spots. Dark saddles in young, obsolete in adults. Light yellowish below.
Distribution Southwest Pacific: east Australia.
Habitat Rocky shores (tidepools), reefs, seagrass beds, 0–140m. Juveniles found in high-energy surge zones.
Behaviour Nocturnal; hides in caves and under ledges by day, feeds at night.
Biology Viviparous, 7–8 pups per litter, usually born in November. Age at maturity about 7 years for males with a maximum age of 19 years; unknown for females. Feed on small fishes, crustaceans, squid and sea anemones.
Status IUCN Red List: Least Concern. Relatively common. Collected for aquaria. Rarely taken in fisheries, extremely hardy and survives release well.

Orectolobidae: wobbegongs

Three genera, 12 species, but additional undescribed species probably present within west Pacific range of the family, from Australia to Japan. Live on the bottom in warm-temperate to tropical continental waters, intertidal to deeper than 200m, often on rocky and coral reefs or on sandy bottom.

Identification Very distinctive, flattened, highly patterned, well-camouflaged sharks. Dermal flaps along sides of broad, flat head, long barbels, short mouth in front of eyes and almost at the very front of the short snout. Heavy jaws, two rows of enlarged, sharp, fang-like teeth in upper jaw, three in lower. Nasoral grooves, circumnarial grooves and flaps, symphysial grooves. Spiracles larger than upward-facing eyes. Two spineless dorsal fins and an anal fin. First dorsal fin origin over pelvic fin bases.

Biology Viviparous with yolk-sac, large litters of 20 or more young. Powerful seabed predators on benthic-dwelling animals (fishes, crabs, lobsters and octopuses), lurking camouflaged by colour patterns and dermal lobes around head, sucking in and impaling prey on large teeth. Uses paired fins to clamber around the bottom, even out of the water.

Status Potentially dangerous – bite if provoked. Some species kept and bred in aquaria. Some are important in fisheries. More than 80% of species are Least Concern in the IUCN Red List of Threatened Species, none are currently assessed as threatened.

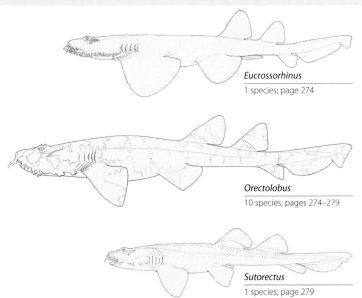

Eucrossorhinus
1 species; page 274

Orectolobus
10 species; pages 274–279

Sutorectus
1 species; page 279

Spotted Wobbegong, *Orectolobus maculatus*, Gold Coast, Australia (p. 277).

Plate 28 ORECTOLOBIDAE I

Very distinctive, flattened, well camouflaged; broad flat head, short mouth in front of eyes almost at short snout end, long barbels with nasoral circumnarial and symphysial grooves, dermal flaps on sides of head, spiracles larger than eyes; first dorsal fin origin over pelvic fin bases, anal fin.

○ **Tasselled Wobbegong** *Eucrossorhinus dasypogon*   page 274

Central Indo-Pacific; inshore coral reefs; 5-50m. Numerous highly branched dermal lobes on head, 'beard' on chin; very broad paired fins; indistinct saddles, variable reticulated pattern of narrow dark lines on light background, scattered enlarged dark blotches at line junctions.

○ **Japanese Wobbegong** *Orectolobus japonicus*   page 276

Northwest Pacific; inshore rocky and coral reefs to 200m. Long branched nasal barbels, 5 dermal lobes below and in front of eye; obvious dark dorsal saddles with spots and blotches and dark (not black), corrugated edges, light background with dark broad reticular lines.

○ **Northern Wobbegong** *Orectolobus wardi*   page 279

Central Indo-Pacific; shallow reefs less than 3m. Unbranched nasal barbels, 2 dermal lobes below and in front of eye, dermal lobes behind spiracle unbranched and broad; simple sombre colour pattern, 3 large dark light-edged rounded saddles, light background with indistinct darker mottling and a few dark spots.

○ **Cobbler Wobbegong** *Sutorectus tentaculatus*   page 279

Southeast Indian Ocean; depth to at least 35m. Rather slender with rows of warty dermal tubercles in rows on back and bases of very low long dorsal fins; simple unbranched nasal barbels, chin smooth, few slender short unbranched dermal lobes form isolated groups in 4–6 pairs; striking pattern of broad dark dorsal saddles with jagged corrugated edges, lighter background with irregular dark spots.

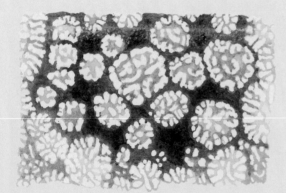
○ *Eucrossorhinus dasypogon*

○ *Orectolobus wardi*

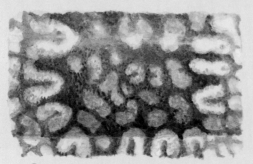
○ *Orectolobus japonicus*

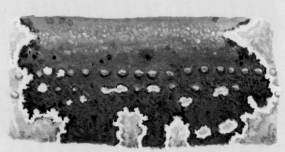
○ *Sutorectus tentaculatus*

○ *Sutorectus tentaculatus*

○ *Eucrossorhinus dasypogon*

○ *Sutorectus tentaculatus*

20cm

○ *Orectolobus japonicus*

○ *Orectolobus wardi*

Plate 29 ORECTOLOBIDAE II

○ **Western Wobbegong** *Orectolobus hutchinsi*   page 275

Southeast Indian Ocean: west Australia; intertidal to about 105m. Long nasal barbels with 1 small branch, 4 dermal lobes below and in front of each eye, lobes behind spiracles weakly or unbranched and slender; strongly contrasting broad dark rectangular dorsal saddles with light spots and deeply corrugated edges, background lighter with numerous broad dark blotches without numerous light O-shaped rings.

○ **Indonesian Wobbegong** *Orectolobus leptolineatus*   page 276

Central Indo-Pacific; below 20m. Nasal barbels branching, dermal lobes complex with 5–8 branches, branching at extremities. Golden-brown, dark brown saddles with irregularly spaced lighter-coloured blotches.

○ **Spotted Wobbegong** *Orectolobus maculatus*   page 277

Southern Australia; intertidal to about 248m or more. Long branched nasal barbels, 6–10 dermal lobes below and in front of each eye; dark back, broad darker dorsal saddles with white O-shaped spots and blotches and corrugated edges, light background with dark broad reticular lines.

○ **Ornate Wobbegong** *Orectolobus ornatus*   page 277

Southwest Pacific: Australia; intertidal to about 100m. Nasal barbels with a few branches, 5 dermal lobes below and in front of each eye, those behind spiracles unbranched or only weakly branched and broad; obvious broad dark dorsal saddles with light spots and conspicuous black corrugated borders, light background with conspicuous dark light-centred spots.

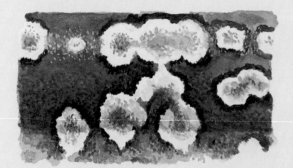

○ *Orectolobus hutchinsi*

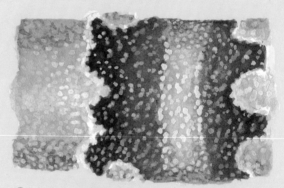

○ *Orectolobus leptolineatus*

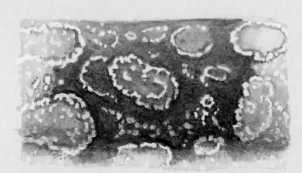

○ *Orectolobus maculatus*

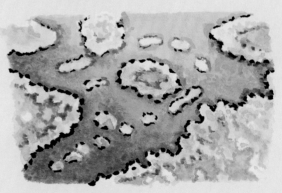

○ *Orectolobus ornatus*

Orectolobus hutchinsi

Orectolobus leptolineatus

50cm

Orectolobus maculatus

Orectolobus ornatus

Plate 30 ORECTOLOBIDAE III

○ **Floral Banded Wobbegong** *Orectolobus floridus*

page 274

Southeast Indian Ocean: southwest Australia; 40–85m. Nasal barbels with a single branch; dermal lobes on sides of head few and unbranched; yellowish above with dark brown bands separated by conspicuous clusters of lighter spots on dorsal surface; head with small black spots; midback with dark brown band.

○ **Gulf Wobbegong** *Orectolobus halei*

page 275

Southern Australia: shallows to 195m. Dermal lobes with multiple branches along head margin; warty tubercles along back with 2 above the eyes; slender nasal barbels with 2 branches; strikingly ornate variegated coloured shark with yellow to grey brown saddles each with paler bluish to white patches and small black spots bordering them.

○ **Dwarf Spotted Wobbegong** *Orectolobus parvimaculatus*

page 278

Southeast Indian Ocean: southwest Australia: 9–135m. Dermal lobes with 6–10 coarse branches; nasal barbels with 1–2 branches; upper surface brownish yellow with dark brown saddles and blotches; white to bluish rings, spots, and reticulations.

○ **Network Wobbegong** *Orectolobus reticulatus*

page 278

Cental Indo-Pacific: Australia; coastal to 20m. Paired dermal lobes in front of eyes and behind spiracles, absent from chin; nasal barbels unbranched; upper surface greyish brown with 3 conspicuous dark saddles, but not lighter margins surrounding saddles.

○ *Orectolobus floridus*

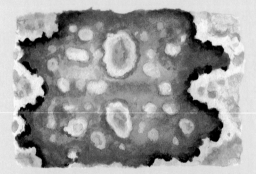

○ *Orectolobus halei*

○ *Orectolobus parvimaculatus*

○ *Orectolobus reticulatus*

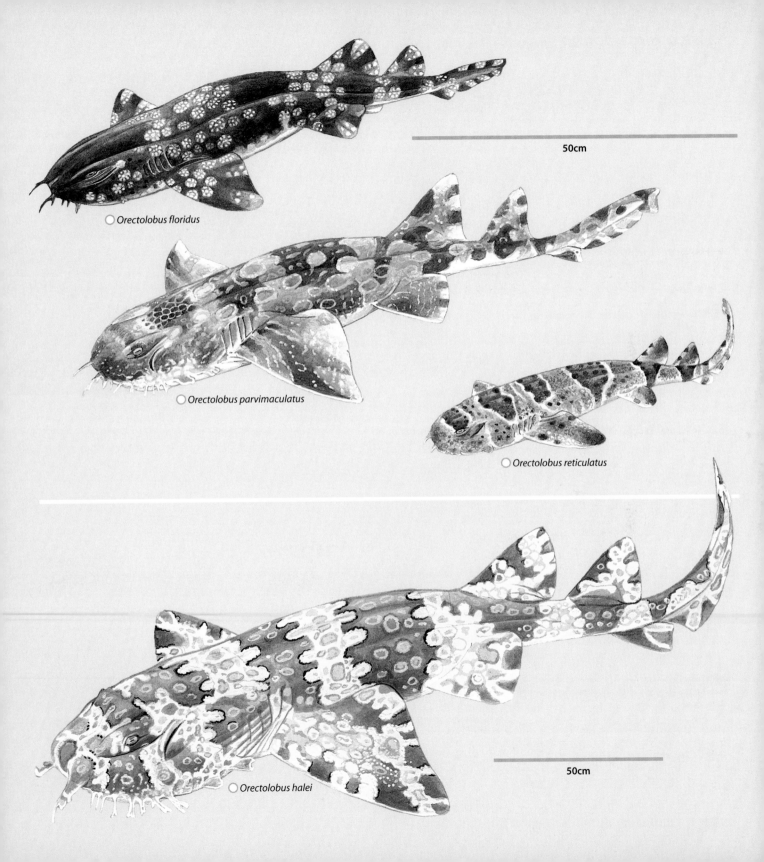

Orectolobus floridus

50cm

Orectolobus parvimaculatus

Orectolobus reticulatus

Orectolobus halei

50cm

TASSELLED WOBBEGONG *Eucrossorhinus dasypogon*

FAO code: **ORE**   Plate page 268

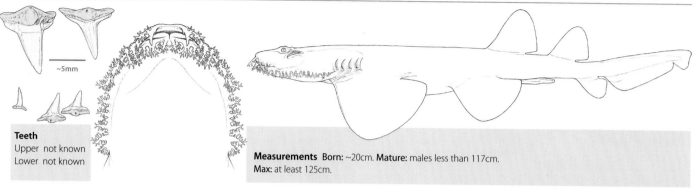

Teeth
Upper not known
Lower not known

Measurements Born: ~20cm. **Mature:** males less than 117cm.
Max: at least 125cm.

Identification Many highly branched dermal lobes on head and 'beard' on the chin. Very broad paired fins. Reticulated pattern of narrow dark lines on a light background. Scattered symmetrical, enlarged dark dots at line junctions. Indistinct saddles.
Distribution Central Indo-Pacific: Indonesia (Wai-geo, Aru), New Guinea, northern Australia.
Habitat Inshore coral reefs (coral heads, channels, reef faces), 5–50m.
Behaviour Nocturnal, possibly solitary. Small home range. Rests by day with curled tail on bottom in caves and under ledges.
Biology Presumably viviparous, nothing else known. Feeds on bottom fishes, possibly invertebrates. Catches nocturnal fishes sharing its caves.
Status IUCN Red List: Least Concern. Part of its Australian range is protected in the Great Barrier Reef Marine Park. Reported to bite divers.

FLORAL BANDED WOBBEGONG *Orectolobus floridus*

FAO code: **OCX**   Plate page 272

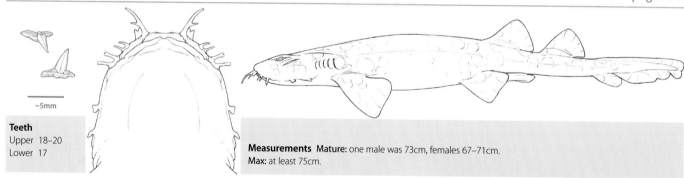

Teeth
Upper 18–20
Lower 17

Measurements Mature: one male was 73cm, females 67–71cm.
Max: at least 75cm.

Identification A small wobbegong. Nasal barbels with a single small branch. Dermal lobes on sides of head few and unbranched; warty tubercles absent. Yellowish brown with dark brown bands separated by conspicuous clusters of lighter spots on dorsal surface of body. Head with small black spots. Midbody has a dark brown band. Ventral surface of body is lighter.
Distribution Southeast Indian Ocean: southwest Australia.
Habitat Inshore reefs and seagrass beds, 40–85m.
Behaviour Unknown.
Biology Viviparous, nothing else known. Diet includes invertebrates and small bony fishes.
Status IUCN Red List: Least Concern. This small, recently described, wobbegong is known from only a few specimens. It is only a minor component of demersal trawl bycatch, unlikely to be retained because of its small size, and likely survives well when discarded.

GULF WOBBEGONG · *Orectolobus halei*

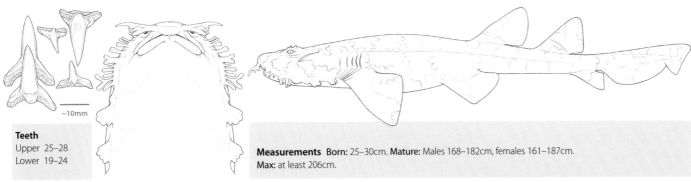

Teeth
Upper 25–28
Lower 19–24

~10mm

Measurements Born: 25–30cm. **Mature:** Males 168–182cm, females 161–187cm. **Max:** at least 206cm.

Identification A large wobbegong. Slender nasal barbels with 2 branches; nasoral and circumnarial grooves present. Dermal lobes with multiple branches along head margin; warty tubercles along back with 2 above eyes. Highly ornate, variegated with yellowish to greyish brown saddles, each with a paler bluish to white patch and bordered with small black spots. A conspicuous white spot is located behind each spiracle.
Distribution Southern Australian coasts, from Queensland to west Australia.
Habitat Coastal inshore reefs down to about 195m depth.
Behaviour Prefer reefs where they can rest in caves, crevices, and under ledges during the day; appear to forage mainly at night. Known to attack divers on occasion.
Biology Viviparous with yolk-sac, 12–47 pups per litter, usually born in spring after 10–11 month gestation. Age at maturity about 16 years. Diet includes bony fishes and small elasmobranchs.
Status IUCN Red List: Least Concern. This species is caught in commercial and recreational fisheries, and sometimes retained, but these fisheries are managed and monitored and the population is stable.

WESTERN WOBBEGONG · *Orectolobus hutchinsi*

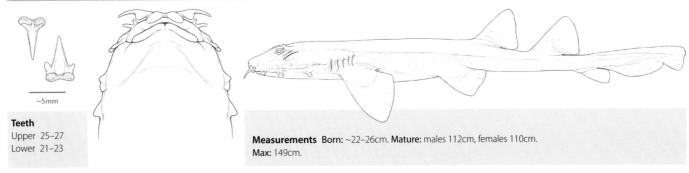

~5mm

Teeth
Upper 25–27
Lower 21–23

Measurements Born: ~22–26cm. **Mature:** males 112cm, females 110cm. **Max:** 149cm.

Identification Long nasal barbels with one small branch; nasoral and circumnarial grooves present. Four dermal lobes below and in front of each eye; dermal lobes behind spiracles unbranched or weakly branched and slender. Strongly contrasting, conspicuous, broad, dark, rectangular dorsal saddles with light spots and deeply corrugated edges (not black edged), separated by lighter areas with numerous broad, dark blotches (without numerous light O-shaped rings).
Distribution Southeast Indian Ocean: west Australia.
Habitat Reefs and seagrass beds, intertidal to about 105m.
Behaviour Unknown.
Biology Viviparous with yolk-sac, 18–29 pups per litter, usually born during winter and spring after 9–11 month gestation. Females give birth every 2–3 years. Age at maturity about 10 years for both sexes. Feeds mainly on bony fishes and octopuses.
Status IUCN Red List: Least Concern. Discarded alive from lobster pots and gill nets.

Orectolobus japonicus   FAO code: **ORJ**   Plate page 268

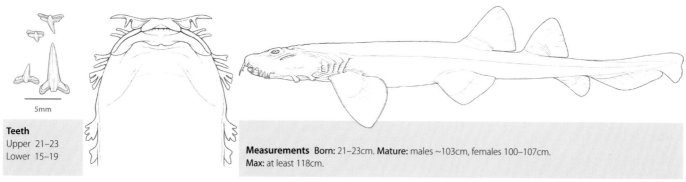

Teeth
Upper 21–23
Lower 15–19

Measurements Born: 21–23cm. **Mature:** males ~103cm, females 100–107cm.
Max: at least 118cm.

Identification Long, branched nasal barbels. Five dermal lobes below and in front of eyes on each side of head. Very obvious colour pattern of broad, dark dorsal saddles with spots and blotches and dark (but not black) corrugated edges separated by lighter areas with dark, broad, reticular lines.
Distribution Northwest Pacific: Japan, Koreas, China, Taiwan and Vietnam (Philippines record is questionable).
Habitat Tropical inshore rocky and coral reefs from intertidal to about 200m.
Behaviour Nocturnal, rarely seen by divers.
Biology Viviparous, up to 20–27 pups per litter born in spring (March–May) in captivity in Japan. Twelve month gestation. Mainly eats benthic fishes, also skates, shark eggcases, cephalopods and shrimp.
Status IUCN Red List: **Least Concern**. Caught by subsistence and artisanal hook and line fisheries and retained for food. Declines are reported in a few locations, but the majority of populations are healthy.

INDONESIAN WOBBEGONG *Orectolobus leptolineatus*   FAO code: **OCX**   Plate page 270

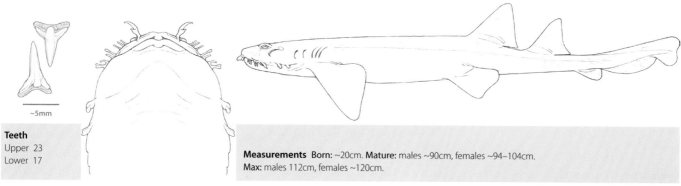

Teeth
Upper 23
Lower 17

Measurements Born: ~20cm. **Mature:** males ~90cm, females ~94–104cm.
Max: males 112cm, females ~120cm.

Identification Nasal barbels branching; nasoral and circumnarial grooves present. Dermal lobes slender, slightly flattened; 5–8 simple to complex lobes, branching at the extremities. Golden brown above with darker brown saddles and lighter coloured blotches irregularly spaced. Ventral surface uniformly lighter in colour.
Distribution Central Indo-Pacific: Indonesia, Malaysian Borneo and Taiwan.
Habitat Little known, but may prefer deeper, cooler water below 20m.
Behaviour Unknown.
Biology Viviparous, a litter of four found in one specimen. Diet unknown, but presumed to include invertebrates and fishes.
Status IUCN Red List: **Not Evaluated**. Currently no information on status.

SPOTTED WOBBEGONG *Orectolobus maculatus*

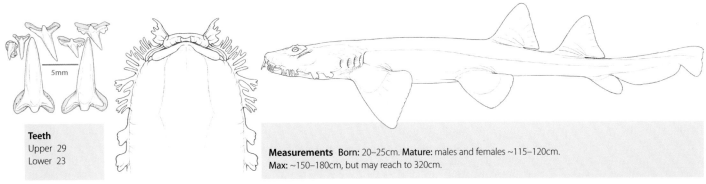

Teeth
Upper 29
Lower 23

Measurements Born: 20–25cm. **Mature:** males and females ~115–120cm.
Max: ~150–180cm, but may reach to 320cm.

Identification Long, branched nasal barbels; nasoral and circumnarial grooves present. 6–10 dermal lobes below and in front of eyes. Dark back; broad, darker dorsal saddles with white O-shaped spots and blotches and corrugated edges, separated by lighter areas with dark, broad, reticular lines.
Distribution Southern half of Australia, from Queensland to west Australia.
Habitat Coral and rocky reefs, bays, estuaries, seagrass, tidepools, under piers, on sand. Intertidal to deeper than 248m.
Behaviour Possibly nocturnal. Sluggish and inactive by day in caves, under overhangs and in channels, singly or in aggregations. May return to daytime resting sites. Can make short trips well above the seabed and climb (with back above water) between tidepools. Males fight in captivity during breeding season (July in New South Wales).
Biology Viviparous, large litters (up to 37 pups), birth occurs in the spring months after 10–11 month gestation. Eats bottom-dwelling invertebrates, bony fishes, sharks and rays.
Status IUCN Red List: Least Concern. Displayed in aquaria. Dangerous bite if provoked.

ORNATE WOBBEGONG *Orectolobus ornatus*

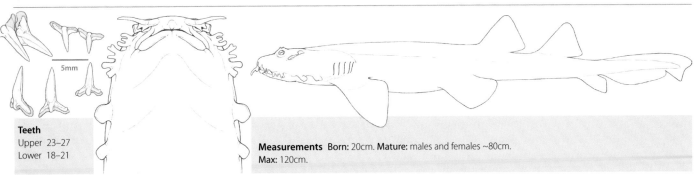

Teeth
Upper 23–27
Lower 18–21

Measurements Born: 20cm. **Mature:** males and females ~80cm.
Max: 120cm.

Identification Nasal barbels with a few branches. Five dermal lobes below and in front of each eye, those behind spiracles unbranched or only weakly branched and broad. Strongly variegated pattern of obvious broad, dark dorsal saddles with light spots and conspicuous black, corrugated borders interspaced with lighter areas and conspicuous dark, light-centred spots.
Distribution Southwest Pacific: east Australia and southern Papua New Guinea.
Habitat Bays, seaweed-covered rock and coral reefs on coast and around offshore islands. Lagoons, reef flats and faces, and reef channels, intertidal to ~100m. Prefers clearer water than *Orectolobus maculatus*.
Behaviour Nocturnal, resting singly and piled in aggregations by day in caves, under ledges and in trenches. Hunts at night.
Biology Viviparous, 2–16 pups per litter, average 9. Birth occurs in the spring after 10–11 month gestation. Feeds on bony fishes, sharks, rays, cephalopods and crustaceans.
Status IUCN Red List: Least Concern. This species is caught in commercial and recreational fisheries, and sometimes retained, but Australian fisheries are managed and monitored and the population is stable.

DWARF SPOTTED WOBBEGONG   *Orectolobus parvimaculatus*

Plate page 272

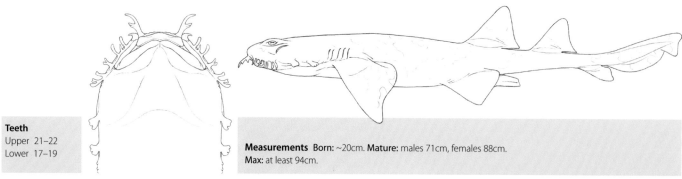

Teeth
Upper 21–22
Lower 17–19

Measurements Born: ~20cm. **Mature:** males 71cm, females 88cm.
Max: at least 94cm.

Identification Nasal barbels slender and with 1–2 branches; nasoral and circumnarial grooves present.
6–10 flattened and coarsely branched dermal lobes. Dorsal surface a brownish yellow with darker brown
saddles and blotches; white to bluish rings, spots and reticulations.
Distribution Southeast Indian Ocean; southwest Australia.
Habitat 9–135m, little else known.
Behaviour Unknown.
Biology Viviparous, nothing else known. Presumably feeds on benthic fishes and invertebrates.
Status IUCN Red List: Least Concern. This small, recently described wobbegong is known from fewer than
50 specimens. It is only a minor component of bycatch in the Western Australian demersal trawl fishery,
unlikely to be retained because of its small size, and likely survives well when discarded.

NETWORK WOBBEGONG   *Orectolobus reticulatus*
FAO code: **OCX**

Plate page 272

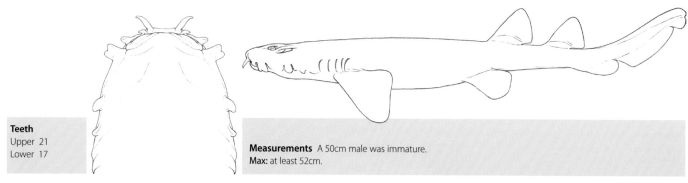

Teeth
Upper 21
Lower 17

Measurements A 50cm male was immature.
Max: at least 52cm.

Identification Nasal barbels simple not branched, flattened; nasoral and circumnarial grooves present.
Dermal lobes broad, paired, in front of eyes and behind spiracles but absent from chin. Dorsal surface
greyish-brown with 3 conspicuous darker saddles on back, without lighter margins surrounding saddles;
smaller darker spots inside each saddle.
Distribution Central Indo-Pacific: Northern Territory, Australia.
Habitat 0–20m, little else known.
Behaviour Unknown.
Biology Unknown. Presumably feeds on benthic fishes and invertebrates.
Status IUCN Red List: Data Deficient. Known from only 4 specimens.

NORTHERN WOBBEGONG *Orectolobus wardi*

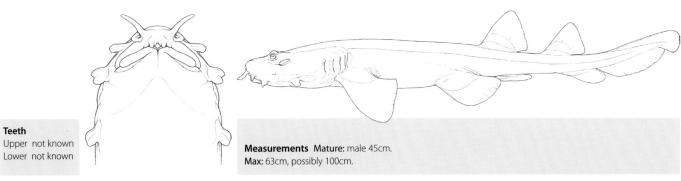

Teeth
Upper not known
Lower not known

Measurements Mature: male 45cm.
Max: 63cm, possibly 100cm.

Identification Unbranched nasal barbels. Two dermal lobes below and in front of each eye; dermal lobes behind spiracles unbranched and broad. Simple, sombre colour pattern: a few dark spots, dusky mottling and 3 large dark, light-edged (ocellate) rounded saddles separated by broad, dusky areas without spots or reticular lines in front of first dorsal fin.
Distribution Central Indo-Pacific: northern Australia and Papua New Guinea, possibly eastern Indonesia.
Habitat Shallow reefs less than 3m deep, often in turbid water.
Behaviour Nocturnal. Inactive (sometimes with head under a ledge) by day.
Biology Viviparous, but nothing else known. Feeds on benthic fishes and invertebrates.
Status IUCN Red List: Least Concern. Not fished in Australia, possibly common.

COBBLER WOBBEGONG *Sutorectus tentaculatus*

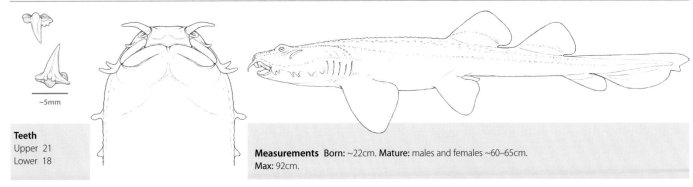

~5mm

Teeth
Upper 21
Lower 18

Measurements Born: ~22cm. **Mature:** males and females ~60–65cm.
Max: 92cm.

Identification Rather slender, less flattened than most wobbegongs; rather narrow head. Simple, unbranched nasal barbels. Chin smooth. A few slender, short, unbranched dermal lobes on sides of head form isolated groups broadly separated from one another, in 4–6 pairs. Rows of large, warty dermal tubercles in rows on back and bases of very low, long dorsal fins (height half base length). Striking colour pattern of broad, dark dorsal saddles with jagged, corrugated edges separated by light areas with irregular dark spots.
Distribution Southeast Indian Ocean: south and west Australia.
Habitat Rocky and coral reefs, in seaweeds. Depth to at least 35m.
Behaviour Unknown.
Biology Viviparous, litters of at least 12; little else known. Diet consists of bony fishes.
Status IUCN Red List: Least Concern. Common, bycatch usually returned alive.

Plate 31 HEMISCYLLIIDAE I – *Chiloscyllium*

Slender, very long tail; mouth well in front of eyes, short barbels, nasoral and circumnarial grooves, large spiracles below eyes; two equal sized unspined dorsal fins, second dorsal fin origin well ahead of anal fin, dorsal and anal fins set far back, notch separates anal and caudal fins; juveniles often different from adults.

○ **Arabian Carpetshark** *Chiloscyllium arabicum*   page 284

Northwest Indian Ocean; coral reefs, lagoons, rocky shores and mangrove estuaries, 3–100m. Prominent ridges on back; thick tail; first dorsal fin origin opposite or just behind pelvic fin insertion, second dorsal base longer than first; adult unpatterned, juveniles with light spots on fins.

○ **Burmese Bambooshark** *Chiloscyllium burmensis*   page 285

Northeast Indian Ocean; 29–33m. No body ridges; thick tail; first dorsal fin origin opposite pelvic fin insertion, long low anal fin; adults with dark fin webs, juvenile colour unknown.

○ **Bluespotted Bambooshark** *Chiloscyllium caerulopunctatum*   page 285

Western Indian Ocean: unknown depth/habitat. Truncated snout; lateral ridges along body; large, angular, very closely spaced dorsal fins; dark grey-brown, numerous light blue spots, no transverse dark bands.

○ **Grey Bambooshark** *Chiloscyllium griseum*   page 286

Indian Ocean and Central Indo-Pacific; inshore, 5–100m. No body ridges; thick tail; first dorsal fin origin over rear; pelvic fin bases, long low anal fin; adult unpatterned, juveniles with dark saddles and transverse bands.

○ **Indonesian Bambooshark** *Chiloscyllium hasselti*   page 286

Central Indo-Pacific; inshore to 12m. No body ridges; slender tail; first dorsal fin origin over rear of pelvic fin base; adults with dusky fins, juveniles with prominent saddles and blotches on fins.

○ **Slender Bambooshark** *Chiloscyllium indicum*   page 287

Indian Ocean; inshore to 90m. Very slender, lateral ridges; dorsal and anal fins set far back on very slender tail, first dorsal fin origin opposite or just behind pelvic fin insertion; adults with small dark spots bars and dashes on light brown background, juveniles with no dark posterior margins to fins.

○ **Whitespotted Bambooshark** *Chiloscyllium plagiosum*   page 287

Indian Ocean and Central Indo-Pacific; inshore to 50m. Lateral ridges on trunk; dorsal and anal fins set far back on very long thick tail, first dorsal fin origin opposite or just behind pelvic fin insertion; numerous light and dark spots on dark body, dark bands and saddles not conspicuously black edged.

○ **Brownbanded Bambooshark** *Chiloscyllium punctatum*   page 288

Central Indo-Pacific; reefs to 85m or more. No body ridges; thick tail; dorsal fins with long free rear tips, first dorsal fin origin over front of pelvic fin base; very light gill slit margins; adults unpatterned, juveniles with dark bands, scattered small dark spots.

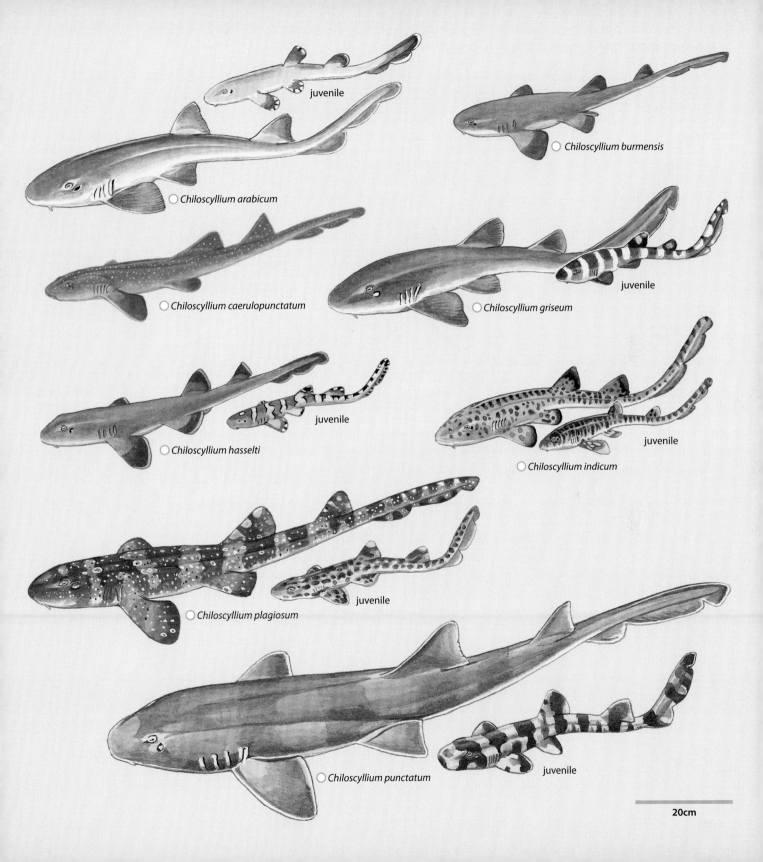

juvenile

Chiloscyllium burmensis

Chiloscyllium arabicum

Chiloscyllium caerulopunctatum

Chiloscyllium griseum

juvenile

Chiloscyllium hasselti

juvenile

Chiloscyllium indicum

juvenile

Chiloscyllium plagiosum

juvenile

Chiloscyllium punctatum

juvenile

20cm

Plate 32 HEMISCYLLIIDAE II – *Hemiscyllium*

○ **Indonesian Speckled Carpetshark** *Hemiscyllium freycineti* page 288

Central Indo-Pacific; shallow coral reefs and seagrass to 10m. Thick tail; small dark spots on snout, large and small dark spots on body, no white spots, moderately large dark epaulette spot without white ring or dark blotches; juveniles with dark paired fins (scattered spots in adults) and broad dark bands under head and encircling tail.

○ **Gale's Epaulette Shark** *Hemiscyllium galei* page 289

Central Indo-Pacific. Reefs, shallow water; 0–25m. Prominent white spots along margin of darker saddles along back, scattered white and dark spots along sides; 7–8 dark oval spots on sides between abdomen and caudal fin.

○ **Papuan Epaulette Shark** *Hemiscyllium hallstromi* page 289

Central Indo-Pacific. Bottom, inshore; to 30m Thick tail; no dark spots on snout, large sparse dark spots on body, moderately large black epaulette spot with white ring and black blotches, juveniles with black webbed paired fins (dusky in adults) and dark bands on tail.

○ **Henry's Epaulette Shark** *Hemiscyllium henryi* page 290

Central Indo-Pacific; rocky outcrops, seagrass to fringing coral reefs 0–30m or deeper. Numerous scattered spots over body including head and fins; distinct double ocelli on sides behind head.

○ **Halmahera Epaulette Shark** *Hemiscyllium halmahera* page 290

Central Indo-Pacific; 0–10m. Coral heads on sandy slope. Light brown, close-set clusters of 2–3 vertically arranged dark spots interspersed with scattered white spots; a few large dark spots on snout and between eyes, 1 spot over pectoral free rear margin with a U-shaped white margin below it.

○ **Michael's Epaulette Shark** *Hemiscyllium michaeli* page 291

Central Indo-Pacific; 0–20m. Fringing and patch reefs. Brilliant leopard pattern of spots covering body with large prominent spot behind and on sides of head; dark blotches on anterior dorsal fin margins; faint saddles on sides of body between pectoral and caudal fins.

○ **Epaulette Shark** *Hemiscyllium ocellatum* page 291

Central Indo-Pacific; coral, very shallow water to 50m. Thick tail; no dark spots on snout, small dark spots on body and unpaired fins, large black epaulette spot with white ring and black blotches; juveniles with black webbed paired fins (fade in adults) and dark bands on tail (light tail surface in adults).

○ **Hooded Carpetshark** *Hemiscyllium strahani* page 292

Central Indo-Pacific; coral, 3–18m. Thick tail; black mask on snout and head, white spots on body and fins over dark saddles and blotches, black epaulette spot partially merged with shoulder saddle, not completely ringed in white, paired fin margins white-spotted on black, dark rings round tail, juveniles no white spots with less bold markings.

○ **Speckled Carpetshark** *Hemiscyllium trispeculare* page 292

Central Indo-Pacific; coral, shallow water to 50m. Thick tail; small dark spots on snout, body and fins; numerous large and small dark spots separated by a reticular light network, no white spots, large black epaulette spot with white ring and 2 curved black blotches, dark saddles encircle back and tail; juvenile coloration unknown.

Hemiscyllium freycineti

juvenile

Hemiscyllium galei

juvenile

Hemiscyllium hallstromi

Hemiscyllium halmahera

Hemiscyllium henryi

Hemiscyllium michaeli

Hemiscyllium ocellatum

juvenile

Hemiscyllium strahani

juvenile

Hemiscyllium trispeculare

20cm

Hemiscylliidae: longtailed carpetsharks

Two Central Indo-Pacific genera, *Chiloscyllium* (eight wide-ranging species) and *Hemiscyllium* (nine species, mainly in the west Pacific, also Indian Ocean: Seychelles). Occur in intertidal pools, very shallow water, rocky and coral reefs close inshore and on sediments, inshore and offshore in bays.

Identification Small (mostly less than 1m) and slender with very long tails, two equal-sized unspined dorsal fins, origin of second well ahead of origin of long, low rounded anal fin, which is separated by a notch from lower caudal fin. Small transverse mouth well in front of dorsolateral eyes, large spiracles below eyes, nasoral and circumnarial grooves, short barbels. Colour patterns of young often different and bolder than adults. *Chiloscyllium* species without a black hood on head or large dark spots on sides of body, mouth closer to eyes than snout tip. *Hemiscyllium* species with spots or hood, nostrils at end of snout, and obvious ridges above eyes. *Hemiscyllium* species are distinguished mainly based on color patterns.

Biology Poorly known. Some (presumably all) are oviparous, laying oval eggcases. Distinct colour patterns of young suggest different habitat preferences from adults. Strong, muscular, leg-like paired fins used to clamber on reefs and in crevices. Large epaulette spots on *Hemiscyllium* species may be eyespots to intimidate predators. Food includes small bottom fishes, cephalopods, shelled molluscs and crustaceans.

Status Taken in multispecies fisheries and as bycatch, which is increasingly retained, sometimes in large numbers. Often common to abundant, until recently, but some species are rare with limited distribution in threatened habitats, and more than 80% of species are now assessed as threatened in the IUCN Red List of Threatened Species. Hardy, attractive and bred in captivity.

Indonesian Speckled Carpetshark, *Hemiscyllium freycineti*, Raja Ampat, Indonesia.

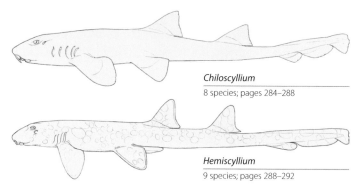

Chiloscyllium
8 species; pages 284–288

Hemiscyllium
9 species; pages 288–292

ARABIAN CARPETSHARK *Chiloscyllium arabicum*

FAO code: **ORA**　　　Plate page 280

Teeth
Upper [26–35]
Lower [21–32]

Measurements Born: less than 10cm. **Mature:** males 55cm, females 52cm. Max: 80cm.

Identification First dorsal fin origin opposite or just behind pelvic fin insertions; second usually with longer base than first. Dorsal and anal fins set far back on very long, thick tail. Prominent ridges on back. Unpatterned but for light spots on juveniles' fins.
Distribution Northwest Indian Ocean: Persian Gulf to India (exact distribution uncertain due to misidentification of the similar Grey Bambooshark *Chiloscyllium griseum*).
Habitat Coral reefs, lagoons, rocky shores, mangrove estuaries; 3–100m, on bottom.
Behaviour Poorly known.
Biology Oviparous, with a single eggcase developing in each uterus. Eggcases are laid on coral reefs with hatching occurring in 70–80 days. Feeds on squids, shelled molluscs, crustaceans and snake eels.
Status IUCN Red List: Near Threatened. Common in summer in Persian Gulf. Has bred in captivity.

BURMESE BAMBOOSHARK   *Chiloscyllium burmensis*   FAO code: **KYL**   Plate page 280

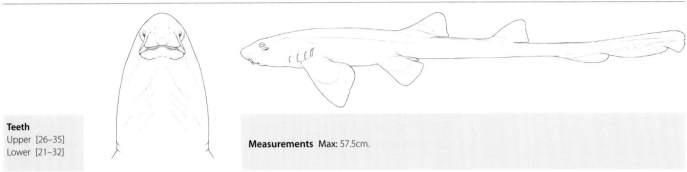

Teeth
Upper [26–35]
Lower [21–32]

Measurements Max: 57.5cm.

Identification Very small eyes. First dorsal fin origin roughly opposite pelvic fin insertions, anal fin long, low, and set far back on very long thick precaudal tail. No lateral ridges on body. Adults with dark fin webs. Juvenile coloration unknown.
Distribution North Indian Ocean: Odisha, northeast coast of India to Burma (Myanmar).
Habitat Shallow inshore habitats 0–30m, predominantly 5–15m, including intertidal pools, rocky and coral reefs, and soft sediments.
Behaviour Unknown.
Biology Virtually unknown, although reproductive mode likely oviparous. Eats small bony fishes.
Status IUCN Red List: **Vulnerable**. Caught throughout its range in a wide range of artisanal and commercial fisheries. Possibly has a refuge in rocky areas. Often formerly discarded; now increasingly landed.

BLUESPOTTED BAMBOOSHARK   *Chiloscyllium caerulopunctatum*   FAO code: **ORP**   Plate page 280

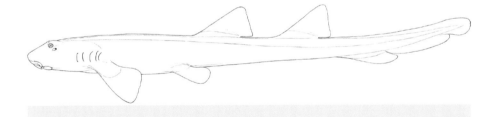

Teeth
Upper unknown
Lower unknown

Measurements Max: 67cm.

Identification Snout tip truncated. Dorsal fins large and angular; interdorsal space very short, less than first dorsal fin base. Lateral ridge along each side of body. Background colour dark grey-brown with numerous light blue spots, but without transverse dark bands.
Distribution West Indian Ocean: Madagascar.
Habitat Unknown.
Behaviour Unknown.
Biology Unknown.
Status IUCN Red List: **Not Evaluated**. Possibly subject to incidental bycatch, but very few specimens have ever been observed.

GREY BAMBOOSHARK *Chiloscyllium griseum* FAO code: **ORR** Plate page 280

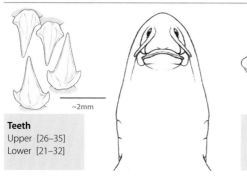

Teeth
Upper [26–35]
Lower [21–32]

~2mm

Measurements Born: less than 12cm. **Mature:** males 45–55cm.
Max: 77cm.

Identification First dorsal fin origin over rear of pelvic fin bases, anal fin low, and set far back on very thick precaudal tail. No lateral ridges on body. Adults often unpatterned. Young with obvious dark saddlemarks and transverse bands, no black edging.
Distribution Indian Ocean and Central Indo-Pacific: Pakistan, India, Sri Lanka, Malaysia, Thailand. Historical records of *Chiloscyllium griseum* from the Philippines, Taiwan, and Japan are misidentifications of *C. punctatum*.
Habitat Inshore, on rocks and in lagoons, 5–100m.
Behaviour Unknown.
Biology Oviparous, lays small oval eggcases on bottom. Probably feeds mainly on invertebrates.
Status IUCN Red List: **Vulnerable**. Common. Bycatch of a wide range of artisanal and commercial gears, including demersal trawl, longline and gillnet. Retained for human consumption in increasingly large quantities. Kept in public aquaria.

INDONESIAN BAMBOOSHARK *Chiloscyllium hasselti* FAO code: **YYL** Plate page 280

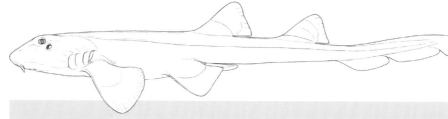

Teeth
Upper [26–35]
Lower [21–32]

Measurements Born: 9–12cm. **Mature:** males 44–54cm.
Max: 61cm.

Identification First dorsal fin origin over rear of pelvic fin bases, anal fin low and set far back on very long thick precaudal tail. No lateral ridges on body. Adults often unpatterned except for dusky fins. Young have prominent saddlemarks (broad dusky patches with conspicuous black edging separated by light areas and blackish spots) and dark blotches on fins.
Distribution Central Indo-Pacific: Myanmar and Andaman Islands east to Vietnam, Indonesia (Sumatra) and possibly the Philippines.
Habitat Probably close inshore, from intertidal pools, sediments, rocky and coral reefs, to 12m.
Behaviour Eggs are attached to benthic marine plants.
Biology Oviparous, eggs hatch in about December.
Status IUCN Red List: **Endangered**. Bycatch of artisanal and commercial fisheries, in demersal trawls, longlines and gillnets. Retained for human consumption in increasingly large quantities. Possibly targeted for aquarium trade.

SLENDER BAMBOOSHARK *Chiloscyllium indicum*

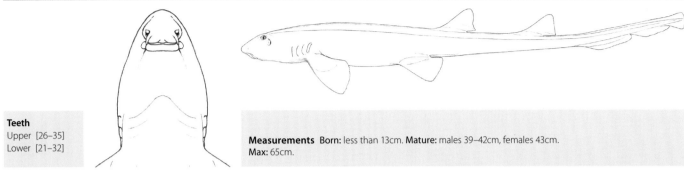

Teeth
Upper [26–35]
Lower [21–32]

Measurements Born: less than 13cm. **Mature:** males 39–42cm, females 43cm.
Max: 65cm.

Identification Very slender body with lateral ridges. First dorsal fin origin opposite or just behind pelvic fin insertions, anal fin long, low and set far back on very long thick precaudal tail. Numerous small dark spots, bars/saddles and dashes on light brown background. No prominent black edges to saddles in juveniles.
Distribution India, Sri Lanka, to about Bangladesh; possibly Arabian Sea, and around Thailand and Sumatra. Records from Solomon Islands and north to China and Japan are different species, possibly *Chiloscyllium plagiosum*.
Habitat Inshore on bottom from intertidal to 90m. Possibly in fresh water (lower Perak River, Malaysia).
Behaviour Little known.
Biology Little known. Oviparous.
Status IUCN Red List: **Vulnerable**. Important in a wide range of inshore artisanal and commercial fisheries, increasingly retained for food (and animal feed in some countries).

WHITESPOTTED BAMBOOSHARK *Chiloscyllium plagiosum*

~2mm

Teeth
Upper [26–35]
Lower [21–32]

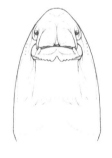

Measurements Born: ~9–12cm. **Mature:** males ~50–60cm, females ~65cm.
Max: 95cm.

Identification First dorsal fin origin opposite or just behind pelvic fin insertions, anal fin set far back on long thick precaudal tail. Lateral ridges on trunk. Body dark with numerous light and dark spots; dark bands and saddles not conspicuously edged with black.
Distribution Indian Ocean and Central Indo-Pacific: India, Indonesia, Malaysia, Singapore, Vietnam, Philippines, Taiwan and Japan.
Habitat Inshore on bottom, reefs in tropics to 50m deep.
Behaviour Nocturnal. Rests in reef crevices by day, feeds at night.
Biology Oviparous, lays about 26 eggcases per season; eggcases take 110–144 days to hatch. Age at maturity 4–5 years both sexes, maximum age 7–14 years. Eats bony fishes and crustaceans.
Status IUCN Red List: **Near Threatened**. Common and important in inshore fisheries. Used for food. Popular aquarium species.

BROWNBANDED BAMBOOSHARK   *Chiloscyllium punctatum*

FAO code: **ORB**   Plate page 280

~5mm

Teeth
Upper 31–33
Lower 30–33

Measurements Eggcases: 11 x 15cm. **Born:** 13–18cm. **Mature:** males ~82cm, females ~87cm. **Max:** 132cm.

Identification First dorsal fin origin over front half of pelvic fin bases, anal fin low and set far back on long thick precaudal tail. No lateral ridges on body. Young with obvious dark bands (not black-edged) and scattered small, dark spots which fade to light brown in adults. Very light gill slit margins.
Distribution Central Indo-Pacific: east India, Southeast Asia, to northern Australia, southern New Guinea, and Japan.
Habitat Coral reefs (intertidal pools, tidal flats, reef faces). Possibly also on soft bottom, offshore to >85m.
Behaviour Hides in crevices and under corals. Can survive out of water for up to half a day.
Biology Oviparous, lays rounded eggcases; may deposit up to 153 eggcases, with eggcases taking about 90 days to hatch. Age at maturity 4.5 years with a maximum age of 12.5 years; in captivity these sharks are known to live to at least 16 years. Feeds on bottom invertebrates and small fishes.
Status IUCN Red List: **Near Threatened**. Common, regularly taken in inshore food fisheries. Displayed and bred in aquaria. May nip if provoked.

INDONESIAN SPECKLED CARPETSHARK   *Hemiscyllium freycineti*

FAO code: **ORF**   Plate page 282

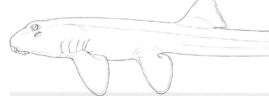

~5mm

Teeth
Upper [26–35]
Lower [21–32]

Measurements **Born:** less than 15cm. **Mature:** males 37–62cm. **Max:** females 69cm.

Identification Dorsal and anal fins set very far back on extremely long, thick tail. Small, dark spots on snout. Large, sparse, dark spots on body, none white; no reticulate pattern. Moderately large, black epaulette spot above pectorals, without white ring or curved black marks around rear half. Dark paired fins with light edges in young, changing to scattered small and large dark spots in adults. Broad, dark bands under head and completely encircling tail in young, lost in light undersides of adults. Dark blotches on anterior edges of dorsal fins.
Distribution Central Indo-Pacific: known only from Raja Ampat Islands, off the island of New Guinea, Indonesia.
Habitat Coral reefs, on sand and in seagrass in shallow water to 10m, but usually in about 2–3m.
Behaviour Hides in reef crevices by day, feeds at night.
Biology Almost unknown.
Status IUCN Red List: **Near Threatened**. Restricted habitat affected by expanding fisheries and pollution.

GALE'S EPAULETTE SHARK   *Hemiscyllium galei*   FAO code: **OCX**   Plate page 282

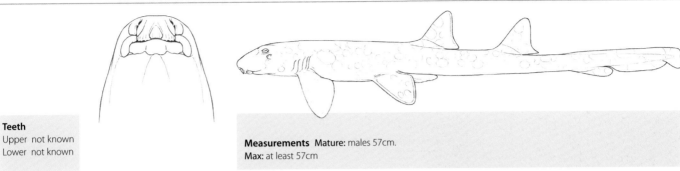

Teeth
Upper not known
Lower not known

Measurements Mature: males 57cm.
Max: at least 57cm

Identification Dorsal and anal fins set very far back on extremely long, thick tail. Interdorsal space shorter than other *Hemiscyllium* species. A line of brilliant white spots along the margin of darker saddles along the back. Scattered white and dark spots along the sides; 7–8 oval dark spots on flanks between abdomen and caudal fin base. Dark spots on dorsal surface of pectoral and pelvic fins.
Distribution Central Indo-Pacific: West Papua, New Guinea.
Habitat Rocky and coral reefs and seagrass beds in shallow water, from intertidal to 10m, occasionally 25m.
Behaviour Hides in reef crevices and overhangs during the day, but is active at night.
Biology Unknown.
Status IUCN Red List: **Vulnerable**. Threatened by artisanal fishing: including hand gleaning and reef flat gill netting, and by habitat loss and degradation. May be targeted for the aquarium trade.

PAPUAN EPAULETTE SHARK   *Hemiscyllium hallstromi*   FAO code: **ORK**   Plate page 282

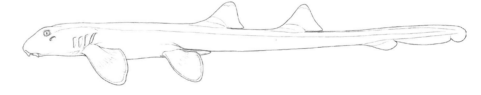

Teeth
Upper [26–35]
Lower [21–32]

Measurements Born: less than 19cm. **Mature:** males 47–64cm.
Max: 77cm.

Identification Dorsal and anal fins set far back on extremely long, thick tail. No dark spots on snout. Dark wide-spaced spots on body, some almost as large or larger than the conspicuous white-ringed black epaulette spot above pectoral fins. Epaulette spot partly ringed (above and behind) by smaller black spots. No white spots or reticular pattern. Paired fins unspotted; black webs and light edges in young fade to dusky in adults. Dark rings around tail in young are lost from uniformly pale underside of adults.
Distribution Central Indo-Pacific: southern Papua New Guinea.
Habitat Inshore seabed, possibly coral reefs, to 30m.
Behaviour Unknown.
Biology Unknown.
Status IUCN Red List: **Vulnerable**. Very limited range affected by fisheries and pollution.

HALMAHERA EPAULETTE SHARK *Hemiscyllium halmahera* FAO code: **OCX** Plate page 282

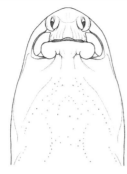

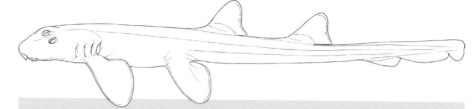

Teeth
Upper not known
Lower not known

Measurements. **Max:** males at least 68cm, females 66cm.

Identification Dorsal and anal fins set very far back on extremely long, thick tail. Body depth at anal fin origin thicker than other *Hemiscyllium* species. Head with a few large dark spots on the snout and between eyes. A pair of large, dark marks on ventral surface, and a large U-shaped dark spot over pectoral free rear margin, with a white margin under its lower half. Overall body coloration along dorsal and lateral surfaces a light brown with close-set, vertically arranged, dense clusters of 2–3 dark spots, interspersed with scattered white spots between dark clusters.
Distribution Central Indo-Pacific: known from islands off the west coast of Halmahera, Indonesia.
Habitat Shallow water, on rocky outcrops, coral reefs and seagrass beds, 0–10m.
Behaviour This shark has been observed resting under coral heads that were sparsely scattered on a steep, black volcanic sand slope.
Biology Unknown.
Status IUCN Red List: Near Threatened. Reported from a small number of locations, where it is threatened by artisanal fishing, collection for the aquarium trade, habitat damage and pollution from mining.

HENRY'S EPAULETTE SHARK *Hemiscyllium henryi* FAO code: **OCX** Plate page 282

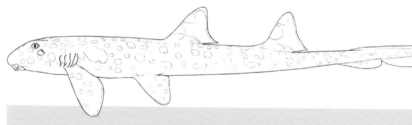

Teeth
Upper not known
Lower not known

Measurements Born: ~20cm. **Mature:** males 60cm, females 82cm.
Max: at least 82cm.

Identification Dorsal and anal fins set very far back on extremely long, thick tail. Body, including head and fins, covered with numerous small scattered spots. Distinct double ocelli marking on sides, just behind head.
Distribution Central Indo-Pacific: West Papua.
Habitat Rocky outcrops and seagrass bed habitats at 0–10m, to fringing tropical coral reefs at 30m or deeper.
Behaviour Nocturnally active, but hides in reef crevices and overhangs during the day.
Biology Unknown.
Status IUCN Red List: Vulnerable. This endemic is known from only six locations within its very restricted range, where it is threatened by artisanal fisheries (it is used for food) and habitat degradation.

MICHAEL'S EPAULETTE SHARK *Hemiscyllium michaeli*

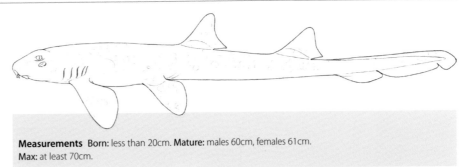

Teeth
Upper not known
Lower not known

Measurements Born: less than 20cm. **Mature:** males 60cm, females 61cm.
Max: at least 70cm.

Identification Dorsal and anal fins set very far back on extremely long, thick tail. Differs from other *Hemiscyllium* species by having a shorter first dorsal fin and anal fin base, and lower anal fin. Brilliant leopard-like pattern of spots covering body, with a large black ocellate spot on side and just behind head; dark blotches on anterior dorsal fin margins. Faint saddles on sides of body between pectoral and caudal fins, becoming darker and more prominent on caudal fin.
Distribution Central Indo-Pacific: Papua New Guinea.
Habitat Fringing and patch tropical coral reefs, rocky outcrops, and seagrass beds at 0–20m.
Behaviour Hides in reef crevices and overhangs during the day, becomes active at night.
Biology Unknown
Status IUCN Red List: **Vulnerable**. A restricted range experiencing habitat degradation and captured for food in artisanal fisheries. Possibly enters aquarium trade.

EPAULETTE SHARK *Hemiscyllium ocellatum*

~2mm

Teeth
Upper 38
Lower 30

Measurements Born: ~14–16cm. **Mature:** males 54–62cm, females ~55–64cm.
Max: 107cm.

Identification Dorsal and anal fins set far back on extremely long, thick tail. No spots on snout. Dark spots on body and unpaired fins much smaller than conspicuous large black epaulette spot (ringed with white, inconspicuous small dark spots behind and below). No white spots or reticular pattern. Pale-margined dark paired fins in young fade in adults; sometimes a few small dark spots on adult paired fins. Dark bands around tail in young; adults with uniform light ventral tail surface.
Distribution Central Indo-Pacific: northern Australia.
Habitat Coral (particularly staghorn) in shallow water and tidepools, sometimes barely submerged, to 50m.
Behaviour More active at dusk and by night. Often feeds at low tide. Crawls, clambers and swims about. Thrashes tail when snout digs in sand. Unafraid of humans, may nip when captured.
Biology Oviparous, eggs hatch in ~115–130 days. Mating occurs from July–December, with eggs being deposited from August to January. Eats worms, crustaceans and small fishes.
Status IUCN Red List: **Least Concern**. Abundant on the Great Barrier Reef.

Teeth
Upper [26–35]
Lower [21–32]

Measurements Mature: males ~60cm, females ~73cm.
Max: 80cm.

Identification Dorsal and anal fins set far back on extremely long, thick tail. Black mask on adult snout and head. Black spots and bands beneath head, not on snout. Black epaulette spot partially merged with shoulder saddle, not completely surrounded by white ring. Dark saddles and blotches on body; many white spots on body and fins. No reticular pattern. Margins of paired fins white-spotted on black. Dark rings around tail. Juvenile markings less bold, no white spots.
Distribution Central Indo-Pacific: northern New Guinea.
Habitat Inshore on coral reef faces and flats, from 3–18m.
Behaviour Nocturnal, hides in crevices and under table corals by day.
Biology Unknown.
Status IUCN Red List: **Vulnerable**. Small, fragmented range is polluted and dynamite-fished.

Teeth
Upper [26–35]
Lower [21–32]

Measurements Mature: males ~53cm.
Max: 79cm.

Identification Dorsal and anal fins set very far back on extremely long, thick tail. Small dark spots on snout; uniformly light under head. Large black epaulette spot with conspicuous white ring and two curved black marks around the posterior half, partly surrounded by smaller black spots. Body and fins covered with numerous small and large dark spots separated by a light reticular pattern. No white spots. Dark saddles on back and tail extend around ventral surface.
Distribution Central Indo-Pacific: northern Australia, possibly Indonesia but confirmation required.
Habitat Coral reefs in shallow water (often under table corals) and in tidepools to 50m.
Behaviour Poorly known.
Biology Oviparous, little else known. Probably feeds on bottom-dwelling invertebrates.
Status IUCN Red List: **Least Concern**. There is little or no fishing for this species in Australian waters, which are thought to represent the majority of this species' range, and some critical habitats lie within protected areas. The possible Indonesian population may be affected by fisheries and habitat degradation.

ZEBRA SHARK *Stegostoma tigrinum*

FAO code: **OSF**   Plate page 294

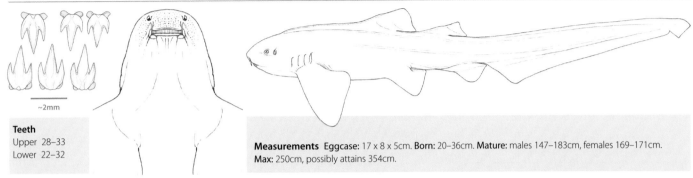

~2mm

Teeth
Upper 28–33
Lower 22–32

Measurements Eggcase: 17 x 8 x 5cm. **Born:** 20–36cm. **Mature:** males 147–183cm, females 169–171cm. **Max:** 250cm, possibly attains 354cm.

Identification Large, slender, flexible, ridged body with a unique banded (juvenile) or variable spotted (adult) pattern. Small barbels; small transverse mouth in front of lateral eyes; large spiracles. First dorsal fin set forwards on back, much larger than second. Anal fin close to tail. Broad caudal fin as long as body. Young are dark brown above, yellowish below, with vertical yellow stripes and spots separating dark saddles. The saddles break up into small brown spots on yellow in sharks 50–90cm long. Spots are more uniformly distributed on large sharks (hence the other English name: Leopard Shark).

Distribution Indian and west Pacific Oceans: tropical continental and insular shelves, East Africa to Japan, New Caledonia and Palau.

Habitat Coral reefs, sands between reefs and offshore sediments, in a narrow band from the intertidal to 62m. Adults and large spotted juveniles rest in coral reef lagoons, channels and faces. Striped young are rarely seen and may occur deeper (over 90m).

Behaviour Poorly known. May rest propped up on pectorals, mouth open, facing current. Usually solitary, aggregations are rare. Sluggish by day, more active at night or when food is present. Can swim strongly and squirm into crevices to search for food. Not wide-ranging and unlikely to disperse far to recolonise depleted areas. The undulating swimming motion of slender newborn Zebra Sharks, combined with their striking banded coloration and long, single-lobed caudal fin, makes them appear remarkably similar at first glance to banded sea snakes (family Elapidae). It has been hypothesised that this is an adaptation that

improves the survival of the most vulnerable life stage of the Zebra Shark, and the first known example of protective mimicry in an elasmobranch.

Biology Oviparous, lays large dark brown or purplish black eggcases, anchored to bottom with fine tufts of fibres. Females of this species exhibit parthenogenesis: they can produce viable eggs that hatch into female pups without needing to be fertilised by a male (see p. 52). Age at maturity 6–8 years for females and 7 years for males, and a maximum age of at least 28 years. Feeds on molluscs, crustaceans, small bony fishes and possibly sea snakes.

Status IUCN Red List: Endangered. Relatively common, but this species' inshore habitat is heavily fished throughout its range except in Australian waters. It is taken in demersal trawls, floating and fixed bottom gillnets and on baited hooks and seen in fish markets in Indonesia, Thailand, Malaysia, Philippines, Pakistan, India, Taiwan and elsewhere. Depleted populations are unlikely to recover rapidly through recolonisation from other areas because of its limited potential for dispersal. Its coral reef habitat is also threatened. Zebra Sharks are not aggressive and can be maintained well in captivity. Some aspects of Zebra Shark morphology, including the longitudinal ridges along the body and numerous internal features, are remarkably similar to those of the Whale Shark (p. 300). Recent molecular research indicates that these two species and the Shorttail Nurse Shark (p. 297) have a close evolutionary relationship. The scientific name *Stegostoma fasciatum* has been in common use for decades, but a recent review of nomenclature shows that the correct (earlier) name should be *Stegostoma tigrinum*.

Zebra Sharks, *Stegostoma tigrinum*, engaging in mating behaviour, Similan Islands, Thailand.

Plate 33 GINGLYMOSTOMATIDAE and STEGOSTOMATIDAE

Head broad and flat, no lateral skin flaps, subterminal mouth in front of eyes, barbels, long nasoral grooves, small spiracles, small gills fifth almost overlaps fourth; precaudal tail much shorter than head and body; two spineless dorsal fins, second dorsal same level and size as anal fin, caudal fin with strong terminal lobe and subterminal notch.

○ **Nurse Shark** *Ginglymostoma cirratum*
page 298

Atlantic Ocean; 0–130m. Long barbels, tiny spiracles; dorsal fins broadly rounded, first dorsal fin larger than second dorsal and anal fins, interdorsal space 5.4–9.5% of TL, caudal fin longer than a quarter of total length; uniform yellow- to grey-brown, young with small dark light ringed ocellar spots and obscure saddles.

○ **Unami Nurse Shark** *Ginglymostoma unami*
page 297

East Pacific; 0–7m. Long barbels, tiny spiracles; dorsal fins broadly rounded, first dorsal fin origin before pelvic fin origin, free rear tip reaches second dorsal fin origin, interdorsal space 3.6–5.6% of TL; adults uniform yellow-brown, young with patterns of small dark spots.

○ **Tawny Nurse Shark** *Nebrius ferrugineus*
page 299

Indian, west and central Pacific Oceans; to 70m or more. Fairly long barbels, tiny spiracles; all fins angular, first dorsal fin larger than second dorsal and anal fins, first dorsal fin base over pelvic fin bases, caudal fin longer than a quarter of total length; uniform shades of brown, depending on habitat.

○ **Shorttail Nurse Shark** *Pseudoginglymostoma brevicaudatum*
page 297

West Indian Ocean; coral reefs to 20m. Short barbels, tiny spiracles; all fins rounded, first dorsal fin about same size as second dorsal and anal fins, caudal fin less than 20% of total length; uniform dark brown.

○ **Zebra Shark** *Stegostoma tigrinum*
page 293

Indian and west Pacific Oceans; to 62m (adult), and >90m (young). Large slender ridged body; small transverse mouth in front of eyes, small barbels, large spiracles; first dorsal fin set forwards on back, first dorsal fin larger than second, anal fin close to caudal fin, broad caudal fin about one-half the total length; distinct spotting in adults, young with unique vertical banding.

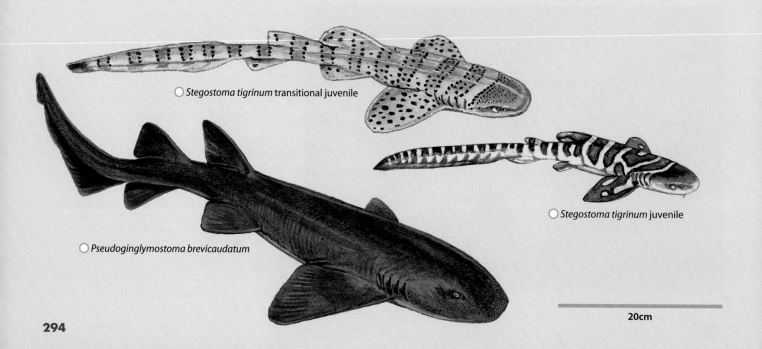

○ *Stegostoma tigrinum* transitional juvenile

○ *Stegostoma tigrinum* juvenile

○ *Pseudoginglymostoma brevicaudatum*

20cm

50cm

○ *Ginglymostoma cirratum*

○ *Ginglymostoma unami*

○ *Nebrius ferrugineus*

○ *Stegostoma tigrinum*

Ginglymostomatidae: nurse sharks

Three genera and four species of subtropical and tropical continental and insular waters, including coral and rocky reefs, sandy areas, reef lagoons and mangrove keys. Depths range from the intertidal and surf zone (sometimes barely covered) to at least 130m. There is at least one more undescribed species, and the genus *Pseudoginglymostoma* may prove to be more closely related to *Rhincodon* and *Stegostoma* than to the other nurse sharks.

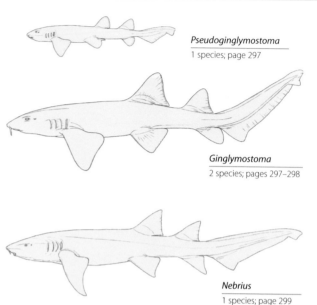

Pseudoginglymostoma
1 species; page 297

Ginglymostoma
2 species; pages 297–298

Nebrius
1 species; page 299

Identification Head broad and flattened, no lateral skin flaps. Snout rounded or truncated. Transverse, subterminal mouth in front of eyes, long nasoral grooves. Nostrils with barbels. Small spiracles behind eyes. Small gill slits, fifth almost overlaps fourth. Two spineless dorsal fins, second level with and about same size as anal fin; latter close to lower caudal fin. Precaudal tail much shorter than head and body. Caudal fin elongated with a strong terminal lobe and subterminal notch but no or very short lower lobe. Unpatterned or with a few dark spots in young.

Biology Viviparous (or possibly so in *Pseudoginglymostoma*). Nocturnal. Social, rests on the seabed in small groups. Cruise and clamber on bottom, mouths and barbels close to bottom, searching for food. Short small mouths with large cavities suck in a variety of bottom invertebrates and fishes, including active reef fish.

Status Larger species are or were formerly common and often caught in local inshore fisheries for food, liver oil and tough leather. Some local extirpations have been reported and those species whose status can be assessed are Threatened. *Ginglymostoma* and *Nebrius* are hardy and can survive well in large aquaria. Not usually aggressive, but can bite hard and hang on if provoked. Popular for dive tourism.

Groups of nocturnal Nurse Sharks, *Ginglymostoma cirratum,* (p. 298) often rest in shallow water on sand or under overhangs by day, moving out at dusk to hunt for food.

SHORTTAIL NURSE SHARK *Pseudoginglymostoma brevicaudatum*

FAO code: **ORX**   Plate page 294

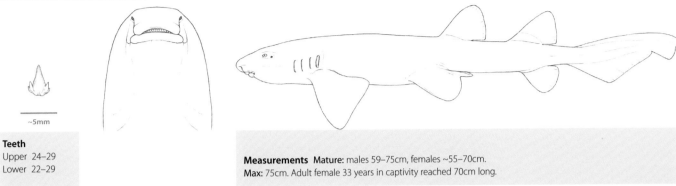

Teeth
Upper 24–29
Lower 22–29

~5mm

Measurements Mature: males 59–75cm, females ~55–70cm.
Max: 75cm. Adult female 33 years in captivity reached 70cm long.

Identification Distinguished from other nurse sharks by very short nasal barbels, two rounded, equal-sized dorsal fins, an anal fin and a short caudal fin (length less than 20% total length). Colour uniformly dark with no spots or markings.
Distribution West Indian Ocean: Mozambique, Tanzania, Kenya and Madagascar, possibly Mauritius and the Seychelles. The Madagascar population may be an undescribed species, and include populations from Mauritius and Seychelles, if confirmed.
Habitat Shallow coral reefs, to 20m.
Behaviour Nocturnal in captivity, otherwise poorly known. Reported to survive for several hours out of water.
Biology Poorly known. Presumed oviparous (egg-laying observed in captivity). A specimen in a public aquarium lived for 33 years.
Status IUCN Red List: **Critically Endangered**. Very poorly known, but most of its limited inshore range is heavily fished, and its coral reef habitat is threatened.

UNAMI NURSE SHARK *Ginglymostoma unami*

FAO code: **GNG**   Plate page 294

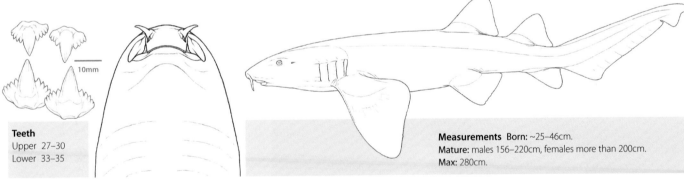

Teeth
Upper 27–30
Lower 33–35

10mm

Measurements Born: ~25–46cm.
Mature: males 156–220cm, females more than 200cm.
Max: 280cm.

Identification Similar in appearance to the Nurse Shark *Ginglymostoma cirratum* but separated by shorter, more robust body trunk; first dorsal fin origin before pelvic fin origins; a shorter interdorsal space (3.6–5.6% of total length); first dorsal fin free rear tip reaches second dorsal fin origin; and posterior end of second dorsal and anal fins reaches to origin of upper and lower caudal fin, respectively. Coloration varies with size: young sharks have a pattern of small dark spots lost with growth; adults are dark to light brownish yellow.
Distribution East Pacific: from the southwest coast of Baja California and the Gulf of California, Mexico, to Peru.
Habitat Nearshore and coastal rocky reefs, including intertidal and bays, to at least 7m and possibly deeper.
Behaviour Unknown.
Biology Viviparous, least 13 pups per litter. Feeds mainly on molluscs, cephalopods, and crustaceans.
Status IUCN Red List: **Not Evaluated**. Only recently distinguished from Ginglymostoma cirratum. This large shark is extremely vulnerable to coastal fisheries, as a bycatch and a target of artisanal and large scale gillnets and longlines, and to habitat degradation. The species is probably depleted, at least in parts of its range.

10mm

Teeth
Upper 30–42
Lower 28–34

Measurements **Born:** 27–30cm. **Mature:** males ~210cm, females 230–240cm.
Max: ~308cm, records of 420cm and larger are likely erroneous.

Identification Long barbels; nasoral grooves. Mouth in front of dorsolateral eyes. Tiny spiracles. Dorsal fins broadly rounded; first much larger than second dorsal and anal fins. Precaudal tail shorter than head and body; caudal fin more than 25% of total length. Similar to *Ginglymostoma unami*, but differs by a longer, more slender body trunk; first dorsal fin origin over or slightly behind pelvic origins; a longer interdorsal space (5.4–9.5% of length); first dorsal fin free rear tip does not reach second dorsal fin origin; and posterior end of second dorsal and anal fins does not reach origin of upper and lower caudal fins, respectively. Adults uniform yellow- to grey-brown; young with small, dark, light-ringed ocellar spots and obscure saddle markings.
Distribution West Atlantic: USA to Gulf of Mexico, Caribbean to southern Brazil. East Atlantic: Cape Verde, Senegal, Cameroon to Gabon (rarely north to France). The east Pacific population has now been described as a different species (*Ginglymostoma unami*).
Habitat Rocky and coral reefs, channels between mangrove keys and sand flats on tropical and subtropical continental and insular shelves, 0–130m, mostly less than 40m.
Behaviour These nocturnal sharks are highly social within a small home range. Nurse Sharks often rest in groups (even in piles) by day in preferred shallow water locations on sand or in caves, which they return to each dawn. They are strong-swimming and more active at night. Their muscular pectoral fins can be used to clamber around on the bottom, the snout is used to root out prey, and the small mouth and large pharynx allows them to suck prey (including sleeping fishes) out of crevices or even to extract conch snails from an intact shell. Courtship and mating behaviour is extremely well-studied. Adult Nurse Sharks always aggregate in the same breeding grounds in very shallow water. Males are present every year, females only on alternate

years. The use of the same breeding site every year enables researchers to tag and take tissue samples from every animal, thus studying populations in considerable detail. Courtship behaviour includes synchronised parallel swimming, sides nearly touching, with one or more males beside or slightly behind and below the female. The successful male (one of a small number of dominant males in the aggregation) bites one of the female's pectoral fins and both roll upside-down on the seabed to mate. Several mating events may take place each day, unless females take refuge in water less than 1m deep, which is too shallow for mating to occur.
Biology Viviparous, 20–30 pups per litter with large yolk-sacs, multiple paternity common within a single litter. Pups are born in late spring/summer after a 5–6 month gestation. Females reproduce every other year. Juvenile nursery areas are located in shallow turtle-grass beds and coral reefs. Male Nurse Sharks mature at 10–15 years, females at 15–20 years. They feed on bottom invertebrates, bony fishes and stingrays.
Status IUCN Red List: **Data Deficient** globally. Historically common in many areas, still common in the USA and Bahamas, but their small home range and aggregating habits make Nurse Sharks highly vulnerable to overexploitation in fisheries. They have been depleted in many regions and reportedly extirpated from southern parts of their former range in the west Atlantic, where they are assessed as **Vulnerable**. There is no information on their status in the east Atlantic. They are fished for their meat, fins, and particularly for their extremely thick skin, which makes good leather. These sharks are docile, popular with divers and hardy in aquaria, but will bite if provoked.

Nurse Sharks, *Ginglymostoma cirratum*, Bahamas.

TAWNY NURSE SHARK   *Nebrius ferrugineus*

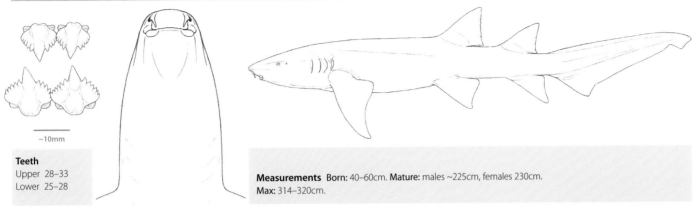

~10mm

Teeth
Upper 28–33
Lower 25–28

Measurements Born: 40–60cm. **Mature:** males ~225cm, females 230cm.
Max: 314–320cm.

Identification Fairly long barbels. Mouth in front of lateral eyes. Tiny spiracles. Large first dorsal fin base over pelvic fin bases. Angular fins. Caudal fin fairly long (more than 25% of total length). Colour may slowly change between shades of brown, depending on habitat. There are a few records from Japan and Taiwan of Tawny Nurse Sharks lacking a second dorsal fin.

Distribution Tropical Indian, west and central Pacific Oceans: southeast Africa to Red Sea and Persian Gulf; India and East Asia, north to Japan; tropical Australia to Palau, Samoa, Marshall Islands and Tahiti.

Habitat On or near bottom in sheltered areas (lagoons, channels and along the edges of coral and rocky reefs) with sand, seagrass and rock crevices, from the intertidal to deeper than 70m, mainly 5–30m. Juveniles are more likely to occur in crevices in shallow lagoons, but adults are more wide-ranging, although likely to return to favoured large crevices and caves.

Behaviour This species is mainly nocturnal, prowling reefs at night searching for prey, which it sucks out of crevices with its muscular pharynx. Tawny Nurse Sharks have a limited home range and often return to the same daytime resting place, where groups may aggregate in the shelter of caves and rocky crevices. They may be sluggish when caught with a hook, or 'spit' streams of water and spin on the line.

Biology Reproductive biology appears to vary in different parts of its range. In Australia, it is viviparous with large litters of about 26 (maximum 32) pups. In Japan, the young feed inside the uterus on large, infertile, yolky eggs (oophagy) and litter size is much smaller (1–4 pups, depending upon competition in the uterus). Preys upon crustaceans, cephalopods (particularly octopuses), sea urchins and reef fishes, occasionally sea snakes.

Status IUCN Red List: **Vulnerable.** Occurs in only a narrow coastal band, which is fished heavily through much of its range (with the exception of Australia, where commercial fishing pressure is low, although this is a popular game fish). Population declines and some local extirpations have been reported in other parts of its range. Small litters and limited dispersal will prevent rapid population recovery following overfishing. Tawny Nurse Sharks are docile and popular with divers, including in Thailand and the Solomon Islands, but may bite if harassed. Hardy when kept in aquaria.

Tawny Nurse Sharks, *Nebrius ferrugineus*, at dusk, Vavuu Atoll, Maldives.

Rhincodontidae: Whale Shark

WHALE SHARK *Rhincodon typus* FAO code: **RHN** Plate page 308

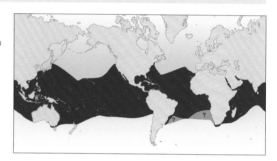

5mm

Teeth
Upper 300+
Lower 300+

Measurements Born: 55–64cm. **Mature:** males at least 600cm, females at least 800cm. **Max:** possibly 1700–2100cm; very large individuals are rarely measured accurately due to their massive size.

Identification Unmistakable, huge filter-feeder with checkerboard pattern of yellow or white spots on grey, bluish or greenish-brown back (white or yellowish underside). Broad, flat head; short snout; huge, transverse, almost terminal mouth in front of eyes. Prominent ridges on body, the lowest ending in a keel on the caudal peduncle. Lunate, unnotched caudal fin.

Distribution Circumglobal, in all tropical and warm temperate seas except the Mediterranean.

Habitat Pelagic, open ocean to close inshore off beaches, coral reefs and islands. Preferred surface water temperature appear to be 21–25°C, but Whale Sharks regularly dive to depths of up to 1928m, where the water is very much colder. The location of pupping and nursery grounds is unknown, although one pregnant female was caught near Taiwan and a few very small individuals have been captured in the Indo-Pacific (India, Philippines).

Behaviour This is a highly migratory shark. Long-distance, long-term movements are undertaken, including one 13,000km journey (in one direction only) made over 37 months. These long-distance migrations are presumably regular occurrences, since very little genetic variation has been found between Whale Sharks in different parts of the world. Tagging and photo-identification have identified regular visits to aggregation

sites, where as many as several hundred Whale Sharks may gather to feed at annual, seasonal or lunar fish and invertebrate spawning events. The high density of plankton produced on these occasions is consumed by suction feeding and gulping, often while the shark hangs vertically in the water. Wild Whale Sharks can readily become used to being hand-fed (see p. 75). Populations appear to segregate by age and sex. Some well-known Whale Shark aggregations used for dive tourism are predominantly of immature males.

Biology Viviparous. The only recorded pregnant female, from Taiwan, had over 300 small pups and eggs in her uterus. Feeds on planktonic crustaceans, fish eggs and small shoaling fishes.

Status IUCN Red List: Endangered. Has been fished, apparently unsustainably, for meat in many areas; harpoon fisheries have caused rapid, steep declines in catches in the Philippines, Taiwan, Maldives and India. Whale Sharks have also been used as a natural fish-aggregating device in some oceanic tuna fisheries, where vessels have encircled them with purse seines (this activity is now widely prohibited by tuna Regional Fisheries Management Organisations). The fins are reportedly not of good quality for soup, but are valued for their great size. Whale Sharks are legally protected in many states, particularly following their protection by international environmental agreements (e.g. CITES, CMS) and Regional Fisheries Management Organisations, and because their economic value for dive tourism has risen hugely in recent decades. A few Whale Sharks are kept on display in very large public aquaria.

Whale Shark, *Rhincodon typus,* feeding, Western Australia.

LAMNIFORMES mackerel sharks

Eight families and 15 species of mainly large, active, pelagic sharks in this order. Families Mitsukurinidae, Carchariidae, Pseudocarchariidae, Megachasmidae and Cetorhinidae have only one living species each. There are two species in Odontaspididae, three in Alopiidae (the threshers) and five in family Lamnidae.

Identification Cylindrical body, two spineless dorsal fins (first originates over abdomen, well in front of pelvic fin origins) and an anal fin. Vertebral axis extends into long upper caudal lobe. Conical head, fairly short snout, five broad gill openings (hind two in front or above pectoral fin origin), large mouth extending behind eyes, nostrils free from mouth, no barbels or grooves. Very small spiracles well behind eyes.

Biology Worldwide geographic distribution, mostly in warm water (some prefer cold) in a wide range of marine habitats, from the intertidal to over 1800m and in the open ocean; none occur in fresh water. Behaviour is highly varied, from slow coastal sharks to fast oceanic swimmers, and top predators to carrion and filter feeders. Some are highly migratory. Many are social and some will hunt cooperatively. Reproduction is viviparous, with oophagy (young eat eggs in the uterus) known in most species where information is available. At least one species also cannibalises other embryos in utero. Diets are varied, from marine mammals, birds and reptiles, to other sharks, rays, bony fishes and invertebrates.

Status Several species are or have been very important in coastal and oceanic commercial and sport fisheries and highly valued for sport, flesh and fins. Others are rare and not often recorded. Some of the larger species occasionally bite people, but are also important for dive ecotourism and film-makers, who view them with or without shark cages. Only one species is commonly kept in aquaria. Two-thirds of species in this order are threatened by overfishing; several are still completely unmanaged and unmonitored, but more than half of the mackerel sharks are now listed in Appendices to major multilateral global and regional environmental agreements, including CITES and CMS (see p. 83).

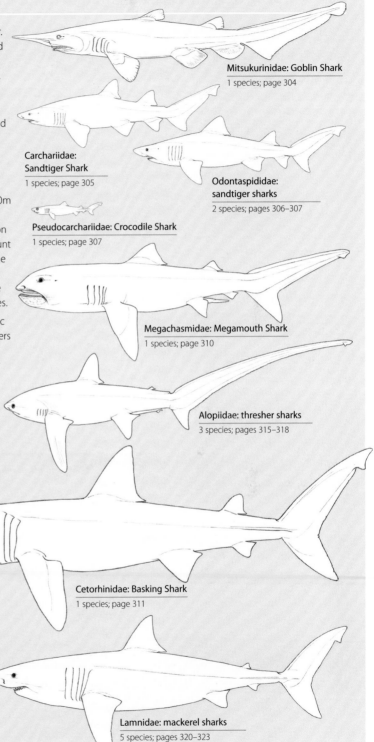

Mitsukurinidae: Goblin Shark
1 species; page 304

Carchariidae: Sandtiger Shark
1 species; page 305

Odontaspididae: sandtiger sharks
2 species; pages 306–307

Pseudocarchariidae: Crocodile Shark
1 species; page 307

Megachasmidae: Megamouth Shark
1 species; page 310

Alopiidae: thresher sharks
3 species; pages 315–318

Cetorhinidae: Basking Shark
1 species; page 311

Lamnidae: mackerel sharks
5 species; pages 320–323

White Shark, *Carcharodon carcharias*, breaching (p. 320).

Plate 34 PSEUDOCARCHARIIDAE, MITSUKURINIDAE, CARCHARIIDAE and ODONTASPIDIDAE

Five species in these four families, each with cylindrical body, conical head, pointed snout, no barbels or grooves, nostrils free from mouth, large mouth extending behind the eyes, very small spiracles well behind eyes; long broad gills in front of pectoral fins, two spineless dorsal fins, an anal fin, vertebral axis extends into upper lobe of asymmetrical caudal fin without keels, upper (no lower) precaudal pits present.

○ **Crocodile Shark** *Pseudocarcharias kamoharai*   page 307

Tropical oceans worldwide; well offshore to at least 590m. Cylindrical, slender, very distinctive; conical head, huge eyes, large mouth with prominent long slender teeth and highly protrusible jaws, very small spiracles and five broad gills; small fins, long upper caudal fin lobe; colour grey or grey-brown, light below with light-edged fins.

○ **Goblin Shark** *Mitsukurina owstoni*   page 304

Patchy in Atlantic, west Indian and Pacific Oceans; deepwater, 100–1300m or more, rarely on surface. Unmistakable soft flabby body; flat elongated snout, large mouth with long-cusped slender teeth and protrusible jaws, very small spiracles, long caudal fin without a lower lobe; colour pinkish white.

○ **Sandtiger Shark** *Carcharias taurus*   page 305

Atlantic, Mediterranean, Indian and west Pacific Oceans; to at least 232m. Large cylindrical heavy body, flattened conical snout, very small spiracles well behind eyes; large mouth extends behind eyes; large slender pointed teeth; long broad gills in front of pectoral fins; large dorsal fins and anal fin of similar size, first dorsal fin closer to pelvic fins than pectoral fins, asymmetrical caudal fin with a short lower lobe; colour light brown, white below, often with scattered dark spots.

○ **Smalltooth Sandtiger** *Odontaspis ferox*   page 306

Worldwide, warm-temperate and tropical; 10–1015m; on or near bottom. Differs from *Carcharias taurus* with long conical snout, fairly large eyes; first dorsal fin closer to pectoral fins than pelvic fins, first dorsal fin larger than second dorsal and anal fins; colour grey or grey-brown above, light below, often with dark spots.

○ **Bigeye Sandtiger** *Odontaspis noronhai*   page 307

Patchy Atlantic, Indian and Pacific Oceans; 35–1000m or more; midwater to near bottom. First dorsal fin larger than second dorsal and anal fins; very large eyes distinguish this species from other sandtigers; colour is uniform dark red-brown to black above and below, with a white blotch on first dorsal fin.

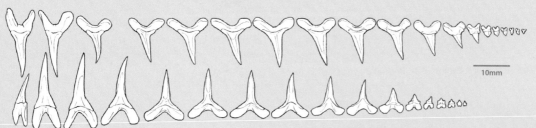

10mm

○ *Mitsukurina owstoni* teeth

10mm

○ *Carcharius taurus* teeth

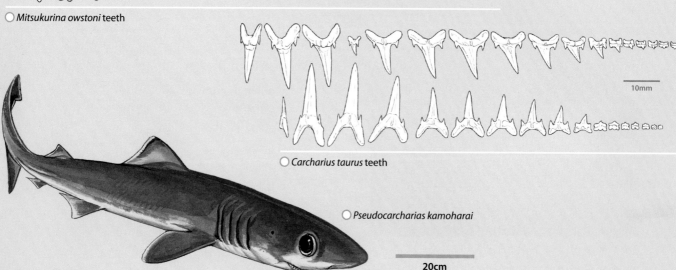

○ *Pseudocarcharias kamoharai*

20cm

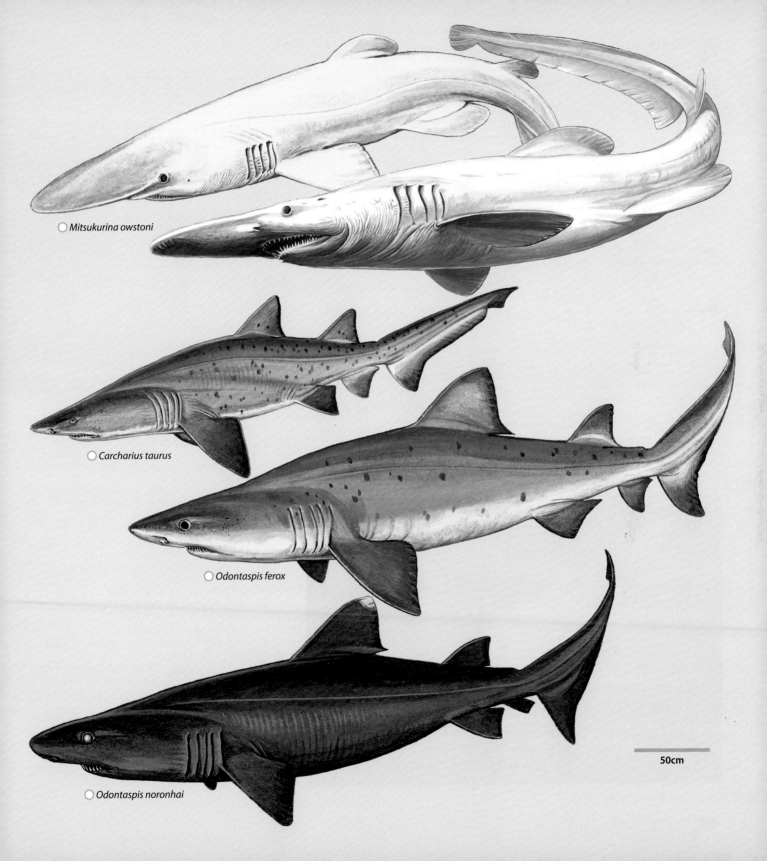

○ *Mitsukurina owstoni*

○ *Carcharius taurus*

○ *Odontaspis ferox*

○ *Odontaspis noronhai*

50cm

GOBLIN SHARK *Mitsukurina owstoni*

FAO code: **LMO**   Plate page 302

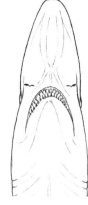

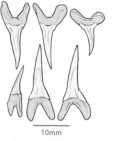

10mm

Teeth
Upper 35–53
Lower 37–46

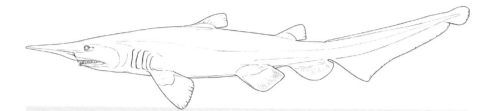

Measurements Born: 80–100cm. **Mature:** males 260–380cm, females more than 420cm.
Max: males 380cm, females 550–620cm.

Identification Body soft, flabby and pinkish-white, with bluish fins. Unmistakable flat, elongated snout; large mouth with protrusible jaws; long-cusped, slender teeth. Very small spiracles. Five broad gills. Long caudal fin without a lower lobe.

Distribution Atlantic, Indian and Pacific Oceans. Patchily distributed.

Habitat Deepwater, outer continental shelves, upper slopes and off seamounts from about 100–1300m. Very rarely comes to the sea surface. The habitat of these sharks, particularly the rarely reported adults, is very poorly known. It has been speculated that, given their soft, flabby body and light pinkish to white colour, Goblin Sharks are more likely to occupy a midwater habitat than to live exclusively on or near the seabed.

Behaviour Its body form suggests that the Goblin Shark is a poor swimmer. The sensitive blade-like snout may be used to detect its prey, which can then be snapped up rapidly when the highly specialised jaws shoot forward. The slender front teeth pictured here suggest a diet of small, soft-bodied fishes and squid, but its back teeth are modified to crush food. At least one individual has been captured at less than 50m depth over water 2000m deep, which supports the theory that they migrate up into the water column.

Biology Very poorly known, since adults are rarely seen or retained for detailed examination. The diet of this shark includes midwater fishes and invertebrates, including squid and crustaceans.

Status IUCN Red List: **Least Concern.** Although Goblin Sharks are considered rare in most places where they have been reported, because they are only a very occasional bycatch of deepwater fisheries, they seem likely to be widely distributed. Most Goblin Sharks captured have been subadults, generally less than 4m total length, taken in seasonal bottom set gill nets. The adults, which can measure up to 6.2m long, presumably occupy areas or habitats that are not fished.

Subadult Goblin Sharks are relatively common in the Tokyo Submarine Canyon between 100–350m deep, where they are caught seasonally in bottom-set nets between October and April. Exceptionally, between 100 and 300 Goblin Sharks were caught by Taiwanese fishermen at a depth of about 600m off the northeast part of the island during a two week period in April 2003. The capture of such a large number of this supposedly rare deepsea shark was unusual; it reportedly followed a strong earthquake. The majority of animals were said to have been males, some up to 4m long. Jaws taken from these specimens were sold for between USD $1,500 and $4,000 each, depending on their size and quality.

Goblin Shark, *Mitsukurina owstoni*.

SANDTIGER SHARK *Carcharias taurus*

FAO code: **CCT**

Plate page 302

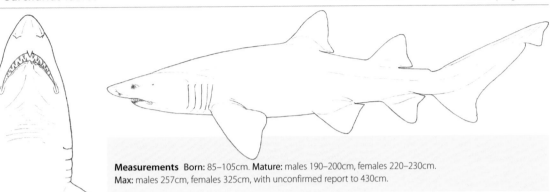

Teeth
Upper 36–54
Lower 32–46

Measurements Born: 85–105cm. **Mature:** males 190–200cm, females 220–230cm.
Max: males 257cm, females 325cm, with unconfirmed report to 430cm.

Alternative names Spotted Raggedtooth, Grey Nurse Shark
Identification Large, heavy, stout body. Flattened, conical snout; long mouth extends behind eyes; large, slender, pointed teeth. Long gill openings in front of pectoral fins. Large dorsal and anal fins similar in size; first dorsal set back and closer to pelvic than to pectoral fins. Asymmetrical tail, short lower lobe. Light brown coloured body, often with scattered darker spots (e.g. in the South African population).
Distribution Atlantic, Mediterranean, Indian and west Pacific Oceans. Coastal warm-temperate and tropical waters. Not recorded in the central and east Pacific.
Habitat Coastal waters, from the surf zone (<1m) to offshore reefs at least 232m deep; mostly 15–25m. Associated with underwater caves, gullies, and reefs. Usually on or near the bottom, occasionally midwater or at surface.
Behaviour A slow but strong swimmer, more active at night. Some populations are highly migratory, moving to cooler water in summer. Uniquely, this species achieves neutral buoyancy by gulping air into its stomach, which enables it to hover virtually motionless in mid water. Sandtigers have the largest brains of all lamnid sharks, and exhibit interesting social behaviours in the wild (where aggregations may number 20–80 sharks) and in captivity. For example, when aggregated they have been observed herding prey fishes to feed and have complex courtship and mating behaviours. Male Sandtigers guard females after mating, which is likely to improve the chance of their offspring surviving to birth (multiple paternity can occur in this species and the largest pups produced by the first eggs fertilised, will kill smaller pups born from later mating events).
Biology The Sandtiger has one of the lowest reproductive rates known among chondrichthyans. It is oophagous, where the developing young feed in utero upon eggs produced by the mother throughout the 9–12 month pregnancy, and in this species also upon the smaller embryos (their siblings and half-siblings – intrauterine cannibalism), until just one precociously toothed pup survives in each uterus. One unborn pup bit an investigating scientist. Two pups are born every other year, as females have a resting year between pregnancies. Males mature at 6–7 years, females at 9–10. Maximum age seen in the wild is 15–17 years, but Sandtigers have lived over 30 years in aquaria. They feed on a wide range of fishes and invertebrates.
Status IUCN Red List: **Vulnerable** globally, many populations seriously depleted. **Critically Endangered** in some regions including Europe, the Mediterranean, and eastern Australia, where it has been a utilised bycatch of commercial and recreational fisheries and targeted by divers for sport. This was the world's first legally protected shark (in Australia), and is now on protected species lists in many countries and in Mediterranean waters. Often seen in public aquaria, it is docile and breeds in captivity. Important for dive ecotourism in South Africa and Australia, but may bite if approached too closely. The Sandtiger's taxonomy, particularly its genus, has regularly been discussed since its scientific discovery and description in the Mediterranean in 1810. It has also been placed in *Eugomphodus*, *Triglochis* and *Odontaspis* (the latter from descriptions of fossil teeth and widely used for all sandtiger sharks for over 100 years). Sandtiger body morphology, however, indicates that they do not belong in the same family or genus as *Odontaspis ferox* and *O. noronhai*; it is now placed into its own family, Carchariidae. Indeed, recent molecular analyses indicate that the Sandtiger Shark is far more closely related, genetically, to the Basking Shark *Cetorhinus maximus* than it is to the two *Odontaspis* species.

Sandtiger Shark, *Carcharias taurus*.

Odontaspididae: deepsea sandtiger sharks

Large heavy-bodied sharks with pointed snouts, upper precaudal pits present but no lower pits, caudal fins asymmetrical and without keels. Two species.

SMALLTOOTH SANDTIGER   *Odontaspis ferox*

FAO code: **LOO**   Plate page 302

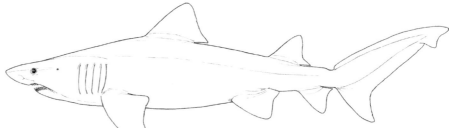

10mm

Teeth
Upper 46–56
Lower 34–48

Measurements Born: ~100–105cm. **Mature:** males 200–250cm, females 300–350cm. **Max:** males 344cm, females 450cm.

Alternative name Bumpytail Raggedtooth

Identification Large, heavy, stout-bodied shark. Long, conical, slightly flattened snout and long mouth. Fairly large eyes. Large first dorsal fin originating over the pectoral fin free rear tips, much larger than the second dorsal and anal fins. Colour is grey, brownish grey, light brown or olive above, often with scattered darker reddish spots over body (e.g. in the South African population), and light below.

Distribution May occur worldwide in warm-temperate and tropical deep water.

Habitat On or near bottom, continental and insular shelves and upper slopes, 10–1015m, possibly epipelagic, with individual caught from 70–500m below surface over water 2000–4000m deep. Sometimes seen near coral reef dropoffs, rocky reefs and gullies. Adults and smaller juveniles (less than 150cm) tend to occur in deeper water (between 300–600m) while intermediate size specimens occur shallower than 150m. Water temperature where these sharks have been caught ranges from 6–20°C.

Behaviour Active offshore swimmer, reported alone and in small groups near reefs and gullies.

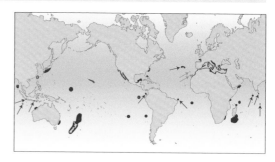

Biology Reproduction poorly known. Presumed viviparous, pups nourished by oophagy. Litter size unknown. Preys on small bony fishes, squid and shrimp.

Status IUCN Red List: Vulnerable globally. Critically Endangered in the northeast Atlantic and Mediterranean due to recent absence from catches and suspected steep declines in its population.

Smalltooth Sandtiger, *Odontaspis ferox*.

BIGEYE SANDTIGER *Odontaspis noronhai*

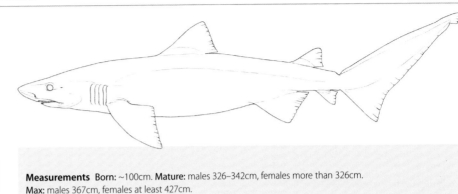

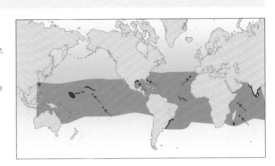

Teeth
Upper 34–43
Lower 37–46

Measurements Born: ~100cm. **Mature:** males 326–342cm, females more than 326cm.
Max: males 367cm, females at least 427cm.

Identification A large, stout shark. Long, bulbous snout and long mouth. Eyes very large. Large first dorsal fin closer to pectoral fins that pelvic fins. Second dorsal fin and anal fins smaller than first dorsal fin. Upper precaudal pit present; no lateral keels on caudal peduncle. Caudal fin asymmetrical with a strong lower lobe. Colour is a uniform blackish, dark chocolate brown or reddish brown, including the ventral surface, without spots; tip of first dorsal fin has a white blotch.
Distribution Atlantic, west Indian and Pacific Oceans. Known from scattered records; may be worldwide in deep warm waters.   **Habitat** Midwater in the open ocean, near bottom on continental and island slopes, 35m to deeper than 1000m. Uniform dark colour suggests this is an oceanic, midwater species.
Behaviour Poorly known. May migrate vertically in mid-ocean (near surface by night, deep water by day).
Biology Unknown. Diet includes bony fishes and cephalopods.
Status IUCN Red List: **Least Concern**. Very little fishing activity occurs in this species' mesopelagic habitat.

Pseudocarchariidae: Crocodile Shark

CROCODILE SHARK *Pseudocarcharias kamoharai*

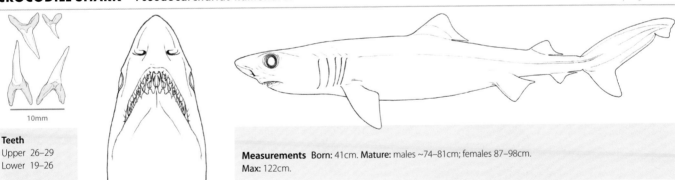

Teeth
Upper 26–29
Lower 19–26

Measurements Born: 41cm. **Mature:** males ~74–81cm; females 87–98cm.
Max: 122cm.

Identification Very distinctive small, slender-bodied oceanic shark with small fins; long upper caudal lobe; conical head with huge eyes; five long, broad gill slits; and prominent, long, slender teeth on highly protrusible jaws. Grey or grey-brown back, lighter below; light-edged fins.
Distribution Worldwide, oceanic tropical waters.
Habitat Usually well offshore, far from land, from surface to at least 590m.
Behaviour Probably a strong, active swimmer. May migrate vertically to surface at night, deeper water by day.
Biology Viviparous, 4 pups per litter feed on unfertilised eggs and possibly cannibalise other young before birth. No known seasonality to the reproductive cycle. Females mature at 5.1 years and reach a maximum age of 13 years. Diet includes midwater bony fishes and cephalopods.
Status IUCN Red List: **Least Concern**. Population has been increasing due to relatively fast growth and possibly the depletion of larger oceanic sharks that predate on and/or compete with this species. Crocodile Sharks are frequently a bycatch in pelagic longline fisheries, but usually discarded.

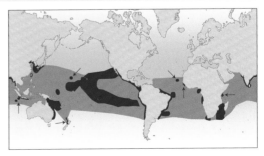

Plate 35 RHINCODONTIDAE, MEGACHASMIDAE and CETORHINIDAE

Each of these three huge plankton feeders is from a different family, and the only species in their family. The Megamouth and Basking Sharks are mackerel sharks of the order Lamniformes. They have a cylindrical body, long head, large gills, a huge mouth that extends behind the eyes with no barbels or grooves, very small spiracles behind eyes, and nostrils free from the mouth. The Whale Shark, however, is a gigantic carpetshark in the order Orectolobiformes and not closely related to the other two plankton feeders. It has a broad flat head with a very short snout, small barbels and long nasoral grooves, and spiracles that are close to and larger than the eyes.

◯ Whale Shark *Rhincodon typus*

page 300

Worldwide, all tropical and warm-temperate seas except Mediterranean; surface to at least 1928m. Unmistakable, huge with prominent ridges on body, lowest terminating in a keel on caudal peduncle; broad flat head, short snout, huge transverse mouth in front of eyes, comparatively small barbels, long nasoral grooves, huge gills, spiracles close to and larger than eyes; 2 spineless dorsal fins, anal fin, caudal fin lunate and unnotched; chequerboard pattern of yellow or white spots and blotches on a grey, blue-grey to green-brown background, white or yellow underside.

◯ Basking Shark *Cetorhinus maximus*

page 311

Worldwide, cold-temperate to tropics; associated with coastal and oceanic fronts; 0–1264m. Unmistakable, very large cylindrical body, strong lateral keels on caudal peduncle; conical head, pointed snout, huge mouth with tiny teeth, nostrils free from mouth, no barbels or grooves, huge gills almost encircle head, very small spiracles well behind eyes; 2 spineless dorsal fins, anal fin, lunate caudal fin; variable colour, darker above often with mottled pattern on back and sides, white blotches under head.

◯ Megamouth Shark *Megachasma pelagios*

page 310

Worldwide in warm-temperate to tropical seas; well offshore; 0–1500m. Unmistakable, large cylindrical body; large long head, short round snout, huge terminal mouth with numerous small hooked teeth, nostrils free from mouth, no barbels or grooves, large gills, very small spiracles behind eyes; 2 spineless dorsal fins, anal fin; grey above, white below, light margins to blackish pectoral and pelvic fins, dark spotting on lower jaw.

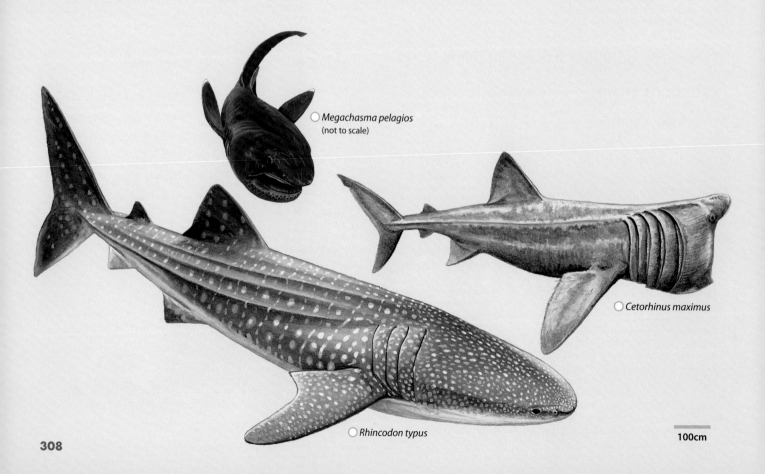

◯ *Megachasma pelagios*
(not to scale)

◯ *Cetorhinus maximus*

◯ *Rhincodon typus*

100cm

308

○ *Rhincodon typus*

○ *Cetorhinus maximus*

○ *Megachasma pelagios*

100cm

MEGAMOUTH SHARK *Megachasma pelagios* FAO code: **LMP** Plate page 308

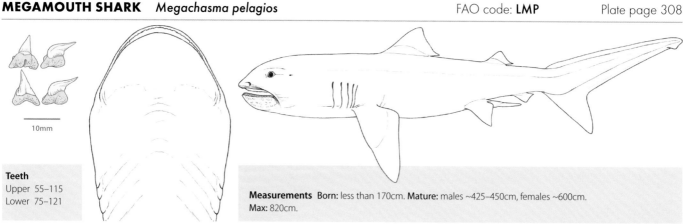

Teeth
Upper 55–115
Lower 75–121

Measurements Born: less than 170cm. **Mature:** males ~425–450cm, females ~600cm.
Max: 820cm.

Identification Unmistakable large, long head with short, rounded snout. Huge terminal mouth extending behind eyes with numerous small, hooked teeth. Large gills. Body grey above (light margins to blackish pectoral and pelvic fins), white below; dark spotting on lower jaw.

Distribution Worldwide, from warm-temperate to tropical seas.

Habitat Oceanic, coastal and offshore, 5–40m on continental shelf, 0–1500m offshore over very deep water.

Behaviour Migrates vertically with krill, rising close to the surface at night, but retreating to deeper water by day. May have luminescent tissue inside mouth to attract prey.

Biology Reproduction unknown, presumed viviparous with oophagy. Feeds on plankton, particularly shrimp, possibly by suction.

Status IUCN Red List: Least Concern. Once thought to be quite rare, this species is relatively common in some areas.

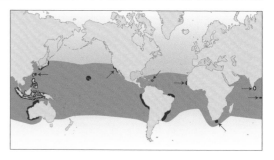

Megamouth Shark, *Megachasma pelagios*, off California, USA.

BASKING SHARK *Cetorhinus maximus*

FAO code: **BSK** Plate page 308

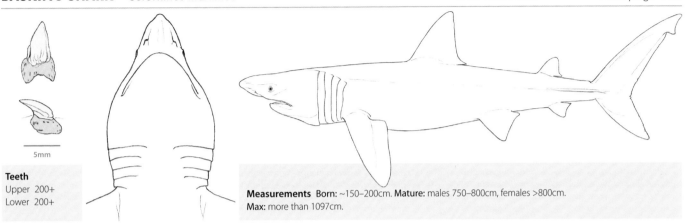

Teeth
Upper 200+
Lower 200+

5mm

Measurements Born: ~150–200cm. **Mature:** males 750–800cm, females >800cm.
Max: more than 1097cm.

Identification Unmistakable, very large shark, second largest after the Whale Shark but easily distinguished from it by a pointed snout, huge subterminal mouth, enormous gills nearly encircling the head, modified gill rakers, strong lateral keels on caudal peduncle, and a huge lunate tail. Colour blackish to grey-brown, grey or blue-grey above and below; ventral surface sometimes lighter, often with irregular white blotches on underside of head and abdomen; mottled on back; lighter linear striping and spots along flanks.

Distribution Worldwide. Near the surface in cold to warm-temperate water, in deep water below the thermocline in tropical and equatorial regions.

Habitat Usually recorded from the coast to the continental shelf edge and slope, often associated with coastal and oceanic fronts. Migrates across ocean basins, and crosses the equator from northern to southern hemisphere at depths of 200–1000m, diving to 1264m, where the water temperature is about 5°C.

Behaviour Usually seen ram-feeding on surface aggregations of zooplankton at coastal or oceanic fronts, sometimes in groups of over 100. Snout tip, dorsal fin and upper caudal fin may all break the surface as the shark swims slowly forward with open mouth, using gill rakers to extract plankton from up to 1.5 million litres of water per hour. Feeds on deepsea zooplankton in winter and offshore. They have been a bycatch of deepsea trawls targeting fish spawning aggregations off New Zealand, presumably while feeding on fish

eggs. Basking Sharks migrate thousands of kilometres across ocean basins and through the tropics, where they remain in cool water below the thermocline. Complex courtship behaviour is reported and breaching is more likely to be observed when sharks are in surface aggregations.

Biology A poorly known plankton-feeder. Huge oil-filled liver provides buoyancy. Presumably oophagous (only 1 litter, of 6 near-term pups, has been reported).

Status IUCN Red List: **Endangered** due to population declines of over 50%. Formerly targeted by harpoon and net fisheries from Norway to Scotland, Ireland and the Bay of Biscay in the North Atlantic, and in the North Pacific off Japan, California and British Colombia (the latter an eradication programme to reduce damage to fishing nets). Liver oil was the main product, traditionally used for lighting, later for engineering oil and cosmetics, also the huge fins and sometimes meat and cartilage. Basking Sharks are now listed in the Appendices of several global and regional environmental agreements (see p. 83), and protected by many range States and the European Union through wildlife legislation or fisheries regulations. Despite protection, there is no sign of recovery for the US and Canadian Pacific stocks, but numbers are reportedly rising in Scottish waters. Observers and photographers can provide sightings records and dorsal fin photographs to scientific research projects like www.baskingshark.org, studying Basking Shark numbers, migrations and life history.

Basking Shark, *Cetorhinus maximus*, Isle of Coll, Scotland.

Plate 36 ALOPIIDAE

Three species. Cylindrical body with very long curving tail, conical head with mouth extending behind eyes, nostrils free from mouth, no barbels or grooves, very small spiracles behind eyes; two spineless dorsal fins, anal fin present.

○ **Pelagic Thresher** *Alopias pelagicus*   page 318

Indian and Pacific Oceans; oceanic, sometimes inshore; 0–300m. Very narrow head with straight forehead and arched profile, no labial furrows, fairly large eyes; long straight broad-tipped pectoral fins, caudal fin upper lobe nearly as long as the rest of the body; deep blue above, white below but no white above pectoral fin bases.

○ **Bigeye Thresher** *Alopias superciliosus*   page 318

Worldwide tropical and temperate seas; oceanic to coastal, to 955m. Distinctive huge eyes extend onto flat-topped head, deep horizontal groove above gills; large, very long narrow pectoral fins; purplish grey or grey-brown above, light grey to white below which does not extend above pectoral fin bases.

○ **Thresher Shark** *Alopias vulpinus*   page 315

Almost worldwide (Indian Ocean distribution not confirmed) in tropical to cold-temperate seas; nearshore to far offshore, to at least 650m. Fairly large eyes, labial furrows; long curved narrow pointed pectoral fins, caudal fin upper lobe nearly as long as the rest of the body; blue-grey to dark grey above, silvery or coppery sides, underside white extending in a patch above pectoral fins, white dot on pectoral fin tips.

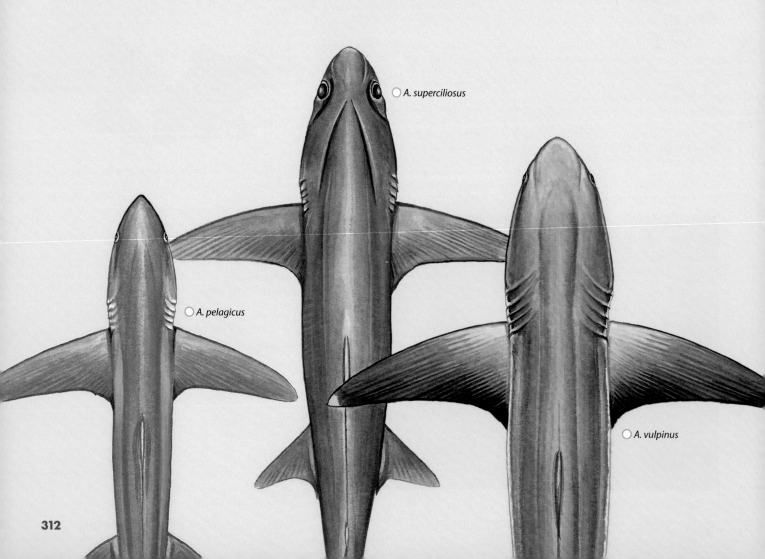

○ *A. superciliosus*

○ *A. pelagicus*

○ *A. vulpinus*

Alopias pelagicus

Alopias superciliosus

Alopias vulpinus

100cm

Alopiidae: thresher sharks

Identification Three species of large-eyed sharks with small mouths, large pectoral, pelvic, and first dorsal fins, tiny second dorsal and anal fins, and elongated, curved whip-like caudal fins almost as long as their bodies. Their large pectoral fins are needed to provide extra lift, thus counterbalancing the upward thrust created by movements of the huge tail.

Biology These species occur from inshore waters (Thresher Shark) to the open ocean in warm tropical seas, where they are specialist feeders on small shoaling fishes and squid. The threshers use their very long tails to herd their prey into tight shoals, which they then stun with tail swats so that they are easier to pick up with their small mouths. Thresher mouth and tooth sizes reflect prey choices: Thresher and Pelagic Thresher Sharks have weaker jaws and smaller teeth and mouths than the Bigeye Thresher, which specialises in slightly larger prey. Fishers report threshers being hooked by the tail – presumably because they attempted to use their tail to stun a baited hook.

Status Because of their similar appearance, these three species are frequently not distinguished in fisheries catch records, but lumped together as 'thresher sharks'. This makes it hard for scientists, who need accurate fisheries data to assess changes in the population status of sharks, to know which species have been most badly affected by the declines reported for the whole family. This is a problem because their very low fecundity makes them particularly susceptible to overfishing and in urgent need of species-specific assessment and management. It is clear, however, that all thresher shark species have been seriously depleted by target and bycatch fisheries for their high value meat and fins (liver oil, cartilage and skin is sometimes also utilised). Every species of thresher shark is assessed as Endangered or Vulnerable in the IUCN Red List of Threatened Species. This is one of the most threatened families of sharks.

Thresher shark fins are particularly valuable and identified by family name in the Asian dried seafood market. In 2000, an estimated 0.5 to 3 million thresher sharks were entering the Hong Kong fin trade annually, greatly exceeding reported fisheries catches, and contributing about 2.3% of all dried shark fins recorded. Fifteen years later, the proportion of thresher shark fins in trade had fallen significantly. The threshers were subsequently listed in CITES Appendix II, to ensure that future international trade in their fins and other products would be legal, sustainable, and traceable. Some RFMOs (IOTC, ICCAT and GFCM) now prohibit or strongly discourage the catch, retention and sale of some or all species of thresher shark within the fisheries that they manage; others may soon follow suit. A few countries, including Spain, protect all thresher shark species.

Pelagic Thresher Shark, *Alopias pelagicus*, being cleaned by a Bluestreak Cleaner Wrasse, *Labroides dimidiatus*, Malapascua Island, Philippines, where this species is important for dive tourism (p. 318).

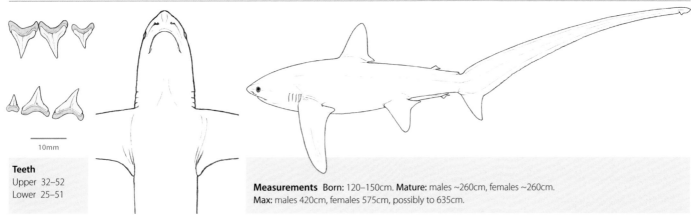

Teeth
Upper 32–52
Lower 25–51

10mm

Measurements Born: 120–150cm. **Mature:** males ~260cm, females ~260cm.
Max: males 420cm, females 575cm, possibly to 635cm.

Identification Labial furrows present. Fairly large eyes on side of head. Upper caudal lobe is about as long as rest of shark. Blue-grey to dark grey above; sides silvery or coppery. Underside white, extending in a patch above the pectoral fins. A white dot on the tips of narrow, pointed pectoral fins.

Distribution Almost worldwide; Indian Ocean records require verification. Oceanic and coastal in tropical to cold-temperate seas.

Habitat From nearshore to far offshore, sea surface to about 400m. Most abundant near land (pups use inshore nurseries) and in temperate waters.

Behaviour Migrates seasonally along the coast. Can breach clean out of the water (one ended up inside a boat after a series of leaps). Slaps the water, herds and stuns small fishes with a tail flick before returning to swallow them, sometimes working cooperatively in schools.

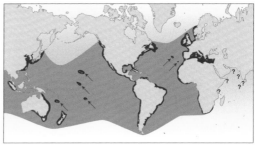

Biology Viviparous, 2–7 (usually 4) pups per litter feed on infertile eggs produced by the mother during the 9 month pregnancy. Males mature at 3–7 years, females at 3–9 years (age varies by region). Maximum known age is 24 years, possibly 50 years. Warm-blooded with a high metabolic rate; threshers use a countercurrent heat exchange system (see p. 33) to maintain a raised body core temperature. They feed on small pelagic schooling fishes, pelagic crabs and cephalopods.

Status IUCN Red List: **Vulnerable** due to declines of 60% in some areas. Threshers are easily overfished for their valuable meat, fins (which were 2–3% of all fins auctioned in Hong Kong, before the introduction of oceanic fisheries and trade regulations), liver oil, and skin; several populations are now seriously depleted. They are also highly prized sports fish. Tail flicks at baited hooks frequently catch threshers by their upper caudal lobe; as ram-ventilators, they suffocate if unable to continue to swim forward. Some divers have been struck by the upper tail lobe.

Thresher Shark, *Alopias vulpinus*, Costa Rica.

Plate 37 LAMNIDAE

Five species. Spindle-shaped body; conical head, fairly short snout, nostrils free from mouth, no barbels or grooves, large mouth extending behind eyes, broad gills, very small spiracles well behind eyes; keels present on caudal peduncle; very small second dorsal and anal fins, nearly symmetrical lunate caudal fin.

○ **White Shark** *Carcharodon carcharias*

page 320

Worldwide except polar seas; coastal and oceanic; 0–1280m. Heavy body; relatively long snout, very black eyes, very long gills; strong keels on caudal peduncle; large first dorsal fin; grey above with sharp demarcation to white below, first dorsal fin with dark free rear tip, black tip to underside of pectoral fins, usually black spot at pectoral fin insertion, old adults often become paler above and off-white below giving rise to a great 'white' shark.

○ **Shortfin Mako** *Isurus oxyrinchus*

page 321

Worldwide in temperate and tropical seas; coastal and oceanic; 0–888m. U-shaped mouth, black eyes, very long gills; strong keels on caudal peduncle; back brilliant blue or purple, paler and more silvery on sides, white below, underside of snout and mouth white in adults (dusky in Azores population 'marrajo criollo'), anterior half of pelvic fins dark, rear and underside white.

○ **Longfin Mako** *Isurus paucus*

page 322

Probably worldwide; possibly epipelagic in deep water; 0–1752m. Differs from *I. oxyrinchus* by a less pointed snout; pectoral fins as long as head and relatively broad-tipped; dusky underside to snout and mouth in adults.

○ **Salmon Shark** *Lamna ditropis*

page 322

North Pacific; cool coastal and oceanic waters to 1864m or more. Heavy body; short snout, long gills; strong keels on caudal peduncle with short secondary keels on caudal fin base; dark grey or blackish above, white below with dusky blotches and dark underside to snout in adults, white patch over pectoral fin bases, first dorsal fin with dark free rear tip.

○ **Porbeagle Shark** *Lamna nasus*

page 323

North Atlantic and cool southern hemisphere waters; inshore to offshore to 1809m. Very similar to *L. ditropis* except for more pointed snout, no white patch above pectoral fins, very distinctive white free tip to first dorsal fin, northern Porbeagle Sharks have no dusky blotches, southern specimens have distinctive dusky hood and dusky blotches similar to *L. ditropis* but these species' ranges do not overlap.

○ *Carcharodon carcharias*

Isurus oxyrinchus

marrajo criollo

Isurus paucus

Lamna ditropis

northern species

Lamna nasus

southern species

100cm

PELAGIC THRESHER *Alopias pelagicus*

FAO code: **PTH**   Plate page 312

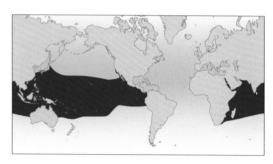

Teeth
Upper 37–45
Lower 37–48

~10mm

Measurements Born: 158–190cm. **Mature:** males 267–277cm, females 280–290cm. **Max:** males 347cm, females 428cm.

Identification Very narrow head with straight forehead and arched profile. No labial furrows. Fairly large eyes. Straight, broad-tipped pectoral fins. Long, curving caudal fin; upper lobe nearly as long as rest of shark. Body deep blue above, white below (no white above pectoral fins).
Distribution Indian Ocean: South Africa to Australia. Pacific Ocean: Tahiti, China, Japan, to southern California, USA and south Peru, including the Galapagos. Absent from the Atlantic.
Habitat Oceanic, wide-ranging, usually offshore, sometimes nearshore on narrow continental shelf, 0–300m. Sometimes near coral reefs, drop-offs and seamounts.
Behaviour Poorly known. Active, strong swimmer. Probably migratory. May repeatedly breach.
Biology Viviparous, 2 pups per litter (1 from each uterus) feed on unfertilised eggs. They do not appear to have a seasonal breeding cycle as pregnant females with embryos at different stages of development are found throughout the year. Males mature at 7–8 years, females 8–13 years; maximum age males ~20 years and females ~29 years. Feed on small fishes and cephalopods.
Status IUCN Red List: Endangered. Threatened by oceanic fisheries and seriously depleted. See p. 314.

BIGEYE THRESHER *Alopias superciliosus*

FAO code: **BTH**   Plate page 312

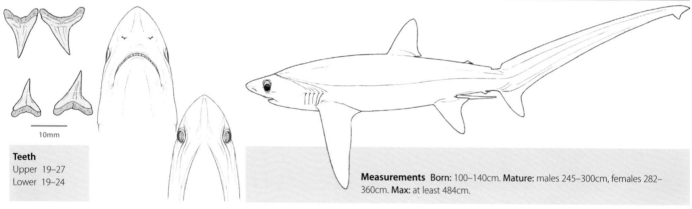

Teeth
Upper 19–27
Lower 19–24

10mm

Measurements Born: 100–140cm. **Mature:** males 245–300cm, females 282–360cm. **Max:** at least 484cm.

Identification Huge eyes extend onto almost flat-topped head. Deep horizontal groove above gills. Very long, narrow pectoral fins. Long, curving upper caudal lobe nearly as long as rest of shark. Body purplish grey or grey-brown above; light grey to white below, not extending above pectoral fins bases.
Habitat Tropical and temperate seas, close inshore to open ocean, surface to 955m, mostly >100m.
Behaviour Uses its tail to stun the pelagic fishes on which it feeds. Discovered visiting a shallow reef fish cleaning station on an Indo-Pacific seamount. This shark migrates vertically from 300–500m during the day to shallower than 100m at night, and is found in areas where the water temperature is 16–25°C.
Biology Viviparous, 2–4 pups per litter. Males mature 9–10 years, females 12–14 years; maximum age 19–20 years. Feeds on small schooling fishes, cephalopods and crustaceans.
Status IUCN Red List: Vulnerable. The slowest rate of population increase of all threshers. Highly threatened by oceanic fisheries; some populations seriously depleted and strictly protected. See p. 314.

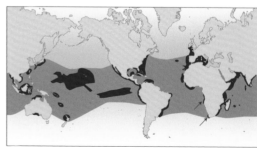

Lamnidae: mackerel sharks

Identification Five species of large spindle-shaped sharks with large teeth, conical heads, long gill openings, and lunate (crescent-shaped) caudal fins with strong caudal keels.

Biology Warm-blooded and viviparous, these species give birth to relatively small litters of large pups (bigger, older females bear larger litters) after a gestation period of around a year. The mother feeds her pups throughout her pregnancy by laying large numbers of infertile eggs. This huge investment of energy in producing large young that will survive well means that females are unable to give birth every year – a long recovery period is needed to rebuild their reserves.

Status Notorious, spectacular, and delicious: this small family includes some of the best known and most economically valuable of large sharks, most of which are classified as Threatened. Everyone who picks up this book will recognise the White Shark, international star of literature, film and ecotourism, and (despite its bad reputation) among the first few shark species to be listed in national, regional and international conservation regulations. The Shortfin Mako and Porbeagle Sharks are not so widely known, although fishers and gourmets appreciate that they produce some of the most delicious, and hence most expensive, shark steaks in the world. While White Shark populations fell mainly due to bycatch and persecution, and their economic value today is mostly derived from non-consumptive use in ecotourism, Shortfin Mako and Porbeagle populations declined steeply because of over-fishing for their delicious meat – particularly in the North Atlantic. They were added to the Appendices of major international biodiversity conservation agreements because scientific advice was not heeded in time and fisheries management measures had been introduced too late to prevent these declines. The other two Lamnid sharks are less well-known. The Longfin Mako is the mysterious, rare and poorly studied cousin of the Shortfin Mako, and the Salmon Shark is the North Pacific 'sister species' to the Porbeagle Shark. An estimated 100,000–150,000 Salmon Shark were caught annually in huge scale oceanic gillnet fisheries during the 1950s and 60s. These 'wall of death' fisheries ended following a United Nations Resolution in 1992 that prohibited the use of nets longer than 2.5km. The result, almost 30 years later, is that the Salmon Shark is showing signs of population recovery; this is the only Lamnid shark species not assessed as threatened in the IUCN Red List – surely a good omen for depleted Porbeagle stocks.

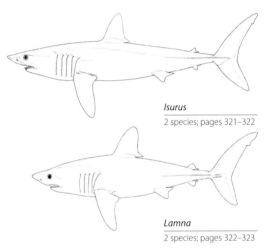

Carcharodon
1 species; page 320

Isurus
2 species; pages 321–322

Lamna
2 species; pages 322–323

Salmon Shark, *Lamna ditropis*, female with mating scar on side and copepod parasites streaming from fins. Port Fidalgo, Prince William Sound, Alaska, USA (p. 322).

Teeth
Upper 23–29
Lower 21–25

20mm

Measurements Born: 107–160cm. **Mature:** males 310–410cm, females 400–500cm. **Max:** males 550cm, females 600–640cm.

Identification Heavy, long-snouted, spindle-shaped body. Huge, flat, triangular, serrated teeth. Very black eyes. Long gill slits. Large first dorsal fin with dark free rear tip. Tiny second dorsal and anal fins. Strong keels on caudal peduncle. Crescent-shaped caudal fin. Very sharp colour change on flanks from greyish back to white underside. Black tip underneath pectoral fins; usually a black spot where rear edge joins body.
Distribution Worldwide. Very wide-ranging in most oceans; among the greatest habitat and geographic range of any fish, tolerating temperatures of 5°–25°C.
Habitat From very shallow water inshore, to the continental shelf and remote oceanic islands, with long periods spent in pelagic habitat while travelling across the mid-ocean at depths of 0–1280m. Most commonly seen in aggregations around rocky reefs near pinniped colonies.
Behaviour An intelligent and inquisitive shark with complex social interactions that minimise conflict within aggregations (females, larger sharks and residents dominate, respectively, males, smaller sharks and new arrivals). Recent research points to complex courtship displays. This very effective predator can breach out of the water when attacking surface-swimming prey, disabling large mammals with a single bite and waiting until they are dead before feeding. Spy-hopping may possibly be used to view or smell for prey. Satellite and genetic studies demonstrate that these sharks are highly migratory, regularly swimming thousands of kilometres to cross and re-cross ocean basins, but also that there is a high degree of

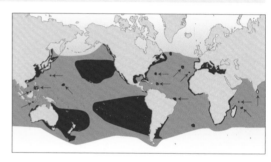

philopatry (homing behaviour) and remarkably little inter-change between populations that aggregate at different coastal sites to feed, even if their migration ranges overlap (p. 56).
Biology Viviparous, litters of 2–17 pups are nourished by unfertilised eggs during ~12 month gestation; there may be a 2–3 year interval between litters to enable females to rebuild their body reserves. Males mature at 9–10 years, females 14–33 years; maximum age is 30–73 years. Feeds on a wide range of prey depending on age, from small fishes to large marine mammals. Warm-blooded, maintaining a high body temperature even in cold water (p. 34).
Status IUCN Red List: **Vulnerable**. There is evidence of stock depletion by bycatch and target fisheries in several parts of the world. However, White Shark is now one of the world's most protected shark species and populations have increased recently in some areas, including where marine mammal populations have rebounded. The species is listed in the Appendices of major global and regional multilateral environmental agreements (see p. 83) and strictly protected by many countries, including Australia, South Africa, Namibia, New Zealand, USA, various Mediterranean and European nations, and several small island States. Teeth, jaw sets and fins are highly prized in trade. White Sharks are extremely valuable to local economies engaged in shark tourism at aggregation sites, particularly in South Africa and Australia. Only kept in aquaria for short periods up to six months. Occasionally bites people, including surfers, and has jumped into boats.

White Shark, *Carcharodon carcharias*, South Africa.

SHORTFIN MAKO *Isurus oxyrinchus*

FAO code: **SMA**   Plate page 316

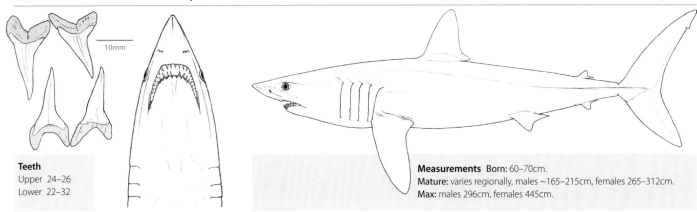

Teeth
Upper 24–26
Lower 22–32

Measurements Born: 60–70cm.
Mature: varies regionally, males ~165–215cm, females 265–312cm.
Max: males 296cm, females 445cm.

Identification Back brilliant blue or purple, underside usually white. Front half of pelvic fins dark, rear and undersides white. Long pointed snout, U-shaped mouth, large blade-like teeth. Underside of snout and mouth white in adults, but dusky in the Azores 'marrajo criollo' population. Pectoral fins shorter than head. Strong keels on caudal peduncle, crescent-shaped tail fin.

Distribution Worldwide in all temperate and tropical oceanic waters from 50°N (60°N in the North Atlantic) to 50°S; close inshore seasonally, particularly where the continental shelf is narrow, but populations in the North and South Atlantic are genetically distinct.

Habitat Coastal and oceanic, 0–888m in water warmer than 16°C (but dives into cold deep water as low as 10°C). Critical habitats are generally unknown, but the western Mediterranean may contain a pupping ground.

Behaviour Possibly the fastest, most muscular and active shark in the world, the Shortfin Mako can swim at speeds of up to 100km/hour (in short bursts) and jump right out of the water. Highly migratory, undertaking very long-distance journeys across ocean basins (including across the Atlantic) and sometimes following warming water bodies towards the poles in summer. Sharks tagged off New Zealand have been recovered from the Marquesas, Tonga, Fiji and New Caledonia; one travelled over 13,000km in 6 months, zipping back and forth between New Zealand and Fiji.

Biology Viviparous, 4–25 (mostly 10 –18, possibly up to 30) pups per litter feed on unfertilised eggs. Larger females have larger litters. Gestation may be 15–18 months with a 3 year reproductive cycle. Mature females

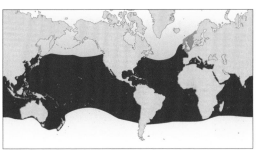

usually give birth in the summer, but nursery grounds are poorly known. Males mature at ~8 years, females at ~21 years; maximum age is 28–32 years, thus females bear only 3–4 litters during their lifetime. Feeds mainly on fishes and squid, but very large sharks may take small cetaceans.

Status IUCN Red List: Endangered. Juveniles under 11 years old (when they become too large to catch) are taken in target and bycatch fisheries worldwide for their highly valuable meat. Shortfin Mako is also an important 'big game' fish and increasingly attractive for dive ecotourism, but may attack if provoked. There is a very high overlap between juvenile habitat and commercial fishing activity, particularly in the Atlantic. Survival of juveniles to adulthood has been very low for decades. The result is that the mature females dying of old age today are not being replaced by the pups that were caught in unmanaged commercial fisheries decades ago. Population declines are reported in all oceans, except possibly the South Pacific. Shortfin Makos are listed in several multilateral environmental agreements, including CITES and CMS (see p. 83). Scientists have advised prohibiting North Atlantic Shortfin Mako catches to allow stock recovery, which would take at least 35 years if fisheries ceased immediately. Some countries have set catch limits, but at the time of writing the tuna RFMOs had been slow to agree and adopt the necessary controls for the high seas fisheries that take unsustainable numbers of Shortfin Mako as a target and utilised bycatch.

Shortfin Mako, *Isurus oxyrinchus*.

MACKEREL SHARKS **LAMNIDAE**   **321**

LONGFIN MAKO *Isurus paucus*

FAO code: **LMA**   Plate page 316

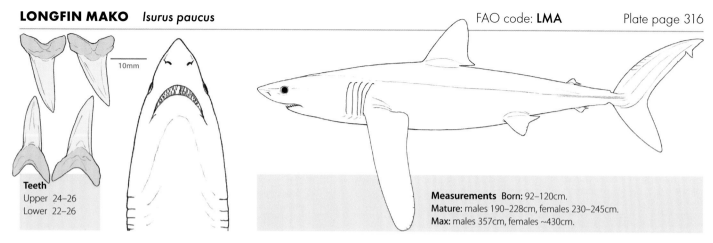

Teeth
Upper 24–26
Lower 22–26

Measurements Born: 92–120cm.
Mature: males 190–228cm, females 230–245cm.
Max: males 357cm, females ~430cm.

Identification Distinguished from *Isurus oxyrinchus* by less pointed snout, pectoral fins as long as head and relatively broad-tipped, and dusky underside to snout and mouth in adults.
Distribution Probably worldwide (poorly recorded); common in oceanic and tropical west Atlantic and possibly central Pacific, rare elsewhere.
Habitat Poorly known. Possibly epipelagic in deep water in the open ocean, 0–1752m.
Behaviour Poorly known. May be a slower swimmer than *I. oxyrinchus*, and appears to occur at greater depths than the Shortfin Mako. **Biology** Very poorly known. Viviparous, with oophagy and intrauterine cannibalism. Litters of 2–8 pups (less fecund than Shortfin Mako). Feeds on fishes and squid. Warm-blooded.
Status IUCN Red List: Endangered. Rarer than the Shortfin Mako and less resilient to over-fishing. No regional fisheries management yet adopted, but listed by CITES and CMS as a 'look-alike' species to Shortfin Mako (see p. 314).

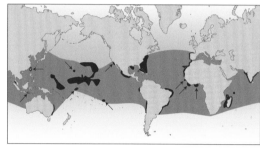

SALMON SHARK *Lamna ditropis*

FAO code: **LMD**   Plate page 316

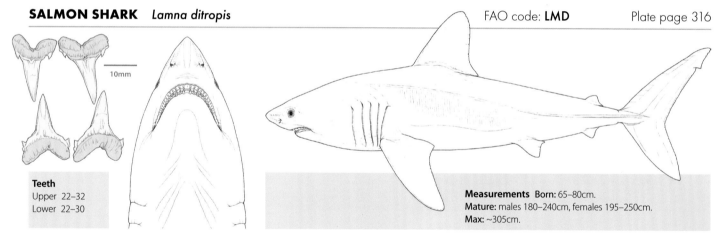

Teeth
Upper 22–32
Lower 22–30

Measurements Born: 65–80cm.
Mature: males 180–240cm, females 195–250cm.
Max: ~305cm.

Identification Heavy body. Short, conical snout. Long gill slits. Strong keels on caudal peduncle, short secondary keels on base. Crescent-shaped caudal fin. Dark grey or blackish. Underside white with dusky blotches and dark underside on snout in adults. First dorsal fin with dark free rear tip; white patch over pectoral fin bases.
Distribution North Pacific (males common in west, females in east).
Habitat Cool coastal and oceanic waters, surface to at least 1864m, but mostly to 300m.
Behaviour Seasonally migratory. Segregate by age and sex: adults move further north; young of the year often occur in lower warmer temperate latitudes. May forage in groups of 30–40.
Biology Viviparous, litters of 2–5 pups born in spring in nursery grounds after a 8–9 month gestation. Reproductive cycle is 2 years, with a resting stage following birth. Males mature about 5 years, females 6–10 years; maximum age 20–30 years. Feed on schooling fishes (salmon, herring, sardines).
Status IUCN Red List: Least Concern. Population recovery followed gillnet bans in the 1990s.

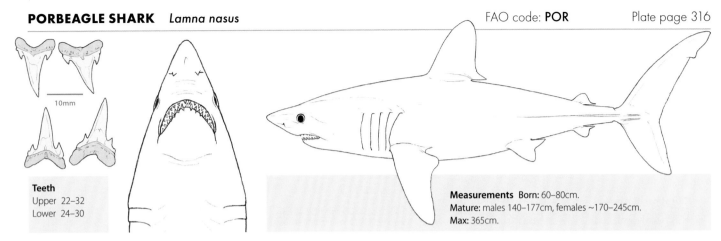

Teeth
Upper 22–32
Lower 24–30

Measurements Born: 60–80cm.
Mature: males 140–177cm, females ~170–245cm.
Max: 365cm.

Identification Very similar to *Lamna ditropis* (the distribution of these species does not overlap), but with a longer, more pointed snout, no white patch above pectoral fin, and distinctive white free rear tip to first dorsal fin. Southern hemisphere Porbeagle Sharks often have dusky markings on underside, as with *L. ditropis*, but North Atlantic sharks have light undersides.

Distribution North Atlantic and southern hemisphere in cool water (2–22°C). Adults prefers 5–10°C, juveniles and neonates are found in warmer water. Not recorded in equatorial seas.

Habitat Inshore to continental offshore fishing banks, 0–200m, occasionally to 1809m in the open ocean. There is a northwest Atlantic pupping ground in the warm waters of the Sargasso Sea and southern hemisphere juveniles are common further north than the adult stocks.

Behaviour Migratory, moves inshore and to the surface in summer, winters offshore in deeper water. There is limited migratory exchange between northeast and northwest Atlantic stocks. Populations are segregated by age (size) and sex, with younger sharks often found in warmer water than adults. Inquisitive, may approach boats and divers, but not dangerous.

Biology Viviparous, 1–5 (average 4) pups per litter. Pups feed on unfertilised eggs. Birth occurs in spring and summer after 8–9 month gestation, with a 1–2 year reproductive cycle. Growth is slightly slower in the northeast than the northwest Atlantic, and northern Porbeagle are significantly larger and faster growing than those in the southern oceans. Males mature at 6–10 years, females at 12–16.5 years; maximum age is

26–60 years in the North Atlantic, up to 65 years for smaller-sized Porbeagle Sharks around New Zealand. Feeds on small teleosts, dogfishes, tope and squid. Warm-blooded and very active; their high body temperature enables both species of *Lamna* to hunt prey in very cold water.

Status IUCN Red List: **Vulnerable** globally. **Critically Endangered** in the northeast Atlantic and Mediterranean, **Endangered** in the northwest Atlantic; these populations were seriously depleted by decades of target and bycatch fisheries for the high value meat (fins are also utilised). Catches in high seas longline fisheries are unknown, but the southwest Atlantic stock is also depleted and catch rates have declined around New Zealand. The status of other southern hemisphere stocks, taken in unknown quantities as bycatch in longline fisheries, is unknown. An important game fish. Listed in several international and regional environmental agreements, including CITES and CMS (see p. 83). A prohibited species or under very restrictive quotas in most northern hemisphere fisheries. Prohibited in Uruguayan fisheries, catch limits set in New Zealand.

Porbeagle Shark, *Lamna nasus*.

CARCHARHINIFORMES   ground sharks

This order is the largest, most diverse and widespread group of sharks. It contains at least 291 species in 10 families: Pentanchidae (deepsea catsharks), Scyliorhinidae (catsharks), Proscylliidae (finback catsharks), Pseudotriakidae (false catsharks), Leptochariidae (Barbeled Houndshark), Triakidae (houndsharks), Hemigaleidae (weasel sharks), Carcharhinidae (requiem sharks), Galeocerdidae (Tiger Shark), and Sphyrnidae (hammerhead sharks). Most are small and harmless to people, but this order also includes some of the largest predatory sharks.

Identification A very wide range of appearances, from strange bottom-dwelling deepsea sharks, to typical large sharks. All have two spineless dorsal fins and an anal fin. A long mouth extends to or behind the eyes, which are protected by nictitating lower eyelids. Nasoral grooves are usually absent (or broad and shallow, when present in a few catsharks). If barbels are present, these are developed from the anterior nasal flaps of nostrils. Largest teeth are distinctly lateral on dental band, not on either side of the symphysis, with no gap or intermediate teeth separating the large anterior teeth from even larger teeth in upper jaw. Intestine usually has a spiral or scroll valve.

Distribution Worldwide, from cold to tropical seas, intertidal to deep ocean, to pelagic in the open ocean. Some are poor swimmers and restricted to small areas of seabed, others are strong, long-distance swimmers and highly migratory.

Biology Highly varied reproductive strategies, also very poorly known in some groups. Many species are oviparous (egg-laying), some depositing eggs on the seabed with embryos developing for up to a year before hatching, others retaining the eggs until close to hatching. More evolutionarily advanced families retain foetuses inside the female, nourished by the yolk-sac or by a placenta, until live young are born.

Status IUCN Red List assessments are available for 98% of Carcharhiniformes; 45% of species are Least Concern (including most deep water species) and 26% are threatened.

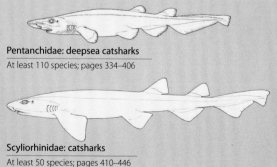

Pentanchidae: deepsea catsharks
At least 110 species; pages 334–406

Scyliorhinidae: catsharks
At least 50 species; pages 410–446

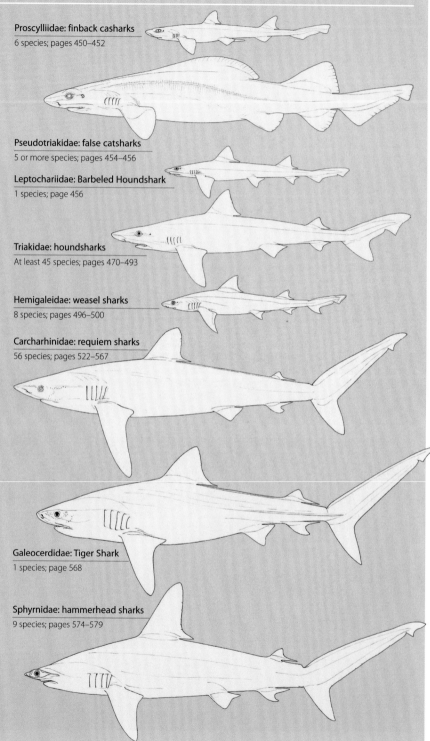

Proscylliidae: finback casharks
6 species; pages 450–452

Pseudotriakidae: false catsharks
5 or more species; pages 454–456

Leptochariidae: Barbeled Houndshark
1 species; page 456

Triakidae: houndsharks
At least 45 species; pages 470–493

Hemigaleidae: weasel sharks
8 species; pages 496–500

Carcharhinidae: requiem sharks
56 species; pages 522–567

Galeocerdidae: Tiger Shark
1 species; page 568

Sphyrnidae: hammerhead sharks
9 species; pages 574–579

Pentanchidae: deepsea catsharks

The deepsea catsharks, family Pentanchidae, had until recently been part of the catshark family, Scyliorhinidae, which was formerly the largest of all the shark families. However, recent taxonomic changes resulted in a split into two: the deepsea catsharks, Pentanchidae, and the catsharks, Scyliorhinidae. As a result, the deepsea catsharks are now the largest shark family, containing 11 genera and at least 110 species. Twelve species in this family are new and presented here for the first time. It is likely these numbers will increase, because new species are continually being discovered and described, as fisheries and research expand into deeper water. Clearly, considerable taxonomic research is still needed to understand these poorly known species. Deepsea catsharks are found worldwide and, as their name implies, mostly live at depths beyond 200m, down to 2200m. Exceptions are the genera *Halaelurus* and *Haploblepharus*, which are mostly shallower-occurring groups. Many of the deepsea catsharks are only known from a small number of individuals. Understandably, therefore, very little is known about the geographic range, status and biology of many of these species.

Identification Mostly small (less than 80cm long; some may mature at about 30cm, a few reach about 90cm); elongated body; two small, spineless dorsal fins (first dorsal base over or behind pelvic bases), and an anal fin. Externally, they are similar to the catsharks (Scyliorhinidae) but are separated from them by the structure of the cranium (skull, or brain case). The deepsea catsharks (Pentanchidae) do not have an internal 'crest' over the orbits of their eyes, whereas the catsharks (Scyliorhinidae) do. The crest, if present, can usually be felt by running your fingers over the eye orbits. Unfortunately, the members of some genera of deepsea catsharks (including the largest genus, demon catsharks, *Apristurus*) are very difficult to tell apart, even for experts. See below.

Biology Mostly oviparous (egg-laying) species, but a few are viviparous and give birth to live young. Eggcases are thick and may or may not have corner tendrils, which are used to attach to the seafloor or invertebrate structures, such as gorgonian corals. Hatching may take up to two or three years, depending on the species. Although no deepsea catsharks are known to make long-distance migrations, several species move vertically hundreds of metres off the bottom into midwater to feed. The newborns of some species inhabit midwater early in life. The deepsea catsharks feed on small bony fishes, cephalopods, crustaceans, and other invertebrates.

Status A large proportion (64%) of these species have been assessed as Least Concern and only 11 (10%) are listed as threatened by the IUCN Red List. This is because the great depths at which they live provide them with a refuge from fishery interactions. Some of the shallow water species are maintained in public aquariums. None are dangerous to humans.

Apristurus clades

We have identified three subgroups, known as clades, of *Apristurus* species that are very similar in appearance, and present the species accounts and plates by clade for this genus. Species in the *A. longicephalus* clade (Plate 40 and pp. 345–347) have an extremely long, slender snout (the distance from the snout tip to nostrils), greater than 6% of its total length; these sharks are sometimes referred to as pinocchio catsharks. The other two clades have snout lengths less than 6% of their total length. Species in the *A. brunneus* clade (Plates 38–39 and pp. 334–344) have relatively slender bodies, with their upper labial furrows longer than the lowers, while those in the *A. spongiceps* clade (Plates 40–41 and pp. 347–353) have a stouter body and upper labial furrows that are equal to or less than the length of the lower labial furrows.

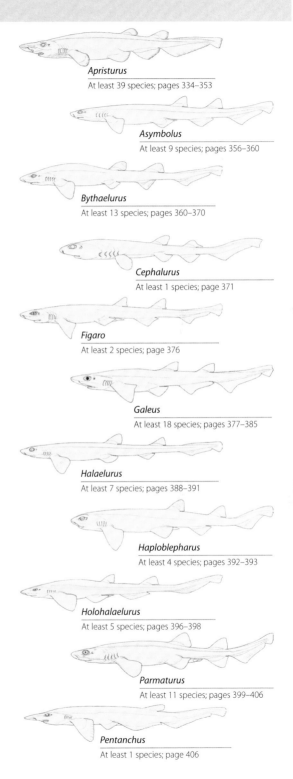

Apristurus
At least 39 species; pages 334–353

Asymbolus
At least 9 species; pages 356–360

Bythaelurus
At least 13 species; pages 360–370

Cephalurus
At least 1 species; page 371

Figaro
At least 2 species; page 376

Galeus
At least 18 species; pages 377–385

Halaelurus
At least 7 species; pages 388–391

Haploblepharus
At least 4 species; pages 392–393

Holohalaelurus
At least 5 species; pages 396–398

Parmaturus
At least 11 species; pages 399–406

Pentanchus
At least 1 species; page 406

Plate 38 PENTANCHIDAE I – *Apristurus* brunneus clade I

Species in the *Apristurus brunneus* clade have relatively slender bodies, short snouts (distance from snout tip to nostrils is less than 6% of total body length), and upper labial furrows that are longer than the lower furrows.

○ **Shortbelly Catshark** *Apristurus breviventralis* _____

Northwest Indian Ocean; insular slopes; 1000–1120m. Elongate, relatively high and angular anal fin; first dorsal fin much smaller than second dorsal fin; pectoral and pelvic fins very close together; medium to dark brown, light brown tongue and palate.

○ **Brown Catshark** *Apristurus brunneus* _____

East Pacific; 33–1341m. On or well above bottom. First dorsal fin same size as second; elongate, relatively high and angular anal fin; dark brown with obvious light posterior margins on fins and upper caudal margin.

○ **Hoary Catshark** *Apristurus canutus* _____

West Atlantic: Caribbean; 521–915m. First dorsal fin much smaller than second; very elongated anal fin; pectoral and pelvic fins fairly close together; dark grey with blackish fin margins.

○ **Flaccid Catshark** *Apristurus exsanguis* _____

Southwest Pacific; 415–1200m. First dorsal fin about the same size as second; very elongated low anal fin; pale grey to pale brown flaccid body.

○ **Humpback Catshark** *Apristurus gibbosus* _____

Central Indo-Pacific to northwest Pacific; 600–913m. First dorsal fin slightly smaller than second; anal fin moderately elongated, relatively high and angular.

○ **Smallbelly Catshark** *Apristurus indicus* _____

West Indian Ocean; 1225–1840m. First dorsal fin larger than second and extends forward as a long low ridge on the back; anal fin moderately elongated, relatively high and angular; paired fins, anal fin and caudal fin all very close together.

○ **Shortnose Demon Catshark** *Apristurus internatus* _____

Northwest Pacific: East China Sea; 670m, possibly 200–1000m. First dorsal fin slightly smaller than second; anal fin moderately elongated, relatively high and angular; dark body, no light fin margins.

○ **Broadnose Catshark** *Apristurus investigatoris* _____

Indian Ocean: Andaman Sea; 1040m. First dorsal fin ⅔ the area of second, extends forward as a low ridge to nearly over pelvic fin origins; anal fin very elongated relatively low and angular.

○ **Japanese Catshark** *Apristurus japonicus* _____

Northwest Pacific; 820–915m. First dorsal fin about the same size as second; anal fin very large moderately elongated and angular; pectoral and pelvic fins widely separated on long abdomen; short snout.

○ **Iceland Catshark** *Apristurus laurussonii* _____

North Atlantic; 560–2060m. First dorsal fin slightly larger than second dorsal and separated by space greater than first dorsal base length; anal fin very large moderately elongated and angular.

○ **Flathead Catshark** *Apristurus macrorhynchus* _____

Northwest Pacific; 220–1140m. First dorsal fin ⅔ size of second dorsal, separated by a space longer than first dorsal base; anal fin very large elongated and angular; light grey-brown above, whitish fins and ventral surface.

○ *Apristurus breviventralis*

○ *Apristurus canutus*

○ *Apristurus brunneus*

○ *Apristurus exsanguis*

○ *Apristurus indicus*

○ *Apristurus gibbosus*

○ *Apristurus internatus*

○ *Apristurus investigatoris*

○ *Apristurus japonicus*

○ *Apristurus macrorhynchus*

○ *Apristurus laurussonii*

20cm

Plate 39 PENTANCHIDAE II – *Apristurus* brunneus clade II

Species in the *Apristurus brunneus* clade have relatively slender bodies, short snouts (distance from snout tip to nostrils is less than 6% of total body length), and upper labial furrows that are longer than the lower furrows.

○ **Broadmouth Catshark** *Apristurus macrostomus*

Central Indo-Pacific to northwest Pacific; 220–1069m. First dorsal fin less than half the size of second and separated from it by a space greater than first dorsal base; anal fin very large, elongated and angular; rear fin margins blackish.

○ **Fleshynose Catshark** *Apristurus melanoasper*

East Atlantic, southwest Pacific, south Indian Oceans; 512–1683m. First dorsal fin same size as second; anal fin high and angular, base shorter than pectoral-pelvic space; slender dark brown body usually with irregularly scattered pale flecks, naked black fin tips.

○ **Smalldorsal Catshark** *Apristurus micropterygeus*

Central Indo-Pacific: South China Sea; 913m. First dorsal fin tiny about ⅑ size of second, originating just behind long pelvic fin bases; anal fin very large, elongated and angular.

○ **Milk-eye Catshark** *Apristurus nakayai*

Central Indo-Pacific; 953–1022m. Very short abdomen about equal to pelvic fin length; first dorsal fin originates well behind pelvic fin insertion; anal fin low, elongated and angular, base longer than pectoral-pelvic space; uniform brownish black, dark tongue and palate, iris is shiny white in life.

○ **Largenose Catshark** *Apristurus nasutus*

East Pacific; 250–950m. First dorsal fin slightly smaller than second dorsal; anal fin very elongated and relatively low; brown grey or grey-blackish body, posterior fin margins pale.

○ **Smallfin Catshark** *Apristurus parvipinnis*

West Atlantic; 600–1220m. First dorsal fin extremely small, originating behind pelvic insertions; anal fin low, elongated and angular, base longer than pectoral-pelvic space; upper caudal margin with fairly prominent crest of enlarged denticles.

○ **Spatulasnout Catshark** *Apristurus platyrhynchus*

Central Indo-Pacific and west Pacific; 400–1080m. Mouth mostly under large eyes; first dorsal fin much smaller than second; large pectoral fins; anal fin low, elongated and angular, base longer than pectoral-pelvic space; light brown or grey to dark brown body with blackish fin margins.

○ **Saldanha Catshark** *Apristurus saldanha*

Southeast Atlantic to southwest Indian Ocean; 344–1009m. Stout body; first dorsal fin slightly larger than second; anal fin very elongated, base 3 x height and relatively low; plain dark slate grey or grey-brown.

○ **Pale Catshark** *Apristurus sibogae*

Central Indo-Pacific; 655m. First dorsal fin much smaller than second dorsal; very large pectoral fins; anal fin large, elongated and angular; body white or reddish white.

○ **South China Catshark** *Apristurus sinensis*

Central Indo-Pacific; 537–1290m. Small dorsal fins, first dorsal about half the size of second; anal fin large, elongated and angular; dark body with no obvious markings. May be a species complex.

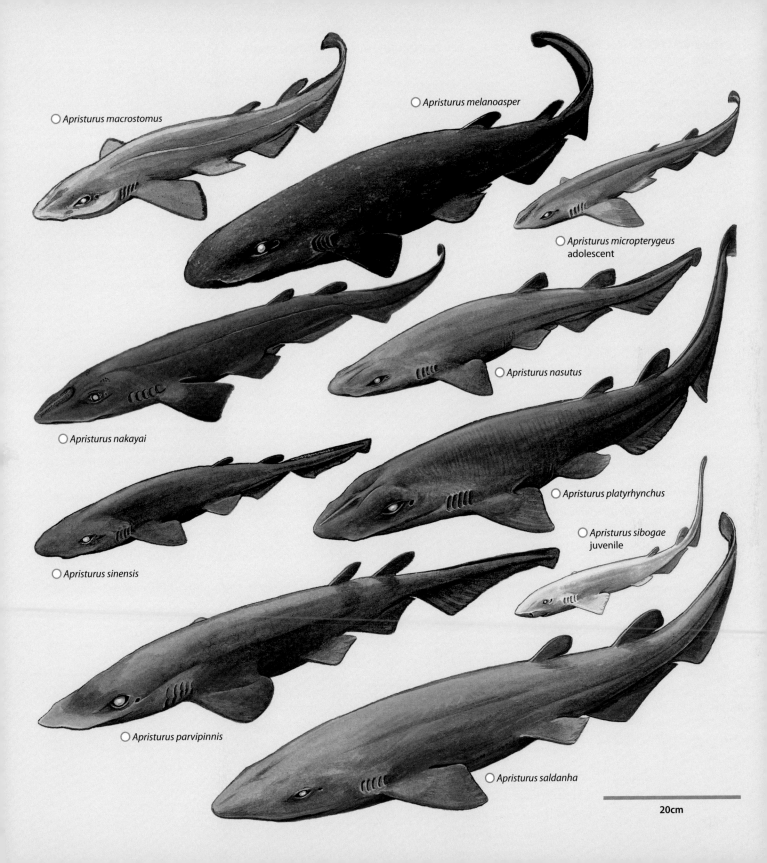

○ *Apristurus macrostomus*

○ *Apristurus melanoasper*

○ *Apristurus micropterygeus*
adolescent

○ *Apristurus nasutus*

○ *Apristurus nakayai*

○ *Apristurus platyrhynchus*

○ *Apristurus sibogae*
juvenile

○ *Apristurus sinensis*

○ *Apristurus parvipinnis*

○ *Apristurus saldanha*

20cm

Apristurus *longicephalus* clade

Species in the *Apristurus longicephalus* clade have an extremely long, slender snout (the distance from the snout tip to nostrils), greater than 6% of its total body length.

○ **Pinocchio Catshark** *Apristurus australis*

page 345

Southwest Pacific; slope and seamounts; 445–1035m. Very long narrow pointed snout; first dorsal fin smaller than second dorsal; anal fin angular, very low, elongated base; throat darker than pale greyish to brown body; fin tips and margins white or with broader pale margins in the northeast Australian population.

○ **Garrick's Catshark** *Apristurus garricki*

page 345

Southwest Pacific; 517–1200m. Snout extremely long, upper labial furrows longer than lowers; first dorsal much smaller than second, both dorsals with conspicuously black denticle-free inner and posterior margins; anal fin elongated, very low; uniformly greyish on head and upper body, becoming pale grey to white on belly.

○ **Longfin Catshark** *Apristurus herklotsi*

page 346

Central Indo-Pacific to northwest Pacific; 423–910m. First dorsal fin about ⅓ the size of the second; anal fin very long low, angular; unusually short abdomen, pelvic and pectoral fins very close together.

○ **Longhead Catshark** *Apristurus longicephalus*

page 346

West Pacific and Indian Oceans; 500–1350m. First dorsal fin slightly smaller than second; anal fin large, angular, elongated; grey-black or dark brown with no conspicuous markings.

○ **Yang's Longnose Catshark** *Apristurus yangi*

page 347

Central Indo-Pacific; 630–786m. Slender body; extremely long snout; short abdomen, less than 10% of body length; dorsal and paired fins bare of denticles and blackish in colour; elongated, very low anal fin; uniform pale brown body, flanks slightly darker than below, mouth cavity dark blackish brown with no denticles.

Apristurus *spongiceps* clade I

Species in the *Apristurus spongiceps* clade have snouts that are less than 6% of total body length, stouter bodies than the other clades, and upper labial furrows that are equal to or less than the length of the lower furrows.

○ **White-bodied Catshark** *Apristurus albisoma*

page 347

Central Indo-Pacific; 935–1564m. First dorsal fin slightly smaller than second; anal fin broadly rounded, high, short base.

○ **Roughskin Catshark** *Apristurus ampliceps*

page 348

Southwest Indian Ocean to southwest Pacific; 800–1503m. First dorsal fin slightly smaller than second; anal fin relatively low, broadly rounded; brownish to black body with scattered paler spots and squiggles; pale caudal fin tip.

○ **White Ghost Catshark** *Apristurus aphyodes*

page 348

Northeast Atlantic; 800–1809m. First dorsal fin about the same size as second; anal fin broadly rounded, high, short base; pale grey body with darker grey edges to some fins.

○ **Bighead Catshark** *Apristurus bucephalus*

page 349

West Indian Ocean: west Australia; continental slope; 920–1140m. First dorsal fin lower than second dorsal; anal fin high and triangular; pleats on throat; greyish brown with black rear margins to anal and caudal fins.

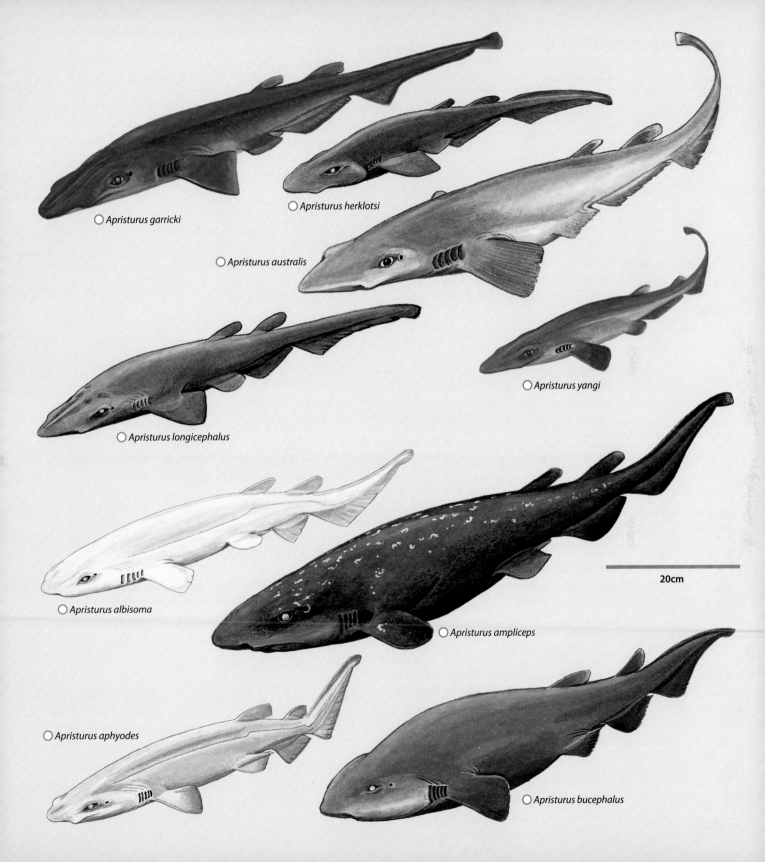

○ *Apristurus garricki*

○ *Apristurus herklotsi*

○ *Apristurus australis*

○ *Apristurus yangi*

○ *Apristurus longicephalus*

○ *Apristurus albisoma*

○ *Apristurus ampliceps*

20cm

○ *Apristurus aphyodes*

○ *Apristurus bucephalus*

Plate 41 PENTANCHIDAE IV – *Apristurus* spongiceps clade II

Species in the *Apristurus spongiceps* clade have snouts that are less than 6% of total body length, stouter bodies than the other clades, and upper labial furrows that are equal to or less than the length of the lower furrows.

○ **Stout Catshark** *Apristurus fedorovi*   page 349

Northwest Pacific; 810–1430m. First dorsal fin slightly smaller than second; anal fin broadly rounded, high, short base.

○ **Longnose Catshark** *Apristurus kampae*   page 350

Northeast and east Pacific; 65–1888m. First dorsal fin about the same size as second; anal fin broadly rounded, high, base length about twice height; precaudal fins may have light edges.

○ **Ghost Catshark** *Apristurus manis*   page 350

North Atlantic, possibly southeast Atlantic; 600–2453m. Distinctively stout body; first dorsal fin slightly smaller than second; anal fin broadly rounded, relatively high; dark grey or blackish body, fin tips whitish in juveniles; sparse lateral trunk denticles, crest of enlarged denticles on caudal fin.

○ **Smalleye Catshark** *Apristurus microps*   page 351

North and southeast Atlantic, southwest Indian Ocean; 700–2200m. Extremely small eyes; first dorsal fin about the same size as second; anal fin broadly rounded, high; stout dark brown to grey-brown to purplish black body, no obvious fin markings; crest of denticles above tail.

○ **Bulldog Catshark** *Apristurus pinguis*   page 351

West Pacific and west Indian Ocean; 858–2057m. Dorsal fins fairly equal-sized; anal fin triangular and high.

○ **Deepwater Catshark** *Apristurus profundorum*   page 352

North Atlantic; 132–1830m. Slender body; dorsal fins of equal size; anal fin triangular and high; brownish body with fuzzy texture due to erect denticles.

○ **Broadgill Catshark** *Apristurus riveri*   page 352

West Atlantic; 622–1500m. Small slender body; first dorsal fin originates in front of pelvic fin insertions and is much smaller than second dorsal; anal fin triangular and high; dark body with unmarked fins.

○ **Spongehead Catshark** *Apristurus spongiceps*   page 353

North central Pacific: Hawaii; 572–1463m. Stout body; pleated and grooved around throat and gills; dorsal fins similar in size; anal fin broadly rounded, relatively high; dark brown body.

○ **Panama Ghost Catshark** *Apristurus stenseni*   page 353

East central Pacific; 915–975m. Slender body; small eyes and wide gill slits; dorsal fins similar in size; anal fin triangular, relatively high, elongated base; blackish body; prominent dorsal crest of denticles on caudal fin.

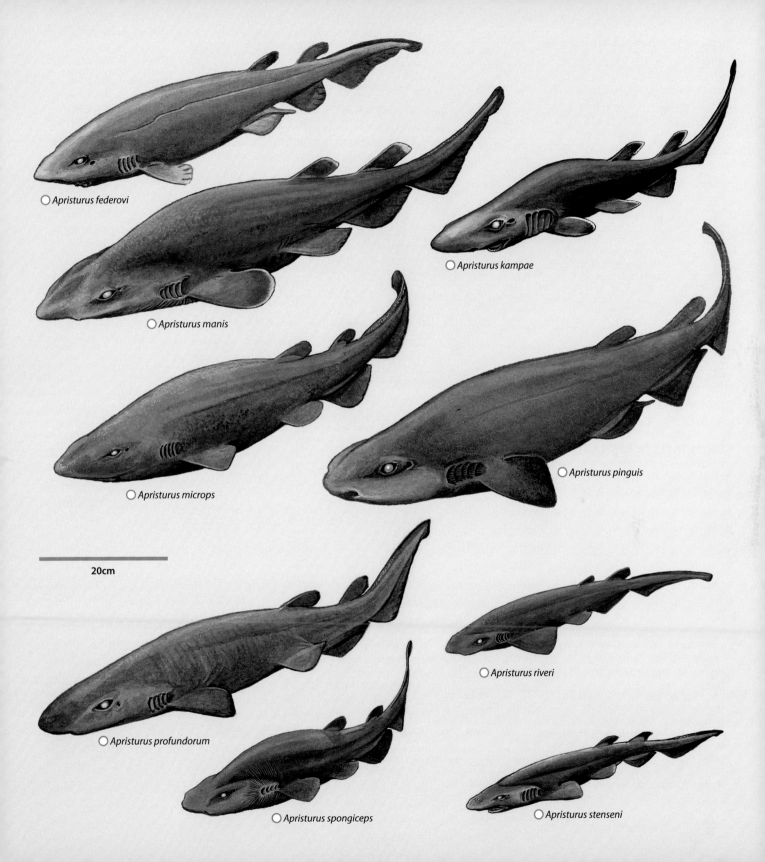

○ *Apristurus federovi*

○ *Apristurus manis*

○ *Apristurus kampae*

○ *Apristurus microps*

○ *Apristurus pinguis*

20cm

○ *Apristurus profundorum*

○ *Apristurus riveri*

○ *Apristurus spongiceps*

○ *Apristurus stenseni*

Spongehead Catshark, *Apristurus spongiceps* (p. 353), filmed in 2002 during an expedition to the northwest Hawaiian Islands. This was only the second record of this species, and the first time one had been seen alive, in its natural habitat at a depth of 1000m next to the Northampton Seamount.

SHORTBELLY CATSHARK *Apristurus breviventralis*

FAO code: **API**   Plate page 326

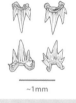
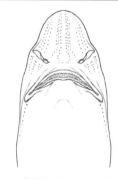

~1mm

Teeth
Upper 73–94
Lower 79–95

Measurements Mature: males 43–49cm.
Max: 49cm.

Identification (**A. brunneus** clade) Body slender, elongate and cylindrical to pelvic fins, becoming laterally compressed behind. Head broad, flattened. Snout moderately long, pointed at tip. Upper labial furrows clearly longer than lowers. First dorsal fin much smaller than second dorsal fin, originating posterior to pelvic fin insertion. Second dorsal fin insertion anterior to anal fin insertion. Anal fin elongate, base length much longer than pectoral-pelvic space, relatively high and angular. Abdominal space between pectoral and pelvic fins very short. Colour a uniform medium to dark brown. Tongue and palate are light brown.
Distribution Northwest Indian Ocean: off the Socotra Islands, Gulf of Aden.
Habitat Deepsea insular slope from 1000–1120m.
Behaviour Unknown.
Biology Oviparous, nothing else known. The species is known from only 9 males; the smallest was immature at 34cm, but the rest, at 43–49cm total length, were mature.
Status IUCN Red List: **Least Concern**. Beyond the depth range of fisheries.

Teeth
Upper 58–74
Lower 48–69

~5mm

Measurements Born: 7–9cm. **Mature:** males 45–55cm, females 48–58cm. Size at maturity varies by region, those in southern portion of their range mature at a smaller size. **Max:** 69cm.

Identification (*A. brunneus* clade) Broad, flattened head. Long snout. Large nostrils. Long, arched mouth extending about opposite front of eyes; very long labial furrows (lower shorter than upper). Gill slits longer than adult eye-length. Dorsal fins equal sized; first origin over pelvic midbases. Small notch between elongate, relatively high, and angular anal fin and elongated caudal fin. Dark brown body, obvious light posterior margins on fins and upper caudal margin.
Distribution East Pacific: southeast Alaska to north Mexico, off Panama, Ecuador, Peru and northern Chile.
Habitat Outer continental shelf and upper slope, 33–1341m, on and well above bottom. This species typically occurs over soft bottom or in rocky areas with high vertical relief. Eggcase deposition sites and nursery areas have been identified off the California coast.
Behaviour Segregation by depth is strong with nearly 90% of subadults occurring below 600m and gravid females mostly occurring between 300–500m.
Biology Oviparous. Pairs of eggcases, ~5 x 2.5cm with long tendrils, laid in spring and summer (Canada; in the southern portion of its range these sharks lay eggcases year-round; may take two years or more to hatch. Feeds on the bottom and in midwater, well off the bottom, on pelagic crustaceans, small shrimp, squid and small fishes.
Status IUCN Red List: Data Deficient. Bycatch in deepsea fisheries.

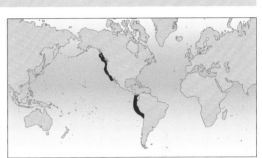

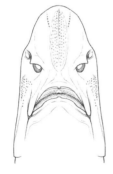

Teeth
Upper >50
Lower >50

Measurements Mostly unknown. **Mature:** 40–45cm.
Max: ~46cm.

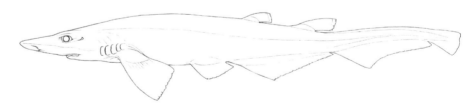

Identification (*A. brunneus* clade) Broad, flattened head. Rounded snout. Large nostrils. Long, arched mouth mostly under eyes; very long labial furrows (uppers reaching upper symphysis, lowers shorter than uppers). Eyes wider than widest gill slit. First dorsal much smaller than second, originating behind pelvic fin insertions. Pectoral and pelvic fins fairly close together. Very elongated anal fin, base 2.5–3 x height, relatively low, separated from elongated caudal fin by small notch. Dark grey with blackish fin margins.
Distribution Caribbean: Straits of Florida, Leeward Islands, Netherlands Antilles, Colombia and Venezuela.
Habitat Insular slopes, 521–915m.
Behaviour Unknown.
Biology Oviparous.
Status IUCN Red List: Least Concern. Poorly known, currently occurs below the depth of fisheries.

FLACCID CATSHARK *Apristurus exsanguis*

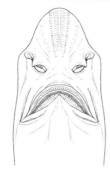

~5mm

Teeth
Upper 48–78
Lower 42–73

Measurements Mature: about 65–70cm (males and females).
Max: 91cm.

Identification (*A. brunneus* clade) Pale grey to pale brown, flaccid body. Broad, flattened head. Moderately elongated snout. Large nostrils. Long, arched mouth extending slightly in front of eyes; very long labial furrows (uppers reaching upper symphysis, lowers shorter than uppers). Dorsal fins similar sized; base of first over rear of pelvic fin bases. Elongated anal fin, low, height less than base length, separated from elongated caudal fin by a small notch.
Distribution Southwest Pacific: widespread around New Zealand and surrounding ridges and islands, including Chatham, Stewart and Campbell Islands.
Habitat Insular slopes, 415–1200m.
Behaviour Unknown.
Biology Oviparous, little else known. Eggcases 6.8 x 2.9cm, lays eggs in large, grooved capsules with long, coiled tendrils.
Status IUCN Red List: Least Concern. Occasionally taken as bycatch, but most of the depth distribution is below 800m, beyond the range of most deepsea fisheries.

HUMPBACK CATSHARK *Apristurus gibbosus*

Teeth
Upper 63–96
Lower 64–98

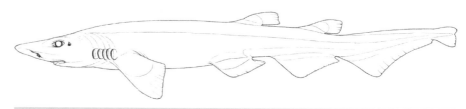

Measurements Mature: males and females ~40–45cm.
Max: males 57cm, females at least 49cm.

Identification (*A. brunneus* clade) Broad, flattened head. Elongated and very broad, spatulate snout. Large nostrils. Long, arched mouth reaching front of eyes; very long labial furrows (uppers reaching upper symphysis, lowers shorter than uppers). Small eyes. Prominent humped back behind very flat head. First dorsal fin base over pelvic fin bases, slightly smaller than second dorsal. Moderately elongated anal fin, relatively high, angular, separated from elongated caudal fin by small notch. Dark (dusky) body, no light fin margins.
Distribution Central Indo-Pacific to northwest Pacific: South China Sea to East China Sea, including Okinawa Trough, also off Taiwan.
Habitat Continental slope, over 600–913m.
Behaviour Oviparous. Eggcases 5–7 x 2–3cm, with fibrous threads on anterior ends and coiled threads on posterior ends, lateral flanges,and thick surface with longitudinal ridges.
Biology Unknown.
Status IUCN Red List: Least Concern. Research vessels have caught a few specimens, none are reported from deepsea trawlers and most of its range is below the depth of fisheries.

SMALLBELLY CATSHARK *Apristurus indicus*

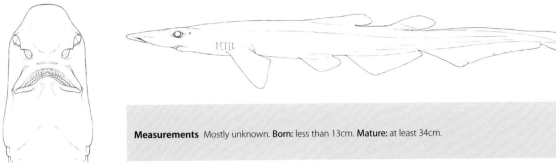

Teeth
Upper 60–62
Lower <60

Measurements Mostly unknown. **Born:** less than 13cm. **Mature:** at least 34cm.

Identification (A. brunneus clade) Broad, flattened head. Elongated snout. Large nostrils. Very short mouth reaching front of eyes; very long labial furrows (uppers reaching upper symphysis, lowers shorter than uppers). Gill slits less than adult eye-length. Two small, spineless dorsal fins, first lower than second and extending forward as long, low ridge on back. Moderately elongated anal fin, relatively high and angular, separated from elongated caudal fin by small notch. Paired fins, anal fin and caudal fin very close together. Brownish or blackish body.
Distribution West Indian Ocean: Somalia, Gulf of Aden, Oman. Records from European waters and west coast of southern Africa are erroneous.
Habitat Continental slopes, deepsea 1225–1840m, on bottom.
Behaviour Unknown.
Biology Unknown.
Status IUCN Red List: Least Concern. Not fished due to depth range. Known from only three female specimens ranging from 13–34cm in length.

SHORTNOSE DEMON CATSHARK *Apristurus internatus*

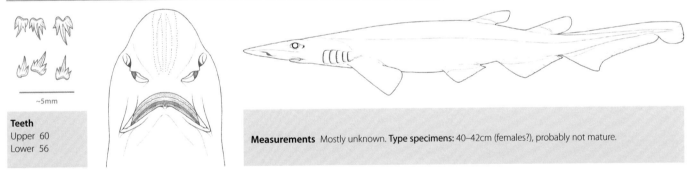

~5mm

Teeth
Upper 60
Lower 56

Measurements Mostly unknown. **Type specimens:** 40–42cm (females?), probably not mature.

Identification (A. brunneus clade) Broad, flattened head. Relatively short and very broad snout. Large nostrils. Long, arched mouth reaching front of eyes or slightly in front of them; very long labial furrows (uppers reaching upper symphysis, lowers shorter). First dorsal base over pelvic fin bases, slightly smaller than second dorsal. Moderately elongated anal fin, relatively high and angular, separated from elongated caudal fin by small notch. Dark body; no light fin edges.
Distribution Northwest Pacific: East China Sea off China (exact locality of type specimens uncertain).
Habitat Continental slope, 670m, possibly 200–1000m.
Behaviour Unknown.
Biology Oviparous. Eggcases 5–7 x 2–3cm with fibrous threads on each corner and long, coiled tendrils on the posterior end.
Status IUCN Red List: Least Concern. Known only from two specimens caught in research surveys; likely also occurs below range of deepsea trawlers. May be a junior synonym of *Apristurus gibbosus*; further research is required to confirm its validity.

BROADNOSE CATSHARK *Apristurus investigatoris*

FAO code: **APV** Plate page 326

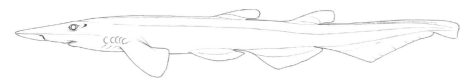

Teeth
Upper 42
Lower 43

Measurements Mostly unknown. **Female type specimen: 26cm.**

Identification (*A. brunneus* clade) Very broad, flattened head. Elongated snout. Large nostrils. Very short, small mouth that extends slightly in front of eyes; very long labial furrows (uppers reaching upper symphysis, lowers about as long as uppers). Gill slits shorter than adult eye-length. First dorsal fin about two-thirds area of second, extending forward as a long, low ridge to nearly over pelvic fin origins. Abdomen short. Very elongated anal fin, relatively low and angular, separated from elongated caudal fin by small notch. Fairly well-developed caudal crest of denticles. Plain medium brown; no prominent fin markings.
Distribution Indian Ocean: Andaman Sea. Records further west, off coast of India, are unconfirmed.
Habitat Continental slope, deepsea 1040m, on bottom.
Behaviour Unknown.
Biology Unknown.
Status IUCN Red List: Least Concern. Known only from two specimens, but most of the population occurs well below the range of deepsea fisheries. A broad-nosed ,short-bodied species similar to this, and to *Apristurus indicus*, occurs south of Madagascar on submarine ridges.

JAPANESE CATSHARK *Apristurus japonicus*

FAO code: **CSJ** Plate page 326

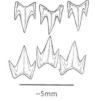

~5mm

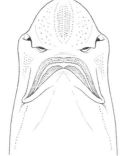

Teeth
Upper 74
Lower 75

Measurements Mature: males at least 62–65cm, females 57–63cm.
Max: 71cm.

Identification (*A. brunneus* clade) Extremely long abdomen (pectoral and pelvic fins widely separated) and short snout (for *Apristurus*). Broad, flattened head. Large nostrils. Mouth large and extending slightly in front of eyes; very long labial furrows (uppers reaching upper symphysis, lowers shorter). Gill slits narrower than adult eye-length. Dorsal fins equal-sized, origin of first over pelvic midbases. Anal fin very large, moderately elongated and angular, separated from elongated caudal fin by small notch. Blackish brown body.
Distribution Northwest Pacific: off Honshu, Japan.
Habitat Slope, near bottom, 820–915m.
Behaviour Unknown.
Biology Oviparous.
Status IUCN Red List: Data Deficient. Reported to be abundant within limited range.

ICELAND CATSHARK *Apristurus laurussonii*

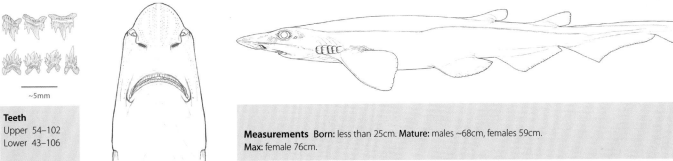

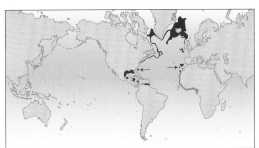

Teeth
Upper 54–102
Lower 43–106

Measurements Born: less than 25cm. **Mature:** males ~68cm, females 59cm.
Max: female 76cm.

Identification (A. brunneus clade) Broad, flattened head. Fairly short snout. Rather broad nostrils. Short, arched mouth extending to anterior ends of eyes; very long labial furrows (uppers reaching upper symphysis, lowers shorter). Small eyes. Narrow gill slits (less than adult eye-length) without prominent medial projections on gill septa. First dorsal fin slightly larger than second; space between greater than first dorsal base. Anal fin very large, moderately elongated and angular, separated from elongated caudal fin by small notch. Dark brown body.

Distribution North Atlantic: from Canary Islands and Madeira to Greenland; in USA from off Massachusetts to Delaware, and to the Gulf of Mexico. Records for the Caribbean from Honduras to Venezuela, and equatorial Africa, should be carefully examined.

Habitat Continental slopes, deepsea (560–2060m), on or near bottom. Temperature range where species caught is 1.7–4.3°C. **Behaviour** Unknown.

Biology Oviparous. Eggcases ~6 x 2–3cm, surface covered with villi-like fibres forming longitudinal striations, with coiled tendrils. Feeds on crustaceans, cephalopods, and small bottom-dwelling fishes.

Status IUCN Red List: Least Concern. A common bycatch in deepsea trawl fisheries, but the majority of its habitat occurs below the depth of these fisheries.

FLATHEAD CATSHARK *Apristurus macrorhynchus*

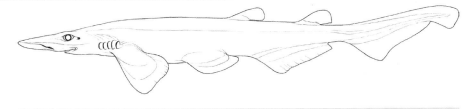

Teeth
Upper 64
Lower 62

Measurements Mostly unknown.
Max: ~67cm (adult female).

Identification (A. brunneus clade) Very broad, flattened head. Rounded, elongated snout. Large nostrils. Short, arched mouth extending to anterior ends of eyes; very long labial furrows (uppers reaching upper symphysis, lowers shorter). Gill slits fairly small (less than adult eye-length). First dorsal originating over last quarter of long pelvic fin bases; two-thirds size of second, separated by a space longer than first dorsal base. Anal fin very large, elongated and angular, separated from elongated caudal fin by small notch Light grey-brown above, whitish below and on fins.

Distribution Northwest Pacific: southeast Honshu, off Japan.

Habitat Insular and continental slopes, on the seabed in deep water, 220–1140m.

Behaviour Unknown.

Biology Oviparous. Lays pairs of eggcases 5–7 x 2–3cm, with long, weak tendrils on each anterior corner and small processes with long, coiled tendrils on posterior corners. Feeds on crustaceans, cephalopods and small bony fishes.

Status IUCN Red List: Data Deficient. Probably discarded from deepsea trawl bycatch.

BROADMOUTH CATSHARK *Apristurus macrostomus*

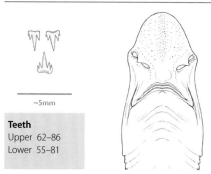
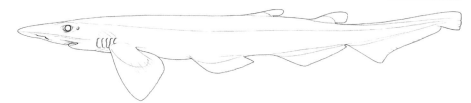

~5mm

Teeth
Upper 62–86
Lower 55–81

Measurements Mature: males 39cm, females 43cm.
Max: males 50cm, females 57cm.

Identification (*A. brunneus* clade) Broad, flattened head. Rounded, elongated snout. Large nostrils. Short, very large, arched mouth extending slightly in front of anterior ends of eyes; very long labial furrows (uppers reaching upper symphysis, lowers shorter than uppers). Gill slits fairly small (less than adult eye-length). First dorsal fin originating well behind pelvic fin bases, less than half size of second; separated from second by a space greater than first dorsal base. Pectoral fins very large. Anal fin very large, elongated and angular, separated from elongated caudal fin by small notch. Possibly dark brown or grey-brown above and below. Rear fin margins blackish.
Distribution Central Indo-Pacific to northwest Pacific: South and East China Seas to Japan, including Philippines, Taiwan and Indonesia. This rare catshark has recently been found to be quite common off north Taiwan.
Habitat Continental slopes from 220–1069m.
Behaviour Unknown. **Biology** Oviparous. Eggcases 7 x 2cm, with fibrous threads on anterior end and coiled tendrils on posterior end, surface smooth, no ridges or striations.
Status IUCN Red List: Least Concern. A retained bycatch in some deepsea fisheries in the north of its range, but the majority of its habitat lies below the depth of these fisheries.

FLESHYNOSE CATSHARK *Apristurus melanoasper*

5mm

Teeth
Upper 59–93
Lower 58–97

Measurements Born: less than 24.7cm. **Mature:** males 62cm, females 59cm.
Max: males 76–79cm, females 73cm.

Identification (*A. brunneus* clade) Slender body. Fairly long, slender, flattened, fleshy snout. Mouth extending to just in front of small eyes; long labial furrows (uppers reaching upper symphysis, lowers shorter). Gill openings slightly smaller than eye diameter. Dorsal fins equal size. Widely separated pectoral and pelvic fins. Anal fin high and angular, base shorter than pectoral-pelvic space, separated from caudal fin by a small notch. Dark brown body. Black, naked fin tips; irregular scattering of pale flecks on most individuals.
Distribution East Atlantic: northeast slope and off Namibia; South Indian Ocean: off Madagascar Ridge, seamounts west of Australia; and southwest Pacific, including southeast Australia and New Zealand.
Habitat Continental slopes and seamounts, 512–1683m.
Behaviour Unknown.
Biology Oviparous. Eggcases 6 x 2cm, posterior ends tapering with 2 very long, tightly coiled tendrils set close together, anterior end with long fibrous threads.
Status IUCN Red List: Least Concern. This species is broadly distributed and much of its range lies below the depth of most fisheries.

SMALLDORSAL CATSHARK *Apristurus micropterygeus*

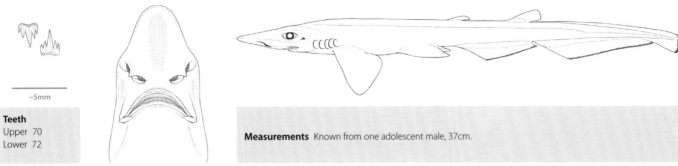

~5mm

Teeth
Upper 70
Lower 72

Measurements Known from one adolescent male, 37cm.

Identification (*A. brunneus* clade) Broad, flattened head. Rounded, rather elongated snout. Large nostrils. Short, arched mouth extending to anterior ends of eyes; very long labial furrows (uppers reaching upper symphysis, lowers shorter). Gill slits fairly small, slightly less than eye-length. First dorsal tiny (about one-ninth size of second), originating just behind long pelvic fin bases. Interdorsal space much greater than first dorsal base. Anal fin very large, elongated and angular, separated from elongated caudal fin by small notch. Possibly grey-brown body; fins not conspicuously marked.
Distribution Central Indo-Pacific: South China Sea, off China.
Habitat Captured at 913m.
Behaviour Unknown.
Biology Unknown.
Status IUCN Red List: **Least Concern**. Known from 1 specimen caught in a research survey. Unlikely to be captured by commercial fisheries, which operate no deeper than 300m in this area.

MILK-EYE CATSHARK *Apristurus nakayai*

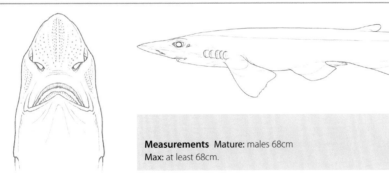

Teeth
Upper 79
Lower 72

Measurements Mature: males 68cm
Max: at least 68cm.

Identification (*A. brunneus* clade) Body slender, elongate and cylindrical before pelvic fins, then becomes laterally compressed further back. Head broad, flattened. Snout relatively long, rounded at tip. Upper labial furrows much longer than lowers. First dorsal fin much smaller than second dorsal fin, originating well posterior to pelvic fin insertion; second dorsal fin insertion slightly anterior to anal fin insertion. Anal fin low, elongated and angular, base longer than pectoral-pelvic space, separated from caudal fin by a small notch. Abdominal space very short, about equal to pelvic fin length. Colour a uniform brownish black. Eye iris is shiny white in life; tongue and palate dark.
Distribution Central Indo-Pacific: known from only three specimens: Coriolis Bank off western New Caledonia; south of New Ireland and off Lae, Papua New Guinea.
Habitat Deepsea slopes from 953–1022m.
Behaviour Unknown.
Biology Presumed oviparous, nothing else known.
Status IUCN Red List: **Least Concern**. Known only from three specimens collected in a region where there are no deepsea fisheries.

LARGENOSE CATSHARK *Apristurus nasutus*

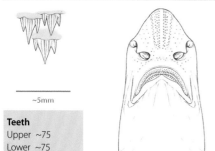

~5mm

Teeth
Upper ~75
Lower ~75

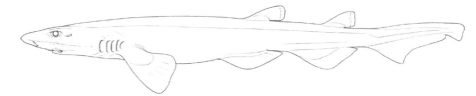

Measurements Mostly unknown. **Mature:** males reach 51–59cm.

Identification (*A. brunneus* clade) Broad, flattened head. Elongated snout. Mouth extending a short distance in front of small eyes; long labial furrows (uppers reaching upper symphysis, lowers shorter). Gill slits shorter than adult eye-length. First dorsal fin slightly smaller than second and originating over pelvic fin midbases. Anal fin very elongated, base 3 x height, relatively low, separated from elongated caudal fin by small notch. Medium brown, grey or blackish grey body. Posterior fin margins pale.
Distribution East Pacific: Costa Rica to central Chile (to Valparaíso ~32°S). Gulf of California, Mexico, records require confirmation.
Habitat Upper continental slopes, on or near bottom, 250–950m.
Behaviour Unknown.
Biology Oviparous. Feeds on deepsea shrimp.
Status IUCN Red List: Least Concern. A rare bycatch in deepsea trawl fisheries, but most of its habitat (particularly in the north) is below these fisheries.

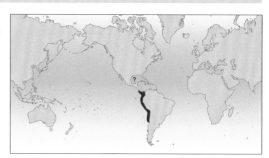

SMALLFIN CATSHARK *Apristurus parvipinnis*

Teeth
Upper 66–90
Lower 66–90

Measurements **Mature:** males to 48cm, females 52cm.
Max: 59cm.

Identification (*A. brunneus* clade) Broad, flattened head. Broad, rounded snout. Mouth mostly under eyes; long labial furrows (uppers reaching upper symphysis, lowers shorter). Eye length much greater than widest gill slit. First dorsal fin extremely small, originating behind pelvic insertions. Anal fin low, elongated and angular, base longer than pectoral-pelvic space, separated from caudal fin by a small notch. Upper caudal margin with fairly prominent crest of enlarged denticles. Grey-brown to blackish.
Distribution West Atlantic: Gulf of Mexico and mainland Caribbean: USA (Florida), Mexico, Honduras, Panama and Colombia. Suriname, French Guiana and off Rio de Janeiro, Brazil.
Habitat Continental slope, on or near bottom, 600–1220m.
Behaviour Unknown.
Biology Oviparous.
Status IUCN Red List: Least Concern. Most of this common species' distribution is below the current depth of fisheries within its wide range.

SPATULASNOUT CATSHARK   *Apristurus platyrhynchus*

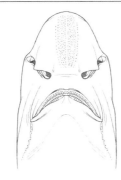

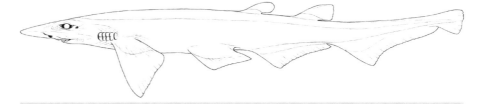

5mm

Teeth
Upper 60–86
Lower 57–87

Measurements Born: less than 28cm. **Mature:** males and females more than 59–60cm.
Max: males 80cm, females 67cm.

Identification (*A. brunneus* clade) Very broad, flattened, rounded snout. Mouth mostly under large eyes (much longer than widest gill slit); long labial furrows (uppers reaching upper symphysis, lowers slightly shorter). First dorsal much smaller than second, originating behind pelvic insertions. Pectoral fins large. High, rounded pelvic fins. Anal fin low, elongated and angular, base longer than pectoral-pelvic space, separated from caudal fin by a small notch. Deep notch near top of caudal fin. Light brown or grey to dark brown with blackish fin margins.
Distribution Central Indo-Pacific and west Pacific: Suruga Bay, Japan southward to the East China Sea (Okinawa Trough); Taiwan; Philippines, South China Sea; Sabah and Borneo, Malaysia; Australia (Queensland, New South Wales, Western Australia) and the Norfolk Ridge.
Habitat Continental and insular slopes, deep water, 400–1080m.
Behaviour Unknown.
Biology Oviparous. Eggcases very long and slender, 9 x 2cm, without coiled tendrils on either end, surface with ridges.
Status IUCN Red List: Least Concern. Poorly known, apparently taken as bycatch in fisheries. *Apristurus acanutus* (China, South China Sea) and *A. verweyi* (Sabah, Borneo) are junior synonyms.

SALDANHA CATSHARK   *Apristurus saldanha*

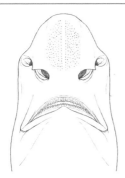

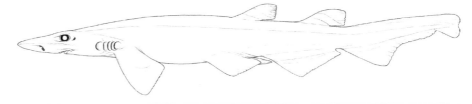

~5mm

Teeth
Upper 44–52
Lower 40–42

Measurements Mature: males 69–74cm, females 70cm.
Max: reported 89cm.

Identification (*A. brunneus* clade) Stout catshark. Long, thick, broad snout. Rather broad nostrils. Mouth not projecting in front of eyes; long labial furrows (uppers reaching upper symphysis, lowers shorter). Fairly large eyes. Gill slits less than adult eye-length. First dorsal slightly smaller or about equal to widely separated second, and originating over pelvic fin midbases or slightly behind them. Large pectoral fins. Anal fin very elongated, base 3 x height, relatively low, separated from elongated caudal fin by small notch. Plain dark slate grey or grey-brown.
Distribution Southeast Atlantic to southwest Indian Ocean: Namibia and South Africa (eastern, western and northern Cape coasts).
Habitat Continental slope, 344–1009m.
Behaviour Unknown.
Biology Oviparous. Eggcases 6–7 x 2–3cm, with rough longitudinal striations or ridges, long coiled tendrils on posterior end, but anterior end lacks tendrils. Eats small bony fishes and cephalopods.
Status IUCN Red List: Least Concern. Most of its distribution occurs beyond current fishing activity.

PALE CATSHARK *Apristurus sibogae*

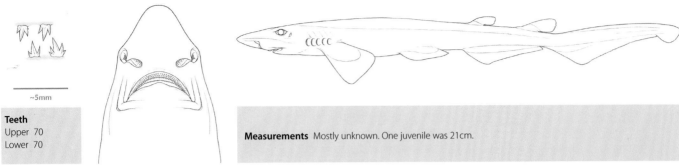

Teeth
Upper 70
Lower 70

Measurements Mostly unknown. One juvenile was 21cm.

Identification (A. brunneus clade) Broad, flattened head. Narrow, pointed snout. Mouth extending well in front of very small eyes (about equal to longest gill slit); long labial furrows (uppers reaching upper symphysis, lowers about same length). First dorsal fin much smaller than second, originating behind pelvic fin insertions. Very large pectoral fins. Small, low pelvic fins. Anal fin large, elongated and angular, separated from elongated caudal fin by small notch. Colour white or reddish white.
Distribution Central Indo-Pacific: Makassar Strait between Borneo and Sulawesi, Indonesia.
Habitat Insular slope, 655m.
Behaviour Unknown.
Biology Oviparous.
Status IUCN Red List: Least Concern. Only known from the holotype, but there are currently no deepsea fisheries in the region.

SOUTH CHINA CATSHARK *Apristurus sinensis*

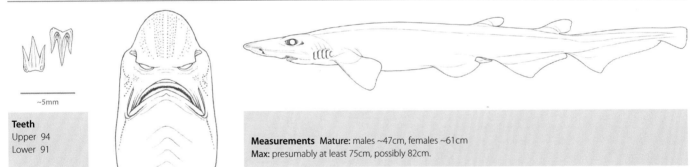

Teeth
Upper 94
Lower 91

Measurements Mature: males ~47cm, females ~61cm
Max: presumably at least 75cm, possibly 82cm.

Identification (A. brunneus clade) Broad, flattened head. Pointed, angular snout. Short mouth not expanded in front of eyes with very long labial furrows (uppers reaching upper symphysis, lowers shorter). Short gill slits (widest much less than eye-length); short medial projections on gill septa. First dorsal fin about half the size of the second, originating over last quarter of pelvic fin bases. Pectoral and pelvic fins well separated. Anal fin large, elongated and angular, separated from elongated caudal fin by small notch. Colour dark with no obvious markings.
Distribution Central Indo-Pacific, possibly southwest Pacific: South China Sea, off China; southeast, east and west Australia, and northern Australia off Ashmore Reef. Records from off New Zealand may be a different species.
Habitat Taken from continental slope, 537–1290m.
Behaviour Unknown.
Biology Oviparous.
Status IUCN Red List: Data Deficient. Possibly a species complex, comprised of 3 different species.

PINOCCHIO CATSHARK *Apristurus australis*

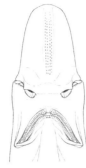

~5mm

Teeth
Upper 50–64
Lower 48–68

Measurements Mature: males ~45–50cm; females 45–55cm. **Max:** 62cm, possibly up to 83cm.

Identification (A. *longicephalus* clade) Slender body. Long, narrow, flattened head. Very long, narrow, pointed snout. Mouth extending opposite small eyes; long labial furrows (uppers reaching upper symphysis, lowers slightly shorter). First dorsal fin smaller than second. Large pectoral fins. Small pelvic fins. Anal fin angular, very low, elongated base, separated from elongated caudal fin by a small notch. Pale greyish brown to brown body. Throat darker than body; tips and borders of fins white or with broader pale margins in the northeast Australian population.
Distribution East coast of Australia. New Zealand form may be a different species.
Habitat Continental slope and seamounts, 445–1035m.
Behaviour Unknown.
Biology Unknown.
Status IUCN Red List: Least Concern. Occasionally caught in trawl fisheries although most of its depth range (>700m) is closed to fishing activities. Relationship with *Apristurus herklotsi* needs assessment.

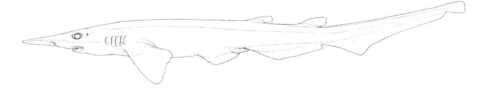

GARRICK'S CATSHARK *Apristurus garricki*

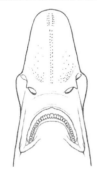

~1mm

Teeth
Upper 37–50
Lower 38–49

Measurements Born: less than 23cm. **Mature:** males 55cm, females 67.5cm. **Max:** 85cm.

Identification (A. *longicephalus* clade) Body elongate, slender. Head relatively narrow, flattened. Snout extremely long, broadly rounded at tip. Upper labial furrows much longer than lowers (uppers almost meet at jaw symphysis). Gills small; first gill slightly smaller than the others. First dorsal fin much smaller than second, its origin behind pelvic fin; dorsal fins inner and posterior margins are conspicuously blackish and bare of denticles. Abdominal space short, less than 10% body length. Anal fin elongated, very low. Colour a uniform greyish on head and upper body, becoming pale grey to white on belly. Eyes, spiracles and gills black around rims; eyes are luminescent green; narrow black margins to pectoral and pelvic fins.
Distribution Southwest Pacific: known only from deep waters surrounding the North Island, New Zealand.
Habitat Occurs along deepsea banks, ridges, and slopes from 517–1200m, but most common below 800m.
Behaviour Unknown.
Biology Oviparous. Eggcases ~14 x 3–4cm, narrow, cylindrical, without coiled tendrils on either end. Diet is unknown, but likely includes small bony fishes, cephalopods and crustaceans.
Status IUCN Red List: Least Concern. A bycatch of deepsea benthic trawls operating in the shallower parts of its range, but its small population also has a refuge from intensive fisheries in deeper waters.

LONGFIN CATSHARK *Apristurus herklotsi*

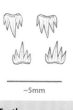

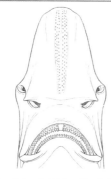

~5mm

Teeth
Upper 44–60
Lower 44–61

Measurements Mature: males ~40–47cm, females 43–48cm.
Max: ~52cm.

Identification (A. *longicephalus* clade) Small body. Broad, flattened head. Greatly elongated, rounded snout. Large nostrils. Short mouth reaching front of eyes; small, closely set teeth; very long labial furrows (uppers reaching upper symphysis, lowers shorter). Short gill slits and incised gill septa. First dorsal fin about one-third size of second, originating in front of or about opposite pelvic insertions. Unusually short abdomen (pectoral and pelvic fins very close together). Anal fin very long low, angular, separated by small notch from long, narrow caudal fin. Plain brownish to blackish brown body.
Distribution Central Indo-Pacific to northwest Pacific: South and East China Seas: China, Philippines, Taiwan, Okinawa Trough, southern Japan (Shikoku).
Habitat Bottom in deep water, 423–910m
Behaviour Unknown.
Biology Oviparous. Eggcases 5 x 1–2cm, without coiled tendrils on either end, anterior end with fibrous threads at each corner; surface with longitudinal ridges. Nothing else known.
Status IUCN Red List: Least Concern. A potential bycatch of deepsea fisheries, but none occur in much of its range and there is no evidence of population decline.

LONGHEAD CATSHARK *Apristurus longicephalus*

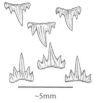

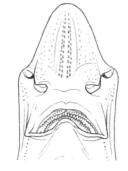

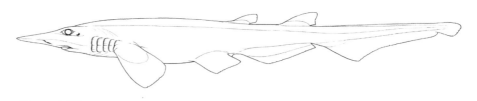

~5mm

Teeth
Upper 35–49
Lower 29–48

Measurements Mature: males 43–49cm, females 47–51cm.
Max: at least 60cm.

Identification (A. *longicephalus* clade) Broad, flattened head. Very long snout (preoral length about 12% total length). Nostrils rather broad. Short, arched mouth extending to anterior ends of eyes; very small, wide-spaced teeth; very long labial furrows (uppers reaching upper symphysis, lowers shorter). Small gill slits (less than adult eye-length). First dorsal fin above pelvic fin base, close to and slightly smaller than second. Pelvic, pectoral and anal fins close together. Anal fin angular, large, elongated, separated from long, narrow caudal fin by a small notch. Grey-black or dark brown; no conspicuous markings.
Distribution West Pacific and Indian Ocean: Japan, East China Sea (Okinawa Trough), China, Taiwan, Philippines, Indonesia (off Java and Sumatra), New Caledonia, northern Australia; Seychelles and Mozambique in Indian Ocean.
Habitat Deep water, probably near bottom, 500–1350m.
Behaviour Unknown.
Biology Oviparous. The first observation of hermaphroditism in a shark (male and female reproductive organs both present) was recorded in this species.
Status IUCN Red List: Least Concern. Although very poorly known, it is wide-ranging and has refuge at depth from most fisheries.

YANG'S LONGNOSE CATSHARK *Apristurus yangi*

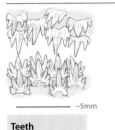

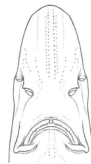

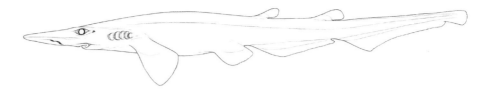

~5mm

Teeth
Upper 52–67
Lower 56–59

Measurements Born: less than 21cm. **Mature:** females 44cm.
Max: at least 44cm.

Identification (*A. longicephalus* clade) Body elongate, slender and cylindrical ahead of pelvic fins. Head narrow, flattened. Snout extremely long, narrowly rounded at tip. Upper labial furrows much longer than lowers. Gills small, all about equal in height. First dorsal fin smaller than second, its origin just behind pelvic fin insertion. Short abdomen, less than 10% body length. Anal fin elongated, very low. Colour a uniform pale brown; sides of body slightly darker than below. Margin of snout tip blackish (more distinct in juveniles); mouth cavity is dark blackish brown and lacks denticles; dorsal, pectoral and pelvic fins bare, with blackish coloration; anterior fin margins darker, most obvious on pectoral fins.
Distribution Central Indo-Pacific: known from 2 specimens off Papua New Guinea.
Habitat Deepsea slopes around Papua New Guinea, from 630–786m.
Behaviour Unknown.
Biology Oviparous, eggcases 6 x 1–2cm, narrow, cylindrical, without coiled tendrils on either end. Size at birth is unknown, but smallest known free-swimming specimen was 21cm total length. Diet includes small bony fishes, but nothing else is known.
Status IUCN Red List: **Least Concern**. Known only from 2 female specimens, 1 adult, 1 juvenile, but there are currently no known deepsea fisheries in this region.

WHITE-BODIED CATSHARK *Apristurus albisoma*

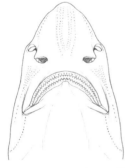

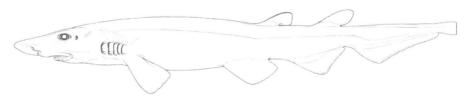

~5mm

Teeth
Upper 56–71
Lower 54–70

Measurements Mature: ~40–50cm.
Max: males 57cm, females 60cm.

Identification (*A. spongiceps* clade) Broad, flattened head. Elongated snout. Moderately large nostrils. Long, arched mouth reaching front of very small eyes; very long labial furrows (uppers reaching upper symphysis, lowers about as long as uppers). First dorsal fin base over pelvic fin bases and slightly smaller than second. Anal fin broadly rounded, high, short base, separated from elongated caudal fin by small notch. Whitish colour.
Distribution Central Indo-Pacific: off New Caledonia and between New Zealand and Australia along the Norfolk Ridge and Lord Howe Rise. A similar species occurs on ridges south of Madagascar.
Habitat Deepsea slope, 935–1564m, possibly on soft bottom.
Behaviour Unknown.
Biology Poorly known. Eats shrimp and cephalopods.
Status IUCN Red List: **Least Concern**. This species' depth range provides a refuge from most fisheries.

ROUGHSKIN CATSHARK *Apristurus ampliceps*

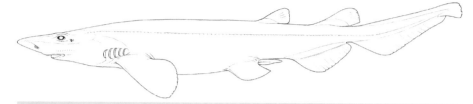

~5mm

Teeth
Upper 57–70
Lower 50–62

Measurements Mature: males 90cm, females 80cm.
Max: at least 90cm.

Identification (*A. spongiceps* clade) Bulky, flattened head. Elongated, broad snout. Large mouth extending well in front of small eyes; large teeth; long labial furrows (uppers reaching upper symphysis, lowers about same length). First dorsal fin slightly smaller than second. Pectoral and pelvic fins well separated. Anal fin relatively low, broadly rounded, base shorter than pectoral-pelvic space, separated from caudal fin by small notch. Dark brown to blackish with small scattered spots and squiggles, caudal fin tip pale. Somewhat similar to *Apristurus manis* in Atlantic.
Distribution Southwest Indian Ocean to southwest Pacific: southern Australia and New Zealand.
Habitat Continental shelf, 800–1503m.
Behaviour Unknown.
Biology Unknown.
Status IUCN Red List: Least Concern. Possibly quite rare. Depth distribution is mostly over 700m, which is outside the range of most fisheries.

WHITE GHOST CATSHARK *Apristurus aphyodes*

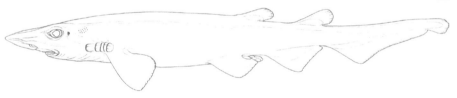

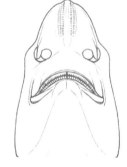

~5mm

Teeth
Upper 56–68
Lowe 49–64

Measurements Born: less than 14cm. **Mature:** ~47–50cm (males and females).
Max: males 55cm, females 53cm.

Identification (*A. spongiceps* clade) Broad, flattened head. Elongated snout. Large nostrils. Long, arched mouth extending well in front of eyes; very long labial furrows (uppers reaching upper symphysis, lowers about as long as uppers). First dorsal fin base over or behind pelvic fin bases, about as large as second. Anal fin broadly rounded, high, short base, separated from elongated caudal fin by small notch. Pale grey, slightly darker grey edges to some fins.
Distribution Northeast Atlantic: Atlantic slope from Iceland to Galicia Bank, Spain.
Habitat Deep slope, 800–1809m, possibly on soft bottom where temperatures are 3.7–9.7°C.
Behaviour Unknown.
Biology Oviparous. Eggcases 5–7 x 2–3cm, vase-shaped and small, horns very short and coiled, without long tendrils. Diet includes crustaceans, cephalopods and small benthic fishes.
Status IUCN Red List: Least Concern. Occurs below the range of most fisheries. Only known from about 30 specimens.

BIGHEAD CATSHARK *Apristurus bucephalus*

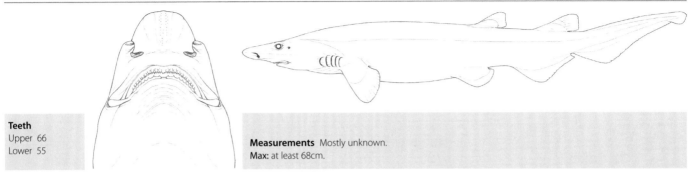

Teeth
Upper 66
Lower 55

Measurements Mostly unknown.
Max: at least 68cm.

Identification (*A. spongiceps* clade) Very stout catshark. Extremely broad, flattened head. Very short snout. Mouth extending well in front of small eyes; long labial furrows (uppers reaching upper symphysis, lowers about same length). Pleats on throat. Fins rather broad and rounded. First dorsal fin lower than second. Pectoral and pelvic fins widely separated. Anal fin high, triangular, base shorter than pectoral-pelvic space, separated from elongated caudal fin by a small notch. Colour uniform greyish brown with black posterior margins to anal and caudal fins.
Distribution Southwest Indian Ocean: west Australia, off Perth.
Habitat Continental slope; collected from 920–1140m.
Behaviour Unknown.
Biology Unknown.
Status IUCN Red List: Data Deficient. Only three specimens recorded from one location.

STOUT CATSHARK *Apristurus fedorovi*

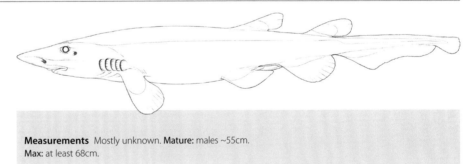

Teeth
Upper 43–61
Lower 40–65

Measurements Mostly unknown. **Mature:** males ~55cm.
Max: at least 68cm.

Identification (*A. spongiceps* clade) Broad, flattened head. Elongated snout. Large nostrils. Long, arched mouth extending well in front of eyes; very long labial furrows (uppers reaching upper symphysis, lowers possibly as long as uppers). Very small eyes. First dorsal slightly smaller or subequal to second, base over pelvic fin bases. Anal fin broadly rounded, high, short base, separated from elongated caudal fin by small notch. Dark brown body.
Distribution Northwest Pacific: northern Japan, including Hokkaido Island and northern Honshu Island and the Emperor Seamount.
Habitat Occurs on deepsea slopes and seamounts from 810–1430m.
Behaviour Unknown.
Biology Oviparous, little else known.
Status IUCN Red List: Data Deficient. A very rarely recorded catshark, with fewer than 30 specimens reported. It may be endemic to northern Japanese waters.

LONGNOSE CATSHARK *Apristurus kampae*

FAO code: **CSZ**   Plate page 332

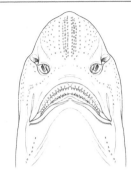

~5mm

Teeth
Upper 49–59
Lower 42–52

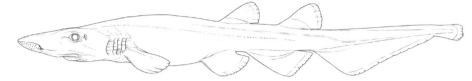

Measurements Born: ~7–14cm. **Mature:** males 49cm, females 49–54cm. **Max:** males 65cm, females 59cm.

Identification (*A. spongiceps* clade) Broad, flattened head. Elongated snout. Large, narrow nostrils. Long, arched mouth extending well anterior to eyes; very long labial furrows (uppers reaching upper symphisis, lowers about as long). Very wide gill slits (longer than adult eye-length). Similar sized dorsal fins, first origin behind pelvic fin insertions. Long pectoral and pelvic fin bases. Anal fin broadly rounded, high, base length about twice height, separated from elongated caudal fin by small notch. Blackish or dark brown to grey body and fins; precaudal fins with light edges or nearly uniform coloration.
Distribution Northeast and east Pacific: USA (Washington) and Mexico, with scattered records through Central America to Peru; Galapagos Islands records likely a different species.
Habitat Outer continental shelf and upper slope, deep water 180–1888m.
Behaviour Unknown.
Biology Oviparous, lays pairs of eggs (one per oviduct). Most gravid females are found between 1000–1200m. Feeds on deep water shrimp, cephalopods and small oceanic bony fishes.
Status IUCN Red List: Data Deficient. Bycatch in deepsea trawls and sablefish traps.

GHOST CATSHARK *Apristurus manis*

FAO code: **APA**   Plate page 332

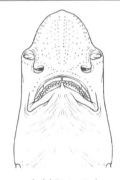

~5mm

Teeth
Upper 59
Lower 52–62

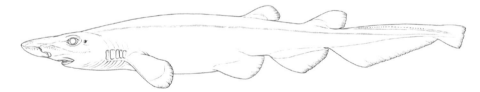

Measurements Born: less than 20cm. **Mature:** males 85cm, females at least 70–76cm. **Max:** 88cm.

Identification (*A. spongiceps* clade) Distinctively stout body strongly tapering in a wedge towards broad, flattened head and elongated snout. Broad nostrils with circular apertures. Very large, long mouth expanded in front of small eyes; enlarged dental bands; very long labial furrows (uppers reaching upper symphisis, lowers about same length). Gill slits shorter than adult eye-length. First dorsal fin slightly smaller than second; first originating over pelvic midbases; interdorsal space greater than first dorsal base. Anal fin broadly rounded, relatively high, separated from elongated caudal fin by small notch; narrow caudal fin with crest of enlarged denticles along dorsal margin. Lateral trunk denticles very sparse. Colour dark grey or blackish; fin tips sometimes whitish (particularly in juveniles).
Distribution North Atlantic, and possibly southeast Atlantic, off South Africa.
Habitat Continental slopes, 600–2453m, most common below 1500m. Eggcases from this species have been found at depths of 2140–2453m.
Behaviour Unknown.
Biology Oviparous. Eggcases 6.3–7.1cm long with both anterior and posterior ends lacking tendrils, surface smooth. Feeds on small fishes, crustaceans and cephalopods.
Status IUCN Red List: Least Concern. A very deep-living catshark that mostly occurs at depths beyond most fishing activity.

SMALLEYE CATSHARK *Apristurus microps*

FAO code: **APX**   Plate page 332

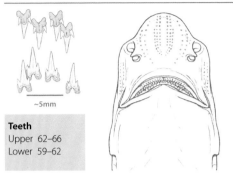

Teeth
Upper 62–66
Lower 59–62

~5mm

Measurements Mature: males 47–51cm, females 47–51cm.
Max: males 61cm possibly 73cm, females 57cm.

Identification (*A. spongiceps* clade) Stout body. Thick, long, broad snout. Mouth projecting forward, well in front of extremely small eyes; long labial furrows (uppers reaching upper symphisis, lowers about same length). Moderately large gill slits (about adult eye-length). Fairly equal-sized dorsal fins, origin of first over rear of pelvic bases; interdorsal space same as first dorsal base. Very short pectoral fins. Anal fin broadly rounded, high, separated from elongated caudal fin by small notch; narrow caudal fin with crest of enlarged denticles along dorsal margin. Dark brown or grey-brown to purplish black body; no obvious fin markings.
Distribution North Atlantic, southeast Atlantic and southwest Indian Ocean off South Africa.
Habitat Continental slopes, 700–2200m.
Behaviour This species is known to migrate far off the bottom into midwater to feed.
Biology Oviparous. Eggcases 4.7–5.2cm long, no tendrils and fine striations along the surface. Eats small bony fishes, shrimp, squid and other small sharks.
Status IUCN Red List: Least Concern. No fisheries value, mostly occurs beyond fishery ranges, and is probably a discarded bycatch from deep trawls.

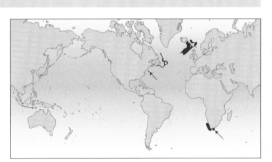

BULLDOG CATSHARK *Apristurus pinguis*

FAO code: **API**   Plate page 332

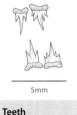

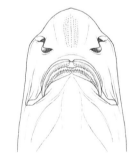

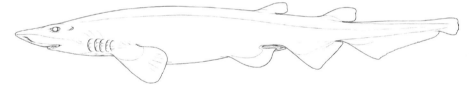

Teeth
Upper 52–70
Lower 50–66

5mm

Measurements Mature: males 57cm, females 50cm.
Max: 70cm (adult male), possibly 84cm.

Alternative name Fat Catshark
Identification (*A. spongiceps* clade) Thick, long, broad snout. Mouth projecting forward, well in front of small eyes; long labial furrows (uppers reaching upper symphisis, lowers about same length). Moderately wide gill slits equal to eye-length in adults). Dorsal fins fairly equal-sized; first dorsal origin over rear of pelvic bases; interdorsal space roughly equal to first dorsal base. Very short pectoral fins. Anal fin triangular, high, separated from elongated caudal fin by small notch. Possibly a crest of denticles on dorsal caudal margin (uncertain). Brown to blackish brown. Slightly lighter above and on fins, often with indistinct blotches; head with irregular lighter blotches.
Distribution West Pacific: East China Sea, Okinawa Trough; Australia, off New South Wales, Victoria, Tasmania, and New Zealand. East Indian Ocean: Australia, off Broken Ridge.
Habitat Captured from 858–2057m.
Behaviour Unknown.
Biology Oviparous. Eggcases 6–7 x 2cm, tapering at each end and lacking tendrils, surface relatively smooth with fine longitudinal striations.
Status IUCN Red List: Least Concern. The majority of this species' range is outside most deepsea fisheries operations.

DEEPWATER CATSHARK *Apristurus profundorum*

FAO code: **APP**   Plate page 332

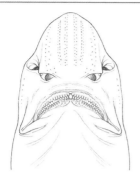
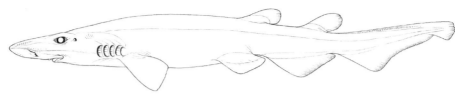

Teeth
Upper 50
Lower 50

Measurements Mature: males and females ~55cm.
Max: 76cm.

Identification (**A. spongiceps clade**) Slender body; erect denticles give skin a fuzzy texture. Thick, flattened snout. Elongated nostrils with narrow apertures. Long, large mouth expanded in front of small eyes with enlarged dental bands; long labial furrows (uppers reaching upper symphysis, lowers about same length). Gill slits rather large (slightly less than adult eye-length). High, rounded fins. Two equal-sized dorsal fins, first originating over pelvic midbases. Interdorsal space slightly greater than first dorsal base. Anal fin triangular, high, separated from elongated caudal fin by small notch. Caudal fin with a crest of enlarged denticles. Brownish body.
Distribution Newfoundland and Labrador, Canada to Delaware, USA, and east to Mid-Atlantic Ridge. East Atlantic record off Mauritania may be a different species, either *Apristurus manis*, *A. microps* or an unknown species.
Habitat Continental slope, 132–1830m.
Behaviour Unknown.
Biology Oviparous.
Status IUCN Red List: Least Concern. A common bycatch in deepsea fisheries, but most of its population lies below the depth range of commercial fishing.

BROADGILL CATSHARK *Apristurus riveri*

FAO code: **CSV**   Plate page 332

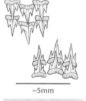

~5mm

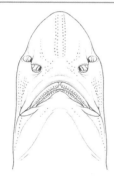

Teeth
Upper 48–58
Lower 38–43

Measurements Mature: males and females ~55cm.
Max: 48cm.

Identification (**A. spongiceps clade**) Small, slender body. Broad, flattened head. Elongated snout. Small nostrils (space between 1.5 x width). Very long mouth expanded in front of eyes; males have much larger conical teeth and much longer and wider mouths and jaws; very long labial furrows (uppers reaching upper symphysis, lowers about same length). Enlarged gill slits, widest nearly adult eye-length. Very small first dorsal fin originating in front of pelvic fin insertions. Anal fin triangular, high, separated from elongated caudal fin by small notch. Dark coloured body with unmarked fins.
Distribution West Atlantic, Gulf of Mexico and Caribbean: Cuba, USA (Florida to Mississippi), Mexico, Honduras, Panama, Colombia, Venezuela and Dominican Republic.
Habitat Continental slopes, on or near bottom, 622–1500m.
Behaviour Unknown.
Biology Oviparous. Eggcases 6 x 2–3cm, long and slender, tendrils absent, fibrous filaments covering case with very fine striations giving it a smooth to the touch texture.
Status IUCN Red List: Least Concern. This poorly known species lives below the depth range of commercial fisheries.

SPONGEHEAD CATSHARK *Apristurus spongiceps*

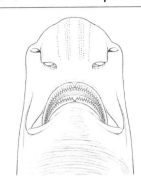
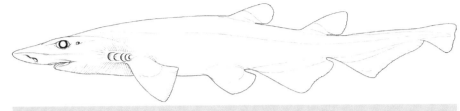

Teeth
Upper 43
Lower 40

Measurements Mostly unknown. Gravid adult female was 51cm.

Identification (**A. spongiceps clade**) Unmistakable dark brown, stout catshark with area around gills and throat covered with grooves and pleats. Broad, flattened head. Rather broad snout and nostrils. Mouth extending well in front of small eyes; long labial furrows (uppers reaching upper symphysis, lowers about same length). Gill slits small, shorter than adult eye-length. Fins high and rounded. Two small (roughly equal sized), spineless dorsal fins. Anal fin broadly rounded, relatively high, separated from elongated caudal fin by small notch.
Distribution North Central Pacific: Hawaiian Islands. Records from Indonesia are of a different species.
Habitat Insular slopes, on or near bottom, 572–1463m.
Behaviour Unknown.
Biology Oviparous, only known specimen was an adult female. Eggcases 7 x 2–3cm, with fibrous covering and no anterior or posterior tendrils.
Status IUCN Red List: Data Deficient. Known from one adult female and one specimen photographed underwater in the Hawaiian Islands (see p. 334).

PANAMA GHOST CATSHARK *Apristurus stenseni*

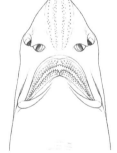
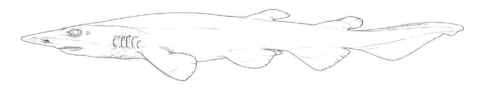

~5mm

Teeth
Upper 28–32
Lower 28–32

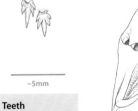

Measurements More than 23cm, possibly to 46cm (adult male).

Identification (**A. spongiceps clade**) Slender body. Broad, flattened head. Elongated snout. Narrow, widely set nostrils. Extremely large mouth extending far in front of eyes; very long labial furrows (uppers reaching upper symphysis, lowers about same length). Very small eyes and very wide gill slits (wider than adult eye-length). Dorsal fins roughly equal size, first originating over pelvic midbases. Distance between pectoral and pelvic fin bases much less than preorbital snout length. Anal fin triangular, relatively high, elongated base, separated from elongated caudal fin by small notch; narrow caudal fin with crest of enlarged denticles along dorsal margin. Blackish body, no obvious markings.
Distribution East Central Pacific: Panama.
Habitat Continental slope, 915–975m.
Behaviour Unknown.
Biology Oviparous.
Status IUCN Red List: Least Concern. Only known from a few specimens collected off Panama in water depths below the current range of commercial fisheries. Records from the Galapagos require confirmation.

Plate 42 PENTANCHIDAE V – *Asymbolus* catsharks

○ **Grey Spotted Catshark** *Asymbolus analis*

page 356

Southwest Pacific; 25–200m. Small dorsal fins behind pelvic fin bases; greyish with obscure saddle-like blotches, widely spaced equal sized dark brown spots, some whitish specks.

○ **Australian Blotched Catshark** *Asymbolus funebris*

page 356

Southeast Indian Ocean; 145m. Dorsal fins behind pelvic fins; brown with dark brown blotches, no small spots, 3 predorsal saddles, bars beneath each dorsal fin and between fins, ventral surface slightly paler.

○ **Starry Catshark** *Asymbolus galacticus*

page 357

Central Indo-Pacific; 235–550m. A very striking small slender catshark with numerous spots and rusty blotches and spots, about 10 faint dusky saddles.

○ **Western Spotted Catshark** *Asymbolus occiduus*

page 357

Southeast Indian Ocean; 98–400m. Yellowish green with 8–9 saddles, equal sized dark brown spots on back and fins (not underside), narrow ridges and often a dark spot below eyes, no white spots.

○ **Pale Spotted Catshark** *Asymbolus pallidus*

page 358

Central Indo-Pacific; 225–444m. Very small catshark; pale yellowish with no saddles, obvious equal sized dark brown spots, usually a pair in front of each dorsal fin and one at centre of each dorsal fin base, but no spots on underside or below eye.

○ **Australian Dwarf Catshark** *Asymbolus parvus*

page 358

Central Indo-Pacific; 59–360m. Very small body, short thick snout, tiny teeth; pale brown with many white spots and lines, faint dark saddles or bands but no dark spots or blotches, no spots on underside.

○ **Orange Spotted Catshark** *Asymbolus rubiginosus*

page 359

Southwest Pacific; 25–290m, possibly to 540m. Pale brown, dark brown spots with orange-brown borders on back and sides, obscure dark saddles separated by clusters of spots, usually a dark mark on edges (not tips) of dorsal fins, no spots on pale underside.

○ **Variegated Catshark** *Asymbolus submaculatus*

page 359

Southeast Indian Ocean; 30–200m. Small catshark, dorsal fins close together, first dorsal base close to pelvic bases; greyish brown with many small black spots, irregular rusty brown saddle-like blotches on back, bluish grey blotches on sides, small black and grey spots on pale grey below head abdomen and tail.

○ **Gulf Catshark** *Asymbolus vincenti*

page 360

Central Indo-Pacific and southwest Pacific; 27–220m. Mottled grey-brown to chocolate, 7–8 saddles, many small faint white spots on back sides and top of fins, pale unspotted underside.

20cm

○ *Asymbolus galacticus*

354

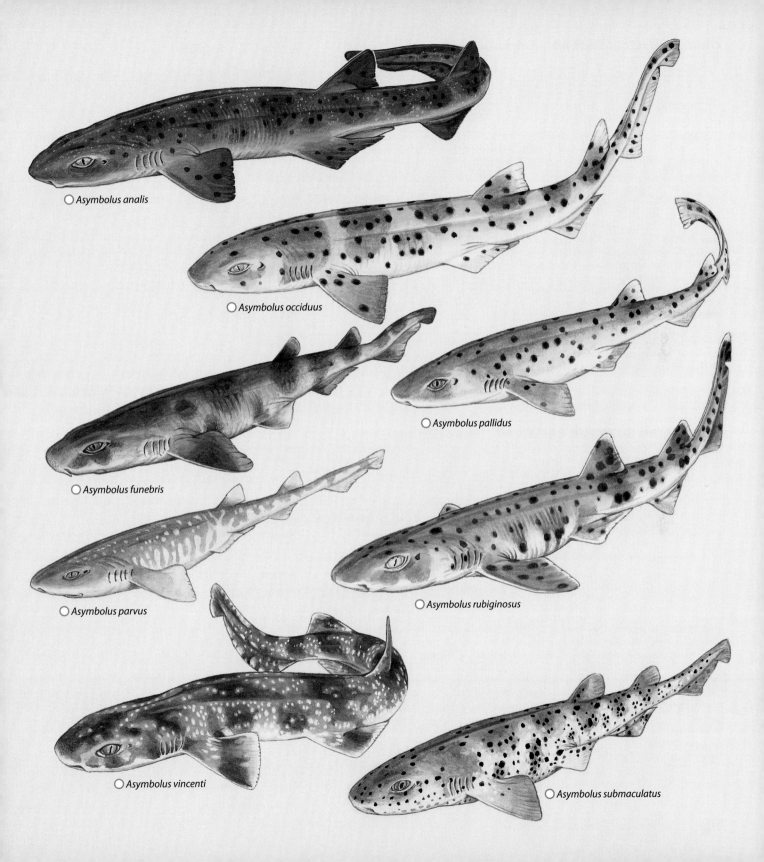

○ *Asymbolus analis*

○ *Asymbolus occiduus*

○ *Asymbolus funebris*

○ *Asymbolus pallidus*

○ *Asymbolus parvus*

○ *Asymbolus rubiginosus*

○ *Asymbolus vincenti*

○ *Asymbolus submaculatus*

GREY SPOTTED CATSHARK *Asymbolus analis*

FAO code: **ASY**   Plate page 354

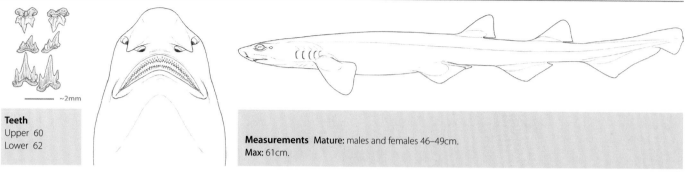

Teeth
Upper 60
Lower 62

~2mm

Measurements Mature: males and females 46–49cm.
Max: 61cm.

Identification Short, slightly flattened, pointed-rounded head. Short, thick snout. Upper teeth exposed; short labial furrows along jaws. Narrow ridges below eyes. Two small dorsal fins behind pelvic bases. Pelvic fin inner margins fused to form apron over claspers in adult males. Anal fin short and angular. Short, broad caudal fin. Colour greyish with evenly sized, widely spaced, dark brown spots and obscure, dark, saddle-like blotches on back and sides; some whitish specks.
Distribution Southwest Pacific: east Australia, southern Queensland to Victoria.
Habitat Bottom, continental shelf, close inshore (25m) to offshore (200m).
Behaviour Unknown.
Biology Oviparous, may be reproductively active year-round. Eggcases 7–8 x 2–3cm, stout, vase-shaped.
Status IUCN Red List: Least Concern. Although this species' range overlaps with considerable trawl fishing effort, six years of monitoring showed no overall catch trend.

AUSTRALIAN BLOTCHED CATSHARK *Asymbolus funebris*

FAO code: **AXM**   Plate page 354

Teeth
Upper 41
Lower not known

Measurements Known from one 44cm female.

Identification Small catshark. Short, slightly flattened, pointed-rounded head. Short, thick snout. Upper teeth exposed; short labial furrows along jaws. Narrow ridges below eyes. Dorsal fins set back behind pelvic fins. Inner margins of pelvic fins presumably fused into apron over adult male claspers. Anal fin short and angular. Short, broad caudal fin. Brown body with large dark brown blotches and saddles (no small spots). Three predorsal saddles; bars beneath each dorsal fin, one between the fins. Ventral surface only slightly paler.
Distribution Southeast Indian Ocean: southwest Australia, off the Recherche Archipelago.
Habitat Outer continental shelf, 145m depth.
Behaviour Unknown.
Biology Unknown.
Status IUCN Red List: Data Deficient. Known from only one specimen.

STARRY CATSHARK *Asymbolus galacticus*

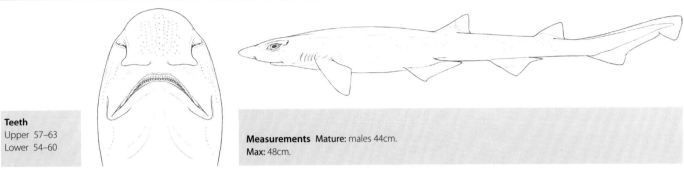

Teeth
Upper 57–63
Lower 54–60

Measurements Mature: males 44cm.
Max: 48cm.

Identification Small, slender. Shortish head; narrow ridges below eyes. First dorsal fin origin over pelvic inserts Light to medium brown catshark with very striking colour pattern. Numerous spots and rusty blotches and spots; about ten faint dusky saddles on flanks; large, dark brown blotches and saddles (no small spots).
Distribution Central Indo-Pacific: off New Caledonia, north of the Norfolk Ridge.
Habitat Continental slope, 235–550m.
Behaviour Unknown.
Biology Oviparous.
Status IUCN Red List: Least Concern. Known only from a few specimens collected by a research vessel.

WESTERN SPOTTED CATSHARK *Asymbolus occiduus*

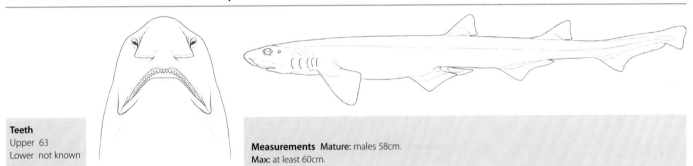

Teeth
Upper 63
Lower not known

Measurements Mature: males 58cm.
Max: at least 60cm.

Identification Short, slightly flattened head. Short, thick snout. Upper teeth exposed; short labial furrows along jaws. Narrow ridges and often a dark spot below eyes. Dorsals behind pelvic fin bases; one dark spot in front of each dorsal fin. Inner pelvic fin margins fused into apron over adult male claspers. Anal fin short and angular. Short, broad caudal fin. Bright yellowish green, with similar-sized brownish black spots and 8–9 distinct saddles (most obvious in juveniles, which have fewer spots). No white spots; no spots on underside.
Distribution Southeast Indian Ocean: southwest Australia.
Habitat Bottom, outer continental shelf, 98–400m.
Behaviour Unknown.
Biology Unknown.
Status IUCN Red List: Least Concern. Most of range unfished.

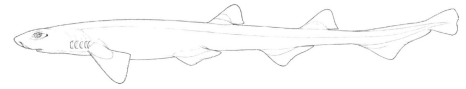

Teeth
Upper 59–66
Lower not known

Measurements Born: 19cm. **Mature:** males 32cm, females 35cm
Max: 47cm.

Identification Very small catshark. Short, slightly flattened head. Short, thick snout. Upper teeth exposed; short labial furrows along jaws. Narrow ridge below eye. Dorsal fins set far back behind pelvics; usually a pair of spots in front of each dorsal fin; one spot at the centre of each dorsal fin base. Inner pelvic fin margins fused into apron over adult male claspers. Anal fin short and angular. Short, broad caudal fin. Pale yellowish with obvious even-sized, dark brown spots (none on underside or beneath eyes). No distinct saddles, bands or white spots.
Distribution Central Indo-Pacific: northeast Australia.
Habitat Bottom, continental shelf, 225–400m.
Behaviour Unknown.
Biology Oviparous with one eggcase in each uterus.
Status IUCN Red List: Least Concern. Unlikely to be caught in commercial fisheries due to low fishing intensity within this species' range.

~2mm

Teeth
Upper 47–59
Lower not known

Measurements Mature: males ~28cm.
Max: ~40cm.

Identification Very small catshark. Short, slightly flattened, pointed-rounded head. Short, thick snout. Tiny teeth; short labial furrows along jaws. Narrow ridges below eyes. Two small dorsal fins set behind pelvic fins. Inner margins of pelvic fins fused over adult male claspers, forming an apron. Anal fin short and angular. Short, broad caudal fin. Colour pale brown with many white spots and lines and faint dark saddles or bands (most obvious on tail). No dark spots or blotches; no spots on underside.
Distribution Central Indo-Pacific: tropical northwest Australia, from just north of Shark Bay to King Sound.
Habitat Bottom, outer continental shelf and upper slope, 59–360m.
Behaviour Unknown.
Biology Oviparous.
Status IUCN Red List: Least Concern. Infrequent bycatch. Thought to survive well when discarded.

ORANGE SPOTTED CATSHARK *Asymbolus rubiginosus*

FAO code: **AXM**   Plate page 354

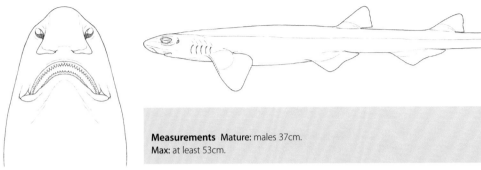

Teeth
Upper 57–58
Lower not known

Measurements **Mature:** males 37cm.
Max: at least 53cm.

Identification Indistinct brownish blotch and ridge below eye. Dorsal fins behind pelvics. Inner margins of pelvic fins fused into apron over adult male claspers. Short, broad caudal fin. Pale brown; many dark brown spots with orange-brown borders on back and sides; flank spots larger but less clear. Usually a dark mark on leading and trailing edges of dorsal fins; obscure dark saddles separated by clusters of spots along spine. No spots on pale underside.
Distribution Southwest Pacific: east Australia, southern Queensland to Tasmania.
Habitat Bottom, continental shelf and upper slope, 25–540m.
Behaviour Unknown.
Biology Oviparous, thought to lay eggs year-round. Eggcases 4–5 x 1.5–2cm, small, stout and vase-shaped.
Status IUCN Red List: Least Concern. Discarded bycatch in some trawl fisheries.

VARIEGATED CATSHARK *Asymbolus submaculatus*

FAO code: **AXM**   Plate page 354

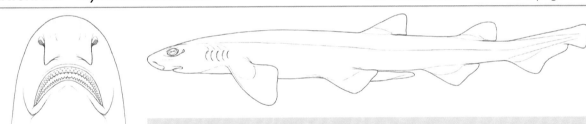

Teeth
Upper 42
Lower not known

Measurements **Mature:** males 38cm.
Max: 44cm

Identification Small catshark. Short, slightly flattened, rounded head. Short, thick snout. Long, arched mouth; large upper teeth exposed; short labial furrows along jaws. Narrow ridges below eyes. Dorsal fins close together with broadly rounded tips; first dorsal base close to pelvic bases. Inner pelvic fin margins fused into an apron over adult male claspers. Anal fin short and angular. Short, broad caudal fin. Colour greyish brown with many small black spots. Darker irregular, rusty brown, saddle-like blotches on back; bluish grey blotches on sides. Small black and grey spots on lower surfaces of head, abdomen and tail, undersides otherwise pale grey.
Distribution Southeast Indian Ocean: southwest Australia.
Habitat Continental shelf, 30–200m, in caves and on ledges.
Behaviour Nocturnal.
Biology Oviparous.
Status IUCN Red List: Least Concern. Rarely seen, unlikely to be caught in fisheries.

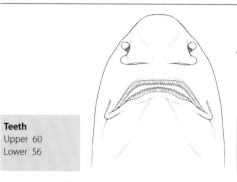
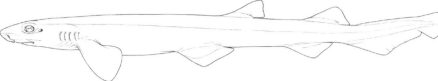

Teeth
Upper 60
Lower 56

Measurements Mature: males 38cm.
Max: at least 61cm.

Identification Short, slightly flattened head. Short, thick snout. Upper teeth exposed; short labial furrows along jaws. Narrow ridges below eyes. Two small dorsal fins behind pelvic fins. Inner pelvic fin margins fused into apron over adult male claspers. Anal fin short and angular. Short, broad caudal fin. Mottled greyish brown or chocolate. 7–8 dark saddles; many small faint white spots. Pale, unspotted underside.
Distribution Southeast Indian Ocean and southwest Pacific: west Australia, south Australia to Victoria, most common in Great Australian Bight.
Habitat Seabed <100m in east, often in seagrass beds. 27–220m in Great Australian Bight.
Behaviour Unknown.
Biology Oviparous, eggcases 5 x 2cm with long filaments, laid in pairs.
Status IUCN Red List: Least Concern. Bycatch in trawl fisheries in only part of its range.

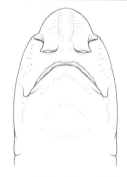
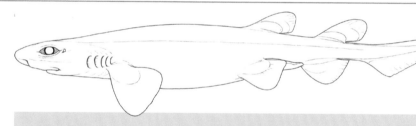

~1mm

Teeth
Upper 70–84
Lower 60–76

Measurements Born: ~12cm. **Mature:** males 40cm, females 39cm.
Max: males 42.2cm, females 44.5cm.

Identification Body firm, stout (slender in juveniles). Snout long, broad and bell-shaped in dorsal view. Mouth roof and tongue densely covered with numerous papillae; labial furrows distinct, lowers distinctly longer than uppers. Anal fin base equal to or less than 1.5 times second dorsal fin base length. Caudal fin dorsal margin lacks prominent crest of comb-like dermal denticles. Coloration a uniform light greyish brown to plain beige above, creamy below, and without patterning or indistinct dark blotches; anterior margin of all fins with very narrow dusky edges.
Distribution Southwest Indian Ocean: only known from the southern end of the Madagascar Ridge at Walters Shoals. Its distribution apparently does not overlap the Dusky Snout Catshark *Bythaelurus naylori*, which is presently only known from a few seamounts along the southwest Indian Ocean Ridge.
Habitat Known only from a few seamounts and at a depth range of 910–1365m.
Behaviour Unknown.
Biology Oviparous, eggcases laid in pairs, one per uterus. Eggcases 6–7 x 2–3cm, small, vase-shaped with very fine striations giving it a smooth texture to the touch; horns at each end short and without tendrils; lateral keels narrow, 1–2mm wide, flat, and without T-shaped flange. Diet uncertain, but likely includes bony fishes, cephalopods and crustaceans.
Status IUCN Red List: Data Deficient. Only recently described and very poorly known, this catshark may be a bycatch of deepsea trawl and longline fisheries.

DUSKY CATSHARK *Bythaelurus canescens*

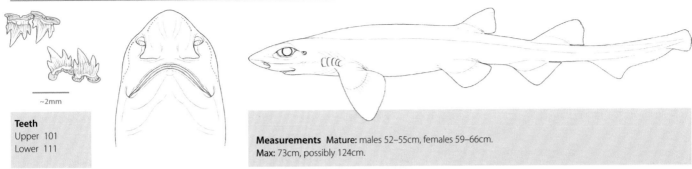

Teeth
Upper 101
Lower 111

~2mm

Measurements Mature: males 52–55cm, females 59–66cm.
Max: 73cm, possibly 124cm.

Identification Fairly large catshark. Short, rounded snout. Long, arched mouth reaches to slightly in front of the large cat-like eyes. Two small dorsal fins; first dorsal base over pelvic fin bases. Anal fin almost as large as second dorsal. Fairly short tail. Coloration plain, dark brown; no markings in adults, young have white fin tips.
Distribution Southeast Pacific: Peru, Chile, Straits of Magellan.
Habitat Deepsea on mud and rock of upper continental slope, 237–1260m.
Behaviour Unknown.
Biology Oviparous, eggcases are vase-shaped, with 12–15 longitudinal striations along surface, long coiled tendrils at each end; eggcases are laid in pairs, one per uterus. Feeds on benthic invertebrates.
Status IUCN Red List: **Vulnerable**. A regular bycatch of several fisheries, including deepsea demersal trawls and longlines which damage the deepwater corals that provide critical egg-laying habitat for this species.

BROADHEAD CATSHARK *Bythaelurus clevai*

Teeth
Upper 62–80
Lower 63–70

~2mm

Measurements Born: ~11–14cm. Mature: males ~28–36cm, females 30–35cm.
Max: ~42cm.

Identification Small catshark. Longish snout, narrow and pointed in side view, broad and bell-shaped from above. Mouth long and arched, reaching past front ends of the small cat-like eyes. Two small dorsal fins; first dorsal base mostly over pelvic fin bases. Anal fin high and triangular, larger than second dorsal. Caudal fin short. Grey above, white below, with few large, conspicuous dark brown blotches and saddles and small spots on sides and upper surface of trunk and tail, but head nearly plain. Pectoral, pelvic, dorsal and anal fins with dark bases and light margins.
Distribution West Indian Ocean: southwest Madagascar, common off Tulear.
Habitat Upper insular slopes, 400–500m.
Behaviour Unknown.
Biology Viviparous, 2 pups per litter. Eats shrimp.
Status IUCN Red List: **Data Deficient**. Very little known about the distribution, life history and fisheries interaction of this species.

Plate 43 PENTANCHIDAE VI – *Bythaelurus* catsharks I

○ **Bach's Catshark** *Bythaelurus bachi*

Southwest Indian Ocean; 910–1365m. Uniform unpatterned light greyish brown to plain beige above, creamy below; anterior fin margins with very narrow dusky edges; no enlarged crest of dermal denticles on upper caudal margin.

○ **Dusky Catshark** *Bythaelurus canescens*

Southeast Pacific; 237–960m. Large, cat-like eyes; uniform dark brown, adults unmarked, young with white fin tips.

○ **Broadhead Catshark** *Bythaelurus clevai*

West Indian Ocean; 400–500m. Small grey catshark, white below, a few large conspicuous dark brown blotches and saddles, small spots on sides and upper surfaces of trunk and tail, not on head; all fins except caudal with dark bases and light margins.

○ **New Zealand Catshark** *Bythaelurus dawsoni*

Southwest Pacific; 50–992m. Light brown or grey body, paler below, white fin tips, dark bands on caudal fin, a line of white spots on sides of small specimens.

○ **Galapagos Catshark** *Bythaelurus giddingsi*

East Pacific; 428–562m. Striking variegated pattern of large white blotches and spots on a grey background.

○ **Bristly Catshark** *Bythaelurus hispidus*

North Indian Ocean; 200–800m. Very small elongated bristly skinned pale brown to whitish catshark, sometimes with faint grey crossbands and white or dusky spots.

○ **Spotless Catshark** *Bythaelurus immaculatus*

Central Indo-Pacific; 534–1020m. Drab yellow-brown, unpatterned, similar to *B. dawsoni* but no white fin tips.

○ **Lollipop Catshark** *Cephalurus cephalus*

East Pacific; 155–927m. Unmistakable tadpole shape; expanded flattened head and gill region; small slender body and tail; very thin-skinned and soft; dorsal fins small, first dorsal slightly ahead of pelvic fin origins.

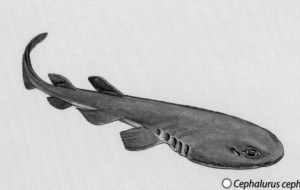

○ *Cephalurus cephalus*

20cm

○ *Bythaelurus bachi*

○ *Bythaelurus canescens*

juvenile

○ *Bythaelurus clevai*

○ *Bythaelurus dawsoni*

○ *Bythaelurus hispidus*

○ *Bythaelurus giddingsi*

○ *Bythaelurus immaculatus*

Plate 44 PENTANCHIDAE VII – *Bythaelurus* catsharks II

○ **Dusky Catshark** *Bythaelurus incanus*   page 368

Central Indo-Pacific; 900–1000m. Very broad flattened head, long soft body; bristly skin; uniformly dark grey-brown, a few pale blotches below.

○ **Mud Catshark** *Bythaelurus lutarius*   page 368

West Indian Ocean; 338–766m. First dorsal smaller than second; plain grey-brown, sometimes with dusky saddles, light below.

○ **Dusky Snout Catshark** *Bythaelurus naylori*   page 369

Southwest Indian Ocean; 752–1443m. Medium to dark brown body with light fin edges and distinctly darker snout; distinctly enlarged prominent crest of comb-like dermal denticles on upper caudal margin.

○ **Error Seamount Catshark** *Bythaelurus stewarti*   page 369

Northwest Indian Ocean; 380–420m. Dark greyish brown with 5–6 indistinct dark blotches on body and tail, beige below with dark brown mottling on head; fins lighten near fin margins; upper caudal margin with slightly enlarged crest of comb-like dermal denticles.

○ **Narrowhead Catshark** *Bythaelurus tenuicephalus*   page 370

West Indian Ocean; 463–550m. Long slender snout broadly pointed at tip, elongate body tapering from first dorsal to caudal fin origin; light to medium brown with 5–6 indistinct dark blotches on trunk and tail, much paler below.

○ **Vivaldi's Catshark** *Bythaelurus vivaldii*   page 370

West Indian Ocean; 628m. Greyish brown above, lighter below, with 8–9 dark broad faint bars dorsally; mouth roof and tongue with few papillae; upper caudal margin with slightly enlarged crest of comb-like dermal denticles.

20cm

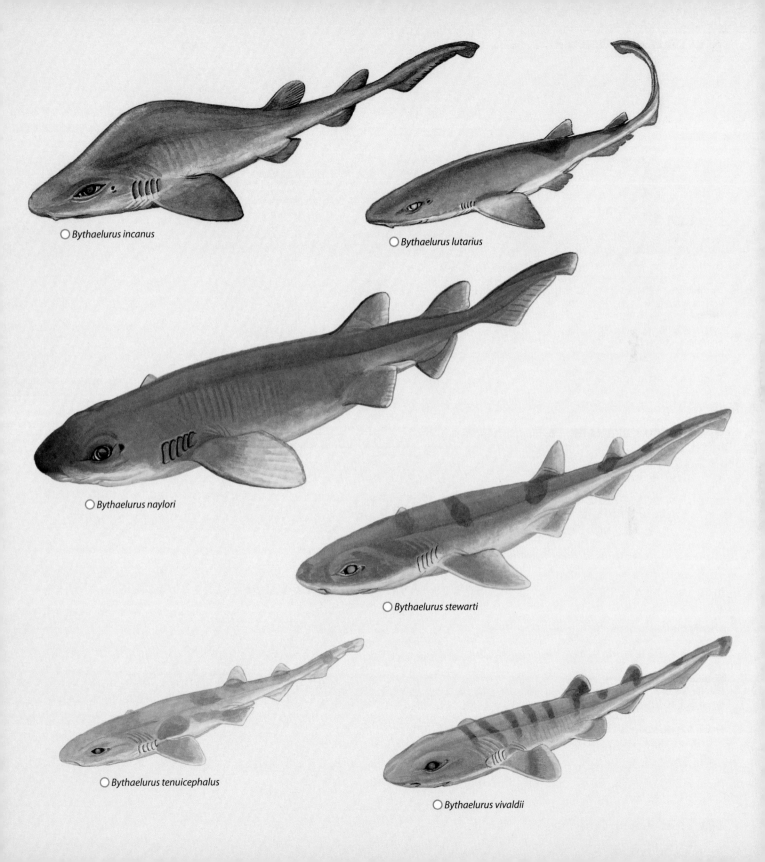

○ *Bythaelurus incanus*

○ *Bythaelurus lutarius*

○ *Bythaelurus naylori*

○ *Bythaelurus stewarti*

○ *Bythaelurus tenuicephalus*

○ *Bythaelurus vivaldii*

NEW ZEALAND CATSHARK *Bythaelurus dawsoni*   FAO code: **HAO**   Plate page 362

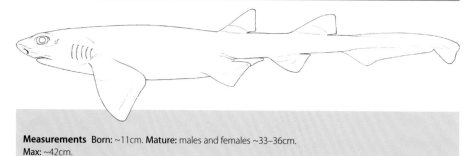

Teeth
Upper 64
Lower 70

Measurements **Born:** ~11cm. **Mature:** males and females ~33–36cm.
Max: ~42cm.

Identification Fairly short body. Broad, flattened head. Elongated, lobate anterior nasal flaps. Long, arched mouth reaches past the front end of the large cat-like eyes. Two dorsal fins; first dorsal base over pelvic bases, second larger. Anal fin short and angular. Light brown or grey body, paler below. A line of white spots on sides of small animals, white fin tips. Dark bands on caudal fin.
Distribution Southwest Pacific: New Zealand, mostly on the Chatham Rise, and around the Auckland and Campbell Islands.
Habitat On or near bottom on the upper slopes of New Zealand and Auckland Islands, 50–992m, most records from 300–600m.
Behaviour Unknown.
Biology Oviparous. Feeds on demersal bony fishes, crustaceans and cephalopods. Sharks larger than 35cm feed more on bony fishes.
Status IUCN Red List: Least Concern. Apparently fairly common in deep water but rarely recorded.

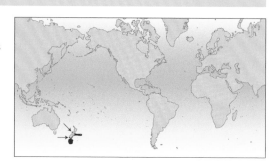

GALAPAGOS CATSHARK *Bythaelurus giddingsi*   FAO code: **CVX**   Plate page 362

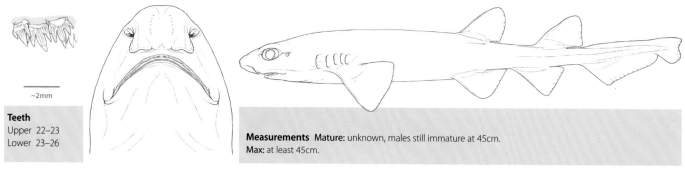

~2mm

Teeth
Upper 22–23
Lower 23–26

Measurements **Mature:** unknown, males still immature at 45cm.
Max: at least 45cm.

Identification Small catshark. Short, flattened, rounded snout. Long, arched mouth reaches past the front end of the cat-like eyes. Two medium-sized, rounded dorsal fins; first dorsal base over pelvic bases. Anal fin about as large as second dorsal. Short tail. Striking variegated pattern of large white spots and blotches on grey background.
Distribution East Pacific: Galapagos Islands.
Habitat Insular slopes, on mud and sand bottoms near lava boulders at 428–562m.
Behaviour Essentially unknown.
Biology Essentially unknown.
Status IUCN Red List: Least Concern. Known only from juveniles – adults may occur in much deeper water. Its steep slope habitat is unsuitable for trawling and there are no fisheries in the Galapagos Marine Reserve.

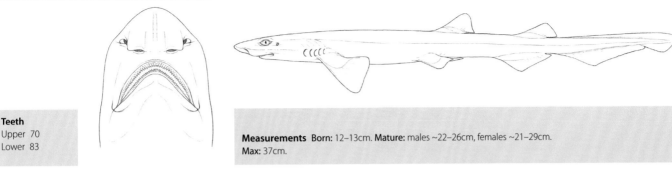

Teeth
Upper 70
Lower 83

Measurements Born: 12–13cm. **Mature:** males ~22–26cm, females ~21–29cm.
Max: 37cm.

Identification Very small, elongated catshark. Short, rounded snout. Long, arched mouth. Two dorsal fins (origin of first over rear of pelvic fin bases) and an anal fin. Bristly skin. Pale brown or whitish coloration, sometimes with faint grey crossbands, white or dusky spots.
Distribution North Indian Ocean: Kenya to Socotra Islands, east to southeast India, Andaman Islands, and Myanmar.
Habitat Bottom on upper continental slopes, 200–800m.
Behaviour Unknown.
Biology Viviparous. Eats small fishes, squid and crustaceans.
Status IUCN Red List: Near Threatened. Apparently common in deep water, overlapping with fisheries in parts of its range. This might be a species complex, containing additional species.

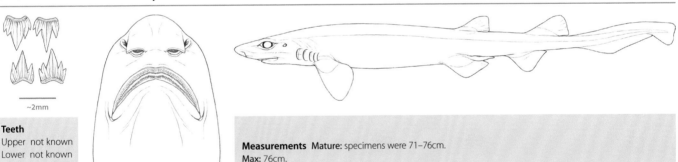

~2mm

Teeth
Upper not known
Lower not known

Measurements Mature: specimens were 71–76cm.
Max: 76cm.

Identification A large, drab, yellowish brown catshark with no markings, similar to *Bythaelurus dawsoni*. Rounded snout. Two dorsal fins; first smaller, over pelvic fin bases; second much larger and over anal fin base. Long abdomen.
Distribution Central Indo-Pacific: South China Sea, ~380–400km east of Hainan Island.
Habitat Bottom, on the continental slope, 534–1020m.
Behaviour Unknown.
Biology Unknown.
Status IUCN Red List: Least Concern. Absence of records from commercial fisheries bycatch suggests that this species is not threatened by the fisheries currently operating in the area.

DUSKY CATSHARK   *Bythaelurus incanus*

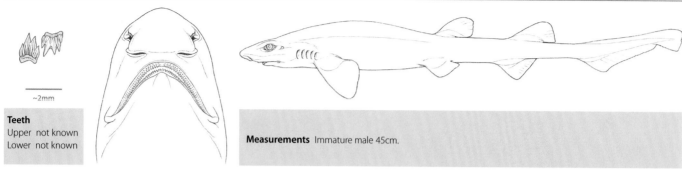

~2mm

Teeth
Upper not known
Lower not known

Measurements Immature male 45cm.

Identification Long, soft body. Very broad, flattened head. Short, rounded snout. Two well separated, similar-sized, tall, rounded dorsal fins; base of first over pelvic bases; second over anal fin. Uniform dark greyish brown bristly skin with a few pale blotches beneath.
Distribution Central Indo-Pacific: northwest Australia.
Habitat Continental slope, 900–1000m.
Behaviour Unknown.
Biology Unknown.
Status IUCN Red List: Data Deficient. Known from only one specimen. Occurs too deep to be caught in fisheries.

MUD CATSHARK   *Bythaelurus lutarius*

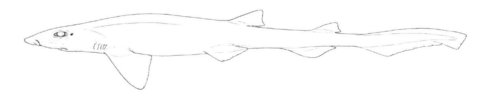

female   male

~2mm

Teeth
Upper 76
Lower 86

Measurements Born: ~10–14cm. Mature: males ~28–31cm, gravid females 29–31cm. Max: ~39cm.

Identification Small catshark. Short, rounded snout. Long, arched mouth reaches past the front end of the cat-like eyes. Two small dorsal fins; base of first roughly over pelvic fin insertions; second (slightly larger) origin over mid base of anal fin. Plain grey-brown, sometimes with dusky saddle bands, light underside.
Distribution West Indian Ocean: Mozambique. Records from Somalia are now classified as a different species, *Bythaelurus vivaldii*, and those from off northern Tanzania are *B. tenuicephalus*.
Habitat Deepsea on tropical continental slopes, on or just above muddy bottom, at 338–766m.
Behaviour Unknown.
Biology Viviparous, 2 pups per litter. Eats cephalopods, small bony fishes and crustaceans.
Status IUCN Red List: Data Deficient. Presumably not taken in fisheries. Formerly thought to occur from Mozambique to Somalia, this species is now known to be endemic to Mozambique.

DUSKY SNOUT CATSHARK *Bythaelurus naylori*   FAO code: **CVX**   Plate page 364

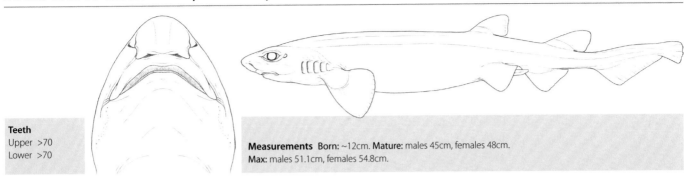

Teeth
Upper >70
Lower >70

Measurements Born: ~12cm. **Mature:** males 45cm, females 48cm.
Max: males 51.1cm, females 54.8cm.

Identification Body stout, flabby, tapering quickly from first dorsal fin to caudal fin origin. Snout short, broad, bluntly rounded. Mouth roof and tongue without numerous papillae; labial furrows distinct, lowers distinctly longer than uppers. Anal fin base equal to or less than 1.5 times second dorsal fin base length. Caudal upper margin with distinctly enlarged prominent crest of comb-like dermal denticles. Coloration a uniform medium to dark brown, with light fin edges, and a distinct dark, dusky to blackish coloured snout.
Distribution Southwest Indian Ocean: only known from the southwest Indian Ridge. The distribution of this species apparently does not overlap with Bach's Catshark *Bythaelurus bachi*, which only occurs on the southern end of the Madagascar Ridge at Walters Shoals.
Habitat Known only from a few seamounts at a depth range of 752–1443m, with one record from 89m.
Behaviour The shallowest record of this species, possibly 89m, came from a midwater trawl.
Biology Oviparous. Small eggcases, 7 x 2cm, vase-shaped, with very fine striations giving it a smooth texture to the touch; horns at each end short and without tendrils; lateral keels ~1mm wide, flat, but without T-shaped flange. Feeds mainly on bony fishes, cephalopods and crustaceans.
Status IUCN Red List: Data Deficient. This poorly known, recently described catshark may be a bycatch of deepsea trawl and longline fisheries around seamounts.

ERROR SEAMOUNT CATSHARK *Bythaelurus stewarti*   FAO code: **CVX**   Plate page 364

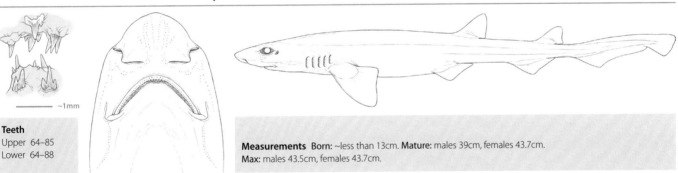

~1mm

Teeth
Upper 64–85
Lower 64–88

Measurements Born: ~less than 13cm. **Mature:** males 39cm, females 43.7cm.
Max: males 43.5cm, females 43.7cm.

Identification Body firm and slender, tapering from first dorsal fin to caudal fin origin. Snout long, broad; tip broadly rounded; strongly bell-shaped in dorsal view. Mouth roof and tongue densely covered with numerous papillae; labial furrows distinct, lowers longer than uppers. Anal fin base more than 1.5 times the second dorsal fin base length. Caudal upper margin with slightly enlarged crest of comb-like dermal denticles. Coloration dark greyish brown above, with 5–6 indistinct dark blotches on trunk and tail. Beige below with dark brown mottling on head, more distinct in larger individuals. Fins similar to body colour, but lightening near margins.
Distribution Northwest Indian Ocean: endemic, known only from the Error Seamount.
Habitat A micro-endemic species restricted to an isolated seamount at a depth range of 380–420m.
Behaviour Unknown.
Biology Unlike most catshark species (which are egg-laying), this shark is live-bearing, yolk-sac viviparous, with a litter size of 2, a single developing embryo in each uterus. Feeds mainly on bony fishes, cephalopods and crustaceans.
Status IUCN Red List: Data Deficient. There are no data on population size or trend, but high seas demersal trawl fisheries target seamounts in the Indian Ocean.

NARROWHEAD CATSHARK *Bythaelurus tenuicephalus*   FAO code: **BZL**   Plate page 364

Teeth
Upper 67–76
Lower 62–64

~1mm

Measurements Mature: males 28cm.
Max: 30cm.

Identification Body firm and elongate, tapering from first dorsal fin to caudal fin origin. Snout long and slender, broadly pointed at tip. Mouth roof and tongue with numerous papillae; labial furrows distinct, lowers distinctly longer than uppers. Anal fin base more than 1.5 times the second dorsal fin base length. Caudal upper margin with slightly enlarged crest of comb-like dermal denticles. Coloration a uniform light to medium brown above with 5–6 indistinct dark blotches on trunk and tail. Lateral trunk anterior to cloaca demarcated with medium brown above and whitish or beige below, with the tail less distinct and gradually fading from brown to whitish or beige.
Distribution West Indian Ocean: off northern Tanzania in the Pemba Channel between Zanzibar and Pemba Island, and off southern Mozambique.
Habitat Deepsea, 463–550m, nothing else known.
Behaviour Unknown.
Biology Reproductive mode unknown. Species is only known from two male specimens, an adult and a juvenile. Diet is unknown, but may include bony fishes, cephalopods and crustaceans. This is one of the smallest known species in this group.
Status IUCN Red List: **Least Concern**. Not seen in bycatch.

VIVALDI'S CATSHARK *Bythaelurus vivaldii*   FAO code: **CVX**   Plate page 364

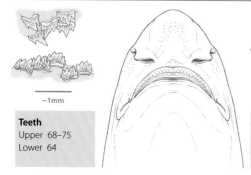

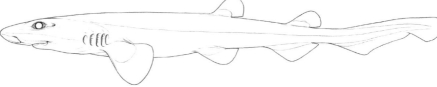

~1mm

Teeth
Upper 68–75
Lower 64

Measurements Mature: females 32.5cm.
Max: 35cm.

Identification Body firm and stout, tapering from first dorsal fin to caudal fin origin. Snout long, broad, bluntly rounded at tip. Mouth roof and tongue with few papillae; labial furrows distinct, lowers distinctly longer than uppers. Anal fin base more than 1.5 times the second dorsal fin base length. Caudal upper margin with slightly enlarged crest of comb-like dermal denticles. Coloration a greyish brown above, lighter ventrally; with 8–9 dark, broad, inconspicuous, transverse bars dorsally, but not reaching the lower half of the body.
Distribution West Indian Ocean: off Somalia.
Habitat Recorded from a depth of 628m.
Behaviour Unknown.
Biology Reproductive mode unknown. Species is only known from two female specimens, an adult and a juvenile. Diet is unknown, but may include bony fishes, cephalopods and crustaceans.
Status IUCN Red List: **Data Deficient**. No information on deepwater fishing pressure in the region.

Teeth
Upper 54–68
Lower 54–68

Measurements Born: ~8–10cm. **Mature:** males ~19cm, females ~24cm.
Max: males at least 25cm, females 37cm.

Identification Unmistakable, very small, unpatterned, tadpole-shaped catshark. Expanded, flattened, rounded head and gill region. Small, slender, very soft, thin-skinned (almost gelatinous) body and tail. Two small dorsal fins and an anal fin; first dorsal origin slightly in front of pelvic fin origins.
Distribution East Pacific: Mexico, southern Baja California, Gulf of California, and the Revillagigedo Archipelago, and possibly south to Peru and Chile.
Habitat Upper continental slope and outermost shelf, on or near bottom, 155–927m. The expanded gills suggests that this species is adapted to survive where oxygen levels are low. In the Gulf of California, it is commonly observed by remote operated vehicles in areas of low dissolved oxygen.
Behaviour Its large head and gill region, which gives it a lollipop shape, may reflect adaptation for life in low oxygen and possibly higher salinity deep-sea basins, where few other predators may be able to survive.
Biology Viviparous, retaining its very thin-walled eggcases in the uterus until the 2 young per litter hatch. Birthing may take place during the summer. Feeds mainly on crustaceans and small bony fishes
Status IUCN Red List: **Least Concern.** A common species in unfished deepsea basins with low oxygen, but very little known. Peruvian and Chilean populations may be a different species.

In 2015, researchers at the Scripps Institution of Oceanography and the Monterey Bay Aquarium Research Institute discovered Lollypop Catsharks thriving in deep basins in the Gulf of California, where oxygen concentrations were less than one percent of those in the surface waters (roughly 1.6 umol/kg). The researchers speculate that the catsharks' large heads and gills may help them absorb oxygen in low-oxygen environments. © 2015 MBARI

Plate 45 PENTANCHIDAE VIII – sawtail catsharks I

○ **Antilles Catshark** *Galeus antillensis*   page 377

West Atlantic: Caribbean Sea; 293–698m. Very similar to *G. arae* but larger. Striking variegated pattern, usually less than 11 dark brown saddle blotches (clearly outlined or obscure), dark bands on tail, usually dark marks on flanks; no black marks on dorsal fin tips, caudal fin may have dusky base and light web; crest of enlarged denticles along upper caudal fin margin.

○ **Roughtail Catshark** *Galeus arae*   page 377

West Atlantic: Gulf of Mexico and continental Caribbean Sea; 292–732m. Very similar to *Galeus antillensis*.

○ **Atlantic Sawtail Catshark** *Galeus atlanticus*   page 378

Northeast Atlantic and Mediterranean; 328–790m. Grey above, whitish below, with dark grey blotches and saddles or narrow vertical bars along body and tail; black mouth cavity; caudal fin with dark margins, not black-tipped; crest of enlarged denticles along upper caudal fin margin.

○ **Blackmouth Catshark** *Galeus melastomus*   page 381

Northeast Atlantic and Mediterranean; 55–2000m. Striking pattern of 15–18 dark saddles, blotches and circular pale-edged spots on back and tail; white-edged fins; crest of enlarged denticles along upper caudal fin margin; blackish mouth cavity.

○ **Southern Sawtail Catshark** *Galeus mincaronei*   page 381

West Atlantic; 130–600m. Striking pattern of 11 oval or circular dark saddles and spots outlined in white on reddish-brown background on back and precaudal tail, whitish below; fins dark but no black tips or white edges; crest of enlarged denticles along upper caudal fin margin.

○ **Mouse Catshark** *Galeus murinus*   page 382

Northeast Atlantic; 380–1300m. Large broadly rounded pelvic fins; body uniformly brown above, slightly paler below, no markings; crest of enlarged denticles along upper caudal fin margin and ventral midline of caudal peduncle; blackish mouth.

○ **Broadfin Sawtail Catshark** *Galeus nipponensis*   page 382

Northwest Pacific; 150–540m. Long space between pelvic fin bases and unusually short anal fin; greatly elongated claspers in mature males; greyish white above, white below, numerous obscure dusky saddles and blotches; black anterior margins to lower caudal lobe and rear tip, dorsal and anal fins; no black fin tips; crest of enlarged denticles along upper caudal fin margin; white mouth lining.

○ **Peppered Catshark** *Galeus piperatus*   page 383

East Pacific; 130–1326m. Black pepper-like spots all over body and tail, with or without variegated white-edged dark saddle blotches; white-edged caudal fin, no black dorsal and caudal fin tips; crest of enlarged denticles along upper caudal fin margin; mouth lining usually dark.

○ **African Sawtail Catshark** *Galeus polli*   page 383

East Atlantic, 159–720m. Usually 11 or fewer well-defined dark saddle blotches outlined in white on back and tail, sometimes uniform dark; no black dorsal and caudal fin tips; crest of enlarged denticles along upper caudal fin margin; mouth lining dark.

○ **Blacktip Sawtail Catshark** *Galeus sauteri*   page 384

Central Indo-Pacific to Northwest Pacific; 60–200m or more. Not patterned, dorsal fins and sometimes caudal fin lobes with prominent black tips; crest of enlarged denticles along upper caudal fin margin; light mouth lining.

○ *Galeus mincaronei*

○ *Galeus melastomus*

20cm

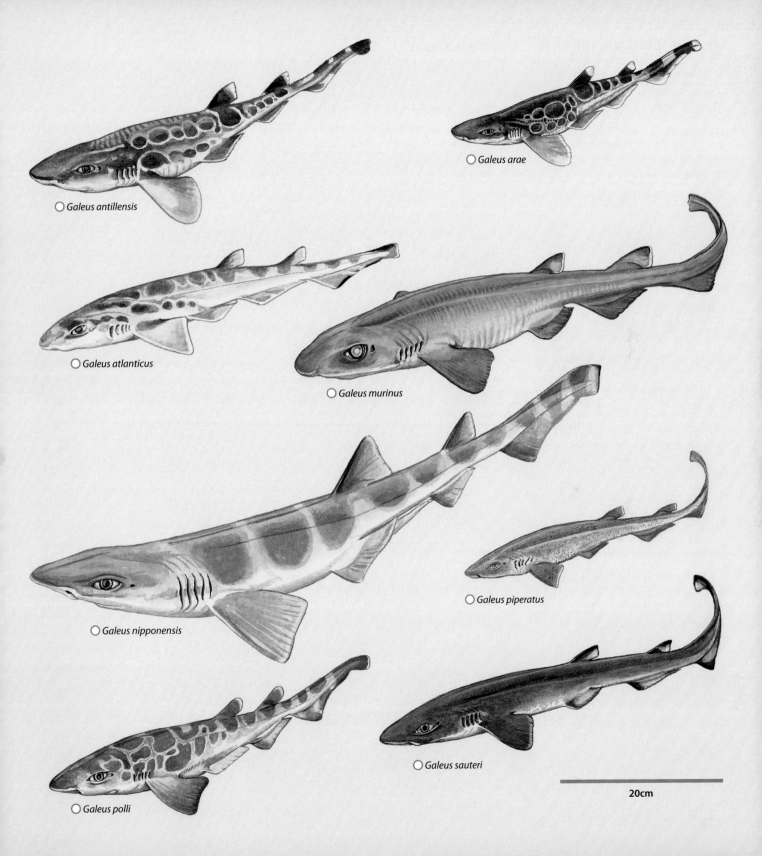

○ *Galeus antillensis*

○ *Galeus arae*

○ *Galeus atlanticus*

○ *Galeus murinus*

○ *Galeus nipponensis*

○ *Galeus piperatus*

○ *Galeus polli*

○ *Galeus sauteri*

20cm

Plate 46 PENTANCHIDAE IX – sawtail catsharks II

○ **Australian Sawtail Catshark** *Figaro boardmani* — page 376

Southwest Indian Ocean to southwest Pacific; 128–823m. Pattern of dark grey-brown saddles and bars, sometimes with white flecks; 3 pale-edged predorsal saddles separated by less distinct narrow bands, a band at and between each dorsal fin and 3 broad postdorsal bands; crests of enlarged denticles along upper caudal fin margin and ventral midline of caudal peduncle.

○ **Northern Sawtail Catshark** *Figaro striatus* — page 376

Central Indo-Pacific; 239–590m. Pale grey-brown; 10–16 predorsal pale-edged saddles and bars, pale-centred bars below dorsal fins; underside flanks and lower fins pale, dorsal and upper caudal fins dusky with pale trailing edges; crests of enlarged denticles along upper and lower caudal fin margins.

○ **Longfin Sawtail Catshark** *Galeus cadenati* — page 378

West Atlantic: Caribbean; 262–549m. Very similar to *Galeus arae* and *G. antillensis* but with longer shallower anal fin.

○ **Corrigan's Catshark** *Galeus corriganae* — page 379

Central Indo-Pacific; 500–742m. Very small slender body, long abdomen; dark to medium grey above, lighter below, 4 distinct dark dorsal saddles, beneath each dorsal fin and on caudal fin; anterior pectoral fin margins black, posterior margins white; mouth dark grey to blackish on roof and tongue; upper caudal fin margin with crest of 2–4 central rows of enlarged denticles between much larger lateral rows.

○ **Gecko Catshark** *Galeus eastmani* — page 379

Northwest Pacific, Central Indo-Pacific; 100–900m. White mouth lining, white-edged dorsal and caudal fins; obscure pattern of saddles and blotches; crest of enlarged denticles along upper caudal fin margin.

○ **Slender Sawtail Catshark** *Galeus gracilis* — page 380

Central Indo-Pacific; 290–470m. Small slender body, pale grey, paler underside, no predorsal markings; 4 dusky saddles, 1 beneath each dorsal fin and 2 on caudal fin; crest of enlarged denticles along upper caudal fin margin.

○ **Longnose Sawtail Catshark** *Galeus longirostris* — page 380

Northwest Pacific; 330–550m. Grey above, whitish below, greyish white mouth lining; young with obscure saddle blotches at dorsal fins and on caudal fin; white-edged dorsal and pectoral fins; no black tips to dorsal and caudal fins; crest of enlarged denticles along upper caudal fin margin.

○ **Phallic Catshark** *Galeus priapus* — page 384

Central Indo-Pacific; 262–830m. Dark grey with 4 predorsal saddles, 1 under each dorsal fin and 2 on caudal fin, ventral surface white; extremely long claspers in adult males; crest of enlarged denticles along upper caudal fin margin.

○ **Dwarf Sawtail Catshark** *Galeus schultzi* — page 385

Central Indo-Pacific; 50–431m. Short rounded snout; obscure dark saddles below dorsal fins and 2 darker bands on tail; no black fin tips; crest of enlarged denticles along upper caudal fin margin.

○ **Springer's Sawtail Catshark** *Galeus springeri* — page 385

West Atlantic: Caribbean Sea; 457–824m. Only *Galeus* with predorsal dark longitudinal stripes outlined in white.

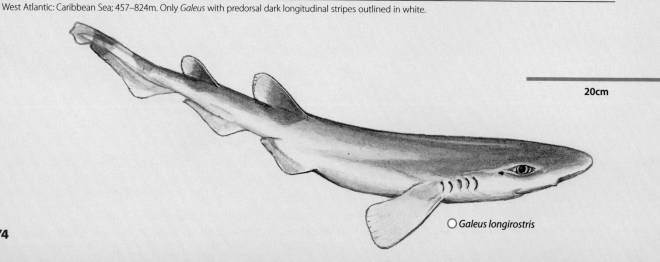

20cm

○ *Galeus longirostris*

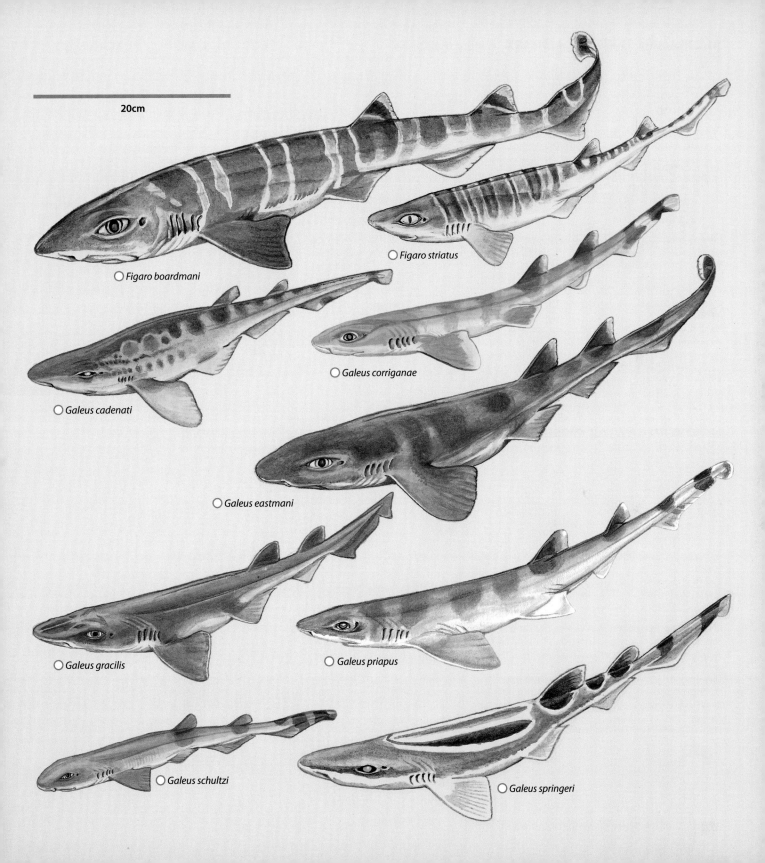

20cm

Figaro boardmani

Figaro striatus

Galeus cadenati

Galeus corriganae

Galeus eastmani

Galeus gracilis

Galeus priapus

Galeus schultzi

Galeus springeri

AUSTRALIAN SAWTAIL CATSHARK *Figaro boardmani*　　　FAO code: **GAB**　　Plate page 374

Teeth
Upper–67
Lower–74

Measurements Mature: males ~40cm, females 40–43cm.
Max: 61cm.

Identification Greyish with a variegated pattern of dark greyish brown saddles and bars. Three broad, pale-edged, predorsal saddles with a narrower, less distinct band between each, a band at and between each dorsal, and three broad bands postdorsally. Bands and saddles sometimes with white flecks. Pale underside. Distinct crests of enlarged dermal denticles along the upper caudal fin margin and along the ventral midline of the caudal peduncle.
Distribution Southwest Indian Ocean to southwest Pacific: Australia, southeast Queensland to west Australia.
Habitat Outer continental shelf and upper slope, on or near bottom, 128–823m.
Behaviour May sometimes aggregate by sex.
Biology Oviparous. Reproductive cycle unknown, but females carrying eggcases in utero during winter have been observed. Feeds on fish, crustaceans and cephalopods.
Status IUCN Red List: Least Concern. Widespread and apparently common.

NORTHERN SAWTAIL CATSHARK *Figaro striatus*　　　FAO code: **CVX**　　Plate page 374

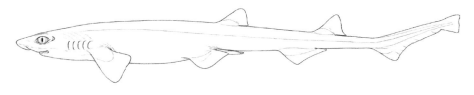

—— ~2mm

Teeth
Upper not known
Lower not known

Measurements Mature: males ~38cm.
Max: at least 42cm.

Identification Pale greyish brown with 10–16 darker, pale-edged saddles and bars in front of first dorsal fin. Bars below dorsal fins with pale centres. Underside and much of sides and lower fins uniformly pale. Dorsal and upper caudal fins dusky, with pale trailing edges. Distinct crests of enlarged dermal denticles along the upper and lower caudal fin margins.
Distribution Central Indo-Pacific: off northeast Australia.
Habitat Continental shelf, bottom from 239–590m.
Behaviour Unknown.
Biology Oviparous.
Status IUCN Red List: Data Deficient. Little fishing activity within small range.

ANTILLES CATSHARK *Galeus antillensis*

FAO code: **GAU** Plate page 372

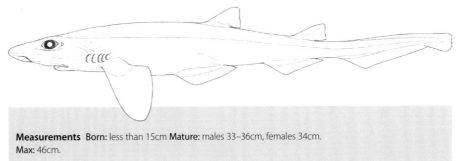

Teeth
Upper 56
Lower 52

Measurements Born: less than 15cm **Mature:** males 33–36cm, females 34cm. **Max:** 46cm.

Identification Larger than *Galeus arae* and with more precaudal vertebrae. Striking pattern of variegated dark brown saddle blotches (usually less than 11) and dark bands on tail; saddles may be clearly outlined or obscure. Usually dark markings on flanks; no black marks on dorsal fin tips; caudal fin tip may have a dusky base and light web. Dark mouth lining. Distinct crest of enlarged dermal denticles along the upper caudal fin margin.
Distribution West Atlantic: Straits of Florida, Caribbean Sea from Hispaniola, Puerto Rico, Jamaica and Leeward Islands to Martinique.
Habitat Deepsea upper insular slopes, on or near bottom, 293–698m; a 150m report is much shallower than other records. Water temperatures where this species occurs range between 4.6–11.1°C (40.3–52°F).
Behaviour May school in large numbers.
Biology Oviparous. Eggcases 5–6 x 1–2cm, with tendrils extending from each corner. Females gravid in late spring and early summer. Feeds mainly on deep water shrimp.
Status IUCN Red List: Least Concern. Common to abundant where it occurs. Formerly considered an island subspecies of *G. arae*.

ROUGHTAIL CATSHARK *Galeus arae*

FAO code: **GAA** Plate page 372

~2mm

Teeth
Upper 59–65
Lower 58–60

Measurements Born: 9cm. **Mature:** males and females ~27–33cm. **Max:** ~33cm.

Identification Similar to *Galeus antillensis*, but smaller and with fewer precaudal vertebrae.
Distribution West Atlantic, including Gulf of Mexico and continental Caribbean Sea: two separate populations, one off USA (North Carolina to the Mississippi Delta), Mexico and Cuba; second off Belize, Honduras, Nicaragua, Costa Rica and adjacent islands.
Habitat Upper continental and insular slopes, on or near bottom, 292–732m.
Behaviour Partly segregates by depth: only adults occur in deep water, adults and juveniles above 450m. May school in large numbers.
Biology Reproduction once thought to be viviparous, but collections of adult female with eggcases confirm that it is oviparous like most other members of this group. Eggcases 5–6 x 1–2cm, with tendrils extending from each corner. Females gravid in late spring and early summer. Feeds mainly on deep water shrimp.
Status IUCN Red List: Least Concern. Common to abundant where it occurs.

ATLANTIC SAWTAIL CATSHARK *Galeus atlanticus*

FAO code: **GHA**   Plate page 372

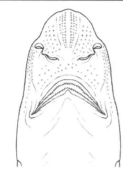

Teeth
Upper 68–73
Lower 73–77

Measurements Born: ~15cm. **Mature:** males 33–42cm, females 37–45cm.
Max: males 42cm, females 46cm.

Identification Long, angular snout; black mouth cavity. Eyes on lateral edges of head. Angular dorsal fins. Large pectoral fins. Anal fin long and low. Distinct crest of enlarged dermal denticles along the upper caudal fin margin. Grey above, whitish below. Dark grey blotches and saddles or narrow vertical bars along body and tail (rarely absent); dorsal fins dusky with light rear webs; caudal fin with dark margin, not black-tipped.
Distribution Northeast Atlantic and Mediterranean: Morocco, Mediterranean Spain (around Straits of Gibraltar), and West Africa from Morocco and Mauritania.
Habitat Continental slopes, 328–790m, but mostly between 500–600m.
Behaviour Essentially unknown.
Biology Oviparous. Apparently with multiple ovipary (nine eggcases in one female), suggesting a short hatching period outside the mother. Eggcases 3–4 x 1cm. Females appear to deposit eggcases year-round.
Status IUCN Red List: Near Threatened. Formerly synonymised with *Galeus melastomus*; possibly mistaken for *G. polli*, but apparently distinct.

LONGFIN SAWTAIL CATSHARK *Galeus cadenati*

FAO code: **GAU**   Plate page 374

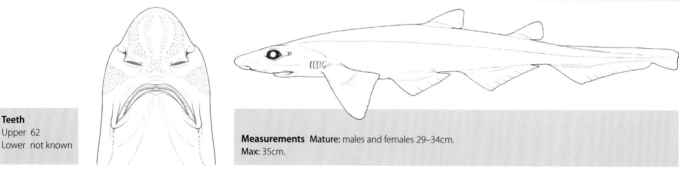

Teeth
Upper 62
Lower not known

Measurements Mature: males and females 29–34cm.
Max: 35cm.

Identification Similar to *Galeus arae* and *G. antillensis* and formerly considered to be a subspecies of *G. arae*. Has a noticeably longer, lower anal fin than either species.
Distribution Southwest Caribbean, off Panama and Colombia.
Habitat Upper continental slopes, 262–549m.
Behaviour Unknown.
Biology Oviparous. Eggcases 5–6 x 1–2cm, with tendrils extending from each corner. Females gravid in late spring and early summer.
Status IUCN Red List: Least Concern. Rare and little-studied.

CORRIGAN'S CATSHARK *Galeus corriganae*

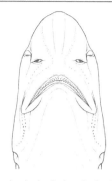
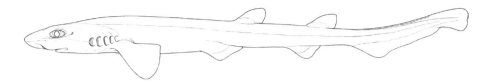

Teeth
Upper 60–69
Lower 54–64

Measurements Born: less than 20cm.
Max: males at least 37cm, females at least 28cm.

Identification Small, very slender-bodied catshark, with a long abdomen. Snout moderately long, rounded at tip. Mouth broadly arched, long; mouth cavity dark grey to blackish on roof and tongue; labial furrows about equal in length. Upper caudal fin margin with crest of 2–4 central rows of denticles between much larger lateral rows; no lower caudal fin margin crest. Coloration dark to medium grey above, lighter below; four distinct dark dorsal saddles, beneath each dorsal fin and on caudal fin; saddles less distinct after preservation. Anterior margin of pectoral fins distinctly black, posterior margin whitish; anterior half of dorsal fins almost entirely dark or dusky.
Distribution Central Indo-Pacific: Papua New Guinea, known only from off the Madang Province and New Britain Province.
Habitat Occurs from 500–742m.
Behaviour Unknown.
Biology Unknown, the only recorded specimens were all immature.
Status IUCN Red List: Least Concern. No deepsea fisheries known in the region.

GECKO CATSHARK *Galeus eastmani*

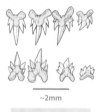

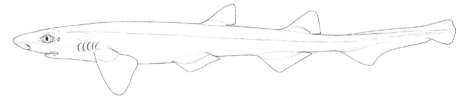

~2mm

Teeth
Upper 34–47
Lower 33–50

Measurements Born: less than 17cm. **Mature:** varies by region, males ~31–36cm, females 35–38cm.
Max: males 40cm, females 45cm, possibly to 50cm. Records to 68cm require confirmation.

Identification Characterised by a white mouth lining and white-edged dorsal and caudal fins, otherwise obscurely patterned, slender body with a fairly small anal fin. Distinct crest of enlarged dermal denticles along the upper caudal fin margin.
Distribution Northwest Pacific to Central Indo-Pacific: Japan, East China Sea, Taiwan, Vietnam, and possibly Malaysia.
Habitat Deepsea, on or near the bottom, 100–900m.
Behaviour Unknown.
Biology Oviparous. Eggs laid in pairs (one per oviduct). In Japanese waters, eggcases are laid between October and April, with a peak in December and January. Feeds mostly on small bony fishes, cephalopods and crustaceans.
Status IUCN Red List: Least Concern. Very common in Japanese waters, of no interest to fisheries.

SLENDER SAWTAIL CATSHARK   *Galeus gracilis*

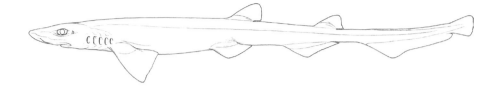

Teeth
Upper 54–57
Lower 54–62

Measurements Mature: males ~33cm, females 32–34cm.
Max: at least 34cm.

Identification Very small, slender catshark. Distinct crest of enlarged dermal denticles along the upper caudal fin margin. Pale grey with four short, dusky saddles (one beneath each dorsal fin and two on the caudal fin). No predorsal markings. Pale underside.
Distribution Central Indo-Pacific: northern Australia.
Habitat Continental slope, on or near bottom, 290–470m.
Behaviour Unknown.
Biology Unknown.
Status IUCN Red List: Data Deficient. Apparently rare. Probably taken as bycatch in trawl fisheries, but of no commercial interest. A similar but distinct species occurs off Indonesia. Records from Papua New Guinea are now known to be *Galeus corriganae*.

LONGNOSE SAWTAIL CATSHARK   *Galeus longirostris*

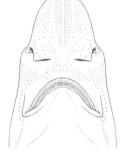

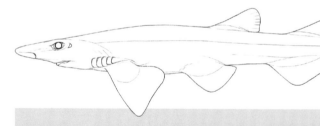

Teeth
Upper 60–70
Lower 60–70

Measurements Mature: males ~66–71cm, females 68–78cm or more.
Max: at least 80cm.

Identification Very long, broadly rounded snout. Greyish white mouth lining. Distinct crest of enlarged dermal denticles along the upper caudal fin margin. Uniform grey above in adults; a few obscure, dusky saddle blotches at dorsal fins and on caudal fin in young. White-edged dorsal and pectoral fins; dorsal and caudal fins without black tips. Whitish below.
Distribution Northwest Pacific: Japan (southern islands).
Habitat On or near bottom, upper insular slopes, 330–550m.
Behaviour Unknown.
Biology Unknown. No adult females yet recorded with eggcases or young.
Status IUCN Red List: Data Deficient. Fairly common where it occurs. Research required on its biology and population, given its limited distribution.

BLACKMOUTH CATSHARK *Galeus melastomus*

FAO code: **SHO**　　Plate page 372

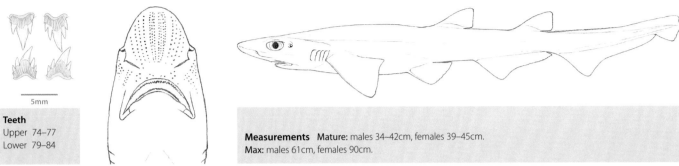

Teeth
Upper 74–77
Lower 79–84

5mm

Measurements Mature: males 34–42cm, females 39–45cm.
Max: males 61cm, females 90cm.

Identification Compressed precaudal tail. Long anal fin reaching or extending past lower caudal origin. Distinct crest of enlarged dermal denticles along the upper caudal fin margin. Striking pattern of 15–18 dark saddles, blotches and circular spots on back and tail. Mouth cavity blackish; white edges to fins.
Distribution Northeast Atlantic and Mediterranean Sea; Faeroes and Norway, south to Senegal and Azores.
Habitat Outer continental shelves and upper slopes, mainly 200–500m, occasionally to 55m and 2000m.
Behaviour Unknown.
Biology Oviparous, up to 13 eggs per female. Eggcases 4–6 x 1.5–3cm. Feeds on benthic invertebrates (e.g. shrimp, cephalopods) and lanternfish.
Status IUCN Red List: Least Concern. Common to abundant where it occurs. Taken as bycatch but low commercial value.

SOUTHERN SAWTAIL CATSHARK *Galeus mincaronei*

FAO code: **GAU**　　Plate page 372

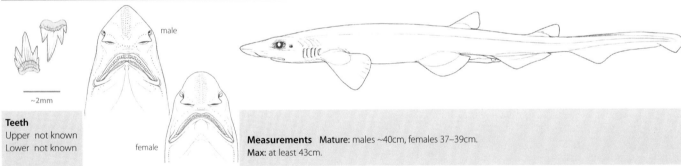

male

female

Teeth
Upper not known
Lower not known

~2mm

Measurements Mature: males ~40cm, females 37–39cm.
Max: at least 43cm.

Identification Long head. Precaudal tail slightly compressed. Moderately long anal fin nearly reaching past lower caudal origin. Distinct crest of enlarged dermal denticles along upper (not lower) caudal fin margin. Reddish brown above, whitish below. Striking pattern of 11 large oval or circular dark saddles and spots outlined in white on back and precaudal tail; fins dark but no black tips or white edges.
Distribution West Atlantic: southern Brazil endemic.
Habitat Deep-reef habitat with gorgonians, corals, sponges, crinoids and brittlestars on upper continental slope, 130–600m.
Behaviour Unknown.
Biology Oviparous, single egg in each oviduct, apparently laid in capsules. Eggcases 5–6 x 4cm, with long filamentous fibres anteriorly and coiled tendrils posteriorly. Often found with *Scyliorhinus haeckelii*.
Status IUCN Red List: Vulnerable. A very narrow range within a heavily fished area may have caused population declines.

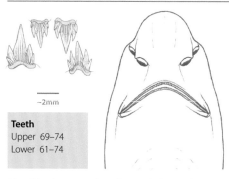

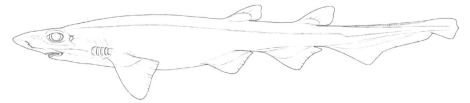

Teeth
Upper 69–74
Lower 61–74

Measurements Born ~8–9cm. **Mature:** males 50–63cm, females 53cm.
Max: males 63cm, females over 53cm.

Identification Large, broadly rounded pelvic fins. Precaudal tail cylindrical, not flattened. Distinct crests of enlarged dermal denticles along the upper caudal fin margin and along the ventral midline of the caudal peduncle. Uniform brown colour (slightly paler below); mouth blackish.
Distribution Northeast Atlantic: west coast of Iceland to Faroe-Shetland Channel, off Scotland, Ireland, France and Spain, and to Morocco and the Western Sahara. Its distribution overlaps other regional *Galeus* species.
Habitat Continental slopes, on or near bottom, 380–1300m.
Behaviour Unknown.
Biology Oviparous. Eggcases 5–6 x 1–2cm, small, slender, without tendrils at either end, but covered in fine fibres producing a rough texture.
Status IUCN Red List: Least Concern. This species small size allows it to escape capture from most fishing gear.

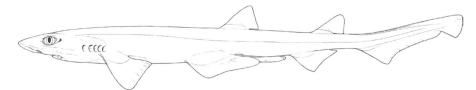

Teeth
Upper not known
Lower not known

Measurements Born 13cm. **Mature:** males 51–62cm, females 55–61cm.
Max: reported 76cm.

Identification Long snout (tip to nostril longer than eye-length); white mouth lining. Mature males have slender, greatly elongated claspers with bases covered by broad extensions of pelvic free rear tips (aprons). Long space between pelvic fin bases. Unusually short anal fin. Distinct crest of enlarged dermal denticles along upper (not lower) caudal fin margin. Greyish white above with numerous obscure dusky saddles and blotches, white below. Black anterior margins to dorsal and anal fins, lower caudal lobe and rear tip; no black fin tips.
Distribution Northwest Pacific: Common in deep water off Japan (southeast Honshu), also Kyushu-Palau Ridge and Taiwan.
Habitat Deep water on bottom, 362–540m.
Behaviour Unknown.
Biology Oviparous, lays pairs of eggs (one per oviduct). Eggcases 9 x 2cm, without tendrils, and with fine longitudinal striations. Feeds mostly on small bony fishes, cephalopods and crustaceans.
Status IUCN Red List: Data Deficient. Common off Taiwan. A similar, smaller species in Philippines and off China is possibly distinct.

PEPPERED CATSHARK *Galeus piperatus* FAO code: **GAP** Plate page 372

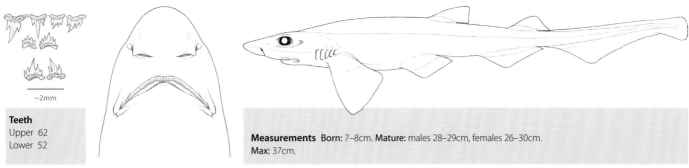

Teeth
Upper 62
Lower 52

~2mm

Measurements Born: 7–8cm. **Mature:** males 28–29cm, females 26–30cm.
Max: 37cm.

Identification Very similar to *Galeus arae* and *G. antillensis* from the west Atlantic. Small, with a distinct crest of enlarged dermal denticles along upper (not lower) caudal fin margin. Body covered with black pepper-like spots, may lack saddle markings or be patterned with variegated, white-edged dark saddle blotches. White-edged caudal fin; no black tips on dorsal or caudal fins. Mouth lining usually dark.
Distribution East Pacific: Mexico, northern Gulf of California; records from Peru need confirmation.
Habitat Bottom in deepsea, 130–1326m.
Behaviour Unknown.
Biology Oviparous.
Status IUCN Red List: Least Concern. Specimens from Peru should be carefully compared with those from the Gulf of California.

AFRICAN SAWTAIL CATSHARK *Galeus polli* FAO code: **GAQ** Plate page 372

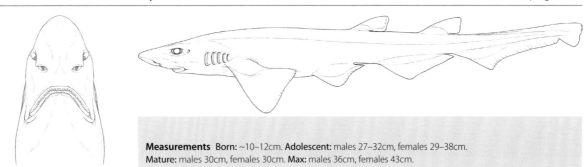

Teeth
Upper 69
Lower 65

Measurements Born: ~10–12cm. **Adolescent:** males 27–32cm, females 29–38cm.
Mature: males 30cm, females 30cm. **Max:** males 36cm, females 43cm.

Identification Fairly small. Distinct crest of enlarged dermal denticles along upper (not lower) caudal fin margin. Usually about 11 or less well-defined, dark grey or blackish grey saddle blotches outlined in whitish on back and tail, sometimes uniform dark above. No black dorsal and caudal fin tips; mouth lining dark.
Distribution East Atlantic: southern Morocco to South Africa (west coast).
Habitat Upper continental slope and outer continental shelf, 159–720m.
Behaviour Unknown.
Biology Viviparous, 5–13 pups per litter; litter size increases with size of the female. Unique in that it bears live young, while most other *Galeus* species lay eggcases. Eats small fishes, squid and shrimp.
Status IUCN Red List: Least Concern. Abundant off Namibia. Only the shallow part of its range is heavily fished, and this catshark is small enough to escape from trawls.

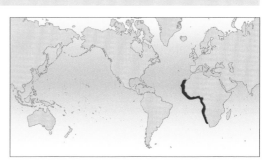

PHALLIC CATSHARK *Galeus priapus*

FAO code: **GAU**   Plate page 374

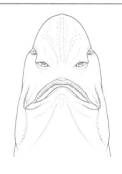

Teeth
Upper 57–60
Lower 60

Measurements Mature: males ~38cm.
Max: 46cm.

Identification A small, slender sawtail catshark. Adult males with extremely long claspers. Distinct crest of enlarged dermal denticles along the upper caudal fin margin. Colour is dark grey with four predorsal saddles, one under each dorsal fin and two saddles on the caudal fin. Ventral surface white.
Distribution Central Indo-Pacific: New Caledonia.
Habitat Mainly on insular slopes, 262–830m.
Behaviour Unknown.
Biology Unknown.
Status IUCN Red List: Least Concern. It is known only from a very few specimens, but range is largely unfished.

BLACKTIP SAWTAIL CATSHARK *Galeus sauteri*

FAO code: **GAI**   Plate page 372

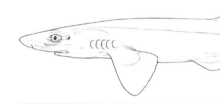

~2mm

Teeth
Upper 70–78
Lower 82

Measurements Mature: males 36–38cm, females 42–45cm.
Max: 50cm.

Identification Caudal peduncle compressed. Distinct crest of enlarged denticles along upper (not lower) caudal fin margin. No pattern of dark saddle blotches; dorsal fins and sometimes upper and lower caudal lobes with prominent black tips; light mouth lining.
Distribution Central Indo-Pacific to northwest Pacific: Taiwan, Philippines.
Habitat Continental shelves, offshore at 60–200m in the Taiwan Strait, possibly deeper elsewhere. Found on soft, muddy bottoms.
Behaviour Unknown.
Biology Oviparous. Eggcases 4 x 1–2cm, smooth and with long tendrils. There does not appear to be a defined breeding season. Feeds on small bony fishes, cephalopods and crustaceans.
Status IUCN Red List: Least Concern. Caught in bottom trawl fisheries, but usually discarded. No evidence of population decline.

DWARF SAWTAIL CATSHARK   *Galeus schultzi*   FAO code: **GAH**   Plate page 374

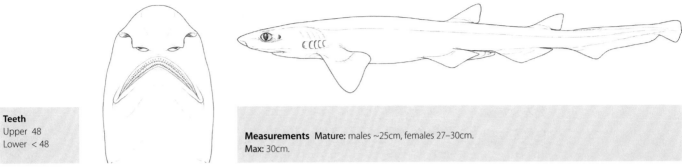

Teeth
Upper 48
Lower < 48

Measurements Mature: males ~25cm, females 27–30cm.
Max: 30cm.

Identification One of the smallest catsharks. Unusually short and rounded snout for a sawtail catshark. Mouth lining light or dusky; very short labial furrows (confined to mouth corners). Distinct crest of enlarged dermal denticles along the upper (not lower) caudal fin margin. Pattern of obscure dark saddle blotches below dorsal fins and two bands on the tail. No black tips to dorsal and caudal fins.
Distribution Central Indo-Pacific: Philippines (off Luzon).
Habitat Mainly on insular slopes, 329–431m. One record on outer shelf, 50m.
Behaviour Unknown.
Biology Poorly known.
Status IUCN Red List: Least Concern. No information on deepsea fisheries in this region, but bycatch is not considered to be a threat.

SPRINGER'S SAWTAIL CATSHARK   *Galeus springeri*   FAO code: **GAU**   Plate page 374

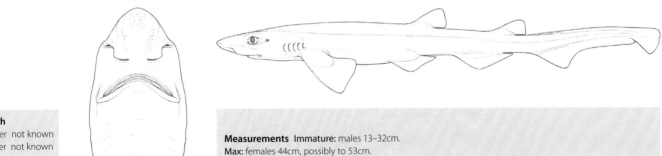

Teeth
Upper not known
Lower not known

Measurements Immature: males 13–32cm.
Max: females 44cm, possibly to 53cm.

Identification Only sawtail catshark with a predorsal pattern of dark longitudinal stripes outlined with white on a dark background. Dark saddles on the tail. Underside white. Distinct crests of enlarged dermal denticles along the upper and lower caudal fin margins.
Distribution West Atlantic: Caribbean Sea; Cuba (north coast), Bahamas, Puerto Rico, Leeward Islands.
Habitat Upper insular slopes, 457–824m.
Behaviour Unknown.
Biology Possibly oviparous.
Status IUCN Red List: Least Concern. A common species that occurs below the depth range of current fishing activities.

Plate 47 PENTANCHIDAE X – *Halaelurus* catsharks

○ **Speckled Catshark** *Halaelurus boesemani*

page 388

West Indian Ocean; 9–157m. Pointed snout, eyes raised above head, gills on upper surface of head; 8 dark saddles separated by narrow bars, dark blotches on dorsal and caudal fins, broad pale edges to pectoral fins; numerous scattered small spots on yellow-brown body, dorsal and caudal fins.

○ **Blackspotted Catshark** *Halaelurus buergeri*

page 389

Central Indo-Pacific and northwest Pacific; 27–100m. Pointed snout, eyes raised above head, gills on upper surface of head; variegated pattern of dusky bands outlined by large black spots on light background.

○ **Lined Catshark** *Halaelurus lineatus*

page 389

West Indian Ocean; 0–290m. Narrow head with upturned knob on snout, eyes raised above head, gills on upper surface of head; about 13 pairs of narrow dark brown stripes outlining dusky saddles and numerous small spots and squiggles over pale brown body, cream below.

○ **Indonesian Speckled Catshark** *Halaelurus maculosus*

page 390

Central Indo-Pacific; 50–80m. Narrowly pointed snout not upturned or knob-like; pale brown with 10 dark brown saddles, 4 predorsal, covered with darker brown spots.

○ **Tiger Catshark** *Halaelurus natalensis*

page 390

Southeast Atlantic and west Indian Ocean; 0–172m. Broad head with pointed upturned snout tip, eyes raised above head, gills on upper surface of head; 10 pairs broad dark brown bars around lighter reddish saddles, no spots.

○ **Quagga Catshark** *Halaelurus quagga*

page 391

North Indian Ocean; 54–300m. Pointed snout, eyes raised above head, gills on upper surface of head; more than 20 dark brown narrow stripes, pairs of bars form saddles under dorsal fins, no spots.

○ **Rusty Catshark** *Halaelurus sellus*

page 391

Central Indo-Pacific; 62–164m. Yellowish brown background, 10 rusty brown saddles outlined by narrow dark brown lines and narrower saddles between them.

Halaelurus boesemani

Halaelurus buergeri

Halaelurus lineatus

Halaelurus maculosus

Halaelurus natalensis

Halaelurus quagga

Halaelurus sellus

20cm

Blackspotted Catshark, *Halaelurus buergeri*, Okinawa Churaumi Aquarium (p. 389).

SPECKLED CATSHARK *Halaelurus boesemani*

FAO code: **HAB**   Plate page 386

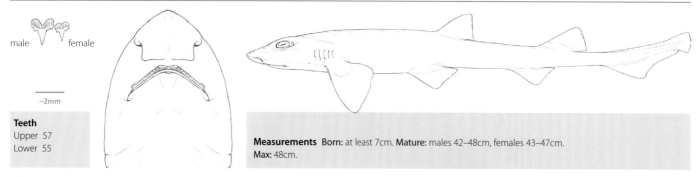

male   female

~2mm

Teeth
Upper 57
Lower 55

Measurements Born: at least 7cm. **Mature:** males 42–48cm, females 43–47cm.
Max: 48cm.

Identification Snout pointed, not upturned. Eyes raised above head. Gills on upper surface of head above mouth. About eight dark saddles separated by narrower bars on yellow-brown back and tail. Dark blotches on dorsal and caudal fins; numerous small, scattered dark spots on body, dorsal and caudal fins. Broad pale bands edge pectoral fins. Pale underside, pelvic and anal fins.
Distribution West Indian Ocean: Somalia, Gulf of Aden and Kenya.
Habitat Continental and insular shelves, 29–157m.
Behaviour Unknown.
Biology Oviparous. Eggcases small, with short tendrils. Up to eight eggcases per oviduct, with a single eggcase in each uterus. Feeds on small bony fishes.
Status IUCN Red List: **Vulnerable**. Only known from a few specimens from an unregulated fishing ground that has been trawled for at least 40 years.

BLACKSPOTTED CATSHARK *Halaelurus buergeri*

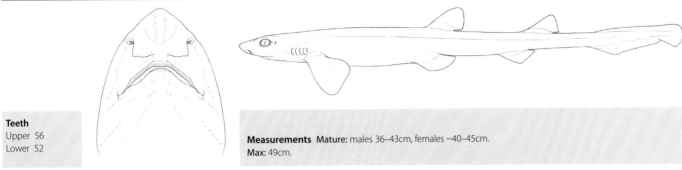

Teeth
Upper 56
Lower 52

Measurements Mature: males 36–43cm, females ~40–45cm.
Max: 49cm.

Alternative name Darkspot Catshark
Identification Snout pointed, not upturned. Eyes raised above head. Gills on upper surface of head above mouth. Variegated pattern with dusky bands outlined by large black spots on a light background.
Distribution Central Indo-Pacific and northwest Pacific: Japan, Korea, Philippines, China, Taiwan, and off northwest Borneo in the South China Sea.
Habitat Continental shelf, 27–100m.
Behaviour Unknown.
Biology Oviparous. Several eggcases are retained in the oviduct and not laid until embryos are advanced and close to hatching.
Status IUCN Red List: Endangered. This species has declined due to intense fishing activity across its entire range.

LINED CATSHARK *Halaelurus lineatus*

female

~2mm

Teeth
Upper 51–60
Lower 47–55

Measurements Born: ~8cm. Mature: males 48–56cm, females 46–52cm.
Max: 56cm.

Identification Narrow head; upturned knob on snout. Eyes raised above head. Gills on upper surface of head above mouth. Pale brown above with about 13 pairs of narrow dark brown stripes outlining dusky saddles; many small spots and squiggles. Cream below.
Distribution West Indian Ocean: South Africa (from East London) to Mozambique.
Habitat Continental shelf, surf line to 290m on rocky and soft bottoms.
Behaviour May segregate by depth or region; most captured off KwaZulu-Natal are pregnant females, adult males and young only rarely caught.
Biology Oviparous. Up to 8 eggcases per oviduct retained until embryos close to hatching (eggs hatch in 23–26 days in captivity). Pregnant females are observed off the KwaZulu-Natal coast in late winter. Eats mostly crustaceans, also bony fishes and cephalopods.
Status IUCN Red List: Least Concern. Common in prawn trawler bycatch. It may have a refuge from fishing pressure over untrawlable rocky reefs.

INDONESIAN SPECKLED CATSHARK   *Halaelurus maculosus*

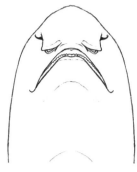
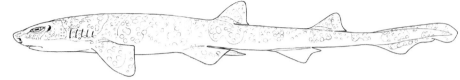

Teeth
Upper 56
Lower 53

Measurements Born: ~8cm. **Adult:** males 40cm, females 48–53cm.
Max: 53cm.

Identification Head with short, narrowly pointed snout, not upturned or knob-like. Pale brown above with about 10 darker brown saddles (four are predorsal); densely covered with darker brown spots, but saddle margins not demarcated by darker lines. Pale yellow to white below.
Distribution Central Indo-Pacific: eastern Indonesia, possibly the Philippines.
Habitat Coastal, possibly associated with shallow reefs, 50–80m.
Behaviour Unknown.
Biology Oviparous, with 6–12 eggcases in uterus. Eats mostly crustaceans, also bony fishes and cephalopods.
Status IUCN Red List: Near Threatened. An occasional retained bycatch in intensive trawl and longline fisheries.

TIGER CATSHARK   *Halaelurus natalensis*

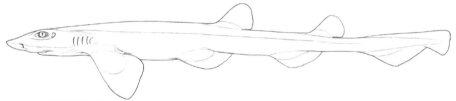

Teeth
Upper 54
Lower 41

~2mm

Measurements Mature: males 35–45cm, females 37–50cm.
Max: females 50cm.

Identification Pointed, upturned snout tip. Eyes raised above broad head. Gills on upper surface of head above mouth. Yellow-brown above with ten pairs of broad, dark brown bars enclosing lighter reddish areas; no spots. Cream below.
Distribution Southeast Atlantic and west Indian Ocean: off Cape Town, South Africa and eastwards to southern Mozambique. Often misidentified with *Halaelurus lineatus*.
Habitat On or near bottom, continental shelf, close inshore to 114m, possibly to 172m, but most records less than 100m.
Behaviour May segregate by size and depth. Offshore trawls mainly take adults.
Biology Oviparous. Eggcases ~4 x 1.5cm. 6–16 (mostly 6–9) eggcases per oviduct, laid when embryos close to hatching. Diet includes small bony fishes, fish offal and crustaceans, also polychaetes, cephalopods and small elasmobranchs.
Status IUCN Red List: Vulnerable. Common, bycaught in prawn trawls and occasionally by shore anglers.

QUAGGA CATSHARK *Halaelurus quagga*

FAO code: **HAQ**   Plate page 386

Teeth
Upper 55–59
Lower 54–55

Measurements Born: ~8cm. **Mature:** males 28–35cm.
Max: 37cm.

Identification Snout pointed but not upturned. Eyes raised above head. Gills on upper surface of head above mouth. Light brown above with more than 20 dark brown, narrow, vertical bars; pairs of bars forming saddles only under dorsal fins. No spots. Lighter below.
Distribution North Indian Ocean: Somalia, India.
Habitat Offshore on continental shelf, on or near bottom, 54–300m.
Behaviour Unknown.
Biology Oviparous. Up to 8 eggcases reported in gravid females.
Status IUCN Red List: Data Deficient. Not important in fisheries.

RUSTY CATSHARK *Halaelurus sellus*

FAO code: **CVX**   Plate page 386

~2mm

Teeth
Upper 55
Lower 46

Measurements Mature: males 34cm, females 38cm.
Max: 42cm.

Identification Snout short, pointed, not upturned or knob-like. Yellowish brown above with ten rusty brown saddles outlined by narrow, dark brown lines, and with narrower saddles between them.
Distribution Central Indo-Pacific: northwest Australia.
Habitat Offshore on continental shelf, on or near bottom, 62–164m.
Behaviour Unknown.
Biology Multiple oviparity with at least 3 eggcases per uterus.
Status IUCN Red List: Least Concern. A small species occasionally caught in bottom trawls.

PUFFADDER SHYSHARK *Haploblepharus edwardsii*

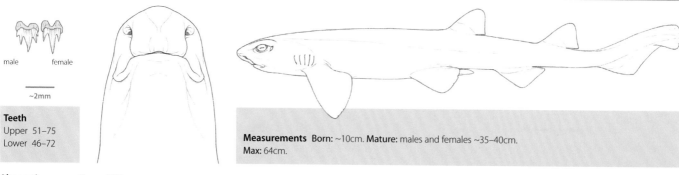

male   female

~2mm

Teeth
Upper 51–75
Lower 46–72

Measurements Born: ~10cm. **Mature:** males and females ~35–40cm.
Max: 64cm.

Alternative name Happy Eddie

Identification Slender body. Stocky, broad head. Very large nostrils; greatly expanded anterior nasal flaps reach mouth. Gill slits on upper sides of body. Pale to dark brown or grey-brown above. Covered with prominent golden-brown or reddish saddles with darker brown margins; many white spots on saddles or between them, mostly smaller or same size as spiracles. White below.

Distribution Southeast Atlantic and southwest Indian Ocean: South Africa (eastern and western Cape).

Habitat Continental shelf, on or near sandy and rocky bottom 0–288m, but mostly 0–90m, closer inshore in west.

Behaviour Apparently social, resting in groups in captivity. Curls up with tail over eyes when captured.

Biology Oviparous. Eggcases ~3.5–5 x 1.5–3cm, laid in pairs (one per oviduct). Eats small bony fishes, fish offal, crustaceans, cephalopods and polychaetes.

Status IUCN Red List: Endangered. Caught by surf anglers, discarded from bottom trawls. Kept in aquaria.

BROWN SHYSHARK *Haploblepharus fuscus*

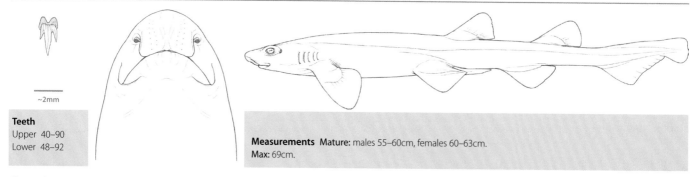

~2mm

Teeth
Upper 40–90
Lower 48–92

Measurements Mature: males 55–60cm, females 60–63cm.
Max: 69cm.

Alternative name Plain Happy

Identification Stocky body in large individuals. Stocky, broad head. Very large nostrils; greatly expanded anterior nasal flaps reach mouth. Gill slits on upper sides of body. Brown above, sometimes with slightly darker, obscure saddles or small white or black spots. White below.

Distribution South Atlantic and west Indian Ocean: South Africa (less than 1000km of coast).

Habitat Inshore on continental shelf, often in shallow, sandy areas, rocky bottom, 0–35m.

Behaviour Curls up with tail over eyes when captured.

Biology Oviparous (lays pairs of eggs). Eats lobsters and bony fishes.

Status IUCN Red List: Vulnerable. Very limited geographic range. It is caught by shore anglers and is exposed to habitat degradation from coastal developments.

NATAL SHYSHARK *Haploblepharus kistnasamyi*

FAO code: **CVX** Plate page 394

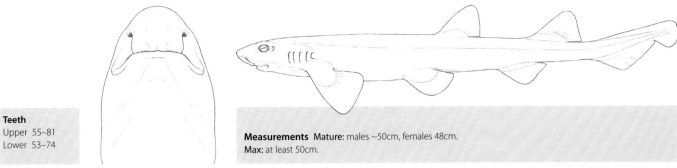

Teeth
Upper 55–81
Lower 53–74

Measurements Mature: males ~50cm, females 48cm.
Max: at least 50cm.

Alternative names Eastern Shyshark or Happy Chappie
Identification Slender body. Stocky, broad, flat head. Very large nostrils; greatly expanded anterior nasal flaps reach mouth. Gill slits on upper sides of body. Body and tail with dark brown, H-shaped dorsal saddle markings; margins conspicuous, dark and dotted with numerous small white spots. Interspaces between saddles and fins with dark mottling on brown background. Underside white.
Distribution Southwest Indian Ocean: South Africa (western Cape to eastern Cape and KwaZulu-Natal).
Habitat Close inshore to 30m on continental shelf, sometimes in surf zone or on rocky reefs.
Behaviour Unknown.
Biology Oviparous. Eggcases unknown.
Status IUCN Red List: **Vulnerable**. Formerly considered identical to *Haploblepharus edwardsii*. Known from only a few specimens.

DARK SHYSHARK *Haploblepharus pictus*

FAO code: **HPP** Plate page 394

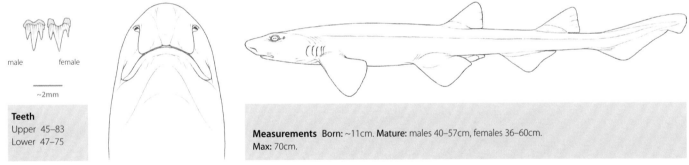

male female

~2mm

Teeth
Upper 45–83
Lower 47–75

Measurements Born: ~11cm. **Mature:** males 40–57cm, females 36–60cm.
Max: 70cm.

Alternative name Pretty Happy
Identification Stocky body. Broad head. Very large nostrils; greatly expanded anterior nasal flaps reach mouth. Gill slits on upper sides of body. Dorsal saddle markings without obvious darker edges, sparsely dotted with large white spots, mostly larger than spiracles and absent between saddles.
Distribution Southeast Atlantic and southwest Indian Ocean: central Namibia to South Africa (East London).
Habitat Continental shelf, kelp forests, rocky inshore reefs and sandy areas, close inshore to ~35m.
Behaviour Curls up with tail over eyes when captured.
Biology Oviparous. Eggcases ~6 x 3cm, with smooth case and long, coiled tendrils at anterior end and short, coiled tendrils on posterior end. One egg laid per oviduct, hatch in ~3.5 months in an aquarium. Eats bony fishes, sea snails, cephalopods, crustaceans, polychaetes and echinoderms, occasionally algae.
Status IUCN Red List: **Least Concern**. Common within limited range. Killed by sports anglers, caught in lobster traps.

Plate 48 PENTANCHIDAE XI – shysharks and Izak catsharks

○ **Puffadder Shyshark** *Haploblepharus edwardsii*

Southeast Atlantic and southwest Indian Ocean; 0–288m. Slender; stocky broad head, very large nostrils with greatly expanded nasal flaps reaching mouth, gills on upper body; prominent golden brown or reddish saddles with darker brown margins, numerous white spots on and between saddles, white below.

○ **Brown Shyshark** *Haploblepharus fuscus*

Southeast Atlantic and southwest Indian Ocean; 0–35m. Stocky; stocky broad head, very large nostrils with greatly expanded nasal flaps reaching mouth, gills on upper body; brown above, sometimes with obscure saddles or small white or black spots, white below.

○ **Natal Shyshark** *Haploblepharus kistnasamyi*

Southwest Indian Ocean; 0–30m. Slender; stocky broad head, very large nostrils with greatly expanded nasal flaps, gills on upper body; body and tail with H-shaped saddles with conspicuous dark margins, numerous small white spots, body and fins with dark mottling, white below.

○ **Dark Shyshark** *Haploblepharus pictus*

Southeast Atlantic and Southwest Indian Ocean; 0–35m. Stocky; broad head, very large nostrils with greatly expanded nasal flaps, gills on upper body; dorsal saddles without obvious darker edges, sparsely dotted with large white spots absent between saddles, white below.

○ **Honeycomb Izak Catshark** *Holohalaelurus favus*

West Indian Ocean; 200–1000m. Broad head; brown body with very distinctive honeycomb pattern of irregular reticulations and spots, few white spots above pectoral fins, ventral surface uniformly grey-brown.

○ **Grinning Izak** *Holohalaelurus grennian*

West Indian Ocean; 238–353m. Head broad with scattered tiny black dots; back with large dark brown spots some fused to form reticulations, blotches and stripes; 'tear'-marks on snout in front of eyes.

○ **Crying Izak Catshark** *Holohalaelurus melanostigma*

West Indian Ocean; 607–658m. Short snout, long mouth, broad head; short angular dorsal fins; slender tail; numerous large dark brown spots, some fused into reticulations blotches and stripes, horizontal 'tear'-marks on snout, tiny black dots beneath head, dark lines and C-shaped mark on bases and webs of dorsal fins.

○ **African Spotted Catshark** *Holohalaelurus punctatus*

West Indian Ocean; 220–420m. Broad head, short snout, long mouth; short angular dorsal fins; slender tail; dense small dark brown spots, a few white spots on back and dorsal fin inserts, sometimes faint saddles; dark C or V-shaped marks on dorsal fin webs, whitish below with tiny black dots underneath head.

○ **Izak Catshark** *Holohalaelurus regani*

Southeast Atlantic and southwest Indian Ocean; 15–1075m. Short snout with long mouth; dorsal fins short and angular; slender tail; dorsal surface covered with dark brown reticulations, bars and blotches over yellowish background; young dark and slender, line of white spots, black bars on tiny fins and very long tail.

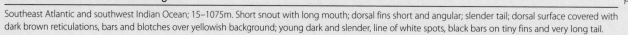

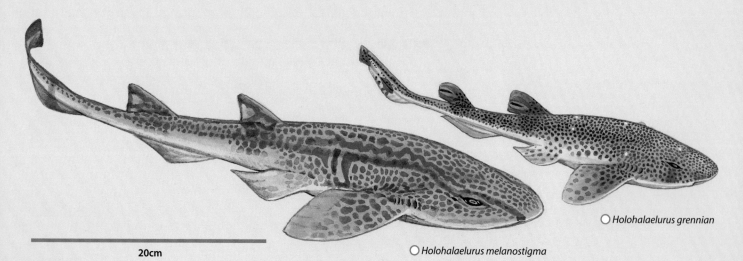

20cm

○ *Holohalaelurus grennian*

○ *Holohalaelurus melanostigma*

○ *Haploblepharus edwardsii*

○ *Haploblepharus fuscus*

○ *Haploblepharus kistnasamyi*

○ *Haploblepharus pictus*

○ *Holohalaelurus favus*

○ *Holohalaelurus punctatus*

○ *Holohalaelurus regani*

20cm

Puffadder Shyshark, *Haploblepharus edwardsii,* False Bay, Western Cape, South Africa (p. 392).

HONEYCOMB IZAK CATSHARK   *Holohalaelurus favus*

FAO code: **CVX**   Plate page 394

male   female

~2mm

Teeth
Upper 65–68
Lower 53–70

Measurements Mature: males ~52cm, females 42cm.
Max: at least 52cm.

Alternative name East African Spotted Izak Catshark, Natal Izak Catshark
Identification A large species of *Holohalaelurus* with a broad head, brown background coloration and a very distinctive honeycomb pattern of irregular reticulations and spots; no white spots above pectoral fins; ventral surface uniformly grey-brown.
Distribution West Indian Ocean: KwaZulu-Natal, South Africa, to southern Mozambique. May have a very limited distribution of only 5 degrees latitude.
Habitat Upper continental slope, 200–1000m. Juveniles may occur deeper than adults.
Behaviour Essentially unknown.
Biology Oviparous.
Status IUCN Red List: Endangered. Historically a common bycatch species in bottom trawls, but very few confirmed records have been reported since the early 1970s.

GRINNING IZAK *Holohalaelurus grennian*

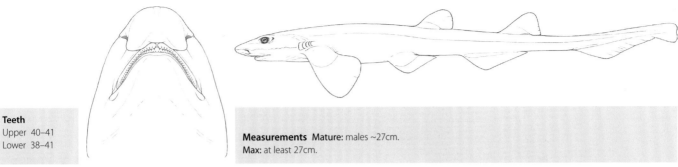

Teeth
Upper 40–41
Lower 38–41

Measurements Mature: males ~27cm.
Max: at least 27cm.

Alternative name East African Spotted Izak
Identification Head very broad; snout short; mouth long. Dorsal fins short and angular. Tail slender. Slightly enlarged, rough denticles on middle of back. Grey-brown above, lighter below. Upper surface with many large, dark brown spots, some fused to form reticulations, blotches and horizontal stripes. Horizontal 'tear'-marks on snout in front of eyes; scattered tiny black dots beneath head; dark lines and dark C-shaped mark on bases and webs of dorsal fins.
Distribution West Indian Ocean: Somalia, Kenya, Tanzania to southern Mozambique and west Madagascar. Presence in South Africa requires confirmation.
Habitat Continental slope, 238–353m.
Behaviour Unknown.
Biology Oviparous.
Status IUCN Red List: Data Deficient. Extremely rare, known only from a few specimens.

CRYING IZAK CATSHARK *Holohalaelurus melanostigma*

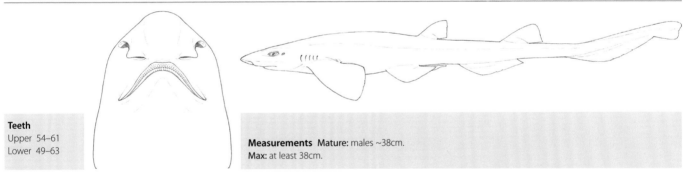

Teeth
Upper 54–61
Lower 49–63

Measurements Mature: males ~38cm.
Max: at least 38cm.

Alternative name Tropical Izak Catshark
Identification Head very broad; snout short; mouth long. Dorsal fins short and angular. Tail slender. Slightly enlarged, rough denticles on middle of back. Grey-brown above, lighter below. Upper surface with many large dark brown spots, some fused to form reticulations, blotches and horizontal stripes. Horizontal 'tear'-marks on snout in front of eyes; scattered tiny black dots beneath head; dark lines and dark C-shaped mark on bases and webs of dorsal fins.
Distribution West Indian Ocean: endemic, known only from north Tanzania, near Pemba Island and south Kenya.
Habitat Continental slope, 607–658m.
Behaviour Unknown.
Biology Oviparous.
Status IUCN Red List: Least Concern. No deepsea fisheries known to operate within its range. Extremely rare, known only from four museum specimens.

AFRICAN SPOTTED CATSHARK   *Holohalaelurus punctatus*

FAO code: **HOP**   Plate page 394

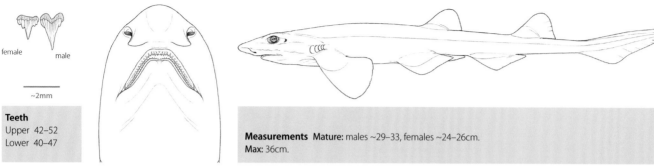

female   male

~2mm

Teeth
Upper 42–52
Lower 40–47

Measurements Mature: males ~29–33, females ~24–26cm.
Max: 36cm.

Identification Head very broad; snout short; mouth long. Short angular dorsal fins. Slender tail. No enlarged, rough denticles on middle of back. Yellow-brown to dark brown upper surface densely covered with small, close-spaced, dark brown spots; faint saddles sometimes present but no reticulations, blotches, horizontal stripes or 'tear'-marks. White spots few and scattered on back and dorsal fin insertions. Highlighted C or V-shaped dark mark often on dorsal fin webs. Whitish below; scattered tiny black dots beneath head.
Distribution Southwest Indian Ocean: South Africa, Mozambique and Madagascar. This is the only *Holohalaelurus* known to occur across the Mozambique Channel.
Habitat Continental shelf and upper slope, 220–420m.
Behaviour Partial sexual segregation (males were more numerous than females off KwaZulu-Natal, in equal numbers off Mozambique).
Biology Oviparous. Eggs laid in pairs. Eats small bony fishes, crustaceans and cephalopods.
Status IUCN Red List: Endangered. Vanished? Formerly common in trawl fisheries and surveys, but rarely seen since the 1970s, and not caught in research trawls for at least 20 years. Entire range heavily fished.

IZAK CATSHARK   *Holohalaelurus regani*

FAO code: **HOR**   Plate page 394

~2mm

Teeth
Upper.–55–72
Lower 27–78

Measurements Born: less than 11cm. **Mature:** males 41cm, females 32cm.
Max: males 69cm, females 62cm.

Identification Very broad head; snout short; mouth long. Dorsal fins short and angular. Tail slender. Enlarged, rough denticles on middle of back. Yellowish to yellow-brown above; upper surface covered with dark brown reticulations, bars and blotches, more spotted in young. No horizontal stripes, white spots, 'tear'-marks or highlighted dark marks on dorsal fins. White below; scattered tiny black dots beneath head. Young dark and slender, with a line of white spots on sides, black bars on tiny fins and very long tail.
Distribution Southeast Atlantic and southwest Indian Ocean: Namibia to KwaZulu-Natal, South Africa; its distribution on the east coast of South Africa is somewhat confused, due to misidentification with other *Holohalaelurus* species.
Habitat Continental shelf and upper slopes from 15–1075m, mainly about 100–300m.
Behaviour At least part of the population migrates inshore in autumn.
Biology Oviparous. Eggcases 3–4 x 1–2cm, with longitudinal striations, a velvety to the touch feel, and long tendrils at each corner. Pairs of eggs laid year-round. Eats small bony fishes, crustaceans, cephalopods, polychaetes, hydrozoans, occasionally kelp.
Status IUCN Red List: Least Concern. Discarded bycatch.

WHITETIP CATSHARK *Parmaturus albimarginatus*

FAO code: **CVX**   Plate page 400

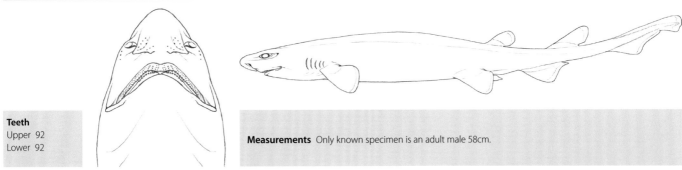

Teeth
Upper 92
Lower 92

Measurements Only known specimen is an adult male 58cm.

Identification A slender, soft-bodied velvety-skinned *Parmaturus* with greatly enlarged denticles on upper and lower lobes of caudal fin. Colour is a uniform brown above, light below, with prominent white-edged fins.
Distribution Central Indo-Pacific: known only from New Caledonia.
Habitat Insular slope, on or near bottom at 590–732m.
Behaviour Unknown.
Biology Unknown.
Status IUCN Red List: Least Concern. Known only from the holotype. Recorded from an unfished area.

WHITE-CLASPER CATSHARK *Parmaturus albipenis*

FAO code: **CVX**   Plate page 400

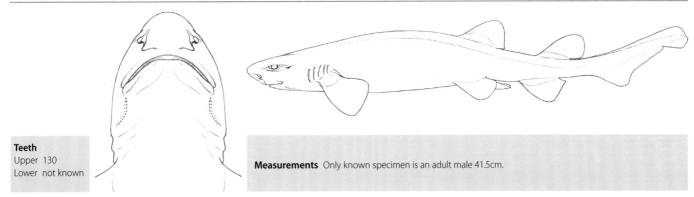

Teeth
Upper 130
Lower not known

Measurements Only known specimen is an adult male 41.5cm.

Identification A slender, soft-bodied velvety-skinned *Parmaturus* with caudal crests present, but not greatly enlarged denticles; crests extending nearly to second dorsal fin. Colour is a uniform brown above, lighter below; white pelvic fins sharply contrast with rest of body colour.
Distribution Central Indo-Pacific: known only from northern New Caledonia.
Habitat Insular slope, on or near bottom at 688–732m.
Behaviour Unknown.
Biology Unknown.
Status IUCN Red List: Least Concern. Known only from the holotype; range not exposed to fisheries.

Plate 49 PENTANCHIDAE XII – filetail catsharks

○ **Whitetip Catshark** *Parmaturus albimarginatus*   page 399

Central Indo-Pacific; 590–732m. Velvety skin with greatly enlarged denticles on upper and lower caudal fin lobes; uniform brown above with prominent white fin edges.

○ **White-clasper Catshark** *Parmaturus albipenis*   page 399

Central Indo-Pacific; 688–732m. Soft velvety skin with caudal crests present, but denticles not greatly enlarged, denticles extending nearly to second dorsal fin; white pelvic fins contrast with brown body.

○ **Brazilian Filetail Catshark** *Parmaturus angelae*   page 402

Southwest Atlantic; 500-600m. Uniform light brown or dark beige, no saddles, except a few small brown spots scattered laterally over body, mouth cavity light to pale greyish yellow.

○ **Beige Catshark** *Parmaturus bigus*   page 402

Central Indo-Pacific; 590–606m. Caudal fin lobes with crests of enlarged denticles; colour uniform pale yellowish brown above, light below.

○ **Campeche Catshark** *Parmaturus campechiensis*   page 403

West Atlantic; 1097m. First dorsal fin slightly smaller than second, first dorsal in front of pelvic fin origin, second dorsal fin same size as anal fin; soft greyish flabby body; dusky on abdomen, around gills and on fin webs.

○ **Velvet Catshark** *Parmaturus lanatus*   page 403

Central Indo-Pacific; 840–855m. Tadpole-shaped with dorsal and ventral crests of enlarged denticles on caudal fin margins; uniform brown velvety textured skin with darker gills and fin margins.

○ **New Zealand Filetail** *Parmaturus macmillani*   page 404

Southwest Pacific; 950–1003m or deeper. First dorsal fin about same size as second, first dorsal origin opposite or just behind pelvic fin origin, second dorsal fin smaller than anal fin; soft flabby grey body, lighter below, fins with dusky webs.

○ **Blackgill Catshark** *Parmaturus melanobranchus*   page 404

Northwest Pacific; 448–1110m. First dorsal fin smaller than second, first dorsal origin well behind pelvic fin origin, second dorsal fin about the same size as anal fin; soft flabby grey to dark brown body, lighter below, blackish gill septa.

○ **Indonesian Filetail Catshark** *Parmaturus nigripalatum*   page 405

Central Indo-Pacific; 170–190m. Slender, flabby, plain brown above, paler below; dorsal pectoral fin surfaces darker than ventral surfaces, all fin edges much darker, claspers and pelvic fins whitish at insertions; mouth roof blackish with darker pores.

○ **Salamander Catshark** *Parmaturus pilosus*   page 405

Northwest Pacific; 358–1177m. First dorsal fin about same size as second, first dorsal origin approximately opposite pelvic fin origin, second dorsal fin much smaller than anal fin; soft flabby body, reddish above, white below, fin webs darker.

○ **Filetail Catshark** *Parmaturus xaniurus*   page 406

Northeast Pacific; 88–1519m. Enlarged gill slits; first dorsal fin same size as second, first dorsal origin just behind pelvic fin origin, second dorsal fin much smaller than anal fin; soft flabby plain dark body, brownish black above, lighter below, fins dark.

○ **Onefin Catshark** *Pentanchus profundicolus*   page 406

Central Indo-Pacific; 673–1070m. Broad round snout with short mouth, short gills with incised gill septa; only species with 5 gills and 1 dorsal fin, very long shallow anal fin; uniformly brownish colour.

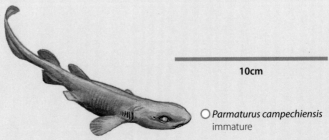

10cm

○ *Parmaturus campechiensis*
immature

20cm

○ *Parmaturus albimarginatus*

○ *Parmaturus angelae*

○ *Parmaturus lanatus* juvenile

○ *Parmaturus macmillani*

○ *Parmaturus nigripalatum*

○ *Parmaturus xaniurus*

○ *Parmaturus albipenis*

○ *Parmaturus bigus*

○ *Parmaturus melanobranchus*

○ *Parmaturus pilosus*

○ *Pentanchus profundicolus*

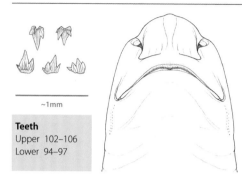

~1mm

Teeth
Upper 102–106
Lower 94–97

Measurements Mature: females 40–42.5cm.
Max: at least 42.5cm.

Identification Body stout but tapering quickly from first dorsal fin to caudal fin. Snout short. Mouth cavity is light to pale greyish yellow; labial furrows poorly developed, lowers longer than uppers. First dorsal fin similar in size and shape to second; first dorsal fin origin slightly anterior to pelvic fin origin; second dorsal fin smaller than anal fin. Caudal crest of enlarged dermal denticles present. Coloration is uniform light brown or dark beige; no saddles, but a few small brown spots scattered laterally over body.
Distribution Southwest Atlantic Ocean: known only from two locations off southern Brazil.
Habitat Continental slope from 500–600m.
Behaviour Unknown.
Biology Oviparous. Eggcases 7.3 x 2.4cm, small, vase-shaped, with very fine striations giving it a smooth texture to the touch; horns at each end short and without tendrils. Feeds on small bony fishes, crustaceans and polychaete worms.
Status IUCN Red List: **Vulnerable**. There are intense, unmanaged, deepsea demersal trawl fisheries throughout this species' range.

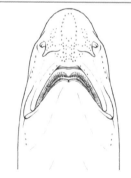

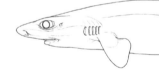

~2mm

Teeth
Upper 120
Lower not known

Measurements Only known specimen is an adult female, 71cm.

Identification A slender, soft-bodied *Parmaturus* with dorsal and ventral crests of enlarged denticles on caudal fin. Colour is a uniform pale yellowish brown above, light below.
Distribution Central Indo-Pacific: known only from northeast Australia.
Habitat Continental slope, on or near bottom at 590–606m.
Behaviour Unknown.
Biology Unknown.
Status IUCN Red List: **Data Deficient**. Known only from the holotype.

CAMPECHE CATSHARK   *Parmaturus campechiensis*   FAO code: **PAH**   Plate page 400

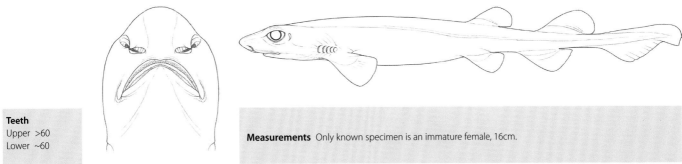

Teeth
Upper >60
Lower ~60

Measurements Only known specimen is an immature female, 16cm.

Identification Soft, flabby body. Short snout. Very small, low anterior nasal flaps. Ridges under eyes. Gills not greatly enlarged. First dorsal fin origin slightly in front of pelvic origins, fin slightly smaller than second dorsal fin. Second dorsal about as large as anal fin and with insertion well behind that of anal fin. Crest of saw-like denticles along top of caudal fin but apparently not below. Greyish body; dusky on abdomen, around gills and on fin webs.
Distribution West Atlantic, Caribbean: collected in Bay of Campeche, Gulf of Mexico.
Habitat Continental slope, on or near bottom, 1097m, possibly to 1378m.
Behaviour Unknown.
Biology Unknown.
Status IUCN Red List: Least Concern. Occurs well below depth of fisheries. Known only from the holotype. Records from Andros Island, Bahamas and Bonaire require confirmation.

VELVET CATSHARK   *Parmaturus lanatus*   FAO code: **CVX**   Plate page 400

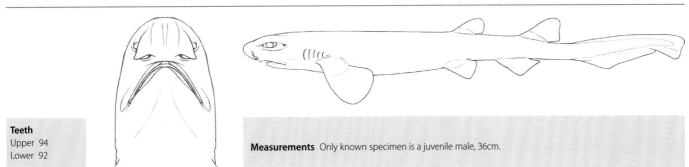

Teeth
Upper 94
Lower 92

Measurements Only known specimen is a juvenile male, 36cm.

Identification A slender, soft-bodied, tadpole-shaped *Parmaturus* with velvety textured skin, dorsal and ventral crests of enlarged denticles on caudal fin margins. Uniform brown with darker gills and fin margins.
Distribution Central Indo-Pacific: known only from Indonesia, near Tanimbar Island, Arafura Sea.
Habitat Continental slope, on or near bottom at 840–855m.
Behaviour Unknown.
Biology Unknown.
Status IUCN Red List: Least Concern. Known from only one specimen (the holotype), but there are currently no deepsea fisheries active within this species' range.

NEW ZEALAND FILETAIL *Parmaturus macmillani*

FAO code: **PAE**   Plate page 400

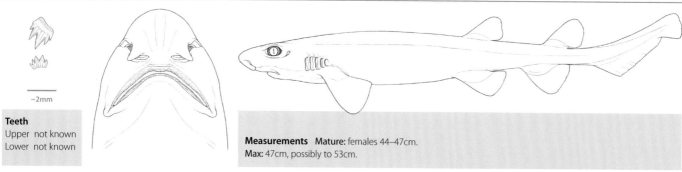

~2mm

Teeth
Upper not known
Lower not known

Measurements Mature: females 44–47cm.
Max: 47cm, possibly to 53cm.

Identification Soft flabby body. Very short, blunt snout. Elongated, lobate anterior nasal flaps. Ridges under eyes. Gills not greatly enlarged. First dorsal fin about as large as second, origin opposite or just behind pelvic origins. Second dorsal fin smaller than anal fin, insertion slightly behind that of anal fin. Crest of saw-like denticles along top of caudal fin but not below. Body grey, lighter below; fins with dusky webs.
Distribution Southwest Pacific: northern New Zealand endemic.
Habitat Captured in deep water, 950–1003m (may extend deeper).
Behaviour Unknown.
Biology Oviparous. Eggcases 5.2 x 3.0cm, lays pairs of large, stout-shelled eggcases (one per oviduct) without tendrils.
Status IUCN Red List: Data Deficient. Known from only a handful of specimens; the species is endemic to northern New Zealand.

BLACKGILL CATSHARK *Parmaturus melanobranchus*

FAO code: **PAV**   Plate page 400

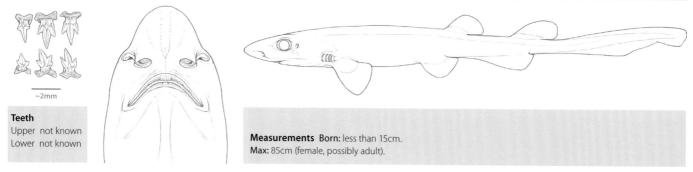

~2mm

Teeth
Upper not known
Lower not known

Measurements Born: less than 15cm.
Max: 85cm (female, possibly adult).

Identification Soft, flabby, plain body. Moderately long, blunt snout. Anterior nasal flaps elongated and pointed. Ridges under eyes. Gills not greatly enlarged. First dorsal fin noticeably smaller than second with origin well behind pelvic fin origins and opposite their mid-bases or insertions. Second dorsal fin about as large as anal fin, insertion well behind insertion of anal fin. Crests of saw-like denticles along top of caudal fin and on preventral caudal margin below. Coloration grey to dark brown above, lighter below; blackish on gill septa.
Distribution Northwest Pacific: China (South China Sea off Hong Kong), Taiwan to Japan (Ryu-Kyu Islands).
Habitat Upper continental and insular slopes, on mud bottom at 448–1110m.
Behaviour Unknown.
Biology Unknown.
Status IUCN Red List: Least Concern. Known from only a few specimens, likely caught in research surveys. Deepsea commercial fisheries appear not to interact with this species.

INDONESIAN FILETAIL CATSHARK   *Parmaturus nigripalatum*

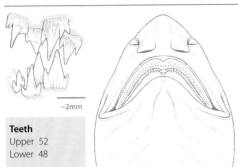

~2mm

Teeth
Upper 52
Lower 48

Measurements Mature: males 54.8cm.
Max: at least 55cm.

Identification Body slender, flabby, soft, slightly depressed anteriorly; tapering from pelvic fin bases to caudal fin. Snout short. Roof of mouth is blackish with darker pores; labial furrows well developed, lowers equal to uppers. First dorsal fin distinctly smaller than second; first dorsal fin origin slightly posterior to pelvic fin insertion. Second dorsal fin smaller than anal fin. Caudal crest of enlarged dermal denticles present. Coloration is a plain medium brown above, slightly paler below. Dorsal surfaces of pectoral fins darker than ventral surfaces; all edges of fins much darker. Claspers and pelvic fins are whitish at insertions.
Distribution Central Indo-Pacific: known only from south of Sumbawa, Indonesia.
Habitat Uncertain, the only known specimen was caught on a longline between 170–190m, which is relatively shallow for members of this genus.
Behaviour May move into shallower water at night based on the only known specimen, caught at night on a longline at a depth of less than 200m.
Biology Unknown, presumed to be oviparous, although the only known specimen was an adult male.
Status IUCN Red List: Data Deficient. Area is heavily fished.

SALAMANDER CATSHARK   *Parmaturus pilosus*

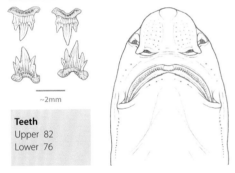

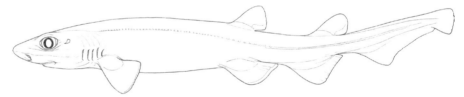

~2mm

Teeth
Upper 82
Lower 76

Measurements Immature males 56cm, females 59–64cm.
Max: at least 64cm.

Identification Soft, flabby, plain body. Moderately long, blunt snout. Anterior nasal flaps elongated and narrowly lobate. Ridges under eyes. Gills not greatly enlarged. Conspicuous crest of saw-like denticles along top of caudal fin and short, inconspicuous crest on preventral caudal margin. First dorsal fin about as large as second, origin about opposite pelvic origins. Second dorsal much smaller than anal fin, with insertion about opposite anal insertion. Coloration reddish above and white below; fin webs darker.
Distribution Northwest Pacific: Japan (southeast Honshu and Riu-Kyu Islands) to Taiwan.
Habitat Upper continental and insular slopes, 358–1177m.
Behaviour Unknown.
Biology Oviparous. Eggcases ~7–8 x 2–3cm with fine striations, smooth to the touch and no tendrils. Large quantities of squalene in liver may maintain neutral buoyancy.
Status IUCN Red List: Data Deficient. Potentially a bycatch of the deepsea trawl fisheries operating within its range, but more information is needed on its life history and population.

FILETAIL CATSHARK *Parmaturus xaniurus*

~2mm

Teeth
Upper 67–71
Lower 78–82

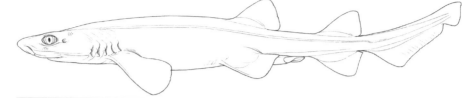

Measurements Born: 7–9cm. **Mature:** males 37–45cm, females 47–50cm.
Max: 61cm.

Identification Soft, flabby body. Moderately long, blunt snout. Large, triangular anterior nasal flaps. Ridges under eyes. Enlarged gill slits. Similar sized dorsal fins, origin of first slightly behind pelvic origins but well in front of midbases. Second dorsal much smaller than anal fin, insertion well in front of anal fin insertion. Crests of saw-like denticles along top of caudal fin, not below. Plain, dark colour; brownish black above, lighter below; fins dark.
Distribution Northeast Pacific: Oregon, USA to Gulf of California, Mexico.
Habitat Outer continental shelf and upper slope, often on or near bottom, 88–1519m, most common between 300–550m; juveniles live in midwater up to 500m above seabed in water over 1000m. Eggcases are deposited in specific areas of high vertical relief at depths of 300–500m.
Behaviour Observed from submersible feeding on moribund lanternfish in almost anoxic conditions.
Biology Oviparous. Eggcases ~7–11 x 2–3cm, with T-shaped lateral flanges and short tendrils. Eggcases are deposited year-round with no defined breeding season. Eats mainly pelagic crustaceans and small bony fishes. Squalene-filled liver may maintain neutral buoyancy. Enlarged gills enable it to thrive in low-oxygen habitats.
Status IUCN Red List: Least Concern. Relatively common. Discarded from bottom trawls and sablefish traps.

ONEFIN CATSHARK *Pentanchus profundicolus*

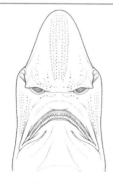

~2mm

Teeth
Upper 61–69
Lower 60–68

Measurements Immature: 38cm. **Mature:** males 51cm.
Max: at least 51cm.

Identification Only species with five gill slits and one dorsal fin. Broadly rounded snout; short mouth. Short gill slits and incised gill septa. Unusually short abdomen (pectoral and pelvic fins very close together). Very long, low anal fin. Long, narrow caudal fin without a crest of denticles. Plain brownish colour.
Distribution Central Indo-Pacific: Philippines, Tablas Straits and Mindanao Sea, east of Bohol.
Habitat Insular slope, on bottom, 673–1070m.
Behaviour Unknown.
Biology Unknown.
Status IUCN Red List: Least Concern. Known from only 2 specimens. Most of its range is likely below the poorly documented deepsea fisheries in the area.

Scyliorhinidae: catsharks

Once the largest shark family, the catsharks now contain about 50 species in seven genera, although considerable taxonomic research is still needed. New species and even new genera are still being discovered and described, as commercial fisheries and research efforts move into deeper water. Catsharks are found worldwide, from tropical to Arctic waters, usually on or near the seabed, from the intertidal to continental slopes, but are often restricted to relatively small ranges. Many are known from very few specimens.

Identification Usually small (less than 80cm long; some may mature at ~30cm, a few reach ~160cm). Elongated body, two small spineless dorsal fins (first dorsal base over or behind pelvic bases) and an anal fin. Long, arched mouth reaches past the front end of the cat-like eyes. The catsharks have an internal 'crest' present on their crania, which can be felt by running your fingers over the eye orbits.

Biology Many species are poorly known. All those for which information is available are oviparous (egg-laying). The eggcases are unique to each species and, if they have been described, can therefore be used to determine the species that laid it. Hatching may take nearly a year. More advanced species retain their eggs until the embryos' development is almost complete, then lay larger numbers of eggs about a month before hatching. A few retain the eggs until the embryos are fully developed and give birth to live young (viviparity). Catsharks are poor swimmers and do not undertake long distance migrations. Some inshore species are nocturnal. They may sleep in groups in crevices by day, moving out to feed at night. They eat benthic invertebrates and small fishes.

Status The majority of these species are assessed as Least Concern (46%) or Data Deficient (28%) in the IUCN Red List; 6% were too recently described to have been assessed by 2020. A few are important in commercial fisheries, and many more are taken as a bycatch. Some are regularly kept and bred in public aquariums. None are dangerous to people.

Akheilos

1 species; page 50

Atelomycterus

At least 6 species; pages 411–413

Aulohalaelurus

At least 2 species; page 414

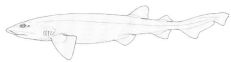

Cephaloscyllium

At least 18 species; pages 415–427

Poroderma

At least 2 species; page 428

Schroederichthys

At least 5 species; pages 429–433

Scyliorhinus

At least 16 species; pages 438–446

Swellshark, *Cephaloscyllium ventriosum* (p. 427).

○ **Ambon Catshark** *Akheilos suwartanai* page 410

Central Indo-Pacific. Medium brown dorsally, pale brown on sides and ventral surface; 2 indistinct saddle bars, 1 each above pectoral and pelvic fins; 3 incomplete ocelli, 1 between pectoral and pelvic fins, and 1 at each dorsal fin base.

○ **New Caledonia Catshark** *Aulohalaelurus kanakorum* page 414

Central Indo-Pacific; to 49m. Fairly slender, thick-skinned, elongated cylindrical body; large equal-sized dorsal fins; dark grey with variegated pattern of large dark close-set blotches on body and fins surrounding numerous large white blotches; white-bordered fins.

○ **Australian Blackspotted Catshark** *Aulohalaelurus labiosus* page 414

Southwest Indian Ocean; to 4m+. Slender, thick-skinned, elongated cylindrical body, narrow head; variegated light grey to yellow-brown, dark saddles and small to large black spots or blotches on sides, back and fins, very few fine white spots; white dorsal caudal and anal fin tips highlighted by dark blotches.

○ **Bali Catshark** *Atelomycterus baliensis* page 411

Central Indo-Pacific. A variegated pattern of 4 well-defined dark brown saddles with scattered blotches of small black spots.

○ **Spotted-belly Catshark** *Atelomycterus erdmanni* page 411

Central Indo-Pacific; 3–62m. Dark background with elaborate pattern of dark brown to black and white blotches and spots, faint dark brown saddles with white spots encircled by 2–4 dark spots; scattered dark spots on ventral surface.

○ **Banded Sand Catshark** *Atelomycterus fasciatus* page 412

Central Indo-Pacific; 27–122m. Slender body, narrow head; triangular dorsal fins much larger than anal fin; light coloured with darker brown saddles, a few scattered small black spots, and sometimes small white spots.

○ **Australian Marbled Catshark** *Atelomycterus macleayi* page 412

Central Indo-Pacific; 0.5–4m. Slender body, narrow head; dorsal fins much larger than anal fin; light grey to grey-brown with darker grey or brown saddles outlined and partly covered in adults by many small black spots, also scattered on flanks; no white spots; hatchlings unspotted.

○ **Coral Catshark** *Atelomycterus marmoratus* page 413

North Indian Ocean and Central Indo-Pacific; 5–100m. Slender, narrow head, dorsal fins much larger than anal fin; dark with no clear saddles, enlarged black spots may merge into dash and bar marks; scattered large white spots on sides back and fin margins.

○ **Whitespotted Sand Catshark** *Atelomycterus marnkalha* page 413

Central Indo-Pacific; 11–74m. Slender, short rounded snout; light grey to brown; 3–4 darker predorsal saddles, 7 poorly defined saddles behind first dorsal fin; vertical rows of small white spots and sparsely scattered black spots.

20cm

○ *Aulohalaelurus kanakorum*

○ *Aulohalaelurus labiosus*

20cm

Akheilos suwartanai

Atelomycterus baliensis

Atelomycterus erdmanni

Atelomycterus fasciatus

Atelomycterus macleayi

Atelomycterus marmoratus

Atelomycterus marnkalha

Coral Catshark, *Atelomycterus marmoratus* (p. 413).

AMBON CATSHARK   *Akheilos suwartanai*

FAO code: **SYX**   Plate page 408

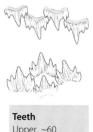

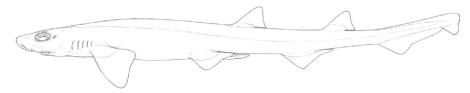

Teeth
Upper ~60
Lower ~53

Measurements Mature: males 53.7cm.
Max: at least 54cm.

Identification Body small, firm, stout, barely tapering from pelvic fin bases to caudal fin origin. Snout short, rounded at tip, distinctly bell-shaped in dorsal view. Nostrils expanded inwards, but not reaching mouth; labial furrows well developed, lowers longer than uppers. Dorsal fins similar in size and shape, first dorsal fin origin over pelvic fin free rear tip. Coloration (based on preserved specimen) is medium brown dorsally, becoming pale brown along sides and ventral surface. Dorsal and lateral surfaces have two indistinct saddles, one each above pectoral and pelvic fins, and three incomplete ocelli laterally, one between pectoral and pelvic fins and one at the base of each dorsal fin; ventral surface without any markings.

Distribution Central Indo-Pacific: Indonesia. Known only from Rumahkay, southwest of Seram Island between Seram and Ambon Islands.

Habitat Unknown.

Behaviour Unknown.

Biology Only known specimen is an adult male. It contained a small bony fish in its stomach. Nothing else known of this species.

Status IUCN Red List: Data Deficient. Fisheries are expanding and fish stocks declining in this region, but impacts on this species are unknown. The only known specimen was found in a museum.

BALI CATSHARK *Atelomycterus baliensis*

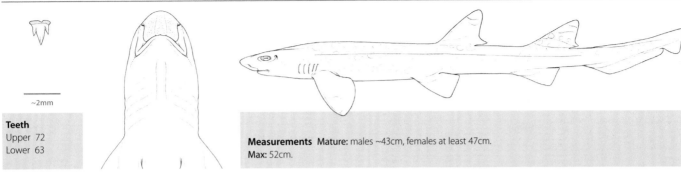

~2mm

Teeth
Upper 72
Lower 63

Measurements Mature: males ~43cm, females at least 47cm.
Max: 52cm.

Identification A small, slender catshark. Relatively short snout. Large nasal flaps, extending nearly to the mouth. Has a variegated pattern of four well-defined, dark brown saddles with scattered blotches of small black spots.
Distribution Central Indo-Pacific: Indonesian endemic, known only from the island of Bali.
Habitat Inshore, likely inhabits crevices and holes on coral and rocky reefs.
Behaviour Unknown.
Biology Oviparous. Presumably feeds on small fishes and crustaceans.
Status IUCN Red List: **Vulnerable**. Destructive fishing practices have likely affected the habitat of this range-restricted species, which is retained when taken as a bycatch in local fisheries. However, it is still regularly seen and its habitat is unsuitable for some fisheries.

SPOTTED-BELLY CATSHARK *Atelomycterus erdmanni*

Teeth
Upper not known
Lower not known

Measurements Mature: males 50cm, females 51cm.
Max: at least 51cm.

Identification Body small, slender, not tapering much. Short snout, rounded at tip. Nasal flaps large, reaching lower jaw; labial furrows long, lowers slightly longer than uppers. Dorsal fins similar in size and shape. Dark background coloration with a complex pattern of dark brown to black and white blotches and spots. Faint dark brown saddles with white spots encircled by 2–4 dark spots; ventral surface with scattered dark spots.
Distribution Central Indo-Pacific: Indonesia. Known only from Lembeh Strait and Bunaken Islands in North Sulawesi, and from off Ambon, Maluku Islands.
Habitat Inshore, from 3–62m on coral and rocky reefs.
Behaviour Unknown.
Biology The only known specimens were an adult male and a female. Occasionally seen by divers.
Status IUCN Red List: **Not Evaluated**. Its reef habitat is unsuitable for some fisheries.

~2mm

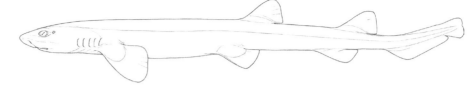

Teeth
Upper 56–73
Lower 50–59

Measurements Mature: males ~33cm, females 35cm.
Max: 45cm.

Identification Slender body. Narrow head. Greatly expanded anterior nasal flaps extending to long mouth; nasoral grooves; very long labial furrows. Dorsal fins broadly triangular, much larger than anal fin; origin of first above rear third of pelvic bases. Brown saddles on a light background; a few scattered small black and sometimes small white spots.
Distribution Central Indo-Pacific: northwest Australia (isolated records from Northern Territory and Queensland may be of a different species).
Habitat Bottom on sand and shelly sand, continental shelf, 27–122m, with the majority occurring shallower than 60m.
Behaviour Unknown.
Biology Oviparous. Eggcases 6.7cm long, laid in pairs (one from each oviduct). Presumably feeds on small fishes and crustaceans.
Status IUCN Red List: Least Concern. Occurs in largely unfished area.

AUSTRALIAN MARBLED CATSHARK *Atelomycterus macleayi*   FAO code: **ATM**   Plate page 408

Teeth
Upper 70
Lower 70

Measurements Born: ~10cm. **Mature:** males ~48cm, females 51cm.
Max: 60cm.

Identification Slender-bodied catshark. Narrow head. Greatly expanded anterior nasal flaps extend to mouth; nasoral grooves; very long labial furrows. Dorsal fins much larger than anal fin; first dorsal origin opposite pelvic fin insertions. Light grey to grey-brown, with darker grey or brown saddles, outlined and partly covered in adults by many small black spots, also scattered on flanks, but no white spots. Hatchling young without spots.
Distribution Central Indo-Pacific: Australia; Western Australia, Northern Territory, possibly Queensland.
Habitat Sand and rock in very shallow water, 0.5–4m.
Behaviour Unknown.
Biology Oviparous. Eggcases 7cm long, laid in pairs.
Status IUCN Red List: Least Concern. Apparently common, no fisheries within its range.

CORAL CATSHARK   *Atelomycterus marmoratus*

FAO code: **ATY**   Plate page 408

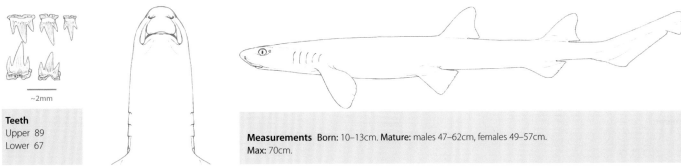

Teeth
Upper 89
Lower 67

~2mm

Measurements Born: 10–13cm. **Mature:** males 47–62cm, females 49–57cm.
Max: 70cm.

Identification Slender body. Narrow head. Greatly expanded anterior nasal flaps extending to long mouth; nasoral grooves; very long labial furrows. Dorsal fins much larger than anal fin; origin of first opposite or slightly in front of pelvic fin insertions. Dark coloration; no clear saddle markings. Enlarged black spots often merge to form dash and bar marks; scattered large white spots on sides, back and fin margins.
Distribution North Indian Ocean, Central Indo-Pacific: Pakistan, India; Sri Lanka to New Guinea, southern China, Philippines, Taiwan and southern Japan.
Habitat Inshore species found in crevices and holes on coral reefs, 5–100m.
Behaviour Nocturnal. Hides among rocks and beneath overhangs by day, emerges to hunt at dusk and by night.
Biology Oviparous. Lays pairs of eggcases, 6–8 x 2–3cm, with no anterior tendrils, posterior end with short tendrils. Has lived for up to 20 years in captivity. Feeds on molluscs, crustaceans and small fishes.
Status IUCN Red List: Near Threatened. A popular aquarium species, because of its small size and beautiful patterning. Hardy and can breed in captivity, but may be aggressive. Common in artisanal fisheries. Its coral reef habitat is threatened in much of its range but offers it a refuge from trawl fisheries.

WHITESPOTTED SAND CATSHARK   *Atelomycterus marnkalha*

FAO code: **SYX**   Plate page 408

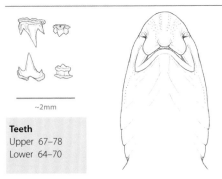

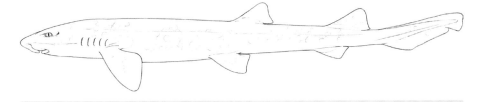

Teeth
Upper 67–78
Lower 64–70

~2mm

Measurements Mature: males 35–39cm, females ~35cm.
Max: 49cm.

Identification A small, slender catshark. Short rounded snout. Very large nasal flaps extending to the mouth. Colour is light grey to brown with 3–4 darker predorsal saddles and seven poorly defined saddles behind first dorsal fin; vertical rows of small white spots and sparsely scattered black spots.
Distribution Central Indo-Pacific: Australia, from Queensland to Northern Territory; also southern Papua New Guinea.
Habitat Sandy to coarse rubble from 11–74m, mostly less than 50m.
Behaviour Unknown.
Biology Oviparous. Feeds on bony fishes, crustaceans and cephalopods.
Status IUCN Red List: Data Deficient. A rarely recorded discarded bycatch of demersal trawl fisheries. Part of its range is protected in the Great Barrier Reef Marine Park.

NEW CALEDONIA CATSHARK *Aulohalaelurus kanakorum* FAO code: **AUK** Plate page 408

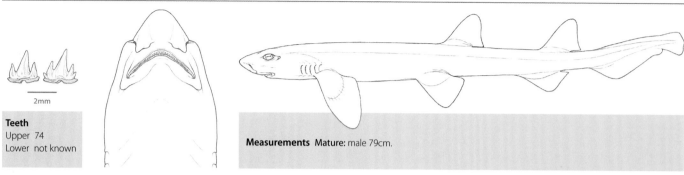

Teeth
Upper 74
Lower not known

Measurements Mature: male 79cm.

Identification Fairly slender, thick-skinned, elongated, cylindrical body. Short, slightly flattened, narrowly rounded head. Long, arched mouth reaches past the front end of the cat-like eyes. Roughly equal-sized dorsal fins; origin of first over pelvic fin insertions. Moderately short, broad caudal fin. Dark grey with a variegated colour pattern of large, dark, close-set blotches on body and fins surrounding numerous large white blotches. White-bordered fins.
Distribution Central Indo-Pacific: New Caledonia, possibly endemic.
Habitat Coral reefs to 49m.
Behaviour Unknown.
Biology Unknown.
Status IUCN Red List: Data Deficient. Presumed endemic. Known from one specimen and two photographs.

AUSTRALIAN BLACKSPOTTED CATSHARK *Aulohalaelurus labiosus* FAO code: **AUL** Plate page 408

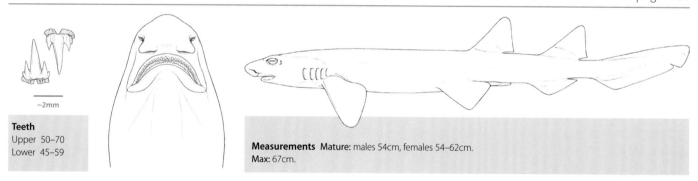

Teeth
Upper 50–70
Lower 45–59

Measurements Mature: males 54cm, females 54–62cm.
Max: 67cm.

Identification Fairly slender, thick-skinned, elongated, cylindrical body. Relatively narrow head; slightly flattened, short, narrowly rounded snout. Equal-sized dorsal fins; origin of first over or slightly in front of pectoral fin insertions. Moderately short, broad caudal fin. Colour is light greyish to yellowish brown, with variegated pattern of small to large black spots or blotches and dark saddles on sides, back and fins; very few small white spots. White fin tips highlighted by dark blotches on dorsal, caudal and anal fins.
Distribution Southwest Indian Ocean: Australia.
Habitat Shallow coastal waters and offshore reefs to at least 4m.
Behaviour Unknown.
Biology Virtually unknown. Probably oviparous.
Status IUCN Red List: Least Concern. Common and unfished within limited range.

WHITEFIN SWELLSHARK *Cephaloscyllium albipinnum*

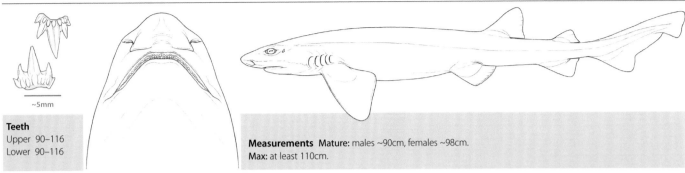

~5mm

Teeth
Upper 90–116
Lower 90–116

Measurements Mature: males ~90cm, females ~98cm.
Max: at least 110cm.

Identification Large, stocky body. Broad head; ridges over eyes. Two dorsal fins, second much smaller than first and over much larger anal fin. Rough skin. Pattern of broad, dark blotches and saddles on medium brownish or greyish background on sides and back (fainter in juveniles). Usually five predorsal bars; fins mostly dark above with pale margins. Pale underneath.
Distribution Southwest Pacific and southeast Indian Ocean: southern Australia, endemic.
Habitat Upper continental slope, 125–555m.
Behaviour Inflates by swallowing water or air.
Biology Oviparous, poorly known. Eggcases ~12 x 10cm, large, flask-shaped and unridged, with long coiled tendrils at each end.
Status IUCN Red List: **Critically Endangered**. Taken in trawl bycatch with significant population declines reported.

COOK'S SWELLSHARK *Cephaloscyllium cooki*

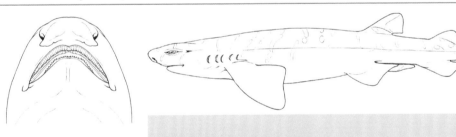

Teeth
Upper 50–61
Lower 49–62

Measurements Mature: males ~29cm.
Max: males at least 30cm, females at least 28cm.

Identification Small, stout-bodied swellshark. Striking colour pattern of dark brown bands edged in white around body and tail forming eight open-centred saddles on a lighter background; first saddle curves forward towards the eyes. Pale spots scattered on snout, fins and inside saddles.
Distribution Central Indo-Pacific: Arafura Sea, off north Australia.
Habitat Upper continental slope, 223–300m.
Behaviour Unknown.
Biology Oviparous.
Status IUCN Red List: **Data Deficient**. Nothing known about life history or fisheries. Only known from about seven specimens.

Plate 51 SCYLIORHINIDAE II – swellsharks I

○ **Whitefin Swellshark** *Cephaloscyllium albipinnum*

Southwest Pacific and southeast Indian Ocean; 125–555m. Large stocky body, broad head, ridges over eyes; medium brownish or greyish background with broad dark blotches and saddles on sides and back fainter in juveniles); usually 5 predorsal bars; fins mostly dark with pale margins and underside.

○ **Draughtsboard Shark** *Cephaloscyllium isabellum*

Southwest Pacific; 690m. Second dorsal much smaller than first; up to 11 dark brown irregular saddles and alternating blotches on sides in checkerboard pattern.

○ **Australian Swellshark** *Cephaloscyllium laticeps*

Southwest Pacific and southeast Indian Ocean; to 60m. Light grey or chestnut background with pattern of dark brown to greyish close-set saddles and blotches; many dark spots, occasional light spots; underside cream, usually with dark stripe down adult belly; no conspicuous light fin margins.

○ **Painted Swellshark** *Cephaloscyllium pictum*

Central Indo-Pacific. A strongly variegated colour pattern with prominent dark blotches saddles and spots.

○ **Flagtail Swellshark** *Cephaloscyllium signourum*

Central Indo-Pacific; 480–700m. Large heavy body; variegated pattern of 9–10 poorly defined dark brown saddles; V-shaped blotch at end of upper caudal lobe.

○ **Balloon Shark** *Cephaloscyllium sufflans*

West Indian Ocean; 40–605m. Light grey with 7 light grey-brown saddles in juveniles, obscure and sometimes absent in adults; pectoral fins dusky above, unspotted below; no obvious light fin margins.

○ **Japanese Swellshark** *Cephaloscyllium umbratile*

Northwest Pacific; 18m to ~590m. Longer snout and more mottled than *C. isabellum*. Pale brown with dark brown saddles and mottling separated by light red-brown areas; scattered small white and dark brown spots; underside lighter, mostly unspotted; fins mottled and spotted, paired fins light below.

○ **Saddled Swellshark** *Cephaloscyllium variegatum*

Central Indo-Pacific; 114–606m. Medium brownish or greyish with obvious dark saddles, usually 5 predorsal; blotches usually absent from sides; fin margins sometimes pale; underside of body pale.

○ **Swellshark** *Cephaloscyllium ventriosum*

East Pacific; mostly 5–40m, up to 457m. Light yellow-brown background, strongly variegated, close-set dark brown saddles and blotches, numerous dark spots, occasional light spots; underside also heavily spotted.

○ *Cephaloscyllium albipinnum*

○ *Cephaloscyllium isabellum*

20cm

○ *Cephaloscyllium laticeps*

○ *Cephaloscyllium pictum*

○ *Cephaloscyllium signourum*

○ *Cephaloscyllium sufflans*

○ *Cephaloscyllium umbratile*

○ *Cephaloscyllium variegatum*

○ *Cephaloscyllium ventriosum*

Plate 52 SCYLIORHINIDAE III – swellsharks II

○ **Cook's Swellshark** *Cephaloscyllium cooki*   page 415

Central Indo-Pacific; 223–300m. A pattern of 8 dark brown saddles outlined in white on greyish brown background; pale spots on snout, fins and inside saddles.

○ **Reticulated Swellshark** *Cephaloscyllium fasciatum*   page 420

West Pacific; 200–450m. Eye ridges; second dorsal much smaller than anal fin; light grey with variegated pattern of dark lines forming open-centred saddles, loops and reticulations and spots on back and sides (no spots in young); spotted underside.

○ **Taiwan Swellshark** *Cephaloscyllium formosanum*   page 420

Central Indo-Pacific. Stout body, short snout; reddish brown above with about 10 dark brownish transverse bands on dorsal and lateral surfaces, numerous white spots; ventral surface lighter to whitish.

○ **Australian Reticulated Swellshark** *Cephaloscyllium hiscosellum*   page 421

Central Indo-Pacific; 294–420m. A striking pattern of dark brown narrow transverse lines forming open-centred saddles and blotches on light background.

○ **Sarawak Swellshark** *Cephaloscyllium sarawakensis*   page 423

West Pacific and Central Indo-Pacific; 82–200m. Dwarf, broad-headed swell shark; light brown with 6 darker brown saddles on back; juveniles have a striking pattern of distinct dark brown polka dots on brownish background.

○ **Indian Swellshark** *Cephaloscyllium silasi*   page 424

North Indian Ocean; 150–500m. Light brown with 7 dark brown saddles; dark blotch over pectoral fin inner margins; no light fin margins; light brown unspotted underside.

○ **Speckled Swellshark** *Cephaloscyllium speccum*   page 424

Central Indo-Pacific; 150–455m. Second dorsal much smaller than first; pale grey with intensely mottled small dark blotches, larger blotches and saddles with small white spots; rounded blotches and white spots below eyes; 3 predorsal saddles.

○ **Steven's Swellshark** *Cephaloscyllium stevensi*   page 425

Central Indo-Pacific; 240–616m. Grey-brown with 8 large dark brown saddles, blotches and numerous scattered smaller white spots.

○ **Narrowbar Swellshark** *Cephaloscyllium zebrum*   page 427

Central Indo-Pacific; 444–454m. Dark brownish to cream background, numerous close-spaced narrow dark bars that do not form rings or saddles, 17–18 predorsal; irregular lines on snout, pale unmarked fins and underside.

○ *Cephaloscyllium cooki*

20cm

○ *Cephaloscyllium fasciatum*

○ *Cephaloscyllium formosanum*

○ *Cephaloscyllium hiscosellum*

○ *Cephaloscyllium sarawakensis*

○ *Cephaloscyllium silasi*

○ *Cephaloscyllium speccum*

20cm

○ *Cephaloscyllium stevensi*

○ *Cephaloscyllium zebrum*

RETICULATED SWELLSHARK *Cephaloscyllium fasciatum*   FAO code: **CPF**   Plate page 418

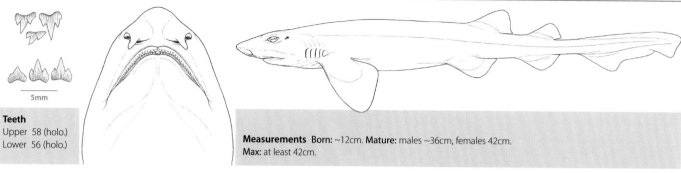

Teeth
Upper 58 (holo.)
Lower 56 (holo.)

Measurements Born: ~12cm. **Mature:** males ~36cm, females 42cm. **Max:** at least 42cm.

Identification Small swellshark with inflatable stomach. Ridges over eyes. Second dorsal fin much smaller than first. Anal fin. Adults have a striking pattern of dark lines forming open-centred saddles, loops, reticulations and spots on light greyish back and sides (spots absent in young). Underside spotted.
Distribution West Pacific: Vietnam, China (Hainan Island), Philippines (Luzon) and Taiwan.
Habitat On or near muddy bottom, uppermost slope, 200–450m.
Behaviour Can expand itself with air or water in an attempt to frighten predators.
Biology Oviparous.
Status IUCN Red List: **Critically Endangered**. This species has a restricted distribution, no refuge from fisheries, and the genus is very sensitive to fishing pressure, even in managed fisheries. Reconstructed shark and ray catches from this region have identified significant population declines across all species.

TAIWAN SWELLSHARK *Cephaloscyllium formosanum*   FAO code: **SYX**   Plate page 418

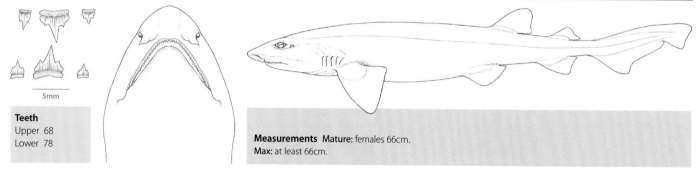

Teeth
Upper 68
Lower 78

Measurements Mature: females 66cm.
Max: at least 66cm.

Identification Body firm, stout and tapering from pelvic fin bases to caudal fin origin. Snout short, rounded at tip; distinctly bell-shaped in dorsal view. Nostrils expanded inwards, but not reaching mouth; labial furrows well developed, lowers longer than uppers. Dorsal fins are dissimilar in size and shape, the first being much larger than the second; first dorsal smaller than pectoral fin, but larger than anal fin. Coloration reddish brown above. About ten dark brownish transverse bands on the upper and sides surfaces of body, with numerous white spots. Lighter to whitish below.
Distribution Northwest Pacific and Central Indo-Pacific: southwest Taiwan and the Philippines.
Habitat Unknown.
Behaviour Unknown.
Biology Only confirmed specimens are an adult female from Taiwan and an immature male from the Philippines. Eggcases with long tendrils on each corner, posterior ends curving inwards, outer cover smooth, lateral margins wavy with ridges.
Status IUCN Red List: **Not Evaluated**. Not recorded in commercial fisheries, so may occur in deeper, unfished waters.

AUSTRALIAN RETICULATED SWELLSHARK   *Cephaloscyllium hiscosellum*   FAO code: **SYX**   Plate page 418

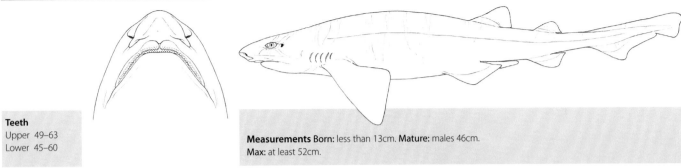

Teeth
Upper 49–63
Lower 45–60

Measurements Born: less than 13cm. **Mature:** males 46cm.
Max: at least 52cm.

Identification Small, stout-bodied swellshark with a striking colour pattern of dark brown, narrow, transverse lines forming open-centered saddles and blotches on a lighter background.
Distribution Central Indo-Pacific: west Australia.
Habitat Upper continental slope, 294–420m.
Behaviour Unknown.
Biology Oviparous. Eggcases smooth, without transverse or longitudal striations or ridges, have been observed with near-term embryos.
Status IUCN Red List: Least Concern. There is very little fishing activity within the range of this swellshark. Previous records of *Cephaloscyllium fasciatum* from Australian waters were of this species.

DRAUGHTSBOARD SHARK   *Cephaloscyllium isabellum*   FAO code: **CPS**   Plate page 416

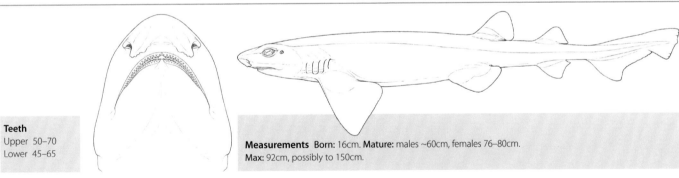

Teeth
Upper 50–70
Lower 45–65

Measurements Born: 16cm. **Mature:** males ~60cm, females 76–80cm.
Max: 92cm, possibly to 150cm.

Identification A large, stocky swellshark with inflatable stomach. Ridges over eyes. Second dorsal fin much smaller than the first. Anal fin. Strongly patterned with up to 11 dark brown, irregular saddles and alternating blotches on its sides in a checkerboard pattern.
Distribution Southwest Pacific: New Zealand endemic.
Habitat Rocky and sandy bottom, shore to 690m, most less than 400m.
Behaviour Can expand body with air or water.
Biology Oviparous. Eggcases 12 x 4cm, laid in pairs, smooth with very long tendrils at each end used for attaching to substrate. Eggcases deposited year-round, with a peak during the summer months. Eats crabs, worms, other invertebrates and probably bony fishes.
Status IUCN Red List: Least Concern. Bycatch in deepwater trawl fisheries, survives discard well.

AUSTRALIAN SWELLSHARK   *Cephaloscyllium laticeps*   FAO code: **CPT**   Plate page 416

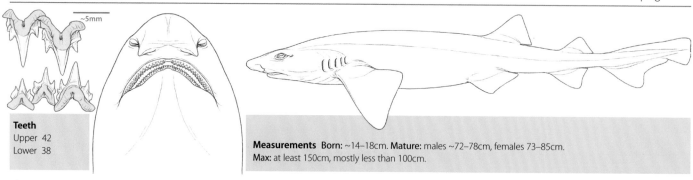

Teeth
Upper 42
Lower 38

Measurements Born: ~14–18cm. **Mature:** males ~72–78cm, females 73–85cm.
Max: at least 150cm, mostly less than 100cm.

Identification Ridges over eyes. Larger first dorsal fin over pelvic fin bases, second over anal fin. Strongly variegated pattern of dark brown or greyish close-set dark saddles and blotches, many dark spots, and occasional light spots on lighter grey or chestnut background. Broad dark stripe between eye and pectoral origin, dark patch below eye. Underside cream, usually with dark stripe down belly in adults. No conspicuous light fin margins.
Distribution Southwest Pacific and southeast Indian Ocean: southern Australia.
Habitat Inshore on continental shelf to at least 60m.
Behaviour Lays ridged, cream-coloured eggcases attached to seaweed and benthic invertebrates.
Biology Oviparous. Eggcases ~13 x 5cm, with distinctive transverse ridges. Females lay year-round, hatching takes place after about 12 months. Feeds on small reef fishes, crustaceans and squid.
Status IUCN Red List: Least Concern. Discarded bycatch from shark gill net fishery; survives well.

PAINTED SWELLSHARK   *Cephaloscyllium pictum*   FAO code: **SYX**   Plate page 416

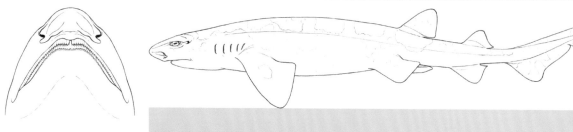

Teeth
Upper 84 (holo.)
Lower 97 (holo.)

Measurements Mature: males ~58–64cm.
Max: males 72cm, females 70cm.

Identification Medium-sized swellshark; moderately stout body. Broadly rounded head. Large nasal flaps expanded laterally, partially overlapping outer lobe but not reaching mouth. Colour pattern strongly variegated with prominent dark blotches, saddles and spots.
Distribution Central Indo-Pacific: eastern Indonesia.
Habitat Known only from several fish markets.
Behaviour Unknown.
Biology Oviparous.
Status IUCN Red List: Data Deficient. This species' depth range and habitat is unknown, but there are intensive fisheries and increasing fishing pressure in this region.

SARAWAK SWELLSHARK *Cephaloscyllium sarawakensis*

FAO code: **SYX**   Plate page 418

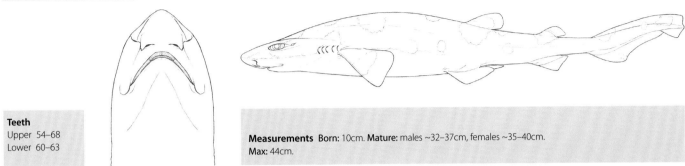

Teeth
Upper 54–68
Lower 60–63

Measurements Born: 10cm. **Mature:** males ~32–37cm, females ~35–40cm.
Max: 44cm.

Identification A small, dwarf swellshark. Broad head. Light brown background colour and six darker brown saddles on back; dark circular to oblong blotches on sides of trunk. Juveniles with a striking pattern of distinct dark brown polka dots on a brownish background.
Distribution West Pacific and Central Indo-Pacific: Taiwan to South China Sea, Vietnam, Hong Kong, Borneo and Malaysia.
Habitat Outer continental shelf from 118–165m.
Behaviour Unknown.
Biology Oviparous. Eggcases 10 x 3cm, smooth, without longitudinal ridges or striations.
Status IUCN Red List: Data Deficient. Known from only a few specimens caught in deepsea trawl nets, but its entire known range is heavily fished, posing a serious threat to this species.

FLAGTAIL SWELLSHARK *Cephaloscyllium signourum*

FAO code: **SYX**   Plate page 416

Teeth
Upper 84
Lower 97

Measurements Max: 74cm.

Identification A large, heavy bodied swellshark. Anterior nasal flaps laterally expanded with overlapping outer lobe, not reaching mouth. Variegated colour pattern of 9–10 poorly defined, dark brown saddles; V-shaped blotch at end of upper caudal lobe. Juveniles pale yellowish with narrow brown bars.
Distribution Central Indo-Pacific: Australia, known only from northeast Queensland.
Habitat Continental slope from 480–700m.
Behaviour Unknown.
Biology Oviparous.
Status IUCN Red List: Data Deficient. Known from only two specimens. Records from outside Australia may be different species.

INDIAN SWELLSHARK *Cephaloscyllium silasi*

FAO code: **CPA**   Plate page 418

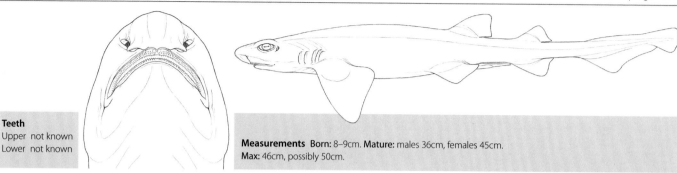

Teeth
Upper not known
Lower not known

Measurements Born: 8–9cm. **Mature:** males 36cm, females 45cm.
Max: 46cm, possibly 50cm.

Identification Dwarf swellshark. Ridges over eyes. First dorsal fin larger, over pelvic fin bases; second much smaller and over anal fin. Pattern of seven moderately broad, dark brown saddles on a light brown background. An obscure darker blotch over the pectoral inner margins; no conspicuous light fin margins. Light brown, unspotted underside.
Distribution North Indian Ocean: India, Sri Lanka and Andaman Islands.
Habitat Seabed on the uppermost continental slope, 150–500m.
Behaviour Unknown.
Biology Oviparous. Eggcases 8–9 x 2–3cm, very long coiled tendrils and a smooth surface. Feeds on crustaceans and cephalopods.
Status IUCN Red List: **Critically Endangered**. This species is a bycatch of the intensive deepsea trawl fisheries that operate across its entire known geographic and depth range.

SPECKLED SWELLSHARK *Cephaloscyllium speccum*

FAO code: **SYX**   Plate page 418

Teeth
Upper 64–75
Lower 75–78

Measurements Mature: males ~64cm.
Max: at least 69cm.

Identification Stocky, short-tailed swellshark; inflatable stomach. Long, arched mouth reaches past the front end of the cat-like eyes; no labial furrows. Ridges over eyes. Two small dorsal fins, second much smaller than the first; first dorsal base over or behind pelvic bases. Anal fin. Pale greyish dorsal surface with intense mottling of small, dark blotches, larger blotches and saddles with small white spots. Rounded blotches and white spots below eyes followed by three predorsal saddles. Fins mostly pale with darker spots and blotches.
Distribution Central Indo-Pacific: tropical northwest Australia.
Habitat Continental shelf and slope, 150–455m.
Behaviour Unknown.
Biology Oviparous.
Status IUCN Red List: **Data Deficient**. Known from only a few specimens. May be rare, but not subject to much fishing pressure.

STEVEN'S SWELLSHARK *Cephaloscyllium stevensi*   FAO code: **SYX**   Plate page 418

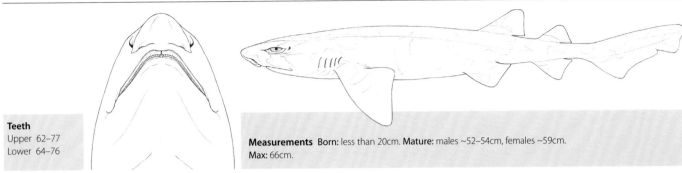

Teeth
Upper 62–77
Lower 64–76

Measurements Born: less than 20cm. **Mature:** males ~52–54cm, females ~59cm.
Max: 66cm.

Identification Moderately stout with a short snout. Colour grey-brown with eight large, dark brown saddles, blotches and numerous scattered smaller white spots.
Distribution Central Indo-Pacific: known only from Papua New Guinea.
Habitat Continental slope, 240–616m.
Behaviour Unknown.
Biology Oviparous.
Status IUCN Red List: Least Concern. There are no known deepsea fisheries operating in the region.

BALLOON SHARK *Cephaloscyllium sufflans*   FAO code: **CPH**   Plate page 416

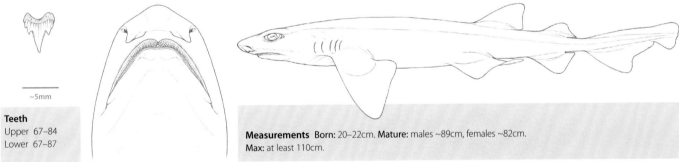

~5mm

Teeth
Upper 67–84
Lower 67–87

Measurements Born: 20–22cm. **Mature:** males ~89cm, females ~82cm.
Max: at least 110cm.

Identification Large swellshark with inflatable stomach. Ridges over eyes. First dorsal fin over pelvic fins; second much smaller and over anal fin. Seven light grey-brown saddles on lighter grey background in juveniles are obscure or absent in adults. Pectoral fins dusky above, unspotted below; no obvious light fin margins.
Distribution West Indian Ocean: South Africa (KwaZulu-Natal), Mozambique, Comoros, Madagascar and possibly Tanzania. A dubious Gulf of Aden record may be a different, smaller species.
Habitat Sand and mud bottom, rocky reefs and areas of steep relief, offshore continental shelf and uppermost slope, 40–605m.
Behaviour Juveniles and adults apparently segregated: immatures are common off KwaZulu-Natal, but not adults. Free eggcases have not yet been found (may occur further north or in deeper water).
Biology Oviparous. Feed on lobsters, shrimp and cephalopods, also teleosts and other elasmobranchs. One specimen was found in the stomach of a coelacanth.
Status IUCN Red List: Near Threatened. Its preference for rocky reefs with vertical faces gives it some refuge from fishing.

JAPANESE SWELLSHARK *Cephaloscyllium umbratile*

FAO code: **CPB** Plate page 416

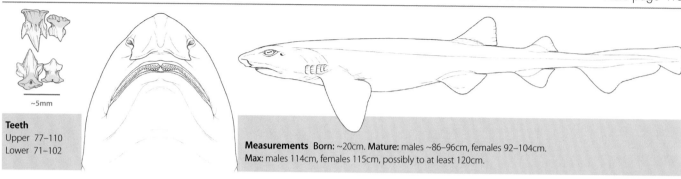

Teeth
Upper 77–110
Lower 71–102

Measurements Born: ~20cm. **Mature:** males ~86–96cm, females 92–104cm. **Max:** males 114cm, females 115cm, possibly to at least 120cm.

Identification Similar to, and formerly synonymised with, *Cephaloscyllium isabellum*, but longer snout and more mottled pattern. Ridges over eyes. Second dorsal fin much smaller than first. Inflatable stomach. Anal fin. Pale brown dorsal surface with dense dark brown mottling and regular saddles separated by lighter reddish brown areas. Scattered small white and dark brown spots. Dorsal and caudal fins mottled and spotted; pectoral and pelvic fins mottled and spotted above, light below; anal fin mottled. Underside lighter, mostly unspotted.
Distribution Northwest Pacific: Japan (wide-ranging), around the Korean Peninsula in the Sea of Japan and Yellow Sea, China and Taiwan.
Habitat Continental shelf from less than 18m to ~590m.
Behaviour Little-known.
Biology Oviparous, eggcase laying appears to occur year-round. Diet includes crustaceans, bony fishes, small sharks and chimaeras.
Status IUCN Red List: **Data Deficient**. A fisheries bycatch throughout its range, but sometimes discarded, may survive well, and is still regularly recorded.

SADDLED SWELLSHARK *Cephaloscyllium variegatum*

FAO code: **SYX** Plate page 416

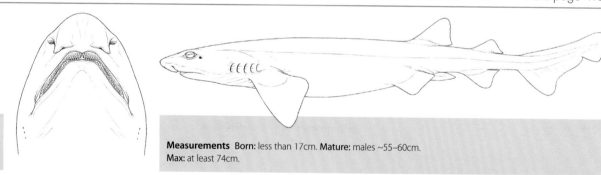

Teeth
Upper 68–82
Lower 68–80

Measurements Born: less than 17cm. **Mature:** males ~55–60cm. **Max:** at least 74cm.

Identification Fairly small swellshark. Slender head; ridges over eyes. Two dorsal fins, close together; first behind pectorals; second much smaller than first and over anal fin. Inflatable stomach. Medium brownish or greyish with obvious dark saddles (usually five in front of dorsal fins). Blotches usually absent from sides; fin margins sometimes pale. Underside of body pale.
Distribution Central Indo-Pacific: east Australia.
Habitat Continental shelf and slope, 114–606m.
Behaviour Unknown.
Biology Oviparous, nothing else known.
Status IUCN Red List: **Near Threatened**. Poorly known, range not heavily fished.

SWELLSHARK *Cephaloscyllium ventriosum*

FAO code: **CPV**   Plate page 416

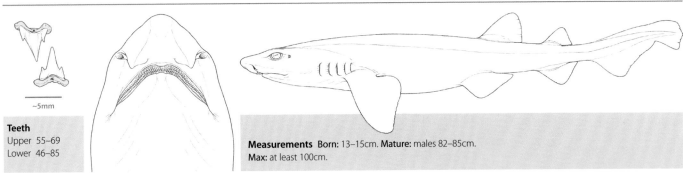

Teeth
Upper 55–69
Lower 46–85

~5mm

Measurements Born: 13–15cm. **Mature:** males 82–85cm.
Max: at least 100cm.

Identification Large swellshark. Ridges over eyes. Two dorsal fins, second much smaller. Anal fin. Strongly variegated with close-set, dark brown saddles and blotches, numerous dark spots and occasional light spots on a lighter yellow-brown background. Underside also heavily spotted.
Distribution East Pacific: California to Mexico, central Chile.
Habitat Continental shelves and upper slopes from inshore to 457m, most common 5–40m. Rocky bottom in kelp beds and other algae.
Behaviour Relatively sluggish and mainly nocturnal. Lies motionless in rocky caves and crevices by day, often in small groups, and swims slowly at night. Inflates stomach when disturbed to wedge itself into crevices.
Biology Oviparous. Eggcases 9–13 x 3–6cm, large, greenish-amber and purse-shaped, very long coiled tendrils, smooth surface, without ridges. Hatch in 7.5–10 months (depending on water temperature). Feeds on fishes and crustaceans.
Status IUCN Red List: **Least Concern**. Not commercially fished. Kept in aquaria.

NARROWBAR SWELLSHARK *Cephaloscyllium zebrum*

FAO code: **SYX**   Plate page 418

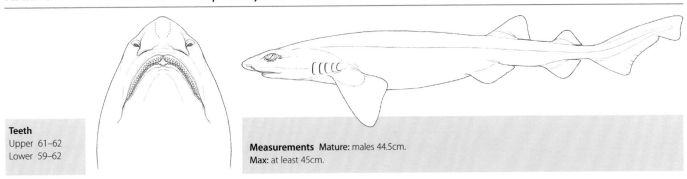

Teeth
Upper 61–62
Lower 59–62

Measurements Mature: males 44.5cm.
Max: at least 45cm.

Identification Ridge over eyes. Two dorsal fins; first set behind pelvic fins; second very much smaller, over anal fin. Inflatable stomach. Very distinctive colour pattern of numerous, narrow, closely spaced, dark bars on a dark brownish to cream background. Bars do not join to form rings or saddles, 17–18 present predorsally. Irregular lines on snout. Uniformly pale fins and underside.
Distribution Central Indo-Pacific: northeast Australia.
Habitat Upper continental slope, 444–454m.
Behaviour Unknown.
Biology Oviparous.
Status IUCN Red List: **Data Deficient**. Known from only a few specimens trawled from one location. Subject to minimal fishing pressure.

PYJAMA SHARK *Poroderma africanum*

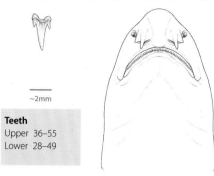

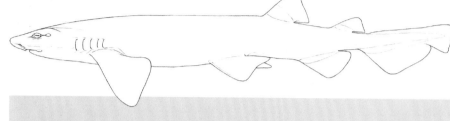

~2mm

Teeth
Upper 36–55
Lower 28–49

Measurements Born: 14–17cm. **Mature:** males 72cm, females 78cm.
Max: 109cm.

Alternative name Striped Catshark.
Identification Unmistakable combination of striking longitudinal stripes, no spots, prominent but short nasal barbels, and dorsal fins (second much smaller) set very far back on body.
Distribution Southeast Atlantic and west Indian Ocean: apparently endemic to South Africa (both Capes, rarely to KwaZulu-Natal).
Habitat Continental shelf to upper slope, from surf zone and intertidal to 108m. On or near bottom in rocky areas, often in caves.
Behaviour Nocturnal, but sometimes active during the day.
Biology Oviparous, lays pairs of eggcases (one per oviduct); one egg hatched in an aquarium after about 5.5 months. Eats bony fishes, hagfish, other small sharks, shark eggcases and a wide range of invertebrates.
Status IUCN Red List: **Least Concern**. Hardy in captivity. Taken by trawlers and anglers, but is quite abundant and likely has high survival when released.

LEOPARD CATSHARK *Poroderma pantherinum*

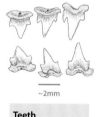

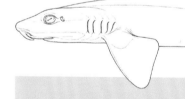

~2mm

Teeth
Upper 36–60
Lower 26–52

Measurements Born: 11cm. **Mature:** males 61cm, females 51cm.
Max: 77cm.

Identification Long nasal barbels reach mouth, which extends behind front of eyes. Dorsal fins set far back, first much larger than second. Striking, leopard-like rosettes of dark spots and lines surrounding light centres, usually arranged in irregular longitudinal rows; variations include numerous, small dense spots to very large, dark spots and partial longitudinal stripes.
Distribution Southeast Atlantic and west Indian Ocean: apparently endemic to South Africa (both Capes, rarely to KwaZulu-Natal).
Habitat Continental to upper slope from surf zone and intertidal to 274m. On or near bottom.
Behaviour Apparently nocturnal.
Biology Oviparous, one egg per oviduct. Feeds on small bony fishes and invertebrates.
Status IUCN Red List: **Least Concern**. Taken by trawlers and anglers, but prefers rocky reef habitat that provides a refuge from bottom trawl fisheries. Hardy when kept in aquaria. The large-spotted variant was formerly considered a separate species, *Poroderma marleyi*.

Pyjama Shark, *Poroderma africanum* (p. 428).

NARROWMOUTH CATSHARK   *Schroederichthys bivius*   FAO code: **SHV**   Plate page 430

Teeth
Upper 46–54
Lower 38–50

~2mm

Measurements Born: 14–20cm. **Mature:** males 53cm, females 40cm. **Max:** males 86cm, females 70cm.

Identification Fairly slim body. Short, narrow, rounded snout. Anterior nasal flaps narrow and lobate. First dorsal origin slightly in front of pelvic insertions. Short caudal fin. 7–8 dark brown saddles on grey-brown back (two conspicuous saddles in interdorsal space); scattered large, dark spots not bordering the saddles; small white spots usually present. Adult males longer and lighter, with much larger teeth and longer, narrower mouths than females; mouths of young even longer and more slender.
Distribution Southeast Pacific and southwest Atlantic: central Chile to southern Brazil.
Habitat Temperate continental shelf and upper slope, 12–359m (mostly <130m), in deeper water in south.
Behaviour Largely unknown.
Biology Oviparous, probably lays pairs of eggs (one per oviduct) in sheltered nursery grounds. Feeds mainly on squat lobsters.
Status IUCN Red List: **Least Concern**. Common where it occurs.

Plate 53 SCYLIORHINIDAE IV – *Poroderma* and *Schroederichthys* catsharks

○ **Pyjama Shark** *Poroderma africanum*

Southeast Atlantic and west Indian Ocean; to 108m. Prominent short nasal barbels; dorsal fins set far back, first dorsal much larger than second; unmistakable pattern of striking longitudinal stripes, continuous from snout to tail; no spots.

○ **Leopard Catshark** *Poroderma pantherinum*

Southeast Atlantic and west Indian Ocean; to 274m. Long nasal barbels reach mouth; dorsal fins set far back, first much larger than second; striking leopard-like rosettes of dark spots and lines surround light centres, usually in irregular rows; variations include many small dense spots, very large spots, partial longitudinal stripes.

○ **Narrowmouth Catshark** *Schroederichthys bivius*

Southeast Pacific to southwest Atlantic; 12–359m. Fairly slim body; short narrow rounded snout, narrow, lobate anterior nasal flaps; short caudal fin; 7–8 dark brown saddles, 2 conspicuous between dorsal fins, scattered large dark spots and usually small white spots.

○ **Redspotted Catshark** *Schroederichthys chilensis*

Southeast Pacific; 1–100m. Moderately slim; short broad rounded snout, very broad wide mouth, anterior nasal flaps broad and triangular; short caudal fin; conspicuous dark saddles, 2 interdorsal; numerous dark spots, few or no white spots.

○ **Narrowtail Catshark** *Schroederichthys maculatus*

West Atlantic: Caribbean; 190–412m. Very slender, elongated trunk and caudal fin; rounded snout, long broad, triangular anterior nasal flaps; juveniles have 6–9 light inconspicuous brown saddles, 3 interdorsal, absent in adults; juveniles and adults with numerous small white spots.

○ **Lizard Catshark** *Schroederichthys saurisqualus*

Southwest Atlantic; 122–500m. Slender trunk and caudal fin; rounded snout, moderately wide mouth, anterior nasal flaps long narrow and lobate; 10 conspicuous dusky saddles on lighter grey or brown background, 4 interdorsal, numerous white spots interspersed with dark spots.

○ **Slender Catshark** *Schroederichthys tenuis*

West Atlantic; 72–450m. Very slender; broad snout, narrow mouth, anterior nasal flaps narrow and lobate; light brown with 7–8 conspicuous dark saddles, 4 interdorsal, outlined by many small dark spots, no white spots.

20cm

◯ *Poroderma pantherinum*

◯ *Poroderma africanum*

◯ *Schroederichthys bivius*

◯ *Schroederichthys chilensis*

◯ *Schroederichthys maculatus*

◯ *Schroederichthys saurisqualus*

◯ *Schroederichthys tenuis*

REDSPOTTED CATSHARK *Schroederichthys chilensis* FAO code: **SHY** Plate page 430

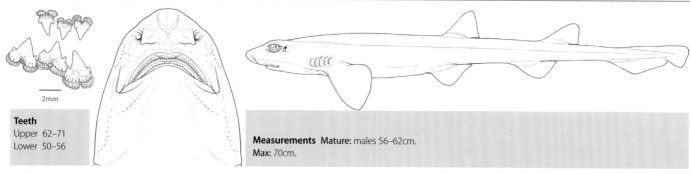

Teeth
Upper 62–71
Lower 50–56

2mm

Measurements Mature: males 56–62cm.
Max: 70cm.

Identification Moderately slim body. Short, broad, rounded snout. Anterior nasal flaps broad and triangular. Very broad mouth. First dorsal origin slightly in front of pelvic insertions. Short caudal fin. Young more slender than adults. Conspicuous dark saddles, including two in the interdorsal space; numerous dark spots that do not border saddles; white spots few or absent.
Distribution Southeast Pacific: South America, off Peru and south-central Chile.
Habitat Temperate inshore continental shelf, on or near bottom, sometimes in very shallow inshore water, 1–100m.
Behaviour Unknown.
Biology Oviparous. Probably lays pairs of eggs (one per oviduct), with tendrils on cases. Feeds on crustaceans and other invertebrates.
Status IUCN Red List: Least Concern. An occasional bycatch in inshore fisheries, usually discarded.

NARROWTAIL CATSHARK *Schroederichthys maculatus* FAO code: **SHU** Plate page 430

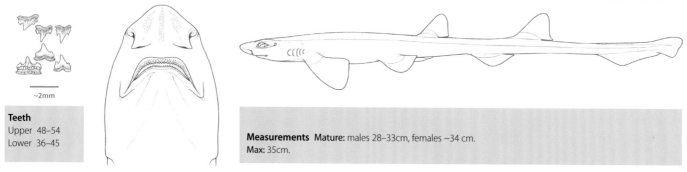

Teeth
Upper 48–54
Lower 36–45

~2mm

Measurements Mature: males 28–33cm, females ~34 cm.
Max: 35cm.

Identification Extremely slender, elongated trunk and caudal fin in juveniles and adults. Rounded snout. Broad, elongate-triangular anterior nasal flaps. Broad mouth. First dorsal origin slightly behind pelvic insertions. Juvenile pattern of 6–9 light, inconspicuous, brown saddles on darker tan to grey background disappears in adults; three saddles in interdorsal space; numerous scattered white spots but no dark spots
Distribution West Atlantic: Central and South America off Honduras, Nicaragua and Colombia, and between the Honduras Bank and Jamaica.
Habitat Shelly or sandy bottom in deep water, 190–412m, tropical outer shelf and upper slope.
Behaviour Unknown.
Biology Oviparous. Probably lays pairs of eggs, with tendrils on cases. Feeds on small bony fishes and squids.
Status IUCN Red List: Least Concern. This species' restricted deep water distribution is largely unfished.

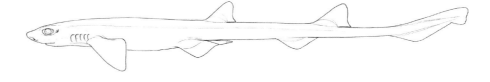

~2mm

Teeth
Upper 56–66
Lower 37–50

Measurements Born: at least 9cm. **Mature:** males 58–59cm, females 55cm.
Max: males 61cm, females 69cm.

Identification Slender, elongated trunk and tail in adults and subadults. Rounded snout. Elongated, narrow, lobate anterior nasal flaps. Moderately wide mouth. First dorsal origin slightly behind pelvic insertions. Ten conspicuous dusky saddles on lighter grey or brown background in adults and subadults, four between dorsal fins; numerous white spots interspersed with dark spots.
Distribution Southwest Atlantic: South America off southern Brazil.
Habitat Outer shelf and upper slope at 122–500m, mostly below 250m on deep reef habitats. Occurs with deepsea gorgonians, hard corals, tube sponges, crinoids, brittle stars and Freckled Catshark *Scyliorhinus haeckelii*.
Behaviour Unknown.
Biology Oviparous. Probably lays pairs of eggs, with tendrils on cases.
Status IUCN Red List: **Vulnerable**. This species, if valid, has a very limited known range.

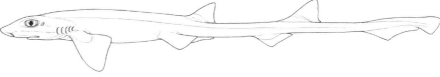

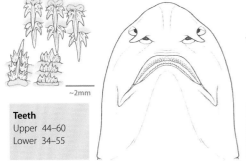

~2mm

Teeth
Upper 44–60
Lower 34–55

Measurements Mature: males 40–47cm, females 37–46cm.
Max: 47cm.

Identification Very slender body. Broad snout. Anterior nasal flaps narrow and lobate. Mouth narrow; males have much larger teeth and a longer, more angular mouth than females (may reach a larger size). First dorsal origin slightly behind pelvic insertions. Colour is light brown with many small, dark spots outlining and scattered between 7–8 conspicuous dark saddles, four between dorsal fins; no white spots.
Distribution West Atlantic: South America, Suriname and northern Brazil (north and just south of the Amazon River mouth).
Habitat On or near bottom, outer continental shelf and upper continental slope, 72–450m.
Behaviour Unknown.
Biology Oviparous, lays eggs in pairs. Eggcases 3.5 x 1.7cm with long tendrils at each end. Eats foraminifera, small fishes, crustaceans, sponges, squids, gastropods and possibly other small sharks.
Status IUCN Red List: **Least Concern**. Captured in deepsea fisheries in the shallower part of its range.

Plate 54 SCYLIORHINIDAE V – *Scyliorhinus* catsharks I

○ **Rio Catshark** *Scyliorhinus cabofriensis*   page 439

Southwest Atlantic; 387–647m. Dorsal surface dark beige with random pattern of various-sized (mostly small) asymmetrical black and white spots, poorly defined saddles without sharp median projections; belly lighter.

○ **Smallspotted Catshark** *Scyliorhinus canicula*   page 439

Northeast Atlantic and west Mediterranean; to 800m. Slender; greatly expanded anterior nasal flaps reach mouth, lower labial furrow only; first dorsal fin larger than second; light background, 8–9 dusky saddles may be unclear, numerous small dark spots, occasionally a few white.

○ **Yellowspotted Catshark** *Scyliorhinus capensis*   page 440

Southeast Atlantic and west Indian Ocean; 26–695m. Anterior nasal flaps small, no nasoral grooves, lower labial furrow only; 8–9 irregular dark grey saddles, numerous small bright yellow spots, no dark spots.

○ **West African Catshark** *Scyliorhinus cervigoni*   page 440

East Atlantic; 45–500m. Very stout; anterior nasal flaps just reach mouth, no nasoral grooves, lower labial furrow only; 8–9 dark saddles centred on dark spots on midline of back, few relatively large and some small scattered dark spots, no white spots.

○ **Comoro Catshark** *Scyliorhinus comoroensis*   page 441

Southwest Indian Ocean; 400–700m. Very large anterior nasal flaps reach mouth, no nasoral grooves, lower labial furrow only; dark bar under eye, bold sharply defined saddles centred on dark spots on midline of back and large blotches on lighter background, numerous small white spots, no small bold dark spots.

○ **Duhamel's Catshark** *Scyliorhinus duhamelii*   page 441

Mediterranean; 43–75m. Beige background with a pattern of scattered various sized dark brown spots forming aggregations and rosettes above; mostly cream without spots below.

○ **Nursehound** *Scyliorhinus stellaris*   page 444

Northeast Atlantic and Mediterranean; 1–380m. Stocky; widely separated anterior nasal flaps do not reach mouth, no nasoral grooves, lower labial furrow only; saddles faint to absent, numerous large to small dark spots over pale background, sometimes white spots, large spots may be irregular, occasionally expand into blotches that totally cover the body.

○ **Dark Freckled Catshark** *Scyliorhinus ugoi*   page 446

West Atlantic; 400–825m. Light brown with darker brown saddles, sometimes with anterior and posterior projections, small light and dark spots covering body, no large white spots; ventrally light brown.

○ *Scyliorhinus capensis*

○ *Scyliorhinus stellaris*

20cm

20cm

○ *Scyliorhinus cabofriensis*

○ *Scyliorhinus duhamelii*

○ *Scyliorhinus canicula*

○ *Scyliorhinus cervigoni*

○ *Scyliorhinus comoroensis*

○ *Scyliorhinus ugoi*

Plate 55 SCYLIORHINIDAE VI – *Scyliorhinus* catsharks II

○ **Boa Catshark** *Scyliorhinus boa*

page 438

West Atlantic: Caribbean; 36–700m. Slender; anterior nasal flaps do not reach mouth, lower labial furrow only; first dorsal fin much larger than second; greyish background, inconspicuous saddles and blotches, outlined by black spots, sometimes in reticulating rows or broken black lines, occasionally a few white spots.

○ **Brownspotted Catshark** *Scyliorhinus garmani*

page 442

Central Indo-Pacific; unknown depth. Stocky; anterior nasal flaps do not reach mouth, no nasoral grooves, lower labial furrow only; first dorsal fin much larger than second; 7 indistinct saddles, scattered large round brown spots, no white.

○ **Freckled Catshark** *Scyliorhinus haeckelii*

page 442

West Atlantic; 35–585m. Slender; anterior nasal flaps do not reach mouth, lower labial furrow only; first dorsal fin much larger than second; 7–8 dusky saddles may be inconspicuous, dark bar under eye, small black spots scattered over back and outline saddles, no light spots.

○ **Whitesaddled Catshark** *Scyliorhinus hesperius*

page 443

West Atlantic: Caribbean; 200–634m. Slender; anterior nasal flaps nearly reach mouth, lower labial furrow only; first dorsal fin much larger than second; 7–8 well-defined dark saddles covered by large white spots, no black spots.

○ **Blotched Catshark** *Scyliorhinus meadi*

page 443

West Atlantic; 146–549m. Stocky; broad head, anterior nasal flaps nearly reach mouth, lower labial furrow only; first dorsal fin much larger than second; 7–8 darker saddles may be obscure, no spots.

○ **Chain Catshark** *Scyliorhinus retifer*

page 444

West Atlantic; 73–754m. Anterior nasal flaps do not reach mouth, lower labial furrow well developed; dorsal fins set well back; black chain patterning outlines faint dusky saddles, no spots.

○ **Cloudy Catshark** *Scyliorhinus torazame*

page 445

Northwest Pacific; close inshore to at least 320m. Slender; narrow head, anterior nasal flaps do not reach mouth, lower labial furrow only; first dorsal fin much larger than second; 6–9 darker saddles, many irregular large dark and light spots on rough dark skin in larger specimens.

○ **Dwarf Catshark** *Scyliorhinus torrei*

page 445

West Atlantic; 180–591m. Small, slender; anterior nasal flaps do not reach mouth, lower labial furrow only; first dorsal fin much larger than second; light brown background with 7–8 darker brown saddles, obscure in adults; many large regularly scattered white spots on back, no black spots.

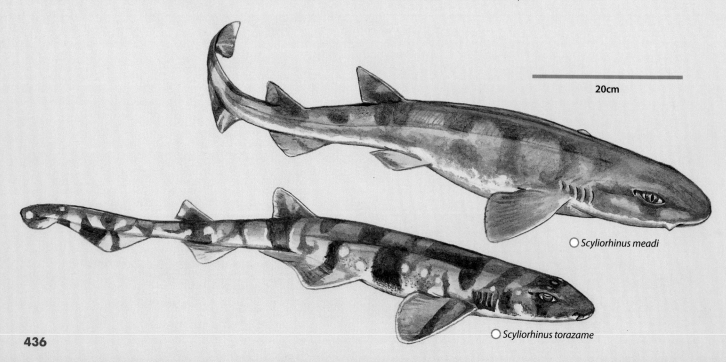

20cm

○ *Scyliorhinus meadi*

○ *Scyliorhinus torazame*

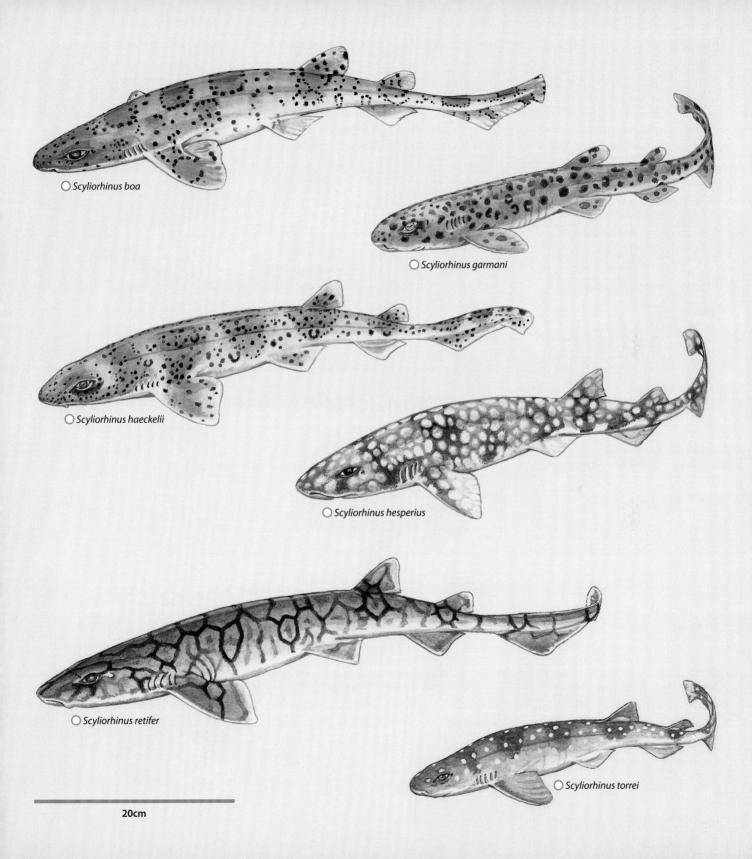

○ *Scyliorhinus boa*

○ *Scyliorhinus garmani*

○ *Scyliorhinus haeckelii*

○ *Scyliorhinus hesperius*

○ *Scyliorhinus retifer*

○ *Scyliorhinus torrei*

20cm

Smallspotted Catshark, *Scyliorhinus canicula* (p. 439).

BOA CATSHARK   *Scyliorhinus boa*   FAO code: **SYA**   Plate page 436

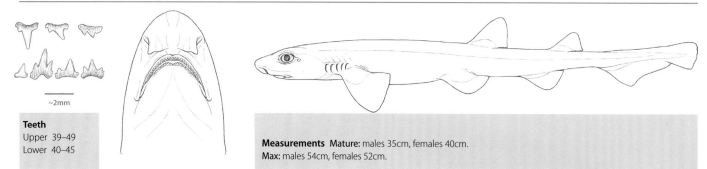

~2mm

Teeth
Upper 39–49
Lower 40–45

Measurements Mature: males 35cm, females 40cm.
Max: males 54cm, females 52cm.

Identification Slender catshark. Small anterior nasal flaps do not reach mouth; no nasoral grooves. Labial furrows on lower jaw only. Second dorsal fin much smaller than first. Colour is greyish with inconspicuous saddles and flank markings outlined by many small, black spots, sometimes in reticulating rows or broken black lines. Sometimes a few white spots; few or no spots inside saddles.

Distribution West Atlantic: Caribbean insular slope off Barbados, Hispanola, Jamaica, Leeward Islands, Windward Islands; continental slope off Nicaragua, Honduras, Panama, Colombia, Venezuela, Suriname, and to southern Brazil and northern Uruguay.

Habitat Continental and insular slopes, on or near bottom, 36–700m, but mostly below 200m.

Behaviour Unknown.

Biology Oviparous.

Status IUCN Red List: Least Concern. Most of this species' range is below the depth of bottom trawl fisheries in the region.

RIO CATSHARK *Scyliorhinus cabofriensis*   FAO code: **SCL**   Plate page 434

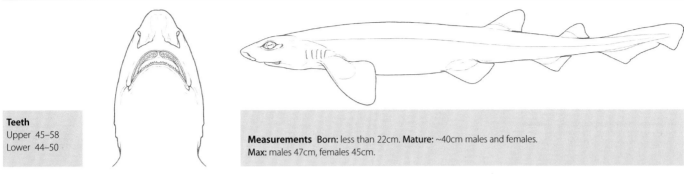

Teeth
Upper 45–58
Lower 44–50

Measurements Born: less than 22cm. **Mature:** ~40cm males and females. **Max:** males 47cm, females 45cm.

Identification Body slender, cylindrical, tapering rapidly from pelvic fin origins to caudal fin origin. Anterior nasal flaps large, covering posterior nasal flaps, extending to just anterior to mouth; nasoral grooves absent. Labial furrows only on lower jaw, uppers absent. First dorsal fin is noticeably larger than second. Background coloration is dark beige, patterned with random, asymmetrical black and white spots of varied sizes, although mostly small; saddles not well defined and without sharp median projections. Lighter below.
Distribution Southwest Atlantic: known only from off Rio de Janeiro state, on the southeast coast of Brazil.
Habitat Continental slope at 387–647m.
Behaviour Unknown.
Biology Oviparous, eggcases unknown. Feeds on small bony fishes and cephalopods.
Status IUCN Red List: Least Concern. Its restricted distribution appears to be unfished.

SMALLSPOTTED CATSHARK *Scyliorhinus canicula*   FAO code: **SYC**   Plate page 434

5mm

Teeth
Upper 40–61
Lower 36–50

Measurements Born: 9–10cm. **Mature:** N Atlantic: males 52–56cm, females 54–60cm. Mediterranean: males 37–40cm, females 37–47cm. **Max:** males 65cm, females 71cm, rarely 80cm, record to 100cm may be *S. stellaris*.

Identification Large, slender body. Greatly expanded anterior nasal flaps reach mouth and cover shallow nasoral grooves. Labial furrows on lower jaw only. Second dorsal much smaller than first. Body covered in numerous small, dark spots on light background; scattered white spots sometimes present; 8–9 dusky saddles may be unclear.
Distribution Northeast Atlantic: Norway and British Isles to western Mediterranean, Morocco, Sahara Republic, and Mauritania to Senegal, Ivory Coast.
Habitat Continental shelves and upper slopes, on sandy and muddy sediment from nearshore to 800m (mostly shallower than 450m).
Behaviour Adults often found in single sex schools; young and hatchlings in shallower water.
Biology Oviparous. Eggcases 4–6 x 2–3cm, long tendrils at each end and fine striations along surface; size varies with female size (smaller in Mediterranean). Females may deposit 40–240 eggcases, in pairs, year-round (mostly November–July) on seaweed. Hatch in 5–11 (mostly 8–9) months. Feeds on small, benthic invertebrates (crustaceans, gastropods, cephalopods, worms) and fishes.
Status IUCN Red List: Least Concern. Taken in many fisheries, retained and discarded, but high discard survival and some populations are stable or increasing. Hardy aquarium species, breeds in captivity.

YELLOWSPOTTED CATSHARK   *Scyliorhinus capensis*   FAO code: **SYP**   Plate page 434

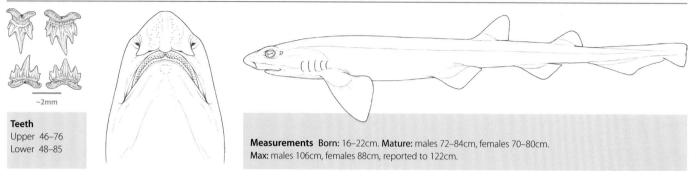

Teeth
Upper 46–76
Lower 48–85

Measurements Born: 16–22cm. **Mature:** males 72–84cm, females 70–80cm.
Max: males 106cm, females 88cm, reported to 122cm.

Identification Fairly large catshark. Small anterior nasal flaps; no nasoral grooves. Labial furrows on lower jaw only. Second dorsal fin much smaller than first. Coloration is grey with 8–9 irregular, darker grey saddles and numerous small bright yellow spots; no dark spots.
Distribution Southeast Atlantic and west Indian Ocean: southern Namibia and South Africa.
Habitat Bottom on continental shelf and upper slope, including soft bottom where it is frequently trawled, 26–695m, mostly 200–400m, deeper in warmer water.
Behaviour Coils up tightly when caught.
Biology Oviparous. Eggcases ~8 x 3cm with long coiled tendrils at each end; deposited in pairs. Feeds on small fishes and many invertebrates.
Status IUCN Red List: Near Threatened. Moderately common on heavily fished offshore banks, discarded from bycatch; population trend appears to be increasing.

WEST AFRICAN CATSHARK   *Scyliorhinus cervigoni*   FAO code: **SYE**   Plate page 434

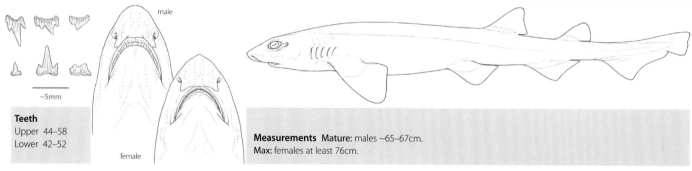

male

female

Teeth
Upper 44–58
Lower 42–52

Measurements Mature: males ~65–67cm.
Max: females at least 76cm.

Identification Very stout. Small anterior nasal flaps barely reach mouth; no nasoral grooves. Labial furrows on lower jaw only. Second dorsal fin much smaller than first. Interdorsal space slightly less than anal base. A few, relatively large and some small, scattered dark spots; 8–9 dusky saddles centred on dark spots on midline of back; no white spots.
Distribution East Atlantic: tropical West Africa from Mauritania to Angola.
Habitat Rocky and mud bottom on continental shelf and upper slope, 45–500m, mostly 150–260m.
Behaviour Unknown.
Biology Oviparous. Eggcases 7–8 x 3cm. Eats bony fishes.
Status IUCN Red List: Data Deficient. Probably taken in trawl fisheries. Has been recorded as *Scyliorhinus stellaris*.

COMORO CATSHARK *Scyliorhinus comoroensis*

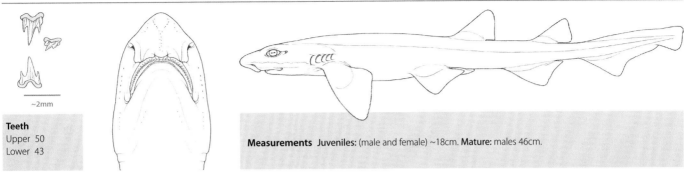

Teeth
Upper 50
Lower 43

~2mm

Measurements Juveniles: (male and female) ~18cm. **Mature:** males 46cm.

Identification Small-bodied. Large anterior nasal flaps reach mouth but no nasoral grooves. Labial furrows on lower jaw only. Second dorsal fin much smaller than first. Patterned with bold, sharply defined, dark grey-brown saddles centred on dark spots on the midline of the back and large blotches on a light grey-brown background; numerous scattered, small white spots in saddles and spaces between them; no small, bold dark spots. Conspicuous dark bar under eye.
Distribution Southwest Indian Ocean: Comoro Islands and northwest coast of Madagascar.
Habitat Insular slopes, on bottom, 400–700m.
Behaviour Photographed in deepsea by a research submersible studying coelacanth *Latimeria chalumnae*.
Biology Probably oviparous.
Status IUCN Red List: Data Deficient. Known only from three specimens. Only recently recorded from Madagascar.

DUHAMEL'S CATSHARK *Scyliorhinus duhamelii*

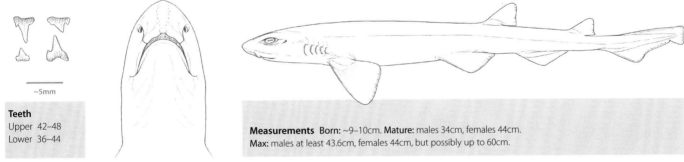

Teeth
Upper 42–48
Lower 36–44

~5mm

Measurements Born: ~9–10cm. **Mature:** males 34cm, females 44cm.
Max: males at least 43.6cm, females 44cm, but possibly up to 60cm.

Identification Body slender, cylindrical, tapering quickly behind pelvic fins to caudal fin origin. Anterior nasal flaps moderate in size, closely spaced, with their borders rounded laterally and posteriorly; extend to the mouth and cover the nasoral grooves. Labial furrows only on lower jaw. First dorsal fin is larger than second. Background coloration is beige, with a pattern of scattered dark brown spots of varied sizes forming aggregations and rosettes on the dorsal and lateral surfaces. Ventrally mostly cream without spots.
Distribution Mediterranean Sea: endemic to the Adriatic.
Habitat Continental shelf from 43–75m.
Behaviour Unknown.
Biology Oviparous, but nothing else known, since the species until recently was misidentified with the Smallspotted Catshark *Scyliorhinus canicula*. Diet is unknown, although as with similar species, likely includes bony fishes, cephalopods and crustaceans.
Status IUCN Red List: Not Evaluated. This species has only recently been resurrected as being a valid species distinct from *Scyliorhinus canicula*.

BROWNSPOTTED CATSHARK   *Scyliorhinus garmani*   FAO code: **SYG**   Plate page 436

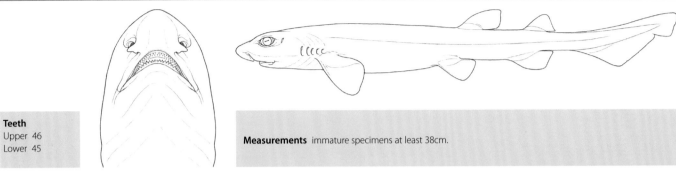

Teeth
Upper 46
Lower 45

Measurements immature specimens at least 38cm.

Identification Stocky body. Anterior nasal flaps do not reach mouth; no nasoral grooves. Labial furrows on lower jaw only. Second dorsal much smaller than first. Anal base shorter than interdorsal space. Patterned with large, scattered, round, brown spots; seven indistinct saddle markings; no white spots.
Distribution Central Indo-Pacific: collected in 'East Indies', possibly Philippines (Negros Island).
Habitat Unknown.
Behaviour Unknown.
Biology Unknown.
Status IUCN Red List: Data Deficient. Described from a few specimens collected over 100 years ago.

FRECKLED CATSHARK   *Scyliorhinus haeckelii*   FAO code: **SYH**   Plate page 436

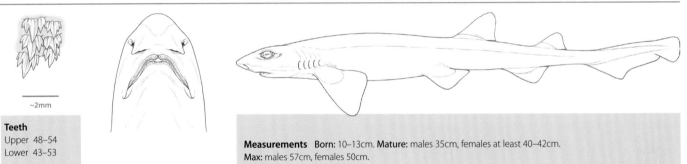

~2mm

Teeth
Upper 48–54
Lower 43–53

Measurements **Born:** 10–13cm. **Mature:** males 35cm, females at least 40–42cm.
Max: males 57cm, females 50cm.

Identification Small, slender-bodied. Small anterior nasal flaps do not reach mouth. Second dorsal fin much smaller than first. Adult males have larger teeth, longer mouths and larger lateral denticles than females. 7–8 dusky (sometimes faint) saddles. Conspicuous dark bar under eye. Very small black spots scattered over back and outlining saddles. No light spots.
Distribution West Atlantic: Venezuela to northern Argentina.
Habitat Continental shelf and upper slope, on or near seabed, 35–585m. On deep-reef habitats (mostly below 250m) off southern Brazil in association with deep water gorgonians, hard corals, tube sponges, crinoids and brittlestars.
Behaviour Unknown.
Biology Oviparous. Eggcases 6–7 x 2–3cm, without longitudinal grooves, laid in pairs on corals and seafans. Feed on cephalopods and bony fishes.
Status IUCN Red List: Data Deficient. Intense fishing pressure throughout its range has depleted many other shark species, but this catshark may have a refuge from fisheries in rough, untrawlable areas.

WHITESADDLED CATSHARK *Scyliorhinus hesperius*

FAO code: **SYU**   Plate page 436

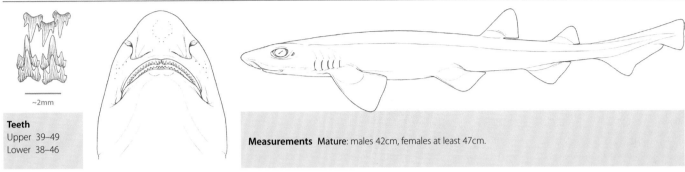

Teeth
Upper 39–49
Lower 38–46

~2mm

Measurements Mature: males 42cm, females at least 47cm.

Identification Fairly small and slender catshark. Small anterior nasal flaps end in front of mouth; no nasoral grooves. Labial furrows on lower jaw only. Second dorsal fin much smaller than first. 7–8 well-defined, dark saddles densely covered with large, closely spaced white spots which sometimes extend to lighter spaces between saddles; no black spots. Conspicuous dark bar under eye.
Distribution West Atlantic: Caribbean; Guatemala, Honduras, Nicaragua, Panama and Colombia.
Habitat Upper continental slope, on or near bottom, 200–634m.
Behaviour Unknown.
Biology Unknown.
Status IUCN Red List: Least Concern. Adult males appear to occupy habitats unsuitable for trawling.

BLOTCHED CATSHARK *Scyliorhinus meadi*

FAO code: **SYM**   Plate page 436

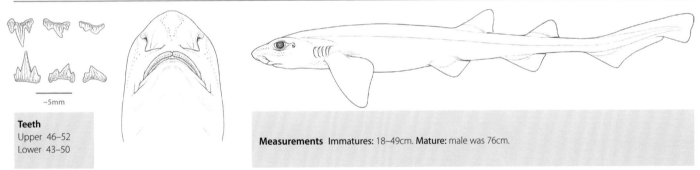

Teeth
Upper 46–52
Lower 43–50

~5mm

Measurements Immatures: 18–49cm. **Mature:** male was 76cm.

Identification Stocky catshark. Broad head. Small anterior nasal flaps end just in front of mouth; no nasoral grooves. Labial furrows on lower jaw only. Second dorsal fin much smaller than first. Colour is dark with 7–8 darker and sometimes obscure saddles; no spots.
Distribution West Atlantic: North Carolina to Florida, USA; Santaren Channel between Cuba and Bahamas Bank.
Habitat Continental slope, on or near bottom, 146–549m.
Behaviour Unknown.
Biology Poorly known, presumably oviparous. Eats cephalopods, shrimp and bony fishes.
Status IUCN Red List: Least Concern. Relatively rare. Records from Mexico (Gulf of Mexico and northern Yucatan Peninsula) were a misidentification; those specimens belong to the genus *Galeus*.

CHAIN CATSHARK *Scyliorhinus retifer*
FAO code: **SYF** Plate page 436

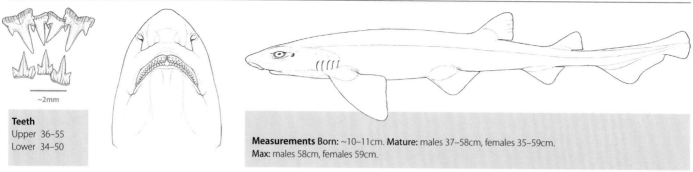

Teeth
Upper 36–55
Lower 34–50

~2mm

Measurements Born: ~10–11cm. **Mature:** males 37–58cm, females 35–59cm.
Max: males 58cm, females 59cm.

Identification Black chain patterning outlining faint dusky saddles, without spots, is unique to this species and *Cephaloscyllium fasciatum*. However, *Scyliorhinus retifer* has well-developed lower labial furrows and dorsal fins set well back.
Distribution West Atlantic: USA (Georges Bank, Massachusetts, to Florida and Texas); Gulf of Mexico and Caribbean, Mexico (Campeche Gulf), Barbados, between Jamaica and Honduras and Nicaragua.
Habitat Outer continental shelf and upper slope, on or near bottom, 73–754m, deeper in the south. May be most common on very rough, rocky untrawlable areas, which are also suitable for egg-laying.
Behaviour Sluggish shark, commonly found resting on bottom. May swallow small pebbles, perhaps for ballast. Females swim rapidly in circles to wrap eggcase tendrils up to 35cm long around seabed projections such as corals.
Biology Oviparous. Eggcases 5–7 x 2–3cm, with longitudinal striations and long coiled tendrils. In captivity eggs laid in pairs every 8–15 days during spring and summer, ~44–52 eggs per year, hatching after about seven months (at 11.7–12.8°C), longer in nursery areas at temperatures down to 7°C. Females caught in the wild, then kept in captivity without males, produced fertilised eggs with viable embryos for seven years.

Young reach 25–30cm after two years in captivity. Feeds on squids, bony fishes, polychaetes and crustaceans.
Status IUCN Red List: Least Concern. Common where it occurs.

NURSEHOUND *Scyliorhinus stellaris*
FAO code: **SYT** Plate page 434

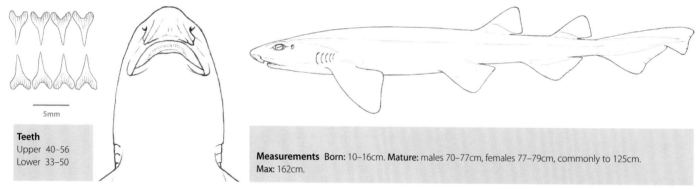

Teeth
Upper 40–56
Lower 33–50

5mm

Measurements Born: 10–16cm. **Mature:** males 70–77cm, females 77–79cm, commonly to 125cm.
Max: 162cm.

Identification Large and stocky. Many large and small black spots, sometimes white spots, over pale background. Saddles very faint or absent. Large spots may be irregular, occasionally expand into large blotches totally covering the body.
Distribution Northeast Atlantic: southern Scandinavia to Mediterranean, Morocco, Mauritania to Senegal. Records further south to Gulf of Guinea and Congo River mouth may be *Scyliorhinus cervigoni*.
Habitat Continental shelf, 1–380m, commonly 20–63m, on rocky or seaweed-covered bottom.
Behaviour Deposit large, thick-walled eggcases with strong tendrils at corners on algae.
Biology Oviparous. Eggcases 10–13 x 3cm, with coiled tendrils at each end and longitudinal striations. Single egg per oviduct laid on algal bottom in subtidal, or extremely low intertidal waters during the spring and summer, may take nine months to hatch. Eats mostly crustaceans, cephalopods, other molluscs, bony fishes and other small sharks.
Status IUCN Red List: Near Threatened. Limited fisheries interest. Less common than *Scyliorhinus canicula*. Size of Mediterranean specimens is on average usually smaller than in the European Atlantic.

CLOUDY CATSHARK *Scyliorhinus torazame*

FAO code: **SYZ**   Plate page 436

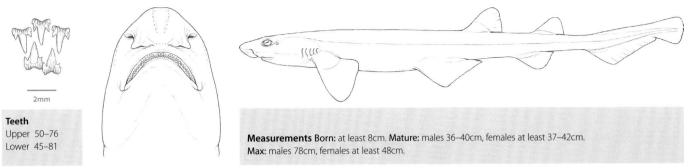

2mm

Teeth
Upper 50–76
Lower 45–81

Measurements Born: at least 8cm. **Mature:** males 36–40cm, females at least 37–42cm. **Max:** males 78cm, females at least 48cm.

Identification Fairly small, slender, narrow-headed shark. Small anterior nasal flaps do not reach mouth; no nasoral grooves. Labial furrows on lower jaw only. Second dorsal fin much smaller than first. Patterned with 6–9 darker saddles and, in larger specimens, many irregular, large dark and light spots on very rough, dark skin.
Distribution Northwest Pacific: Japan (Hokkaido and Honshu to Okinawa), Korea, China and Taiwan.
Habitat Continental shelf and upper slope, close inshore to at least 320m.
Behaviour Eggcases deposited year round in a nursery or hatching ground.
Biology Oviparous. Eggcases 5–6 x 2–3cm, translucent with smooth surface, one egg per oviduct. Feeds on crustaceans, cephalopods and small bony fishes.
Status IUCN Red List: Least Concern. A common discarded bycatch of demersal trawl, gillnet, and longline fisheries in Japanese waters.

DWARF CATSHARK *Scyliorhinus torrei*

FAO code: **SYI**   Plate page 436

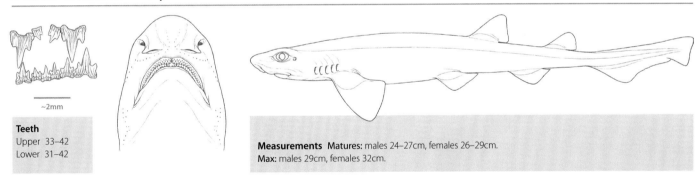

~2mm

Teeth
Upper 33–42
Lower 31–42

Measurements Matures: males 24–27cm, females 26–29cm. **Max:** males 29cm, females 32cm.

Identification Very small, slender catshark. Small anterior nasal flaps do not reach mouth; no nasoral grooves. Labial furrows on lower jaw only. Second dorsal fin much smaller than first. Coloration is light brown with 7–8 darker brown saddles (obscure in adults) and many large, regularly scattered white spots on back; no black spots.
Distribution West Atlantic: Florida Straits, Bahamas, northern Cuba, Virgin Islands.
Habitat Upper continental slope, on or near bottom, 229–550m, mostly below 366m.
Behaviour Unknown.
Biology Poorly known. Diet includes squids and cuttlefish.
Status IUCN Red List: Least Concern. Occurs outside the depth range of current fishing activities.

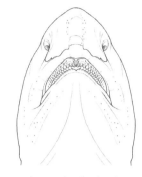

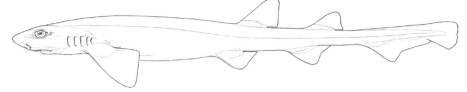

Teeth
Upper 47–56
Lower 45–53

Measurements Mature: males ~45cm, females ~47cm.
Max: males 53cm, females 63cm.

Identification Body stout, depressed on head region, tapering quickly from pelvic fin origins to caudal fin origin. Anterior nasal flaps large, widely spaced, covering posterior nasal flaps; nasoral grooves absent. Lower labial furrows short, uppers absent. First dorsal fin slightly larger than second. Coloration is light brown with darker brown saddles, occasionally with anterior and posterior projections. Small light and dark spots, but no large white spots, cover body. Light brown below.

Distribution West Atlantic: from southern Santa Catarina state, southeast Brazil, to Barbados, Caribbean Sea.

Habitat Continental slope at 400–825m. Usually associated with coral formations, where this species deposit their eggcases.

Behaviour Unknown.

Biology Oviparous. Eggcases 15 x ~6cm, surface smooth to the touch; horns at each end short and without tendrils. Feeds on small bony fishes, cephalopods and, in one specimen, the eggcase of a hagfish.

Status IUCN Red List: **Least Concern**. Poorly known, but its deep slope distribution is currently unfished.

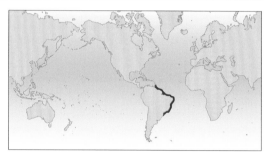

Nursehound, *Scyliorhinus stellaris* (p. 444).

Proscylliidae: finback catsharks

A small family containing three genera, with six described species included here: *Ctenacis* (one species), *Eridacnis* (three species) and two species of *Proscyllium*.

Identification Dwarf to small sharks (adults 16–65cm) with narrowly rounded head and rounded or subangular snout, no deep groove in front of elongated, cat-like eyes. Rudimentary nictitating eyelids. No barbels or nasoral grooves, internarial space less than 1.3 times nostril width, long angular arched mouth reaching past anterior ends of eyes, small papillae on palate and edges of gill arches, labial furrows very short or absent. First dorsal fin base short and well ahead of pelvic fin bases, but closer to pelvic bases than pectoral bases, no precaudal pits, caudal fin without a strong lower lobe or lateral undulations on its dorsal margin. Body and fin colour are usually variegated, but *Eridacnis* species have plain bodies and striped caudal fins.

Biology Most are viviparous, except for the oviparous *Proscyllium habereri*, and feed on small fishes and invertebrates.

Status Poorly known deepsea sharks of outer continental and insular shelves and upper slopes, on or near bottom, 50–766m. Most are assessed as Least Concern in the IUCN Red List, due to their deepwater refuge from fishing pressure. Disjunct distribution, mostly in the Central Indo-Pacific but one species in tropical northwest Atlantic.

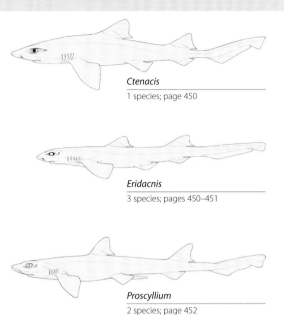

Ctenacis
1 species; page 450

Eridacnis
3 species; pages 450–451

Proscyllium
2 species; page 452

Graceful Catshark, *Proscyllium habereri* (p. 452).

Plate 56 PROSCYLLIIDAE, PSEUDOTRIAKIDAE and LEPTOCHARIIDAE

○ **Harlequin Catshark** *Ctenacis fehlmanni*

page 450

Northwest Indian Ocean; 70–300m. Stoutish; nictitating eyelids, anterior nasal flaps do not reach large triangular mouth, very short labial furrows; large red-brown saddle blotches, smaller round spots and vertical bars, spots on fins.

○ **Cuban Ribbontail Catshark** *Eridacnis barbouri*

page 450

West Atlantic; 400–650m. Slender; nictitating eyelids, anterior nasal flaps do not reach large triangular mouth, very short labial furrows; anal fin ⅔ dorsal fin height; light greyish brown body, faint dark bands on ribbon-like caudal fin, light edges to dorsal fins.

○ **Pygmy Ribbontail Catshark** *Eridacnis radcliffei*

page 451

Indian Ocean and west Pacific; 71–766m. Extremely small, slender; nictitating eyelids, anterior nasal flaps do not reach broad triangular mouth, no labial furrows; anal fin less than ½ dorsal fin height; dark brown body, prominent dark bands on ribbon-like caudal fin, blackish marks on dorsal fins.

○ **African Ribbontail Catshark** *Eridacnis sinuans*

page 451

Southwest Indian Ocean; 180–480m. Slender; nictitating eyelids, anterior nasal flaps do not reach triangular mouth, short labial furrows; anal fin ½ dorsal fin height; grey-brown body, faint dark bands on ribbon-like caudal fin, light margins on dorsal fins.

○ **Graceful Catshark** *Proscyllium habereri*

page 452

West Pacific and Central Indo-Pacific; 50–320m. Slender; large eyes, nictitating eyelids, anterior nasal flaps nearly reach triangular mouth; indistinct dusky saddle blotches on body and fins, small to large dark brown spots, occasionally small white spots.

○ **Magnificent Catshark** *Proscyllium magnificum*

page 452

Northeast Indian Ocean; 141–300m. Similar to *Proscyllium habereri* but more variegated pattern of large and small spots, including spots and a curved bar forming 'clown face' pattern below dorsal fins.

○ **Slender Smoothhound** *Gollum attenuatus*

page 454

Southwest Pacific; 129–975m. Slender; very long angular snout, nictitating eyelids, anterior nasal flaps short, triangular mouth, short labial furrows, small spiracles; relatively short subtriangular first dorsal fin, second dorsal slightly larger than first; body unpatterned grey above, light below.

○ **Sulu Gollumshark** *Gollum suluensis*

page 454

Central Indo-Pacific; ~730m. Shorter snout than other gollumshark; dark grey-brown above lighter below, dorsal, pectoral and pelvic fins with paler posterior margins, terminal caudal lobe light with dark to black edge.

○ **False Catshark** *Pseudotriakis microdon*

page 456

Patchy worldwide, except east Pacific; 100–2430m. Stocky soft body; short bell-shaped snout, elongated cat-like eyes with nictitating eyelids, very large spiracles, anterior nasal flaps short, huge angular mouth, short labial furrows; unpatterned dark brown to blackish.

○ **Eastern Dwarf False Catshark** *Planonasus indicus*

page 455

North Indian Ocean; 200–1000m. Stocky flabby body; fairly long bell-shaped snout, cat-like eyes, nictitating eyelids, very large mouth, short labial furrows uppers shortest; low rounded first dorsal fin, second higher, longer, and narrow tipped; uniformly brownish black.

○ **Pygmy False Catshark** *Planonasus parini*

page 455

Northwest Indian Ocean; 560–1120m. Stocky soft body; short bell-shaped snout, elongated cat-like eyes with nictitating eyelids, anterior nasal flaps short, angular mouth, short labial furrows; relatively short subtriangular first dorsal fin, second dorsal fin larger; grey-brown body, darker fins.

○ **Barbeled Houndshark** *Leptocharias smithii*

page 456

East Atlantic; 5–75m. Slender body; horizontal oval eyes, internal nictitating eyelids; nostrils with slender barbels, long arched mouth, very long labial furrows; light grey-brown, lighter below.

50cm

○ *Pseudotriakis microdon*

Ctenacis fehlmanni

Eridacnis barbouri

Eridacnis radcliffei

Eridacnis sinuans

Proscyllium haberei

Proscyllium magnificum

Gollum attenuatus

50cm

Gollum suluensis

Planonasus parini

Planonasus indicus

Leptocharias smithii

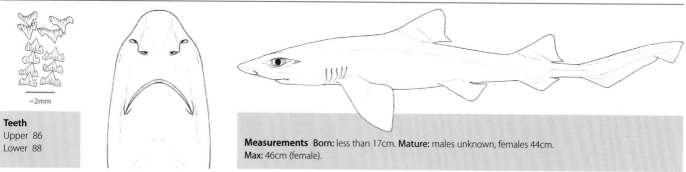

Teeth
Upper 86
Lower 88

~2mm

Measurements Born: less than 17cm. **Mature:** males unknown, females 44cm.
Max: 46cm (female).

Identification   Rather stout body and tail. Short anterior nasal flaps not reaching mouth. Large, triangular mouth; very short labial furrows. Nictitating eyelids. Has a unique colour pattern of large, reddish brown, irregular dorsal saddle blotches, interspersed with smaller, round spots and vertical bars, and spots on fins.
Distribution   Northwest Indian Ocean: off Somalia and Oman.
Habitat   Outer continental shelf and upper slopes, 70–300m.
Behaviour   Unknown.
Biology   Viviparous, with yolk-sac. One adult female had a single developing embryo, but little else known. Smallest free-swimming individual was 17cm. Large mouth, small teeth and large pharynx with gill rakers suggest a diet of very small invertebrates. Unidentified crustaceans were found in a single individual.
Status   IUCN Red List: **Least Concern**. Known from only a few specimens, this little known deepsea shark occurs in an area where no bottom trawling takes place.

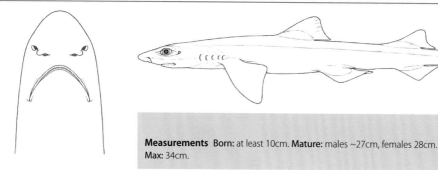

Teeth
Upper 55–68
Lower 63–76

Measurements Born: at least 10cm. **Mature:** males ~27cm, females 28cm.
Max: 34cm.

Identification   Very small, slender. Short anterior nasal flaps do not reach mouth; no nasoral grooves or barbels. Triangular mouth; very short but well-developed labial furrows. Nictitating eyelids. Anal fin about two-thirds of dorsal fin heights. Long, narrow, ribbon-like caudal fin. Light greyish brown with light edges on two dorsal fins and faint dark banding on caudal fin.
Distribution   West Atlantic: Florida Straits and off north coast of Cuba.
Habitat   Upper continental and insular slopes, on the bottom, 400–650m.
Behaviour   Unknown.
Biology   Viviparous, 2 young per litter.
Status   IUCN Red List: **Least Concern**. This deepwater species occurs below the current maximum depth of fisheries in the region.

PYGMY RIBBONTAIL CATSHARK *Eridacnis radcliffei*

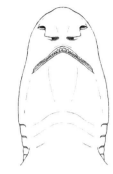

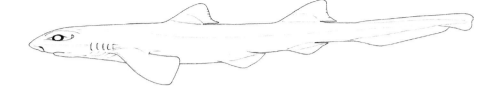

Teeth
Upper 72-78
Lower 65–77

Measurements Born: ~10–11cm. **Mature:** males 16–19cm, females 14–16cm.
Max: males 26cm, females 24cm.

Identification Extremely small shark. Two spineless dorsal fins. Anal fin less than half height of dorsals.
Long, narrow, ribbon-like caudal fin. Dark brown colour; dorsal fins with blackish markings; caudal fin with
prominent dark banding.
Distribution Patchy, Indian Ocean and west Pacific: Tanzania, Gulf of Aden, India (Gulf of Mannar and Bay
of Bengal), Sri Lanka, Andaman Islands, Vietnam, Philippines and Taiwan.
Habitat Mud bottoms on upper continental and insular slopes, outer shelves, 71–766m.
Behaviour Unknown.
Biology Viviparous, 1–2 very large young per litter. Females pregnant year-round in Indian waters, with a
peak in December and January where 76–87% of the female population is gravid. Probably fast-growing.
Feeds mainly on small bony fishes and crustaceans, also squid.
Status IUCN Red List: Least Concern. A bycatch of deepsea trawl fisheries in parts of its range, unfished
elsewhere and in deeper water. This is one of the smallest known shark species.

AFRICAN RIBBONTAIL CATSHARK *Eridacnis sinuans*

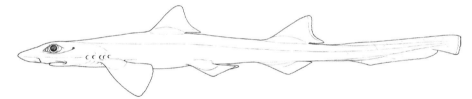

Teeth
Upper 67
Lower 77

Measurements Born: 15–17cm. **Mature:** males 29–30cm, females 37cm.
Max: 37cm.

Identification Slender dwarf catshark. Fairly long snout. Short anterior nasal flaps that do not reach
triangular mouth. Nictitating eyelids. Long, narrow, ribbon-like caudal fin. Grey-brown with faint dark
banding on caudal fin and light margins on dorsal fins.
Distribution Southwest Indian Ocean: South Africa, Mozambique and Tanzania.
Habitat Upper continental slope and outer shelf, 180–480m.
Behaviour Sexes apparently segregate by area or depth: mostly males are (or were) taken off KwaZulu-
Natal.
Biology Viviparous, 2 young per litter. Feeds on small bony fishes, crustaceans and cephalopods.
Status IUCN Red List: Least Concern. Only part of its range is trawled. Species is not utilised and other
areas are unfished.

GRACEFUL CATSHARK *Proscyllium habereri*

FAO code: **CPY** Plate page 448

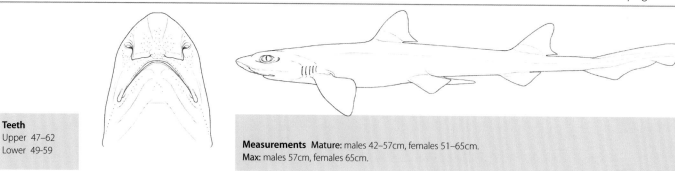

Teeth
Upper 47–62
Lower 49-59

Measurements Mature: males 42–57cm, females 51–65cm.
Max: males 57cm, females 65cm.

Identification Slender body. Large anterior nasal flaps nearly reaching triangular mouth, which extends past eyes. Large eyes with nictitating eyelids. Rather long tail with broad caudal fin. Small to large dark brown spots, sometimes small white spots and indistinct dusky saddle blotches on body and fins.
Distribution West Pacific and Central Indo-Pacific: northwest Java, Vietnam, China, Taiwan, Korea, Ryukyu Islands and southeast Japan.
Habitat Continental and insular shelves, 50–320m.
Behaviour Unknown.
Biology Oviparous, 1 egg per uterus. Feed on bony fishes, crustaceans and cephalopods.
Status IUCN Red List: Vulnerable. Entire range is intensively fished by industrial and artisanal fisheries, which have driven declines in elasmobranch populations, but this small-bodied species appears to be relatively resilient to fishing pressure.

MAGNIFICENT CATSHARK *Proscyllium magnificum*

FAO code: **CVX** Plate page 448

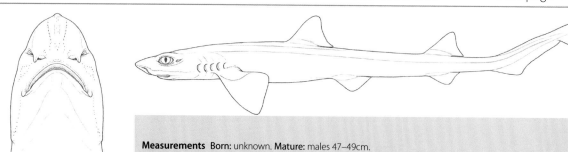

Teeth
Upper 80+
Lower 80+

Measurements Born: unknown. **Mature:** males 47–49cm.
Max: 49cm.

Identification Similar to *Proscyllium habereri*, but with a more variegated pattern of small and large reddish brown spots and dots on lighter background, including clusters of two small round spots above a large upcurved spot and an intermediate small spot forming 'clown faces' below dorsal fins.
Distribution Northeast Indian Ocean: Bay of Bengal off Andaman Islands and Andaman Sea off Myanmar (Burma).
Habitat Near edge of outer continental shelf, known from 141–300m.
Behaviour Unknown.
Biology Unknown.
Status IUCN Red List: Not Evaluated. Originally described from five specimens off Myanmar, several more specimens have been caught in deep water off the Andaman Islands and Myanmar.

Pseudotriakidae: false catsharks

This poorly known family of deepsea sharks contains at least five species in three genera, with possibly one or two undescribed species.

Identification Small to large sharks (adults 56–295cm) with narrowly rounded head and more or less elongated bell-shaped snout, a deep groove in front of elongated, cat-like eyes. Rudimentary nictitating eyelids. No barbels or nasoral grooves, internarial space over 1.5 times nostril width, long angular arched mouth reaching past anterior ends of eyes, no papillae inside mouth and none on edges of gill arches, labial furrows short but always present. First dorsal fin more or less elongated, base closer to pectoral fin bases than to pelvic fin bases, no precaudal pits, caudal fin with a weak lower lobe or none, no lateral undulations on its dorsal margin. Colour usually plain grey to brown or blackish (white spots and fin margins on some *Gollum* species).

Biology Viviparous as far as known, at least three species are known to be oophagous, with foetuses eating nutritive eggs. Probably prey upon small fishes and invertebrates.

Status Poorly known deepsea sharks of outer continental and insular shelves and slopes, on or near the bottom, 100–2430m. Most are assessed as Least Concern, because all or part of their range is currently deeper than fisheries activities. The large *Pseudotriakis microdon* is wide-ranging, but small species have restricted distributions in west Indian Ocean and west Pacific. Absent from the east Pacific.

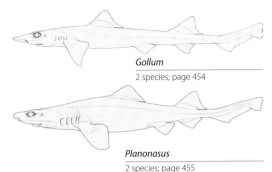

Gollum
2 species; page 454

Planonasus
2 species; page 455

Pseudotriakis
1 species; page 456

False Catshark, *Pseudotriakis microdon* (p. 456).

SLENDER SMOOTHHOUND *Gollum attenuatus*

FAO code: **CPG**   Plate page 448

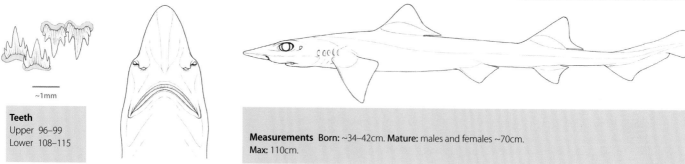

Teeth
Upper 96–99
Lower 108–115

~1mm

Measurements Born: ~34–42cm. **Mature:** males and females ~70cm.
Max: 110cm.

Alternative name Gollumshark
Identification Small, with slender body and tail. Very long, angular snout (bell-shaped in dorsoventral view, wedge-shaped laterally). Short anterior nasal flaps. Angular mouth extending behind eyes; numerous small teeth; short labial furrows. Elongated, cat-like eyes; nictitating eyelids. Small spiracles. Relatively short, subtriangular first dorsal fin; second as high or slightly higher. Colour greyish and unpatterned.
Distribution Southwest Pacific: New Zealand and surrounding seamounts and ridges.
Habitat Outermost continental shelf, upper slope and adjacent seamounts, 129–975m, most abundant 400–600m.
Behaviour Unknown.
Biology Viviparous and oophagous, 1–3 (usually 2) pups per litter. Diet includes small teleosts, cephalopods and crustaceans.
Status IUCN Red List: Least Concern. Taken as bycatch, but occurs mostly where fishing effort is low or prohibited.

SULU GOLLUMSHARK *Gollum suluensis*

FAO code: **CVX**   Plate page 448

Teeth
Upper 94
Lower 81

~1mm

Measurements Mature: males ~58cm.
Max: at least 65cm (female).

Identification Distinguished from other species of gollumshark by shorter snout. Colour plain dark greyish-brown above, lighter grey below; no conspicuous white markings on body or fins. Narrow light posterior margins on dorsal, pectoral and pelvic fins. Web of terminal caudal lobe light, edged with dark grey or blackish.
Distribution Central Indo-Pacific: Sulu Sea, Philippines, off Palawan Island.
Habitat Unknown. Deepsea on upper continental slope or upper shelf, known from ~730m.
Behaviour Unknown.
Biology Unknown.
Status IUCN Red List: Least Concern. Known only from a very few specimens and part of the population likely lives below the depth of most fisheries.

EASTERN DWARF FALSE CATSHARK *Planonasus indicus*

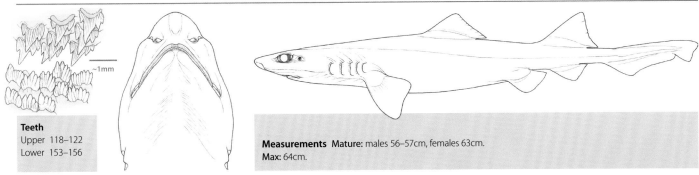

Teeth
Upper 118–122
Lower 153–156

Measurements Mature: males 56–57cm, females 63cm.
Max: 64cm.

Identification Small, stocky, flabby body. Moderately elongate snout, tip bluntly to broadly rounded (bell-shape in dorsal-ventral view, wedge-shaped laterally). Mouth very large and broadly rounded; roof of mouth and tongue without oral papillae; short labial furrows, lowers longer than uppers. Relatively large, cat-like eyes; nictitating membrane. First dorsal fin low; tip broadly rounded; origin behind pectoral fin free rear tips. Second dorsal fin higher; tip narrow; base longer than first dorsal fin. Colour, including fins, uniform dark brownish black, without any distinct blotches, mottling, patterning, spots or stripes. First dorsal fin without white free rear tip.
Distribution North Indian Ocean: off Kochi, Kerala, southern India, Sri Lanka and possibly the Maldives.
Habitat Occurs along continental and insular slopes from 200–1000m, with most records between 300–600m.
Behaviour All known specimens have come from fisheries targeting gulper sharks (*Centrophorus* spp.), suggesting they co-occur in a similar environment.
Biology Oophagous, with 2 pups per litter, 1 embryo develops per uterus. Diet includes bony fishes, cephalopods, and possibly crustaceans.
Status IUCN Red List: Data Deficient. A very poorly known bycatch in some deepsea fisheries.

PYGMY FALSE CATSHARK *Planonasus parini*

Teeth
Upper 110–115
Lower 120

Measurements Mature: males unknown (over 34cm), females unknown (over 53cm).
Max: at least 53cm.

Identification Small, moderately stocky shark with soft body and tail. Short snout (bell-shaped in dorsoventral view, wedge-shaped laterally). Short anterior nasal flaps. Angular mouth extending behind eyes; numerous small teeth; short labial furrows. Elongated cat-like eyes; nictitating eyelids. Small spiracles. Relatively short, subtriangular first dorsal fin; second dorsal fin higher. Unpatterned, dark grey-brown body with darker fins.
Distribution Northwest Indian Ocean: Arabian Sea, off Socotra Island. Specimens reported from India and the Maldives are a different species, *Planonasus indicus*.
Habitat Continental and insular slopes, 560–1120m.
Behaviour Unknown.
Biology Poorly known; only 3 specimens, 2 immature females (39cm and 53cm) and an immature male 34cm recorded.
Status IUCN Red List: Least Concern. Occurs deeper than any known fisheries in the region.

FALSE CATSHARK *Pseudotriakis microdon*

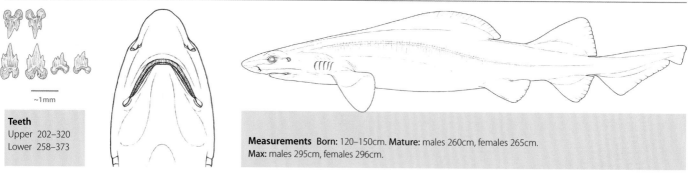

~1mm

Teeth
Upper 202–320
Lower 258–373

Measurements Born: 120–150cm. **Mature:** males 260cm, females 265cm. **Max:** males 295cm, females 296cm.

Identification Large, with stocky, bulky, soft body and tail. Short snout (bell-shaped in dorsoventral view, wedge-shaped laterally). Short anterior nasal flaps. Huge, angular mouth extending behind eyes; numerous small teeth; short labial furrows. Elongated cat-like eyes; nictitating eyelids. Very large spiracles, about as long as eyes. Long, low, keel-like first dorsal fin; second much higher. Dark brown to blackish, unpatterned.
Distribution Patchy worldwide, except in the east Pacific.   **Habitat** Deepsea floor on continental and insular slopes, occasionally continental shelves and shallower near submarine canyons, 100–2430m.
Behaviour Large body cavity, soft fins, skin and musculature indicates an inactive and sluggish shark of virtually neutral buoyancy. Seemingly uncommon or rare wherever it occurs, although they are known to congregate at times. Photographed feeding on a fish in deep water.
Biology Viviparous, with oophagy, 2 pups per litter. No reproductive seasonality; females with embryos at different developmental stages co-occur in birthing areas. Feeds on fishes, including other elasmobranchs, cephalopods, and crustaceans.
Status IUCN Red List: Least Concern. Wide-ranging but only a sporadic bycatch of deepsea fisheries.

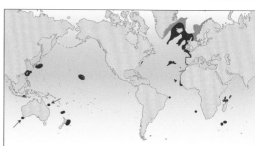

Leptochariidae: Barbeled Houndshark

A single genus and species that was previously placed in the families Triakidae and Carcharhinidae, however its morphology shows it is distinct from these two families. Endemic to the east Atlantic.

BARBELED HOUNDSHARK *Leptocharias smithii*

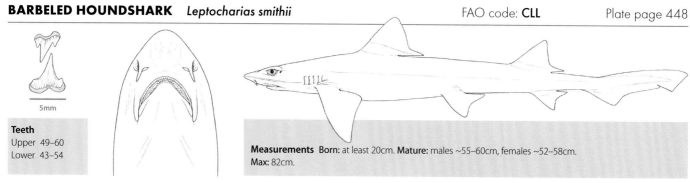

5mm

Teeth
Upper 49–60
Lower 43–54

Measurements Born: at least 20cm. **Mature:** males ~55–60cm, females ~52–58cm. **Max:** 82cm.

Identification Small, slender shark. Nostrils with slender barbels. Long, arched mouth reaching past anterior ends of eyes; small cuspidate teeth (males have greatly enlarged anterior teeth); labial furrows very long. Horizontally oval eyes; internal nictitating eyelids. Light grey-brown colour.
Distribution East Atlantic: Mauritania to Angola.
Habitat Continental shelf, near bottom, 5–75m. Especially common on mud off river mouths.
Behaviour Unknown. Males may use greatly enlarged front teeth in courtship and copulation.
Biology Viviparous, with 7 pups per litter, and a gestation period of at least 4 months; unique spherical placenta. Pregnant females occur July–October off Senegal. Feeds on crustaceans, small bony fishes and skates.
Status IUCN Red List: Near Threatened. Taken in many fisheries as a utilised bycatch throughout its restricted habitat; status needs further study.

Triakidae: houndsharks

One of the largest families of sharks, with over 45 species distributed worldwide in warm and temperate coastal seas. Most species occur in continental and insular waters, from the shoreline and intertidal to the outermost shelf, often close to the bottom, with many in sandy, muddy and rocky inshore habitats, enclosed bays and near river mouths. A few deep water species range down continental slopes to great depths, possibly exceeding 2000m. Many are endemic with a very restricted distribution.

Identification Small to medium-sized sharks with two medium to large spineless dorsal fins, the first dorsal base well ahead of pelvic bases, and an anal fin. Horizontally oval eyes with nictitating eyelids, no nasoral grooves, anterior nasal flaps not barbel-like (except in *Furgaleus*), a long, angular or arched mouth reaching past the front of the eyes, moderate to very long labial furrows. Caudal fin without a strong lower lobe or lateral undulations on its dorsal margin. Some (e.g. *Mustelus*) can be very hard to identify without vertebral counts and several species are undescribed.

Biology Some species are very active and swim almost continuously, others can rest on the bottom, and many swim close to the seabed. Some are most active by day, others at night. Houndsharks are viviparous, either with or without a yolk-sac placenta, bearing litters from 1–2, to 52 pups. They feed mainly on bottom and midwater invertebrates and bony fishes; some largely take crustaceans, some mainly fishes, and a few primarily cephalopods; none feed on birds and mammals.

Status Houndsharks are generally fairly common and some are very abundant in coastal waters, where they are fished extensively for their meat, liver oil and fins (e.g. *Galeorhinus* and *Mustelus*). Some of the smaller coastal species reproduce rapidly and can support well-managed fisheries (42% of species are assessed as Least Concern by IUCN). Others have a history of stock collapse, where fisheries are or have been unmonitored and unregulated, and require very careful management (36% are threatened globally, including 13% Critically Endangered). Some species are extremely rare. None are harmful to people.

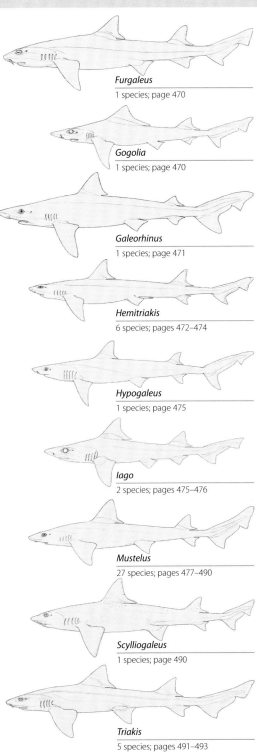

Furgaleus
1 species; page 470

Gogolia
1 species; page 470

Galeorhinus
1 species; page 471

Hemitriakis
6 species; pages 472–474

Hypogaleus
1 species; page 475

Iago
2 species; pages 475–476

Mustelus
27 species; pages 477–490

Scylliogaleus
1 species; page 490

Triakis
5 species; pages 491–493

Spotted Gully Shark, *Triakis megalopterus* (p. 492).

Plate 57 TRIAKIDAE I – *Furgaleus, Gogolia, Hemitriakis* and *Hypogaleus*

○ **Whiskery Shark** *Furgaleus macki*

Southeast Indian Ocean; 0–220m. Stocky almost humpbacked; conspicuous subocular ridges, dorsolateral eyes, anterior nasal flaps form barbels, very short arched mouth; grey with variegated dark blotches or saddles that fade with age, lighter below.

○ **Sailback Houndshark** *Gogolia filewoodi*

Central Indo-Pacific; 73m. Preoral snout long; first dorsal fin very large and triangular (length about the same as caudal fin); grey-brown above, lighter below.

○ **Deepwater Sicklefin Houndshark** *Hemitriakis abdita*

Central Indo-Pacific; 224–402m. Slender; snout long and parabolic, small anterior nasal flaps, arched mouth; falcate dorsal pectoral and anal fins, first dorsal fin same size as second; grey-brown with dark stripe under snout, distinct white dorsal, pectoral and anal fin tips, juveniles with prominent dark bars and saddles.

○ **Striped Topeshark** *Hemitriakis complicofasciata*

Northwest Pacific; inshore to ~90–100m. Slender; snout long and blunt, prominent subocular ridge, small anterior nasal flaps, moderately arched mouth; alcate dorsal, pectoral and anal fins; first dorsal fin higher than second; greyish-brown above with vague saddles; caudal fin with white posterior margins; whitish below, no stripe under snout; near term young distinctly patterned with dark O-shaped spots, bars and saddles which fade with age.

○ **Sicklefin Houndshark** *Hemitriakis falcata*

Central Indo-Pacific; 110–200m. Slender; snout long and blunt, small anterior nasal flaps, arched mouth; strongly falcate dorsal pectoral and anal fins, first dorsal fin same size as second, first dorsal fin origin over to behind pectoral fin rear tip; grey-brown with no stripe under snout, distinct white dorsal fin tips, juveniles with saddles and large spots.

○ **Indonesian Houndshark** *Hemitriakis indroyonoi*

Central Indo-Pacific and east Indian Ocean; 60m or more. Slender elongated body; small anterior nasal flaps, broadly arched mouth, adults plain grey with prominent white fin tips, newborns with dark blotches and bars.

○ **Japanese Topeshark** *Hemitriakis japanica*

Northwest Pacific; inshore to over 100m. Snout moderately long and parabolic, slit-like eyes with subocular ridge, short anterior nasal flaps, broadly arched mouth; first dorsal fin about same size as second, first dorsal fin origin over to behind pectoral fin rear tip, dorsal fins much larger than anal fin; conspicuous white-edged fins.

○ **Whitefin Topeshark** *Hemitriakis leucoperiptera*

Central Indo-Pacific; coastal to ~48m. Snout moderately long and parabolic; moderately long eyes with subocular ridge, small anterior nasal flaps, broadly arched mouth; strongly falcate fins with conspicuous white edges, first dorsal fin larger than second, which is larger than anal fin, first dorsal fin origin over to behind pectoral fin inner margin rear tip.

○ **Blacktip Topeshark** *Hypogaleus hyugaensis*

Indian Ocean and west Pacific; 40–480m. Fairly slender; long broadly pointed snout, large oval eyes with subocular ridges, small anterior nasal flaps, arched mouth; second dorsal fin smaller than first, larger than anal fin, relatively short terminal caudal lobe; bronzy to grey-brown above lighter below, dusky upper and lower caudal fin tips especially in young.

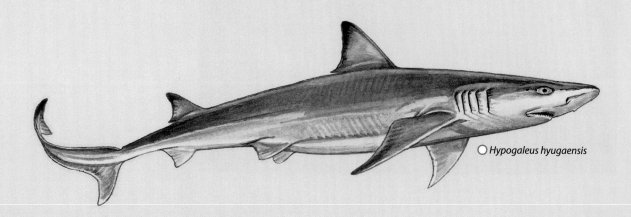

○ *Hypogaleus hyugaensis*

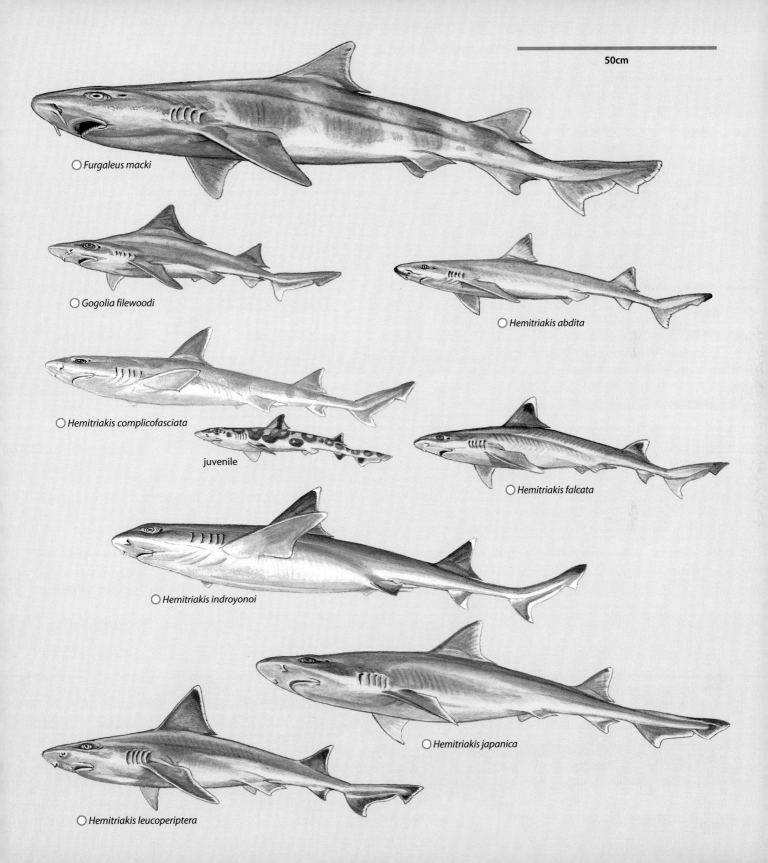

○ *Furgaleus macki*

50cm

○ *Gogolia filewoodi*

○ *Hemitriakis abdita*

○ *Hemitriakis complicofasciata*

juvenile

○ *Hemitriakis falcata*

○ *Hemitriakis indroyonoi*

○ *Hemitriakis japanica*

○ *Hemitriakis leucoperiptera*

Plate 58 TRIAKIDAE II – *Iago, Scylliogaleus* and *Triakis*

○ **Longnose Houndshark** *Iago garricki* page 475

Southwest Pacific and Central Indo-Pacific; 250–477m. Slender; first dorsal fin very slightly larger than second dorsal, first dorsal fin origin over pectoral fin inner margins; grey-brown above, becoming lighter below; dorsal fins with conspicuous black tips; dorsal, pectoral and caudal fin posterior margins pale.

○ **Bigeye Houndshark** *Iago omanensis* page 476

North Indian Ocean; 92–2195m. Slender; large eyes, large gills; dorsal fins small, first dorsal fin origin over pectoral fin bases, small lower caudal lobe; grey-brown, lighter below, sometimes with darker dorsal fin posterior margins.

○ **Flapnose Houndshark** *Scylliogaleus quecketti* page 490

Southwest Indian Ocean; 0–73m. Short blunt snout, fused anterior nasal flaps large, cover mouth; first dorsal fin same size as second, second dorsal fin much larger than anal fin; grey above, cream below, newborns with white posterior margins to dorsal anal and caudal fins.

○ **Sharpfin Houndshark** *Triakis acutipinna* page 491

Southeast Pacific; 50–200m. Snout broadly rounded, widely separated anterior nasal flaps do not reach mouth, long upper labial furrows reach lower symphysis; narrow fins, pectoral fins falcate, first dorsal fin with abrupt vertical posterior margin; no spots or bands.

○ **Spotted Houndshark** *Triakis maculata* page 491

East Pacific; 10–200m. Stout; snout broadly rounded, widely separated lobate anterior nasal flaps do not reach mouth, long upper labial furrows reach lower symphysis; broad fins, pectoral fins falcate with straight posterior margins, first dorsal fin with backward sloping posterior margin; many black spots (some unspotted).

○ **Spotted Gully Shark** *Triakis megalopterus* page 492

Southeast Atlantic; 0–50m. Broad blunt snout, widely separated small lobate anterior nasal flaps do not reach mouth, upper labial furrows do not reach lower symphysis; large broad fins, pectoral fins falcate with concave posterior margin, first dorsal almost vertical; grey-bronze above, usually many small black spots.

○ **Banded Houndshark** *Triakis scyllium* page 492

Northwest Pacific; 30–150m. Short snout broadly rounded, widely separated lobate anterior nasal flaps do not reach mouth, long upper labial furrows reach lower symphysis; pectoral fins almost triangular, first dorsal almost vertical; sparsely scattered small black spots may be absent, dusky saddles in young.

○ **Leopard Shark** *Triakis semifasciata* page 493

Northeast Pacific; 0–156m. Snout broadly rounded, widely separated anterior nasal flaps do not reach mouth, upper labial furrows reach lower symphysis; pectoral fins falcate; unique striking colour pattern of saddles and spots on tan to greyish background, fading to whitish below.

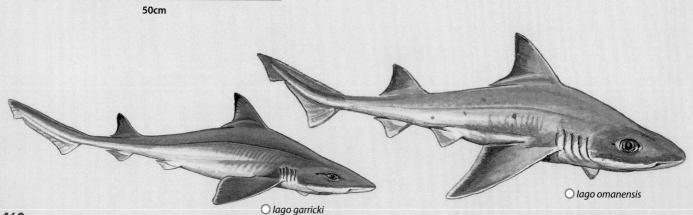

50cm

○ *Iago garricki*

○ *Iago omanensis*

○ *Scylliogaleus quecketti*

○ *Triakis acutipinna*

○ *Triakis maculata*

○ *Triakis megalopterus*

50cm

○ *Triakis scyllium*

○ *Triakis semifasciata*

Plate 59 TRIAKIDAE III – *Galeorhinus* and *Mustelus* I

○ **Tope** *Galeorhinus galeus*

Almost worldwide: temperate; 0–800m. Slender; long conical snout, no obvious subocular ridges, small anterior nasal flaps, large arched mouth; first dorsal fin much larger than second, which is same size as anal fin; very long terminal caudal lobe; grey above, white below, black fin markings in young, sometimes a few dusky spots.

○ **Starry Smoothhound** *Mustelus asterias*

Northeast Atlantic and Mediterranean; to 200m. Slender; rounded snout, nostrils closer together than similar species in region; unfringed dorsal fins, pectoral and pelvic fins fairly small; grey or grey-brown, only European smoothhound with numerous white spots.

○ **Dusky Smoothhound** *Mustelus canis*

West Atlantic; 0–808m. Slender; short head and snout, large close-set eyes, nostrils widely spaced, upper labial furrows longer than lowers; unfringed dorsal fins, caudal fin deeply notched; grey, usually unspotted above, white below; newborns with dusky-tipped dorsal and caudal fins, juveniles with white fin margins.

○ **Smoothhound** *Mustelus mustelus*

East Atlantic; 0–800m. Short head and snout, large close-set eyes, nostrils widely spaced, upper labial furrows slightly longer than lowers; unfringed dorsal fins; grey to grey-brown, lighter below, usually unspotted, occasionally dark spots in very large individuals.

○ **Whitespotted Smoothhound** *Mustelus palumbes*

Southeast Atlantic, southwest Indian Ocean; 0–443m. Nostrils relatively widely spaced, upper labial furrows longer than lowers; unfringed dorsal fins, relatively large pectoral and pelvic fins (larger than *Mustelus asterias*, *M. manazo* and *M. mustelus*); grey to grey-brown, only South African smoothhound with white spots.

○ **Blackspot Smoothhound** *Mustelus punctulatus*

East Atlantic and Mediterranean; <200m. Short narrow head and snout, large eyes, nostrils close together, upper labial furrows slightly longer than lowers; prominent fringed dorsal fins; body grey above, light below, sides usually distinctly black-spotted; often mistaken for *Mustelus mustelus*.

○ **Narrownose Smoothhound** *Mustelus schmitti*

Southwest Atlantic; 2–195m. Short head, snout moderately long, nostrils close together, upper labial furrows much longer than lowers; dorsal fins prominently fringed giving dark frayed appearance; grey, white-spotted; easily distinguished from other white-spotted *Mustelus*.

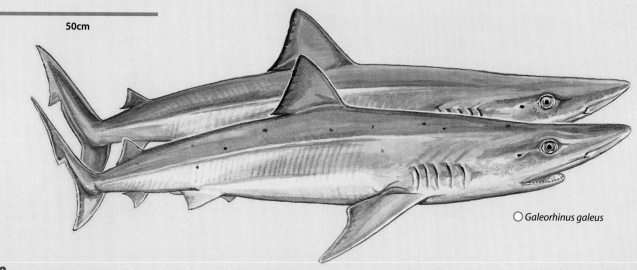

50cm

○ *Galeorhinus galeus*

○ *Mustelus asterias*

○ *Mustelus canis*

○ *Mustelus mustelus*

○ *Mustelus palumbes*

○ *Mustelus punctulatus*

○ *Mustelus schmitti*

50cm

Plate 60 TRIAKIDAE IV – *Mustelus* II

○ **Gummy Shark** *Mustelus antarcticus*

Southeast Indian Ocean to southwest Pacific; to 350m. Snout relatively long, large nostrils widely spaced, long upper labial furrows slightly longer than lowers; unfringed dorsal fins, first dorsal fin slightly larger than second, relatively small pectoral and pelvic fins; bronze to grey-brown, white spotted, rarely with some black spots.

○ **Grey Smoothhound** *Mustelus californicus*

Northeast Pacific; 0–265m. Short narrow head, fairly small eyes, nostrils widely spaced, short mouth, upper labial furrows same length as lowers; triangular dorsal fins, lower caudal lobe poorly developed; uniform grey, no spots, light below.

○ **Sharptooth Smoothhound** *Mustelus dorsalis*

East Pacific; 20–200m. Long acutely pointed snout, very small wide-set eyes, large nostrils moderately spaced, upper labial furrows longer than lowers; broadly triangular first dorsal fin; uniform grey or grey-brown, no spots, light below.

○ **Spotless Smoothhound** *Mustelus griseus*

Northwest Pacific; inshore to 131m, possibly 300m. Short head and snout, small eyes, nostrils widely spaced, upper labial furrows about same length as lowers; unfringed dorsal fins, lower caudal lobe semifalcate; uniform pale grey or grey-brown, no spots, light below.

○ **Brown Smoothhound** *Mustelus henlei*

East Pacific; 0–281m or more. Short head, moderately long snout, large close-set eyes, nostrils widely spaced, upper labial furrows longer than lowers, caudal peduncle long, dark broadly fringed dorsal fins; body usually iridescent bronze-brown, occasionally grey, no spots, white below.

○ **Rig** *Mustelus lenticulatus*

Southwest Pacific; close inshore to 1000m. Short head, moderately long snout, fairly large wide-set eyes, nostrils widely spaced, long upper labial furrows longer than lowers; unfringed dorsal fins, relatively large pectoral and pelvic fins; grey or grey-brown, white-spotted (only one in range), light below.

○ **Starspotted Smoothhound** *Mustelus manazo*

Northwest Pacific to Central Indo-Pacific; 1–360m. Short head, moderately long snout, large close-set eyes, nostrils fairly close-set, upper labial furrows much longer than lowers; unfringed dorsal fins, relatively small pectoral and pelvic fins; grey to grey-brown, many white spots (only white-spotted *Mustelus* in range), light below.

○ **Speckled Smoothhound** *Mustelus mento*

Southeast Pacific; 16–50m. Short blunt angular snout, fairly small moderately spaced eyes; nostrils widely spaced, upper labial furrows slightly longer than lowers; caudal peduncle short, unfringed dorsal fins; grey to grey-brown, white-spotted, light below.

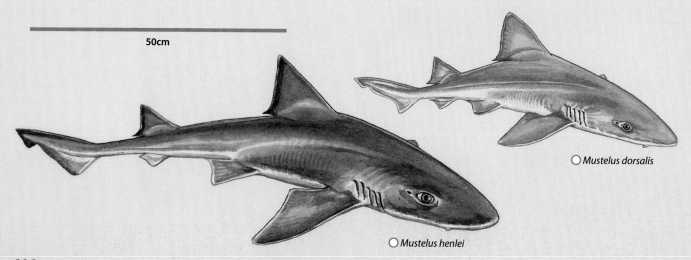

50cm

○ *Mustelus dorsalis*

○ *Mustelus henlei*

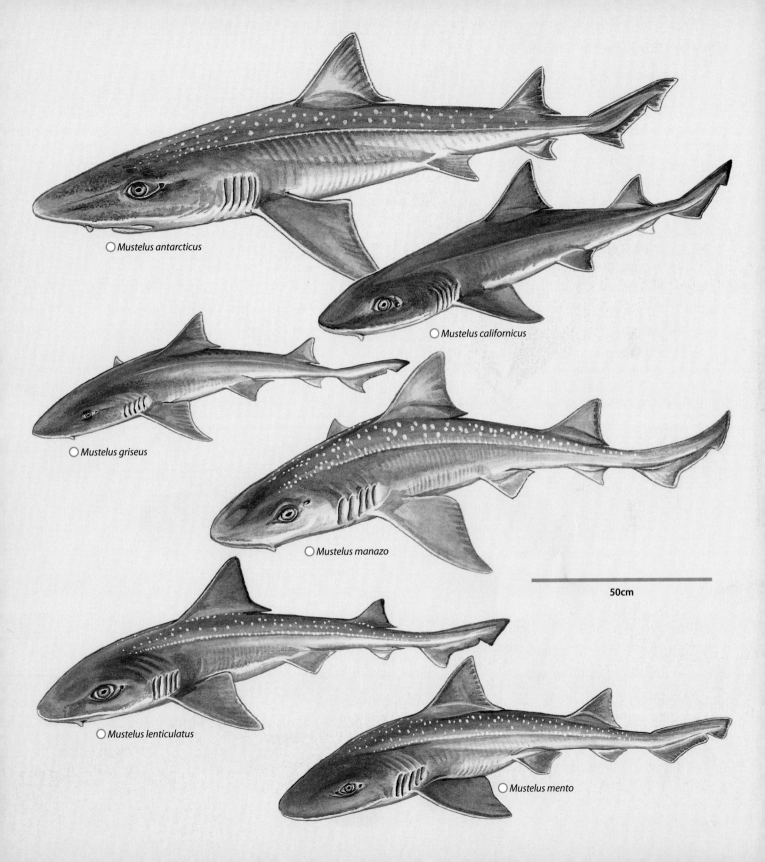

○ *Mustelus antarcticus*

○ *Mustelus californicus*

○ *Mustelus griseus*

○ *Mustelus manazo*

50cm

○ *Mustelus lenticulatus*

○ *Mustelus mento*

Plate 61 TRIAKIDAE V – *Mustelus* III

○ **Striped Smoothhound** *Mustelus fasciatus*   page 480

Southwest Atlantic; 10–500m. Long head, long angular pointed snout, very small eyes, nostrils widely spaced, upper labial furrows much longer than lowers; caudal peduncle short; broadly triangular unfringed dorsal fins; vertical dark bars in young, unspotted.

○ **Smalleye Smoothhound** *Mustelus higmani*   page 481

West Atlantic; close inshore to 1463m. Long acutely pointed snout, very small wide-set eyes, nostrils widely spaced, upper labial furrows same length as lowers; unfringed dorsal fins, falcate first dorsal fin only slightly larger than second; uniform grey or grey-brown, no spots, light below.

○ **Sicklefin Smoothhound** *Mustelus lunulatus*   page 482

East Pacific; 9–200m. Distinguished from similar *Mustelus californicus* by more pointed snout, more widely separated eyes, shorter mouth, shorter upper labial furrows and strongly falcate dorsal fins; uniform grey to grey-brown, no spots, light below.

○ **Venezuelan Dwarf Smoothhound** *Mustelus minicanis*   page 484

West Atlantic; 71–183m. Short head and snout, very large close-set eyes, nostrils widely spaced, upper labial furrows slightly longer than lowers; unfringed dorsal fins, lower caudal lobe poorly developed; uniform grey, no spots, light below, newborns with dusky dorsal and caudal fin tips.

○ **Arabian Smoothhound** *Mustelus mosis*   page 484

North and west Indian Ocean; 20–250m. Short head and snout, large fairly close-set eyes, nostrils widely spaced, upper labial furrows same length as lowers; unfringed dorsal fins, lower caudal lobe semifalcate; uniform grey or grey-brown, no spots, light below; in South Africa: white first dorsal fin tip, black second dorsal and caudal fin tips.

○ **Narrowfin Smoothhound** *Mustelus norrisi*   page 485

West Atlantic; close inshore to 100m occasionally 260m. Short narrow head, relatively large eyes, nostrils close-set, relatively long mouth, upper labial furrows equal to or slightly longer than lowers; strongly falcate fins; uniform grey, no spots, light below, newborns with dusky dorsal and caudal fin tips.

○ **Gulf of Mexico Smoothhound** *Mustelus sinusmexicanus*   page 488

West Atlantic; 36–229m. Short head and snout, large close-set eyes, nostrils widely spaced, upper labial furrows longer than lowers; unfringed dorsal fins, large first dorsal fin, lower caudal lobe moderately expanded; uniform grey, no spots, light below, newborns with dusky dorsal and caudal fin tips.

○ **Humpback Smoothhound** *Mustelus whitneyi*   page 489

East to southeast Pacific; 16–211m. Stocky, almost humpbacked; fairly long head and snout, fairly large eyes, nostrils widely spaced, upper labial furrows much longer than lowers; dark fringed dorsal fins, lower caudal lobe very slightly falcate; uniform grey, no spots, light below.

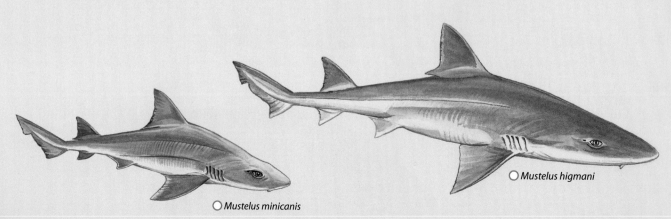

○ *Mustelus higmani*

○ *Mustelus minicanis*

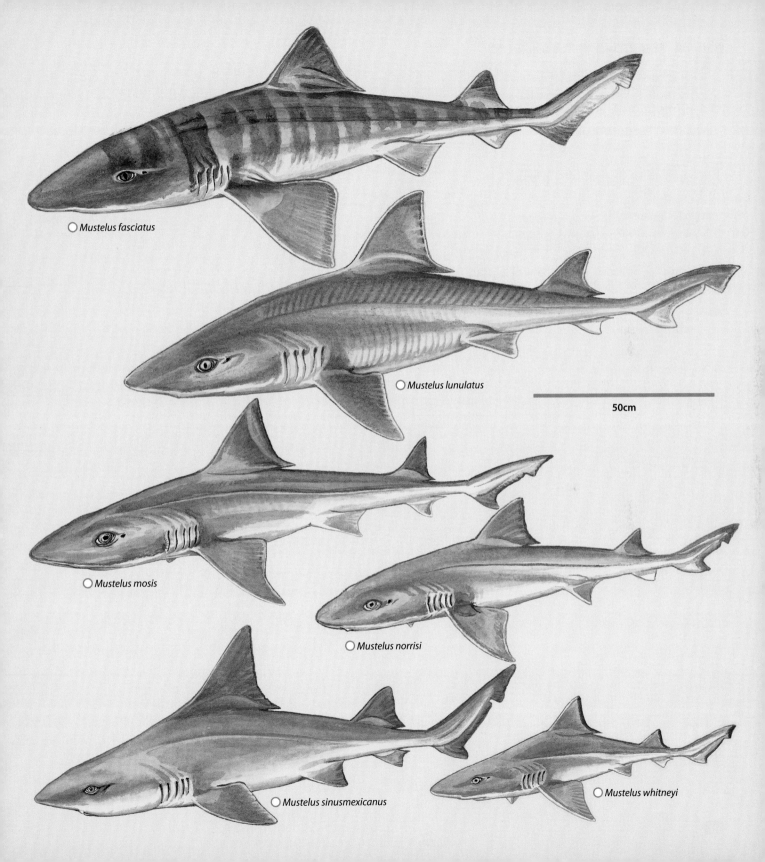

○ *Mustelus fasciatus*

○ *Mustelus lunulatus*

50cm

○ *Mustelus mosis*

○ *Mustelus norrisi*

○ *Mustelus sinusmexicanus*

○ *Mustelus whitneyi*

Plate 62 TRIAKIDAE VI – *Mustelus* IV

○ **White-margin Fin Houndshark** *Mustelus albipinnis*

Northeast Pacific; 30–281m. A slender-bodied houndshark distinguished from other species by a combination of uniformly dark grey-brown above, lighter below, and with distinctive white fin edges on the posterior margins of the dorsal, pectoral, anal and caudal fins.

○ **Australian Grey Smoothhound** *Mustelus ravidus*

Central Indo-Pacific; 100–300m. Slender; lower labial furrows longer than uppers; deeply notched caudal fin; uniform bronze to grey-brown, without spots.

○ **Western Spotted Gummy Shark** *Mustelus stevensi*

Central Indo-Pacific; 121–735m. Upper labial furrows longer than lowers; deeply notched caudal fin; bronzy, spotted.

○ **Whitefin Smoothhound** *Mustelus widodoi*

Central Indo-Pacific; 60–120m. Large slender-bodied smoothhound shark, without spots; uniform grey above, lighter below, a broad white margin on the first dorsal fin and a distinct black margin on the second dorsal fin apex.

○ **Andaman Smoothhound** *Mustelus andamanensis*

Andaman Sea; less than 100m. Slender-bodied smoothhound shark, without spots; uniform greyish brown above, lighter below, dorsal fins dusky becoming blackish in upper half and without white margins, .caudal fin with narrow white margin and black tip.

50cm

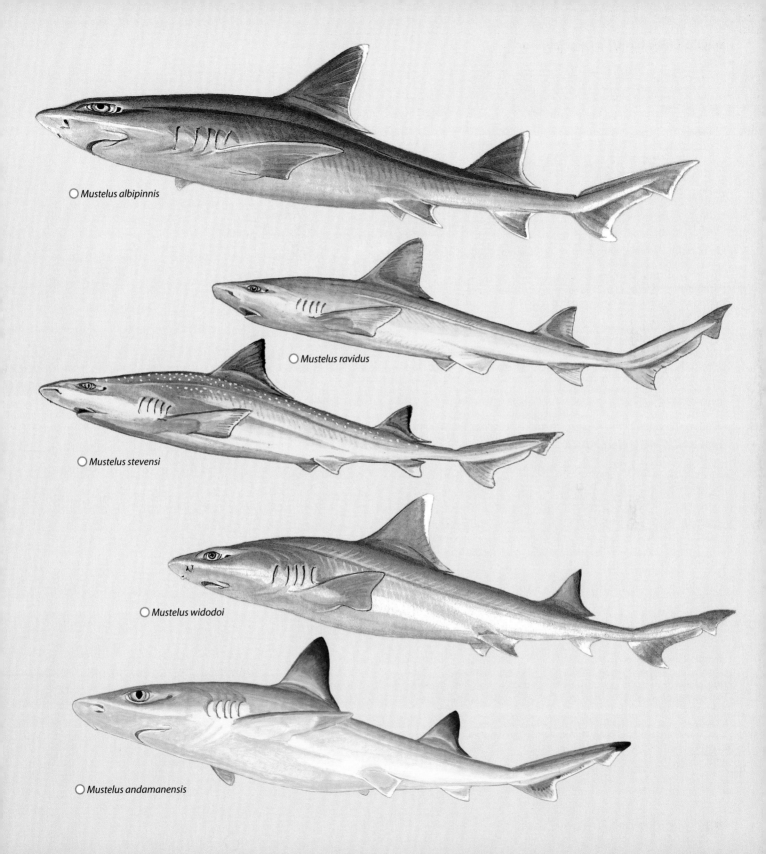

○ *Mustelus albipinnis*

○ *Mustelus ravidus*

○ *Mustelus stevensi*

○ *Mustelus widodoi*

○ *Mustelus andamanensis*

WHISKERY SHARK *Furgaleus macki*

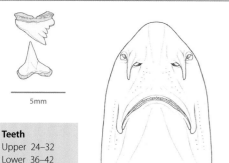

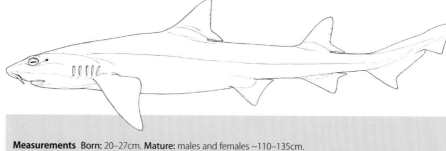

Teeth
Upper 24–32
Lower 36–42

Measurements Born: 20–27cm. **Mature:** males and females ~110–135cm.
Max: 160cm.

Identification Only houndshark with anterior nasal flaps forming slender barbels. Stocky, almost humpbacked. Obvious ridges below dorsolateral eyes. Very short, arched mouth. Grey above, with variegated dark blotches or saddles that fade with age. Light below.
Distribution Southeast Indian Ocean: endemic to south and west Australia.
Habitat Shallow temperate continental shelf, on or near bottom on rock, seagrass and in kelp from 0–220m.
Behaviour Active swimmer.
Biology Viviparous. No yolk-sac placenta, 4–29 pups per litter (average 19) with gestation lasting about 7–9 months and birth occurring in the late winter and early spring. Age at first maturity ~4.5 years for males and 6.5 years for females; maximum age is 10.5–11.5 years. These sharks are specialised feeders on octopus, and other cephalopods, and to a much lesser extent bony fishes.
Status IUCN Red List: Least Concern. Stock was depleted by target gillnet fishery to less than 30% of original size in 1960–70s, but the fishery is now well managed, with the population fairly stable since mid-1980s and now increasing.

SAILBACK HOUNDSHARK *Gogolia filewoodi*

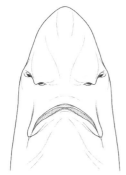

Teeth
Upper 40–41
Lower 35

Measurements Born: ~22cm. **Mature:** A female measured 74cm.

Identification Small, grey-brown houndshark distinguished by huge, triangular first dorsal fin (about same length as caudal fin) and long preoral snout (1.6–1.7 times mouth width).
Distribution Central-Indo Pacific: one site known, northern Papua New Guinea.
Habitat Only known specimen from 73m on continental shelf, probably near bottom.
Behaviour Unknown.
Biology Viviparous. Pregnant female was carrying 2 pups.
Status IUCN Red List: Data Deficient. Presumably endemic; recent surveys in Papua New Guinea have failed to find any new records of this species.

Teeth
Upper 30–41
Lower 31–46

10mm

Measurements Vary between regions. **Born:** 30–40cm. **Mature:** males ~120–170cm, females 130–185cm. **Max:** males 175cm, females 195cm.

Alternative name School, Soupfin, Liver-oil or Vitamin Shark

Identification Slender, long-nosed houndshark without obvious anterior nasal flaps or subocular ridges. Large, arched mouth; small blade-like teeth. Second dorsal fin much smaller than first and about as large as anal fin. Extremely long terminal caudal lobe (about half the dorsal caudal margin). Greyish above, white below; black markings on fins of young.

Distribution Almost worldwide in temperate waters except the northwest Atlantic and northwest Pacific.

Habitat Most abundant in cold to warm-temperate continental shelf seas, from the surfline and very shallow water to well offshore, not oceanic, often near the bottom, 0–800m.

Behaviour This is an active, strong swimmer, capable of travelling 35 miles (56km) per day. Populations in higher latitudes make long-distance migrations between their winter and summer feeding and breeding grounds – up to 1400km in the southwest Atlantic. Other populations migrate shorter distances seasonally, inshore and offshore. This species also makes vertical diurnal movements, from deep water by day to shallower water at night. Although generally considered to be a species of the continental shelf, tag returns have identified long oceanic journeys: across the Tasman Sea from Australia to New Zealand (particularly females), and over 2500km from the UK to the Canary Islands, also to the Mediterranean, Azores, Norway and north of Iceland. Genetic studies indicate that the Pacific Ocean is a barrier to their migration, but there is high connectivity between the populations in South Africa and Australia. Tope usually occur in small schools, partly segregated by size and sex. Pregnant females move from offshore feeding grounds into shallow bays and estuaries to give birth. Juveniles remain in nursery grounds for up to 2 years (but may move into deeper water in winter), then join schools of immature animals elsewhere.

Biology Viviparous with yolk-sac. Reproductive parameters, such as litter size and gestation, vary by region. Litter size, 6–52 pups, increases with the size of the mother. Age at maturity has not been studied in all regions, but has a very low biological productivity because it matures late (females at 11–17 years, males

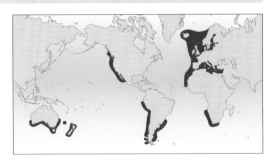

~9–13 years), there is a 12 month gestation, and the females may have a 1–2 year resting period between pregnancies, so only give birth every 2–3 years. Maximum age is possibly 60 years. This species is mainly an opportunistic feeder on bony fishes, also invertebrates. Predators include other large sharks and probably marine mammals.

Status IUCN Red List: **Critically Endangered** globally, due to depletion in unmanaged target gillnet and longline fisheries in many parts of its range, while bycatch and damage to inshore nursery ground habitat are also of concern. Several regional populations have been assessed; this species is depleted to a level qualifying as **Critically Endangered** in the southwest Atlantic (where the population has been heavily fished throughout its range and may be on the brink of vanishing), Australia and South Africa. It has been listed as **Vulnerable**, but is now known to have declined to **Endangered** status in European waters. It has also declined, but not so steeply, in New Zealand, where important nursery grounds likely replenish the population fished in Australian waters. Depletion in the northeast Pacific has resulted in it being listed under Canada's Species At Risk Act. *G. galeus* probably has more vernacular names than any other shark because it has been targeted by so many fisheries in different localities for its meat (which is popular in many parts of the world), liver oil (the Oil or Vitamin Shark) and fins (Soupfin Shark) – and it is now also targeted by sports anglers. *G. galeus* fisheries are regulated in Australia, New Zealand, EU and UK and the species is largely protected in Canada and the Mediterranean, where the General Fisheries Council for the Mediterranean requires the release of *G. galeus* caught in bottom-set nets, longlines and tuna traps. This species is listed in Appendix II of the Convention on Migratory Species.

Tope, *Galeorhinus galeus*.

DEEPWATER SICKLEFIN HOUNDSHARK   *Hemitriakis abdita*

FAO code: **TRK**   Plate page 458

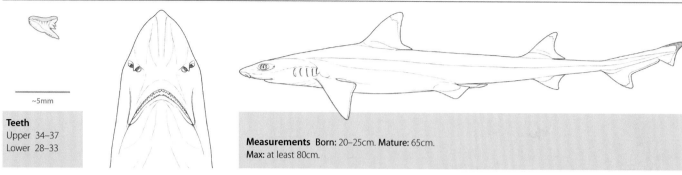

~5mm

Teeth
Upper 34–37
Lower 28–33

Measurements Born: 20–25cm. Mature: 65cm.
Max: at least 80cm.

Identification Slender houndshark. Rather long, parabolic snout. Small anterior nasal flaps. Arched mouth; small blade-like teeth. Dorsal fins similar-sized. Dorsal, pectoral and anal fins falcate. Greyish brown. Snout with dark stripe on underside; dorsal, pectoral and anal fins with distinct white tips. Juveniles have prominent dark bars and solid saddles on body and fins.
Distribution Central Indo-Pacific: Australia, possibly New Caledonia.
Habitat Deep water on upper continental slope, 224–402m.
Behaviour Unknown.
Biology Unknown.
Status IUCN Red List: Data Deficient. Known from only a few specimens from a small area.

STRIPED TOPESHARK   *Hemitriakis complicofasciata*

FAO code: **TRK**   Plate page 458

5mm

Teeth
Upper 31–37
Lower 28–33

Measurements Born: ~22cm. **Mature:** males 76–84cm, females 81–93cm.
Max: 93cm.

Identification Slender houndshark with short, blunt snout, prominent subocular ridge, small nasal flaps, moderately arched mouth, long upper labial furrows, short lowers; dorsal, pectoral and anal fins falcate; first dorsal fin higher than second. Adults and adolescents greyish brown above, with faint saddles, no stripe on ventral snout surface, white below; apex and posterior margin of dorsal, pectoral, pelvic, anal and caudal fins with white tips and posterior margins. Young juveniles with striking pattern of complicated bars, lines, spots and rings over body; coloration fades with growth.
Distribution Northwest Pacific: Ryukyu Islands, Japan to southwest Taiwan. **Habitat** Inshore to ~90–100m.
Behaviour Unknown.   **Biology** Viviparous without yolk-sac placenta; bears 5–8 live young.
Status IUCN Red List: Vulnerable. This inshore endemic is a retained bycatch of trawl, longline and gillnet fisheries, but may now be benefiting from decreasing fishing pressure and some inshore trawl bans in parts of its limited range.

newborn / juvenile

SICKLEFIN HOUNDSHARK *Hemitriakis falcata*

FAO code: **TRK**   Plate page 458

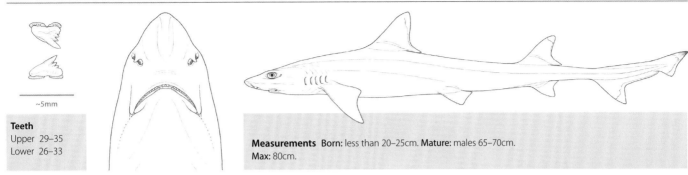

Teeth
Upper 29–35
Lower 26–33

~5mm

Measurements Born: less than 20–25cm. **Mature:** males 65–70cm.
Max: 80cm.

Identification Slender-bodied. Small anterior nasal flaps. Arched mouth; small blade-like teeth. Dorsal fins similar-sized; first dorsal origin over pectoral insertions, in front of pectoral free rear tips. Dorsal, pectoral and anal fins strongly falcate in adults. Greyish brown body. No stripe below snout; dorsal fins with distinct white tips. Juveniles and subadults less than 50cm have bold saddles and large spots with solid centres on body and fins.
Distribution Central Indo-Pacific: northwest Australia.
Habitat Outer continental shelf, 110–200m.
Behaviour Unknown.
Biology Unknown.
Status IUCN Red List: **Least Concern**. Very limited range is largely unfished.

INDONESIAN HOUNDSHARK *Hemitriakis indroyonoi*

FAO code: **TRK**   Plate page 458

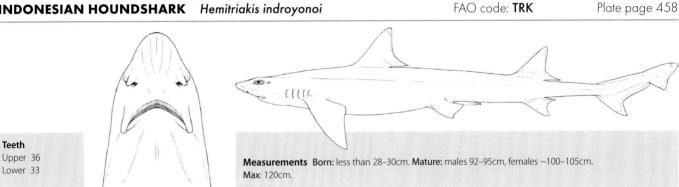

Teeth
Upper 36
Lower 33

Measurements Born: less than 28–30cm. **Mature:** males 92–95cm, females ~100–105cm.
Max: 120cm.

Identification Slender houndshark. Short narrow pointed snout. Small nasal flaps. Broadly arched mouth, long upper labial furrows, shorter lowers. Large oval eyes. Dorsal, pectoral and anal fins strongly falcate; first dorsal fin slightly larger than second, origin behind pectoral fin rear tip. Adults greyish brown above, becoming lighter below, with prominent white tipped fins; newborns with dark blotches and bars, but fades with growth.
Distribution Central Indo-Pacific and east Indian Ocean: east Indonesia, Bali and Lombok Islands, and off the Andaman and Nicobar Islands, Andaman Sea.
Habitat Outer continental shelf at 60m or more.
Behaviour Unknown.
Biology Viviparous, without yolk-sac placenta; litters from 6–11. Diet mostly consists of small demersal fishes and crustaceans.
Status IUCN Red List: **Not Evaluated**. This species has yet to be assessed.

JAPANESE TOPESHARK   *Hemitriakis japanica*

FAO code: **THJ**   Plate page 458

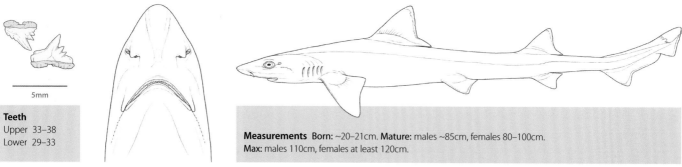

5mm

Teeth
Upper 33–38
Lower 29–33

Measurements Born: ~20–21cm. **Mature:** males ~85cm, females 80–100cm.
Max: males 110cm, females at least 120cm.

Identification Fairly low, slit-like eyes above ridges. Moderately long, parabolic snout. Short anterior nasal flaps. Broadly arched mouth; small blade-like teeth. Fins not strongly falcate in adults. First dorsal origin over or behind pectoral free rear tips (except in newborns). Anal fin much smaller than dorsals. Uniform grey-brown above, becoming lighter below, with conspicuous white-edged fins.
Distribution Northwest Pacific: China, Taiwan, Korea and Japan.
Habitat Temperate to subtropical continental shelf, close inshore to over 100m offshore.
Behaviour Unknown.
Biology Viviparous, no yolk-sac placenta; 8–22 pups per litter (average 10), increasing with size of female. In East China Sea, mating occurs June–September (mostly June–August) and pupping in June–August (mainly June). Feeds on small fishes, cephalopods and crustaceans.
Status IUCN Red List: Least Concern. A common, utilised fisheries bycatch, with no evidence of population declines.

WHITEFIN TOPESHARK   *Hemitriakis leucoperiptera*

FAO code: **THL**   Plate page 458

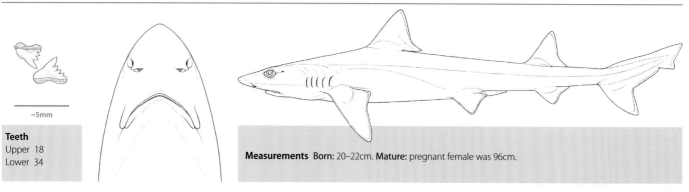

~5mm

Teeth
Upper 18
Lower 34

Measurements Born: 20–22cm. **Mature:** pregnant female was 96cm.

Identification Moderately elongated dorsolateral eyes with prominent subocular ridges. Moderately long, parabolic snout. Nostrils with short anterior nasal flaps. Broadly arched mouth; small blade-like teeth. Fins strongly falcate. First dorsal origin over pectoral inner margins. Second dorsal smaller, but much larger than anal fin. Snout without dark stripe on underside; fins with conspicuous white edges.
Distribution Central Indo-Pacific: Philippines.
Habitat Coastal waters to 48m.
Behaviour Unknown.
Biology Viviparous, 12 pups per litter.
Status IUCN Red List: Endangered. Two specimens collected over 75 years ago; a few additional specimens caught in the past 15 years.

BLACKTIP TOPESHARK   *Hypogaleus hyugaensis*   FAO code: **THH**   Plate page 458

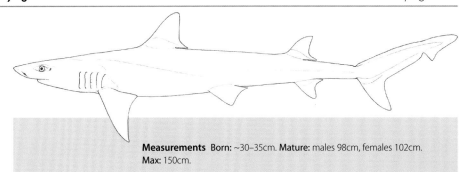

Teeth
Upper 46–51
Lower 41–48

5mm

Measurements Born: ~30–35cm. **Mature:** males 98cm, females 102cm.
Max: 150cm.

Alternative name Pencil Shark
Identification Fairly slender, medium-sized houndshark. Long, broadly pointed snout. Blade-like teeth.
Large, oval eyes. Second dorsal fin smaller than first but larger than anal fin. Relatively short terminal caudal
lobe (less than half dorsal caudal margin). Bronzy to grey-brown (lighter underside) with dusky dorsal and
upper caudal fin tips, particularly in young.
Distribution Indian Ocean and west Pacific: South Africa, Mozambique, Tanzania, Kenya, Australia, Taiwan
and Japan (Persian Gulf records may be based on *Paragaleus randalli*).
Habitat Tropical and warm-temperate continental shelves and uppermost slope, 40–230m, occasionally to
480m, near bottom.
Behaviour Unknown.
Biology Viviparous, yolk-sac placenta; 2–15 (average 10) pups per litter. Gestation is up to 12 months,
possibly followed by a 12 month resting period. Pups in December in South African waters and in February
off Australia after estimated 10–11 month gestation. Feeds on bony fishes and cephalopods.
Status IUCN Red List: Least Concern. Patchy distribution and generally low abundance; bycatch in fisheries.

LONGNOSE HOUNDSHARK   *Iago garricki*   FAO code: **TIK**   Plate page 460

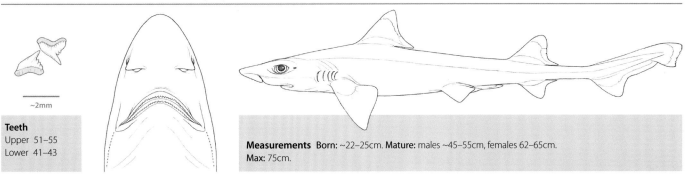

Teeth
Upper 51–55
Lower 41–43

~2mm

Measurements Born: ~22–25cm. **Mature:** males ~45–55cm, females 62–65cm.
Max: 75cm.

Identification Slender houndshark. Long narrow snout; small blade-like teeth. First dorsal origin situated
well forward (over pectoral fin inner margins). Second dorsal nearly as large as first. Poorly developed lower
caudal lobe. Greyish brown above, becoming lighter below; dorsal fins with conspicuous black tips, posterior
margins pale; pectoral and caudal fin posterior margins pale.
Distribution Central Indo-Pacific: New Hebrides, Vanuatu, Papua New Guinea, Solomon Islands, northern
Australia, Philippines.
Habitat Deep water tropical upper continental and insular slopes, 250–477m.
Behaviour Unknown.
Biology Viviparous, yolk-sac placenta; 4–5 pups per litter. Eats cephalopods.
Status IUCN Red List: Least Concern. Only rarely taken by fisheries.

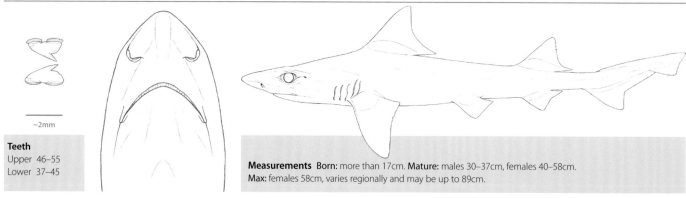

Teeth
Upper 46–55
Lower 37–45

~2mm

Measurements Born: more than 17cm. **Mature:** males 30–37cm, females 40–58cm.
Max: females 58cm, varies regionally and may be up to 89cm.

Identification Slender-bodied. Small, blade-like teeth. Large gill slits, width of longest nearly equal to length of large eyes. Small dorsal fins, origin of first set far forward over pectoral fin bases. Small lower caudal lobe. Uniform grey-brown above, lighter below; sometimes with darker dorsal fin margins.
Distribution North Indian Ocean: Red Sea, Gulf of Oman, Pakistan, southwest India, Sri Lanka, possibly Bay of Bengal and Myanmar (Burma). Sharks similar to *Iago omanensis* in Bay of Bengal may be distinct. The holotype of *I. mangalorensis*, reported from the Gulf of Aden/southwest India, is lost and may have been *I. omanensis*.
Habitat On or near bottom, continental shelf and slope, less than 92–1000m, or more, possibly to 2195m in Red Sea. Often found in warm, oxygen-poor conditions.
Behaviour Enlarged gills may permit survival in warm, low-oxygen, and probably hypersaline waters.
Biology Viviparous, yolk-sac placenta; 2–10 pups per litter. Reproductive parameters and maximum size may vary across its range. Eats bony fishes and cephalopods.
Status IUCN Red List: **Least Concern.** This species has a relatively wide depth and geographic range.

Bigeye Houndshark, *Iago omanensis*.

WHITE-MARGIN FIN HOUNDSHARK   *Mustelus albipinnis*   FAO code: **SDV**   Plate page 468

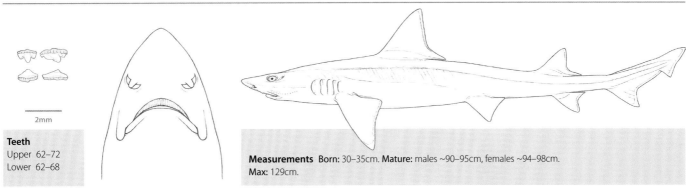

2mm

Teeth
Upper 62–72
Lower 62–68

Measurements Born: 30–35cm. **Mature:** males ~90–95cm, females ~94–98cm. **Max:** 129cm.

Identification A slender-bodied houndshark distinguished from other species by its coloration: a combination of uniformly dark grey-brown above, lighter below, and with distinctive white fin edges on the posterior margins of the dorsal, pectoral, anal and caudal fins.
Distribution Northeast Pacific: Gulf of California, Mexico possibly ranging to Ecuador and the Galapagos Islands.
Habitat A deepsea *Mustelus* species, usually occurring from 30–281m. Tends to inhabit areas of rock and other hard-bottom substrates.
Behaviour Unknown.
Biology Viviparity without yolk-sac, 3–23 pups per litter (average 16). Mostly eats crustaceans and small fishes.
Status IUCN Red List: **Least Concern**. A trawl bycatch in part of its distribution, but the deeper part of this species' range is mostly unfished.

GUMMY SHARK   *Mustelus antarcticus*   FAO code: **CTU**   Plate page 464

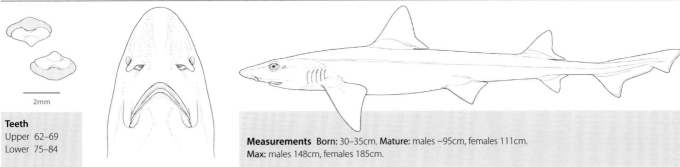

2mm

Teeth
Upper 62–69
Lower 75–84

Measurements Born: 30–35cm. **Mature:** males ~95cm, females 111cm. **Max:** males 148cm, females 185cm.

Identification Slender houndshark. Wide internarial space. Angular mouth; flat 'pavement' of crushing teeth; long upper labial furrows. Second dorsal fin nearly as large as first; fin margins not frayed. Relatively small pectoral and pelvic fins. Bronze to greyish brown; white-spotted, rarely with some black spots; pale underside.
Distribution Southeast Indian Ocean to southwest Pacific: Australia, where it is the only *Mustelus* found in temperate waters.
Habitat On or near bottom, mainly on continental shelf, shore to about 80m, also upper slope to 350m.
Behaviour No major migrations. Some large females move long distances, but no defined seasonal migration patterns.
Biology Viviparity without yolk-sac; 1–57 pups per litter (depending on female size, average 14). Gestation 11–12 months. Females give birth annually or on alternate years in shallow coastal nurseries. Age at maturity ~4–5 years, with longevity 16 years. Eats crustaceans, marine worms and small fishes.
Status IUCN Red List: **Least Concern**. Abundant, a high-productivity endemic harvested for its meat throughout range. Management protects nursery grounds and large mature females; stocks have recovered from earlier depletion by fisheries. *Mustelus walkeri* is now known to be this species.

STARRY SMOOTHHOUND *Mustelus asterias*

FAO code: **SDS**   Plate page 462

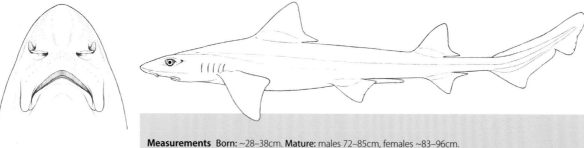

Teeth
Upper 30–41
Lower 31–46

Measurements Born: ~28–38cm. **Mature:** males 72–85cm, females ~83–96cm.
Max: 140cm.

Identification Large, slender-bodied smoothhound. Snout moderately long and bluntly angular in lateral view; narrower internarial space than similar regional species. Unfringed dorsal fins. Pectoral and pelvic fins relatively small. Only European smoothhound with numerous small, white spots on grey or grey-brown flanks and back (no dark spots or bars).
Distribution Northeast Atlantic: British Isles, North Sea to Canaries and Western Sahara; Mediterranean Sea.
Habitat Continental and insular shelves, on or near sand and gravel, intertidal to ~200m.
Behaviour Swims actively in captivity. May migrate inshore in summer. Specialist feeder on crustaceans.
Biology Viviparous, no yolk-sac placenta; 6–35 per litter, increasing with maternal size. Gives birth inshore in summer after ~1 year gestation. Males mature at 4–5 years, females 6 years; maximum age ~12 years for males, 20 years females.
Status IUCN Red List: **Least Concern** globally (assessment out of date and over-optimistic). **Near Threatened** in northeast Atlantic waters, and **Vulnerable** in the Mediterranean where significant declines have been caused by intensive fisheries. They are taken by anglers in some regions, and kept in aquaria.

GREY SMOOTHHOUND *Mustelus californicus*

FAO code: **CTN**   Plate page 464

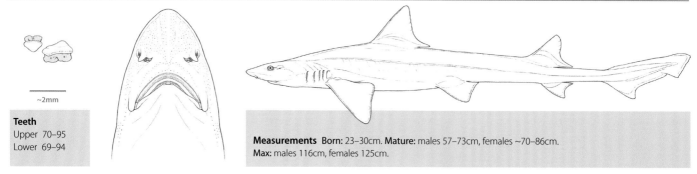

Teeth
Upper 70–95
Lower 69–94

Measurements Born: 23–30cm. **Mature:** males 57–73cm, females ~70–86cm.
Max: males 116cm, females 125cm.

Identification Short, narrow head. Broad internarial space. Short mouth with pavement of flat teeth; equal-sized labial furrows. Fairly small eyes. Triangular dorsal fins, first closer to pelvics than pectorals. Poorly developed lower caudal lobe. Unpatterned, uniform grey above; light below.
Distribution Northeast Pacific: northern California to Gulf of California.
Habitat Warm-temperate to tropical inshore and offshore waters, 0–265m, from shallow muddy bays to the edge of the continental shelf.
Behaviour Summer visitor to north-central California waters, resident further south.
Biology Viviparous, 2–16 pups per litter; born after 10–12 month gestation. Age at maturity 1–3 years; maximum age at least 9 years. Feeds mostly on crabs.
Status IUCN Red List: **Least Concern**. Common where it occurs. Important in fisheries in the southern half of its range.

DUSKY SMOOTHHOUND   *Mustelus canis*   FAO code: **CTI**   Plate page 462

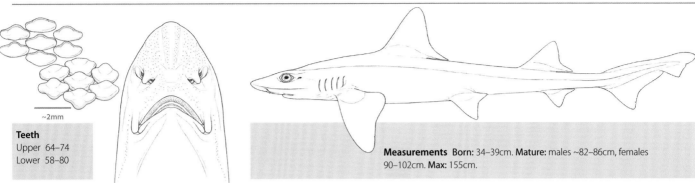

Teeth
Upper 64–74
Lower 58–80

~2mm

Measurements Born: 34–39cm. **Mature:** males ~82–86cm, females 90–102cm. **Max:** 155cm.

Identification Large, slender-bodied. Short head and snout. Broad internarial space. Low, flat pavement teeth; upper labial furrows longer than lowers. Large ,closely spaced eyes. Dorsal fins unfringed. Deeply notched caudal fin. Usually unspotted grey above, white below. Newborn young with dusky-tipped dorsal and caudal fins.
Distribution West Atlantic: Canada to Argentina, in several widely separated, discrete populations, coastal to well offshore.
Habitat Continental shelf populations prefer mud and sand, 0–18m, but occur to 200m, rarely on upper slopes to 808m.
Behaviour Very active shark, constantly patrolling for food that can be located even when hidden. Northern population migrates inshore and north in summer, south and offshore in winter. Not territorial in captivity, but larger animals are dominant.
Biology Viviparous, with yolk-sac placenta, bearing 4–20 pups per litter (average 10) after 10 month gestation. Matures in 2–7 years, with a maximum age of 10–16 years. Feeds mostly on crustaceans and bony fishes.
Status IUCN Red List: **Near Threatened**. Abundant where it occurs. Largest females are fished heavily by long-line and gill net, declines reported. Kept in public aquaria.

SHARPTOOTH SMOOTHHOUND   *Mustelus dorsalis*   FAO code: **CTD**   Plate page 464

Teeth
Upper not known
Lower not known

~2mm

Measurements Born: 21–23cm. **Mature:** males and females ~43cm. **Max:** 66cm.

Identification Small, fairly slender-bodied. Long, acutely pointed snout. Large, moderately spaced nostrils. High-cusped teeth arranged in a pavement; upper labial furrows longer than lowers. Very small, widely set eyes. Dorsal fins unfringed, broadly triangular. Lanceolate lateral trunk denticles. Unspotted plain grey or grey-brown, light below.
Distribution East Pacific: Baja California (Mexico) to Peru.
Habitat Tropical, inshore continental shelf, 20–200m.
Behaviour Unknown.
Biology Viviparous, yolk-sac placenta; 4 pups per litter. Eats mantis shrimps and other crustaceans.
Status IUCN Red List: **Vulnerable**. Possibly uncommon compared with other houndsharks in region, but likely taken in intensive, largely unregulated, fisheries that have recorded declining landings of 'tollo' (aggregated species of *Mustelus*). These are used for meat.

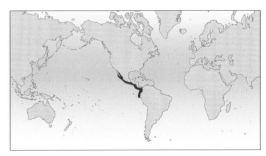

STRIPED SMOOTHHOUND *Mustelus fasciatus*   FAO code: **CTF**   Plate page 466

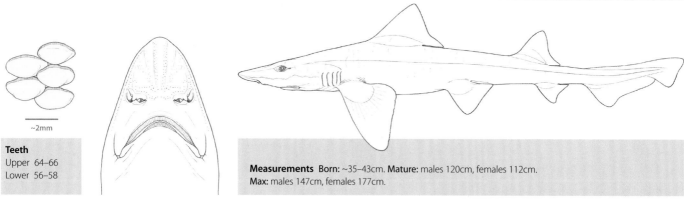

Teeth
Upper 64–66
Lower 56–58

~2mm

Measurements Born: ~35–43cm. **Mature:** males 120cm, females 112cm.
Max: males 147cm, females 177cm.

Identification Fairly stocky. Very long head and long, angular, acutely pointed snout. Wide internarial space. Teeth arranged in pavement without cusps, crowns broadly rounded; upper labial furrows much longer than lowers. Eyes very small. Broadly triangular, unfringed dorsal fins. Short caudal peduncle. Vertical dark bars on unspotted body, at least in young. Very distinctive, closest to *Mustelus mento*, but has a longer and more angular head.
Distribution Southwest Atlantic: southern Brazil, Uruguay, northern Argentina.
Habitat Temperate continental shelf and upper-most slope, on bottom inshore and offshore, possibly from intertidal to 70m, rarely 10–500m off southern Brazil.
Behaviour Unknown.
Biology Viviparous, with placenta; 6–12 pups per litter, increasing with female size, with an annual reproductive cycle and gestation period of 11–12 months. Eats mostly crustaceans and fishes.
Status IUCN Red List: **Critically Endangered**. Uncommon to rare within its very limited range, where intense fisheries occur.

SPOTLESS SMOOTHHOUND *Mustelus griseus*   FAO code: **CTE**   Plate page 464

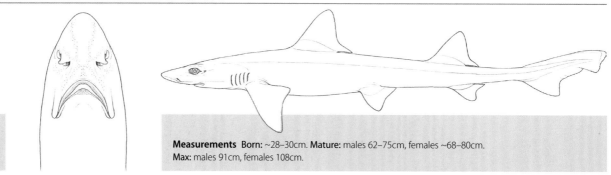

Teeth
Upper 71
Lower 71

Measurements Born: ~28–30cm. **Mature:** males 62–75cm, females ~68–80cm.
Max: males 91cm, females 108cm.

Identification Moderate-sized. Short head and snout. Wide internarial space. Low-crowned teeth with weak cusps; upper labial furrows equal or slightly shorter than lowers. Small eyes. Unfringed dorsal fins. Semifalcate lower caudal lobe. Plain grey or grey-brown, light below.
Distribution Northwest Pacific: Japan, Koreas, Taiwan, China, Vietnam.
Habitat Bottom, inshore to at least 131m, possibly to 300m.
Behaviour Unknown.
Biology Viviparous, yolk-sac placenta; 2–20 pups per litter, larger females bear more. Gestation 10 months in Japanese waters (mating in July, birth April–May). Feeds mostly on crustaceans.
Status IUCN Red List: **Data Deficient**. Common where it occurs. Important in fisheries off Japan, China and Taiwan.

BROWN SMOOTHHOUND   *Mustelus henlei*

FAO code: **CTK**   Plate page 464

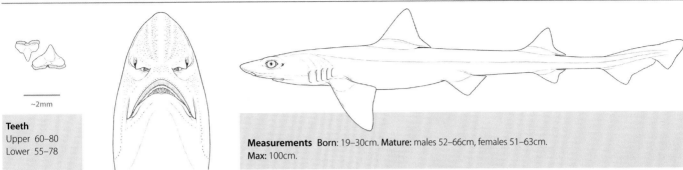

Teeth
Upper 60–80
Lower 55–78

~2mm

Measurements Born: 19–30cm. **Mature:** males 52–66cm, females 51–63cm.
Max: 100cm.

Identification Slender-bodied. Short head. Moderately long snout. Wide internarial space. High-cusped 'pavement' teeth; upper labial furrows longer than lowers. Large, close-set eyes. Trailing dorsal fin edges appear broadly frayed with dark margins of bare ceratotrichia. Long caudal peduncle. Unspotted, usually iridescent bronzy-brown (occasionally greyish); white below.
Distribution East Pacific: USA (Washington State) to Peru.
Habitat Continental shelf, intertidal to at least 281m. Most abundant in enclosed, shallow, muddy bays. Most cold-tolerant of the three *Mustelus* spp. occurring north to Washington State.
Behaviour Active agile swimmer in captivity, mainly patrolling over bottom, sometimes in midwater and at surface. Often rests on seabed. One tagged individual migrated about 160km in three months.
Biology Viviparous, 1–21 pups per litter; number of pups increases with size of the female. Gestation is 10–11 months; annual reproductive cycle. Age at maturity 2–3 years, maximum age 7–13 years. Crustaceans most important prey, also polychaete worms and fishes.
Status IUCN Red List: Least Concern. Common to abundant where it occurs. Heavily fished. Kept in large aquaria.

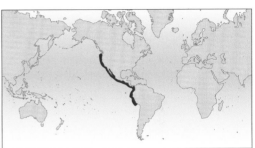

SMALLEYE SMOOTHHOUND   *Mustelus higmani*

FAO code: **CTJ**   Plate page 466

Teeth
Upper 66–78
Lower 62–69

~2mm

Measurements Born: 20–29cm. **Mature:** males ~43–48cm, females 43–48cm.
Max: males 69cm, females 88cm.

Identification Small-bodied. Long, acutely pointed snout. Wide internarial space. Fairly low-cusped teeth; upper labial furrows same length as lowers. Very small, widely separated eyes. Unfringed, falcate dorsal fins, second almost as large as first. Plain grey or grey-brown above, no spots; light below.
Distribution Tropical west Atlantic: northern Gulf of Mexico to Brazil.
Habitat Continental shelf and upper slope. Close inshore to well offshore (130m) on mud, sand and shell seabed off South America. Deep continental slope (at least 1463m) in Gulf of Mexico.
Behaviour Partial sexual segregation occurs.
Biology Viviparous, yolk-sac placenta; 1–7 pups per litter (usually 3–5). Feeds mainly on crustaceans, occasionally on bony fishes and squid.
Status IUCN Red List: Endangered. A target and bycatch of largely unmanaged artisanal and commercial fisheries throughout its range. Landings have declined despite continued high fishing effort, suggesting population collapse in parts of its range.

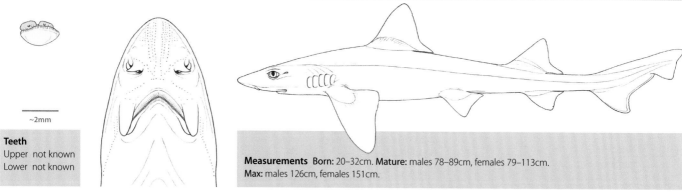

Teeth
Upper not known
Lower not known

~2mm

Measurements Born: 20–32cm. **Mature:** males 78–89cm, females 79–113cm.
Max: males 126cm, females 151cm.

Identification Large-bodied. Short head. Moderately long snout. Wide internarial space. Long upper labial furrows, longer than lowers. Fairly large, wide-set eyes. Dorsal fin margins unfrayed. Relatively large pectoral and pelvic fins. Grey or grey-brown, white-spotted (only *Mustelus* in range); lighter below.
Distribution Southwest Pacific: New Zealand (five management stocks).
Habitat Cold-temperate insular shelf and slope, close inshore to 1000m, but uncommon below 250m.
Behaviour Schooling species, move inshore to summer feeding and mating grounds. Segregates by size and sex: immatures school separately from largely single-sex adult schools. Females migrate further than males. Young occupy shallow coastal areas, including bays and estuaries.
Biology Viviparous, 2–37 pups per litter (increasing with female size); about an 11 month gestation. Fast-growing, males mature at 4–6 years and females 5–8 years, longevity at least 20 years or more. Eats crustaceans, particularly crabs.
Status IUCN Red List: Least Concern. Abundant, commercially fished. Stocks have rebuilt since fishery quotas introduced.

~2mm

Teeth
Upper 72–102
Lower 71–106

Measurements Born: 28–35cm. **Mature:** males 70–92cm, females 97–103cm.
Max: possibly 175cm.

Identification Large-bodied. Distinguished from *Mustelus californicus* by more pointed snout, more widely separated eyes, shorter mouth, strongly falcate fins, and shorter upper labial furrows. Unspotted grey or grey-brown above, light below.
Distribution East Pacific: southern California (possibly only in warm summers) to northern Peru.
Habitat Warm-temperate to tropical continental shelf, close inshore to well offshore, from 9–200m.
Behaviour Unknown.
Biology Viviparous, with 6–19 pups per litter after a 11 month gestation. Feeds mostly on crustaceans, with a change in diet with growth: young sharks feed almost exclusively on crabs while adults have a more varied diet that includes squids.
Status IUCN Red List: Least Concern. Common to abundant where it occurs. Important in longline fisheries for meat in the Gulf of California. Often misidentified with other regional smoothhound sharks.

STARSPOTTED SMOOTHHOUND   *Mustelus manazo*

FAO code: **MTZ**   Plate page 464

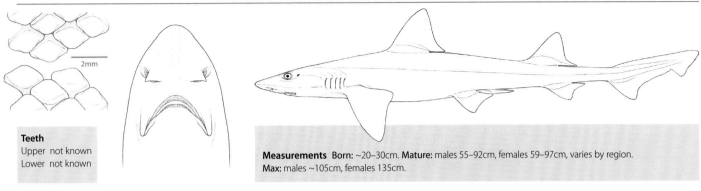

Teeth
Upper not known
Lower not known

Measurements Born: ~20–30cm. **Mature:** males 55–92cm, females 59–97cm, varies by region. **Max:** males ~105cm, females 135cm.

Identification Medium-sized. Short head. Moderately long snout. Fairly narrow internarial space. Upper labial furrows much longer than lowers. Large, close-set eyes. Unfringed dorsal fins. Relatively small pectoral and pelvic fins. Grey to grey-brown, with many white spots; light below. Only white-spotted smoothhound in its range.
Distribution Northwest Pacific: southern Siberia, Japan, Koreas, China, Taiwan and southeast Asia. Possibly west Indian Ocean: Kenya, Tanzania and Madagascar.
Habitat Temperate to tropical continental shelf, intertidal to close inshore, on mud and sand, 1–360m.
Behaviour Unknown.
Biology Viviparous, no yolk-sac placenta; 1–22 pups per litter (mostly 2–6, average ~5) increasing with size of mother, born in spring after ~10–12 month gestation. Adults mate in summer. Fairly fast-growing, age at maturity varies by region, females 3–7 years and males 2–6 years, maximum age 9–17 years. Feeds on mostly bottom-dwelling invertebrates, particularly crustaceans.
Status IUCN Red List: Data Deficient. Generally abundant where it occurs. Important in longline fisheries for meat off Japan, also China and the Koreas. Records of this smoothhound from the west Indian Ocean may be a different species.

SPECKLED SMOOTHHOUND   *Mustelus mento*

FAO code: **MTE**   Plate page 464

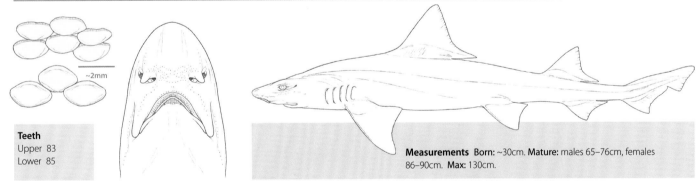

Teeth
Upper 83
Lower 85

Measurements Born: ~30cm. **Mature:** males 65–76cm, females 86–90cm. **Max:** 130cm.

Identification Stocky body. Short, bluntly angular snout. Wide internarial space. A crushing 'pavement' of broadly rounded, uncusped teeth; upper labial furrows slightly longer than lowers. Fairly small, moderately spaced eyes. Unfringed dorsal fins. Short caudal peduncle. White-spotted grey to grey-brown, light below; vertical dark bands only in young. Fairly similar to *Mustelus fasciatus* but with a shorter, more rounded head.
Distribution Southeast Pacific: Galapagos Islands, Peru, Chile, Juan Fernandez Island.
Habitat Temperate continental and insular shelves, inshore and offshore, 16–50m.
Behaviour Unknown.
Biology Viviparous, without yolk-sac placenta; 7 pups per litter.
Status IUCN Red List: Critically Endangered. This and other species of 'tollo' (*Mustelus*) are heavily exploited by artisanal and some industrial fisheries and used for meat. Landings records illustrate 'boom-and-bust' fisheries for this species, which is now seriously depleted.

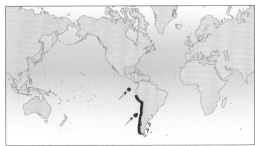

VENEZUELAN DWARF SMOOTHHOUND *Mustelus minicanis*

FAO code: **SDV**   Plate page 466

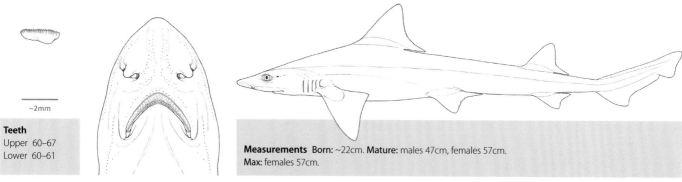

Teeth
Upper 60–67
Lower 60–61

~2mm

Measurements Born: ~22cm. **Mature:** males 47cm, females 57cm.
Max: females 57cm.

Identification Small, stout body. Short head and snout. Wide internarial space. Pavement of low-crowned teeth with weak cusps; upper labial furrows somewhat longer than lowers. Very large, fairly close-set eyes. Unfringed dorsal fins. Poorly developed lower caudal lobe. Uniform grey, without spots. Newborns have dusky-tipped dorsal and caudal fins.
Distribution West Atlantic: Colombia and Venezuela.
Habitat Offshore outer continental shelf, 71–183m.
Behaviour Unknown.
Biology Poorly known. Viviparous (placental), 1 known litter of 5 pups.
Status IUCN Red List: **Endangered**. Possibly rare or uncommon; only 9 specimens recorded. Possibly caught in offshore trawl fisheries.

ARABIAN SMOOTHHOUND *Mustelus mosis*

FAO code: **MTM**   Plate page 466

Teeth
Upper 72–83
Lower 68–77

Measurements Born: 26–28cm. **Mature:** males 65–78cm, females 73–106cm.
Max: 150cm.

Alternative name Hardnose Smoothhound
Identification Large, fairly slender-bodied. Short head and snout. Broad internarial space. Low-crowned teeth with weak cusps; labial furrows similar length. Large, fairly close-set eyes. Unfringed dorsal fins. Semifalcate lower caudal lobe. Unspotted grey or grey-brown, light below. White-tipped first dorsal and black-tipped second dorsal and caudal fins in South Africa.
Distribution Indian Ocean: Red Sea, Persian Gulf, India, Pakistan and Sri Lanka in north; Kenya to KwaZulu-Natal and South Africa. The only *Mustelus* in most of its range.
Habitat Continental shelf, bottom inshore and offshore, some on coral reefs, 20–250m.
Behaviour Unrecorded.
Biology Viviparous, with yolk-sac placenta; 2–16 pups per litter. Eats small bottom-dwelling fish, molluscs and crustaceans.
Status IUCN Red List: **Near Threatened**. Common where present. Fished for food off Pakistan and India. Does well in captivity. Often misidentified with other houndshark species.

SMOOTHHOUND *Mustelus mustelus*

FAO code: **SMD**   Plate page 462

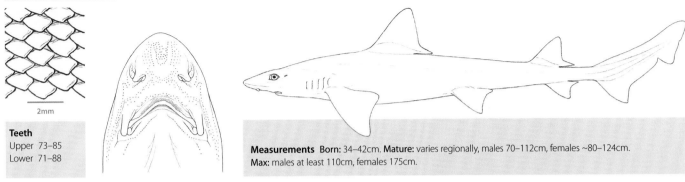

2mm

Teeth
Upper 73–85
Lower 71–88

Measurements Born: 34–42cm. **Mature:** varies regionally, males 70–112cm, females ~80–124cm. **Max:** males at least 110cm, females 175cm.

Identification A large, usually unspotted houndshark. Short head and snout. Internarial space >1.5 times nostril width. Upper labial furrows slightly longer than lowers. Unfringed dorsal fins. Moderately large pectoral and pelvic fins. Uniform grey or grey-brown above, lighter below, with no white spots or dark bars; very large individuals may have a few scattered dark spots.
Distribution Temperate east Atlantic: UK to Mediterranean, Morocco, Canaries, possibly Azores, Madeira. Angola to South Africa, including Indian Ocean coast.
Habitat Continental shelves and upper slopes, usually 5–50m, often in intertidal but occasionally to at least 800m.
Behaviour Prefers swimming near bottom, but sometimes found in midwater.
Biology Viviparous, yolk-sac placenta; 4–18 pups per litter after a 9–11 month gestation; larger females have more pups. Age at maturity ~9 years for males, 10–11 years for females; maximum age is 25 years. Primarily feeds on crustaceans, also cephalopods and bony fishes.
Status IUCN Red List: **Vulnerable** globally and regionally in European waters and the Mediterranean. Still common to abundant, despite depletion in fisheries. Particularly important in European, Mediterranean and West African fisheries. Taken in bottom trawls, fixed nets and line gear. Used fresh, frozen, dried-salted and smoked for food; liver for oil, and for fishmeal. Also taken by sports fishers. Kept in aquaria.

NARROWFIN SMOOTHHOUND *Mustelus norrisi*

FAO code: **MTR**   Plate page 466

~2mm

Teeth
Upper 58–65
Lower 57–60

Measurements Born: 29–37cm. **Mature:** males 76–81cm, females ~76–87cm. **Max:** males 118cm, females 123cm.

Identification Fairly large-bodied. Short, narrow head. Narrow internarial space. Long mouth; upper labial furrows equal to or slightly longer than lowers. Relatively large eyes. Strongly falcate fins. Grey in colour, unspotted. Dorsal and caudal fin tips dusky in newborns. Smaller, slenderer and narrower-headed than *Mustelus canis*.
Distribution West Atlantic: USA (Gulf of Mexico); southern Caribbean coast of Colombia and Venezuela; southern Brazil.
Habitat Continental shelf, sand and mud bottom, close inshore to 100m and occasionally to 260m, mostly less than 55m.
Behaviour Segregates by size and sex off Florida: adult males are inshore in winter. May be migratory in Gulf of Mexico, moving inshore to shallower than 55m in winter, offshore in other seasons.
Biology Viviparous, yolk-sac placenta; 7–14 pups per litter. Eats crabs and shrimp, also small fishes.
Status IUCN Red List: **Data Deficient**. Common where it occurs. No fisheries information.

WHITESPOTTED SMOOTHHOUND   *Mustelus palumbes*   FAO code: **MUP**   Plate page 462

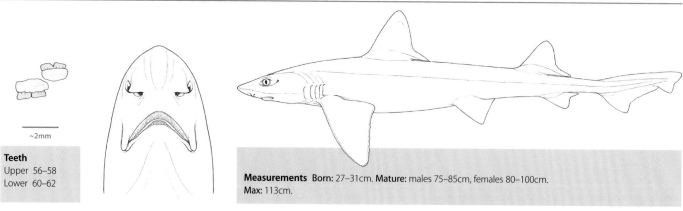

Teeth
Upper 56–58
Lower 60–62

Measurements Born: 27–31cm. **Mature:** males 75–85cm, females 80–100cm.
Max: 113cm.

Identification Large-bodied. Relatively broad internarial space. Upper labial furrows longer than lowers. Unfringed dorsal fins. Relatively large pectoral and pelvic fins (larger than those of *Mustelus asterias, M. manazo* and *M. mustelus*). Uniform grey to grey-brown; usually white-spotted (the only white-spotted smoothhound in southern African waters).
Distribution Southeast Atlantic, southwest Indian Ocean: Namibia, South Africa, southern Mozambique.
Habitat Continental shelf and upper slope, nearshore to 443m, but mostly below 70m, on or near sand and gravel bottom.
Behaviour Unknown.
Biology Viviparous, no yolk-sac placenta; 3–15 pups per litter (average 7). Eats crabs and other crustaceans.
Status IUCN Red List: Least Concern. Common offshore, and taken as bycatch in some fisheries, but usually discarded. Occasionally taken by sports anglers, but far less common than the Smoothhound *Mustelus mustelus*. Population appears to have increased over the past three decades.

BLACKSPOT SMOOTHHOUND   *Mustelus punctulatus*   FAO code: **MPT**   Plate page 462

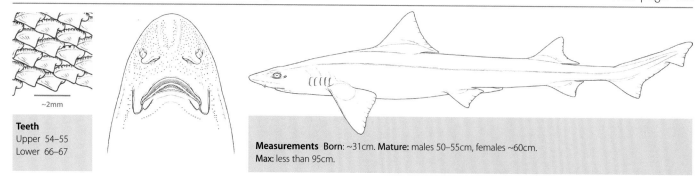

Teeth
Upper 54–55
Lower 66–67

Measurements Born: ~31cm. **Mature:** males 50–55cm, females ~60cm.
Max: less than 95cm.

Identification A poorly known, slender-bodied, smoothhound shark. Short, narrow head and snout. Internarial space <1.5 times nostril width. Upper labial furrows slightly longer than lowers. Large eyes. Fringed dorsal fins. Colour grey above with distinct black spots along lateral sides of trunk, lighter below. Often misidentified as more common *Mustelus mustelus*.
Distribution East Atlantic: western Sahara, Mediterranean.
Habitat Inshore continental shelf, on bottom, <200m.
Behaviour Unknown.
Biology Little known, presumably viviparous. Probably feeds on crustaceans.
Status IUCN Red List: Data Deficient. Presumably fished for food, but recorded as *M. mustelus*.

AUSTRALIAN GREY SMOOTHHOUND   *Mustelus ravidus*   FAO code: **SDV**   Plate page 468

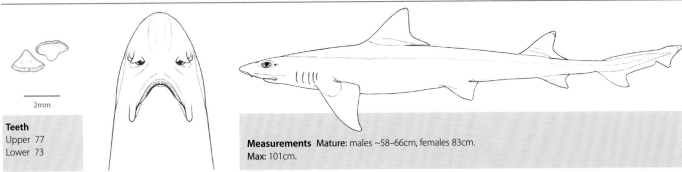

2mm

Teeth
Upper 77
Lower 73

Measurements Mature: males ~58–66cm, females 83cm.
Max: 101cm.

Identification Slender houndshark. Snout relatively long. Wide internarial space. Upper labial furrows shorter than lowers. Rather high-cusped crushing teeth arranged in a 'pavement'. Unfringed dorsal fins, first higher than second; pectoral fins moderately large, acutely pointed at anterior margin apex. Deeply notched caudal fin. Uniform unspotted bronzy colour above, becoming white below.
Distribution Central Indo-Pacific: northern and western tropical Australia.
Habitat Deep continental shelf, 100–300m.
Behaviour Unknown.
Biology Largely unknown. Bears 6–24 pups (average 18) per litter, therefore presumably fairly productive.
Status IUCN Red List: **Least Concern**. Taken as fisheries bycatch, but much of range unfished.

NARROWNOSE SMOOTHHOUND   *Mustelus schmitti*   FAO code: **SDP**   Plate page 462

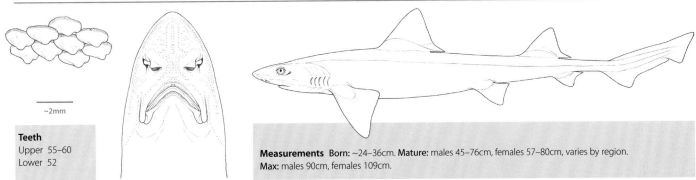

~2mm

Teeth
Upper 55–60
Lower 52

Measurements Born: ~24–36cm. **Mature:** males 45–76cm, females 57–80cm, varies by region.
Max: males 90cm, females 109cm.

Identification Fairly slender smoothhound. Short head, moderately long snout. Narrow internarial space. Upper labial furrows longer than lowers. Dorsal fins prominently fringed on posterior margin giving it a dark frayed appearance, first dorsal fin origin well over pectoral fin inner margins and much larger than second. Grey above with white spots, lighter below Easily distinguished from other white-spotted *Mustelus*.
Distribution Southwest Atlantic: southern Brazil to southern Argentina.
Habitat Offshore continental shelf, 2–195m.
Behaviour Migrates seasonally from Brazil in winter to Uruguay and Argentina in summer.
Biology Viviparous, without placenta, 1–14 pups (average 8) per litter; gestation 11–12 months. Age at maturity 3–6 years males, 4–7 years females, with maximum age 9 years males, 16 years females. Eats crabs, other crustaceans, many other invertebrates and small benthic fishes.
Status IUCN Red List: **Critically Endangered**. Formerly an important commercial fish species, intensively exploited throughout its range, including in breeding and nursery grounds, and now severely depleted.

GULF OF MEXICO SMOOTHHOUND *Mustelus sinusmexicanus*   FAO code: **SDV**   Plate page 466

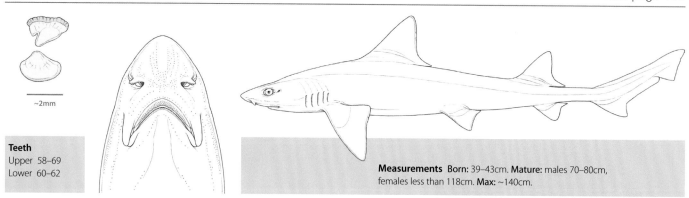

Teeth
Upper 58–69
Lower 60–62

~2mm

Measurements Born: 39–43cm. **Mature:** males 70–80cm,
females less than 118cm. **Max:** ~140cm.

Identification Large-bodied. Short head and snout. Broad internarial space. Pavement of high-crowned teeth (unlike sympatric *Mustelus canis* and *M. norrisi*); upper labial furrows longer than lowers. Large, close-set eyes. Unfringed dorsal fins; first dorsal large. Lower caudal lobe non-falcate but strong and moderately expanded. Colour grey, unspotted. Young have dusky-tipped dorsal and caudal fins.
Distribution West Atlantic: Gulf of Mexico (USA and Mexico).
Habitat Offshore continental shelf and upper slope, 36–229m, mostly 42–91m. Not in shallow water.
Behaviour Unknown.
Biology Apparently placental viviparous, 8 pups per litter. Poorly known.
Status IUCN Red List: Data Deficient. Recently described endemic, confused with *M. canis*.

WESTERN SPOTTED GUMMY SHARK *Mustelus stevensi*   FAO code: **SDV**   Plate page 468

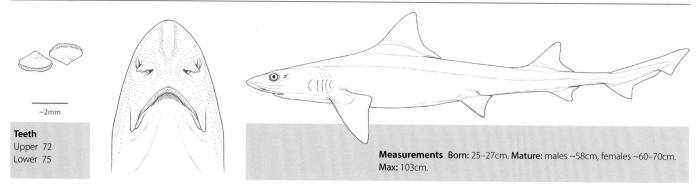

Teeth
Upper 72
Lower 75

~2mm

Measurements Born: 25–27cm. **Mature:** males ~58cm, females ~60–70cm.
Max: 103cm.

Identification Slender-bodied. Snout relatively long, tip narrowly rounded. Upper labial furrow longer than lowers. Flattened low-cusped teeth arranged in a 'pavement'. Greyish brown above with numerous white spots on adults (juveniles often plain without spots); paler below; dorsal and caudal fins with dusky or black tips, posterior caudal fin margin not pale-edged. Very similar to cold-water *M. antarcticus*, but range does not overlap.
Distribution Central Indo-Pacific: northern tropical Australia, Java, Bali, Lombok and Myanmar (Burma).
Habitat Deep continental shelf, 121–402m, possibly to 735m.
Behaviour Unknown.
Biology Viviparous, 4–17 pups per litter. Feeds on crustaceans (mainly crabs), fishes and cephalopods.
Status IUCN Red List: Least Concern. Probably widespread. Irregular bycatch in fisheries, but most of range is unfished.

HUMPBACK SMOOTHHOUND *Mustelus whitneyi*

FAO code: **MUW**   Plate page 466

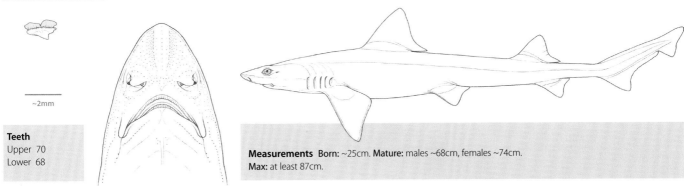

Teeth
Upper 70
Lower 68

Measurements Born: ~25cm. **Mature:** males ~68cm, females ~74cm.
Max: at least 87cm.

Identification Rather stocky (almost humpbacked). Fairly long head and snout. Wide internarial space. Strongly cuspidate teeth; upper labial furrows much longer than lowers. Fairly large eyes. A dark margin of bare ceratotrichia on trailing edges of dorsal fins, giving them a broadly frayed appearance. Short caudal peduncle. Lower caudal lobe very slightly falcate. Colour grey, unspotted; lighter below.
Distribution East to southeast Pacific: Panama to southern Chile.
Habitat Continental shelf, 16–211m, most common 70–100m. Prefers rocky bottom around islands.
Behaviour Unknown.
Biology Viviparous, 5–10 pups per litter. Eats crabs, mantis shrimp and small bony fishes.
Status IUCN Red List: Critically Endangered. This and other species of 'tollo' (*Mustelus*) are heavily exploited by artisanal and some industrial fisheries and used for meat. Landings records illustrate former 'boom-and-bust' fisheries for this species.

WHITEFIN SMOOTHHOUND *Mustelus widodoi*

FAO code: **SDV**   Plate page 468

Teeth
Upper 73
Lower 69

Measurements Born: less than 31cm. **Mature:** males 83–89cm, females over 92cm.
Max: at least 110cm.

Identification A large, slender-bodied, smoothhound shark. Relatively short snout. Wide internarial space,. Upper labial furrows shorter than lowers. Teeth cuspidate with low rounded cusp. Dorsal fin posterior margins unfringed. Pectoral fins moderately large, falcate, acutely pointed at tip. Uniform grey above without spots, lighter below. Broad white margin on the first dorsal fin; distinct black margin on the second dorsal fin apex; terminal caudal lobe also with distinct black tip.
Distribution Central Indo-Pacific: Indonesia: possibly endemic to eastern Indonesia.
Habitat Inshore, usually 60–120m.
Behaviour Unknown.
Biology Viviparous. Eats crustaceans and small bony fishes.
Status IUCN Red List: Data Deficient. Very little information other than it is seen in low numbers at fish markets.

ANDAMAN SMOOTHHOUND   *Mustelus andamanensis*   FAO code: **SDV**   Plate page 468

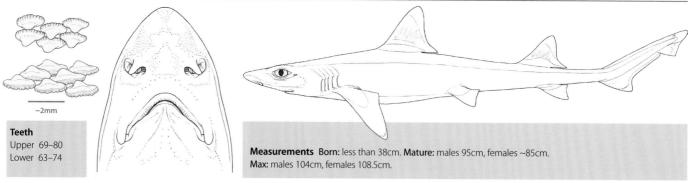

Teeth
Upper 69–80
Lower 63–74

Measurements Born: less than 38cm. **Mature:** males 95cm, females ~85cm.
Max: males 104cm, females 108.5cm.

Identification Slender-bodied. Snout narrowly rounded at tip. Upper labial furrows shorter than lower furrows. Teeth similar in both jaws; flattened, low-cusped and arranged in pavement. Externally, looks similar to other *Mustelus* species. Greyish brown above, lighter below, white spots never present, dorsal fins dusky at bases, upper half becoming blackish, without white margins, caudal fin with narrow white margin and a black tip.
Distribution Known only from the Andaman Sea, from Phuket, Thailand to Yangon, Myanmar (Burma) and the Andaman Islands.
Habitat Caught at depths of less than 100m.
Behaviour Adults and juveniles appear to segregate by size and maturity. Adult females group separately from adult males.
Biology Unknown.
Status IUCN Red List: Not Evaluated. This species was only recently described.

FLAPNOSE HOUNDSHARK   *Scylliogaleus quecketti*   FAO code: **TSK**   Plate page 460

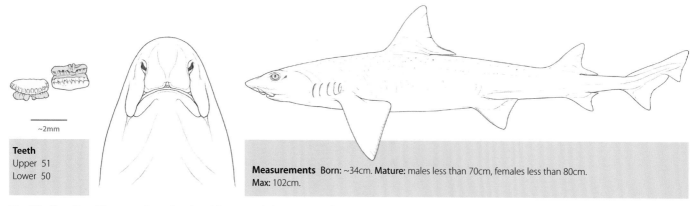

Teeth
Upper 51
Lower 50

Measurements Born: ~34cm. **Mature:** males less than 70cm, females less than 80cm.
Max: 102cm.

Identification Short, blunt snout. Large, fused nasal flaps expanded to cover mouth; nasoral grooves. Small, blunt pebble-like teeth. Second dorsal fin as large as first and much larger than anal fin. Grey above, cream below. Newborns with white rear edges on dorsal, anal and caudal fins.
Distribution Southwest Indian Ocean: South Africa.
Habitat Inshore continental shelf, at the surfline and close offshore, 0–73m.
Behaviour Poorly known. The species appears to have very localised movement patterns.
Biology Viviparous with litters of 2–4 (usually 2–3) pups after a gestation of 9–10 months. Feeds primarily on crustaceans (including lobsters), also squid.
Status IUCN Red List: Vulnerable. Restricted distribution is heavily fished; targeted for shark meat export trade.

SHARPFIN HOUNDSHARK *Triakis acutipinna*

FAO code: **TTA** Plate page 460

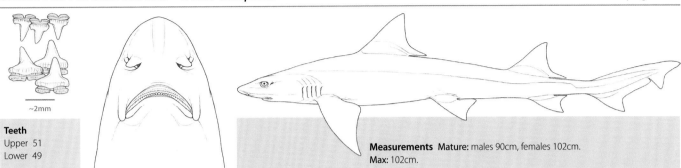

Teeth
Upper 51
Lower 49

Measurements Mature: males 90cm, females 102cm.
Max: 102cm.

Identification Short, broadly rounded snout. Widely separated anterior nasal flaps do not reach the mouth. Teeth not blade-like; long upper labial furrows reach the lower symphysis of the mouth. Narrow fins. First dorsal with an abruptly vertical posterior margin. Pectoral fins narrowly falcate. Colour description from preserved specimen: uniform greyish brown above, lighter below, no spots or bands.
Distribution Southeast Pacific: Ecuador.
Habitat Tropical continental waters from 50–200m.
Behaviour Unknown.
Biology Unknown.
Status IUCN Red List: Endangered. Rare, known from two specimens caught in heavily fished waters.

SPOTTED HOUNDSHARK *Triakis maculata*

FAO code: **TTM** Plate page 460

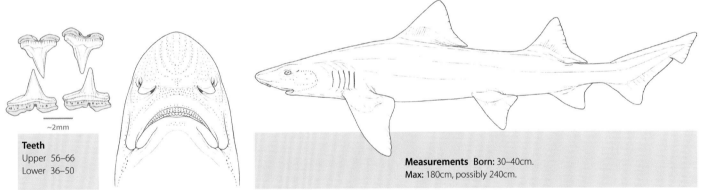

Teeth
Upper 56–66
Lower 36–50

Measurements Born: 30–40cm.
Max: 180cm, possibly 240cm.

Identification Very stout (for a houndshark). Short, broadly rounded snout. Widely separated, lobate anterior nasal flaps do not reach mouth. Teeth not blade-like; long upper labial furrows reaching lower symphysis of mouth. Broad fins. First dorsal fin with a backwards-sloping rear edge. Pectoral fins broadly falcate. Greyish above, usually with many black spots; some are unspotted (plain females may have spotted young), lighter below.
Distribution East Pacific: Peru to northern Chile, Galapagos Islands.
Habitat Inshore temperate continental shelf, 10–200m.
Behaviour Unknown.
Biology Poorly known. Viviparous, without yolk-sac placenta; one record of 14 pups per litter.
Status IUCN Red List: Critically Endangered. This and other species of 'tollo' (*Mustelus*) are heavily exploited by artisanal and some industrial fisheries off Peru and Chile, where they are used for food. Landings records illustrate former 'boom-and-bust' fisheries for this species, which is now seriously depleted.

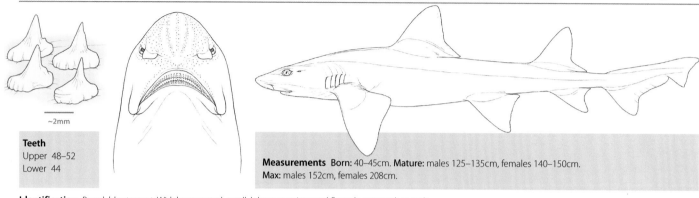

Teeth
Upper 48–52
Lower 44

~2mm

Measurements Born: 40–45cm. **Mature:** males 125–135cm, females 140–150cm. **Max:** males 152cm, females 208cm.

Identification Broad, blunt snout. Widely separated, small, lobate anterior nasal flaps do not reach mouth. Large mouth; small, pointed teeth; upper labial furrows do not reach lower symphysis. Large, broad fins. First dorsal almost vertical. High interdorsal ridge. Pectorals falcate with concave posterior margin. Short, heavy caudal peduncle. Grey-bronze, usually many black spots; white below. No or few spots in young (some adults also plain).

Distribution Southeast Atlantic: southern Angola to South Africa.   **Habitat** Shallow inshore to surfline, prefers sandy shores and rocks and crevices in shallow bays, 0–50m, usually less than 10m.

Behaviour Schools in summer, often many pregnant females present. Actively patrol very close to bottom in captivity, sometimes in midwater but rarely in the open.

Biology Viviparous, no yolk-sac placenta, 5–15 pups per litter (average 9–10); gestation is ~19–21 months, with a reproductive cycle of 2–3 years between pregnancies. Eats crabs, bony fishes, small sharks; larger sharks consume more fish, while smaller sharks feed mostly on crustaceans.

Status IUCN Red List: Least Concern. Locally common but range is heavily fished. Caught by sports anglers and commercial fishers, but low commercial value. Hardy in captivity.

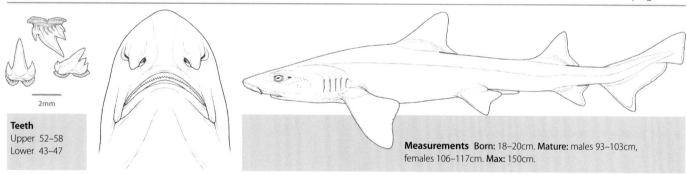

2mm

Teeth
Upper 52–58
Lower 43–47

Measurements Born: 18–20cm. **Mature:** males 93–103cm, females 106–117cm. **Max:** 150cm.

Identification Fairly slender-bodied. Short, broadly rounded snout. Widely separated, lobate anterior nasal flaps do not reach mouth. Partly blade-like teeth; long upper labial furrows reach lower symphysis. Relatively narrow fins. First dorsal rear margin almost vertical. Pectoral fins triangular in adults. Brownish grey above, sparsely scattered, small, black spots and broad, dusky saddles in young; spots fading (sometimes absent) in adults.

Distribution Northwest Pacific: Russia (Peter the Great Bay), Japan, Koreas, China, Taiwan. Possibly Central Indo-Pacific: Philippines uncertain.

Habitat Continental and insular, close inshore, on or near bottom, 30–150m. Often in estuaries and shallow bays, on sand, seaweed and eelgrass flats.

Behaviour Seldom gregarious, but some gather in seabed resting areas.

Biology Viviparous, no yolk-sac placenta; 10–24 pups per litter (average 9–10). Males mature at 5–6 years, females at 6–7 years, with maximum age for males 15 years and females 18 years. Eats crustaceans and other invertebrates, and small fishes.

Status IUCN Red List: Least Concern. Common to abundant where it occurs. Often fished.

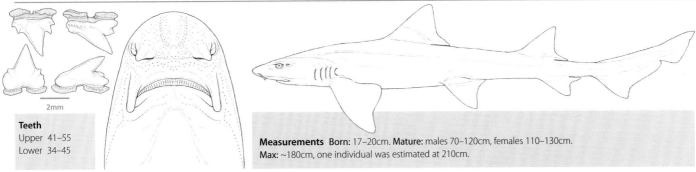

Teeth
Upper 41–55
Lower 34–45

2mm

Measurements Born: 17–20cm. **Mature:** males 70–120cm, females 110–130cm.
Max: ~180cm, one individual was estimated at 210cm.

Identification Snout broadly rounded. Widely separated anterior nasal flaps do not reach mouth. Upper labial furrows reach lower symphysis. Pectoral fins falcate. A unique pattern of striking black saddle marks and spots on a pale tan to greyish background, fading to whitish below. The centres of the saddle marks become lighter in adults.

Distribution Northeast Pacific: from southern Washington (USA) to the Gulf of California (Mexico). The latter population may be isolated; few Leopard Sharks are taken in the southern Gulf.

Habitat Cool to warm-temperate waters on the inshore and offshore continental shelf. Apparently well-adapted to poorly oxygenated waters, and most common on or near the seabed in bays and estuaries, from the intertidal to 20m. It may also occur on open coasts and around offshore islands, and down to a depth of 156m. Females give birth in water less than 1m deep, including over eelgrass beds.

Behaviour This is an active, strong-swimming shark, although it may also be seen resting on sand among rocks. Leopard Sharks form large, nomadic schools (sometimes associating with smoothhound sharks *Mustelus californicus* and *M. henlei*, North Pacific Spiny Dogfishes *Squalus suckleyi* and Bat Rays *Myliobatis californicus*). Most have a small home range, but some have been recorded travelling up to 150km. They may follow the tide in and out to feed on shallow mudflats. In captivity, Leopard Sharks form a loose social hierarchy: larger individuals maintain their dominant status by gently nipping the pectoral fins of smaller animals.

Biology Viviparous, no yolk-sac placenta, with a gestation of ~10–12 months. 1–37 pups per litter, increasing with the size of the mother. Age at maturity is 7–15 years (females older than males) and they may live for 30 years. Although sharks of the same age may vary in size, the older they are, the slower they grow: one shark tagged when 125cm long had only reached 129cm when recaptured 12 years later. Small sharks eat crabs and other benthic invertebrates, nipping siphons off clams and sucking worms out of the sediment, but large sharks may consume fishes and even other small sharks.

Status IUCN Red List: Least Concern. This is one of the most common sharks along the Pacific coast of North America and may be abundant where exploitation levels are low. There have been small fisheries for

Leopard Sharks in California, mostly by sports anglers (the meat is good to eat, although older sharks may contain high levels of mercury) and for the aquarium trade, but this species is generally well managed. Bag limits and a minimum landing size are used to regulate the Californian sports fishery and reduce harvesting of small sharks for the aquarium trade. Virtually nothing is known about the Leopard Shark in the southern part of its range, in Mexico, where the population in the Gulf of California may be isolated from the population further north; however, this species makes up only a very small part of the small shark fishery off Baja California.

Leopard Shark, *Triakis semifasciata*.

Plate 63 HEMIGALEIDAE

○ **Hooktooth Shark** *Chaenogaleus macrostoma*

page 496

North Indian Ocean to Central Indo-Pacific; 0–160m. Fairly long angular snout, very long mouth, protruding extremely long hooked lower teeth, large eyes with nictitating eyelids, small spiracles, gills at least 2 x eye-length; second dorsal ⅔ size of first; light grey or bronzy, second dorsal and posterior caudal fin tips occasionally black.

○ **Australian Weasel Shark** *Hemigaleus australiensis*

page 497

Central Indo-Pacific; 1–170m. Very similar to *H. microstoma* but has different coloration; unmarked first dorsal fin, dark second dorsal tip and margin, dark caudal fin posterior tip, no white spots.

○ **Sicklefin Weasel Shark** *Hemigaleus microstoma*

page 497

North Indian Ocean and Central Indo-Pacific; 0–170m. Fairly long rounded snout, small spiracles, very short arched mouth, gills short; dorsal pectoral and pelvic fins and lower caudal lobe strongly falcate, second dorsal fin ⅔ size of first, same as anal fin; light grey or bronzy, dorsal fins with light tips and margins, occasionally white spots on sides.

○ **Snaggletooth Shark** *Hemipristis elongata*

page 498

West Pacific and Indian Ocean; 1–132m. Long broadly rounded snout, large curved saw-edged upper teeth, hooked lower teeth protrude from mouth, gills more than 3 x eye-length; strongly curved concave fins, second dorsal fin ⅔ size of first, anal fin smaller and further back; light grey or bronzy, dusky second dorsal tip and caudal fin posterior tip.

○ **Whitetip Weasel Shark** *Paragaleus leucolomatus*

page 499

West Indian Ocean; 1–20m. Long snout, large oval eyes, fairly long mouth, gills ~2 x eye-length; dorsal, pectoral fins and lower caudal lobe not falcate, dorsal fins similar in height; dark grey, white below, prominent white tips and margins on most fins except black-tipped second dorsal, broad dusky patches beneath snout.

○ **Atlantic Weasel Shark** *Paragaleus pectoralis*

page 499

East Atlantic; 1–100m. Moderately long snout, large oval eyes, short small mouth, gill slits less than 1.5 x eye-length; second dorsal fin ⅔ size of first and ahead of much smaller anal fin origin; striking long yellow bands on light grey or bronze background, white below, unpatterned fins.

○ **Slender Weasel Shark** *Paragaleus randalli*

page 500

North Indian Ocean; to 18m. Snout with narrowly rounded tip, gills same length as large lateral eyes, small spiracles, long mouth; concave fins, second dorsal fin ⅔ size of first and origin slightly ahead of smaller anal fin origin; grey to grey-brown, light below, dark fins with inconspicuous light posterior margins, underside of snout with narrow black lines.

○ **Straight-tooth Weasel Shark** *Paragaleus tengi*

page 500

Central Indo-Pacific to northwest Pacific; inshore to 20m. Moderately long rounded snout, large lateral eyes, small spiracles, short arched mouth, gill slits about 1.2–1.3 x eye-length; second dorsal fin ⅔ the size of first and slightly ahead of smaller anal fin origin; light grey, no prominent markings.

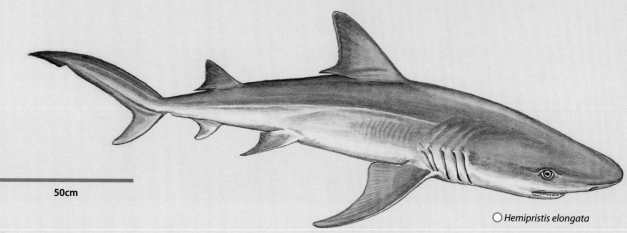

50cm

○ *Hemipristis elongata*

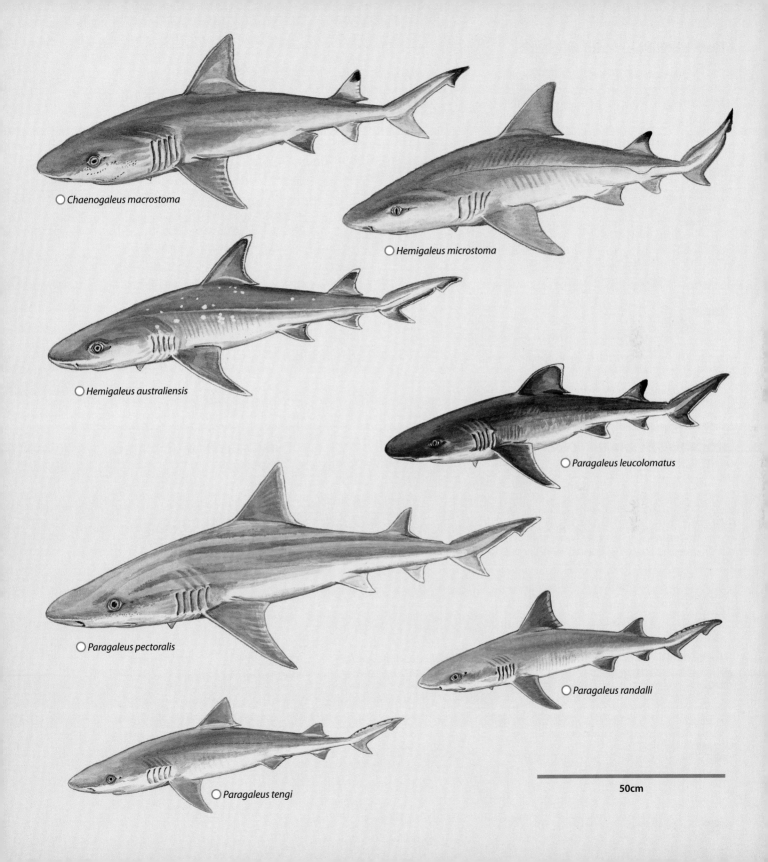

○ *Chaenogaleus macrostoma*

○ *Hemigaleus microstoma*

○ *Hemigaleus australiensis*

○ *Paragaleus leucolomatus*

○ *Paragaleus pectoralis*

○ *Paragaleus randalli*

○ *Paragaleus tengi*

50cm

Hemigaleidae: weasel sharks

There are four genera in this family, containing eight described species. *Chaenogaleus* and *Hemipristis* contain one species each, *Hemigaleus* two species, and *Paragaleus* at least four, although there is probably at least a fifth undescribed species in the latter.

Identification Small to moderate-sized sharks with horizontal, oval eyes with nicitating eyelids, small spiracles, long labial furrows, precaudal pits, spiral intestinal valves, and large second dorsal fins. Caudal fin with a strong lower lobe and wavy dorsal edge.

Biology Live-bearing (viviparous), with yolk-sac placenta. Some are specialist feeders on cephalopods, others have a highly varied diet.

Status Worldwide, in fossil record, now restricted to shallow tropical continental and insular shelf waters of the east Atlantic and Indo-West Pacific. Common and important in inshore fisheries, but over 60% of species are depleted to the extent that they have been assessed by IUCN as threatened.

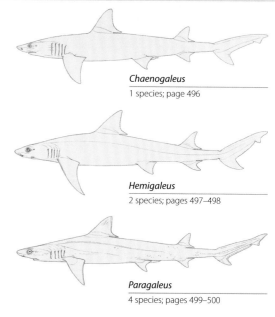

Chaenogaleus
1 species; page 496

Hemigaleus
2 species; pages 497–498

Paragaleus
4 species; pages 499–500

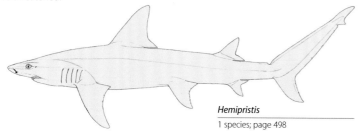

Hemipristis
1 species; page 498

HOOKTOOTH SHARK *Chaenogaleus macrostoma*

FAO code: **HCM**   Plate page 494

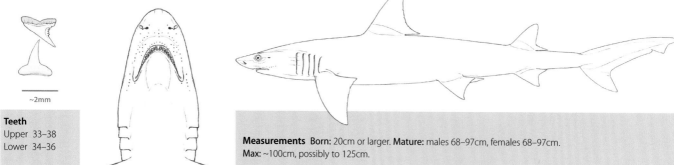

~2mm

Teeth
Upper 33–38
Lower 34–36

Measurements Born: 20cm or larger. **Mature:** males 68–97cm, females 68–97cm. **Max:** ~100cm, possibly to 125cm.

Identification Small, slender shark. Fairly long, angular snout. Extremely long, hooked, smooth-edged lower teeth protrude from very long mouth. Large, lateral eyes with nicitating eyelids. Small spiracles. Gill slits at least twice eye-length. Second dorsal about two-thirds size of first; origin opposite or slightly ahead of smaller anal fin origin. Light grey or bronzy colour, often without prominent markings. Black tip to second dorsal and terminal lobe of caudal fin.
Distribution North Indian Ocean to Central Indo-Pacific: Persian Gulf to Indonesia, China and Taiwan.
Habitat Continental and insular shelves, 0–160m.
Behaviour Unknown.
Biology Viviparous, 4 pups per litter. Poorly known.
Status IUCN Red List: **Vulnerable**. Commonly caught in fisheries.

AUSTRALIAN WEASEL SHARK   *Hemigaleus australiensis*   FAO code: **CVX**   Plate page 494

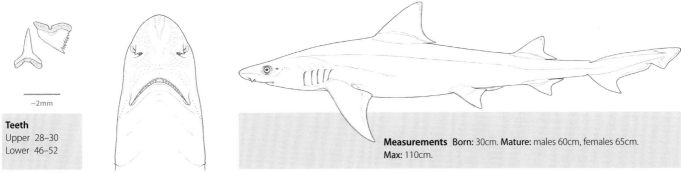

~2mm

Teeth
Upper 28–30
Lower 46–52

Measurements Born: 30cm. **Mature:** males 60cm, females 65cm.
Max: 110cm.

Identification Similar to *Hemigaleus microstoma*, with pelvic and dorsal fins, and lower caudal lobe strongly falcate. Has different coloration (second dorsal and caudal fins have dark margins and tips), vertebral counts and tooth counts.
Distribution Central Indo-Pacific: northern Australia and Papua New Guinea.
Habitat Continental shelf, on or near bottom, 1–170m.
Behaviour Unknown.
Biology Litters of 1–19 pups (average 8) after 6 month gestation, with possibly 2 pregnancies per year. Specialist feeder on cephalopods, also takes crustaceans.
Status IUCN Red List: Least Concern. With relatively low levels of fisheries throughout most of its range this species is able to withstand fishing pressure.

SICKLEFIN WEASEL SHARK   *Hemigaleus microstoma*   FAO code: **HEH**   Plate page 494

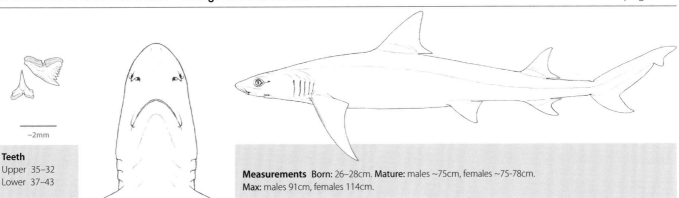

~2mm

Teeth
Upper 35–32
Lower 37–43

Measurements Born: 26–28cm. **Mature:** males ~75cm, females ~75-78cm.
Max: males 91cm, females 114cm.

Identification Small, slender shark. Fairly long, rounded snout. Very short, arched mouth. Eyes with nictitating eyelids. Short gills. Paired fins, dorsal fins and lower caudal lobe strongly falcate. Second dorsal about two-thirds size of first, opposite origin of equal-sized anal fin. Colour light grey or bronzy. Dorsal fins with light margins and tips, sometimes white spots on sides; black tip to second dorsal and terminal lobe of caudal fin.
Distribution North Indian Ocean and Central Indo-Pacific: Tanzania to Red Sea, Gulf of Aden, Sea of Oman, India to Philippines, China and Taiwan.
Habitat Continental shelf, on or near bottom, 0–170m.
Behaviour Unknown.
Biology Viviparous with yolk-sac placenta, with litters of 2–4 (average 3); gestation 6 months, possibly 2 pregnancies per year. Poorly known, probably a specialist feeder on cephalopods, but also feeds on crustaceans and echinoderms.
Status IUCN Red List: Vulnerable. Heavily fished, but apparently reproduces quite rapidly with fast population growth rates.

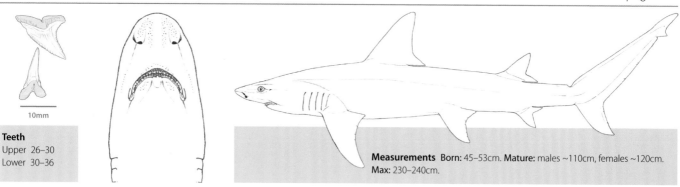

Teeth
Upper 26–30
Lower 30–36

10mm

Measurements Born: 45–53cm. **Mature:** males ~110cm, females ~120cm. **Max:** 230–240cm.

Identification A slender-bodied shark. Long, broadly rounded snout. Large, curved, saw-edged upper teeth; hooked lower teeth that protrude from mouth. Large, lateral eyes with nictitating eyelids. Small spiracles. Long gill slits (more than 3 x eye-length). Strongly curved, concave fins. Second dorsal fin two-thirds size of first and ahead of smaller anal fin origin. Light grey or bronzy colour without prominent markings. Tip of second dorsal and terminal lobe of caudal fin sometimes with a dusky blotch, more prominent in juveniles than adults.

Distribution West Pacific and Indian Ocean: South Africa to northern Australia, Philippines, China and Taiwan.

Habitat Continental and insular shelves, 1–132m.

Behaviour Poorly known.

Biology Litter 2–11 pups (average 6) , increasing with female size; gestation 7–8 months, possibly breeds in alternate years. Matures in about 2–3 years, with a maximum age of 15 years. Feeds on cephalopods and fishes.

Status IUCN Red List: **Vulnerable**. Important, intensively fished commercial species with declining stocks reported.

Snaggletooth Shark, *Hemipristis elongata*, southern Mozambique.

WHITETIP WEASEL SHARK *Paragaleus leucolomatus*

FAO code: **HEC**   Plate page 494

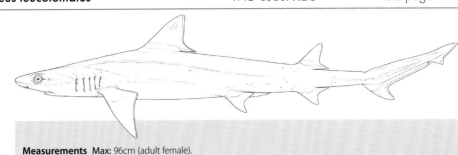

Teeth
Upper 28
Lower 30

Measurements Max: 96cm (adult female).

Identification Slender shark with a long snout. Fairly long mouth; small, serrated upper teeth; erect-cusped lower teeth. Large, oval eyes with nictitating eyelids. Small spiracles. Gills about twice eye-length. Dorsal and pelvic fins and lower caudal lobe not falcate. Dorsal fins unequal in height, first high, apically narrow, larger than second, origin about opposite free rear tips of pectorals, second dorsal origin slightly ahead of anal fin origin. Dark grey colour. Prominent white tips and margins on most fins, except abrupt black tip on second dorsal. Underside of snout with broad dusky patches, otherwise white below.
Distribution West Indian Ocean: South Africa, southern Mozambique, and northern Madagascar; possibly Sudan, but requires confirmation.
Habitat Coastal and tropical, in shallow water 1–20m.
Behaviour Unknown.
Biology Viviparous with yolk-sac placenta; 2 pups per litter.
Status IUCN Red List: **Vulnerable**. Long known from a single record from 1984, but in recent years more specimens have been observed off South Africa and southern Mozambique.

ATLANTIC WEASEL SHARK *Paragaleus pectoralis*

FAO code: **HEI**   Plate page 494

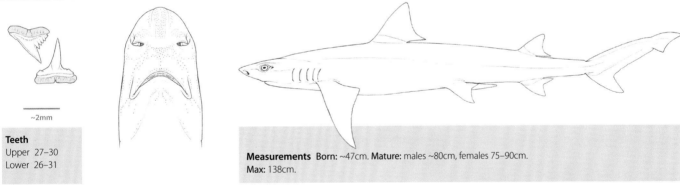

Teeth
Upper 27–30
Lower 26–31

Measurements Born: ~47cm. **Mature:** males ~80cm, females 75–90cm.
Max: 138cm.

Identification Slender-bodied. Moderately long snout. Short, small mouth; small, serrated upper teeth; erect-cusped lower teeth. Large, oval eyes with nictitating eyelids. Small spiracles. Gill slits less than 1.5 x eye-length. Second dorsal fin two-thirds the size of first, origin slightly in front of anal fin origin. Striking yellow, longitudinal bands on light grey or bronze background; unpatterned fins. White below.
Distribution East Atlantic: Cape Verde Islands and Mauritania to Angola; possibly south to northern Namibia, possibly north to Morocco. One northwest Atlantic record (1906), but the location of that specimen may be erroneous.
Habitat Tropical shelf, surfline to 100m.
Behaviour Specialist feeder on cephalopods, also small fishes.
Biology Viviparous with 1–4 (mostly 2) pups per litter in May–June in Senegal.
Status IUCN Red List: **Data Deficient**. Common where it occurs.

SLENDER WEASEL SHARK *Paragaleus randalli* FAO code: **IEI** Plate page 494

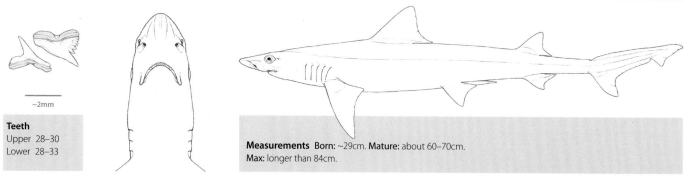

Teeth
Upper 28–30
Lower 28–33

~2mm

Measurements Born: ~29cm. **Mature:** about 60–70cm.
Max: longer than 84cm.

Identification Snout with narrowly rounded tip. Long mouth and relatively deep lower jaw. Large, lateral eyes with nictitating eyelids. Small spiracles. Gills about equal to eye-length. Concave fins. Second dorsal two-thirds the area of first; origin slightly ahead of smaller anal fin origin. Grey to grey-brown in colour, light below. Fins mostly dark with inconspicuous, light posterior margins; no obvious white or black tip. Snout with a pair of narrow black lines but no dark patches on underside.
Distribution North Indian Ocean and northwest Pacific: Arabian Gulf, Gulf of Oman, India, Sri Lanka and Taiwan. A somewhat similar *Paragaleus*, possibly a distinct species, occurs off Myanmar.
Habitat Inshore, in shallow water (down to 18m) on the continental shelf.
Behaviour Unknown.
Biology Viviparous with yolk-sac placenta. Two pups per litter (1 in each uterus).
Status IUCN Red List: Near Threatened. Poorly known and frequently misidentified throughout its range, this inshore species occurs in areas with intense fishing pressure and habitat loss.

STRAIGHT-TOOTH WEASEL SHARK *Paragaleus tengi* FAO code: **HEN** Plate page 494

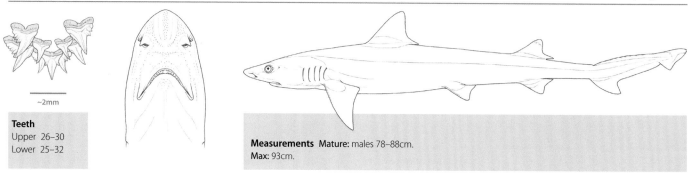

Teeth
Upper 26–30
Lower 25–32

~2mm

Measurements Mature: males 78–88cm.
Max: 93cm.

Identification Small, slender shark. Rounded, moderately long snout. Rather short, arched mouth; lower teeth not prominently protruding. Large, lateral eyes with nictitating eyelids. Small spiracles. Moderate-sized gill slits (about 1.2–1.3 times eye-length in adults, less in young). Light grey colour with no prominent markings.
Distribution Confirmed from Taiwan, Hong Kong, and Penang, Malaysia. Records from southern China, Vietnam, and Gulf of Thailand require confirmation. Records from southern Japan are erroneous.
Habitat Inshore to ~20m.
Behaviour Unknown.
Biology Viviparous with yolk-sac placenta.
Status IUCN Red List: Endangered. Depleted due to heavy fishing pressure throughout its shallow inshore distribution, and affected in some areas by habitat damage. Possibly extirpated from part of its former range.

In a remarkable example of cooperative feeding by several species of apex pelagic predator, a bait ball of scad provides a meal for Yellowfin Tuna *Thunnus albacares* and several Carcharhinid shark species.

Carcharhinidae: requiem sharks

The requiem shark family (Carcharhinidae) contains 11 genera and at least 56 species worldwide. The name 'requiem' is said to be derived from the ancient Norman French word 'reschignier', which means to bare teeth, or grimace.

Requiem sharks are members of one of the largest, most important shark families, and include many common and wide-ranging species. These sharks are dominant (in terms of their biodiversity, abundance and biomass) in tropical continental shelf and offshore habitats, but some species are also found in subtropical and warm-temperate seas. Several members of this family are closely associated with coral reefs and oceanic islands, while other species range far into ocean basins. One pelagic requiem shark, the Blue Shark, has one of the greatest geographic ranges of any shark, or indeed any marine vertebrate, ranging from high latitude, cool-temperate waters into the tropics. A few other species are found in temperate waters and some even at great depths, but none are truly specialised deepwater sharks (compared with many species of the families Squalidae and Pentanchidae). Some requiem shark species occur in freshwater rivers and lakes. Although members of other families may enter river mouths and ascend rivers for a short distance, the little-known Central Indo-Pacific river sharks, *Glyphis* species, and the broadly distributed Bull Shark appear to be the only living sharks that can inhabit fresh water for extended periods. The Bull Shark has a wide range in tropical and warm temperate rivers and lakes around the world and is remarkable for the apparent ease with which it can move from saline to freshwater conditions and back again (see p. 35).

Identification Although some species are relatively small (65–100cm), most are medium to large in size, with maximum lengths ranging to nearly 400cm. They have a long, arched mouth with blade-like teeth (often broader in the upper jaw), most with short labial furrows (except *Rhizoprionodon*), and no nasoral grooves or barbels. Their eyes are usually round (to horizontal) with internal nictitating eyelids, usually no

Whitetip Reef Sharks, *Triaenodon obesus*, (p. 500), hunt for reef fishes on a rocky-bottomed reef at night near Cocos Island, Costa Rica.

Bull Shark, *Carcharhinus leucas* (p. 468), at Pinnacles, Mozambique.

spiracles. They have two dorsal fins and one anal fin, the first dorsal fin is medium to large in size with its base located well in front of the pelvic fin bases, the second dorsal is usually much smaller. There are precaudal pits, a caudal fin with a strong lower lobe and lateral undulations on its upper margin. Some requiem sharks have an interdorsal ridge along their backs (e.g. Caribbean Reef Shark, Silky Shark); others lack this feature (e.g. Blacktip Reef Shark, Bull Shark) – checking for its presence or absence between the two dorsal fins is very important when identifying requiem sharks of the genus *Carcharhinus*. The bodies of most requiem sharks are unpatterned (particularly in *Carcharhinus*). The extremely rare river sharks (genus *Glyphis*), the only truly freshwater shark species, are very difficult to distinguish without tooth and vertebral counts. There may be at least one undescribed *Glyphis* species.

Biology All Carcharhinidae are viviparous: they give birth to litters of live young, ranging in size from just one or two pups, to 135 in the Blue Shark. Their reproductive mode is placental viviparity (see p. 51), in which the embryo's empty yolk-sac becomes attached to the wall of the uterus, and the mother continues to feed her growing pups until birth. Nutrient is usually provided to the litter in the form of uterine secretions, but some species continuously produce unfertilised eggs throughout the pregnancy for their pups to eat. The Pelagic Thresher only produces one fertilised egg for each uterus, followed by many infertile eggs to feed her large twins. Other species produce several fertile eggs and give birth to several pups. Large litters, with many umbilical cords, come with a risk of cord entanglement. The Blacktip Shark is one of the requiem sharks that mitigates this risk by subdividing each uterus into several compartments, or crypts, one for each pup. The age of a newborn pup can be estimated by examining how far the open umbilical scar between its pectoral fins has healed – this usually takes a few weeks; it soon vanishes completely. Producing and carrying around litters of large pups requires a huge input from mature female requiem sharks. As a result, some species need to take one or two years off between litters, so that they can replenish their energy reserves before the next pregnancy (see Lemon Shark life cycle, p.505).

Requiem sharks are active, strong swimmers, occurring singly or in small to large schools. Some species are 'ram-ventilators', which need to swim continually to force oxygenated water through their gills, while others are capable of resting motionless for extended periods on the bottom. Many are more active at night, or during dawn and dusk, than during the daytime. Some are solitary or may socialise in small groups, some are social schooling species. These sharks are major predators, feeding on a wide variety of prey including bony fishes, elasmobranchs, cephalopods, and crustaceans, as well as sea birds, turtles, sea snakes, marine mammals, benthic

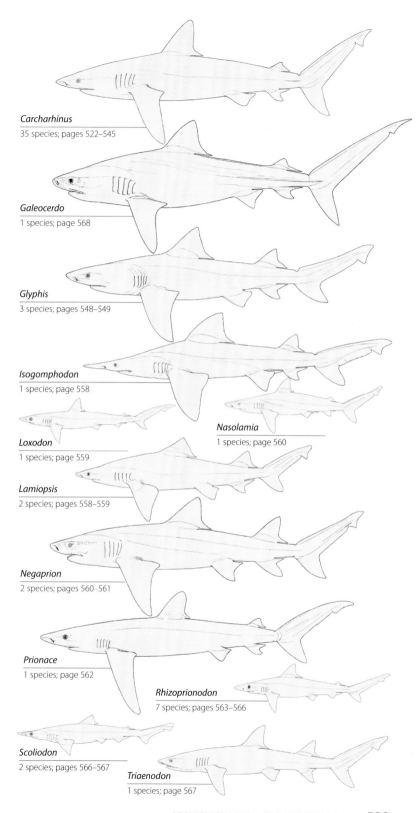

Carcharhinus
35 species; pages 522–545

Galeocerdo
1 species; page 568

Glyphis
3 species; pages 548–549

Isogomphodon
1 species; page 558

Loxodon
1 species; page 559

Nasolamia
1 species; page 560

Lamiopsis
2 species; pages 558–559

Negaprion
2 species; pages 560–561

Prionace
1 species; page 562

Rhizoprionodon
7 species; pages 563–566

Scoliodon
2 species; pages 566–567

Triaenodon
1 species; page 567

invertebrates, and carrion. Smaller species and young animals tend to specialise on a fairly narrow selection of prey, but larger species and adults will feed on a wider range of prey items.

At least some requiem sharks exhibit specialised behavioural displays when confronted by divers or other sharks, which may be indicative of aggressive or defensive threat. There is a clear hierarchical dominance between certain species that occur together: Oceanic Whitetip Sharks are dominant over Silky Sharks of the same size, which in turn can dominate Grey Reef Sharks; Galapagos Sharks are dominant over Blacktip Sharks but subordinate to the Silvertip. Danger signs which may appear if a requiem shark has been approached too closely or cornered in shallow water, include a hunched back, lowered pectoral fins, jerky movements, jaw gaping and S-shaped swimming. These signal danger because they indicate that an attack in self-defence is likely to follow. Other sharks recognise the signs and will respond accordingly to avoid a bite; swimmers and divers who might encounter sharks need to learn them!

Status Requiem sharks are by far the most important family in tropical and warm temperate coastal and oceanic shark fisheries, including commercial, subsistence and sports fishing. During the past decade, the Carcharhinidae have contributed one third of all shark landings reported to FAO (and a substantial amount of the 48% of global shark landings that are not identified at least to family level). Blue Shark alone has provided almost 25% of the global total shark catch, and 70% of all reported Carcharhinids; this is likely the world's most heavily fished shark and has been extraordinarily resilient (so far) to intensive oceanic fisheries. Other requiem sharks reported in large numbers in global fisheries include the biologically more vulnerable Silky Shark and easily recognisable threshers.

These species are mainly utilised for food (meat) and for fins, although liver oil, cartilage and skin may also be marketed. Many requiem sharks are consumed locally, but their fins almost invariably enter international trade. Identifying these fins produces a far better picture of global shark landings and trends than is possible from the incomplete catch data reported to FAO (see p. 69). Almost 50% of the 76 species recently identified in the international fin trade through Hong Kong were members of the family Carcharhinidae. The oceanic Blue and Silky Sharks contributed 35% and over 5%, respectively, of all shark fins sampled in 2014–15; a complex of four blacktip species another 2%, Bull and Spinner Sharks ~1% each, and unidentified requiem sharks 4%. The fins of Dusky and Oceanic Whitetip Sharks, threshers, and even the small fins of sharpnose shark species are also regularly seen. Overall, the requiem sharks likely contribute at least 50% of all shark fins identified in international trade. There has been remarkable stability in the species composition of the Hong Kong fin trade over time, with one notable exception: The Sandbar Shark was common in auctioned fins 20 years ago, but has been rare in recent surveys. This is attributed to the closure of two major shark fisheries, off Western Australia and the USA Atlantic coast.

Their importance in fisheries has had serious consequences for these species' conservation status: over half of requiem sharks are assessed as threatened (16% Critically Endangered, 14% Endangered, 23% Vulnerable) in the IUCN Red List of Threatened Species; 23% are Near Threatened (including Blue Shark), and only 20% are Least Concern. The latter mostly include smaller more fecund species of *Carcharhinus*, several *Rhizoprionodon* species, and

some Australian endemics. Notably, the newly described Lost Shark (p. 539), which has not been recorded since 1934, had the unfortunate distinction in 2020 of becoming the first shark species to be assessed as 'Critically Endangered – Possibly Extinct'.

Not many people can readily identify requiem sharks, even under the best of circumstances, which means that this family is likely under-represented in reports of shark attacks. Despite this, the Bull Shark is third on the list of shark species most implicated in unprovoked fatal and non-fatal shark attacks (after the non-Carcharhinid White and Tiger Sharks) – this shark is particularly dangerous because it occurs in inshore and brackish-to-freshwater habitats that are also heavily used by people and is still relatively common. Its offshore distribution and dwindling numbers means that it is rare for the large and inquisitive Oceanic Whitetip Shark to come into contact with swimmers, or it might rival the Bull Shark for numbers of attacks. In total, 15 species of requiem sharks have been reported to attack people and boats, seven species with at least one fatal result. While the Blacktip (which mistakenly nips swimmers as it chases small schooling fishes along the Florida coast), Spinner, Lemon and Blacktip Reef Sharks have been responsible for several unprovoked bites, none of these have resulted in fatalities.

Several requiem sharks, including Bull, Sandbar and Silvertip sharks, are popular and impressive on public display and some species have bred in large aquaria. These and some others (including dangerous species such as Bull Shark, and several reef sharks) have also gained popularity as attractive subjects for viewing underwater by divers. A few (particularly Lemon Shark and several *Carcharhinus* species) are the subject of intensive research activity.

Right: The life cycle of the Lemon Shark (p. 561), illustrated here, is representative of many requiem and coastal sharks. Pregnant females go back to the sheltered, shallow nurseries where they were born, to give birth to their own litter. They return to pup again at two-year intervals (some species return every year, while a few return every three years). The newborn pups usually remain in their nursery for several years, until they have grown big enough to risk encountering the larger predators found in deeper water habitats. Their home range expands and diet changes as they grow and move into different habitats. These sharks usually start to migrate long distances once they mature, at about 14 years old, although falling winter temperatures may force pups away from nursery grounds into warmer water. Aggregations of large juveniles and adults can be seen in summer, when they move into higher latitudes as water temperatures rise, before returning to warmer water as winter approaches.

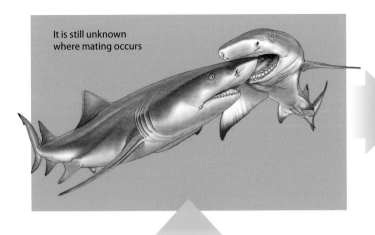

It is still unknown where mating occurs

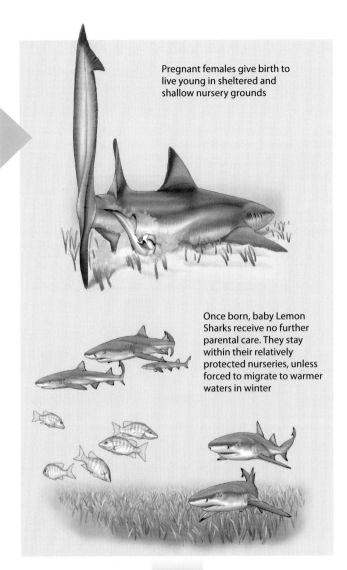

Pregnant females give birth to live young in sheltered and shallow nursery grounds

Once born, baby Lemon Sharks receive no further parental care. They stay within their relatively protected nurseries, unless forced to migrate to warmer waters in winter

Lemon Sharks migrate seasonally to stay in their preferred water temperature, visit rich hunting grounds and possibly to congregate and mate

Upon reaching maturity, Lemon Sharks are found in deeper water

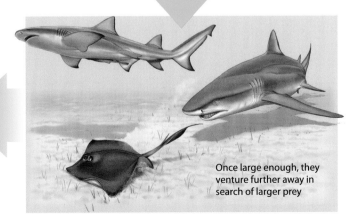

Once large enough, they venture further away in search of larger prey

Plate 64 CARCHARHINIDAE I and GALEOCERDIDAE

○ **Tiger Shark** *Galeocerdo cuvier*

page 568

Worldwide: warm seas; 0–1136m. Huge; very short broad bluntly rounded snout, eyes fairly large, big mouth with very long upper labial furrows, large spiracles; low caudal keels; interdorsal ridge very prominent, moderately broad semi-falcate pectoral fins, first dorsal fin more than 2.5 times height of second, both with very long rear tips; grey above with vertical black to dark grey bars and blotches, bold in young fading in adults, white below.

○ **Bull Shark** *Carcharhinus leucas*

page 535

Worldwide: warm seas, large rivers and lakes; 1–164m. Massive body, large thick head; very short broad bluntly rounded snout, small eyes, triangular saw-edged upper teeth, upper labial furrows very short, no spiracles; weak caudal keels; no interdorsal ridge, large angular pectoral fins, broad triangular first dorsal fin less than 3.2 times height of second, both dorsal fins with short rear tips; grey to grey-brown, fin tips dusky but only conspicuously so in young, white below.

○ **Oceanic Whitetip Shark** *Carcharhinus longimanus*

page 537

Worldwide: warm oceans; 0–1082m. Large; short bluntly rounded snout, small eyes, upper labial furrows very short, no spiracles; weak caudal keels; interdorsal ridge low, long paddle-like pectoral fins, huge rounded first dorsal fin much larger than second; grey or brown, first dorsal fin pectoral fins and sometimes caudal fin tips conspicuous mottled white, juveniles with black tips on some fins and black patches on caudal peduncle, white below.

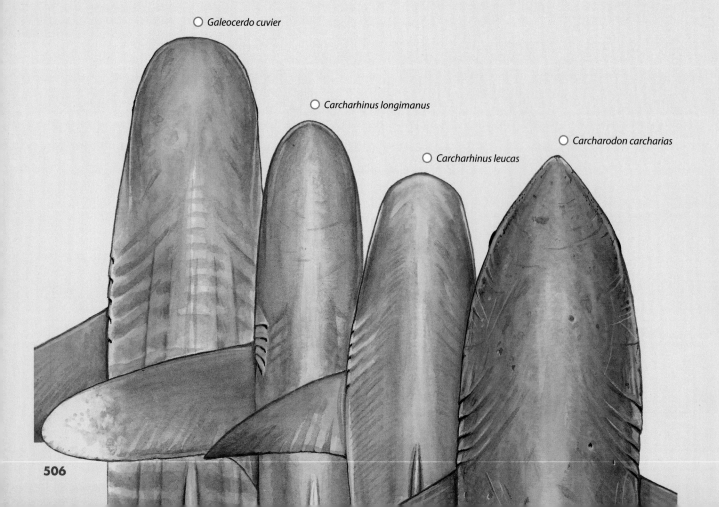

○ *Galeocerdo cuvier*

○ *Carcharhinus longimanus*

○ *Carcharodon carcharias*

○ *Carcharhinus leucas*

Carcharhinus longimanus

Carcharhinus leucas

Galeocerdo cuvier

50cm

Plate 65 CARCHARHINIDAE II

○ **Silky Shark** *Carcharhinus falciformis*   page 531

Worldwide: warm seas; 0–500m. Large, slim; fairly long flat rounded snout, large eyes, small mouth; no caudal keels; narrow interdorsal ridge, long narrow pectoral fins (shorter in young), first dorsal behind pectoral fins, second dorsal low with greatly elongated inner margin and rear tip; dark grey brown or nearly black, inconspicuous pale flank band, white below, all fins except first dorsal with inconspicuous dusky tips.

○ **Blue Shark** *Prionace glauca*   page 562

Worldwide; 0–1000m. Large, graceful, slim; long conical snout, large eyes, no spiracles, small mouth; weak caudal keels; no interdorsal ridge, long narrow scythe-shaped pectoral fins, first dorsal fin well behind pectoral fins closer to pelvic fin base, second dorsal fin less than a third the size of first; usually dark blue back with bright blue flanks and sharply demarcated white below.

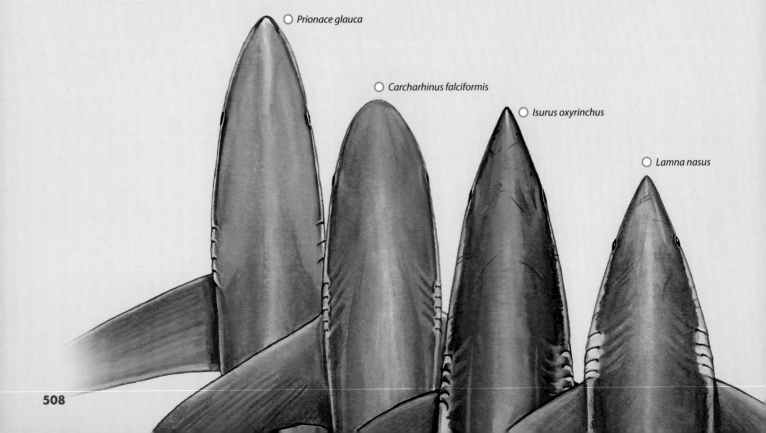

○ *Prionace glauca*

○ *Carcharhinus falciformis*

○ *Isurus oxyrinchus*

○ *Lamna nasus*

Prionace glauca

Carcharhinus falciformis

50cm

Plate 66 CARCHARHINIDAE III

○ **Silvertip Shark** *Carcharhinus albimarginatus*   page 523

Tropical, Indian and Pacific Oceans; 0–800m. Moderately long broadly rounded snout, eyes round; interdorsal ridge present, pectoral fins with narrow tips, first dorsal fin apex narrowly rounded or pointed; dark grey occasionally bronze-tinged above, white below, faint white lateral band, striking white tips and posterior margins on all fins except small black second dorsal.

○ **Grey Reef Shark** *Carcharhinus amblyrhynchos*   page 524

Indian and Pacific Oceans; 0–275m. Moderately long broadly rounded snout, eyes usually round; no interdorsal ridge, pectoral fins narrow and falcate, first dorsal fin apex narrowly rounded or pointed; body grey above, white below, first dorsal fin plain or irregularly to prominently white-edged, obvious broad black posterior margin to entire caudal fin, blackish tips to other fins.

○ **Blacktip Reef Shark** *Carcharhinus melanopterus*   page 539

Indo-West Pacific, Mediterranean; 0–100m. Short broadly rounded snout, horizontally oval eyes; no interdorsal ridge, pectoral fins narrow and falcate, first dorsal fin apex rounded, second dorsal fin large with short rear tip; brown-grey above, white below, brilliant black fin tips highlighted by white.

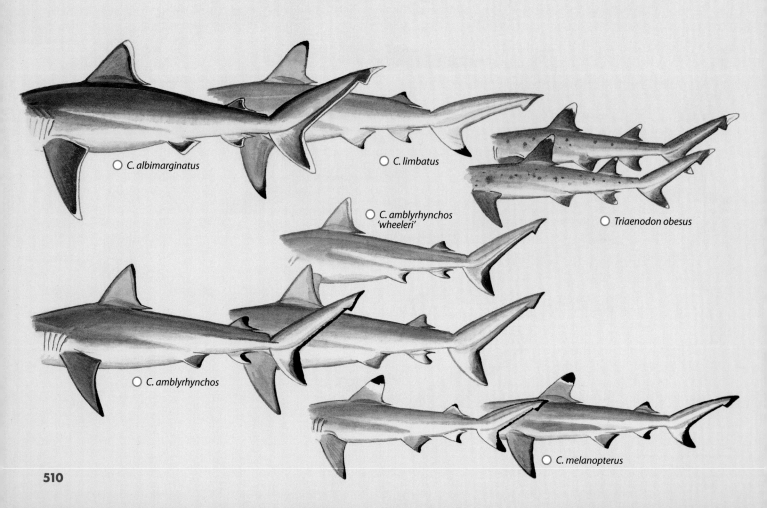

○ *C. albimarginatus*

○ *C. limbatus*

○ *C. amblyrhynchos 'wheeleri'*

○ *Triaenodon obesus*

○ *C. amblyrhynchos*

○ *C. melanopterus*

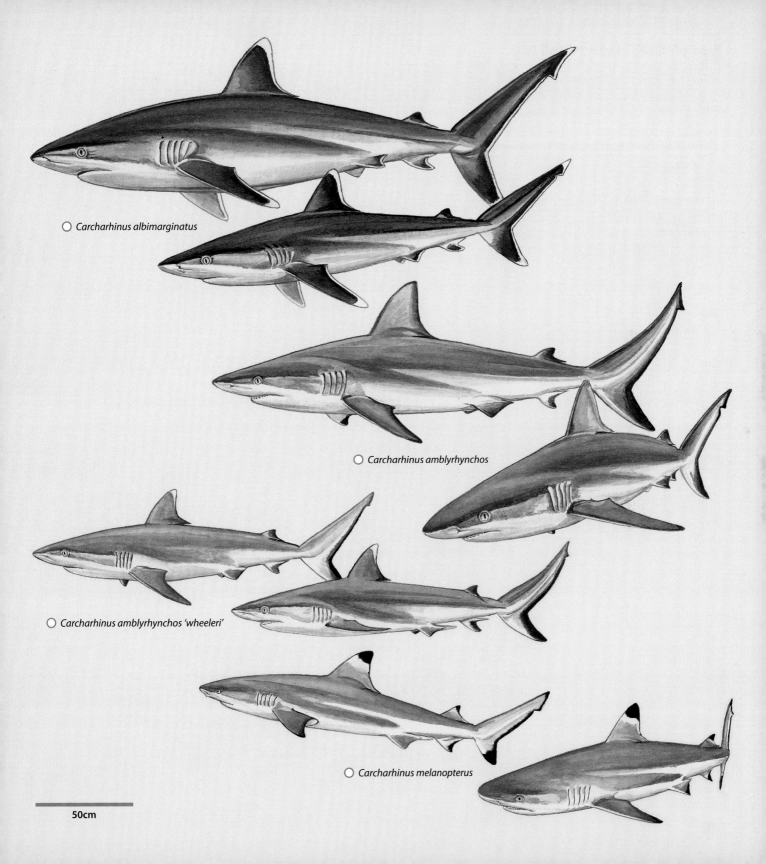

○ *Carcharhinus albimarginatus*

○ *Carcharhinus amblyrhynchos*

○ *Carcharhinus amblyrhynchos 'wheeleri'*

○ *Carcharhinus melanopterus*

50cm

Plate 67 CARCHARHINIDAE IV

○ **Graceful Shark** *Carcharhinus amblyrhynchoides*   page 525

West Indian Ocean to Central Indo-Pacific; 0–75m. Tubby; fairly short wedge-shaped pointed snout, serrated upper tooth cusps, fairly large eyes, large gills; no interdorsal ridge, moderately large pectoral fins, large triangular first dorsal fin, both with short rear tips; grey with conspicuous white flank band, often black-tipped fins but less prominently marked than *Carcharhinus leiodon*.

○ **Sicklefin Lemon Shark** *Negaprion acutidens*   page 560

Tropical Indian and Pacific Oceans; 0–90m. Stocky; broad blunt snout, fairly small eyes, moderately large gills; no interdorsal ridge, very similar to *Negaprion brevirostris*, but dorsal pectoral and anal fins usually more falcate, first dorsal fin about same size as second; yellow-brown above, white below.

○ **Whitetip Reef Shark** *Triaenodon obesus*   page 567

Pacific and Indian Oceans; 1–330m. Small slender; very short broad snout, oval eyes; no interdorsal ridge, fairly broad triangular pectoral fins, first dorsal fin behind pectoral fins, large second dorsal fin; grey-brown, lighter below, occasionally scattered dark spots on sides, brilliant very conspicuous white tips on first dorsal and terminal caudal fin.

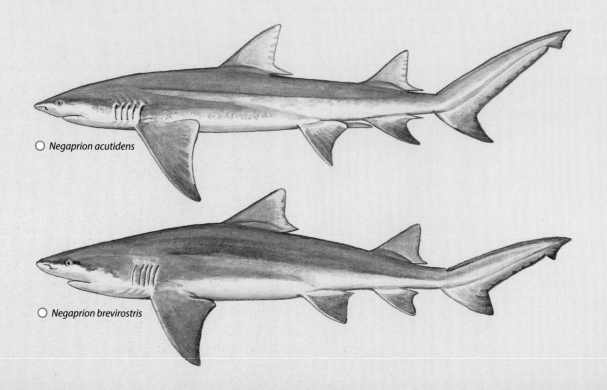

○ *Negaprion acutidens*

○ *Negaprion brevirostris*

○ *Carcharhinus amblyrhynchoides*

○ *Triaenodon obesus*

○ *Negaprion acutidens*

50cm

Plate 68 CARCHARHINIDAE V

○ **Galapagos Shark** *Carcharhinus galapagensis* page 532

Worldwide; 0–286m. Fairly long broad snout, fairly large eyes, low anterior nasal flaps; low interdorsal ridge, large semifalcate pectoral fins, moderately large first dorsal with short free rear tip, first dorsal originating over inner margin of pectoral fins; grey-brown, most fins with inconspicuous dusky tips, inconspicuous white flank stripe, white below.

○ **Dusky Shark** *Carcharhinus obscurus* page 540

Possibly worldwide in warm seas; 0–500m. Broad rounded snout, triangular saw-edged upper teeth, fairly large eyes, low poorly developed anterior nasal flaps; low interdorsal ridge, curved pectoral fins, falcate first dorsal fin, first dorsal origin over to ahead of pectoral fin free rear tips; grey to bronzy, most fins with inconspicuous dusky tips, inconspicuous white flank stripe, white below.

○ **Sandbar Shark** *Carcharhinus plumbeus* page 542

Possibly worldwide in warm seas; 1–280m. Moderately long rounded snout, fairly large eyes, low short anterior nasal flaps; interdorsal ridge present, large pectoral fins, very large erect first dorsal fin, origin over to slightly ahead pectoral fin inserts; grey-brown to bronzy, tips and posterior margins of fins often inconspicuously dusky, inconspicuous white flank stripe, white below.

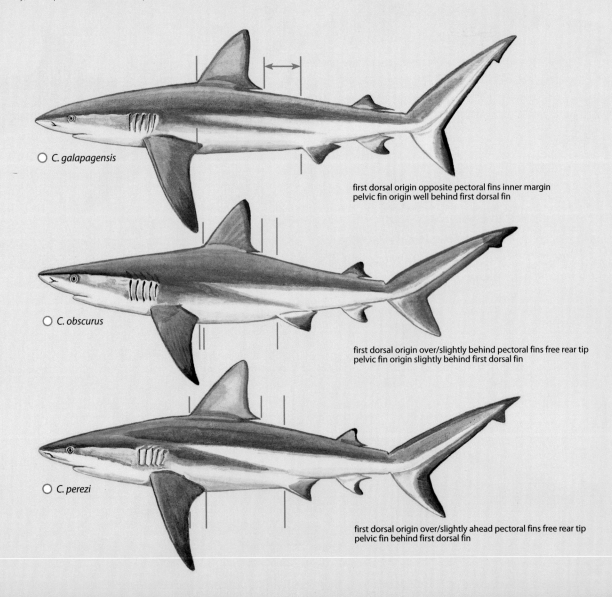

○ *C. galapagensis*

first dorsal origin opposite pectoral fins inner margin
pelvic fin origin well behind first dorsal fin

○ *C. obscurus*

first dorsal origin over/slightly behind pectoral fins free rear tip
pelvic fin origin slightly behind first dorsal fin

○ *C. perezi*

first dorsal origin over/slightly ahead pectoral fins free rear tip
pelvic fin behind first dorsal fin

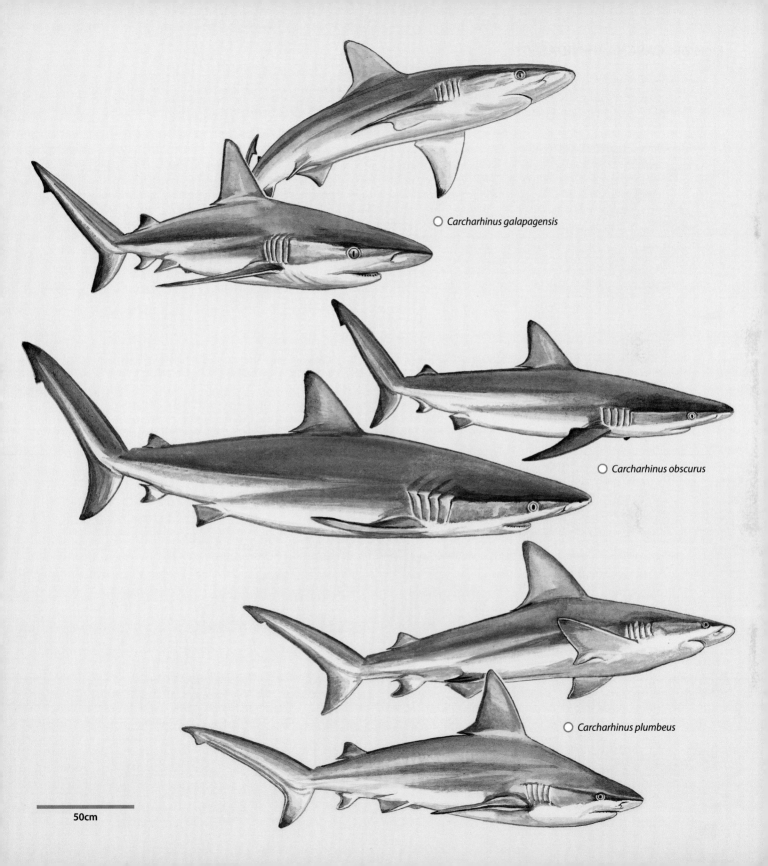

○ *Carcharhinus galapagensis*

○ *Carcharhinus obscurus*

○ *Carcharhinus plumbeus*

50cm

Plate 69 CARCHARHINIDAE VI

○ **Bignose Shark** *Carcharhinus altimus* page 522

Probably worldwide in warm seas; 0–1000m. Heavy cylindrical body; large long broad snout, circular moderately large eyes, long anterior nasal flaps, upper labial furrows short and inconspicuous, moderately long gills; prominent high interdorsal ridge, large straight pectoral and dorsal fins; grey occasionally bronzy above with obscure dusky fin tips except pelvic fins, inconspicuous white flank stripe, white below.

○ **Spinner Shark** *Carcharhinus brevipinna* page 528

Atlantic, Indian and Pacific Oceans, Mediterranean; 0–200m. Slender; long narrow pointed snout, small circular eyes, short relatively inconspicuous anterior nasal flaps, prominent labial furrows, long gills; no interdorsal ridge, small pectoral and first dorsal fins, both dorsal fins with short free rear tips; adults and juveniles (but not the young) have conspicuous black fin tips on pectoral, second dorsal, anal and caudal lower lobe, but absent on pelvic fins; grey body, inconspicuous white flank stripe, white below.

○ **Blacktip Shark** *Carcharhinus limbatus* page 536

Widespread: warm seas; 0–140m. Stout; long narrow pointed snout, small circular eyes, low triangular anterior nasal flaps, upper labial furrows short and inconspicuous, long gills; no interdorsal ridge, moderately large falcate pectoral fins, high first dorsal fin, second dorsal moderately large; grey to grey-brown, black fin tips to pectoral second dorsal and terminal caudal fins, sometimes pelvic and rarely anal fins, usually black edges on first dorsal apex and upper caudal lobe, conspicuous white flank stripe, white below.

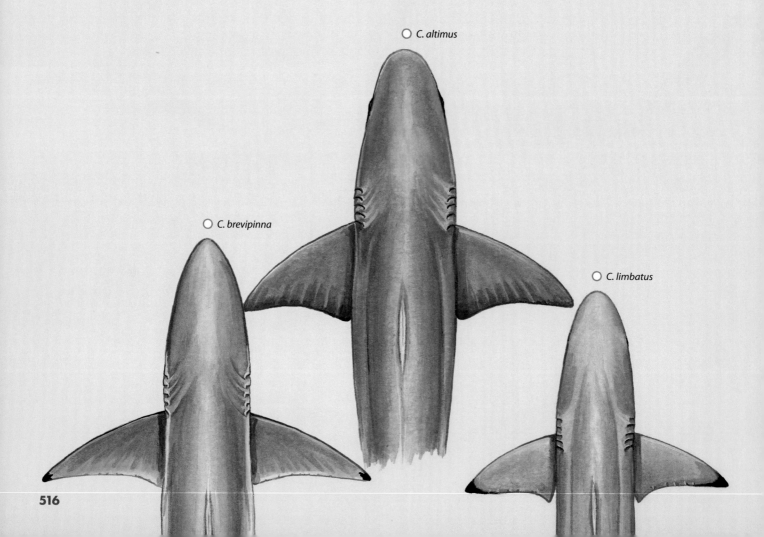

○ C. altimus

○ C. brevipinna

○ C. limbatus

○ *Carcharhinus altimus*

○ *Carcharhinus brevipinna*

juvenile

○ *Carcharhinus limbatus*

50cm

Plate 70 CARCHARHINIDAE VII

○ **Bronze Whaler** *Carcharhinus brachyurus*   page 527

Indo-Pacific, Atlantic and Mediterranean; close inshore–145m. Broad bluntly pointed snout; no interdorsal ridge, long pectoral fins, small dorsal fins with short free rear tips; olive-grey to bronzy, most fins with inconspicuous darker margins and dusky tips, fairly prominent white flank stripe, white below.

○ **Caribbean Reef Shark** *Carcharhinus perezi*   page 541

West Atlantic; 0–378m. Bluntly rounded snout; interdorsal ridge present, large narrow pectoral fins, small first dorsal fin and moderately large second dorsal with short free rear tip; dark grey or grey-brown, underside of pectoral pelvic anal fins and lower caudal lobe dusky but not prominently marked, inconspicuous white flank stripe, white below.

○ **Lemon Shark** *Negaprion brevirostris*   page 561

Atlantic and east Pacific; 0–120m. Short-nosed; no interdorsal ridge, dorsal, pectoral and pelvic fins weakly falcate, first dorsal fin about the same as second; pale yellow-brown, no fin markings, no flank stripe, light below.

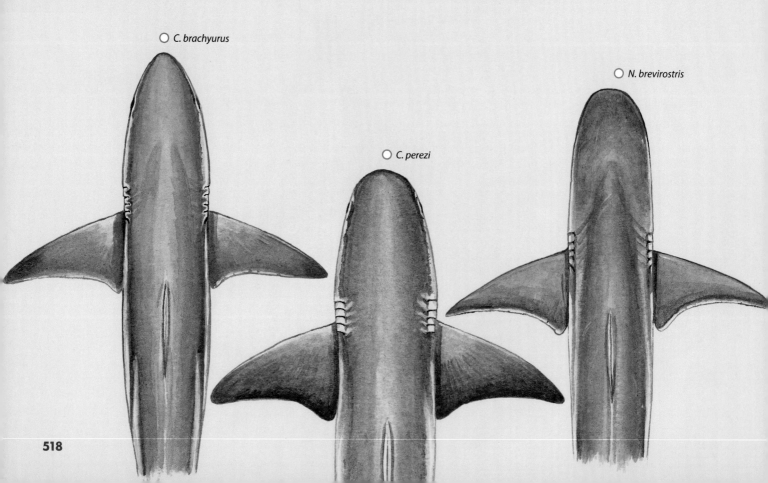

○ *C. brachyurus*

○ *C. perezi*

○ *N. brevirostris*

Carcharhinus brachyurus

Carcharhinus perezi

Negaprion brevirostris

50cm

PLATE 71 CARCHARHINIDAE VIII

○ **Blacknose Shark** *Carcharhinus acronotus*   page 522

West Atlantic; 3–>100m. Moderately long rounded dark-tipped snout; no interdorsal ridge, small pectoral fins, small first dorsal fin, second moderately large, both dorsal fins with short rear tips, second dorsal fin origin approximately over anal fin origin, second dorsal and terminal caudal fin tips dark.

○ **Finetooth Shark** *Carcharhinus isodon*   page 534

West Atlantic; 0–20m. Long pointed snout; fairly large eyes, very long gills; no interdorsal ridge, small pectoral fins, dorsal fins with short rear tips, first small, second moderately large with origin over posterior of anal fin origin; dark blue-grey, no prominent fin markings, inconspicuous white flank stripe, white below.

○ **Smoothtooth Blacktip Shark** *Carcharhinus leiodon*   page 534

Northwest Indian Ocean; 0–40m. Similar to *Carcharhinus amblyrhynchoides* but with smooth and erect-cusped upper teeth; short bluntly pointed snout; moderately large eyes, long gill slits; no interdorsal ridge, small pectoral fins, fairly large dorsal fins with short free rear tips, very conspicuous black tips on all fins.

○ **Night Shark** *Carcharhinus signatus*   page 543

Atlantic; 0–600m. Long pointed snout; large eyes, short inconspicuous upper labial furrows; small pectoral fins, interdorsal ridge present, first dorsal small with moderately long free rear tip, second dorsal low with long free rear tip, first dorsal fin origin over pectoral fins; grey-brown with no conspicuous fin markings, white below.

○ **Australian Blacktip Shark** *Carcharhinus tilstoni*   page 545

Central Indo-Pacific; 0–150m. Similar to *Carcharhinus limbatus*; long snout; fairly large eyes; no interdorsal ridge, moderately large pectoral fins, dorsal fins with short free rear tips, first dorsal large, second dorsal moderately large, first dorsal fin origin approximately over pectoral fin insertions; grey to bronzy with black-tipped fins, pelvic and anal fins occasionally plain, pale flank stripe, pale below.

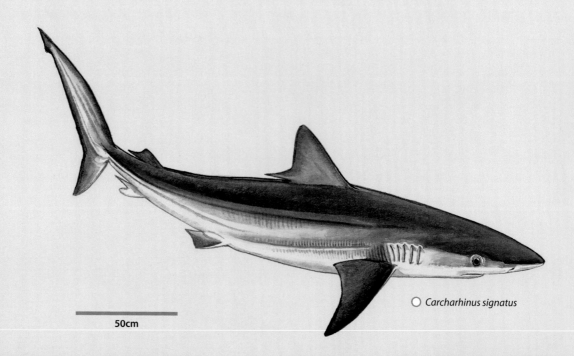

○ *Carcharhinus signatus*

50cm

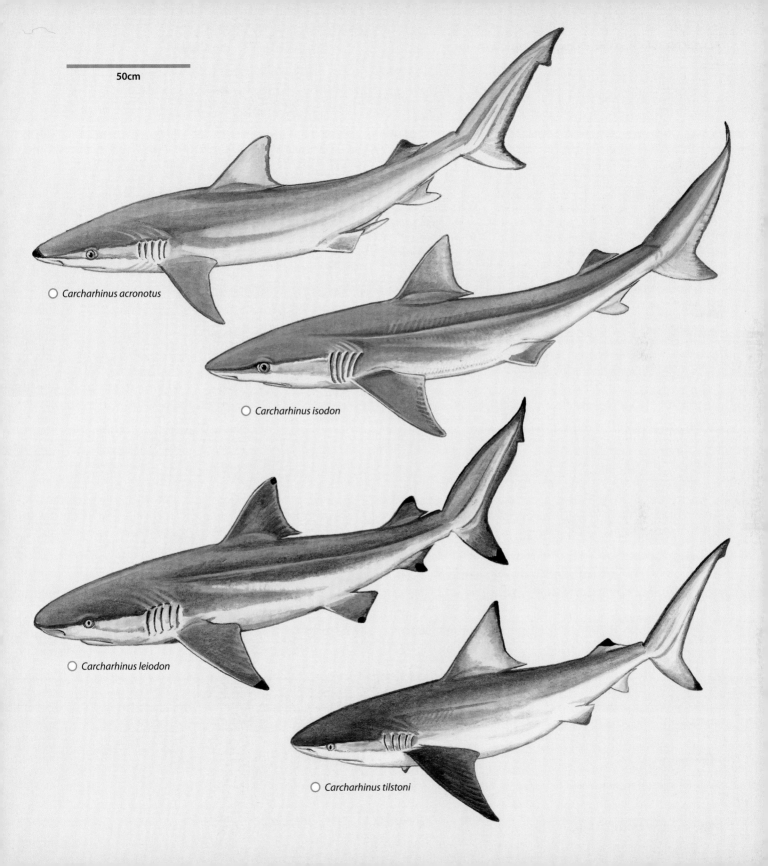

50cm

○ *Carcharhinus acronotus*

○ *Carcharhinus isodon*

○ *Carcharhinus leiodon*

○ *Carcharhinus tilstoni*

BLACKNOSE SHARK *Carcharhinus acronotus*

FAO code: **CCN**   Plate page 520

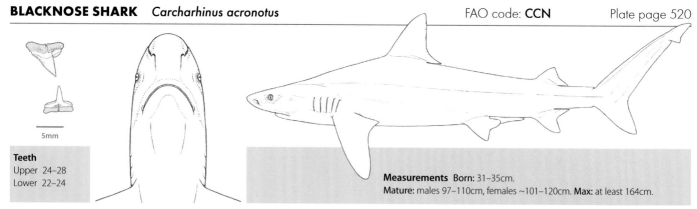

Teeth
Upper 24–28
Lower 22–24

5mm

Measurements Born: 31–35cm.
Mature: males 97–110cm, females ~101–120cm. **Max:** at least 164cm.

Identification Moderately long, rounded snout. Short, inconspicuous upper labial furrows. Moderately large eyes. Short gills. Small first dorsal fin, second moderately large (but still half the size of the first), both dorsal fins with short rear tips, second dorsal fin origin approximately over anal fin origin. No interdorsal ridge. Small pectoral fins. Grey in colour with dark tip to snout. Second dorsal and upper caudal fin tips dark.
Distribution West Atlantic: southern USA (Virginia) to southern Brazil, including Caribbean.
Habitat Coastal continental and insular shelves, mainly over sand, shell and coral, 3m to over 100m.
Behaviour Performs 'hunch' display (back arched, caudal fin lowered, head raised) when threatened. Migrates short distances seasonally. A small, harmless shark that is often eaten by other larger sharks.
Biology Viviparous, yolk-sac placenta, 1–6 pups per litter. Reproduces annually after 9–11 month gestation. Age at maturity varies by region, but ranges from 2–6.6 years; longevity ~10–19 years. Feeds on small fishes.
Status IUCN Red List: Near Threatened (out of date). Caught in large numbers for food. Kept in public aquaria.

BIGNOSE SHARK *Carcharhinus altimus*

FAO code: **CCA**   Plate page 516

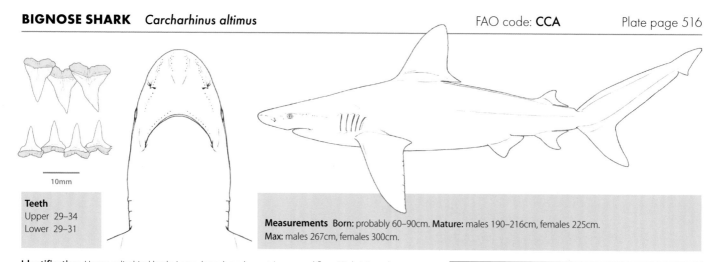

Teeth
Upper 29–34
Lower 29–31

10mm

Measurements Born: probably 60–90cm. **Mature:** males 190–216cm, females 225cm.
Max: males 267cm, females 300cm.

Identification Heavy, cylindrical body. Large, long, broad snout. Long nasal flaps. High, triangular, saw-edged upper teeth; short, inconspicuous upper labial furrows. Fairly large, circular eyes. Moderately long gills. Large, straight dorsal and pectoral fins. Prominent, high interdorsal ridge. Greyish (sometimes bronzy) above. Inconspicuous, faint white flank marking and dusky fin tips (except for pelvics). White below.
Distribution Probably worldwide in most tropical and warm waters, but patchy.
Habitat Offshore, on deep continental and insular shelf edge and uppermost slopes, sometimes on surface to at least 1000m. Young occupy shallower waters, up to 25m.
Behaviour Unknown.
Biology Viviparous, 3–15 pups per litter. Eats bony fishes, other sharks, stingrays and cuttlefish.
Status IUCN Red List: Near Threatened. Bycatch of deep pelagic longlines, occasionally bottom trawls.

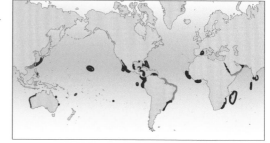

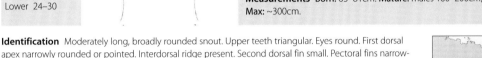

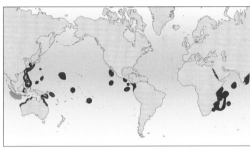

Teeth
Upper　26–30
Lower　24–30

10mm

Measurements Born: 63–81cm. **Mature:** males 160–200cm, females 160–199cm.
Max: ~300cm.

Identification Moderately long, broadly rounded snout. Upper teeth triangular. Eyes round. First dorsal apex narrowly rounded or pointed. Interdorsal ridge present. Second dorsal fin small. Pectoral fins narrow-tipped. Dark grey, sometimes bronze-tinged; faint white band on sides. Very striking white tips and trailing edges on all fins except for black second dorsal. White below.

Distribution Tropical, Indian and Pacific Oceans, wide-ranging but with a fragmented distribution. Unconfirmed in the west Atlantic.

Habitat Continental shelf, offshore islands, coral reefs and offshore banks; also inside lagoons, near drop offs and offshore, from the surface to depths of 800m. Young occur among reefs in shallower water closer to the shore, moving into deeper water as they grow. Adults are more wide-ranging, but this is not an oceanic species.

Behaviour Ranges vertically from the surface to the seabed. It is often seen following boats, but is not thought to disperse widely between sites. This species is more aggressive than, and dominant over,

Galapagos Sharks *Carcharhinus galapagensis* and Blacktip Sharks *C. limbatus*. Adults are often quite heavily scarred, presumably from interactions with other Silvertip Sharks.

Biology Viviparous, with a yolk-sac placenta. 1–11 pups per litter, usually 5–6, which are born in summer after a gestation of about one year. Feeds on a variety of midwater and bottom fishes, including smaller sharks and eagle rays and octopus.

Status IUCN Red List: Vulnerable. This is a large, slow-growing shark that is widely taken as bycatch and also targeted for its large fins and meat. Some very remote Silvertip Shark populations have reportedly been fished out by shark fin fisheries, but artisanal fisheries use most parts of this shark, including its liver, jaws and cartilage. Silvertip Sharks are highly vulnerable to overfishing, particularly as they appear not to disperse widely between sites; recolonisation of areas that have been fished out is, therefore, unlikely to occur rapidly. This bulky shark is often bold, verging on aggressive, when it encounters divers. It is potentially dangerous if present in large numbers and attracted by bait, when it is reported to become very aggressive; there is at least one confirmed incident of an unprovoked Silvertip Shark bite. Caution is therefore advised when this shark is encountered underwater.

Silvertip Sharks, *Carcharhinus albimarginatus*.

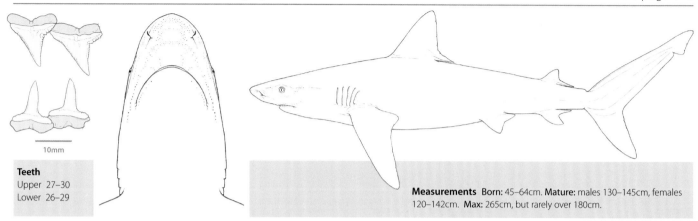

Teeth
Upper 27–30
Lower 26–29

10mm

Measurements Born: 45–64cm. **Mature:** males 130–145cm, females 120–142cm. **Max:** 265cm, but rarely over 180cm.

Identification Moderately long, broadly rounded snout. Narrow, serrated upper teeth. Eyes usually round. First dorsal fin apex narrowly rounded or pointed. No interdorsal ridge. Second dorsal fin small and high, with short free rear tip, origin about over anal fin origin. Pectoral fins moderately large, narrow and falcate. Grey above, white below. First dorsal plain or irregularly to prominently white-edged; obvious broad, black posterior margin to entire caudal fin; blackish tips to other fins.

Distribution Indian and Pacific Oceans, less common in east Pacific.

Habitat Continental and insular shelves and adjacent oceanic waters, coastal-pelagic and inshore, 0–140m. Common on coral reefs, often in deeper areas near drop-offs (including fringing reefs), in atoll passes, and shallow lagoons adjacent to strong currents. Prefers the leeward sides of small, low coral islands. The Grey Reef Shark is found in deeper water than the Blacktip Reef Shark *Carcharhinus melanopterus* (which occupies shallow flats), but is commonly found on the flats if *C. melanopterus* are absent. Young favour shallower water than adults.

Behaviour This is an active, strong-swimming, highly social species, which aggregates by day in or near reef passes or lagoons, and becomes even more active at night, when aggregations disperse. Groups of juveniles are often seen in shallow water, probably in pupping and nursery grounds. Pregnant females may also be seen in large groups in shallow warm water. Grey Reef Sharks often cruise near the seabed but will visit the surface to investigate food sources and may venture several kilometres offshore before returning to their home site. One animal tagged on a remote reef in the Coral Sea undertook a 250km round trip to the Great Barrier Reef and back. Groups of this highly inquisitive species may approach divers closely in seldom-dived areas, but they soon disperse and seldom reappear, except at a distance, during repeated dives. Baiting may therefore be used to attract this species. In the absence of food, if approached too closely, cornered or startled, Grey Reef Sharks may perform a threat display that is intended to intimidate other large animals. This consists of an exaggerated swimming pattern with the back arched, the head lifted, the pectoral fins lowered and the head and tail wagging in broad sweeps. Horizontal spiral swimming or figure-of-eight loops may also be performed. If the perceived threat is not frightened off by this display, it may finish by biting and fleeing.

Biology Viviparous, with a yolk-sac placenta, bearing 1–6 pups per litter after ~a 12–14 month gestation. Age at maturity ranges from ~6–11 years and longevity is 15–25 years, depending on the region. This species feeds mostly off the seabed on small bony reef fishes, also cephalopods (squids and octopus) and crustaceans.

Status IUCN Red List: Endangered. The Grey Reef Shark was one of the most abundant Indo-Pacific reef sharks (along with Blacktip Reef Shark and Whitetip Reef Shark *Triaenodon obesus*). Formerly common in clear tropical coastal waters and oceanic atolls, it is now depleted in many regions because of intensive and largely unmanaged fisheries, degradation of its restricted inshore habitat (due to climate change, destructive fishing practices and poor water quality), site fidelity, and low productivity (small litter size and relatively late age at maturity). Its fins are regularly recorded in small quantities in international trade. The Grey Reef Shark is far more valuable protected for dive tourism than it is for fisheries. In Palau, where fewer than 10,000 visiting divers are attracted to the islands to dive with sharks, individual sharks are estimated to have an annual value approaching US$180,000, and a lifetime value of US$1.9 million to the tourism industry. However, this species can be aggressive (particularly when spear-fishing is occurring) and may bite if cornered, harassed, or excited by the presence of food stimuli. Although most dive encounters with Grey

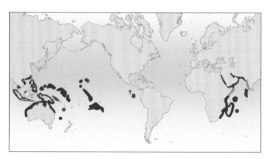

Reef Sharks take place without incident, these animals should be treated with respect. The western Indian Ocean population is sometimes referred to as *Carcharhinus wheeleri*.

Grey Reef Shark, *Carcharhinus amblyrhynchos*.

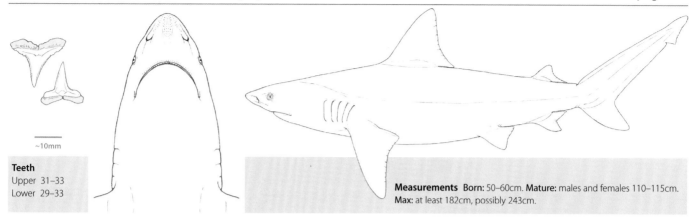

Teeth
Upper 31–33
Lower 29–33

~10mm

Measurements Born: 50–60cm. **Mature:** males and females 110–115cm.
Max: at least 182cm, possibly 243cm.

Identification Fairly large, tubby shark. Fairly short, pointed, wedge-shaped snout. Fairly large eyes. Large gill slits. Large, triangular first dorsal, second dorsal fin small and high, origin over or anterior to anal fin origin, dorsals with short free rear tips. No interdorsal ridge. Moderately large pectoral fins. Grey with conspicuous white flank mark. Often with black-tipped fins.
Distribution West Indian Ocean to Central Indo-Pacific: Somalia to Philippines and northern Australia.
Habitat Coastal-pelagic on continental and insular shelves, 0–75m. **Behaviour** Unknown.
Biology Poorly known. Viviparous, litters of 1–9, average 3; birth occurs after a 9–10 month gestation. Probably feeds mostly on fishes, along with cephalopods and crustaceans.
Status IUCN Red List: **Near Threatened** (out of date). **Vulnerable** in north Indian Ocean. Taken and landed as bycatch in commercial fisheries throughout range. One of four similar black-tipped shark species that together contribute ~4% of shark fins traded through Hong Kong.

PIGEYE SHARK *Carcharhinus amboinensis* FAO code: **CCF** Plate page 550

Teeth
Upper 23–27
Lower 23–25

10mm

Measurements Born: ~60–72cm.
Mature: males ~195–210cm, females 198–223cm. **Max:** 280cm.

Alternative name Java Shark. **Identification** Large with massive, thick head; very short, broad, blunt snout; large, triangular, saw-edged upper teeth. Small eyes. High, erect, triangular first dorsal (over three times second dorsal height); dorsals with short free rear tips. No interdorsal ridge. Large, angular pectoral fins. Greyish above, underside white. Fin tips dusky, but not strikingly marked.
Distribution West Pacific and Indian Ocean: South Africa to Australia. East Atlantic: Senegal, Gambia, Sierra Leone, Ivory Coast, Ghana and possibly Nigeria.
Habitat Continental and insular shelves, inshore near surf line and along beaches, 0–100m.
Behaviour Less common where *Carcharhinus leucas* are common; possibly competitive exclusion?
Biology Viviparous, litters of 6–13. Gestation is ~12 months, birth usually occurs in late spring or early summer. Feeds on bottom fishes, crustaceans and molluscs.
Status IUCN Red List: **Data Deficient** (out of date). Often misidentified with the Bull Shark, *Carcharhinus leucas*. Shark fisheries are intensifying in its range and it is regularly observed in the fin trade.

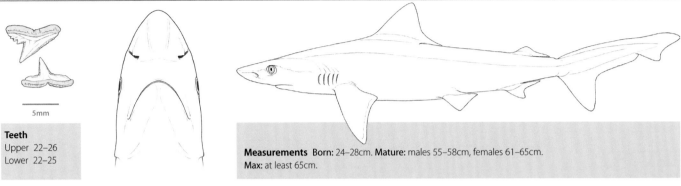

Teeth
Upper 22–26
Lower 22–25

5mm

Measurements Born: 24–28cm. **Mature:** males 55–58cm, females 61–65cm.
Max: at least 65cm.

Identification Small-bodied. Long, pointed snout; enlarged pores alongside corners of large mouth. Large eyes. Small dorsal fins with short free rear tips; origin of second about over anal midbase. No interdorsal ridge. Small pectoral fins. Brown above, white below. Black spot beneath snout tip; dusky to blackish markings on the second dorsal and upper caudal tip; paired fins and anal fin with inconspicuous light edges.
Distribution Central Indo-Pacific: confirmed only from northwest Borneo (eastern Malaysia), China, and possibly Indonesia; records from Java and the Philippines cannot be confirmed. Its distribution is now likely more restricted.
Habitat Tropical inshore/coastal. Rare.
Behaviour Unknown.
Biology Viviparous, yolk-sac placenta, with litters of 6. Little else known
Status IUCN Red List: **Endangered** (out of date). Recently reported in small numbers at fish markets in Borneo, where previously it had not been seen since 1937.

Teeth
Upper 25–30
Lower 23–28

~10mm

Measurements Born: 35–40cm. **Mature:** males ~80–91cm, females ~85–101cm. **Max:** 150cm.

Identification Medium-sized. Short, bluntly rounded snout. Nipple-like anterior nasal flaps. Short labial furrows. Horizontally oval eyes. Moderately large gills. No interdorsal ridge. Fairly large second dorsal with a short free rear tip. Grey to light brownish with conspicuous white band on flank. Black dorsal and caudal fin edges (no obvious black blotch on first dorsal fin); black tips on caudal lobes and pectoral fins. White below.
Distribution Central Indo-Pacific: northern Australia to southern Papua New Guinea and the Solomon Islands.
Habitat Shallow inshore water on continental and insular shelves, coral reefs and in estuaries, 0–20m; may range into deeper water.
Behaviour Reported skittish and timid when accosted by people (hence its name).
Biology Viviparous, 1–6 pups per litter (average 4) after 8–11 months gestation; in some tropical regions it may give birth twice annually. Reproductive cycle is two years. Age at maturity is 4 years for males and 5 years for females, with a maximum age of 12–16 years. Eats small fishes and, to a lesser extent, crustaceans and cephalopods.
Status IUCN Red List: **Least Concern**. Relatively common in northern Australia, this is a small shark with relatively fast growth; the Australian fishery is well-managed.

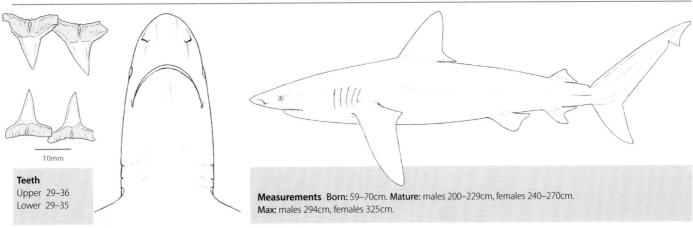

Teeth
Upper 29–36
Lower 29–35

10mm

Measurements Born: 59–70cm. **Mature:** males 200–229cm, females 240–270cm.
Max: males 294cm, females 325cm.

Alternative name Copper Shark

Identification A large shark. Broad, bluntly pointed snout. Narrow, bent cusps on upper teeth. Dorsal fins with short free rear tips, first dorsal moderately large, second dorsal small, relatively low, origin over or slightly posterior to anal origin. Usually no interdorsal ridge. Long pectoral fins. Olive-grey to bronzy above with a fairly prominent white band on flank. Most fins with inconspicuous darker edges and dusky to black tips, not boldly marked. White below.

Distribution Most warm-temperate waters in Indian, Pacific and Atlantic Oceans, and the Mediterranean.

Habitat Close inshore to at least 145m offshore.

Behaviour Active, seasonally migratory in at least part of its range but very little exchange between adjacent regional populations. Large numbers follow winter sardine run off KwaZulu-Natal, South Africa. Nursery grounds in inshore bays and coastal waters.

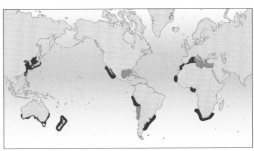

Biology Viviparous, yolk-sac placenta. Gestation 12–16 months with possibly a one year resting period between pregnancies; litters of 13–24 pups every other year. Males mature at ~13–16 years and females ~16–20 years, maximum age is ~25–31 years. Feeds on bony fishes, elasmobranchs and cephalopods.

Status IUCN Red List: **Vulnerable**. This shark is considered Vulnerable due to its exceptionally slow growth rate, large size, inshore habitat vulnerable to damage, and the ease of which it is captured. Caught for food and by sports anglers. Potentially harmful to swimmers and divers and sometimes aggressive during encounters with divers.

Bronze Whalers, *Carcharhinus brachyurus*, feeding on a baitball of the sardine, *Sardinops sagax*, during the annual Sardine Run off the east coast of South Africa.

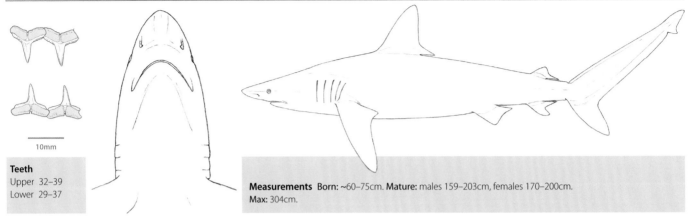

Teeth
Upper 32–39
Lower 29–37

10mm

Measurements Born: ~60–75cm. **Mature:** males 159–203cm, females 170–200cm. **Max:** 304cm.

Identification Slender body. Long, narrow, pointed snout. Short, relatively inconspicuous anterior nasal flaps. Small, narrow-cusped teeth; prominent labial furrows (longer than any other *Carcharhinus*). Small, circular eyes. Long gill slits. Small first dorsal fin, second dorsal moderately large; both with short free rear tips. No interdorsal ridge. Small pectoral fins. An inconspicuous white band on flanks. Adults and juveniles (but not the young) have conspicuous black fin tips on pectoral, second dorsal, anal and caudal lower lobe; absent on pelvic fins. Underside white.

Distribution Cosmopolitan in the warm-temperate to tropical Atlantic, Indian and Pacific Oceans, and Mediterranean Sea. Not recorded from the central and east Pacific.

Habitat Coastal-pelagic on continental and insular shelves, common close inshore (in depths of less than 30m) off beaches, in bays and off river mouths, to at least 200m, from the sea surface to the bottom. It is less common in a pelagic habitat offshore. Nursery grounds are located in very shallow nearshore areas.

Behaviour This is a very active, schooling shark, and primarily a fish-eater, especially schooling fishes such as the migrating schools of mackerel with which it is often associated. Its English name is derived from its unusual feeding behaviour: the Spinner Shark swims rapidly upwards through fish schools with its mouth open, spinning along its horizontal axis and snapping in all directions, then shoots, still spinning, out of the water at the end of its feeding run. These aerial displays may include up to three rotations before the shark falls back into the water, often on its back. This species no doubt participates in feeding frenzies when

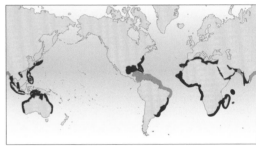

schooling fishes are present, but will also congregate to eat trash fish dumped off trawlers. The Spinner Shark is highly migratory in the Gulf of Mexico (and possibly elsewhere): the sharks move inshore during spring-summer to feed and breed, then possibly head south and into deeper water in winter. There is a certain amount of segregation by age and sex; the young prefer lower water temperatures than the adults.

Biology Viviparous, with a yolk-sac placenta. 3–15 pups per litter (litter size increases with female size) born after an 11–15 month gestation; thus there is a two year reproductive cycle. Young Spinner Sharks are fast-growing (in the northwest Atlantic). Age at maturity is 8–10 years in females (younger in males) and the maximum age is 17–19 years or possibly older. Although this species is primarily a fish-eater, it also takes stingrays and cephalopods.

Status IUCN Red List: Vulnerable. Common, but its inshore distribution is vulnerable to fishing pressure and habitat degradation. The Spinner Shark is taken in target commercial and recreational fisheries, and is important in multi-species fisheries, but likely under-reported in landings due to misidentification as Blacktip Shark. The meat is valuable (it is often sold as 'Blacktip Shark' in the USA because of consumer preference for the latter species), and the fins enter international trade. The skin may also be used for leather and the liver for oil, particularly when taken in artisanal fisheries. This species has bitten swimmers, but is probably of very limited threat to people because of its feeding habits.

A Spinner Shark, *Carcharhinus brevipinna,* finishing a feeding run through a school of fish.

PACIFIC SMALLTAIL SHARK *Carcharhinus cerdale* FAO code: **CWZ** Plate page 550

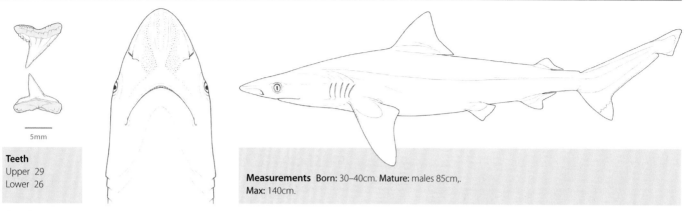

Teeth
Upper 29
Lower 26

5mm

Measurements Born: 30–40cm. **Mature:** males 85cm,.
Max: 140cm.

Identification Very similar to *Carcharhinus porosus*. Small shark. Long, pointed snout. Short labial furrows. Large circular eyes. Large falcate first dorsal fin. No interdorsal ridge. Small second dorsal (origin over anal fin midbase). Small pectoral fins. Deeply notched anal fin. Grey above with inconspicuous white flank band. Tips of dorsal, pectoral and caudal fins frequently dusky or blackish, not conspicuously so. Light below.
Distribution East Pacific: Gulf of California to Peru, not on offshore islands.
Habitat Shallow continental shelf and estuaries, near muddy bottom, from close inshore to at least 40m.
Behaviour Unknown.
Biology Viviparous, but otherwise poorly known. Possibly eats bony fishes, small sharks, crabs and shrimp.
Status IUCN Red List: **Critically Endangered**. Artisanal gillnet, longline and industrial trawl fisheries operate throughout its shallow coastal range. Common until 1980s, but extirpated from Mexico 20 years ago. A greatly depleted bycatch of Colombia's shrimp trawl fishery. Harmless to people.

COATES'S SHARK *Carcharhinus coatesi* FAO code: **CWZ** Plate page 554

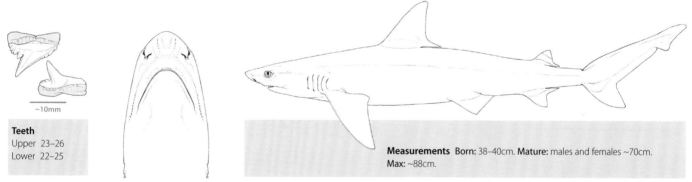

Teeth
Upper 23–26
Lower 22–25

~10mm

Measurements Born: 38–40cm. **Mature:** males and females ~70cm.
Max: ~88cm.

Identification Small, slender shark. Snout moderately long and narrowly rounded. Teeth oblique, blade-like and coarsely serrated. First dorsal fin moderately high, weakly falcate; origin just behind pectoral fin free rear tips. Interdorsal ridge present. Greyish brown with bronze hues above. Second dorsal fin with clearly defined prominent black tip not extending onto fin base; other fins lack any dark marks. White below.
Distribution Central Indo-Pacific: northern Australia, from Shark Bay (western Australia) to Fraser Island (Queensland), and Papua New Guinea.
Habitat Inshore on continental and insular shelves, from the surfline to ~123m.
Behaviour Although common, this coastal shark is poorly known.
Biology Viviparous, 1–3 pups per litter (usually 2). Matures in possibly 2 years with a maximum age of 6.5 years. Feeds mostly on bony fishes, but also cephalopods and crustaceans.
Status IUCN Red List: **Least Concern**. The species has only recently been separated from the very similar-looking *Carcharhinus sealei*.

WHITECHEEK SHARK   *Carcharhinus dussumieri*   FAO code: **CCD**   Plate page 550

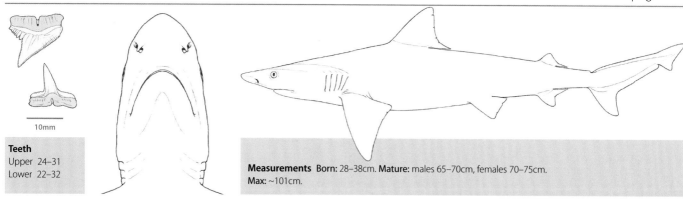

Teeth
Upper 24–31
Lower 22–32

Measurements Born: 28–38cm. **Mature:** males 65–70cm, females 70–75cm.
Max: ~101cm.

Identification Small shark. Moderately long, rounded snout. Small, widely spaced nostrils. Oblique-cusped, serrated teeth; short inconspicuous upper labial furrows. Fairly large, horizontally oval eyes. Small, triangular first dorsal fin; both dorsals with short free rear tips. Interdorsal ridge usually absent. Small, semifalcate pectoral fins. Colour grey to grey-brown with inconspicuous light flank stripe. Black or dusky spot only on second dorsal; other fins with pale trailing edges. Often confused with *Carcharhinus tjutjot*.
Distribution North Indian Ocean: from at least the Arabian/Persian Gulf to the east coast of India and Sri Lanka. Full extent of range unknown due to previous misidentification with *C. tjutjot*.
Habitat Tropical, inshore, continental and insular shelves, 0–100m. **Behaviour** Poorly known.
Biology Viviparous, yolk-sac placenta, normally 2 pups per litter, exceptionally up to 4. Feeds on small fishes, cephalopods and crustaceans.
Status IUCN Red List: **Endangered**. Common, but heavily fished for meat and sometimes fins in most of range. Large declines and some local extinctions are documented over the past 15 years. Harmless to people.

CREEK WHALER   *Carcharhinus fitzroyensis*   FAO code: **CCZ**   Plate page 554

Teeth
Upper 28–30
Lower 26–30

Measurements Born: ~45–55cm. **Mature:** males ~80–88cm, females 90–100cm. **Max:** 139cm.

Identification Fairly large shark. Long, parabolic snout. Lobate anterior nasal flaps. Narrow teeth; short labial furrows. Moderately large, round eyes. Short gill slits. Broad, triangular fins. First dorsal origin over pectoral fin free rear tips; second about over anal origin. No interdorsal ridge. Bronze to greyish-brown above, without any conspicuous fin or body markings. Light below.
Distribution Central Indo-Pacific: northern Australia.
Habitat Mainly inshore, from intertidal to at least 40m. Bays appear to be important habitat as nursery grounds for newborn sharks.
Behaviour Unrecorded.
Biology Viviparous, 1–7 pups per litter, born annually after 7–9 month gestation. Age at maturity for males is 3–5 years, females 5 years, and maximum age is 13 (males and females). Eats mainly small fishes, some crustaceans.
Status IUCN Red List: **Least Concern**. A small, relatively productive shark whose population can tolerate the small numbers caught in Australian inshore gillnet fisheries.

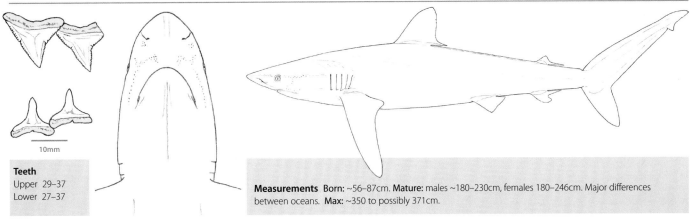

Teeth
Upper 29–37
Lower 27–37

Measurements **Born:** ~56–87cm. **Mature:** males ~180–230cm, females 180–246cm. Major differences between oceans. **Max:** ~350 to possibly 371cm.

Identification Large, slim shark. Fairly long, flat, rounded snout. Small jaws; oblique-cusped, serrated upper teeth. Large eyes. First dorsal behind pectoral fins. Narrow interdorsal ridge. Second dorsal low with greatly elongated inner margin and free rear tip. Long, narrow pectoral fins. No caudal keels. The English name comes from the smooth hide, covered with small, tightly packed, overlapping denticles. Dark grey to grey-brown or nearly blackish above, with inconspicuous pale flank band. Fin markings inconspicuous, tips dusky except for first dorsal. White below.

Distribution Worldwide in tropical seas.

Habitat Oceanic and epipelagic, surface to at least 500m. Commonest in water less than 200m near continental and insular shelf edge and over deep water reefs and seamounts; also in the open sea, occasionally inshore to 18m. Longline catches are more abundant offshore near land than in the open ocean.

Behaviour An active, swift, bold, inquisitive and sometimes aggressive shark that displays an interesting range of behaviours. 'Hunching' with back arched, head raised and caudal fin lowered, may be a threat to divers that approach too closely. When in groups, Silky Sharks have been seen 'tilting' (presenting their full lateral profile towards each other), gaping their jaws and puffing out their gills. Sometimes they suddenly charge straight upwards, veer away just before reaching the surface then glide back down to deeper water. Silky Sharks segregate by size: young in offshore nursery areas on the shelf edge and oceanic banks, and associated with drifting objects including fish aggregating devices (FADs) set by tuna purse seiners. Sub-adults and adults are further offshore, often associated with tuna schools. Twice as many adult males than females are caught on tuna longlines. Maldivian fishers believe that Silky Sharks aggregate tuna and make them easier to catch (all sharks are protected in Maldivian waters). Silky Sharks may also be found in schools of Scalloped Hammerheads and follow marine mammals. Juvenile Pilot Fish *Naucrates doctor* often ride the pressure wave ahead of their snout.

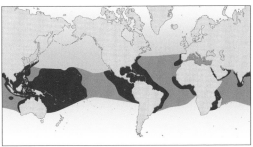

Biology Viviparous, yolk-sac placenta. Litters of 2–18 pups (average 5–10); females possibly have a 2 year reproductive cycle and do not associate with tuna while pregnant. Pupping occurs in summer in some regions, but there is no obvious seasonality in others. Age at maturity ranges from 5–15 years, depending on region. Longevity is estimated at up to 36 years, but few large adults are caught. Primarily eats fish, including small tunas, also cephalopods and pelagic crabs.

Status IUCN Red List: Vulnerable. Serious declines are reported in several regions. There is huge juvenile Silky Shark mortality under entangling FADs, which dangle nets below them. Researchers calculated that Indian Ocean Silky Sharks had a 29% chance of surviving to one year old, 9% to two years, and 3% to three years because so many die unseen beneath FADs. Although frequently misidentified or completely unreported in fisheries records, Silky Sharks are the secondmost widely caught and traded (for their fins) shark after the Blue Sharks; the fins of 0.5 to 1.5 million per year Silkies entered international trade, from bycatch in oceanic longliners and purse seiners, and artisanal fisheries. It is also utilised for meat. The retention of Silky Shark bycatch is prohibited in several oceanic fisheries (see p. 83) and FAD Management Plans are reducing bycatch mortality and ghost fishing (see p. 68). The species is also listed in CITES and CMS. This shark is regarded as dangerous and potentially harmful to people, partly because of its size and aggressive behaviour, but is rarely encountered except on offshore reefs near deep water and no serious incidents have been attributed to it. It is important for ecotourism in the Red Sea, where divers have photographed adults on offshore reefs and found them cooperative and unaggressive.

Silky Sharks, *Carcharhinus falciformis*, Republic of Cuba, Caribbean.

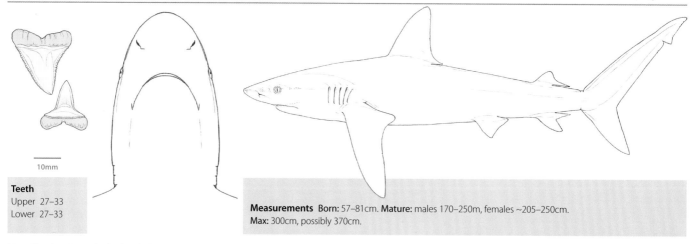

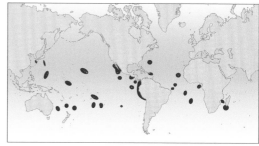

Teeth
Upper 27–33
Lower 27–33

10mm

Measurements Born: 57–81cm. **Mature:** males 170–250m, females ~205–250cm. **Max:** 300cm, possibly 370cm.

Identification Large-bodied. Fairly long, broadly rounded snout. Low anterior nasal flaps. Large, erect teeth. Fairly large eyes. Moderately large first dorsal fin with short free rear tip; originates over inner margin of pectoral fins. Low interdorsal ridge. Large, semifalcate pectoral fins. Brownish grey above with inconspicuous white band on flank. No conspicuous fin markings, but most fin tips are dusky (not black or white). White below.

Distribution Worldwide, patchy, mainly around warm-temperate to tropical oceanic islands.

Habitat This species is partly defined by its preferred habitat: in clear water over rugged coral or rocky bottom around offshore islands, seamounts, and occasionally the edge of the continental shelf. However, this is a coastal, pelagic, non-oceanic, species, occurring from shallow inshore (1m) to well offshore (0–286m). Juvenile nursery grounds are usually in depths of less than 25m, sometimes very shallow water, adults are found further offshore. The Galapagos Shark is often found in areas of strong currents.

Behaviour A locally common to abundant shark that often swims within a few metres above the bottom, but will readily come to the surface to feed or investigate disturbances. They are found in aggregations, but apparently do not form coordinated schools. Inquisitive and sometimes aggressive, Galapagos Sharks perform 'hunch' displays near divers (with an arched back, raised head, lowered caudal and pectoral fins) while twisting and rolling, which may be followed by biting. During group interactions, Galapagos Sharks are dominant over Blacktip Sharks but subordinate to Silvertip Sharks of equal size.

Biology Viviparous, with a yolk-sac placenta, bearing 4–16 pups per litter. Their reproductive life history is not well studied, but females probably only breed every 2–3 years, with mating occurring from winter to spring. Around Bermuda, 95% of Galapagos Sharks caught are newborns, suggesting that this is a nursery ground. Estimated age at maturity is 6–8 years for males and 6.5–9 years for females. However, this species is very close, genetically and morphologically, to the Dusky Shark (see below and p. 540); if their life history is also similar, these estimates are far too short, and these sharks will have a significantly slower life history cycle and lower resistance to overfishing than previously thought. Galapagos Sharks feed mainly on bottom fishes.

Status IUCN Red List: Least Concern. Common or abundant in pelagic waters and around islands and seamounts, this is one of the most common sharks around the Hawaiian Islands, and its population in large parts of its Pacific range is stable. Large areas of its habitat are now protected in very large, offshore island marine protected areas. Elsewhere, however, it is often heavily fished; some populations around Central America and in the mid-Atlantic have reportedly been extirpated. This species can be a nuisance to divers and anglers because of its inquisitiveness and occasional aggressiveness; it occasionally bites people and one fatality has been reported. Recent molecular genetic analyses have found virtually no difference between the Galapagos Shark and the Dusky Shark (p. 540), which are primarily differentiated by their habitat (insular and continental, respectively) and precaudal vertebral counts; these two species, as currently identified, may prove to be the insular/oceanic and onshore/continental forms of the same species.

Galapagos Sharks, *Carcharhinus galapagensis*.

PONDICHERRY SHARK *Carcharhinus hemiodon*

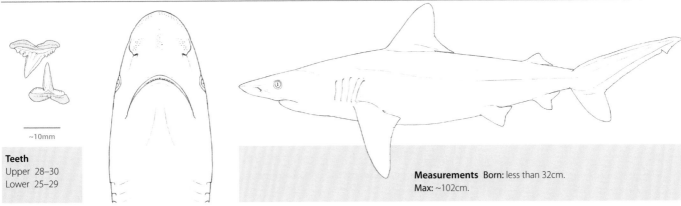

Teeth
Upper 28–30
Lower 25–29

~10mm

Measurements Born: less than 32cm.
Max: ~102cm.

Identification Probably small-bodied. Fairly long, pointed snout. Small, widely spaced nostrils. Upper labial furrows short and inconspicuous. Fairly large eyes. Fairly large first dorsal and moderately large second dorsal, both with short free rear tips. An interdorsal ridge. Small pectoral fins. No preanal ridge. Grey above, with conspicuous white band on flank. Black tips to pectorals, second dorsal and lower caudal lobe. White below.
Distribution Indian and Pacific Oceans: Oman to eastern India, and to Indonesia, Malaysia and possibly China.
Habitat Coastal, on tropical continental and insular shelves; 10–150m. Unconfirmed reports from river mouths and fresh water upstream.
Behaviour Unknown. **Biology** Virtually unknown.
Status IUCN Red List: Critically Endangered. Very rare, known from ~20 museum specimens collected from heavily fished sites. Last seen in 1979.

HUMAN'S WHALER SHARK *Carcharhinus humani*

Teeth
Upper 24–26
Lower 22–25

5mm

Measurements Born: 35–45cm. **Mature:** males 75cm, females 75cm.
Max: 94.6cm.

Identification Small, slender whaler shark. Long, narrowly rounded snout. Moderately high first dorsal fin originating just anterior to the pectoral fin free rear tip. A weak interdorsal ridge. Second dorsal fin less than one-half the height of the first, originating about opposite the anal fin origin. Pale brownish to grey above. Second dorsal fin has a black blotch on the upper one to two thirds of the fin, strongly demarcated from the ground colour, and not extending onto the upper body; most of the other fins are whitish along the outer margins. Whitish below.
Distribution West Indian Ocean: Kuwait to the Socotra Islands, and south to the KwaZulu-Natal coast of South Africa; off Madagascar, Seychelles, and possibly other islands in the region.
Habitat Coastal, from close inshore off sandy beaches down to ~40m. One specimen was found at the surface over water 1260m deep off Madagascar, but close to shallow water.
Behaviour Apparently common year-round in South African waters; otherwise a very poorly known species.
Biology Viviparous, 1–2 pups per litter. Feeds mainly on bony fishes, cephalopods and crustaceans.
Status IUCN Red List: Data Deficient. This little-known species' range overlaps significantly with inshore fisheries, but population trends are unknown.

FINETOOTH SHARK *Carcharhinus isodon*

FAO code: **CCO**   Plate page 520

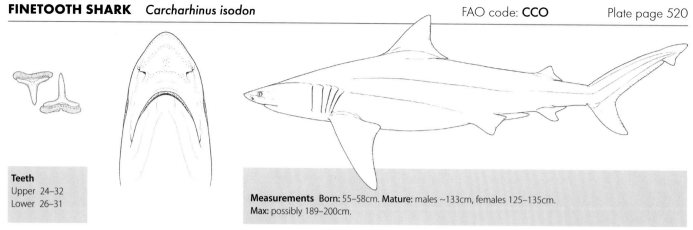

Teeth
Upper 24–32
Lower 26–31

Measurements Born: 55–58cm. **Mature:** males ~133cm, females 125–135cm.
Max: possibly 189–200cm.

Identification Small-bodied. Moderately long, pointed snout. Erect, smooth or irregularly serrated teeth; short, inconspicuous upper labial furrows. Fairly large eyes. Very long gill slits. Small first dorsal; moderately large second dorsal with origin over posterior of anal fin origin; both dorsals with short free rear tips. No interdorsal ridge. Small pectoral fins. Dark bluish grey above; inconspicuous white band on flank. No prominent fin markings. White below.
Distribution West Atlantic: United States from North Carolina to the Gulf of Mexico; and Brazil from Sao Paulo to Catarina State, with a few scattered records from Guyana and Trinidad.
Habitat Warm-temperate to tropical inner continental shelf, intertidal to ~20m.
Behaviour Very active. Large aggregations migrate seasonally with changing water temperature.
Biology Viviparous, yolk-sac placenta. 1–6 pups per litter (average 2–4) in May–June, two-year reproductive cycle. Age at maturity ~4 years, maximum age ~8 years. Feeds on small fishes, also shrimp.
Status IUCN Red List: **Least Concern** (out of date). Locally common, caught throughout range.

SMOOTHTOOTH BLACKTIP SHARK *Carcharhinus leiodon*

FAO code: **CCJ**   Plate page 520

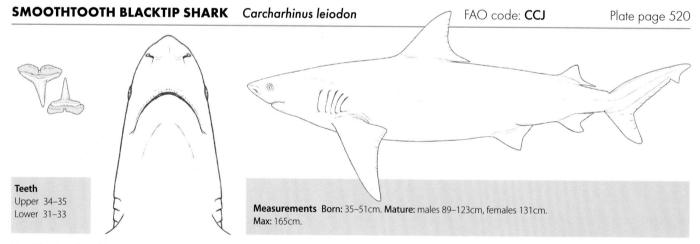

Teeth
Upper 34–35
Lower 31–33

Measurements Born: 35–51cm. **Mature:** males 89–123cm, females 131cm.
Max: 165cm.

Identification Short, bluntly pointed snout. Smooth, erect-cusped upper teeth. Fairly large eyes. Long gill slits. Fairly large first dorsal and moderately large second dorsal, both with short free rear tips. No interdorsal ridge. Small pectoral fins. Conspicuous black tips on all fins. Similar to *Carcharhinus amblyrhynchoides*, which has serrated upper tooth cusps and less prominently marked fin tips.
Distribution Northwest Indian Ocean: Gulf of Aden to the Persian Gulf.
Habitat Inshore to a maximum depth of 30–40m. Water temperature is 19–30℃.
Behaviour Unknown.
Biology Viviparous with litters of 4–6. Diet includes small bony fishes.
Status IUCN Red List: **Endangered**. Originally described in 1985 based on one specimen collected in 1902; it was 2009 before any additional specimens were discovered.

BULL SHARK *Carcharhinus leucas*

FAO code: **CCE**   Plate page 506

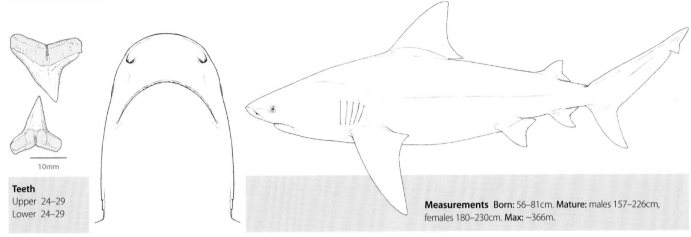

Teeth
Upper 24–29
Lower 24–29

10mm

Measurements Born: 56–81cm. **Mature:** males 157–226cm, females 180–230cm. **Max:** ~366m.

Identification A massive, solid, broad-headed shark. Very short, broad, bluntly-rounded snout. Triangular, saw-edged upper teeth; upper labial furrows very short. Small eyes. No spiracles. Broad, triangular first dorsal fin (less than 3.2 times the second dorsal fin height); both dorsal fins with short free rear tips, No interdorsal ridge. Large, angular pectoral fins. Inconspicuous caudal keels. Greyish above. Fin tips dusky, but not conspicuously marked except in juveniles. Underside white.

Distribution Worldwide, in tropical and subtropical waters and occasionally fresh water.

Habitat Usually close inshore (1–30m) in hypersaline lagoons, bays, river mouths, passages between islands, coastal canals, near wharves and along surf line; but also many thousand kilometres upstream in warm rivers (often in very turbid water) and freshwater lakes. Found to depth of 164m off coasts. Apparently even newborn Bull Sharks are euryhaline, and juveniles commonly migrate into fresh water. They prefer a water temperature of 26°C or more and migrate seasonally as temperatures change. For example, in the west Atlantic, some Bull Sharks move northwards along the US coast in summer, then as the water cools, they retreat southwards, back to their tropical stronghold.

Behaviour Bull Sharks usually cruise slowly near the seabed in water less than 20m deep, travelling an average of 5–6km per day, but are agile and quick when chasing and attacking prey. Young sharks may be seen spinning out of the water. They have small eyes, as sight is not very useful for hunting prey in turbid water conditions. Females often have courtship scars, but males with fighting scars are rare. These sharks prefer a coastal environment, and pregnant females give birth to their pups in estuaries, a vital inshore habitat for Bull Sharks, which has been badly damaged by a wide range of human activities.

Their preferred distribution, close to the coast and in fresh water, also means that Bull Sharks are likely to come into regular contact with people. This, combined with their large size, massive jaws with disproportionately large teeth, indiscriminate appetite and propensity to take fairly large prey, makes the Bull Shark one of the world's three most dangerous shark species, particularly in turbid tropical waters. (The other two, measured by the numbers of recorded attacks on people, are the White and Tiger Shark, but because Bull Sharks are not as easy to identify, bites from this species are probably under-recorded and the Bull Shark may actually be the world's most dangerous shark.)

Biology Viviparous, with a yolk-sac placenta. Litters of 1–13 pups are usually born in estuaries and rivers, after an estimated 10–12 month gestation. They grow quite fast for the first 5 years of life, averaging 15–20cm per year, then slow to ~10cm per year from 6–10 years, 5–7cm per year for years 11–16, and 4–5cm per year once they have reached maturity, at 15–20 years old. Maximum age is ~32 years, but may be up to 50 years. Bull Sharks eat a very broad range of food, from bony fishes, elasmobranchs and invertebrates, to sea turtles, birds, dolphins, whale offal, and terrestrial mammals. Abattoir offal, and fish and other animals scavenged from fishing gear, are readily taken, but this shark is less likely than Tiger Sharks to swallow inedible garbage.

Status IUCN Red List: Near Threatened (out of date). Their nearshore and estuarine habitat is extremely vulnerable to human impacts; habitat loss and damage have affected some populations. Bull Sharks are taken as a fisheries bycatch throughout their range, and populations decline rapidly if targeted. They are caught mainly with longlines, hook-and-line gear and gillnets, and utilised fresh, fresh–frozen or smoked for human consumption. The hide is used for leather, fins for shark-fin soup (almost 2% of fins identified in Hong Kong are from Bull Sharks), and the liver for oil and vitamins (Bull Shark liver oil has a high vitamin content). This shark is

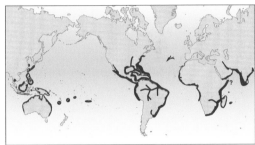

notorious as being responsible for several fatal and nonfatal bites on people, particularly swimmers and bathers. Large animals are, therefore, targeted by bather protection nets and drum lines. Bull Sharks are very popular for dive ecotourism in areas with good visibility; divers often encounter this species without incident, although they must be treated with respect. Bull Sharks are a popular sports angling trophy and do well in captivity (they have lived in aquaria for over 20 years).

Bull Shark, *Carcharhinus leucas*.

REQUIEM SHARKS **CARCHARHINIDAE** 535

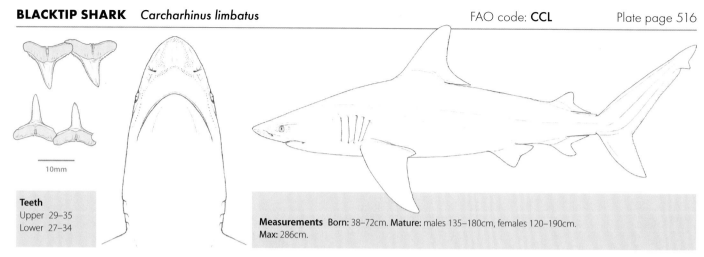

Teeth
Upper 29–35
Lower 27–34

10mm

Measurements Born: 38–72cm. **Mature:** males 135–180cm, females 120–190cm. **Max:** 286cm.

Identification A fairly large, stout shark. Long, narrow, pointed snout. Low, triangular anterior nasal flaps. Narrow-cusped, erect upper teeth; short, inconspicuous upper labial furrows. Small eyes. Long gill slits. High first dorsal fin. No interdorsal ridge. Grey to grey-brown above, with conspicuous white band on the flank; white below. Fin tips of pectorals, second dorsal and lower caudal lobe usually black, sometimes also pelvic and anal fins (but anal fins are usually plain). There are usually black edges on the first dorsal apex and upper caudal lobe, but some adults lack black tips.

Distribution Widespread in tropical and subtropical seas.

Habitat The Blacktip Shark occurs on continental and insular shelves, usually close inshore (off river mouths, in estuaries, shallow muddy bays, saline mangrove swamps, island lagoons and coral reef drop-offs), to offshore in water at least 140m deep. It tolerates reduced salinities, but not fresh water.

Behaviour Often segregated by age and sex; pregnant females are seasonally migratory. Aggregates most strongly in the early summer when the sharks are youngest, suggesting that they are seeking refuge in numbers from predators (which are mostly larger sharks). Pregnant females move inshore to drop their young in nursery and pupping grounds. This is a very active, fast swimmer, often occurring in large schools near the surface. Blacktip Sharks may leap out of the water and rotate up to three times around their axis before dropping back into the sea at the end of a feeding run on small schooling fishes, but do so less often than Spinner Sharks. Hunting activity peaks at dawn and dusk. Blacktip Sharks enter a feeding frenzy on highly concentrated food sources. Small individuals of this shark have approached divers, apparently out of 'curiosity', but circled them at a distance without closing.

Biology Viviparous, with a yolk-sac placenta. 1–10 pups per litter (commonly 4–7) are born in inshore nursery grounds, in alternate years, after a 10–12 month gestation. Females have one functional ovary and

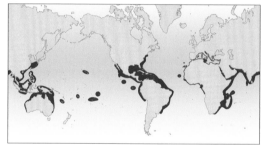

two functional uteri; each uterus is separated into compartments containing a single embryo. Maturity occurs at 4–7 years old, with a maximum age of 12 years. Mainly eats fishes, also cephalopods and crustaceans.

Recent molecular studies indicate that the Blacktip Shark may actually represent a species-complex, as different regional populations appear to be quite genetically distinct from each other.

Fossil teeth belonging to this species have been dated to the Early Miocene.

Status IUCN Red List: **Near Threatened** (out of date). Its inshore distribution is heavily fished and susceptible to habitat degradation. This is an important commercial and sports fishery species, with valuable meat and fins. It is utilised fresh, fresh-frozen, or dried-salted for food, the hides for leather, liver oil for vitamins (it has a high vitamin content), and carcasses for fishmeal. This species is hard to distinguish from a few other blacktipped sharks in Hong Kong fin markets; combined, they comprise 4% of fins identified. Blacktip Sharks are among the most important species in the northwest Atlantic shark fishery, second only to the Sandbar Shark. The meat is considered superior to that of the latter, so Sandbar and other requiem shark meat is often sold under the name 'Blacktip Shark' in the United States. Very few serious bites on people are reported. It is not potentially harmful unless stimulated by food (but may harass spear-fishers when fish speared). Important for dive tourism.

The United States and Australia are the only two countries that (at time of writing) manage fisheries catching Blacktip Sharks.

Blacktip Shark, *Carcharhinus limbatus.*

OCEANIC WHITETIP SHARK *Carcharhinus longimanus*

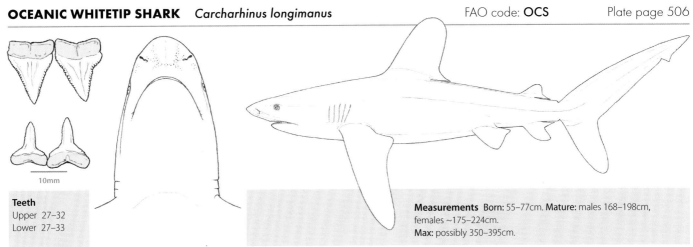

Teeth
Upper 27–32
Lower 27–33

Measurements Born: 55–77cm. **Mature:** males 168–198cm, females ~175–224cm.
Max: possibly 350–395cm.

Identification Large, stocky shark. Snout bluntly rounded. Upper teeth triangular. Small eyes. Huge, rounded first dorsal fin. Interdorsal ridge present. Long, paddle-like pectoral fins. Inconspicuous caudal keels. Grey or brownish above, white below. First dorsal fin and pectoral fins with obvious white-mottled tips; juveniles have black tips on some fins and black patches or saddles on the caudal peduncle.
Distribution Worldwide. Formerly the most abundant warm-water oceanic-pelagic shark.
Habitat Oceanic-epipelagic (occasionally coastal), far offshore in the open ocean from the surface to 1082m in temperatures of 18–28°C. Occasionally in shallow water (37m) off oceanic islands or where the continental shelf is very narrow.
Behaviour Slow-moving but active by day and at night, cruising slowly at or near the surface with huge pectoral fins outspread. Very inquisitive, aggressive and persistent, especially when competing for food with Silky Sharks and sometimes when investigating divers. Some size and sexual segregation reported.
Biology Viviparous, yolk-sac placenta, 1–15 pups per litter (increasing with female size) after ~1 year gestation. Age at maturity varies regionally, but in general is 4–9 years, with a maximum age of 11–25 years.

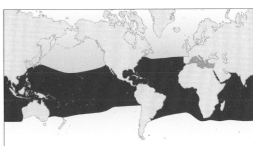

Mainly feeds on oceanic bony fishes and cephalopods, also stingrays, sea birds, turtles, marine gastropods, crustaceans, marine mammal carrion and garbage.
Status IUCN Red List: Critically Endangered. Declines reported of up to 99% in the northwest and central Atlantic, 90% in the central Pacific, and 60–70% elsewhere. This species was originally the most widespread and common warm-water oceanic-pelagic shark, but its very inquisitive nature makes it highly catchable. This, combined with a low reproductive capacity, makes it extremely vulnerable to depletion due to bycatch mortality in tuna and billfish fisheries, and directed oceanic shark fin fisheries. The Oceanic Whitetip Shark's unmistakable, huge fins have a very high value in international trade (they are known as Liu Qiu by fin traders). Demand from the international fin trade has driven fisheries mortality and still poses a major threat to this species, despite shark finning bans and its prohibited status in all oceanic fisheries managed by the tuna Regional Fisheries Management Organisations. In 2000, Liu Qiu comprised slightly less than 2% of all shark fins identified in Hong Kong shark fin auctions. In 2013 the species was listed in Appendix II of CITES, which requires trade to be legal, from sustainable sources, and reported to CITES (p. 83). Subsequently, Oceanic Whitetip fins comprised 1% of samples in the Hong Kong retail market. In addition to legal trade reported to CITES, some large illegal shipments of this species' fins are have been reported. While the Oceanic Whitetip Shark is now rarely encountered in the water, but has been bitten swimmers and boats.

Oceanic Whitetip Shark, *Carcharhinus longimanus*.

HARDNOSE SHARK *Carcharhinus macloti*

FAO code: **CCM**   Plate page 556

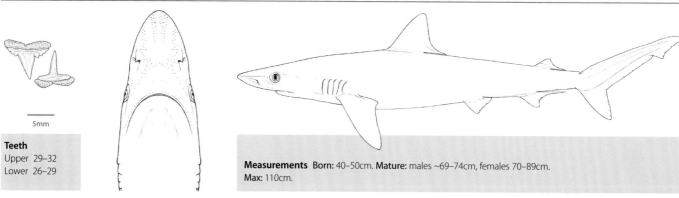

5mm

Teeth
Upper 29–32
Lower 26–29

Measurements Born: 40–50cm. **Mature:** males ~69–74cm, females 70–89cm.
Max: 110cm.

Identification Small, slender shark. Long, pointed snout; oblique-cusped, smooth-edged upper teeth. Fairly large eyes. Dorsal fins small (second very low) with extremely long free rear tips. No interdorsal ridge. Grey or grey-brown above, with inconspicuous light flank marks. Fins light-edged but not conspicuously marked. White below. Only *Carcharhinus* with a hypercalcified rostrum (felt if snout pinched).
Distribution Indian and west Pacific Oceans, from Tanzania to South Korea and northern Australia.
Habitat Continental and insular shelves, from inshore shallows to depth of 200m offshore.
Behaviour Forms large, sexually segregated aggregations.
Biology Viviparous, yolk-sac placenta. 1–2 (usually 2) pups per litter after ~1 year gestation followed by 1 year resting before next breeding cycle. Limited age data, but a shark recaptured 10 years after being tagged as a mature adult suggests longevity is 15–20 years. Eats mostly small fishes, also cephalopods and crustaceans.
Status IUCN Red List: Near Threatened (out of date). Fished throughout range, heavily in southern Asia, population likely reduced.

SMALLTAIL SHARK *Carcharhinus porosus*

FAO code: **CCR**   Plate page 550

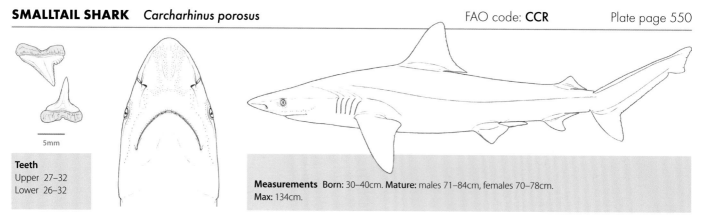

5mm

Teeth
Upper 27–32
Lower 26–32

Measurements Born: 30–40cm. **Mature:** males 71–84cm, females 70–78cm.
Max: 134cm.

Identification Small shark. Long, pointed snout. Short labial furrows. Large, circular eyes. Large, falcate first dorsal fin. No interdorsal ridge. Small second dorsal (origin over anal midbase). Small pectoral fins. Deeply notched anal fin. Grey above, with inconspicuous white flank band. Tips of pectoral, dorsal and caudal fins frequently dusky or blackish, not conspicuously so. Light below.
Distribution West Atlantic: northern Gulf of Mexico to southern Brazil, but not around Caribbean Islands. East Pacific: continental coast, Gulf of Mexico to Peru, not on offshore islands.
Habitat Shallow continental shelf and estuaries, near muddy bottom, from close inshore to at least 84m.
Behaviour Unknown.
Biology Viviparous, yolk-sac placenta, 2–9 (average 6) pups per litter. Gestation is ~12 months. Age at maturity is ~6 years for males and females, with a maximum age of 12 years. Eats bony fishes, small sharks, crabs and shrimp.
Status IUCN Red List: Critically Endangered. Catches have fallen very steeply or ceased in the intensive artisanal and industrial fisheries operating throughout this species' inshore range. Harmless to people.

BLACKTIP REEF SHARK *Carcharhinus melanopterus*

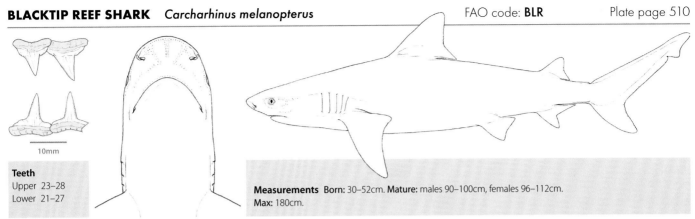

Teeth
Upper 23–28
Lower 21–27

Measurements **Born:** 30–52cm. **Mature:** males 90–100cm, females 96–112cm. **Max:** 180cm.

Identification Medium-sized shark. Short, bluntly rounded snout. Narrow-cusped teeth. Horizontally oval eyes. Rounded first dorsal fin apex. No interdorsal ridge. Largish second dorsal fin with short free rear tip. Narrow, falcate pectoral fins. Brownish grey above. Brilliant black fin tips highlighted by white. Underside white.
Distribution West Pacific and Indian Oceans, with a few records from the east Mediterranean (introduced through Suez Canal from the Red Sea).
Habitat Very shallow water on coral reefs and reef flats, also near reef drop-offs, 0–100m, rarely offshore or in brackish water. **Behaviour** Strong, active swimmer, dorsal fins above surface in very shallow water, alone or in small groups (not strongly schooling).
Biology Viviparous, yolk-sac placenta, 2–4 pups per litter (usually 4) after possible 8–16 month gestation. Eats small fishes and invertebrates.
Status IUCN Red List: **Vulnerable.** A formerly common Indo-Pacific reef shark. Steep population declines in some areas, due to unmanaged industrial and artisanal target and bycatch inshore fisheries. Probably also threatened by reef habitat deterioration. Important for aquarium display, and for dive tourism where still common. Very occasionally bites people swimming or wading on reefs, but more cautious when encountering divers.

LOST SHARK *Carcharhinus obsoletus*

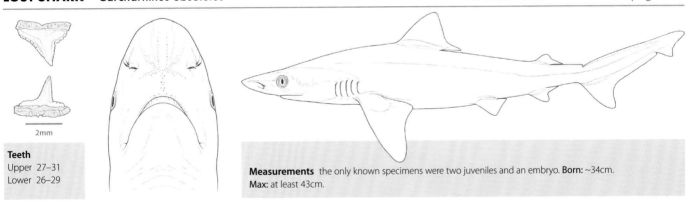

Teeth
Upper 27–31
Lower 26–29

Measurements the only known specimens were two juveniles and an embryo. **Born:** ~34cm. **Max:** at least 43cm.

Identification A small, slender whaler shark. Short, rounded snout. Relatively small first dorsal fin originating just behind the pectoral fin base and over the inner fin margin; a much smaller, low, second dorsal fin, origin about opposite midbase of anal fin. No interdorsal ridge. Pale grey, becoming lighter along flanks and below. Fins without any distinct dark or light markings, blotches or spots (but coloration is known only from specimens in preservative for over 85 years).
Distribution Central Indo-Pacific: known from the South China Sea, including the Gulf of Thailand, Vietnam, and Malaysian Borneo. The full extent of its distribution is uncertain.
Habitat Coastal inshore, but nothing else known. **Behaviour** Unknown.
Biology Viviparous, nothing else known.
Status IUCN Red List: **Critically Endangered (Possibly Extinct).** This species is known from only 3 specimens, a late term embryo and 2 juvenile females. The last record of this species was from 1934; none have been reported since then.

Teeth
Upper 29–33
Lower 27–37

Measurements Born: 69–100cm. **Mature:** males 265–280cm, females 257–310cm. **Max:** 420cm.

Identification Large-bodied shark. Broadly rounded snout. Low, poorly developed anterior nasal flaps. Triangular, saw-edged upper teeth. Fairly large eyes. Large, falcate first dorsal fin. Low interdorsal ridge. Curved, moderate-sized pectoral fins. Grey to bronzy above, with inconspicuous white band on flank. Most fin tips dusky but not boldly marked. Underside white.

Distribution Possibly worldwide in tropical and warm-temperate shelf waters.

Habitat Continental and insular shelves, from the shoreline to adjacent oceanic waters, 0–500m. Avoids estuaries. Often follows ships offshore.

Behaviour Dusky Sharks are highly migratory with changing temperatures; they move into higher latitudes in subtropical and temperate regions during the warmer summer months, then retreat as these waters cool in winter. Adult female Dusky Sharks also move close inshore during the summer months to give birth, and then immediately move back offshore (minimising the risk of cannibalism). Newborns segregate in the nursery areas where they are born. The young tend to remain in nearshore waters, often forming large schools and feeding aggregations. As they grow, they move into an offshore habitat. Juveniles and adults may undertake different, partially sexually-segregated, seasonal movements north-south and inshore-offshore.

Biology Viviparous, with a yolk-sac placenta, with 2–18 pups per litter (number varies between regions but is not correlated with size of mother). Females have a 2–3 year reproductive cycle, with mating occurring

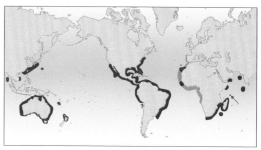

every other year. Females move inshore to give birth after an estimated 16–22 month gestation. Maturity varies regionally, ~17–24 years old, with a maximum age of 34–53 years. Bony fishes are the primary prey, followed by elasmobranchs and crustaceans, with other species also taken. Other large sharks eat young Dusky Sharks.

Status IUCN Red List: **Endangered**. This is one of the most vulnerable of all vertebrates to exploitation, because it reproduces so slowly. It is difficult to manage or protect, as it is taken in mixed species fisheries and has a high mortality rate when taken as bycatch. It is also a popular sport fish (on the east coast of the USA, recreational fisheries drove the initial decline in this species). Although the Dusky Shark is now protected in US waters and numbers of juveniles are increasing, the adult population is still declining. Dusky Sharks are the subject of ecotourism diving in some areas where they occur, but caution is necessary as these large sharks can be very aggressive and have been implicated in attacks on swimmers, surfers and divers. Young Dusky Sharks are kept in aquaria. This species is closely related to the Galapagos Shark (p. 532), and they are difficult to distinguish from each other except by habitat, internal anatomy, and analysis of nuclear DNA (p. 60). Earlier molecular studies had suggested that these two species may in fact be the one and the same, with one form (*C. galapagensis*) being found far from continental landmasses and the other (*C. obscurus*) occurring in association with islands, and the continental shelf edge and upper slope. However, they are now known to be distinct with no current evidence of hybridisation between the two species, including where they occur together around Norfolk Island, Australia.

Dusky Shark, *Carcharhinus obscurus*, Wild Coast, Eastern Cape, South Africa.

CARIBBEAN REEF SHARK *Carcharhinus perezi*

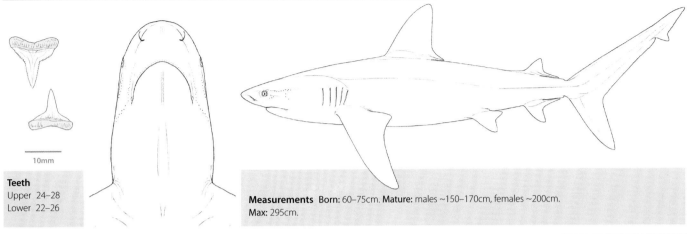

Teeth
Upper 24–28
Lower 22–26

10mm

Measurements Born: 60–75cm. **Mature:** males ~150–170cm, females ~200cm.
Max: 295cm.

Identification Large reef shark. Short, bluntly rounded snout. Narrow teeth. Small first dorsal and moderately large second dorsal with short free rear tips. Interdorsal ridge present. Large, narrow pectoral fins. Dark grey or grey-brown above, with inconspicuous white band on flanks. Undersides of paired fins, anal and lower caudal lobe dusky but not prominently marked. White below.
Distribution West Atlantic: North Carolina (USA), throughout Caribbean and south to Brazil.
Habitat Commonest Caribbean coral reef shark, near bottom and near drop-offs on outer reef to at least 378m. On hard bottom (including calcareous algae) and on mud off river deltas in Brazil.
Behaviour Poorly studied. Can lie motionless on the bottom, pumping water over gills with its pharynx. Can be closely approached while lying 'sleeping' in caves or in the open.
Biology Viviparous, yolk-sac placenta, 3–6 pups per litter born after ~1 year gestation every 2 years. There may be a pupping ground off the north coast of Brazil. Eats bony fishes.
Status IUCN Red List: **Near Threatened** (out of date). Common and quite heavily fished for human consumption, hides, oil and fins but far more valuable for dive tourism. Often encountered by divers without incident; only rarely bites people (some during feeding to attract sharks to divers). Kept in a few large aquaria, where it has given birth.

Caribbean Reef Shark, *Carcharhinus perezi*, Bahamas.

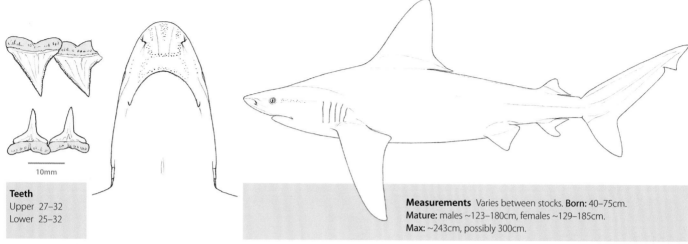

Teeth
Upper 27–32
Lower 25–32

10mm

Measurements Varies between stocks. **Born:** 40–75cm.
Mature: males ~123–180cm, females ~129–185cm.
Max: ~243cm, possibly 300cm.

Identification Stout-bodied. Moderately long, rounded snout. High, triangular, saw-edged upper teeth. Very large, erect first dorsal fin with origin over/slightly ahead of large pectoral fins insert. Interdorsal ridge present. Grey-brown or bronzy above, with inconspicuous white band on flanks. Tips and posterior edges of fins often dusky, but no obvious markings. Underside white.
Distribution Widespread in tropical and warm-temperate waters worldwide.
Habitat Common in bays, harbours and at river mouths, also offshore in adjacent deep water and on oceanic banks. Usually near bottom, 20–55m, ranging 1–280m.
Behaviour The Sandbar Shark feeds slightly more actively at night. Some stocks migrate seasonally, often in large schools, as water temperatures change. Young form mixed-sex schools on shallow coastal nursery grounds, then move into deeper, warmer water in winter. Adults are segregated from juveniles (reducing cannibalism) and the sexes are also usually separate, except when mating in spring and summer. Bite marks found on females during the mating season are caused by males, who follow and bite the female in the back until they swim upside down, then mate with both claspers. Multiple paternity is found in Sandbar Shark litters.

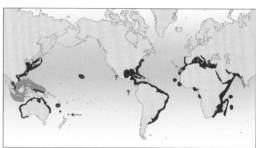

Biology A large, slow-growing, late maturing and low-fecundity coastal shark. Viviparous, with yolk-sac placenta. Litter size and gestation varies by region, 1–14 pups per litter (5–12 common) increasing with size of female, after a gestation of 8–12 months. Females give birth every 2nd or 3rd year. Age at maturity varies by region, but in general 8–14 years for males, and 7.5–16 years for females; maximum age is 19–25 years. The Sandbar Shark is one of the slowest growing, latest maturing of known sharks (making it very vulnerable to overfishing), but it may grow and mature more rapidly in captivity. Feeds mainly on small bottom fishes, also some molluscs and crustaceans.
Status IUCN Red List: **Vulnerable** (out of date). The Sandbar Shark is an important component of target coastal shark net and line fisheries and a bycatch in most areas where it occurs, but catches are largely unrecorded. It has been severely overfished in the northwest Atlantic, where meat and fins are extremely valuable. Declines are reported from shark nets off South African beaches. Sandbar Shark fins are large and highly prized by Hong Kong fin traders. This species formerly comprised at least 2–3% of fins in international trade, but more recently less than 0.25%; this decline is likely due to seriously depleted stocks and declining catches – few range states have improved their management of this species. This is a hardy and spectacular species in public aquaria and reproduces in captivity.

Sandbar Shark, *Carcharhinus plumbeus*, Hawaii.

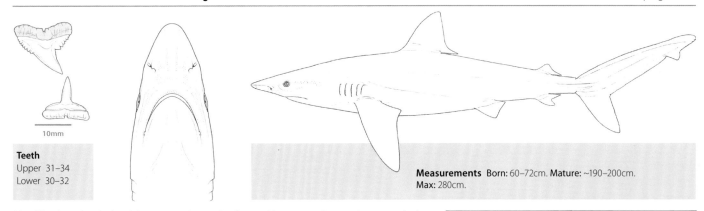

Teeth
Upper 31–34
Lower 30–32

10mm

Measurements Born: 60–72cm. **Mature:** ~190–200cm.
Max: 280cm.

Identification Slim shark with long, pointed snout. Small jaws; oblique-cusped, serrated upper teeth; short, inconspicuous upper labial furrows. Large eyes. Front of first dorsal fin over pectoral fins; both dorsals low with elongated free rear tips. Interdorsal ridge. Small pectoral fins. Grey-brown above, white below. No conspicuous fin markings.

Distribution Atlantic: tropical and warm-temperate waters. USA to Argentina in west; West Africa, from Senegal to Namibia, in east.

Habitat Deep water coastal and semi-oceanic, on or along outer continental and insular shelves and along upper slopes. Prefers 50–100m, ranges 0–600m.

Behaviour Active schooling shark, migrating vertically into shallower water at night. Possibly seasonal geographic migrations.

Biology Viviparous, yolk-sac placenta, 4–18 (usually 12–18) pups per litter. Age at maturity is ~8 years for males, 10 years for females; maximum age is at least 17 years, possibly 31 years. Feeds on small, active bony fishes, squid, and shrimp.

Status IUCN Red List: **Vulnerable** (out of date). Formerly very common in Caribbean fisheries, now apparently rare. Harmless to people.

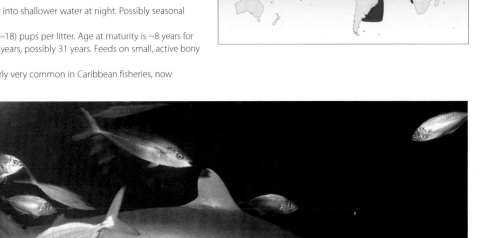

Night Shark,
Carcharhinus signatus,
Praia do Forte, Brazil.

BLACKSPOT SHARK *Carcharhinus sealei* FAO code: **CCI** Plate page 556

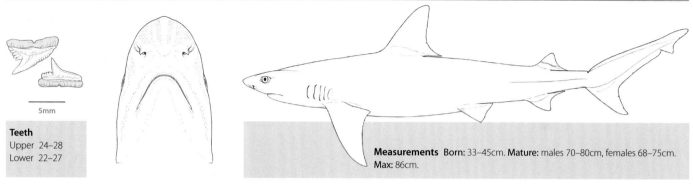

5mm

Teeth
Upper 24–28
Lower 22–27

Measurements Born: 33–45cm. **Mature:** males 70–80cm, females 68–75cm.
Max: 86cm.

Identification Small, slender shark. Long, rounded snout. Oblique-cusped teeth. Large, oval eyes. No or very low interdorsal ridge. Grey or tan above with inconspicuous light stripes on flanks, lighter below. Conspicuous black or dusky tip on second dorsal fin; other fins with pale trailing edges, no dark marks.
Distribution Central Indo-Pacific. Former records from the west Indian Ocean are now known to be Human's Whaler Shark *Carcharhinus humani*.
Habitat Coastal, continental and insular shelves, from surf line and intertidal to 40m. Not off river mouths.
Behaviour Unknown.
Biology Viviparous, yolk-sac placenta, 1–2 pups per litter born in spring after ~9 month gestation. Fast-growing, matures at ~1 year, maximum age at least 5 years. Eats small fishes, squid and prawns.
Status IUCN Red List: **Near Threatened** (out of date). Occurs in intensively fished coastal areas, declining in some areas. Life history information on this species needs to be re-examined, since some of it was based on Human's Whaler Shark *C. humani*.

SPOTTAIL SHARK *Carcharhinus sorrah* FAO code: **CCQ** Plate page 554

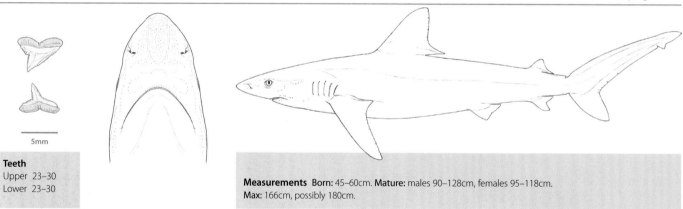

5mm

Teeth
Upper 23–30
Lower 23–30

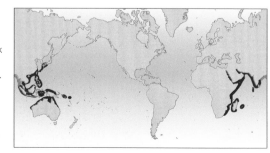

Measurements Born: 45–60cm. **Mature:** males 90–128cm, females 95–118cm.
Max: 166cm, possibly 180cm.

Identification Small, spindle-shaped shark. Long, rounded snout. Long, narrow, nipple-like anterior nasal flaps. Oblique-cusped, serrated teeth; short, inconspicuous labial furrows. Large, circular eyes. Moderately short gills. Interdorsal ridge present. Very low, elongated second dorsal fin and lower caudal lobe. Medium-grey above with conspicuous white band on flank. First dorsal plain or black edged; large, conspicuous black tips on pectoral, second dorsal and lower caudal fins. White below.
Distribution West Pacific and Indian Ocean: South Africa to China, northern Australia and Solomon Islands.
Habitat Continental and insular shelves, shallow water and around coral reefs, usually 20–50m, ranging 0–140m. **Behaviour** Young occur in quiet, shallow water, segregated from adults.
Biology Viviparous, yolk-sac placenta, 1–8 pups (average 3–6) per litter, after 10 month gestation. Age at maturity is ~2–3 years with a maximum age of 5–7 years. Feeds on bony fishes and octopus.
Status IUCN Red List: **Near Threatened**.(out of date). Fished heavily in part of range and fairly common in fin trade, but fisheries not monitored. **Vulnerable** in the north Indian Ocean due to intensive fisheries there.

AUSTRALIAN BLACKTIP SHARK *Carcharhinus tilstoni*

FAO code: **CCU**

Plate page 520

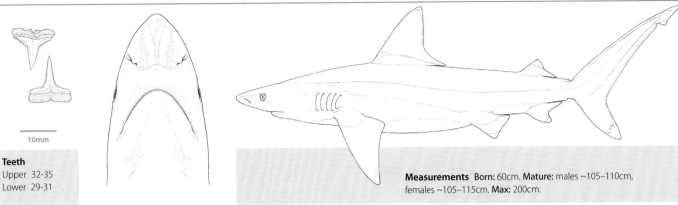

10mm

Teeth
Upper 32–35
Lower 29–31

Measurements Born: 60cm. **Mature:** males ~105–110cm, females ~105–115cm. **Max:** 200cm.

Identification Medium-sized shark. Long snout. Slender, erect, serrated teeth. First dorsal origin about over pectoral fin insertions. No interdorsal ridge. Distinguished from *Carcharhinus limbatus* by vertebral counts. Bronzy to grey above, with pale flank stripe. Black-tipped fins (pelvic and anal fins may be plain). Pale below.
Distribution Central Indo-Pacific: tropical Australia.
Habitat Continental shelf waters, close inshore to ~150m, midwater or near surface.
Behaviour Often in large aggregations.
Biology Viviparous, 1–6 pups (average 3) per litter born in January after ~10 month gestation. Matures in 3–4 years, with a maximum age of 8–15 years. Feeds on bony fishes and cephalopods.
Status IUCN Red List: Least Concern. Fast growth rates, early maturity and relatively high fecundity make it fairly resilient to fisheries. Catches have been reduced under management.

INDONESIAN WHALER SHARK *Carcharhinus tjutjot*

FAO code: **CWZ**

Plate page 554

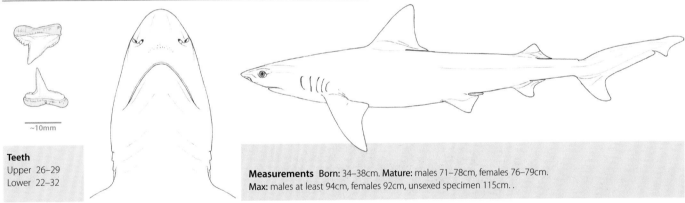

~10mm

Teeth
Upper 26–29
Lower 22–32

Measurements Born: 34–38cm. **Mature:** males 71–78cm, females 76–79cm.
Max: males at least 94cm, females 92cm, unsexed specimen 115cm. .

Identification Snout moderately long, rounded. Small, widely spaced nostrils. Short upper labial furrows. Fairly large, horizontally oval eyes. Low dorsal fins, short free rear tips; not falcate in adults. Moderate to strong interdorsal ridge present. Falcate pectorals. Greyish to pale brown shark; demarcation between darker above and lighter below from snout to tail. Black blotch (white below) on second dorsal; other fins unmarked with paler margins.
Distribution West Pacific and Central Indo-Pacific: Indonesia to Taiwan; previously misidentified as the Whitecheek Shark *Carcharhinus dussumieri*.
Habitat Inshore, usually shallower than 100m, but otherwise unknown. **Behaviour** Poorly known.
Biology Viviparous, usually with 2 pups (rarely 4). No defined reproductive season; females pup year-round. Feeds on bony fishes and, to a lesser extent, cephalopods and crustaceans.
Status IUCN Red List: Vulnerable. Common as bycatch landed into fish markets in Indonesia and Taiwan. Restricted shallow inshore habitat and small litter size suggests that rising fishing pressure may be a threat.

Plate 72 CARCHARHINIDAE IX – *Glyphis* sharks

○ **Ganges Shark** *Glyphis gangeticus*

West Pacific and Indian Ocean; 0–50m. Large, stocky; short broadly rounded snout, tiny eyes; no interdorsal ridge, longitudinal upper precaudal pit, first dorsal fin origin over posterior third of pectoral fin base, second dorsal about half height of first, deeply notched anal fin; grey above with no conspicuous markings, white below.

○ **New Guinea River Shark** *Glyphis garricki*

Central Indo-Pacific; brackish fresh water in rivers and adjacent marine waters. Large, slender; flat-headed, short broadly rounded snout, tiny eyes; no interdorsal ridge, longitudinal upper precaudal pit, first dorsal fin origin over posterior third of pectoral fin base, second dorsal large about two-thirds the of height first, anal fin with deep notch; grey above with dusky fin margins, white below.

○ **Speartooth Shark** *Glyphis glyphis*

Central Indo-Pacific; freshwater rivers, estuaries and adjacent coastal waters. Large, stocky; short broadly rounded snout, tiny eyes; no interdorsal ridge, longitudinal upper precaudal pit, first dorsal fin origin over rear pectoral fin base, second dorsal large about three-fifths height of first, anal fin with deep notch; grey-brown above with no conspicuous markings, white below.

○ **Daggernose Shark** *Isogomphodon oxyrhynchus*

West Atlantic; 4–40m. Medium-sized; unmistakable extremely long flat pointed snout, tiny circular eyes, short prominent labial furrows; no interdorsal ridge, large paddle-shaped pectoral fins, first dorsal fin origin over pectoral fins, second dorsal fin about half the size of first, notched anal fin; uniform grey or yellow-grey above, light below.

○ *G. gangeticus*   ○ *G. garricki*   ○ *G. glyphis*   ○ *I. oxyrhynchus*

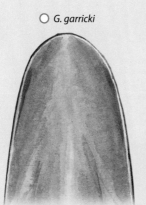

Glyphis gangeticus

Glyphis garricki

Glyphis glyphis

juvenile

50cm

Isogomphodon oxyrhynchus

River sharks

Identification of *Glyphis* species

The Indo-West Pacific river sharks (genus *Glyphis*) are rare, poorly known and difficult to identify. Specimens and good photographs of whole animals, measurements, jaws and teeth and, ideally, vertebral counts (see table) are essential for confirmation. Only three species in the genus are valid; previously five species had been named, but two have been synonymised with *Glyphis gangeticus* based on new molecular and morphological data. Unidentified *Glyphis* sharks are also reported from rivers in Malaysian Borneo and central Kalimantan, these may be another species or hitherto unknown populations of *Glyphis gangeticus*.

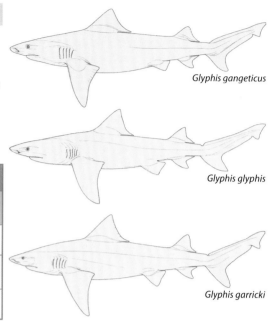

Glyphis gangeticus

Glyphis glyphis

Glyphis garricki

SPECIES	TOOTH ROWS	VERTEBRAL COUNTS			
		total	precaudal	monospondylous precaudal	caudal
Glyphis gangeticus	30–37/31–34	169	80	50	89
Glyphis glyphis	probably 26–29/27–29 (Papua New Guinea)	213–222		69–73	90 (type specimen)
Glyphis garricki	32–34/32–34	137–151	79–83	44–50	65–68

GANGES SHARK *Glyphis gangeticus*   FAO code: **CGA**   Plate page 546

Teeth
Upper 30–37
Lower 31–34

20mm

Measurements Born: 56–61cm. Mature: males ~178cm.
Max: at least 275cm, possibly larger.

Identification Large, stocky shark. Short, broadly rounded snout. Upper teeth with high, broad, serrated, triangular cusps; first few lower front teeth with weakly serrated cutting edges and low cusplets on crown foot. Tiny eyes. First dorsal origin over rear third of pectoral bases; second dorsal about half height of first. No interdorsal ridge. Anal fin with deeply notched posterior margin. Longitudinal upper precaudal pit. Grey above, white below, without conspicuous markings.
Distribution West Pacific and north Indian Ocean: Indus River outside Karachi, Pakistan, to Bangladesh, Myanmar, Thailand and Borneo.
Habitat Fresh water in rivers, possibly estuaries and seasonally brackish inshore marine waters to ~50m.
Behaviour Unknown.
Biology Viviparous, but nothing else known.
Status IUCN Red List: Critically Endangered. Habitat heavily fished and degraded. Originally known from three 19th century museum specimens, but several recent records have confirmed and extended its range. Man-eating reputation may be confused with that of *Carcharhinus leucas*.

NEW GUINEA RIVER SHARK *Glyphis garricki*

FAO code: **RSK**   Plate page 546

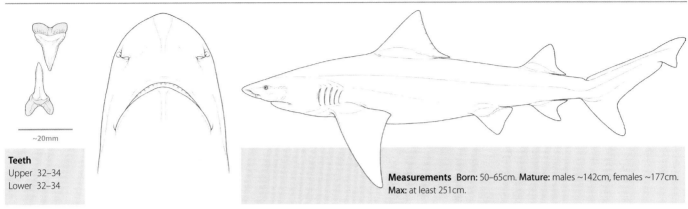

Teeth
Upper 32–34
Lower 32–34

~20mm

Measurements Born: 50–65cm. **Mature:** males ~142cm, females ~177cm. **Max:** at least 251cm.

Identification Large, slender, rather flat-headed shark. Short, broadly rounded snout. Upper teeth with high, broad, serrated, triangular cusps; first few lower front teeth with long, hooked, protruding cusps with serrated cutting edges confined to spear-like tips, and no cusplets. Tiny eyes. First dorsal origin over rear third of pectoral bases; second dorsal about two-thirds height of first. No interdorsal ridge. Anal fin with deeply notched posterior margin. Longitudinal upper precaudal pit. Grey above, without conspicuous body markings. Dusky fin margins. White below.
Distribution Central Indo-Pacific: northern Australia, Papua New Guinea.
Habitat Turbid and brackish fresh water in rivers and adjacent coastal marine waters.
Behaviour Unknown. **Biology** Viviparous, with litter size of at least 9; birth may occur in the spring. Feeds on other elasmobranchs, especially stingrays and bony fishes.
Status IUCN Red List: Critically Endangered. Very rare, population extremely small and fragmented. Captured in artisanal fisheries. Surveys using environmental DNA have produced new records.

SPEARTOOTH SHARK *Glyphis glyphis*

FAO code: **CGG**   Plate page 546

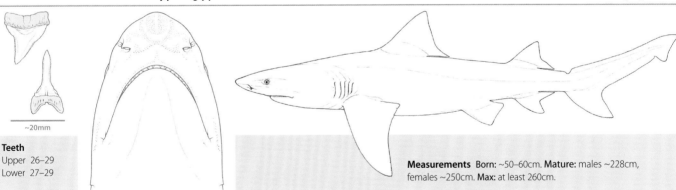

Teeth
Upper 26–29
Lower 27–29

~20mm

Measurements Born: ~50–60cm. **Mature:** males ~228cm, females ~250cm. **Max:** at least 260cm.

Identification Large, stocky shark. Short, broadly rounded snout. Upper teeth with high, broad, serrated, triangular cusps; first few lower front teeth with long, hooked, protruding cusps, with unserrated or serrated cutting edges confined to spear-like tips, no cusplets. Tiny eyes. First dorsal origin over rear ends of pectoral bases; second dorsal rather large (about three-fifths height of first). No interdorsal ridge. Anal fin with deeply notched posterior margin. Longitudinal upper precaudal pit. Grey above, white below, without conspicuous markings.
Distribution Central Indo-Pacific: Australia – Ord River (western Australia) to Wenlock River (Queensland) – and southern Papua New Guinea. Records outside this area need confirmation, as they may be based on other *Glyphis* species. **Habitat** Occurs in turbid freshwater rivers, estuaries and adjacent coastal waters.
Behaviour Unknown. **Biology** Viviparous, with litters of 6–7 pups. In northern Australia, newborns are common in some rivers from October to December. Feeds on bony fishes and crustaceans.
Status IUCN Red List: Endangered. Known for over a century from a single museum specimen, it appears to be restricted to river systems in northern Australia and Papua New Guinea (the latter habitats face significant development and exploitation pressure). Extremely rare.

Plate 73 CARCHARHINIDAE X

○ **Pigeye Shark** *Carcharhinus amboinensis*

West Pacific, Indian Ocean and east Atlantic; 0–100m. Large; massive thick head, very short broad blunt snout, small eyes, short labial furrows; no interdorsal ridge, large angular pectoral fins, high erect triangular first dorsal fin, second dorsal fin less than a third the size of first, both with short free rear tips; grey above, white below, fin tips dusky not conspicuous marked.

○ **Pacific Smalltail Shark** *Carcharhinus cerdale*

East Pacific; 0–40m. Very similar to *Carcharhinus porosus*. Small; long pointed snout, large circular eyes, short labial furrows; no interdorsal ridge, large falcate first dorsal fin, small second dorsal fin with origin over anal fin midbase, small pectoral fins, deeply notched anal fin; grey above, light below, inconspicuous white flank band, pectoral dorsal and caudal fin tips frequently dusky or blackish.

○ **Whitecheek Shark** *Carcharhinus dussumieri*

North Indian Ocean; 0–100m. Small; moderately long rounded snout, fairly large oval eyes, inconspicuous upper labial furrows; triangular first dorsal, short free rear tips to dorsal fins, interdorsal ridge usually absent, semifalcate pectoral fins; grey to grey-brown, inconspicuous light flank stripe, white below, black or dusky spot on second dorsal, pale posterior margins to other fins. Often confused with *Carcharhinus tjutjot*.

○ **Pondicherry Shark** *Carcharhinus hemiodon*

Indian and Pacific Oceans; 10–150m. Moderately long pointed snout, fairly large eyes, short inconspicuous upper labial furrows; interdorsal ridge, small pectoral fins, fairly large first dorsal fin, short free rear tips to both dorsals; grey above with black tips to pectorals second dorsal and caudal fins, conspicuous white flank stripe, white below.

○ **Human's Whaler Shark** *Carcharhinus humani*

West Indian Ocean; 0–40m. Small, slender; long narrowly rounded snout; weak interdorsal ridge, moderately high first dorsal fin originates just before pectoral fin free rear tip, second dorsal less than half height of firs; pale brownish grey body, whitish below, black blotch on upper part of second dorsal not extending onto body, most other fins whitish on outer margins.

○ **Lost Shark** *Carcharhinus obsoletus*

Central Indo-Pacific. Small, slender; short rounded snout; no interdorsal ridge, fairly small first dorsal fin originating just behind the pectoral fin base, much smaller low second dorsal fin, origin about opposite midbase of anal fin; preserved specimens pale grey, lighter along flanks and below; fins unmarked.

○ **Smalltail Shark** *Carcharhinus porosus*

West Atlantic and east Pacific; 0–84m. Small; long pointed snout, large circular eyes, short labial furrows; no interdorsal ridge, small pectoral fins, large falcate first dorsal fin, small second dorsal with origin over anal fin midbase, deeply notched anal fin; grey above, light below, pectoral dorsal and caudal fin tips frequently dusky, not conspicuous marked, inconspicuous white flank band.

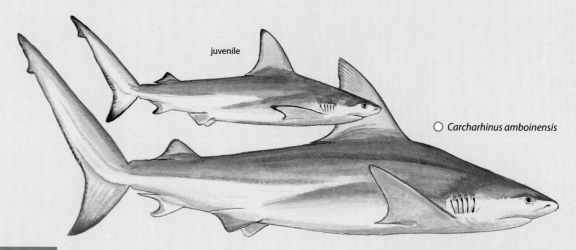

juvenile

○ *Carcharhinus amboinensis*

50cm

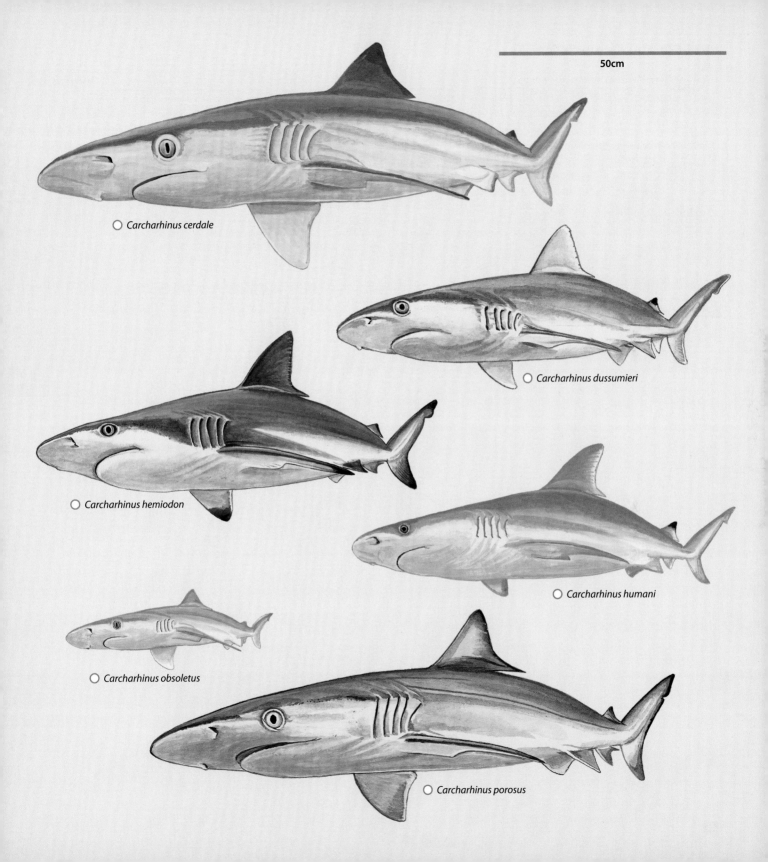

○ *Carcharhinus cerdale*

○ *Carcharhinus dussumieri*

○ *Carcharhinus hemiodon*

○ *Carcharhinus humani*

○ *Carcharhinus obsoletus*

○ *Carcharhinus porosus*

50cm

Plate 74 CARCHARHINIDAE XI

○ **Milk Shark** *Rhizoprionodon acutus*   page 563

East Atlantic, Mediterranean and Indo-West Pacific; 1–200m. Long narrow snout, small widely spaced nostrils, large eyes, only requiem shark in range with long upper and lower labial furrows; interdorsal ridge present, small pectoral fins, first dorsal fin origin well behind pectoral fin origin, small low second dorsal behind larger anal fin; bronze to grey, most fin tips slightly pale, dark dorsal and terminal caudal fin tips in juveniles and occasionally adults, white below.

○ **Pacific Sharpnose Shark** *Rhizoprionodon longurio*   page 564

East Pacific; 0–27m. Long snout, small widely spaced nostrils, large eyes, long upper and lower labial furrows; interdorsal ridge present, small pectoral fins, first dorsal fin origin usually over or slightly in front pectoral fin free rear tips, second dorsal origin well behind larger anal fin origin, long preanal ridge; grey or grey-brown, light-edged pectoral fins, dorsal fin tips dusky, white below.

○ **Caribbean Sharpnose Shark** *Rhizoprionodon porosus*   page 565

West Atlantic; 1–500m. Small; long snout, fairly large eyes, long labial furrows; no interdorsal ridge, first dorsal fin origin usually over or slightly behind pectoral fin free rear tips, small second dorsal fin origin over anal fin midbase; brown or grey-brown above, white below, white posterior pectoral fin margins, dorsal and caudal fin posterior margins blackish; occasionally white spots along sides.

○ **Atlantic Sharpnose Shark** *Rhizoprionodon terraenovae*   page 566

Northwest Atlantic; 0–280m. Similar to *Rhizoprionodon porosus*; small; long snout, fairly large eyes, long upper labial furrows; no interdorsal ridge, first dorsal fin origin usually over or slightly in front of pectoral fin free rear tips, second dorsal fin small origin over anal fin midbase inserts; grey to grey-brown above, white below, with small white spots on sides in larger specimens, pectoral fins with white margins, dorsal fin tips dusky.

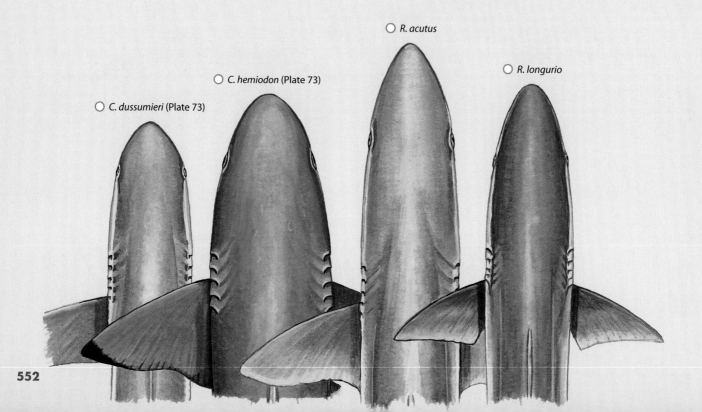

○ *R. acutus*

○ *R. longurio*

○ *C. hemiodon* (Plate 73)

○ *C. dussumieri* (Plate 73)

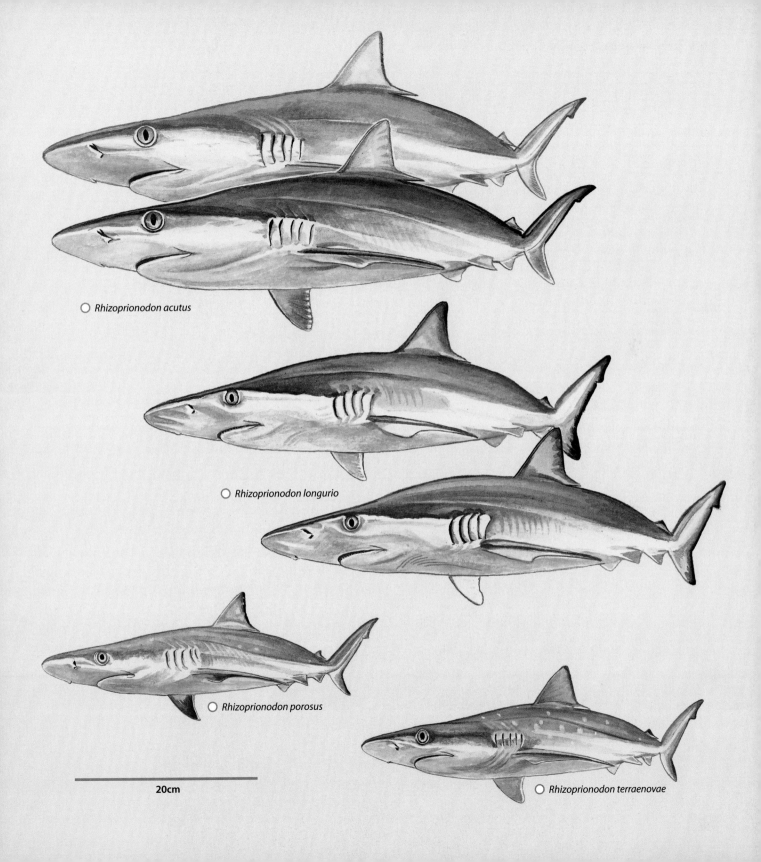

Rhizoprionodon acutus

Rhizoprionodon longurio

Rhizoprionodon porosus

20cm

Rhizoprionodon terraenovae

Plate 75 CARCHARHINIDAE XII

○ **Nervous Shark** *Carcharhinus cautus*

page 526

Central Indo-Pacific; 0–20m. Short bluntly rounded snout, horizontally oval eyes, nipple-like anterior nasal flaps, short labial furrows; no interdorsal ridge, second dorsal fin with short free rear tip; grey to light brown, conspicuous white flank stripe, white below, black dorsal and caudal fin margins, black-tipped pectoral fins.

○ **Coates's Shark** *Carcharhinus coatesi*

page 529

Central Indo-Pacific; 0–123m. Small; moderately long and narrowly rounded snout; first dorsal moderately high, weakly falcate, origin just behind pectoral fin free rear tip; interdorsal ridge; greyish brown above, white below, fins unmarked except for clearly defined black upper one-third of second dorsal fin. Only recently distinguished from *Carcharhinus sealei*.

○ **Indonesian Whaler Shark** *Carcharhinus tjutjot*

page 545

West Pacific and Central Indo-Pacific; <100m. Moderately long, rounded snout; small, widely spaced nostrils; short upper labial furrows; fairly large, horizontally oval eyes; low dorsal fins, short free rear tips; moderate to strong interdorsal ridge present; falcate pectorals; greyish to pale brown above with prominent black fin blotch on second dorsal fin; lighter below.

○ **Creek Whaler** *Carcharhinus fitzroyensis*

page 530

Central Indo-Pacific; 0–40m. Long parabolic snout, round eyes, lobate anterior nasal flaps, short labial furrow, short gills; no interdorsal ridge, broad triangular fins, first dorsal origin over pectoral fin free rear tips, second dorsal approximately over anal fin origin; grey-brown to bronze, lacks conspicuous markings, light below.

○ **Spottail Shark** *Carcharhinus sorrah*

page 544

West Pacific and Indian Ocean; 0–140m. Long rounded snout, large round eyes, long narrow nipple-like anterior nasal flaps, short inconspicuous labial furrow, moderately short gills; interdorsal ridge; very low long second dorsal fin; medium-grey body with conspicuous white flank stripe, white below, large conspicuous black tips on pectoral, second dorsal and lower caudal fins, first dorsal plain or black-edged.

○ **Broadfin Shark** *Lamiopsis temmincki*

page 558

North Indian Ocean; <100m. Moderately long snout nearly equal to mouth width, slightly heart-shaped upper teeth, smooth lower teeth, small round eyes, short broadly triangular anterior nasal flaps, short labial furrows, fifth gill slit about half the size the first; no interdorsal ridge, very broad triangular pectoral fins, anal fin posterior margin approximately straight; light grey or tan, lacks any conspicuous markings, light below.

○ **Borneo Broadfin Shark** *Lamiopsis tephrodes*

page 559

Central Indo-Pacific; <100m. Similar in appearance to *Lamiopsis temmincki* except that upper teeth are triangular and lower teeth have fine serrations.

○ **Whitenose Shark** *Nasolamia velox*

page 560

East Pacific; 15–192m. Very long conical snout, large round eyes, very large close-set nostrils, very short labial furrows; no interdorsal ridge, moderately broad triangular pectoral fins, first dorsal fin much larger than second, anal fin slightly larger than second dorsal; grey-brown to light brown, prominent black spot outlined with white on upper snout, light below.

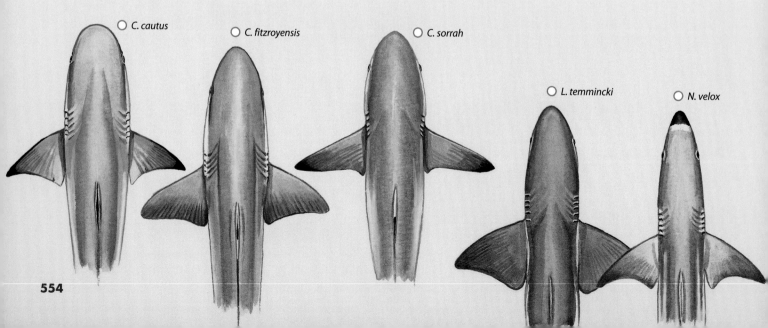

○ *C. cautus* ○ *C. fitzroyensis* ○ *C. sorrah* ○ *L. temmincki* ○ *N. velox*

○ *Carcharhinus cautus*

○ *Carcharhinus coatesi*

○ *Carcharhinus fitzroyensis*

○ *Carcharhinus sorrah*

○ *Carcharhinus tjutjot*

○ *Lamiopsis tephrodes*

○ *Lamiopsis temmincki*

50cm

○ *Nasolamia velox*

Plate 76 CARCHARHINIDAE XIII

○ **Borneo Shark** *Carcharhinus borneensis*

Central Indo-Pacific; coastal inshore. Long pointed snout, large eyes; no interdorsal ridge, small pectoral fins, small dorsal fins with short free rear tips; brown with dark markings on second dorsal and upper caudal fin tips, pectoral pelvic and anal fins with light margins, black spot beneath snout tip.

○ **Hardnose Shark** *Carcharhinus macloti*

West Pacific and Indian Ocean; 0–200m. Long pointed hard snout, moderately large eyes; no interdorsal ridge, small dorsal fins, second dorsal very low with very long free rear tip; grey to grey-brown, fins with light margins, inconspicuous light flank stripe.

○ **Blackspot Shark** *Carcharhinus sealei*

Central-Indo Pacific; 0–40m. Small, slender; long rounded snout, large oval eyes; no or low interdorsal ridge, small pectoral and first dorsal fins; grey or tan with conspicuous dusky to black tip on second dorsal fin, other fins with light margins, light flank stripe.

○ **Sliteye Shark** *Loxodon macrorhinus*

West Pacific and Indian Ocean; 7–120m. Long narrow snout, large eyes with posterior notch; no to rudimentary interdorsal ridge, first dorsal fin much larger than second, which is behind the anal fin and has a long free rear tip; grey to brown, black margin on first dorsal and caudal fins, other fins with light margins.

○ **Brazilian Sharpnose Shark** *Rhizoprionodon lalandii*

West Atlantic; 0–70m. Long snout, large eyes, widely spaced nostrils, long upper and lower labial furrows; small second dorsal fin with origin far behind anal origin; dark grey to grey-brown, light below, light margins to pectoral fins and dusky dorsal fins.

○ **Grey Sharpnose Shark** *Rhizoprionodon oligolinx*

Tropical west Pacific and Indian Ocean; 0–36m. Very small shark; long snout, large eyes, no spiracles, small widely spaced nostrils, short labial furrows; small second dorsal fin with origin far behind anal fin origin, long anal ridge; grey to bronzy, inconspicuous dusky fin margins.

○ **Australian Sharpnose Shark** *Rhizoprionodon taylori*

Central Indo-Pacific; 0–110m. Very similar to *Rhizoprionodon oligolinx*; dorsal and caudal fins with dark margins, terminal caudal fins dark, other fins with light margins.

○ **Spadenose Shark** *Scoliodon laticaudus*

West and north Indian Ocean; 10–75m. Unmistakable very long flattened spade-like snout, small eyes; no interdorsal ridge, short broad triangular pectoral fins, first dorsal fin much larger than second; bronzy-grey with no conspicuous markings.

○ **Pacific Spadenose Shark** *Scoliodon macrorhynchos*

West Pacific Ocean; coastal. Appearance very similar to *S. laticaudus*, main difference is a longer second dorsal fin to anal fin length.

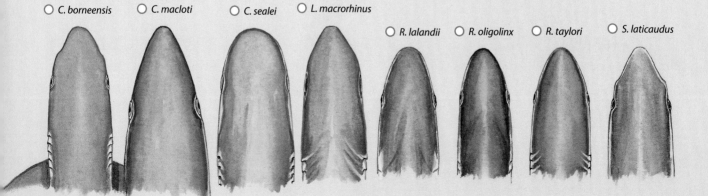

○ *C. borneensis*   ○ *C. macloti*   ○ *C. sealei*   ○ *L. macrorhinus*   ○ *R. lalandii*   ○ *R. oligolinx*   ○ *R. taylori*   ○ *S. laticaudus*

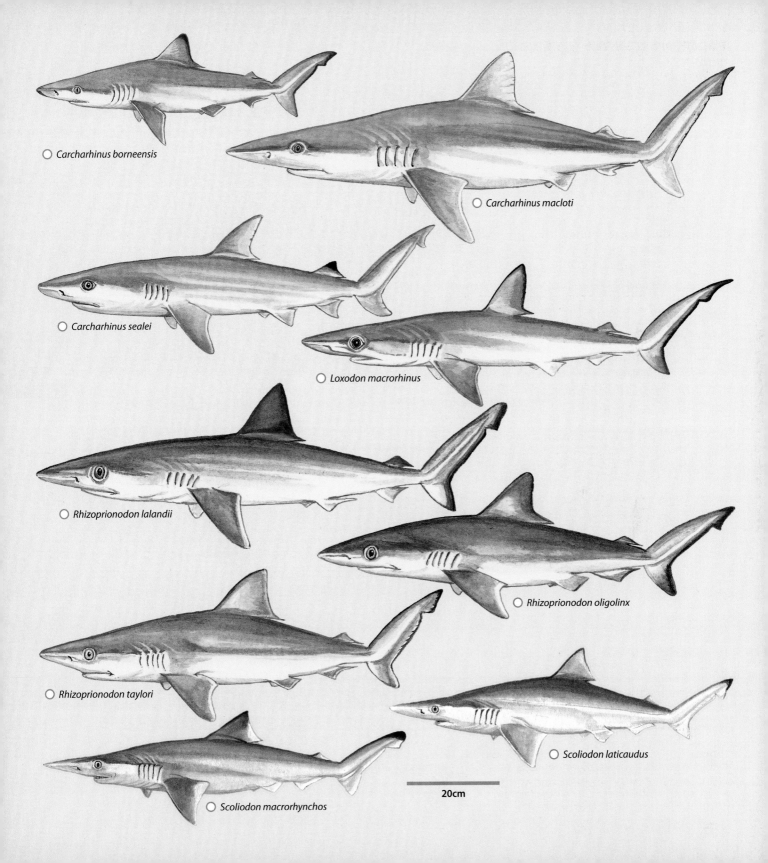

○ *Carcharhinus borneensis*

○ *Carcharhinus macloti*

○ *Carcharhinus sealei*

○ *Loxodon macrorhinus*

○ *Rhizoprionodon lalandii*

○ *Rhizoprionodon oligolinx*

○ *Rhizoprionodon taylori*

○ *Scoliodon laticaudus*

○ *Scoliodon macrorhynchos*

20cm

DAGGERNOSE SHARK *Isogomphodon oxyrhynchus*

FAO code: **CIO**   Plate page 546

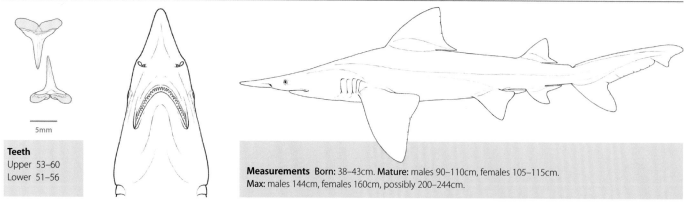

Teeth
Upper 53–60
Lower 51–56

5mm

Measurements Born: 38–43cm. **Mature:** males 90–110cm, females 105–115cm.
Max: males 144cm, females 160cm, possibly 200–244cm.

Identification Unmistakable, extremely long, flat, pointed snout. Small, spike-like teeth in both jaws; upper teeth serrated; short, prominent labial furrows. Tiny, circular eyes. First dorsal fin origin over large paddle-shaped pectoral fins. No interdorsal ridge. Second dorsal fin about half the size of first. Notched anal fin. Unpatterned grey or yellow-grey shark, light below.
Distribution West Atlantic: South America, from Venezuela, Trinidad and Tobago to Brazil's Amazon coast.
Habitat Turbid water in estuaries, mangroves, river mouths and over shallow banks, 4–40m.
Behaviour Migrates inshore in dry season, offshore in rainy season (apparently intolerant of reduced salinities). Females found in deeper waters than males. Long snout and small eyes may be adaptations for life in turbid water.   **Biology** Viviparous, yolk-sac placenta, 2–8 pups per litter born at start of rainy season after ~1 year gestation. Possible 2 year birth cycle. Maturity is between 5–7 years with a maximum age of 7–12 years. Feeds on small schooling fishes.   **Status** IUCN Red List: Critically Endangered. Populations declining steeply due to bycatch in fisheries and habitat degradation.

BROADFIN SHARK *Lamiopsis temmincki*

FAO code: **LMT**   Plate page 554

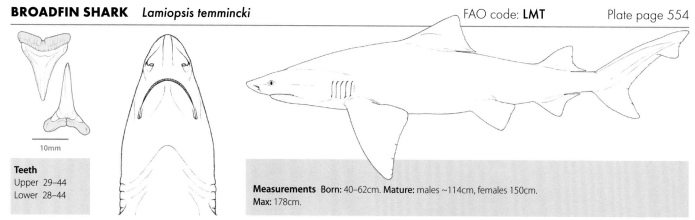

Teeth
Upper 29–44
Lower 28–44

10mm

Measurements Born: 40–62cm. **Mature:** males ~114cm, females 150cm.
Max: 178cm.

Identification Small, rather stocky shark. Moderately long snout, length nearly equal to mouth width. Short, broadly triangular anterior nasal flaps. Upper teeth serrated with broad, triangular cusps; lower teeth smooth, hooked, narrow-cusped; short labial furrows. Small, round eyes. Fifth gill about half height of first. Second dorsal fin nearly as large as first. No interdorsal ridge. Very broad, triangular pectoral fins. Anal fin posterior margin nearly straight. Longitudinal upper precaudal pit. Light grey or tan above, light below. No prominent markings.
Distribution Northern Indian Ocean: scattered from Pakistan to India, occurrence in the Bay of Bengal requires confirmation.
Habitat Inshore continental shelf, less than 100m deep.   **Behaviour** Unknown.
Biology Viviparous, 4–8 pups per litter (usually 6) born before the monsoon. Possibly 8-month gestation. Feeding habits may include small bony fishes and invertebrates.
Status IUCN Red List: Endangered (out of date). Rare. Taken in fisheries, trends unknown. West Pacific records are of the closely related *Lamiopsis tephrodes*.

BORNEO BROADFIN SHARK *Lamiopsis tephrodes*   FAO code: **RSK**   Plate page 554

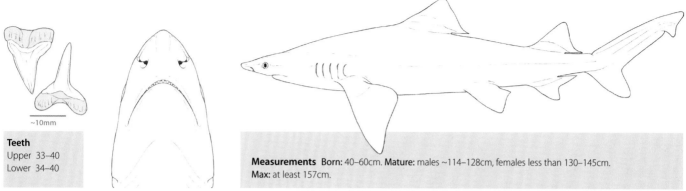

Teeth
Upper 33–40
Lower 34–40

~10mm

Measurements Born: 40–60cm. **Mature**: males ~114–128cm, females less than 130–145cm.
Max: at least 157cm.

Identification Small, rather stocky shark. Very similar to *Lamiopsis temmincki* but has triangular upper teeth and finely serrated lower teeth (*L. temmincki* has more heart-shaped upper teeth and smooth-edged lower teeth). Slate grey above, light below. No prominent markings.
Distribution Central Indo-Pacific: Indonesia, southeast Asia, and possibly off southern China. Exact distribution requires confirmation, due to previous misidentification with *L. temmincki*.
Habitat Inshore continental shelf, <100m deep.
Behaviour Unknown.
Biology Viviparous, with 4–8 pups per litter (usually 8), but little else known due to misidentification with *L. temmincki*. Possibly feeds on small bony fishes and invertebrates.
Status IUCN Red List: **Not Evaluated**. Taken in fisheries, trends unknown.

SLITEYE SHARK *Loxodon macrorhinus*   FAO code: **CLD**   Plate page 556

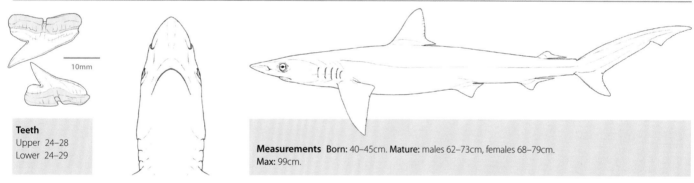

Teeth
Upper 24–28
Lower 24–29

10mm

Measurements Born: 40–45cm. **Mature**: males 62–73cm, females 68–79cm.
Max: 99cm.

Identification Small, very slim shark. Long, narrow snout. Small, smooth-edged, oblique-cusped teeth; short labial furrows. Big eyes with posterior notches. Small, low second dorsal fin with very long free rear tip, behind larger anal fin. Interdorsal ridge absent or rudimentary. Grey to brownish above. Fins with inconspicuous light rear edges; black margin on first dorsal and caudal fins. White below.
Distribution West Pacific and Indian Ocean.
Habitat Continental and insular shelves in shallow, clear water, 7–120m.
Behaviour Unreported.
Biology Viviparous, yolk-sac placenta, 2–4 pups per litter. Age at maturity is 1.4 years for males and 1.9 years for females. Eats small bony fishes, shrimp and cuttlefish.
Status IUCN Red List: **Least Concern** (out of date). Commonly caught in fisheries. A fast-growing coastal shark able to sustain reasonable fishing pressure.

WHITENOSE SHARK   *Nasolamia velox*

FAO code: **CNX**   Plate page 554

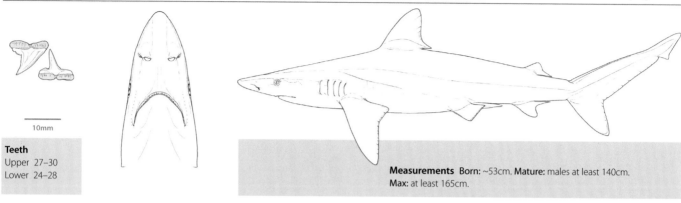

Teeth
Upper 27–30
Lower 24–28

Measurements Born: ~53cm. **Mature:** males at least 140cm.
Max: at least 165cm.

Identification Slender shark. Very long, conical snout. Very large, close-set nostrils separated by a space only slightly greater than the nostril width. Very short labial furrows. Moderately large, round eyes. First dorsal fin much larger than second. No interdorsal ridge. Moderately broad, triangular pectoral fins. Anal fin slightly larger than second dorsal. Grey-brown to light brown above, light below. A prominent black spot outlined with white on upper snout tip.
Distribution East Pacific: Southern California, USA, to Peru.
Habitat Continental shelf, inshore and offshore, usually 15–24m or less, occasionally to 192m.
Behaviour Unknown.
Biology Viviparous, yolk-sac placenta, 5 pups per litter. Feeds on small bony fishes, cephalopods and crabs.
Status IUCN Red List: Endangered. Taken in gillnet, longline and trawl fisheries for food and fishmeal, but is now extremely rare across much of its former range.

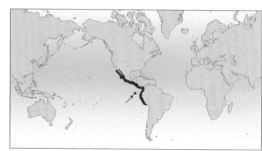

SICKLEFIN LEMON SHARK   *Negaprion acutidens*

FAO code: **NGA**   Plate page 512

Teeth
Upper 29–32
Lower 28–30

Measurements Born: 45–80cm. **Mature:** both sexes ~220–240cm.
Max: 310cm.

Identification Big, stocky shark. Broad, blunt snout. Narrow, smooth-cusped teeth in both jaws. Fairly small eyes. Moderately large gills. Second dorsal fin about as large as first. No interdorsal ridge. Yellowish above, white below. Very similar to *Negaprion brevirostris*, but dorsal, pectoral and anal fins usually more falcate.
Distribution Tropical Indian and Pacific Oceans. Widespread.
Habitat Inshore, on or near bottom, 0–90m. Prefers turbid, still water in bays, estuaries, sandy plateaus, outer reef shelves and reef lagoons. Young sharks are found on very shallow reef flats, dorsal fins exposed.
Behaviour Sluggish, swimming slowly near or resting on the bottom. May surface when stimulated by food, but shy and reluctant to approach divers. Young sharks are more inquisitive. This shark has a high site fidelity, with very little movement outside lagoons or atolls in the west Indian Ocean.
Biology Viviparous, 1–14 pups (average 9) per litter after ~10–11 month gestation. Appears to have a 2 year reproductive cycle. Feeds on bottom bony fishes and stingrays.
Status IUCN Red List: Vulnerable (out of date). Heavily fished throughout range outside Australia. Hardy in captivity. Valuable for dive tourism. Potentially harmful, can be aggressive when provoked.

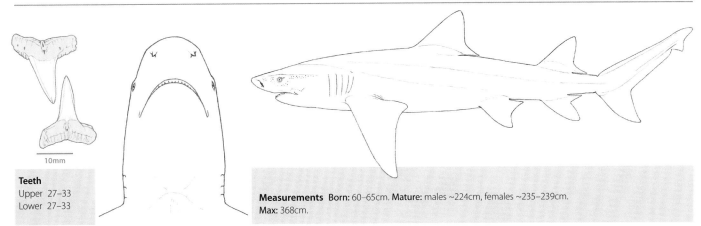

Teeth
Upper 27–33
Lower 27–33

10mm

Measurements Born: 60–65cm. **Mature:** males ~224cm, females ~235–239cm.
Max: 368cm.

Identification Big, stocky, short-nosed shark. Narrow, smooth-cusped teeth in both jaws. Dorsal, pectoral and pelvic fins weakly falcate; usually has less falcate fins than *Negaprion acutidens*. Second dorsal about as large as first. No interdorsal ridge. Pale yellow-brown above, light below. No conspicuous markings.
Distribution Atlantic: widespread in tropical east and west. East Pacific: Mexico to Ecuador.
Habitat Inshore and coastal, from surface and intertidal to at least 120m. Usually around coral keys, mangrove fringes, piers and docks, on sand or coral mud, in saline creeks, enclosed sounds or bays, and river mouths. This species is adapted to low-oxygen, shallow-water environments, and may even travel short distances upriver.
Behaviour The Lemon Shark occurs alone or forms loose aggregations of up to 20 individuals, based on size and sex. These sharks are most active at dawn and dusk. They have been seen congregating near docks and fishing piers during the night, returning to deep water during the day. They can rest on the bottom. Pups have high site fidelity; they have small home ranges within shallow, sheltered nursery grounds where they remain for several years, with their range expanding as they grow. Adults may undertake long migrations, possibly including to deeper waters at the onset of winter; sometimes at or near the surface in the open ocean. The shallow nursery grounds are only visited by adults when pregnant females enter to give birth. Although the ecology and behaviour of Lemon Sharks is extremely well-researched, both in the wild and in captivity, aspects of their behaviour and life history remain a mystery. Genetic research suggests

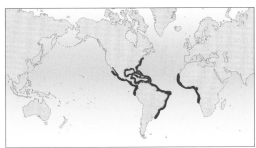

either that adult sharks travel very long distances to mate, or some regional stocks that are now geographically widely separated were relatively recently linked. Over 80% of Lemon Shark litters have been found to have more than one father (in one case a single litter had up to five different fathers) indicating that Lemon Sharks aggregate to mate.
Biology Viviparous, with a yolk-sac placenta. Mating occurs in shallow water during the spring months. 4–17 pups per litter are born in summer after a 10–12 month gestation; females may then take nearly a year off before mating again. Lemon Sharks are slow-growing, maturing at 11–13 years old, with longevity of ~27–30 years. They mainly eat fishes, also crustaceans and molluscs. Small Lemon Sharks are preyed upon by larger sharks, which is why they spend the first few years of their life in shallow, sheltered nursery grounds.
Status IUCN Red List: **Near Threatened**. Although Lemon Sharks are widespread and still relatively common, some of their shallow, coastal nursery grounds are subject to habitat deterioration. There is also evidence of some population depletion by largely unmanaged commercial and artisanal fisheries for fins, meat and leather in the east Pacific and west Atlantic. Lemon Sharks are also caught in recreational fisheries. They are valuable for dive tourism and usually well-behaved, but are potentially harmful as they can be aggressive when provoked. They are a popular public aquarium exhibit.

Lemon Shark, *Negaprion brevirostris*.

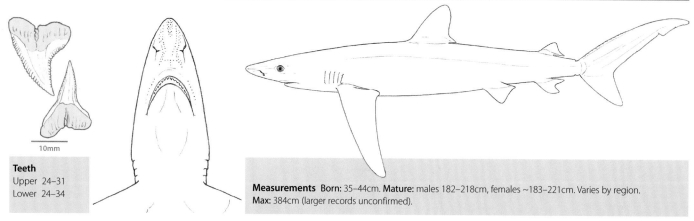

Teeth
Upper 24–31
Lower 24–34

Measurements Born: 35–44cm. **Mature:** males 182–218cm, females ~183–221cm. Varies by region. **Max:** 384cm (larger records unconfirmed).

Identification Slim, graceful shark. Long, conical snout. Curved, saw-edged, triangular upper teeth. Large eyes. No spiracles. Long, narrow, scythe-shaped pectoral fins well in front of first dorsal fin. No interdorsal ridge. Mature females often bear bite wounds (mating scars). Dark blue back, bright blue flanks and sharp demarcation to white underside.

Distribution Worldwide in temperate and tropical oceanic waters (temperature 7–25℃, preferably 12–20℃, latitude 60°N to 50°S). Possibly the most wide-ranging of sharks.

Habitat Oceanic and pelagic, usually off the edge of the continental shelf, 0–1000m (deeper in warmer waters). Migrations often follow major trans-oceanic currents. Occasionally venture inshore at night, particularly around oceanic islands or where the continental shelf is narrowest. Nursery areas offshore.

Behaviour Cruises slowly at the surface, tips of dorsal and caudal fins out of the water and long pectoral fins extended. Most active in early evening and at night, when they may move inshore. Forms large aggregations (where still sufficiently abundant) to feed on shoals of prey or carrion. Have been seen biting at floating objects. Known to harass spearfishers. May circle swimmers, boats and divers for some time before approaching and biting. Highly migratory, with complex movements related to prey availability and reproductive cycles. Segregates by age, sex and reproductive phase: juveniles, subadults, mature sharks and pregnant females are usually found in separate areas, with adult males and females meeting only briefly to mate. Moves seasonally to higher latitudes where prey is more abundant in productive oceanic convergence or boundary zones. Frequent vertical excursions made into deep water or to the thermocline, returning regularly to the surface (possibly to prevent body cooling). Tagging studies have demonstrated that Atlantic Blue Sharks undertake numerous transatlantic migrations, swimming slowly with the major current systems. Pacific Blue Sharks may migrate up to 9200km.

Biology Viviparous, yolk-sac placenta, 4–135 pups per litter (usually 25–35) born in spring and summer after 9–12 month gestation. In European waters, pups remain in offshore nursery area until ~130cm long, when they begin to migrate with other sharks of the same age and sex. Males mature at 4–6 years, females

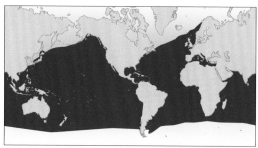

at 5–7 years. Mature females may breed annually, or on alternate years. Longevity ~20 years. Feeds on relatively small prey: usually squid and pelagic fishes, also invertebrates, bottom-dwelling fish and small sharks, sometimes seabirds taken at the surface.

Status IUCN Red List: Near Threatened. This is the most heavily fished shark in the world. Although the meat is relatively low value in many markets compared with that of other shark species, the large, valuable fins of Blue Sharks are retained and carcasses may be discarded at sea if shark finning bans are not enforced. There have been reports of a substantial decline in catch rates and reductions in sightings frequency in some regions (it is **Critically Endangered** in the Mediterranean), but there are insufficient data to assess all regional population declines. Based upon the estimated number of Blue Shark fins that entered the international fin trade in 2000 (when Blue Shark contributed ~17% of the global shark fin trade), over ten million of these sharks were being fished annually, mainly as bycatch in pelagic longline fisheries for tunas and billfish. Fifteen years later, reported catches had tripled and the species now contributes an estimated 34–64% of the international fin trade, while the contribution of depleted and protected species has declined. This level of capture mortality possibly exceeds the maximum levels that Blue Shark populations can sustain, particularly in the Atlantic where longline vessels and Blue Sharks usually occupy the same areas, and is likely to lead to stock depletion. This species is becoming valuable for offshore dive tourism in temperate waters. It is potentially harmful, having been responsible for a few fatal bite incidents on people, but often timid.

Blue Shark, *Prionace glauca*.

MILK SHARK *Rhizoprionodon acutus*

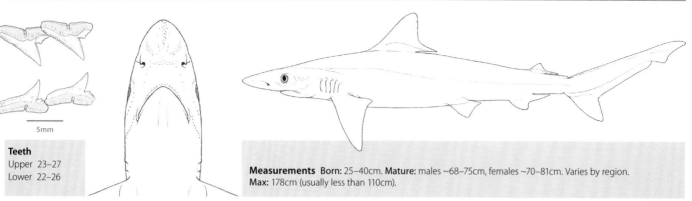

Teeth
Upper 23–27
Lower 22–26

Measurements Born: 25–40cm. **Mature:** males ~68–75cm, females ~70–81cm. Varies by region. **Max:** 178cm (usually less than 110cm).

Identification Small-bodied. Long, narrow snout. Small, wide-spaced nostrils. Oblique, narrowly triangular, smooth-edged teeth; only requiem shark in its range with long upper and lower labial furrows. Big eyes. Small, low second dorsal fin well behind larger anal fin. Bronze to greyish above. Most fin tips slightly pale; juvenile dorsal and upper caudal fin tips dark (sometimes also in adults). White below.
Distribution Indian Ocean, west Pacific, east Atlantic, Mediterranean (Gulf of Taranto off Italy).
Habitat Continental shelf, midwater to near bottom, 1–200m. Often off sandy beaches, sometimes in estuaries.
Behaviour Unknown.
Biology Viviparous, yolk-sac placenta, 1–8 pups per litter (usually 2–5), ~1 year gestation. Matures at 2 years, but may be up to 5–6 years in West Africa, with a maximum age of 8–9 years. Feeds mainly on bony fishes. Eaten by larger sharks.
Status IUCN Red List: **Vulnerable**. Heavily fished, but productive, common and widespread.

BRAZILIAN SHARPNOSE SHARK *Rhizoprionodon lalandii*

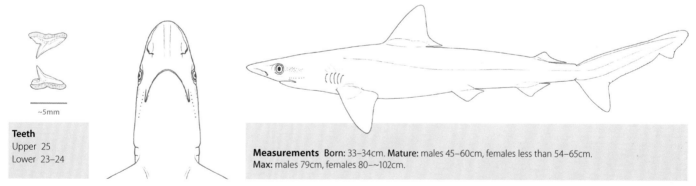

Teeth
Upper 25
Lower 23–24

Measurements Born: 33–34cm. **Mature:** males 45–60cm, females less than 54–65cm. **Max:** males 79cm, females 80–~102cm.

Identification Long snout. Small, widely spaced nostrils. Only requiem shark in its range with long upper and lower labial furrows, small second dorsal fin with origin far behind anal origin, long anal ridges, and apex of pectoral fin – when fin is laid back against body – falling in front of first dorsal midbase. Dark grey or grey-brown above, light below. Light margins to pectorals and dusky dorsal fins.
Distribution West Atlantic: Panama to southern Brazil.
Habitat Littoral, continental shelf, 0–70m, on sand and mud, not normally in lagoons and estuaries.
Behaviour Unknown.
Biology Viviparous, yolk-sac placenta, 1–5 pups per litter. Summer mating season, with winter births in coastal nursery areas. Eats small bony fishes, shrimp and squid.
Status IUCN Red List: **Vulnerable**. Steep declines reported in landings from intensive, largely unmanaged fisheries that operate throughout this species' range. Meat is consumed locally.

PACIFIC SHARPNOSE SHARK   *Rhizoprionodon longurio*

FAO code: **RHU**   Plate page 552

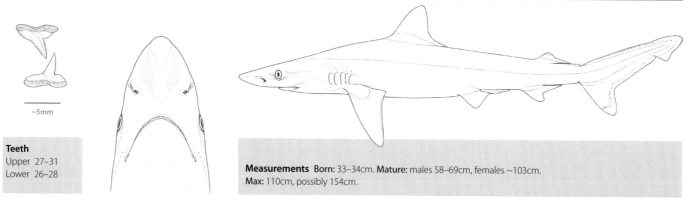

Teeth
Upper 27–31
Lower 26–28

Measurements Born: 33–34cm. **Mature:** males 58–69cm, females ~103cm.
Max: 110cm, possibly 154cm.

Identification Only east Pacific requiem shark with long labial furrows and second dorsal fin origin well behind anal fin origin. Long snout. Small, widely spaced nostrils. Large eyes. No spiracles. Grey or grey-brown above. Light-edged pectoral fins; dusky-tipped dorsal fins. White below.
Distribution East Pacific: southern California to Peru.
Habitat Littoral on continental shelf, from intertidal to at least 27m.
Behaviour Unknown.
Biology Viviparous, litters of 2–5 pups. Feeds mainly on small bony fishes and crustaceans.
Status IUCN Red List: **Vulnerable**. Locally very abundant, but intensive fisheries throughout its range have caused population declines.

GREY SHARPNOSE SHARK   *Rhizoprionodon oligolinx*

FAO code: **RHX**   Plate page 556

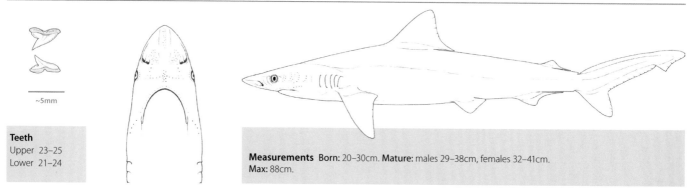

Teeth
Upper 23–25
Lower 21–24

Measurements Born: 20–30cm. **Mature:** males 29–38cm, females 32–41cm.
Max: 88cm.

Identification Very small shark. Long snout. Small, widely spaced nostrils. Small, oblique, narrow-cusped teeth in both jaws; short labial furrows. Large eyes. No spiracles. Second dorsal fin origin far behind anal fin origin. Grey or brownish grey to bronzy above, pale below. Inconspicuous dusky fin edges.
Distribution Tropical west Pacific and north Indian Oceans.
Habitat Littoral, continental and insular shelves, inshore and offshore to at least 36m.
Behaviour Unknown.
Biology Viviparous, yolk-sac placenta, 3–5 pups per litter. Feeds on small bony fishes, cephalopods and crustaceans.
Status IUCN Red List: **Least Concern** (out of date). Heavily fished, but abundant and productive.

CARIBBEAN SHARPNOSE SHARK   *Rhizoprionodon porosus*   FAO code: **RHR**   Plate page 552

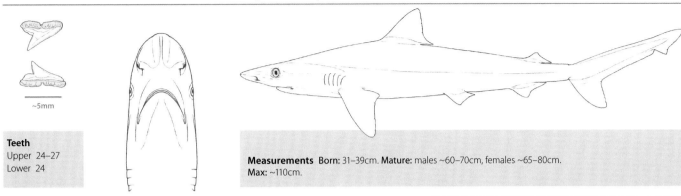

Teeth
Upper 24–27
Lower 24

~5mm

Measurements Born: 31–39cm. **Mature:** males ~60–70cm, females ~65–80cm.
Max: ~110cm.

Identification Small shark. Long snout. Small, wide-spaced nostrils. Serrated teeth; long labial furrows.
Fairly large eyes. No spiracles. Brown or grey-brown above, white below, white posterior pectoral fin margins,
dorsal and caudal fin posterior margins blackish; occasionally white spots along sides.
Distribution West Atlantic: Caribbean and tropical South America.
Habitat Usually close inshore on continental and insular shelves, also offshore to 500m.
Behaviour Unknown.
Biology Viviparous, yolk-sac placenta, 2–8 pups per litter. 10–11 month gestation, born in spring or early
summer off southern Brazil. Matures in ~2 years, with a maximum age of 5 years for males and 8 years for
females. Mostly eats small bony fishes, also invertebrates.
Status IUCN Red List: **Least Concern** (out of date). Very common.

AUSTRALIAN SHARPNOSE SHARK   *Rhizoprionodon taylori*   FAO code: **RHY**   Plate page 556

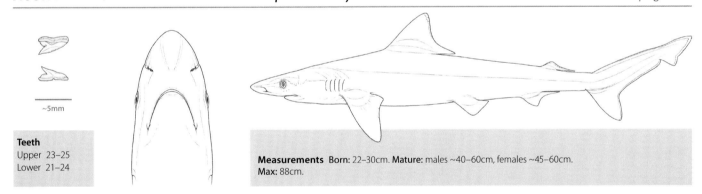

Teeth
Upper 23–25
Lower 21–24

~5mm

Measurements Born: 22–30cm. **Mature:** males ~40–60cm, females ~45–60cm.
Max: 88cm.

Identification Small shark, very similar to *Rhizoprionodon oligolinx*. Bronze to greyish above, pale below.
Not conspicuously marked, but dorsal and caudal fin margins and upper caudal fin tip dark; other fins light-
edged.
Distribution Central Indo-Pacific: tropical northern Australia and Papua New Guinea.
Habitat Tropical inshore continental shelf from 0–110m.
Behaviour Unknown.
Biology Viviparous, yolk-sac placenta, 1–10 pups per litter, with a gestation of 11–12 months. Matures in
~one year, with a maximum age of 7 years. Feeds mostly on small fishes, cephalopods and crustaceans.
Status IUCN Red List: **Least Concern**. Taken in bycatch but abundant and one of the most productive
species of shark known, growing very rapidly, maturing after one year with females producing litters each
year.

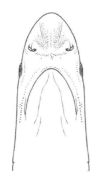

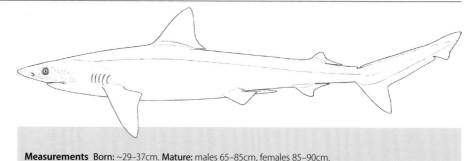

~5mm

Teeth
Upper 24–27
Lower 24–27

Measurements Born: ~29–37cm. **Mature:** males 65–85cm, females 85–90cm. **Max:** at least 113cm.

Identification Very similar to *Rhizoprionodon porosus*. Small-bodied. Long snout. Long upper labial furrows. Fairly large eyes. First dorsal fin origin usually over or slightly in front of pectoral fin free rear tips. No interdorsal ridge. Small second dorsal fin origin over anal fin midbase inserts. Grey to grey-brown above with small light spots on sides of large specimens. Pectoral fins with white margins; dorsals with dusky tips. White below.
Distribution Northwest Atlantic: New Brunswick, Canada, to Mexico and Honduras.
Habitat Coastal in enclosed bays, sounds, harbours, and marine to brackish estuaries, from intertidal to ~280m, usually less than 10m. Often close to surf zone off sandy beaches.
Behaviour Migrates seasonally in Gulf of Mexico: offshore in winter, inshore in spring.
Biology Viviparous, yolk-sac placenta, 1–7 pups per litter, (usually 4–6, increasing with female size) born inshore in spring-summer after 10–11 month gestation. Age at maturity ~2.4–3.9 years, maximum age 10 years. Feeds mainly on small bony fishes.
Status IUCN Red List: **Least Concern** (out of date). Abundant and able to sustain fairly intensive fishing pressure.

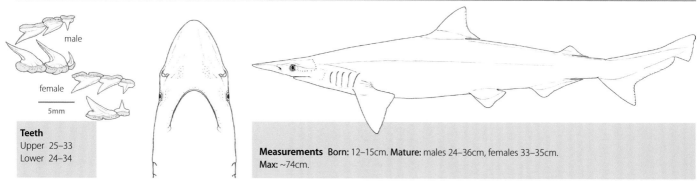

male

female

5mm

Teeth
Upper 25–33
Lower 24–34

Measurements Born: 12–15cm. **Mature:** males 24–36cm, females 33–35cm. **Max:** ~74cm.

Identification Small, stocky shark. Unmistakable, very long, flattened, spade-like snout. Small, smooth-edged, blade-like teeth. Small eyes. Rear tip of first dorsal over pelvic fin midbases. No interdorsal ridge. Second dorsal much smaller and originating well behind origin of larger anal fin. Short, broad, triangular pectoral fins. Bronzy-grey colour with no conspicuous markings.
Distribution North and west Indian Ocean: Bay of Bengal, India to Tanzania.
Habitat Close inshore, often in rocky areas and lower reaches of large tropical rivers, from 10–75m.
Behaviour Occurs in large schools.
Biology Viviparous, unusual columnar placenta and long umbilical cord (egg yolks far too small to nourish pups). Breed year-round, 1–14 pups per litter. Matures at 1–2 years, longevity 5–6 years. Eats small pelagic schooling and bottom-living bony fishes, also shrimp and cuttlefish.
Status IUCN Red List: **Near Threatened** (out of date). Abundant but very heavily fished. Previously confused with the very similar looking *Scoliodon macrorynchus*.

PACIFIC SPADENOSE SHARK *Scoliodon macrorhynchos*   FAO code: **RSK**   Plate page 556

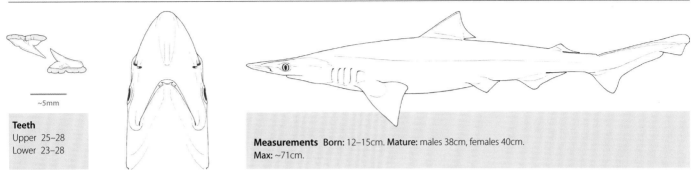

Teeth
Upper 25–28
Lower 23–28

Measurements Born: 12–15cm. **Mature:** males 38cm, females 40cm.
Max: ~71cm.

Identification Small, stocky shark. Very similar in appearance to *Scoliodon laticaudus*, the main morphological difference appears to be the longer second dorsal fin to anal fin length: 6–9.1% total length versus 4.6–6.2% total length for *S. laticaudus*. No interdorsal ridge. Bronzy-grey colour with no conspicuous markings
Distribution West Pacific Ocean: western Indonesia to Japan.
Habitat Close inshore, often in rocky areas and lower reaches of large tropical rivers.
Behaviour Occurs in large schools.
Biology Viviparous, but other life history characteristics little known due to confusion with *S. laticaudus*. Most likely eats small bony fishes, shrimp and cuttlefish.
Status IUCN Red List: Near Threatened. Abundant and common, but very heavily fished across most of its range in intensive industrial trawl and artisanal fisheries. Previously confused with the very similar-looking *S. laticaudus*.

WHITETIP REEF SHARK *Triaenodon obesus*   FAO code: **TRB**   Plate page 512

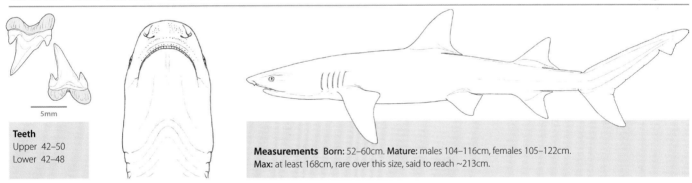

Teeth
Upper 42–50
Lower 42–48

Measurements Born: 52–60cm. **Mature:** males 104–116cm, females 105–122cm.
Max: at least 168cm, rare over this size, said to reach ~213cm.

Identification Small, slender shark. Very short, broad snout. Oval eyes. First dorsal fin behind pectorals. No interdorsal ridge. Second dorsal nearly as large as first. Greyish brown above, underside lighter; sometimes scattered dark spots on sides. Brilliant, very conspicuous, white tips on first dorsal and upper caudal fins.
Distribution Pacific and Indian Oceans. Widespread.
Habitat Continental shelves and island terraces. Usually on or near bottom in crevices or caves in coral reefs, and in coral lagoons in shallow clear water, 8–40m but ranging from 1–330m.
Behaviour Often seen resting on the bottom, in caves and under ledges in coral and on sand by day. More active during slack tide and at night. Small home range occupied for months or years. Social but not territorial; shares home range without conflict. Specialises in capturing bottom prey in crevices, holes and caves in coral heads and ledges, located by scent and sound, sometimes hunting in packs. Readily attracted to bait and may be hand-fed by divers; rarely aggressive.
Biology Viviparous, 1–5 pups per litter (commonly 2–3) after at least 5 month gestation. Matures at 7–8 years, longevity at least 19 years, possibly up to 25 years. Feeds on bony fishes and cephalopods.
Status IUCN Red List: Vulnerable. Steep declines in abundance reported in some parts of its range, where it is caught in intensive industrial and small scale fisheries and may be retained for its fins, flesh, and other parts; elsewhere it remains abundant. Some of its coral reef habitat is threatened by climate change, destructive fishing practices and poor water quality. Popular for dive tourism, kept in captivity.

Galeocerdidae: Tiger Shark

A recently resurrected family with a single genus and species. The Tiger Shark has always been an outlier in the Carcharhinidae due to distinct morphological and biological differences. The only known species is also distinct genetically from the other Carcharhinidae members based on molecular studies. Details on the family and genus are given below under the species account.

TIGER SHARK *Galeocerdo cuvier* FAO code: **TIG** Plate page 506

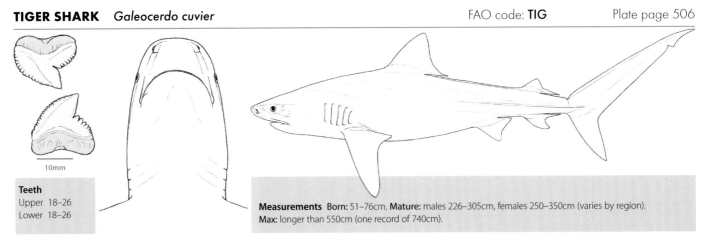

Teeth
Upper 18–26
Lower 18–26

Measurements Born: 51–76cm. **Mature:** males 226–305cm, females 250–350cm (varies by region). **Max:** longer than 550cm (one record of 740cm).

Identification Huge shark. Broad, bluntly rounded snout. Big mouth with large, saw-edged, cockscomb-shaped teeth; very long upper labial furrows. Large spiracles. Long free rear tip to first dorsal fin is almost half of the fin height. Prominent interdorsal ridge. Low caudal keels. Grey above with vertical black to dark grey bars and spots; stripes are bold in young but fade in adults. White below.

Distribution Worldwide, in temperate and tropical seas.

Habitat On or near continental and insular shelves, from the surface and intertidal to 1136m. Found from turbid areas with a high runoff of fresh water, including estuaries and harbours, to clear-water coral atolls and lagoons. Travels long distances visiting different habitats.

Behaviour Apparently nocturnal, large Tiger Sharks may move inshore at night to very shallow water and return to deeper areas by day. Smaller Tiger Sharks may be more active by day. They are strong swimmers, undertaking very long-distance movements that are often very unpredictable. It has been suggested that its irregular movements and very short residency periods at each location are a tactic to take local populations of prey species by surprise; the sharks move on rapidly when their prey become wary and harder to catch. More regular seasonal migrations may be associated with mating and pupping, or timed to the availability of young and inexperienced prey. For example, normally solitary Tiger Sharks aggregate annually to feed on fledging albatross chicks at French Frigate Shoals, 500 miles from the main Hawaiian Islands in the Pacific. Long-term tagging studies have shown that while some Tiger Sharks appear to have a large semiresident home range, covering perhaps 100km of coastline, others travel many thousands of kilometres in a year. A Tiger Shark tagged off Ningaloo Reef, western Australia, swam as far north as Semba Island, Indonesia, then south to the Great Australian Bight during the summer, before returning to Ningaloo a year later.

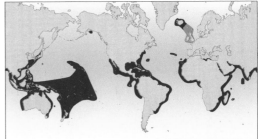

Biology Unlike Carcharhinids, this is a viviparous species with a yolk-sac (no placenta). Tiger Sharks bear very large litters (10–82 pups, average 26–33) that are born in spring to early summer. Mating occurs in spring (sometimes before pregnant females have given birth), so gestation takes slightly over a year (15–16 months) and litters are born every 2 years or less. They are fast-growing: age at maturity varies by region, but is between 4–13 years; their longevity is at least 20–22 years, but may be up to 27–37 years. Sometimes called a 'garbage can with fins', the Tiger Shark eats bony fishes, elasmobranchs (including Tiger Shark pups), sea turtles, sea snakes, marine iguanas, seabirds, marine mammals, jellyfish, carrion and rubbish.

Status IUCN Red List: Near Threatened. This is (or was formerly) a common large shark but has been caught regularly in target and non-target fisheries; its large fins are high value although not very abundant in Hong Kong markets, and it is easy to hook. Its meat is not considered particularly good, but liver oil and skin are high quality. Some population declines have been reported, but survival of juvenile Tiger Sharks increases when adults are depleted. It is assessed as **Vulnerable** in the north Indian Ocean due to intense fisheries and a decline in larger individuals. This shark is potentially harmful; it is a target of shark control programmes and, because it has been responsible for fatal attacks on people, is sometimes the focus of short-term local eradication programmes although the shark responsible had probably already left the scene. Tiger Sharks can, however, often be unaggressive when encountered underwater and are valuable for dive tourism.

Tiger Shark, *Galeocerdo cuvier.*

Sphyrnidae: hammerhead sharks

A small family containing two genera, with nine described species. Genus *Eusphyra* contains one very distinctive species, and there are eight described species in genus *Sphyrna*. A cryptic species of *Sphyrna* was identified through genetic analysis of tissue samples, but is virtually indistinguishable from the Scalloped Hammerhead.

Identification Unmistakable hammer-shaped heads. These function as a submarine-like bow plane to improve manoeuvrability, and increase sensory capacity by enhancing stereoscopic vision and the sharks' ability to triangulate sources of scent and electromagnetic signals.

Biology Live-bearing (viviparous), with yolk-sac placenta. Feed on bony fishes, smaller sharks, rays, cephalopods and invertebrates, but not on marine mammals or other large vertebrates.

Status Found worldwide in tropical and warm-temperate seas, on or adjacent to continental and insular shelves and seamounts, from the surface to at least 1043m, sometimes in large schools. Target and bycatch fisheries have depleted many populations. The fins are particularly valuable (making up over 6% of large shark fins identified in international trade during the early 2000s, rising to over 8% in 2014–15) and hammerheads die very quickly when hooked or entangled; live release of bycatch is therefore unusual. With the exception of one Data Deficient species, the entire family is now assessed as threatened in the IUCN Red List and 56% are Critically Endangered; this is a higher exposure to risk of extinction than any other shark family. Concern over hammerhead population declines has resulted in the Scalloped Hammerhead and the other two similar large species (Great and Smooth Hammerheads)

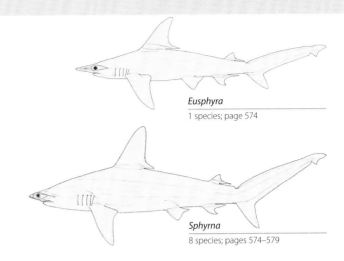

Eusphyra
1 species; page 574

Sphyrna
8 species; pages 574–579

being listed in Appendix II of CITES and CMS. Large hammerhead species are also protected in the Mediterranean and prohibited species in some oceanic tuna fisheries (although developing countries may continue to take them for domestic consumption). The largest hammerheads are important for dive tourism in a few locations. They have occasionally bitten divers and swimmers, but most are shy and very difficult to approach, not aggressive.

Smooth Hammerhead, *Sphyrna zygaena*, Baja California Sur, Mexico (p. 579).

Plate 77 SPHYRNIDAE I

○ **Winghead Shark** *Eusphyra blochii*   page 574

Central Indo-Pacific; 0–127m. Immense wing-shaped head with very long narrow blades, width between eyes about half of total length; first dorsal fin origin over pectoral fin bases, further forward than other hammerheads, upper precaudal pit longitudinal not crescentic; body brown above, white below.

○ **Scalloped Bonnethead** *Sphyrna corona*   page 574

East Pacific; 0–100m. Moderately broad anterior arched mallet-shaped head, medial and lateral indentations on anterior edge and transverse posterior margin, no prenarial grooves, snout moderately long about ⅖ head width, small strongly arched mouth; first dorsal fin free rear tip over pelvic fin insertion, anal fin posterior margin approximately straight, upper precaudal pit transverse crescentic; grey above, white below extending to back of head.

○ **Carolina Hammerhead** *Sphyrna gilberti*   page 575

Northwest Atlantic; coastal to at least 100m. A cryptic species, externally indistinguishable from the Scalloped Hammerhead *Sphyrna lewini*. Originally identified through molecular analysis.

○ **Scoophead Shark** *Sphyrna media*   page 575

West Atlantic and east Pacific; 0–100m. Moderately broad anterior arched mallet-shaped head, weak medial and lateral indentations on anterior edge and transverse posterior margin, no prenarial grooves, snout moderately short about ⅓ the head width, moderately large broadly arched mouth; first dorsal fin free rear tip over pelvic fin insert, upper precaudal pit transverse crescentic; grey-brown above, light below.

○ **Bonnethead Shark** *Sphyrna tiburo*   page 578

West Atlantic and east Pacific; 0–90m. Unique very narrow smooth shovel-shaped head, no indentations, no prenarial grooves, snout moderately long about ⅖ the head width, broadly arched mouth; first dorsal fin free rear tip anterior to pelvic fin origins, anal fin posterior margin shallowly concave, upper precaudal pit tranverse crescentic; grey or grey-brown, often with spots, light below.

○ **Smalleye Hammerhead** *Sphyrna tudes*   page 579

West Atlantic; 5–80m. Broad anterior arched mallet-shaped head, deep medial and lateral indentations on anterior edge none on posterior margin, no prenarial grooves, snout short less than a ⅓ the head width, moderately large broadly arched mouth; first dorsal fin free rear tip over pelvic fin insertion, anal fin posterior margin moderately concave, upper precaudal pit transverse crescentic; grey-brown to golden, light below.

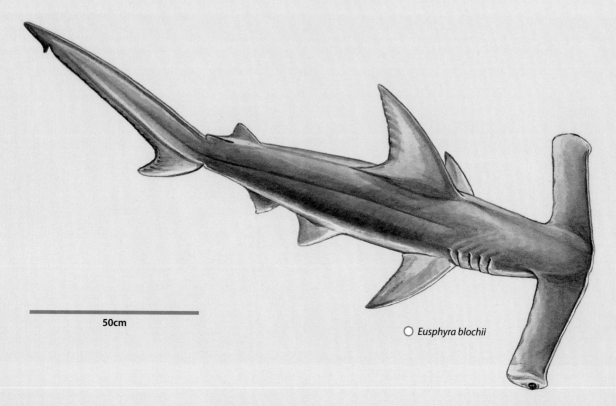

50cm

○ *Eusphyra blochii*

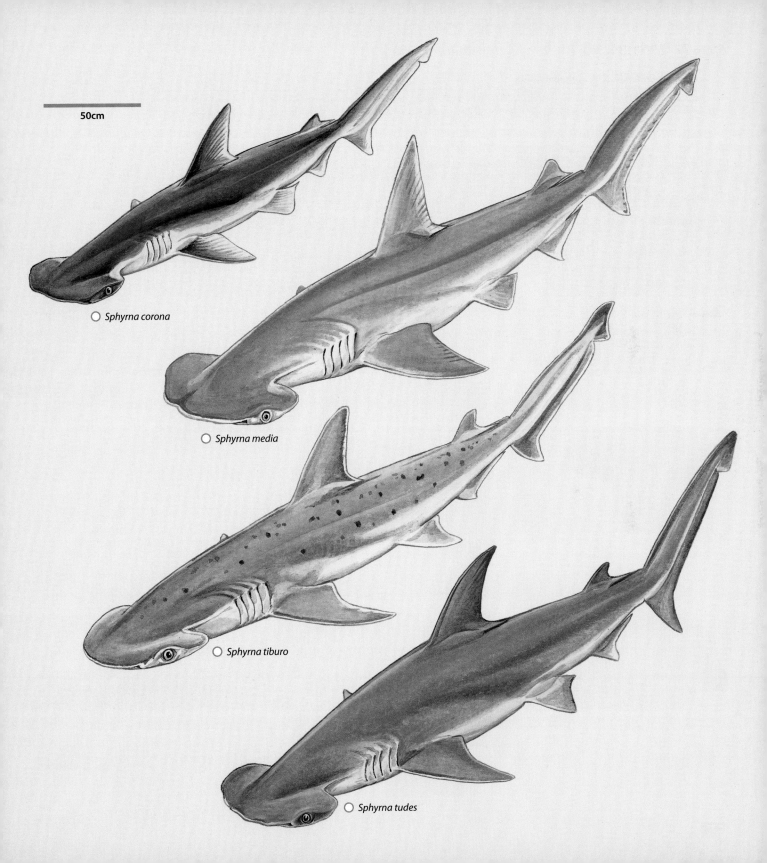

50cm

○ *Sphyrna corona*

○ *Sphyrna media*

○ *Sphyrna tiburo*

○ *Sphyrna tudes*

Plate 78 SPHYRNIDAE II – large hammerhead sharks

○ **Scalloped Hammerhead** *Sphyrna lewini*   page 576

Worldwide; 0–1043m. Broad narrow-bladed head with arched anterior margin, prominent median indentation and lateral indentations, well-developed prenarial grooves, short snout ⅕–⅓ of the head width, broadly arched mouth; first dorsal fin moderately high, second dorsal fin low less than anal fin height; upper precaudal pit transverse crescentic; light grey or bronzy with dusky pectoral fin tips and dark blotch on lower caudal lobe, white below. Indistinguishable externally from *Sphyrna gilberti* in the west Atlantic.

○ **Great Hammerhead** *Sphyrna mokarran*   page 577

Worldwide; 0–300m. Broad narrow-bladed head with nearly straight anterior margin, prominent median indentation and lateral indentations, no or weakly developed prenarial grooves, snout short less than ⅓ the head width, broadly arched mouth; first dorsal fin very high and falcate, second dorsal fin about equal to anal fin height; upper precaudal pit transverse crescentic; light grey or grey-brown, fins unmarked, white below.

○ **Smooth Hammerhead** *Sphyrna zygaena*   page 579

Worldwide; 0–200m. Broad narrow-bladed head with broadly arched anterior margin, no median indentation but prominent lateral indentations, well-developed prenarial grooves, snout short ⅕ to less than ⅓ the head width, broadly arched mouth; first dorsal fin moderately high, second dorsal fin low, less than anal fin height; upper precaudal pit transverse crescentic; olive-grey or dark grey-brown, undersides of pectoral fin tips dusky, white below.

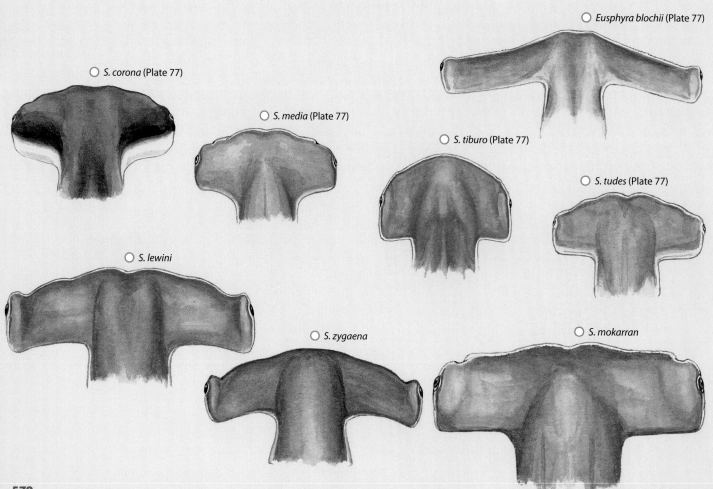

○ *Eusphyra blochii* (Plate 77)

○ *S. corona* (Plate 77)

○ *S. media* (Plate 77)

○ *S. tiburo* (Plate 77)

○ *S. tudes* (Plate 77)

○ *S. lewini*

○ *S. zygaena*

○ *S. mokarran*

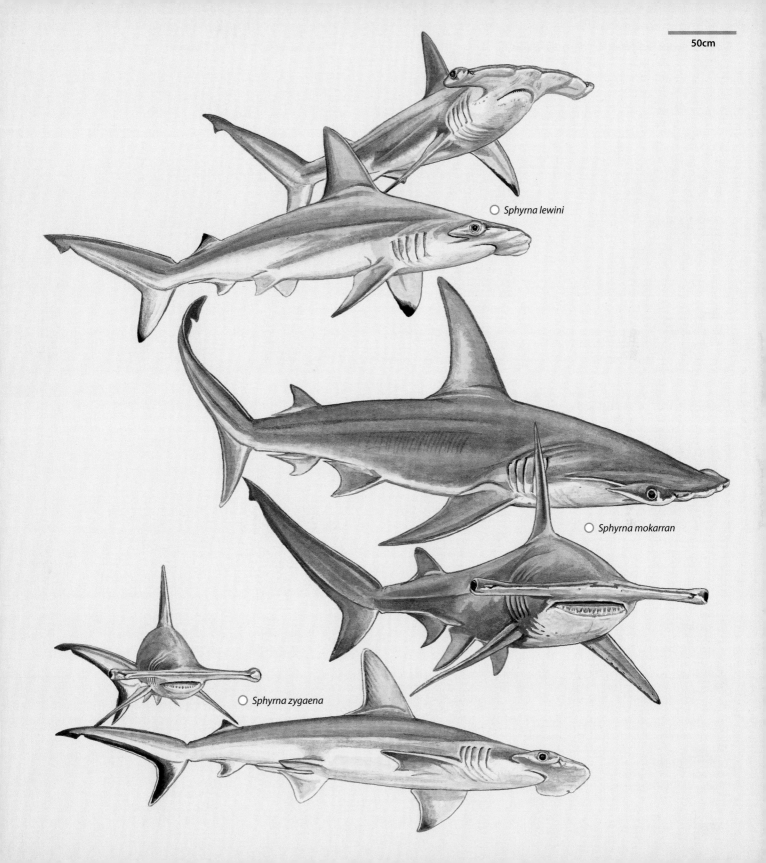

50cm

Sphyrna lewini

Sphyrna mokarran

Sphyrna zygaena

WINGHEAD SHARK *Eusphyra blochii*

FAO code: **EUB**   Plate page 570

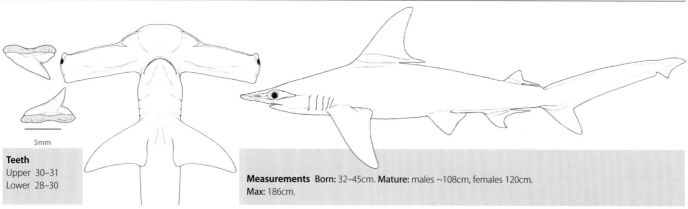

Teeth
Upper 30–31
Lower 28–30

5mm

Measurements Born: 32–45cm. **Mature:** males ~108cm, females 120cm.
Max: 186cm.

Identification Medium-sized, slender body. Unmistakable, immense, wing-shaped head with very long narrow blades; width between eyes about half of total length. First dorsal fin origin over pectoral fin bases, further forward than other hammerheads. Upper precaudal pit is longitudinal, not crescent-shaped. Coloured brown above, white below.
Distribution West Pacific and north Indian Ocean: Persian Gulf to Australia and China.
Habitat Shallow water, continental and insular shelves to 127m.
Behaviour Pregnant females reportedly fight each other. Feeds on small fishes, but also on cephalopods and crustaceans.
Biology Viviparous, with yolk-sac placenta, 6–25 (average 11) pups per litter after 8–11 month gestation. Maturity is reached at 5.5 years for males, 7.2 years for females, with a maximum age of 21 years.
Status IUCN Red List: Endangered. Heavily fished in much of range except in Australian waters, where it is Least Concern because it is such a small component of the commercial catch. Not known to bite people.

SCALLOPED BONNETHEAD *Sphyrna corona*

FAO code: **SSN**   Plate page 570

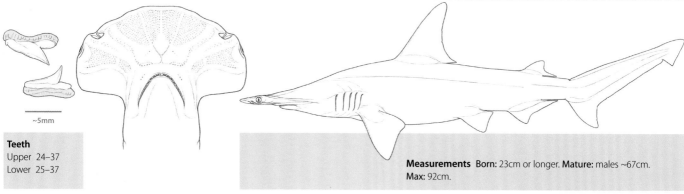

Teeth
Upper 24–37
Lower 25–37

~5mm

Measurements Born: 23cm or longer. **Mature:** males ~67cm.
Max: 92cm.

Identification Small-bodied. Moderately broad, anteriorly arched, mallet-shaped head with medial and lateral indentations on its anterior edge and transverse posterior margins; no prenarial grooves; snout rather long, about ~two-fifths of the head width. Small, strongly arched mouth. Free rear tip of first dorsal fin over pelvic insertions. Posterior margin of anal fin nearly straight. Transverse, crescentic upper precaudal pit. Grey above. White below, extending to back of head.
Distribution East Pacific: Gulf of California to Peru.
Habitat Continental shelf, mostly a coastal species, 0–100m.
Behaviour Unknown.
Biology Viviparous, possibly 2 pups per litter.
Status IUCN Red List: Critically Endangered. Captured throughout its range by a wide variety of intensive, unmanaged commercial and artisanal fisheries; formerly rare, but recently no longer recorded at all.

CAROLINA HAMMERHEAD *Sphyrna gilberti*

FAO code: **SPN**   Plate page 570

Teeth
Upper 32–34
Lower 30–31

Measurements Born: ~40cm.

Identification Large hammerhead, externally indistinguishable from the Scalloped Hammerhead *Sphyrna lewini* (p. 576). The only subtle morphological difference is the head length is >20% of the precaudal length and the number of precaudal vertebrae is ≤91. Coloration uniform grey to brown dorsally, fading to white ventrally, with ventral pectoral fin surfaces variably white to dusky, and lower caudal lobe dusky to black. Molecular data have identified this cryptic species.
Distribution Northwest Atlantic: southeastern USA ,off South Carolina. Possibly also Panama and Brazil.
Habitat Uncertain due to misidentification with the Scalloped Hammerhead *S. lewini*, but likely occupies a similar habitat, from inshore coastal waters to at least 100m.
Behaviour Unknown.   **Biology** Viviparous, but nothing else known due to misidentification with the Scalloped Hammerhead. Diet probably includes bony fishes, cephalopods and crustaceans.
Status IUCN Red List: Data Deficient. Distribution, depth range and impact of fisheries unknown.

SCOOPHEAD SHARK *Sphyrna media*

FAO code: **SPE**   Plate page 570

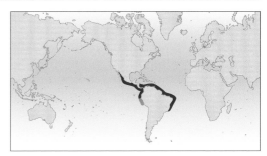

~5mm

Teeth
Upper 24–37
Lower 25–37

Measurements Born: 34cm or less. **Mature:** males ~90–100cm, females 100–133cm.
Max: 150cm.

Identification Small-bodied. Moderately broad, anteriorly arched, mallet-shaped head with weak medial and lateral indentations on anterior edge and transverse posterior margins; no prenarial grooves; snout rather short, about one-third of head width. Moderately large, broadly arched mouth. Free rear tip of first dorsal fin over pelvic insertions. Transverse, crescentic upper precaudal pit. Grey-brown above, light below.
Distribution West Atlantic: Panama to southern Brazil. East Pacific: Gulf of California to Ecuador, probably northern Peru.
Habitat Continental shelf, 0–100m. Mangroves may be important habitat for juveniles.
Behaviour Unknown.   **Biology** Nothing known.
Status IUCN Red List: Critically Endangered. This small shark was common or abundant during the 1970s, but has since been subject to intense and largely unmanaged fishing pressure, and the loss or degradation of mangrove habitats. Its former range has declined in some areas and recent records are very sparse.

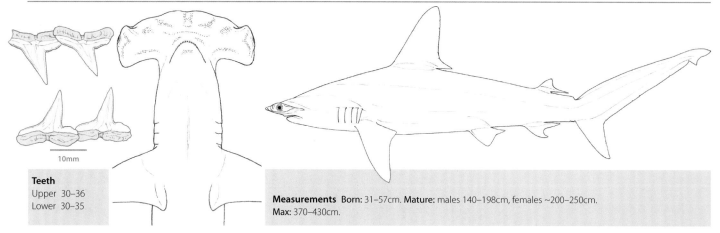

Teeth
Upper 30–36
Lower 30–35

10mm

Measurements Born: 31–57cm. **Mature:** males 140–198cm, females ~200–250cm. **Max:** 370–430cm.

Identification Large shark. Broadly arched, narrow-bladed head with a central notch and two smaller lateral indentations at the front of the head. First dorsal moderately high. Light grey or bronzy above, white below. Dusky or black-tipped pectoral fins; dark blotch on lower caudal lobe. The cryptic species *Sphyrna gilberti*, which is only distinguishable from the Scalloped Hammerhead at molecular level, is apparently widely distributed in the west Atlantic.

Distribution Worldwide, a coastal-pelagic semi-oceanic species of warm-temperate and tropical waters. Populations from different regions are genetically distinct.

Habitat Over continental and insular shelves and adjacent deep water, from the surface to deeper than 1043m, often close inshore and in enclosed bays and estuaries. Juveniles mainly occur in shallow inshore areas, subadults in deeper water, while adults aggregate further offshore around seamounts. Scalloped Hammerheads tend to remain in colder waters below the thermocline during the summer, and closer to the surface in the cooler months.

Behaviour A seasonally migratory, schooling shark. Large schools, primarily females, aggregate around seamounts and offshore islands by day, then disperse alone or in small groups to feed at night. Tagged female Scalloped Hammerheads move fairly regularly between Malpelo Island, Cocos Island and the Galapagos (some 600–700km apart in the east Pacific), but return to the Central American coast to pup in

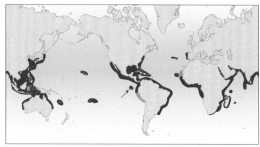

the same inshore nursery grounds in shallow bays. Females tend not to disperse widely to other areas, but males undertake much longer-distance dispersions.

Biology Viviparous, females give birth to 12–41 pups per litter after an 8–12 month gestation, which is followed by a one year resting period. Growth parameters vary between populations. Scalloped Hammerheads may grow more slowly and attain a smaller size in cooler waters than in the tropics. Males mature at ~10 years, females ~13–15 years, with a maximum age of up to 35 years. Scalloped Hammerheads eat bony fishes, sharks, rays and invertebrates; populations of some of these prey species have reportedly increased in size in areas where Scalloped Hammerhead populations were depleted.

Status IUCN Red List: Critically Endangered. Adult and pups of this formerly common, widespread species have been extremely heavily fished in many regions, offshore and in their nursery grounds. Globally, the population has declined by over 80%. Hammerhead fins are highly valued in the international fin trade because of their size and high needle (ceratotrichia) count. Fins of Scalloped and Smooth Hammerheads combined (they cannot be distinguished by eye) made up 4.4% of fins traded in the Hong Kong market in 2003. By 2014–15, Scalloped Hammerhead fins were the third most abundant species and comprised 4% of fins identified using genetic analyses in Hong Kong, with an additional 3.4% from Smooth Hammerheads. The Scalloped Hammerhead is listed in Appendix II of CMS and CITES, protected in the Mediterranean, and is a prohibited species in the Atlantic and east Pacific pelagic fisheries managed by ICCAT and IATTC. Schooling females are highly valued for dive tourism.

Scalloped Hammerheads, *Sphyrna lewini*.

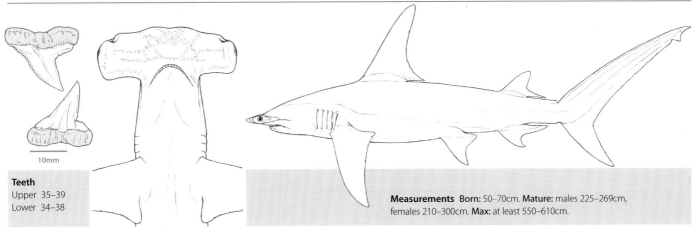

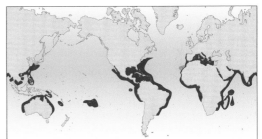

Teeth
Upper 35–39
Lower 34–38

10mm

Measurements Born: 50–70cm. **Mature:** males 225–269cm, females 210–300cm. **Max:** at least 550–610cm.

Identification Very large hammerhead with a notch at the centre of head. First dorsal very high and curved. Second dorsal and anal fins high, with deeply concave rear margins. Light grey or grey-brown above. Fins unmarked. White below.

Distribution Worldwide, tropical seas.

Habitat Coastal-pelagic and semi-oceanic, over continental shelves, island terraces, in passes and lagoons of coral atolls and on coral reefs, close inshore to well offshore, 1–300m or more.

Behaviour Nomadic and seasonally migratory.

Biology Viviparous, with 6–42 pups per litter after ~7–11 month gestation. Age at maturity varies by region, but usually 5–6 years, with a maximum age of 42 years for males and 44 years for females. Varied prey; apparently prefers stingrays and other batoids, groupers and marine catfishes.

Status IUCN Red List: Critically Endangered. Never abundant, populations have undergone significant declines due to exploitation in target and bycatch fisheries. This species' very high value fins comprise 0.8% of fins recorded in Hong Kong during 2014–15, a much smaller proportion than Scalloped and Smooth Hammerheads. Important for dive tourism in some areas. May occasionally bite people.

Great Hammerhead, *Sphyrna mokarran*.

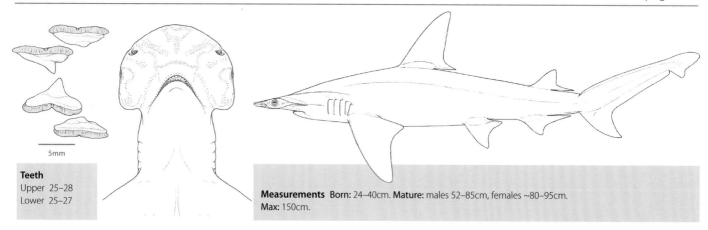

Teeth
Upper 25–28
Lower 25–27

5mm

Measurements Born: 24–40cm. **Mature:** males 52–85cm, females ~80–95cm.
Max: 150cm.

Identification Small shark. Unique, very narrow, smooth, shovel-shaped head. First dorsal free rear tip in front of pelvic fin origins. Shallowly concave posterior anal fin margin. Grey or grey-brown above, often with small dark spots. Light below.

Distribution West Atlantic: Rhode Island, USA, to Brazil. East Pacific: southern California to Ecuador.

Habitat Continental and insular shelves, over mud and sand, on coral reefs, in estuaries, shallow bays and channels, 0–90m (mainly 10–25m). Pupping females are most common in shallow water.

Behaviour Migratory and social. Often in groups of 3–15, rarely alone. Behaviour well studied and complex.

Biology Viviparous, with 4–21 pups per litter. Male and female maturity varies by region, but around 2–3 years; maximum age ~8–12 years. Mainly eats crustaceans, also bivalves, octopus and small fishes.

Status IUCN Red List: Endangered. Once abundant, now rare to absent in the Atlantic and Pacific due to intensive fisheries operating across most of its range. Low fishing mortality only in USA waters. Kept in public aquaria.

Bonnethead Shark, *Sphyrna tiburo*.

SMALLEYE HAMMERHEAD *Sphyrna tudes*

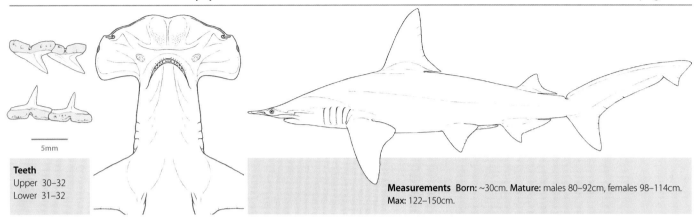

Teeth
Upper 30–32
Lower 31–32

5mm

Measurements Born: ~30cm. **Mature:** males 80–92cm, females 98–114cm. **Max:** 122–150cm.

Identification Broad, arched, mallet-shaped head, deeply indented in front, straight behind; short snout. Fairly large, broadly arched mouth. Free rear tip of first dorsal fin over pelvic insertions. Transverse, crescentic upper precaudal pit. Grey-brown to golden above, light below.
Distribution West Atlantic: Venezuela to Brazil. Old records from Mediterranean Sea and West Africa were based on *Sphyrna lewini* and not this species.
Habitat Continental shelf, to 80m or more.   **Behaviour** Nursery grounds on shallow, muddy bays.
Biology Viviparous, 5–19 pups per litter, 10 month gestation. Eats small bony fishes, newborn Scalloped Hammerheads, small crustaceans and squid.
Status IUCN Red List: Critically Endangered. Target or bycatch in intensive and largely unmanaged commercial and artisanal fisheries throughout its inshore range, which have caused population depletions and local extinctions.

SMOOTH HAMMERHEAD *Sphyrna zygaena*

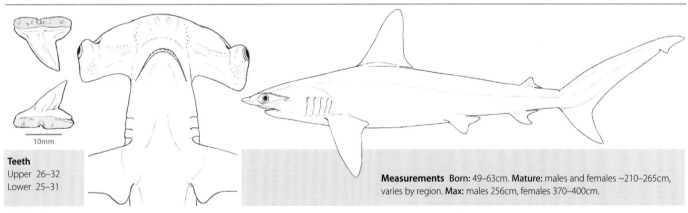

Teeth
Upper 26–32
Lower 25–31

10mm

Measurements Born: 49–63cm. **Mature:** males and females ~210–265cm, varies by region. **Max:** males 256cm, females 370–400cm.

Identification Large hammerhead. Curved head without a notch at the centre. First dorsal fin moderately high. Second dorsal fin low, less than anal fin height. Olive-grey or dark grey-brown above. Undersides of pectoral fin tips dusky. White below.
Distribution Worldwide in tropical and temperate waters (tolerates cooler water than other hammerheads).
Habitat Continental and insular shelves, from close inshore to well offshore, 0–200m, possibly to 500m.
Behaviour Young sharks (up to 1.5m) sometimes occur in huge migrating schools.
Biology Viviparous, with 20–50 pups per litter after a 10–11 month gestation. Age at maturity ~15 years, with a maximum age of 24–25 years. This is the most productive large hammerhead. Feeds on bony fishes, small sharks, skates and stingrays.
Status IUCN Red List: Vulnerable. Target and bycatch of commercial and artisanal fisheries throughout range. Fourth most abundant species in Hong Kong fin trade (2014–15). May be confused with *Sphyrna lewini* in tropics. Largely unmanaged outside northwest Atlantic, Australia and New Zealand. Protected in Mediterranean, prohibited in ICCAT and IATTC pelagic fisheries; listed in CITES and CMS Appendix II.

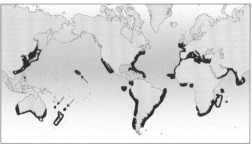

Appendix I: Glossary

See also Topography section on pp. 13–15.

Abdominal ridges or keels. In some sharks, paired longitudinal dermal ridges that extend from the bases of the pectoral fins to the pelvic fin bases.

Abyss. The deep sea bottom, ocean basins or abyssal plain descending from 2000m to about 6000m (see p. 586).

Abyssal plain. The extensive, flat, gently sloping or nearly level region of the ocean floor from about 2000m to 6000m depth (see p. 586).

Anal fin. A single fin on the ventral surface of the tail between the pelvic fins and caudal fin of some sharks, absent in batoids, dogfish, sawsharks, angelsharks, and some chimaeras.

Annular rings or annuli. In a vertebral centrum in cross section, rings of calcified cartilage separated by uncalcified cartilage that occupy the intermedialia only, or concentric rings that cross both the intermedialia and basalia.

Anterior. Forward, in the [longitudinal] direction of the snout tip.

Anterior margin. In precaudal fins, the margin from the fin origin to its apex.

Anterior nasal flap. A flap on the front edges of the nostrils, that serves to partially divide the nostril into incurrent and excurrent apertures or openings.

Apex (plural, **apices**). In precaudal fins, the distal tip, which can be acutely pointed to broadly rounded.

Apical. In oral teeth, towards the tip of the crown or cusp. Can also be used as indicating direction towards the apex or tip of a fin, fin-spine, etc.

Aplacental yolk-sac viviparity. A reproductive mode where the maternal adult gives birth to live young which are primarily nourished by the yolk in their yolk-sac. The yolk is gradually depleted and the yolk-sac reabsorbed until the young are ready to be born. Often referred to as ovoviviparity.

Aplacental viviparity. A reproductive mode where the maternal adult gives birth to live young which do not have a yolk-sac placenta.

Barbels. Long conical paired dermal lobes on the snouts of sharks, which may serve to locate prey. Sawsharks have barbels on the underside of the snout in front of the nostrils (as in sturgeon), but most barbelled sharks have them associated with the nostrils.

Base. In precaudal fins, the proximal part of the fin between the origin and insertion, extending distally, and supported by the cartilaginous fin skeleton. In the caudal fin, that thickened longitudinal part of the fin enclosing the vertebral column and between the epaxial and hypaxial lobes or webs of the fin. In oral teeth, the proximal root and crown foot, in apposition to the distal cusp. In denticles, the proximal anchoring structures, often with four or more lobes, holding the denticles in the skin.

Bathyal. Benthic habitats from 200m to 2000m depth (see p. 586).

Bathypelagic zone. The sunless zone of the water column, usually beyond 1,000m depth. Some oceanic sharks may transit the epipelagic, mesopelagic and bathypelagic zones to the bottom while migrating vertically (see p. 586).

Batoid. A flat elasmobranch fish, e.g. skate or ray, with the pectoral fins fused to the sides of the head and the gill openings on the ventral surface. Species of the order Rajiformes: the sawfish, sharkray, wedgefish, guitarfish, thornrays, panrays, electric rays, skates, stingrays, stingarees, butterfly rays, eagle rays, cownose rays and devilrays.

Benthic. Referring to organisms that are bottom-dwelling.

Benthopelagic. Species occurring on the bottom as well as in pelagic habitats.

Bioluminscence. Light produced by biochemical means in some organisms.

Biomass. The total quantity or weight of all plant and animal material in a given area or volume.

Bivalves. Large class of molluscs that have a hard calcareous shell made of two parts or 'valves', these are usually bilaterally symmetrical.

Boreal. Cold northern regions.

Blade. In oral teeth, an arcuate, convex-edged section of the cutting edge of the crown foot, without cusplets.

Body ridges. Elongated longitudinal dermal ridges on the sides of the trunk and precaudal tail in certain carpet sharks (Orectolobiformes); present in the Whale, Zebra, and some bamboosharks.

Body. Can refer to an entire shark, sometimes restricted to the trunk and precaudal tail.

Brackish. Waters within the mid ranges of salinity, having more salinity than freshwater, but not as much as seawater, for example estuarine water.

Branchial arches. The paired cartilaginous arches that support the gills.

Breach. A complete or almost complete leap above the water's surface followed by a splash on re-entry.

Bycatch. The part of a catch taken incidentally in addition to the target species towards which fishing effort is directed. In a broad context, this includes all non-targeted catch including byproduct, discards and other interactions with gear.

Calcified cartilage. Shark skeletons are formed of hyaline cartilage or gristle, but this is often reinforced with layers of calcified cartilage, cartilage impregnated with a mineral, hydroxyapatite, similar to that of bone.

Calcification. The process by which calcium and carbonate ions are combined to form calcerous skeletal materials.

Cannibal vivipary. See Uterine cannibalism.

Capsule length. Length of an eggcase excluding the horns or tendrils.

Carcharhinoid. A ground shark, a member of the order Carcharhiniformes, and including the catsharks, false catsharks, finbacked catsharks, barbeled houndsharks, houndsharks, weasel sharks, requiem sharks, and hammerheads.

Cartilaginous fishes. Members of the class Chondrichthyes.

Caudal crest. A prominent saw-like row of enlarged pointed denticles along the dorsal caudal margin and sometimes along the ventral caudal margin of the caudal fin. Found in certain sharks, especially some catsharks.

Caudal fin. The fin on the end of the tail in shark-like fishes, lost in some batoids.

Caudal keels. A dermal keel on each side of the caudal peduncle that may extend onto the base of the caudal fin, and may, in a few sharks, extend forward as a body keel to the side of the trunk.

Caudal peduncle. That part of the precaudal tail extending from the insertions of the dorsal and anal fins to the front of the caudal fin.

Centrum (plural, **centra**). Spool-shaped, partially or usually fully calcified structures articulated together to form the vertebral column.

Cephalopods. A member of the class Mollusca, cephalopods includes the octopus, squids and nautilus.

Ceratotrichia. Slender soft or stiff filaments of an elastic protein, superficially resembling keratin or horn (from the Greek keratos, horn, and trichos, hair). Ceratotrichia run in parallel and radial to the fin base and support the fin webs. The prime ingredient of shark-fin soup.

Cetacean. Referring to Cetacea.

Cetacea. An order of marine mammals with modified forelimbs, no visible hind limbs and horizontal tail fin, which comprised of whales, dolphins and porpoises.

Chimaera or **Chimaeroid**. A chimaera, ratfish, silver shark, ghost shark, spookfish, or elephant fish, a member of the order Chimaeriformes, subclass Holocephali.

Chondrichthyan. Referring to the class Chondrichthyes.

Chondrichthyes. The class Chondrichthyes (from Greek chondros, cartilage, and ichthos, fish), a major taxonomic group of aquatic, gill-breathing, jawed, finned vertebrates with primarily cartilaginous skeletons, one to seven external gill openings, oral teeth in transverse rows on their jaws, and mostly small, toothlike scales or dermal denticles. Chondrichthyes include the living elasmobranchs and holocephalans and their numerous fossil relatives, and also can be termed shark-like fishes or simply sharks.

Chondrocranium. See neurocranium.

Circumglobal. Occurring around the world.

Circumnarial fold. A raised semicircular, lateral flap of skin around the incurrent aperture of a nostril, in heterodontoids, orectoloboids, and a few batoids, defined by a circumnarial groove.

Circumnarial groove. A shallow groove defining the lateral bases of the circumnarial folds.

Circumtropical. Occurring around the tropical regions of the world.

CITES. Convention on International Trade in Endangered Species of Fauna and Flora. An international agreement which aims to ensure that international trade in specimens of wild fauna and flora does not threaten the survival of species. Appendix I of CITES lists species threatened with extinction. Appendix II includes 'species not necessarily threatened with extinction, but in which trade must be controlled in order to avoid utilisation incompatible with their survival'. www.cites.org

Claspers. Paired copulatory organs present on the pelvic fins of male cartilaginous fishes, for internal fertilisation of eggs.

Class. One of the taxonomic groups of organisms, containing related orders; related classes are grouped into phyla.

Classification. The ordering of organisms into groups on the basis of their relationships, which may be by similarity or common ancestry.

Cloaca. The common chamber at the rear of the body

cavity of elasmobranchs through which body wastes and reproductive products including sperm, eggs, and young pass, to be expelled to the outside through a common opening or vent.

Common name. The informal vernacular name for an organism, which may vary from location to location.

Concave. An outline or surface that curves inwards.

Continental rise. The gently sloping base of the continental shelf made up of sediment deposits; see also Rise (see p. 586).

Continental shelf. The gently sloping, shelf-like part of the seabed adjacent to the coast extending to a depth of about 200m (see p. 586).

Continental slope. The often steep, slope-like part of the seabed extending from the edge of the continental shelf to a depth of about 2000m (see p. 586).

Convex. An outline or surface that curves outwards.

Crustacean. A mainly aquatic arthropod taxon of the class Crustacea, typically with a hard shell covering the body and includes crabs, lobsters, crayfish, shrimps, prawns, krill and barnacles.

Cusp. A usually pointed large distal projection of the crown. A primary cusp is situated on the midline of the crown foot. Multicuspid refers to oral teeth or denticles with more than one cusp. In lateral trunk denticles, the posterior ends of the crown may have medial and lateral cusps, sharp or blunt projections associated with the medial and lateral ridges.

Cusplet. As with a cusp, but a small projection in association with a cusp, and usually mesial and distal but not medial on the crown foot.

Cutting edge. In oral teeth, the compressed sharp longitudinal ridge on the mesodistal edges of the crown.

Demersal. Occurring or living near or on the bottom of the ocean (cf. pelagic).

Dermal denticle. A small tooth-like scale found in cartilaginous fishes, covered with enameloid, with a core and base of dentine and usually small and often close-set to one another and covering the body. A few non-batoid sharks, many batoids, and chimaeroids generally have them enlarged and sparse or reduced in numbers.

Dermal lobes. In wobbegongs, family Orectolobidae, narrow or broad-based, simple or branched projections of skin along the horizontal head rim and on the chin.

Diphycercal. A caudal fin with the vertebral axis running horizontally into the fin base, which is not elevated.

Distal. In any direction, at the far end of a structure. In oral teeth, used in a special sense for structures on the teeth towards the posterolateral mouth corners or rictuses. See apical and basal.

Dorsal fin spine. A small to large enameloid-covered, dentine-cored spine located on the anterior margins of one or both of the dorsal fins, found on bullhead sharks (Heterodontiformes), many dogfish sharks, fossil (but not living) batoids, chimaeroids, but lost entirely or buried in the fin bases of other shark-like fishes.

Dorsal. Upwards, in the vertical direction of the back. See ventral.

Dorsal fin. A fin located on the trunk or precaudal tail or both, and between the head and caudal fin. Most sharks have two dorsal fins, some batoids one or none.

Dorsal lobe. See upper lobe.

Dorsal margin. In the caudal fin, the margin from the upper origin to its posterior tip. Usually continuous, in angel sharks (Squatiniformes) with their hypocercal, superficially inverted caudal fins, it is subdivided. See Squatinoid caudal fin.

Drop-off. Steep or sheer underwater cliff or precipice.

Echinoderms. Referring to members of the Echinodermata.

Echinodermata. An entirely marine invertebrate phylum with a five-part symmetry, water-based vascular system and tube feet; the diverse group includes brittlestars, starfish, sea urchins, sea cucumbers and sea lilies

Ecosystem. The living community of different species, interdependent on each other, together with their non-living environment.

EEZ. Exclusive Economic Zone. A zone under national jurisdiction (up to 200-nautical miles wide) declared in line with the provisions of 1982 United Nations Convention of the Law of the Sea, within which the coastal State has the right to explore and exploit, and the responsibility to conserve and manage, the living and nonliving resources.

Eggcase. A stiff-walled elongate-oval, rounded rectangular, conical, or dart-shaped capsule that surrounds the eggs of oviparous sharks, and is deposited by the female shark on the substrate. It is analogous to the shell of a bird's egg and is made of protein, which is a type of collagen that superficially resembles horn or keratin. Eggcases often have pairs of tendrils or hornlike structures on their ends, or flat flanges on their sides or spiral flanges around their lengths, which anchor the cases to the bottom. As the egg travels from the ovaries into the oviducts and through the nidamental glands, the eggcase is secreted around it and the egg is fertilised. Live-bearing sharks may retain eggcases, and these vary from being rigid and similar to those of oviparous sharks to soft, bag like, degenerate and membranous. Soft eggcases may disintegrate during the birth cycle.

Elasmobranch. Referring to the subclass Elasmobranchii.

Elasmobranchii. One of the two subclasses of the class Chondrichthyes. Elasmobranchii are the shark-like fishes, the other subclass Holocephali are the chimaeras. The Elasmobranchii includes all the living sharks, batoids, and a host of fossil species. They differ from holocephalans in having five to seven pairs of gill openings, open to the exterior and not covered by a soft gill cover; separate oral teeth not formed as tooth plates; a rigid spined or spineless first dorsal fin and usually a relatively short, rigid, spined or spineless second dorsal fin.

Embryo. An earlier development stage of the young of a live-bearing shark, ranging from nearly microscopic to moderate-sized but not like a miniature adult. See foetus.

Enameloid. The shiny hard external coating of the crowns of shark oral teeth, superficially similar to enamel in land vertebrates.

Endemic. A species or higher taxonomic group of organisms that is only found in a given area. It can include national endemics found in a river system or along part or all of the coast of a given country, but also regional endemics, found off or in adjacent countries with similar habitat, but not elsewhere.

Endemism. The ecological condition of a species being restricted in geographical distribution to an area or region. See endemic.

Epibenthic zone. The area of the ocean just above and including the sea bottom, from shallow seas to deep abysses, epibenthic sharks live on or near the bottom (see p. 586).

Epipelagic zone. That part of the oceans beyond the continental and insular shelves, in oceanic waters, from the surface to the limits of where most sunlight penetrates, about 200m. Also known as the sunlit sea

or 'blue water'. Most epipelagic sharks are found in the epipelagic zone, but may penetrate the mesopelagic zone (see p. 586).

Euryhaline. An organism that is able to tolerate a large variance in salinity.

Euselachian. Referring to the Euselachii.

Euselachii. The cohort Euselachii (Greek Eu, true, good or original, and selachos, shark or cartilaginous fish), the spined or 'phalacanthous' sharks, including the modern sharks or Neoselachii, and fossil shark groups including the hybodonts, the ctenacanths, and the xenacanths, all primitively with anal fins and having two dorsal fins with fin spines.

Excretion. The separation and removal of waste material from an organism usually applied to nitrogen waste normally in the form of urea or ammonia.

Excurrent apertures. The posterior and ventrally facing openings of the nostrils, which direct water out of the nasal cavities and which are often partially covered by the anterior nasal flaps. These are usually medial on the nostrils and posteromedial to the incurrent apertures, but may be posterior to the incurrent apertures only.

Extant. Applies to a species that is still living.

Extinct. Applied to a species or group that has no living representatives.

Eye notch. A sharp anterior or posterior indentation in the eyelid, where present cleanly dividing the upper and lower eyelids.

Eyespots or ocelli. Large eye-like pigment spots located on the dorsal surface of the pectoral fins or bodies of some sharks including rays, angel sharks, and some bamboo sharks, possibly serving to frighten potential enemies.

Falcate. Sickle-shaped.

Family. One of the taxonomic groups of organisms, containing related genera; related families are grouped into orders.

FAO. United Nations Food and Agriculture Organisation.

Fauna. The community of animals peculiar to a region, area, specified environment or period.

Fecundity. A measure of the capacity of the maternal adult to produce young.

Filter-feeding. A form of feeding whereby suspended food particles are extracted from the water using the gill rakers.

Fin base. See base.

Fin insertion. See insertion.

Fin origin. See origin

Fin web. The usually thin, compressed part of the fin, distal to the base, that is supported by ceratotrichia alone (in aplesodic fins) or by ceratotrichia surrounding expanded fin radials or by radials only (plesodic fins).

Finning. The practice of slicing off a shark's valuable fins and discarding the body at sea.

First dorsal fin. The anteriormost dorsal fin of two, ranging in position from over the pectoral fin bases to far posterior on the precaudal tail.

FL. Fork length. A standard morphometric measurement used for sharks, from the tip of the snout to the fork of the caudal fin.

Flank. Sides of the shark excluding the head and tail regions.

Foetus. A later development stage of the unborn young of a live-bearing shark, that essentially resembles a small adult. Term foetuses are ready to be born, and generally have oral teeth and denticles erupting, have a colour pattern (often more striking than adults) and, in ovoviviparous sharks, have their yolk-sacs resorbed.

Free rear tips. The pectoral, pelvic, dorsal, and anal fins all have a movable rear corner or flap, the free rear tip, that is separated from the trunk or tail by a notch and an inner margin. In some sharks the rear tips of some fins are very elongated.

Fusiform. Spindle- or torpedo-shaped, elongated with tapering ends.

Galeomorph. Referring to the Galeomorphii.

Galeomorphii. The neoselachian superorder Galeomorphii, including the heterodontoid, lamnoid, orectoloboid, and carcharhinoid sharks.

Gastropod. Largest class of molluscs, many marine species have a hard calcareous shell but others do not have shells and are commonly known as sea slugs.

Generation. Measured as the average age of parents of newborn individuals within the population. Where generation length varies under threat, the more natural, i.e. pre-disturbance, generation length should be used.

Genus (plural, **genera**). One of the taxonomic groups of organisms, containing related species; related genera are grouped into families. The first of two scientific names assigned to each species.

Gestation period. The period between conception and birth in live-bearing animals.

Gill openings or **slits**. In elasmobranchs, the paired rows of five to seven transverse openings on the sides or underside of the head for the discharge of water through the gills. Chimaeras have their four gill openings hidden by a soft gill cover and discharge water through a single external gill opening.

Gill raker denticles. In the basking shark (Cetorhinidae), elongated denticles with hairlike cusps arranged in rows on the internal gill openings, which filter out planktonic organisms.

Habitat. The locality or environment in which an animal lives.

Hadal. The benthic zone of the deep trenches, 6000 to about 11,000m, from which no cartilaginous fishes have been observed or recorded (see p. 586).

Hadopelagic zone. The pelagic zone inside the deep trenches, 6000 to about 11,000m, from which no chondrichthyans have been observed or recorded (see p. 586).

Head. That part of a cartilaginous fish from its snout tip to the last (or first and only in chimaeras) gill slits.

Heterocercal. A caudal fin with the vertebral axis slanted dorsally into the fin base, which is also dorsally elevated.

Heterodontoid. A bullhead shark, horn shark, or Port Jackson shark, a member of the order Heterodontiformes, family Heterodontidae.

Heterodonty. In oral teeth, structural differences between teeth in various positions on the jaws, between teeth in the same position during different life stages, or between teeth in the same positions in the two sexes.

Hexanchoid. A cowshark or frilled shark, members of the order Hexanchiformes, and including the sixgill sharks, sevengill sharks, and frilled sharks.

Holocephalan. Referring to the Holocephali.

Holocephali. The subclass Holocephali (from Greek holos, entire, and kephalos, head), the living chimaeras and their numerous fossil relatives, a major subdivision of the class Chondrichthyes. The name is in reference to the fusion of the upper jaws or palatoquadrates to the skull in all living species and in many but not all fossils. The living holocephalans include three families in the order Chimaeriformes. The living species differ from

elasmobranchs in having four pairs of gill openings covered by a soft gill cover and with a single pair of external gill openings, oral teeth fused and reduced to three pairs of ever-growing tooth plates, an erectile first dorsal fin with a spine and a long, low spineless second dorsal.

Holotype. The single specimen designated as the type of a species by the original author at the time the species name and description was published.

Homodonty. In oral teeth, structural similarity between teeth in various positions on the jaws, between teeth in the same position during different life stages, or between teeth in the same positions in the two sexes.

Hypaxial web. The entire fin web below the vertebral column (vertebral axis) and the caudal base.

Hypercalcified structures. Parts of the skeleton that have developed extremely dense calcified cartilage, primarily during growth and maturation, which sometimes swell to knobs that distort and engulf existing cartilaginous structures. The rostrum of the Salmon Shark (Lamna ditropis) is a particularly impressive hypercalcified structure.

Hypocercal. A caudal fin with the vertebral axis slanted ventrally into the fin base, which is also ventrally depressed. Found only in angelsharks (Squatiniformes) among living sharks.

ICES. International Council for the Exploration of the Seas.

Ichthyologist. See ichthyology.

Ichthyology. The branch of zoology devoted to the study of fish.

Incurrent apertures. The anterior and ventrally facing openings of the nostrils, which direct water into the nasal cavities. These are usually lateral on the nostrils and anterolateral to the excurrent apertures, but may be anterior to the excurrent apertures only.

Indo-Pacific. An area covering the Indian Ocean and the western Pacific Ocean.

Inner margin. In precaudal fins including the pectoral, pelvic, dorsal and anal fins, the margin from the fin insertion to the rear tip.

Insertion. The posterior or rear end of the fin base in precaudal fins. See Origin.

Inshore. Shallow waters seaward side of the surf zone.

Insular shelf. See Shelf.

Interdorsal ridge. A ridge of skin on the midback of sharks, in a line between the first and second dorsal fins; particularly important in identifying grey sharks (genus Carcharhinus, family Carcharhinidae).

Intermediate teeth. Small oral teeth between the laterals and anteriors of the upper jaw, found in most lamnoids.

Intertidal zone. Shoreline between high and low tide marks that is diurnally exposed to the air by tidal movement (see p. 586).

Intestinal valve. A dermal flap inside the intestine, protruding into its cavity or lumen, and of various forms in different cartilaginous fishes. Often formed like a corkscrew or augur. See spiral, ring and scroll valves.

Invertebrate. Animals without a backbone.

IPOA-Sharks. International Plan of Action for the Conservation and Management of Sharks.

Island slope. See slope.

ITQ. Individual Transferable Quota. a catch limit or quota (a part of the Total Allowable Catch) allocated to an individual fisher or vessel owner which can either be harvested or sold to others.

IUCN. The World Conservation Union. A union of sovereign

states, government agencies and non-governmental organisations. www.iucn.org.

Jaws. See mandibular arch.

Juvenile. The life stage between hatching from egg or birth to sexual maturity and adulthood.

Keel. See caudal keel.

Labial cartilages. Paired internal cartilages that support the labial folds at the lateral angles of the mouth. Living neoselachians typically have two pairs of upper labial cartilages, the anterodorsal and posterodorsal labial cartilages, and one pair of ventral labial cartilages, but these are variably reduced and sometimes absent in many sharks.

Labial folds. Lobes of skin at the lateral angles of the mouth, usually with labial cartilages inside them, separated from the sides of the jaws by pockets of skin (labial grooves or furrows).

Labial furrows or **labial grooves**. Grooves around the mouth angles on the outer surface of the jaws of many cartilaginous fishes, isolating the labial folds. Primitively there is a distinct upper labial furrow above the mouth corner and a lower labial furrow below it.

Lagoon. Area of relatively shallow water that is partially or completely separated from the open water by a land barrier.

Lamnoid. A mackerel shark, a member of the order Lamniformes, and including the sandtiger sharks, Goblin Shark, Crocodile Shark, Megamouth Shark, thresher sharks, Basking Shark, and the makos, Porbeagle, Salmon Shark, and White Shark.

Lateral. Outwards, in the transverse direction towards the periphery of the body. See medial.

Lateral keel. See caudal keel.

Lateral line. A sensory canal system of pressure sensitive cells, that run along the sides of the body, often branching at the head, which detect water movements, disturbances and vibrations.

Lateral ridges. Reinforced ridges along the side of the body, one of which is often an extension of the caudal keel.

Lateral teeth. Large broad-rooted, compressed, high-crowned oral teeth on the sides of the jaws between the anteriors and laterals.

Lateral trunk denticle. A dermal denticle from the dorsolateral surface of the back below the first dorsal fin base.

Littoral zone. That part of the oceans over the continental and insular shelves, from the intertidal to 200m (see p. 586).

Live-bearing. A mode of reproduction in which female sharks give birth to young sharks, which are miniatures of the adults. See vivipary.

Longevity. The maximum expected age, on average, for a species or population in the absence of human-induced or fishing mortality.

Longline fisheries. Commercial fisheries that use a long line with baited hooks attached at intervals.

Lower lobe. In the caudal fin, the expanded distal end of the preventral and lower postventral margins, defined by the posterior notch of the caudal fin.

Lower origin. In the caudal fin, the anteroventral beginning of the hypaxial or lower web of the caudal fin, at the posterior end of the anal-caudal or pelvic-caudal space.

Lower postventral margin. In the caudal fin, the lower part of the postventral margin of the hypaxial web, from the ventral tip to the posterior notch.

Mandibular arch. The paired primary jaw cartilages of

sharks, including the dorsal palatoquadrates and the ventral Meckel's cartilages.

Medial. Inwards, in the transverse direction towards the middle of the body. See lateral.

Medial lobe. Small lobe on anterior nasal flap of nostril.

Medial teeth. Small oral teeth, generally symmetrical and with narrow roots, in one row at the symphysis and often in additional paired rows on either side of the symphysial one.

Mesopelagic zone. That part of the oceans beyond the continental and insular shelves, in oceanic waters, from about 200–1000m, the twilight zone where little light penetrates (see p. 586).

Migratory. The systematic (as opposed to random) movement of individuals from one place to another, often related to season and breeding or feeding. Knowledge of migratory patterns helps to manage shared stocks and to target aggregations of fish.

Molariform. In oral teeth, referring to a tooth with a broad flat crown with low cusps or none, for crushing hard-shelled invertebrate prey.

Monospecific. Genus containing only one known extant species.

MPA. Marine Protected Area. Any area of the intertidal or subtidal terrain, together with its overlying water and associated flora, fauna, historical and cultural features, which has been reserved by law or other effective means to protect part or all of the enclosed environment.

MSY. Maximum sustainable yield. The largest theoretical average catch or yield that can continuously be taken from a stock under existing environmental conditions without causing the stock to become depleted (assuming that removals and natural mortality are balanced by stable recruitment and growth).

Multiple ovipary. A mode of egg-laying or ovipary in which female sharks retain several pairs of cased eggs in the oviducts, in which embryos grow to advanced developmental stages. When deposited on the bottom (in captivity) the eggs may take less than a month to hatch. Found only in the scyliorhinid genus *Halaelurus*, with some uncertainty as to whether the eggs are normally retained in the oviducts until hatching. Eggs laid by these sharks may be abnormal, unusual, or an alternate to ovovivipary. The Whale Shark (*Rhincodon typus*) may have multiple retention of eggcases; near-term foetuses have been found in their uteri and eggcases with developing foetuses have been collected on the bottom.

Nape. Back of the neck region.

Nares. See nostrils.

Nasal aperture. On the neurocranium, an aperture in the anteroventral surface or floor of each nasal capsule, through which the nostril directs water into and out of the nasal organ.

Nasal curtain. Greatly expanded anterior nasal flaps, that reach the mouth to form the base of the nasoral grooves, helping direct water flow from the nostrils to the mouth.

Nasal flap. One of a set of dermal flaps associated with the nostrils, and serving to direct water into and out of them, including the anterior, posterior, and mesonarial flaps.

Nasoral grooves. Many bottom-dwelling, relatively inactive sharks have nasoral grooves, shallow or deep grooves on the ventral surface of the snout between the excurrent apertures and the mouth. The nasoral grooves are covered by expanded anterior nasal flaps that reach the mouth, and form water channels that allow the respiratory current to pull water by partial pressure into

and out of the nostrils and into the mouth. This allows the shark to actively irrigate its nasal cavities while sitting still or when slowly moving. Nasoral grooves occur in heterodontids, orectoloboids, chimaeroids, some carcharhinoids and most batoids.

Neoselachian. Referring to the Neoselachii.

Neoselachii. From Greek *neos*, new, and *selachos*, shark. The modern sharks, the subcohort Neoselachii, consisting of the living elasmobranchs and their immediate fossil relatives. See Euselachii.

Neotype. A specimen, not part of the original type series for a species, which is designated by a subsequent author, particularly if the holotype or other types have been destroyed, were never designated in the original description, or are presently useless.

Neritic province. A body of water from the shoreline to 200m deep that includes the continental shelf (see p. 586).

Neurocranium. In sharks, a box-shaped complex cartilaginous structure at the anterior end of the vertebral column, containing the brain, housing and supporting the nasal organs, eyes, ears, and other sense organs, and supporting the visceral arches or splanchnocranium. Also termed chondrocranium, chondroneurocranium, or endocranium.

Nictitating lower eyelid. In the ground sharks (order Carcharhiniformes), a movable lower eyelid that has special posterior eyelid muscles that lift it and, in some species, completely close the eye opening.

Nictitating upper eyelid. In parascylliid orectoloboids, the upper eyelid has anterior eyelid muscles that pull it down and close the eye opening, analogous to the nictitating lower eyelids of carcharhinoids.

Nomenclature. In biology, the application of distinctive names to groups of organisms.

Non-target species. Species which are not the subject of directed fishing effort (cf. target catch), including the bycatch and byproduct.

Nostrils. The external openings of the cavities of the nasal organs, or organs of smell.

Nursery ground. Area (often inshore and in sheltered waters and with abundant food organisms) where newborn sharks live and feed and grow to a certain size in their life cycle, then move elsewhere.

Oceanic. Referring to organisms inhabiting those parts of the oceans beyond the continental and insular shelves, over the continental slopes, ocean floor, sea mounts and abyssal trenches. The open ocean.

Oceanic ridge. Part of the ocean floor where new material is added to plate margins by magma eruptions forming an ever expanding parallel line of deepsea ridges.

Oceanic seamount. See seamount.

Oceanic trench. The deepest part of the ocean floor, typically formed when one tectonic plate slides under another (see p. 586).

Ocellus (plural **ocelli**). Eye-like marking in which the central colour is bordered in a full or broken ring of another colour.

Ocular. Of or associated with the eye.

Offshore. Waters that are over continental shelves and slope edges (see p. 586).

Olfactory. Parts of the body that are associated with the sense of smell.

Oophagy. Egg-eating, a mode of live-bearing reproduction employing uterine cannibalism; early foetuses deplete their yolk-sacks early and subsist by eating nutritive eggs produced by the mother.

Known in several lamnoid sharks, the carcharhinoid family Pseudotriakidae, and in the orectoloboid family Ginglymostomatidae (*Nebrius ferrugineus*).

Order. One of the taxonomic groups of organisms, containing related families; related orders are grouped into classes.

Orectoloboid. A carpet shark, a member of the order Orectolobiformes, including barbelthroat carpet sharks, blind sharks, wobbegong sharks, bamboosharks, epaulette sharks, nurse sharks, Zebra Shark, and Whale Shark.

Origin. The anterior or front end of the fin base in all fins. The caudal fin has upper and lower origins but no insertion. See insertion.

Ovipary or **oviparity.** A mode of reproduction in which female sharks deposit eggs enclosed in oblong or conical egg-cases on the bottom, which hatch in less than a month to more than a year, producing young sharks which are miniatures of the adults.

Ovoviviparity. See aplacental yolk-sac vivipary.

Ovovivipary. Generally equivalent to yolk-sac vivipary, live-bearing in which the young are nourished primarily by the yolk in the yolk-sac, which is gradually depleted and the yolk-sac resorbed until the young are ready to be born. Sometimes used to cover all forms of aplacental vivipary, including cannibal vivipary.

Paired fins. The pectoral and pelvic fins.

Parasite. An organism that lives in or on another organism, the host, and obtains nutrients from the host at it's expense.

Paratype. Specimens other than the holotype designated as paratypes by the author at the time of a new species description. If not designated they become non-type or voucher specimens.

Pectoral fins. A symmetrical pair of fins on each side of the trunk just behind the head and in front of the abdomen. These are present in all cartilaginous fishes and correspond to the forelimbs of a land vertebrate (a tetrapod or four-footed vertebrate).

Pelagic. Referring to organisms that live in the water column, not on the sea bottom.

Pelvic fin. A symmetrical pair of fins on the sides of the body between the abdomen and precaudal tail which correspond to the hindlimbs of land vertebrate (a tetrapod or four-footed vertebrate). Also, ventral fins.

Pharynx. Part of the gut that links the mouth to the oesophagus.

Photophores. Conspicuously pigmented small spots on the bodies of most lantern sharks (family Etmopteridae) and some kitefin sharks (family Dalatiidae). These are tiny round organs that are covered with a conspicuous dark pigment (melanin) and produce light by a low-temperature chemical reaction.

Phylum. In animal taxonomy it is one of the major groupings dividing organisms into general similarities and comprises of superclasses, classes and all other lower groupings.

Placenta. See yolk-sac placenta.

Placental vivipary. Live-bearing in which the young develop a yolk-sac placenta, which is apparently confined to the carcharhinoid sharks.

Placoid scale. See dermal denticle.

Poikilothermic. An organism whose body temperature varies in accordance with the temperature of its surroundings.

Population. A group of individuals of a species living in a particular area. (This is defined by IUCN (2001) as the total number of mature individuals of the taxon, with subpopulations defined as geographically or otherwise distinct groups in the population between which there is little demographic or genetic exchange (typically one successful migrant individual or gamete per year or less).)

Pores, pigmented. In a few sharks and skates, the pores for the lateral line and ampullae of Lorenzini are conspicuously black-pigmented, and look like little black specks.

Posterior. Rearwards, in the longitudinal direction of the caudal fin tip or tail filament. Also caudad.

Posterior margin. In precaudal fins, the margin from the fin apex to either the free rear tip (in sharks with distinct inner margins) or the fin insertion (for those without inner margins).

Posterior nasal flaps. Low flaps or ridges arising on the posterior edges of the excurrent apertures of the nostrils.

Posterior notch. In the caudal fin, the notch in the postventral margin dividing it into upper and lower parts.

Posterior tip. The posteriormost corner or end of the terminal lobe of the caudal fin.

Postventral margin. In the caudal fin, the margin from the ventral tip to the subterminal notch of the caudal fin. See lower and upper postventral margins.

Preanal ridges. A pair of low, short to long, narrow ridges on the midline of the caudal peduncle extending anteriorly from the anal fin base.

Precaudal fins. All fins in front of the caudal fin.

Precaudal pit. A depression at the upper and sometimes lower origin of the caudal fin where it joins the caudal peduncle.

Precaudal tail. That part of the tail from its base at the vent to the origins of the caudal fin

Predorsal ridge. A low narrow ridge of skin on the midline of the back anterior to the first dorsal fin base.

Predorsal space. Space between snout tip and dorsal fin(s).

Prenarial. In front of the nostrils (nares).

Preventral margin. In the caudal fin, the margin from the lower origin to the ventral tip of the caudal fin.

Pristiophoroid. A saw shark, order Pristiophoriformes, family Pristiophoridae.

Productivity. Relates to the birth, growth and mortality rates of a fish stock. Highly productive stocks are characterised by high birth, growth and mortality rates and can usually sustain higher exploitation rates and, if depleted, could recover more rapidly than comparatively less productive stocks.

Protrusible. Capable of being thrust forward, term normally applied to jaws.

Proximal. In any direction, at the near end of a structure.

Pupping ground. Area favoured for giving birth and depositing young.

Ray. See Batoid.

Red List of Threatened Species. Listing of the conservation status of the world's flora and fauna administered by IUCN. www.redlist.org

Reef channel. Channels that develop in the reef linking the water in the reef flat and lagoons to the open sea.

Reef face. Seaward face to the coral reef, often known as 'drop-off' and the most productive area of the reef (see p. 586).

Reef flat (Reef terrace). Horizontal part of the reef extending from the shore to, the reef crest, usually only a few metres down from the seas surface (see p. 586).

Ring valve. A type of spiral intestinal valve in which the

valve turns are very numerous and short resembling a stack of washers.

Rise. The transitional and less steep bottom zone from the lower slope to the abyss or ocean floor, between 2250m and 4500m. The rise can be divided into upper (2250–3000m), middle (3000–3750m) and lower (3750–4500m) rises. Few sharks are known from the rise, and those mostly from the upper rise. See Abyss, Hadal, shelf and slope (see p. 586).

Rostral keel. In squaloids, a large vertical plate on the underside of the rostrum and internasal septum, sometimes reduced, and with the cavities of the subnasal fenestrae on either side of the keel.

Rostrum. The cartilaginous anteriormost structure that supports the prenasal snout including lateral line canals and masses of ampullae. It is absent in a few nonbatoid sharks and in many batoids.

Row. In teeth, a single replicating line of teeth, running parallel to the jaw, which includes functional teeth and their replacements, derived from one tooth-producing area on the jaw.

Saddle. Darker dorsal marking that extends downwards either side of the shark but does not meet on the ventral surface.

Sargasso Sea. Central body of water surrounded by the North Atlantic circulation gyre.

Scientific name. The formal binomial name of a particular organism, consisting of the genus and specific names; a species only has one valid scientific name.

Scroll valve. A type of spiral intestinal valve in requiem and hammerhead sharks in which the valve has uncoiled and resembles a rolled-up bib or scroll.

Seagrass. The collective name for marine flowering plants that are found in shallow waters in mud and sand.

Seamount. A large isolated elevation in the open ocean, characteristically of conical form, that rises at least 1000m from the ocean floor; often a productive area for deepwater fisheries (see p. 586).

Second dorsal fin. The posteriormost dorsal fin of two in cartilaginous fishes, not always present, ranging in position from over the pelvic fin bases to far posterior on the precaudal tail.

Secondary caudal keels. Low horizontal dermal keels on the ventral base of the caudal fin in mackerel sharks (Lamnidae) and some somniosids.

Secondary lower eyelid. The eyelid below or lateral to the nictitating lower eyelid, separated from it by a subocular groove or pocket, and, in many carcharhinoids with internal nictitating lower eyelids, functionally replacing them as lower eyelids.

Seine netting. A fishing method using nets to surround an area of water where the ends of the nets are drawn together to encircle the fish (includes purse seine and Danish seine netting).

Semi-pelagic. Occasionally living on the bottom but is found mainly far off the bottom in the water column above.

Series. In teeth, a line of teeth running from the front of the jaw inward.

Serrations. In oral teeth, minute teeth formed by the cutting edge of the crown that enhance the slicing abilities of the teeth.

Shark. Generally used for cylindrical or flattened cartilaginous fishes with five to seven external gill openings on the sides of their heads, pectoral fins that are not attached to the head above the gill openings, and a large, stout tail with a large caudal fin; that is, all

living elasmobranchs except the rays or batoids. Living sharks in this sense are all members of the Neoselachii, the modern sharks and rays. Rays are essentially flattened sharks with the pectoral fins attached to their heads, while living chimaeras are also called ghost sharks or silver sharks. Hence shark is used here in an alternate and broader sense to include the rays and chimaeras.

Shelf, continental and insular. The sloping plateau-like area along the continents and islands between the shoreline and approximately 200m depth. It is roughly divided into inshore (intertidal to 100m), and offshore (100–200m) zones. The shelves have the greatest diversity of cartilaginous fishes. See Abyss, rise and slope (see p. 586).

Single ovipary. A mode of egg-laying or ovipary in which female sharks produce encased eggs in pairs, which are not retained in the oviducts and are deposited on the bottom. Embryos in the eggcases are at an early developmental stage, and take a few months to over a year to hatch. Found in almost all oviparous cartilaginous fishes.

Slope, continental and insular. The precipitous bottom zone from the edge of the outer shelf down to the submarine rise, between 200–2250m. The slope can be divided into upper (200–750m), middle (750–1500m) and lower (1500–2250m) slopes, of which the upper and middle slope has the highest diversity of deepwater benthic sharks. See Abyss, rise and shelf.

Snout. That part of a cartilaginous fish in front of its eyes and mouth, and including the nostrils.

Species. A group of interbreeding individuals with common characteristics that produce fertile (capable of reproducing) offspring and which are not able to interbreed with other such groups, that is, a population that is reproductively isolated from others; related species are grouped into genera.

Spiracle. A small to large opening between the eye and first gill opening of most sharks and rays, representing the modified gill opening between the jaws and hyoid (tongue) arch. Lost in chimaeras and some sharks.

Squalene. A long-chain oily hydrocarbon present in the liver oil of deepwater cartilaginous fishes. It is highly valued for industrial and medicinal use.

Squaloid. A dogfish shark, a member of the order Squaliformes, including bramble sharks, spiny dogfish, gulper sharks, lanternsharks, viper sharks, rough sharks, sleeper sharks, kitefin sharks, and cookiecutter sharks.

Squalomorph. Referring to the Squalomorphii.

Squalomorphii. The neoselachian superorder Squalomorphii, including the hexanchoid, squaloid, squatinoid, pristiophoroid, and batoid sharks.

Squatinoid. An angelshark, order Squatiniformes, family Squatinidae.

Squatinoid caudal fin. Angelsharks (Squatiniformes) are unique among living sharks in having hypocercal caudal fins that resemble inverted caudal fins of ordinary sharks. The dorsal margin is subdivided into a predorsal margin from the upper origin to its dorsal tip (analogous to the preventral margin and ventral tips in ordinary sharks), a postdorsal margin (like the postventral margin) from the dorsal tip to its supraterminal notch (similar to the subterminal notch), and a short supraterminal margin and large ventral terminal margin (similar to the subterminal and terminal margins) between the supraterminal notch and the ventral tip of the caudal. The ventral margin has a preventral margin forming a ventral lobe with the ventral tip and the ventral terminal margin.

SSC. Species Survival Commission**.** One of six volunteer commissions of IUCN.

SSG. Shark Specialist Group (part of the IUCN Species Survival Commission network).

Stock. A group of individuals of a species, which are under consideration from the point of view of actual or potential utilisation, and which occupy a well defined geographical range independent of other stocks of the same species. A stock is often regarded as an entity for management and assessment purposes.

Subcaudal keel. In a few dogfish sharks (family Centrophoridae), a single longitudinal dermal keel on the underside of the caudal peduncle.

Sublittoral zone (subtidal zone). The benthic region from low water mark to the outer limit of the continental shelf, usually about 200m deep (see p. 586).

Subquadrate. Nearly square.

Subterminal margin. In the caudal fin, the margin from the subterminal notch to the ventral beginning of the terminal margin.

Subterminal mouth or **ventral mouth**. Mouth located on the underside of the head, behind the snout.

Subterminal notch. On the caudal fin of most sharks, the notch in the lower distal end of the caudal fin, between the postventral and subterminal margins, and defining the anterior end of the terminal lobe.

Subtriangular. Nearly triangular.

Subtropical region. The intermediate region between the tropical and temperate zones.

Supraorbital crest. On the neurocranium, an arched horizontal plate of cartilage forming the dorsal edge of the orbit on each side; it arises from the medial orbital wall and the cranial roof and extends horizontally from the preorbital process to the postorbital process. It is apparently primitive for shark-like fishes but is variably reduced or absent in some living elasmobranchs.

Supraorbital or **brow ridge**. A dermal ridge above each eye, particularly well-developed in heterodontoids and some orectoloboids.

Sympatric. Different species which inhabit the same or overlapping geographic areas.

Symphyseal or **symphysial groove**. A longitudinal groove on the ventral surface of the lower jaw of some orectoloboid sharks, extending posteriorly from the lower symphysis.

Symphysial teeth. Larger oral teeth in one row on either side of the symphysis, distal to medials or alternates where present. Symphysials are broader than medials and usually have asymmetrical roots.

Symphysis. The midline of the upper and lower jaws, where the paired jaw cartilages articulate with each other.

Syntype. Two or more specimens used and mentioned in an original description of a species, where there was no designation of a holotype or a holotype and paratype(s) by the describer of the species.

Systematics. Scientific study of the kinds and diversity of organisms, including relationships between them.

Tail. That part of a cartilaginous fish from the cloacal opening or vent (anus in chimaeroids, which lack a cloaca) to the tip of the caudal fin or caudal filament, and including the anal fin, usually the second dorsal fin when present, and caudal fin.

Target catch. The catch which is the subject of directed fishing effort within a fishery; the catch consisting of the species primarily sought by fishers.

Taxon (plural **taxa**). A taxonomic group at any level in a classification. Thus the taxon Chondrichthyes is a class with two taxa as subclasses, Elasmobranchii and Holocephali, and the taxon *Galeorhinus*, a genus, has one taxon as a species, *G. galeus*.

Taxonomy. Often used as a synonym of systematics or classification, but narrowed by some researchers to the theoretical study of the principles of classification.

Temperate. Two circumglobal bands of moderate ocean temperatures usually ranging between 10° and 22°C at the surface, but highly variable due to currents and upwelling. Including the north temperate zone between the Tropic of Cancer, 23°27′N latitude, to the Arctic Circle, 66°30′N; and the south temperate zone between the Tropic of Capricorn, 23°27′S latitude, to the Antarctic Circle, 66°30′N (see p. 586).

Terminal lobe. In the caudal fin of most non-batoid sharks and at least one batoid, the free rear wedge-shaped lobe at the tip of the caudal fin, extending from the subterminal notch to the posterior tip.

Terminal margin. In the caudal fin, the margin from the ventral end of the subterminal margin to the posterior tip.

Terminal mouth. Mouth located at the very front of the animal. Most cartilaginous fishes have subterminal mouths, but some species (Viper Shark, wobbegongs, angelsharks, frilled sharks, Whale Shark, Megamouth Shark and mantas) have it terminal or nearly so.

Thorn. In many batoids, most angel sharks and the Bramble Shark, enlarged, flat conical denticles with a sharp, erect crown and a flattened base (which may grow as the shark grows).

TL. **Total length**. A standard morphometric measurement for sharks and some batoids, from the tip of snout or rostrum to the end of the upper lobe of the caudal fin.

Transverse. Across the long access of the body.

Tropical. Circumglobal band of warm coastal and oceanic water, usually above 22°C at the surface (but varying because of currents and upwelling), between the latitudes of 23°27′N (Tropic of Cancer) and 23°27′S (Tropic of Capricorn) and including the Equator (see p. 586).

Truncate. Blunt, abbreviated.

Trunk. That part of a cartilaginous fish between its head and tail, from the last gill openings to the vent, including the abdomen, back, pectoral and pelvic fins, and often the first dorsal fin.

Umbilical cord. A modified yolk stalk in placental viviparous sharks, carrying nutrients from the placenta to the foetus.

Undescribed species. An organism not yet formally described by science and so does not yet a have a formal binomial scientific name. Usually assigned a letter or number designation after the generic name, for example, *Gollum* sp. A is an undescribed species of false catshark belonging to the genus *Gollum*.

Unpaired fins. The dorsal, anal, and caudal fins.

Upper eyelid. The dorsal half of the eyelid, separated by a deep pocket (conjunctival fornix) from the eyeball. The upper eyelid fuses with the eyeball and the pocket is lost in all batoids.

Upper lobe. In the caudal fin, the entire fin including its base, epaxial and hypaxial webs but excepting the ventral lobe.

Upper origin. In the caudal fin, the anterodorsal beginning of the epaxial or upper web of the caudal fin, at the posterior end of the dorso-caudal space.

Upper postventral margin. In the caudal fin, the upper part of the postventral margin of the hypaxial web, from the posterior notch to the subterminal notch.

Uterine cannibalism or **cannibal vivipary**. A mode of reproduction in which foetuses deplete their yolk-sacks early and subsist by eating nutritive eggs produced by the mother (see oophagy) or first eat smaller siblings and then nutritive eggs (see adelphophagy).

Vent. The opening of the cloaca on the ventral surface of the body between the inner margins and at the level of the pelvic fin insertions.

Ventral fin. See pelvic fin.

Ventral lobe. See lower lobe

Ventral margin. In the caudal fin, the entire ventral margin from lower origin to posterior tip, either a continuous margin or variably subdivided into preventral, postventral, subterminal and terminal margins.

Ventral tip. In the caudal fin, the ventral apex of the caudal fin.

Ventral. Downwards, in the vertical direction of the abdomen. See dorsal.

Vertebra, plural vertebrae. A single unit of the vertebral column, including a vertebral centrum and associated cartilages that form arches and ribs.

Vertebral axis. That part of the vertebral column inside the base of the caudal fin.

Vertebral column. The entire set or string of vertebrae, or 'backbone', of a shark, from the rear of the chondrocranium to the end of the caudal base. Living elasmobranchs range from having as few as 60 vertebrae (some squaloids of the family Dalatiidae) to as many as 477 vertebrae (thresher sharks).

Viviparity. A reproductive mode where the maternal adult gives birth to live young. Encompasses aplacental viviparity and placental viviparity.

Vivipary. Used in two ways in recent literature, as being equivalent to placental vivipary only, that is for carcharhinoid sharks with a yolk-sac placenta; or for all forms of live-bearing or aplacental vivipary.

Yolk-sac. Almost all sharks start embryonic development somewhat like a chicken, as a large spherical yolky egg inside an elongated shell, the eggcase. A small disk of dividing cells represents the pre-embryo or blastula atop the huge yolk mass. The blastula expands around the sides and ventral surface of the yolk mass, and differentiates into an increasingly shark-like embryo, the yolk sac or bag-like structure containing the yolk, and a narrow tubular yolk stalk, between the abdomen of the embryo and the yolk-sac.

Yolk-sac placenta. An organ in the uterus of some ground sharks (order Carcharhiniformes), formed from the embryonic yolk-sac of the embryo and maternal uterine lining, through which maternal nutriment is passed to the embryo. It is analogous to the placenta of live-bearing mammals. There are several forms of yolk-sac placentas in carcharhinoid sharks, including entire, discoidal, globular, and columnar placentas (see Compagno, 1988).

Yolk-sac vivipary. Live-bearing in which the young are nourished primarily by the yolk in the yolk-sacs, which is gradually depleted and the yolk-sacs resorbed until the young are ready to be born

Appendix II: Oceans and seas

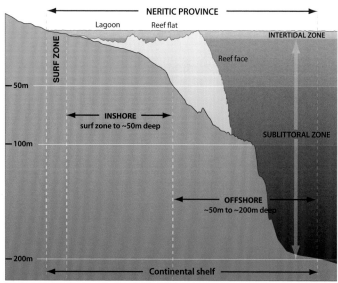

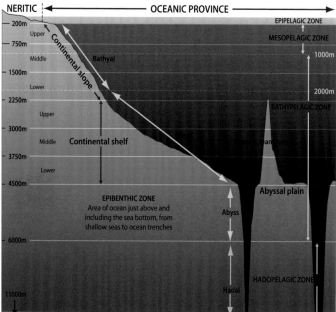

Figure 62: (above) continental shelf zones, including reef zones.

Figure 63: (right) oceanic zones.

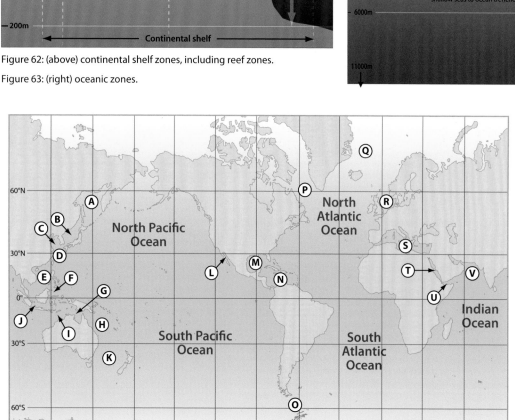

A Sea of Okhotsk 1,583,000km²; 859m

B Sea of Japan 978,000km²; 1752m

C Yellow Sea 380,000km²; 44m

D East China Sea 1,249,000km²; 188m

E South China Sea 3,500,000km²; 1212m

F Celebes Sea 280,000km²

G Arafura Sea 650,000km²

H Coral Sea 4,791,000km²; 2394m

I Timor Sea 610,000km²; 406m

J Java Sea 310,000km²; 46m

K Tasman Sea 2,300,000km²

L Gulf of California 160,000km²; 818m

M Gulf of Mexico 1,600,000km²; 1615m

N Caribbean Sea 2,754,000km²; 2200m

O Scotia Sea 900,000km²; 3500m

P Labrador Sea 841,000km²; 1898m

Q Greenland Sea 1,205,000km²; 1444m

R North Sea 750,000km²; 95m

S Mediterranean Sea 970,000km²; 1500m

T Red Sea 438,000km²; 490m

U Gulf of Aden 530,000km²; 1800m

V Arabian Sea 3,862,000km²; 4652m

Figure 64: Large bodies of water referred to in this book; figures refer to sea area and mean depth.

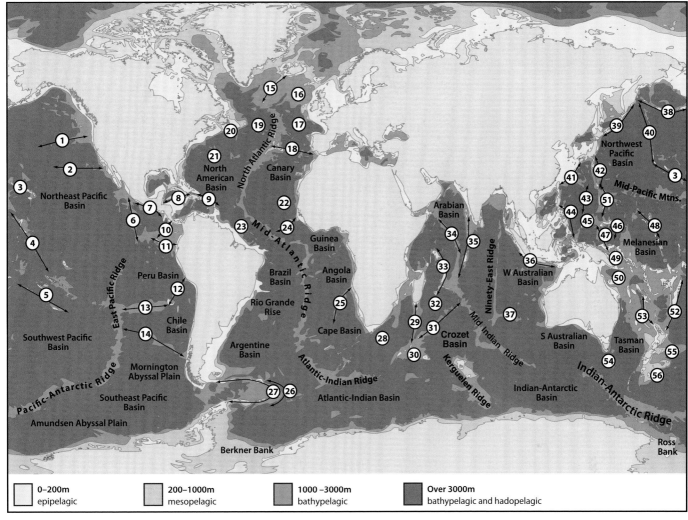

| 0–200m epipelagic | 200–1000m mesopelagic | 1000–3000m bathypelagic | Over 3000m bathypelagic and hadopelagic |

1	Mendocino Seascarp	17	NE Atlantic Basin– 5943m	33	Mascarene Ridge– 8m	49	New Britain Trench– 9140m
2	Murray Seascarp– 412m	18	Azores-Cape St. Vincent Range	34	Carlsberg Ridge	50	Coral Sea Basin
3	Hawaiian Ridge	19	Newfoundland Basin– 4685m	35	Maldive Ridge	51	Mariana Trench– 11022m
4	Line Islands	20	Grand Banks– 69m	36	Java Ridge & Trench– 7125m	52	Kermadec & Tonga Trench– 10047m
5	Austral Ridge	21	Bermuda Rise	37	W Australian Ridge	53	Norfolk Is. Ridge & Trough
6	East Pacific Rise	22	Cape Verde Plateau	38	Aleutian Trench– 782 m	54	Tasman Plateau– 770m
7	Mid. America Trench– 6662m	23	Guiana Basin	39	Kuril Trench– 10542 m	55	Chatham Rise
8	Cayman Trench– 7536m	24	Sierra Leone Rise	40	Emperor Seamount Chain	56	New Zealand Plateau
9	Puerto Rico Trench– 8742m	25	Walvis Ridge– 24m	41	Ryukyu Trench– 7181m		
10	Cocos Ridge	26	S Sandwich Trench– 8325m	42	Japan Trench– 10374m		
11	Carnegie Ridge	27	Scotia Ridge	43	Kyush-Palau Ridge		
12	Nazca Ridge	28	Agulhas Plateau	44	Philippine Trench– 10497m		
13	Sal y Gomez Ridge	29	Madagascar Ridge– 18m	45	Palau Trench– 8054m		
14	Challenger Fracture Zone	30	Crozet Plateau	46	E Caroline Basin– 7208m		
15	Reykjanes Ridge– 550m	31	SW Indian Ridge	47	New Guinea Rise		
16	Rockall Bank	32	Madagascar Basin– 6400m	48	Marshall Islands		

Figure 65: Major oceanic physical features.

Appendix III: Field observations

Many readers will use this book for identifying sharks that they see while diving or fishing and will probably mainly come across fairly common and widely distributed species. Even so, it is worth keeping good records of your observations because scientific knowledge of even some of the most widespread species is lacking. Anyone may be able to contribute important new information, particularly if they are in a part of the world where little research is underway.

Before handling a live shark, please remember that these animals live supported by water. Their internal organs (gut, liver etc.) do not have the surrounding protection and support provided by the ribs and abdominal muscles of land animals. They can, therefore, be extremely easily damaged if lifted out of the water without good abdominal support (i.e. in a sling). Please take as many measurements as possible in the water alongside the boat, don't take your shark out to hug it, and try to ensure that it is released in good condition! You may need to spend some time helping the shark to get oxygenated water flowing over the gills again, if it has been motionless for a while. Pushing or pulling it gently through the water (headfirst, naturally) should help.

Several codes of conduct and good practice guidance for the safe handling of sharks, that aim to minimise damage to both the catch and the angler, are available online. E.g. www.rac-spa.org/sites/default/files/doc_fish/gl_shark_ray_en.pdf and www.sharktrust.org/pages/faqs/category/angling-project.

Fish markets, particularly, are some of the best hunting grounds for poorly known or even completely unknown sharks. You may have to get up extremely early in the morning to see the fish as they are landed and before they are chopped up and sold and are advised to wear old clothes and shoes that can be washed (or thrown away) afterwards, but the experience can be really memorable. With luck, stall-holders will be enthusiastic and helpful if you show a keen interest in their stock.

Where possible the following information should be recorded, whether from live sharks in the water, on hook and line alongside a boat, onboard or onshore, or from dead specimens.

i. Name and address of observer or collector

ii. Date, time, location, habitat and water depth (where available).

iii. Other relevant observations (e.g. behaviour).

iv. Photographs of the whole specimen, particularly if it is an uncommon record or outside its usual geographic range, following the guidance on p. 591.

v. Measurements. Anglers like to record weights, but since these are very variable, depending upon time of year, point in the reproductive cycle etc., scientists prefer to record length as well, if not instead of weight. Total length should be recorded as a point-to-point distance, not over the curve of the body (which overestimates length). Precaudal length, from tip of snout to the tail origin, is another important measurement and can be easier to record than total length. Other useful measurements are illustrated in the following pages.

vi. Record whether male (a photograph of the claspers may be useful in judging maturity) or female, and whether there are any signs of pregnancy.

vii. If the specimen is dead, remove and dry a strip of teeth from each of the upper and lower jaws (label them!) or keep the entire jaws. The latter should be pinned to a board to dry in order to prevent distortion. If it is also possible to keep and dry the fins and the vertebral column, these can be very useful in confirming identification of species that are difficult to identify. DNA can also be extracted from dried tissue for scientific studies.

viii. If the specimen is dead, small and apparently unusual, it can be useful to keep the whole animal. Freezing is the easiest way to do this in the short term. In the longer term, it will be necessary to fix and preserve the specimen, using formalin and alcohol, but these procedures are usually undertaken in museums because of the difficulties of storing safely these toxic and flammable chemicals. For more information, see Compagno (2001) www.fao.org/3/x9293e/x9293e.pdf.

Remember, that many of the species described in this field guide may well turn out to be two or more very similar species. They may be distinguished by careful examination and measurements, by differences in size (they may mature at quite different lengths), by their use of different habitats and prey species, or because of different (but sometimes overlapping) distribution. Confusion between such similar species makes it very hard to understand their distribution, habitat, life history and other biological characters. Good records can help to overcome these problems.

Records of sharks, particularly of unusual species or species recorded outside their usual range, should be sent to the relevant national or state museum, or fishery department, of the country in whose waters they were recorded. County biological record centres may be interested in receiving UK records. At minimum, a good photograph, ideally the specimen itself, should accompany unusual records, particularly if it is the first record from that country or region, but please contact the curator with details before sending in any specimens (particularly if these are large). While every country needs a national fish collection for reference purposes, to help train its fisheries staff and researchers and for the use of visiting scientists, and should be offered new specimens, some institutes may not have the necessary facilities to curate and keep them in good condition. In these cases, or if more than one important specimen has been collected, it may be necessary to send them to one of the major international fish collections as well, or instead.

These are the key measurements recorded by scientists who are describing shark species. They may vary slightly between individuals of a single species but together help to differentiate between similar species.

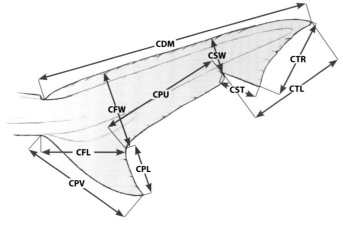

CDM Caudal fin dorsal margin
CPU Caudal fin upper postventral margin
CFW Caudal fin fork width
CSW Caudal fin subterminal width
CFL Caudal fin fork length
CPV Caudal fin preventral margin
CPL Caudal fin lower postventral margin
CST Caudal fin subterminal margin
CTR Caudal fin terminal margin
CTL Caudal fin terminal lobe

Figure 66. Caudal fin measurements.

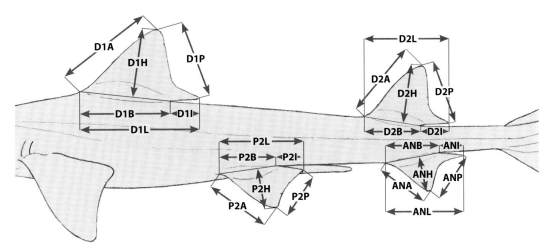

D1A	First dorsal fin anterior margin	**D2A** Second dorsal fin anterior margin	**P2A** Pelvic fin anterior margin	**ANA** Anal fin anterior margin
D1P	First dorsal fin posterior margin	**D2P** Second dorsal fin posterior margin	**P2P** Pelvic fin posterior margin	**ANP** Anal fin posterior margin
D1H	First dorsal fin height	**D2H** Second dorsal fin height	**P2H** Pelvic fin height	**ANH** Anal fin height
D1B	First dorsal fin base	**D2B** Second dorsal fin base	**P2B** Pelvic fin base	**ANB** Anal fin base
D1I	First dorsal fin inner margin	**D2I** Second dorsal fin inner margin	**P2I** Pelvic fin inner margin	**ANI** Anal fin inner margin
D1L	First dorsal fin length	**D2L** Second dorsal fin length	**P2L** Pelvic fin length	**ANL** Anal fin length

Figure 67. Dorsal, pelvic and anal fin measurements.

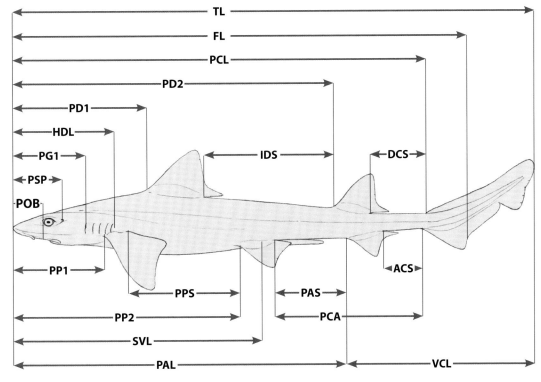

TL	Total length
FL	Fork length
PCL	Precaudal fin length
PD2	Pre-second dorsal fin length
PD1	Pre-first dorsal fin length
HDL	Head length
PG1	Prebranchial length
PSP	Prespiracular length
POB	Pre-orbital length
IDS	Interdorsal space
DCS	Dorsal fin insert to caudal fin upper origin
PP1	Pre-pectoral fin length
PP2	Pre-pelvic fin length
SVL	Snout to vent
PAL	Pre-anal fin length
PPS	Pectoral fin insert to pelvic fin origin
PAS	Pelvic fin insert to anal fin origin
ACS	Anal fin insert to caudal fin lower origin
PCA	Pelvic fin insert to caudal fin lower origin
VCL	Vent to caudal fin posterior tip

Figure 68. Main longitudinal measurements.

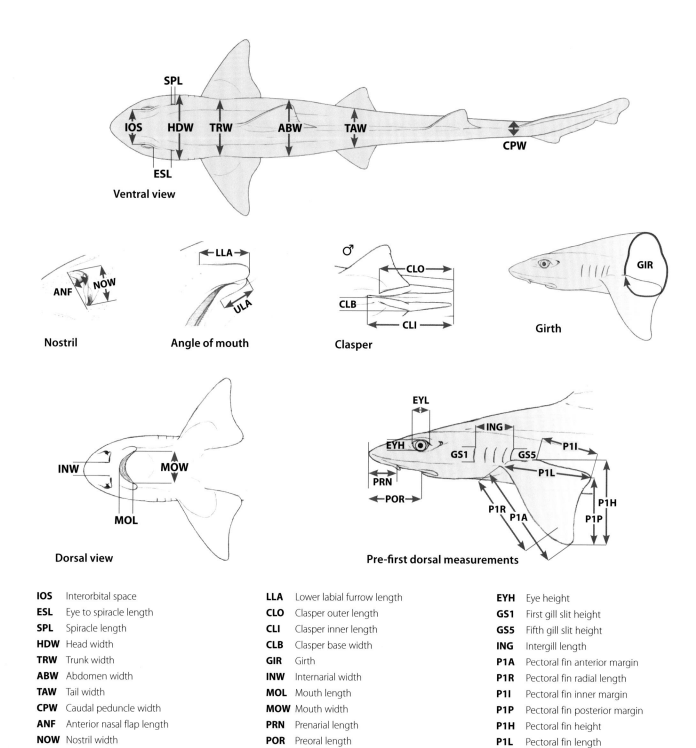

SPL

IOS **HDW** **TRW** **ABW** **TAW**

CPW

ESL

Ventral view

ANF **NOW**

Nostril

LLA

ULA

Angle of mouth

♂

CLO

CLB

CLI

Clasper

GIR

Girth

INW **MOW**

MOL

Dorsal view

EYL

EYH **ING**

GS1 **GS5** **P1I**

PRN **P1L**

POR **P1H**

P1R **P1A** **P1P**

Pre-first dorsal measurements

IOS	Interorbital space	**LLA**	Lower labial furrow length	**EYH**	Eye height
ESL	Eye to spiracle length	**CLO**	Clasper outer length	**GS1**	First gill slit height
SPL	Spiracle length	**CLI**	Clasper inner length	**GS5**	Fifth gill slit height
HDW	Head width	**CLB**	Clasper base width	**ING**	Intergill length
TRW	Trunk width	**GIR**	Girth	**P1A**	Pectoral fin anterior margin
ABW	Abdomen width	**INW**	Internarial width	**P1R**	Pectoral fin radial length
TAW	Tail width	**MOL**	Mouth length	**P1I**	Pectoral fin inner margin
CPW	Caudal peduncle width	**MOW**	Mouth width	**P1P**	Pectoral fin posterior margin
ANF	Anterior nasal flap length	**PRN**	Prenarial length	**P1H**	Pectoral fin height
NOW	Nostril width	**POR**	Preoral length	**P1L**	Pectoral fin length
ULA	Upper labial furrow length	**EYL**	Eye length		

Figure 69. Other measurements.

Taking photographs

If possible try and place a ruler or other measuring scale alongside the specimen; if no ruler is available, then some other object to show a size relationship. A handwritten label that includes a number, the date, location, other relevant capture information and includes the photographer's name is desirable. A plain coloured or an neutral background contrasting with the specimen's colour is best.

Take photographs in lateral view and in total length, and dorsal and ventral views, if possible with the fins erected and spread. Add close-ups of details, e.g. lateral and ventral view of head to gill openings or to origin of pectoral fins, mouth-nasal region, the jaws with dentition, individual fins, interdorsal ridge and colour marks or patterns. Close-ups of teeth are also helpful, especially for the sharks of the genus *Carcharhinus*.

Lateral view, total length.

Trunk fin markings.

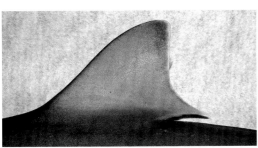

Ventral view, head to gill openings.

Dorsal view, head and pectoral fins.

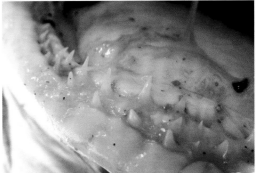

Upper and lower teeth.

Close-up of first dorsal fin.

Appendix IV: Fin identification

This guide covers the fins of many shark species that are listed in various regional and international agreements, including CITES Appendix II, and/or that are common in the international dried fin trade. These are: Porbeagle Shark, Oceanic Whitetip, Scalloped Hammerhead, Smooth Hammerhead, Great Hammerhead, Blue Shark and Shortfin Mako.

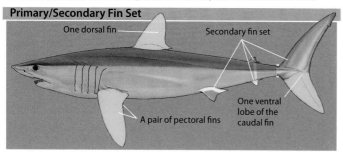

Primary/Secondary Fin Set

One dorsal fin

Secondary fin set

A pair of pectoral fins

One ventral lobe of the caudal fin

The following pages focus on dried, unprocessed first dorsal fins because these are the most easily identified of the traded fins for the species covered here. Caution is suggested when using this guide to identify dorsal fins less than 20cm across the base, to avoid possible mis-identification of samples from very small sharks. Pectoral fins descriptions help to confirm identification.

The key characteristics that can be used to separate the first dorsal fins of the above seven species from other types of shark fins in trade are described. Porbeagle and Oceanic Whitetip first dorsal fins can be rapidly identified to the species level based on the diagnostic white markings detailed here. The first dorsal fins of the three largest hammerhead sharks, as a group, which are the only hammerhead species that are common in trade, can also rapidly be separated from all other large sharks using two simple measurements that describe their characteristic shape (much taller than they are broad) and colour (dull brown or light grey). Identification of hammerhead shark fins to species requires examination of dorsal and pectoral fin sets or genetic testing. Blue and Shortfin Mako Shark fins are also described.

How to use this guide

Step 1. Distinguish first dorsal fins from other highly-valued traded fins: pectoral fins and lower caudal lobes (see below).

Step 2. Look for white first dorsal fin markings, and use the flowchart to identify either Porbeagle or Oceanic Whitetip Sharks or exclude many species with black fin markings.

Step 3. Take several simple measurements (opposite) to help identify hammerhead first dorsal fins, which are much taller than they are broad and are dull brown or light grey.

Step 4. Check for Blue and Shortfin Mako Shark fins.

STEP 1   **Distinguish first dorsal fins from pectoral fins and lower caudal lobes**

a. Check the fin colour on each side

Dorsal fins are the same colour on both sides (see both side views below). Pectoral fins are darker above (dorsal view) than underneath (ventral view), see below.

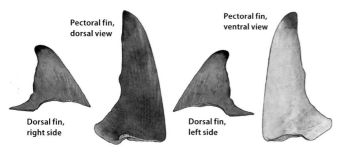

Pectoral fin, dorsal view

Pectoral fin, ventral view

Dorsal fin, right side

Dorsal fin, left side

---

b. Check the base of the fin

Dorsal fins (D) have a continuous row of closely spaced cartilaginous blocks running along almost the entire fin base. When looking at a cross section of the base of a lower caudal lobe (LC1), there is typically only a yellow, 'spongy' material called ceratotrichia, which is the valuable part of the lower caudal lobe. In some lower caudal lobes (LC2) there may be a small number of the cartilaginous blocks, but they are usually widely spaced and/or occur only along part of the fin base. Usually the lower caudal lobe has been cut along its entire base when removed from the shark; in contrast, dorsal fins frequently have a free rear tip that is fully intact.

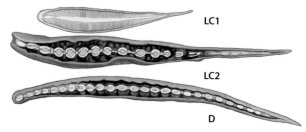

LC1

LC2

D

STEP 2   **Identify Porbeagle and Oceanic Whitetip first dorsal fins**

A STOP in the flowchart below indicates that the fin is not from a species covered in this guide.

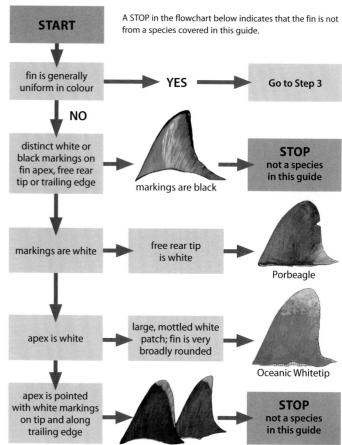

START

fin is generally uniform in colour → YES → Go to Step 3

NO

distinct white or black markings on fin apex, free rear tip or trailing edge → markings are black → **STOP** not a species in this guide

markings are white → free rear tip is white → Porbeagle

apex is white → large, mottled white patch; fin is very broadly rounded → Oceanic Whitetip

apex is pointed with white markings on tip and along trailing edge → **STOP** not a species in this guide

Take fin measurements

1 Measure fin origin to apex (O–A) with a flexible tape measure.
2 Measure the fin width (W) at the halfway point of O–A (i.e. if O–A is 10cm, measure W at 5cm along O–A).
3 Divide O–A by W (O–A/W).

Origin to apex and fin width (measured from leading edge to trailing edge) are the landmarks found to be the most useful for species identification purposes, as measurements based on fin height, fin base and free rear tip are often too variable and dependent on cut and condition of the fin.

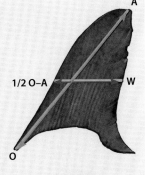

1/2 O–A · W

O · A

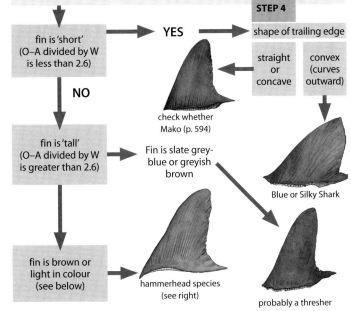

fin is 'short' (O–A divided by W is less than 2.6)

YES

STEP 4

shape of trailing edge

straight or concave

convex (curves outward)

NO

check whether Mako (p. 594)

fin is 'tall' (O–A divided by W is greater than 2.6)

Fin is slate grey-blue or greyish brown

Blue or Silky Shark

fin is brown or light in colour (see below)

hammerhead species (see right)

probably a thresher

Distinguishing hammerhead dorsals from other tall fins (mako and thresher sharks)

Thresher and hammerhead shark first dorsal fins are very tall and slender, mako slightly shorter. Thresher and mako fins are slate grey or dark greyish brown. Great Hammerhead first dorsal fins have a distinctive curved shape and are much lighter. Scalloped and Smooth Hammerhead first dorsal fins are similar in shape to thresher shark dorsal fins, but much lighter in colour and usually light brown not grey. Shortfin Mako Shark first dorsal fins are dark greyish-brown or slate grey, with a steep-angled leading edge, smooth texture and short free rear tip.

Shortfin Mako
Isurus oxyrinchus

Thresher
Alopias vulpinus

Great Hammerhead
Sphyrna mokarran

Scalloped Hammerhead
Sphyrna lewini

Distinguishing hammerhead dorsals from other tall fins (guitarfish and Blacktip Sharks)

Dorsal fins that are tall and slender and dull brown or light greyish-brown are probably one of three species of hammerhead sharks: Great *Sphyrna mokarran*, Scalloped *S. lewini* or Smooth *S. zygaena*. See descriptions on the following page.

Tall dorsal fins can also come from several species of guitarfish or Blacktip Sharks *Carcharhinus limbatus*. In guitarfish first dorsal fins, cartilaginous blocks do not extend across the entire fin base (Image **A**). In hammerheads, these cartilaginous blocks are present along almost the entire fin base (Image **A**). Guitarfish dorsal fins also exhibit a glossy sheen (Image **B**), and some species also have white spots, unlike the dull brown, uniform coloration of hammerhead dorsal fins.

Some Blacktip Shark first dorsal fins exhibit O–A/W that is close to or slightly greater than 2.5. However, they often (but not always) have a black spot on the dorsal fin apex, and the fin has a glossy appearance that is unlike the dull surface of the hammerheads (Image **C**).

If fins from a single shark have been kept together in sets and are available for comparison, Blacktip Shark pectoral fins are also longer and more slender than the short, broad fins of the hammerheads (Image **D**).

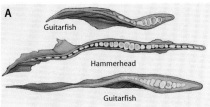

A

Guitarfish

Hammerhead

Guitarfish

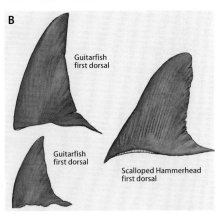

B

Guitarfish first dorsal

Guitarfish first dorsal

Scalloped Hammerhead first dorsal

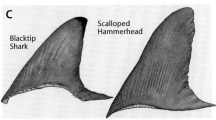

C

Blacktip Shark

Scalloped Hammerhead

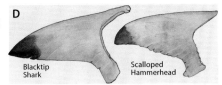

D

Blacktip Shark

Scalloped Hammerhead

PORBEAGLE SHARK *Lamna nasus*   pp. 316 and 323

First dorsal fin: dark blue/black to dark greyish brown, rounded apex with white patch on lower trailing edge onto free rear tip.

First dorsal fin

Dorsal view

Pectoral fins

Ventral view

Pectoral fins: short, rounded at apex; ventral surface has dusky coloration from apex throughout midsection of fin and along leading edge.

Conservation Status: CITES Appendix II and other international and regional agreements.

OCEANIC WHITETIP *Carcharhinus longimanus* pp. 506 and 537

First dorsal fin: large and broadly rounded paddle-like); mottled white colour at apex.

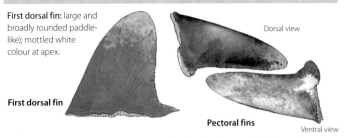

First dorsal fin

Dorsal view

Pectoral fins

Ventral view

Pectoral fins: long, broadly rounded at apex; dorsal surface has mottled white colour at apex, ventral surface is typically white but can have mottled brown coloration.
- mottled white colour also present on caudal fin (upper and lower lobe).
- very small juveniles may have mottled black coloration on dorsal, pectoral and caudal fins.

Conservation Status: CITES Appendix II, prohibited in many Atlantic and Pacific pelagic fisheries (see p. 83).

SCALLOPED HAMMERHEAD *Sphyrna lewini* pp. 570 and 574

First dorsal fin: tall, flattening out toward apex; straight to moderately curved trailing edge (similar to Smooth Hammerhead, less slender than Great Hammerhead first dorsal fin).

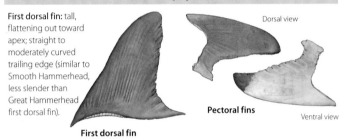

Dorsal view

Pectoral fins

Ventral view

First dorsal fin

Pectoral fins: short and broad with black tips visible at the apex on ventral side.

Conservation Status: CITES Appendix II and prohibited by some international and regional fisheries agreements (see p. 83).

SMOOTH HAMMERHEAD *Sphyrna zygaena* pp. 572 and 579

First dorsal fin: tall, sloping more at apex; moderately curved trailing edge (similar to Scalloped Hammerhead, less slender than Great Hammerhead first dorsal fin).

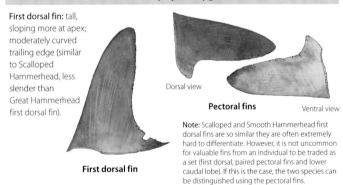

Dorsal view

Pectoral fins

Ventral view

Note: Scalloped and Smooth Hammerhead first dorsal fins are so similar they are often extremely hard to differentiate. However, it is not uncommon for valuable fins from an individual to be traded as a set (first dorsal, paired pectoral fins and lower caudal lobe). If this is the case, the two species can be distinguished using the pectoral fins.

First dorsal fin

Pectoral fins: short and broad with faint to no markings on ventral side.

Conservation Status: CITES Appendix II and a prohibited species in some regional fisheries agreements (see p. 83).

GREAT HAMMERHEAD *Sphyrna mokarran* pp. 572 and 577

First dorsal fin: tall, slender from leading edge to trailing edge; elongated and pointed at apex.

Dorsal view

Ventral view

Pectoral fins

Dorsal fin

Note: Small to moderate-sized *S. mokarran* first dorsal fins are similar to those of the Winghead Shark *Eusphyra blochii* (p. 574), but the latter are extremely rare in trade and only landed in the northern Indo-Pacific.

Pectoral fins: Pointed apex, moderately curved along trailing edge with dusky colour at apex on ventral side and often along trailing edge.

Conservation Status: CITES Appendix II and a prohibited species in some regional fisheries agreements (see p. 83).

BLUE SHARK *Prionace glauca* pp. 508 and 562

First dorsal fin: dark grey-blue to greyish brown; rounded apex, trailing edge convex, leading edge with rather a shallow angle; moderate free rear tip. Silky Shark dorsal is same shape but light grey; its pectoral fins are much shorter.

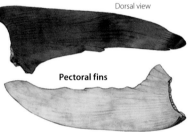

Dorsal view

First dorsal fin

Pectoral fins

Ventral view

Pectoral fins: dorsal surface dark grey-blue or dark greyish brown, ventral white and unmarked; long and slender from leading edge to trailing edge, radial cartilage visible under skin.

Conservation Status: no catch limits (in 2020); the most abundant species in fin trade (see p. 83).

SHORTFIN MAKO *Isurus oxyrinchus* pp. 316 and 321

First dorsal fin: dark greyish-brown or slate grey; very erect steep-angled leading edge, moderately straight trailing edge, short free rear tip, thick base; smooth texture.

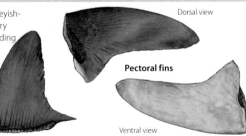

Dorsal view

Pectoral fins

Ventral view

First dorsal fin

Pectoral fins: dorsal surface dark greyish brown or slate grey, white margin on edge of free rear tip, ventral surface white unmarked; rounded apex, moderately short and broad from leading edge to trailing edge.

Conservation Status: CITES Appendix II and a prohibited species in some regional fisheries agreements (see p. 83).

For terminology used in this section, see p. 580. This information is reproduced from *Identifying Shark Fins: Oceanic Whitetip, Porbeagle and Hammerheads* (Stony Brook University/Pew Environment group, D.L. Abercrombie and D.D. Chapman authors) and Abercrombie *et al.* 2013. For more information on fin identification, see **www.fisheries.noaa.gov/national/international-affairs/shark-conservation** and **www.fao.org/ipoa-sharks/tools/software/isharkfin/en/**

Appendix V: Tooth identification

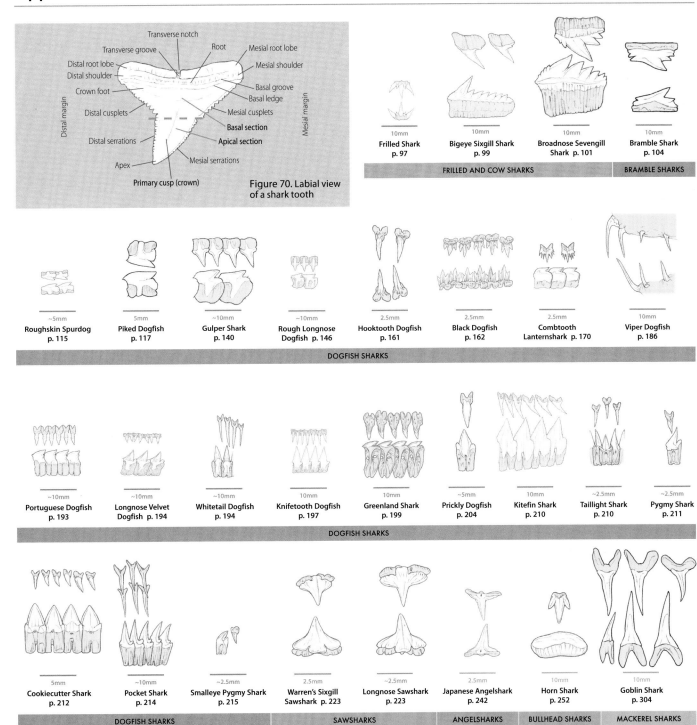

Figure 70. Labial view of a shark tooth

Transverse notch
Transverse groove
Root
Mesial root lobe
Distal root lobe
Mesial shoulder
Distal shoulder
Crown foot
Basal groove
Basal ledge
Distal margin
Distal cusplets
Mesial cusplets
Mesial margin
Basal section
Distal serrations
Apical section
Apex
Mesial serrations
Primary cusp (crown)

Frilled Shark p. 97
Bigeye Sixgill Shark p. 99
Broadnose Sevengill Shark p. 101
Bramble Shark p. 104

FRILLED AND COW SHARKS · **BRAMBLE SHARKS**

Roughskin Spurdog p. 115
Piked Dogfish p. 117
Gulper Shark p. 140
Rough Longnose Dogfish p. 146
Hooktooth Dogfish p. 161
Black Dogfish p. 162
Combtooth Lanternshark p. 170
Viper Dogfish p. 186

DOGFISH SHARKS

Portuguese Dogfish p. 193
Longnose Velvet Dogfish p. 194
Whitetail Dogfish p. 194
Knifetooth Dogfish p. 197
Greenland Shark p. 199
Prickly Dogfish p. 204
Kitefin Shark p. 210
Taillight Shark p. 210
Pygmy Shark p. 211

DOGFISH SHARKS

Cookiecutter Shark p. 212
Pocket Shark p. 214
Smalleye Pygmy Shark p. 215
Warren's Sixgill Sawshark p. 223
Longnose Sawshark p. 223
Japanese Angelshark p. 242
Horn Shark p. 252
Goblin Shark p. 304

DOGFISH SHARKS · **SAWSHARKS** · **ANGELSHARKS** · **BULLHEAD SHARKS** · **MACKEREL SHARKS**

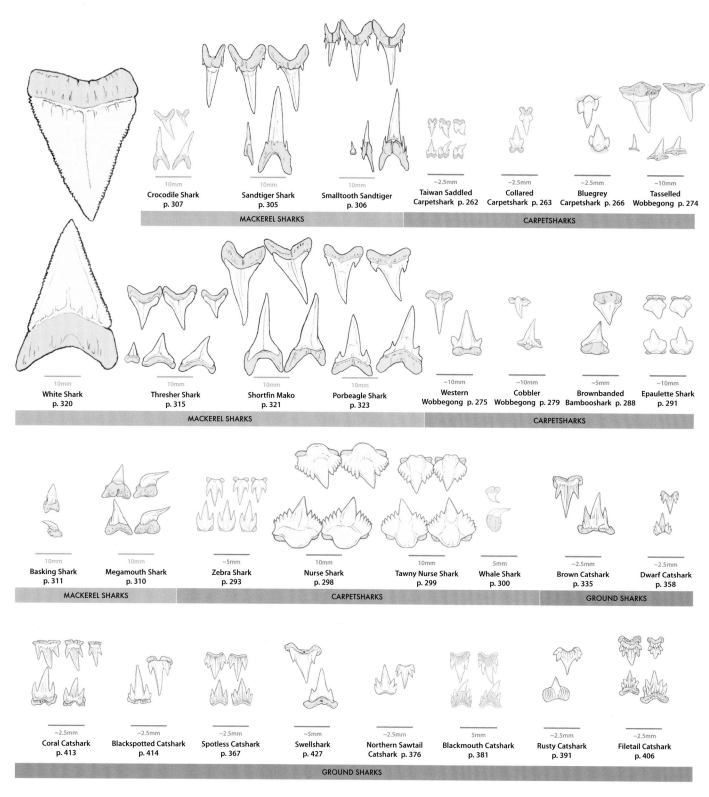

Crocodile Shark
p. 307

10mm

MACKEREL SHARKS

Sandtiger Shark
p. 305

10mm

Smalltooth Sandtiger
p. 306

10mm

Taiwan Saddled
Carpetshark p. 262

~2.5mm

Collared
Carpetshark p. 263

~2.5mm

Bluegrey
Carpetshark p. 266

~2.5mm

CARPETSHARKS

Tasselled
Wobbegong p. 274

~10mm

White Shark
p. 320

10mm

MACKEREL SHARKS

Thresher Shark
p. 315

10mm

Shortfin Mako
p. 321

10mm

Porbeagle Shark
p. 323

10mm

Western
Wobbegong p. 275

~10mm

Cobbler
Wobbegong p. 279

~10mm

Brownbanded
Bambooshark p. 288

~5mm

CARPETSHARKS

Epaulette Shark
p. 291

~10mm

Basking Shark
p. 311

10mm

MACKEREL SHARKS

Megamouth Shark
p. 310

10mm

Zebra Shark
p. 293

~5mm

Nurse Shark
p. 298

10mm

CARPETSHARKS

Tawny Nurse Shark
p. 299

10mm

Whale Shark
p. 300

5mm

Brown Catshark
p. 335

~2.5mm

GROUND SHARKS

Dwarf Catshark
p. 358

~2.5mm

Coral Catshark
p. 413

~2.5mm

Blackspotted Catshark
p. 414

~2.5mm

Spotless Catshark
p. 367

~2.5mm

Swellshark
p. 427

~5mm

Northern Sawtail
Catshark p. 376

~2.5mm

Blackmouth Catshark
p. 381

5mm

Rusty Catshark
p. 391

~2.5mm

Filetail Catshark
p. 406

~2.5mm

GROUND SHARKS

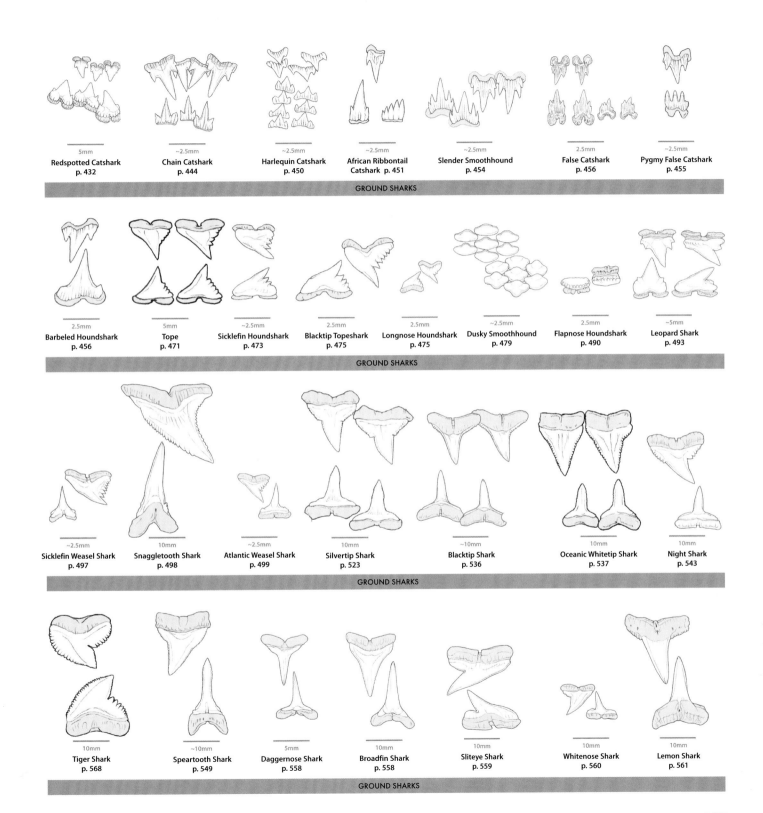

Redspotted Catshark
p. 432

Chain Catshark
p. 444

Harlequin Catshark
p. 450

African Ribbontail Catshark p. 451

Slender Smoothhound
p. 454

False Catshark
p. 456

Pygmy False Catshark
p. 455

GROUND SHARKS

Barbeled Houndshark
p. 456

Tope
p. 471

Sicklefin Houndshark
p. 473

Blacktip Topeshark
p. 475

Longnose Houndshark
p. 475

Dusky Smoothhound
p. 479

Flapnose Houndshark
p. 490

Leopard Shark
p. 493

GROUND SHARKS

Sicklefin Weasel Shark
p. 497

Snaggletooth Shark
p. 498

Atlantic Weasel Shark
p. 499

Silvertip Shark
p. 523

Blacktip Shark
p. 536

Oceanic Whitetip Shark
p. 537

Night Shark
p. 543

GROUND SHARKS

Tiger Shark
p. 568

Speartooth Shark
p. 549

Daggernose Shark
p. 558

Broadfin Shark
p. 558

Sliteye Shark
p. 559

Whitenose Shark
p. 560

Lemon Shark
p. 561

GROUND SHARKS

2.5mm

Blue Shark
p. 562

10mm

Milk Shark
p. 563

~5mm

Atlantic Spadenose Shark
p. 566

10mm

Whitetip Reef Shark
p. 567

5mm

Winghead Shark
p. 574

10mm

Scalloped Hammerhead
p. 576

10mm

Great Hammerhead
p. 577

GROUND SHARKS

Sources for the introduction

Anderson, S.D. *et al.* 2011. Long-term individual identification and site fidelity of white sharks, *Carcharodon carcharias*, off California using dorsal fins. *Marine Biology*. DOI: 10.1007/s00227-011-1643-5.

Boussarie *et al.* 2018. Environmental DNA illuminates the dark diversity of sharks. *Science Advances* 02 May 2018:Vol. 4, no. 5, eaap9661. DOI: 10.1126/sciadv. aap9661.

Brazeau, M.D. *et al.* 2020. Endochondral bone in an Early Devonian 'placoderm' from Mongolia. *Nat Ecol Evol* 4, 1477–1484.

Cagua, E.F. *et al.* 2014. Whale shark economics: A valuation of wildlife tourism in South Ari atoll, Maldives. *PeerJ*, 2, e515.

Cardeñosa, D. *et al.* 2018. CITES-listed sharks remain among the top species in the contemporary fin trade. *Conserv. Lett.* 43, e12457-7.

Cardeñosa, D. *et al. 2019*. Small fins, large trade: a snapshot of the species composition of low-value shark fins in the Hong Kong markets. *Animal Conservation*. doi:10.1111/acv.12529

Chapman, D.D. *et al.* 2015. There and back again: a review of residency and return migrations in sharks, with implications for population structure and management. *Annu. Rev. Mar.* Sci. 7, 547–570.

Cisneros-Montemayor, A. *et al.* 2013. Global economic value of shark ecotourism: implications for conservation. *Oryx*. Available on CJO2013. doi:10.1017/ S0030605312001718.

Clarke, S.C. *et al.* 2006a. Identification of Shark Species Composition and Proportion in the Hong Kong Shark Fin Market Based on Molecular Genetics and Trade Records. *Conservation Biology*. Volume 20, No. 1, 201–211.

Clarke, S.C. *et al.* 2006b. Global estimates of shark catches using trade records from commercial markets. *Ecology Letters*. 9: 1115–1126.

Dent, F. and Clarke, S. 2015. *State of the global market for shark products*. FAO Fisheries and Aquaculture Technical Paper No. 590, Rome, FAO, 187 pp.

Diop, M. and Dossa, J. 2011. *30 years of shark fishing in West Africa*. FIBA, PCRM, CSRP/SRFC.

Dulvy, N.K. *et al.* 2008. You can swim but you can't hide: the global status and conservation of oceanic pelagic sharks and rays. *Aquatic Conservation*. 18:459– 482.

Dulvy, N.K. *et al.* 2014. Extinction risk and conservation of the world's sharks and rays. *eLIFE* 2014;3:e00590. DOI: 10.7554/eLife.00590.

FAO. 2020. The State of World Fisheries and Aquaculture 2020. In brief. Sustainability in action. Rome. https://doi.org/10.4060/ca9231en

Ferter, K. *et al.* 2013. Unexpectedly high catch-and-release rates in European marine recreational fisheries: implications for science and management. *ICES Journal of Marine Science* 70 (7), 1319-1329.

Fields, A.T. *et al.* 2017. Species composition of the international shark fin trade assessed through a retail-market survey in Hong Kong. *Conserv. Biol.* 32, 376– 389.

Fowler, S. and Seret, B. 2010. *Shark Fins in Europe: Implications for reforming the EU finning ban*. European Elasmobranch Association and IUCN Shark Specialist Group.

Fowler, S.L. and Valenti, S. 2007. *Review of migratory Chondrichthyan Fishes. CMS Technical Report Series* No. 15. IUCN–The World Conservation Union / Convention on Migratory Species (CMS).

Fowler, S.L. *et al.* (eds). 2002. Elasmobranch Biodiversity, Conservation and Management. Occasional Paper of the IUCN SSC No. 25. IUCN, Switzerland, and UK. x + 256 pp.

Fowler, S.L. *et al.* 2003. *The conservation status of Australasian chondrichthyans*: Report of the IUCN Shark Specialist Group Australia and Oceania Regional Red List Workshop. IUCN Shark Specialist Group Newbury, UK, and the University of Queensland.

Fowler, S.L. *et al.* 2005. *Sharks, Rays and Chimaeras: The Status of the Chondrichthyan Fishes*. IUCN SSC Shark Specialist Group. IUCN, Gland, Switzerland and Cambridge, UK. x + 461pp.

Freire, K.M.F, *et al.* 2020. Estimating Global Catches of Marine Recreational Fisheries. *Frontiers in Marine Science* 7. https://www.frontiersin.org/article/10.3389/ fmars.2020.00012.

Gallagher, A.J. and Hammerschlag N. 2011. Global shark currency: the distribution, frequency, and economic value of shark ecotourism. *Curr Issues Tourism*. 8:797– 812.

Gallagher, A.J. and Huveneers, C. 2018. Emerging challenges to shark-diving tourism. *Marine Policy* 96:9-12. DOI: 10.1016/j.marpol.2018.07.009

Gallagher, A.J. *et al.* (2015). Biological effects, conservation potential, and research priorities of shark diving tourism. Biological Conservation, 184, 365–379.

Haas, A.R. *et al.* 2017. The contemporary economic value of elasmobranchs in The Bahamas: Reaping the rewards of 25 years of stewardship and conservation. *Biological Conservation*, 207, 55–63. https://doi.org/10.1016/j.biocon. 2017.01.007.

Heupel M.R. *et al.* 2014. Sizing up the ecological role of sharks as predators. *Mar Ecol Prog Ser* 495:291-298. https://doi.org/10.3354/meps10597

Huveneers, C. *et al.* 2017. The economic value of shark-diving tourism in Australia. *Reviews in Fish Biology and Fisheries*, Vol. 27, No. 3, pp. 665-680.

Hyder, K. *et al.* 2018. Recreational sea fishing in Europe in a global context— Participation rates, fishing effort, expenditure, and implications for monitoring and assessment. *Fish and Fisheries* 19 (2), 225-243.

Hyder, K. *et al.* 2020. Marine recreational fisheries—current state and future opportunities. *ICES Journal of Marine Science* 77 (6), 2171–2180.

IPBES 2019. *Global assessment report on biodiversity and ecosystem services of the Intergovernmental Science-Policy Platform on Biodiversity and Ecosystem Services*. IPBES Secretariat, Bonn, Germany.

IPBES 2019. IBPES Global Assessment Report Key Statistics and Facts on Oceans and Fishing. https://www.facebook.com/IPBES/videos/361549754465850

Jabado *et al.* 2017. *The conservation status of sharks, rays, and chimaeras in the Arabian Sea and adjacent waters*. Environment Agency – Abu Dhabi, UAE and IUCN Species Survival Commission Shark Specialist Group, Vancouver, Canada. 236 pp.

Jorgensen, S.J. *et al.* 2012. Eating or Meeting? Cluster Analysis Reveals Intricacies of White Shark (*Carcharodon carcharias*) Migration and Offshore Behavior. *PLoS ONE*. 7(10): e47819.

Kempster R.M. *et al.* 2016. How Close is too Close? The Effect of a Non-Lethal Electric Shark Deterrent on White Shark Behaviour. *PLoS ONE* 11(7): e0157717. doi:10.1371/journal.pone.0157717

Kohler, N.E. and Turner, P.A. 2019. Distributions and Movements of Atlantic Shark Species: A 52-Year Retrospective Atlas of Mark and Recapture Data. *Marine Fisheries Review*. doi: https://doi.org/10.7755/MFR.81.2.1

MacNeil, M.A. *et al.* 2020. Global status and conservation potential of reef sharks. *Nature* 583, 801–806. https://doi.org/10.1038/s41586-020-2519-y

Mustika P.L.K. *et al.* 2020. The Economic Value of Shark and Ray Tourism in Indonesia and Its Role in Delivering Conservation Outcomes. *Front. Mar. Sci.* 7:261. doi: 10.3389/fmars.2020.00261

Naylor, G.J.P. *et al.* 2012. A DNA sequence-based approach to the identification of shark and ray species and its implications for global elasmobranch diversity and parasitology. *Bull. Am. Mus. Nat. Hist.* 367, 262 pp.

Okes, N. and Sant, G. 2019. *An overview of major global shark traders, catchers and species*. TRAFFIC. https://www.traffic.org/publications/reports/major-global- shark-traders-catchers-and-species/

Pacoureau, N. *et al.* 2021. Half a century of global decline in oceanic sharks and rays. *Nature*, 589(7843), 567-571.

Pinhal, D. *et al.* 2012. Cryptic hammerhead shark lineage occurrence in the western South Atlantic revealed by DNA analysis. *Mar Biol.* 159:829–836.

Potts, W.M. *et al.* 2020. What constitutes effective governance of recreational fisheries? – A global review. *Fish and Fisheries* 21: 91–103. https://doi. org/10.1111/faf.12417

Queiroz, N. *et al.* 2019. Global spatial risk assessment of sharks under the footprint of fisheries. *Nature* 572, 461–466 (2019). https://doi.org/10.1038/s41586-019-1444-4

Roberts, C.M. 2002. Deep impact: The rising toll of fishing in the deep sea. *Trends in Ecology and Evolution* 17(5):242-245. DOI: 10.1016/S0169-5347(02)02492-8

Schaeffer, B. 1967. Comments on elasmobranch evolution. In: Gilbert, P.W. *et al.* (Eds). *Sharks, skates and rays.* The Johns Hopkins University Press, Baltimore, MD, USA.

Tickler, D. *et al.* 2018. Far from home: Distance patterns of global fishing fleets. *Science Advances* 4, eaar3279 (2018).

Vianna, G.M.S. *et al.* 2010. *Wanted Dead or Alive? The relative value of reef sharks as a fishery and an ecotourism asset in Palau.* Australian Institute of Marine Science and University of Western Australia, Perth, Australia.

Vianna, G.M.S. *et al.* 2010. Socio-economic value and community benefits from shark-diving tourism in Palau: A sustainable use of reef shark populations. *Biological Conservation*, 145(1), 267–277.

Vianna, G.M.S. *et al.* 2011. *The socio-economic value of the shark-diving industry in Fiji.* Australian Institute of Marine Science. University of Western Australia. Perth (26pp).

Vianna, G.M.S. *et al.* 2018. Shark-diving tourism as a financing mechanism for shark conservation strategies in Malaysia. *Marine Policy*, 94, 220–226. https://doi.org/10.1016/j.marpol.2018.05.008.

Vincent, A.C.J. *et al.* 2013. The role of CITES in the conservation of marine fishes subject to international trade. *Fish and Fisheries*. DOI: 10.1111/FAF.12035.

White, W.T. and Last P.R. 2012. A review of the taxonomy of chondrichthyan fishes: a modern perspective. *Journal of Fish Biology*, 80, 901–917.

Worm, B. *et al.* 2013. Global catches, exploitation rates, and rebuilding options for sharks. *Marine Policy*. 40:194–204.

Zimmerhackel, J.S. *et al.* 2018. Evidence of increased economic benefits from shark-diving tourism in the Maldives. *Marine Policy*, https://doi.org/10.1016/j.marpol.2018.11.004

Zimmerhackle, J.S. *et al.* 2018. How shark conservation in the Maldives affects demand for dive tourism. *Tourism Management*. https://doi.org/10.1016/j.tourman.2018.06.009

Further reading

Abel, D.C. and Grubbs, R.D. 2020. *Shark Biology and Conservation: Essentials for Educators, Students, and Enthusiasts.* The Johns Hopkins University Press, Baltimore, MD, USA.

Camhi, M.D., Pikitch E.K. and Babcock E.A. (Eds). 2007. *Sharks of the Open Ocean: Biology, Fisheries and Conservation.* Blackwell Publishing, Oxford, UK.

Carrier, J.C., Musick J.A. and Heithaus M.R. (Eds). 2010. *Sharks and their relatives II: biodiversity, adaptive physiology, and conservation.* CRC Press, Boca Raton, USA.

Carrier, J.C., Musick J.A. and Heithaus M.R. (Eds). 2012. *The biology of sharks and their relatives.* 2nd Edition. CRC Press, Boca Raton, USA.

Castro, J.I. 2011. *The sharks of North America.* Oxford University Press, USA.

Castro, J.I. 2020. *Genie: the life and recollections of Eugenie Clark.* Save Our Seas Foundation, Geneva, Switzerland.

Dipper, F. 2016. *The Marine World: A natural history of ocean life.* Wild Nature Press. UK.

Eilperin, J. 2011. *Demon fish: Travels through the hidden world of sharks.* Pantheon. New York, USA.

Hamlett, W. (Ed.). 1999. *Sharks, skates, and rays: the biology of elasmobranch fishes.* The Johns Hopkins University Press, Baltimore, MD, USA.

Hamlett, W. (Ed.). 2005. *Reproductive Biology and Phylogeny of Chondrichthyes: Sharks, Rays, and Chimaeras. Vol. 3: Reproductive Biology and Phylogeny.* Science Publishers, Inc. Enfield, NH, USA.

Klimley, A.P. 2013. *The Biology of Sharks and Rays.* University of Chicago Press. USA.

Stafford-Deitsch, J. 2015. *Shark Doc, Shark Lab: the life and work of Samuel Gruber.* Save Our Seas Foundation, Geneva, Switzerland.

Further identification guides

Abercrombie, D.L., Cardeñosa, D. and Chapman, D.D. 2018. *Genetic Approaches for Identifying Shark Fins and Other Products: A Tool for International Trade Monitoring and Enforcement.* Abercrombie and Fish, Marine Biological Consulting, Suffolk County, New York, USA.

Abercrombie, D.L., Chapman, D.D., Gulak, S.J.B. and Carlson, J.K. 2013. Visual Identification of Fins from Common Elasmobranchs in the Northwest Atlantic Ocean. NMFS-SEFSC-643, 51 pp.

Compagno, L.J.V. 1988. *Sharks of the Order Carcharhiniformes.* Princeton University Press, Princeton, New Jersey, USA.

Compagno, L.J.V., Ebert D.A. and Smale M.J. 1989. *Guide to the sharks and rays of southern Africa.* Struik Publishers, Cape Town, South Africa.

Ebert, D.A. 2003. *Sharks, Rays, and Chimaeras of California.* University of California Press, California, USA.

Ebert, D.A. 2013. *Deep-sea Cartilaginous Fishes of the Indian Ocean. Volume 1. Sharks.* FAO Species Catalogue for Fishery Purposes. No. 8, Vol. 1. Rome. FAO. 256 p.

Ebert, D.A. 2014. *On board guide for the identification of pelagic sharks and rays of the Western Indian Ocean.* Food and Agriculture Organization, SmartFish Programme, Indian Ocean Commission. 109 p.

Ebert, D.A. 2015. *Deep-sea Cartilaginous Fishes of the Southeastern Atlantic Ocean.* FAO Species Catalogue for Fishery Purposes. No. 9. Rome. FAO. 251 p.

Ebert, D.A. 2016. *Deep-sea Cartilaginous Fishes of the Southeastern Pacific Ocean.* FAO Species Catalogue for Fishery Purposes. Rome, Italy. FAO, 241 p.

Ebert, D.A. and Dando, M. 2020. *Field Guide to Sharks, Rays and Chimaeras of Europe and the Mediterranean.* Wild Nature Press, Princeton University Press, Princeton, New Jersey, USA.

Ebert, D.A. and Dando, M. In prep. *Field Guide to Sharks, Rays and Chimaeras of East Coast of North America.* Wild Nature Press, Princeton University Press, Princeton, New Jersey, USA.

Ebert, D.A. and Dando, M. In prep. *Field Guide to Sharks, Rays and Chimaeras of West Coast of North America.* Wild Nature Press, Princeton University Press, Princeton, New Jersey, USA.

Ebert, D.A. and Mostarda, E. 2015. *Identification guide to the deep-sea cartilaginous fishes of the Southeastern Atlantic Ocean.* FishFinder Programme, FAO, Rome. 70 p.

Ebert, D.A. and Mostarda, E. 2016. *Identification guide to the deep-sea cartilaginous fishes of the Southeastern Pacific Ocean.* FishFinder Programme, FAO, Rome, Italy, 53 p.

Ebert, D.A. and Stehmann M.F.W. 2013. *Sharks, batoids, and chimaeras of the North Atlantic.* FAO Species Catalogue for Fishery Purposes. No. 7. FAO, Rome, Italy.

Ebert, D.A., Fowler, S. and Dando, M. 2014. *An illustrated pocket guide to sharks of the world.* Wild Nature Press, UK.

Fricke, R., Eschmeyer, W.N., and Van der Laan, R. (eds) 2021. Eschmeyer's catalogue of fishes on-line. California Academy of Sciences: San Francisco. Available from: http://researcharchive.calacademy.org/research/ichthyology/catalog/fishcatmain.asp

Jabado R.W. and Ebert D. 2015. *Sharks of the Arabian Seas: an identification guide.* International Fund for Animal Welfare (IFAW), Dubai, UAE. 240 pp.

Last, P.R. and Stevens J.D. 2009. *Sharks and rays of Australia.* CSIRO, Australia.

Nakabo, T. (Ed.). 2002. *Fishes of Japan with pictorial keys to the species.* English edition. Tokai University Press, Tokyo.

FAO has published regional and taxonomic guides to many commercially-important fish species, including sharks, rays and chimaeras. Download from www.fao.org/ipoa-sharks/background/sharks/en/

Scientific societies, research and conservation bodies

American Elasmobranch Society (AES): www.elasmo.org

Angelshark Project (Northeast Atlantic and Mediterranean): www.angelsharkproject.com

Associação Portuguesa para o Estudo e Conservação de Elasmobrânquios (APECE)/ Portuguese Association for the Study and Conservation of Elasmobranchs: www.apece.pt

Association Pour l'Etude et la Conservation des Sélaciens (APECS, France): www.asso-apecs.org

Bimini Biological Field Station (Bahamas): www.biminisharklab.com

Blue Resources Trust/Sri Lanka Elasmobranch Project: www.blueresources.org

Blue Ventures: www.blueventures.org

Deutsche Elasmobranchier-Gesellschaft e.V./German Elasmobranch Society: www.elasmo.de

Dutch Shark Society: www.dutchsharksociety.org

Elasmoproject (Indian Ocean and West Africa): www.elasmoproject.com

European Elasmobranch Association (umbrella organisation for member bodies): www.eulasmo.org

Fundación Colombiana para la investigación y conservación de tiburones y rayas: www.squalus.org

Gruppo Ricercatori Italiani sugli Squali, Razze e Chimere (GRIS)/Group of Italian Researchers on Sharks, Rays and Chimaeras: www.griselasmo.wixsite.com/ griselasmo

HAI Norge/Norwegian Shark Alliance: www.hainorge.org

Irish Elasmobranch Group (IEG): www.irishelasmobranchgroup.com

iSea, Greece: www.isea.com.gr

IUCN Shark Specialist Group: www.iucnssg.org

Japanese Society for Elasmobranch Studies: www.jses.info

Manta Trust (global Mobulid ray research and conservation): www.mantatrust.org.

MarAlliance (Central America and Caribbean): www.maralliance.org

Marine Megafauna Foundation (research and conservation in the Americas, Western Indian Ocean, Southeast Asia): www.marinemegafaunafoundation.org

MarViva (Pacific coast of Central America): www.marviva.net

Nederlandse Elasmobranchen Vereniging/Dutch Elasmobranch Society: www.elasmobranch.nl

Oceania Chondrichthyan Society (Asia-Pacific and Indian Ocean): www.oceaniasharks.org.au

Project AWARE Foundation: www.projectaware.org

Shark Advocates: www.sharkadvocates.org

Shark Alliance: www.sharkalliance.org

Shark Foundation, Switzerland: www.shark.ch

Shark Research Institute: www.sharks.org

Shark Specialist Group (part of IUCN Species Survival Commission): www.iucnssg.org

Shark Trust: www.sharktrust.org

SharkLab, Malta: www.sharklab-malta.org

Sharks in Israel: www.sharks.org.il

Sociedade Brasileira para o Estudo de Elasmobrânquios: www.sbeel.org.br

South African Shark Conservancy: www.sharkconservancy.org

SUBMON, Spain: www.submon.org

TRAFFIC International (wildlife trade monitoring network): www.traffic.org/what-we-do/species/sharks-and-rays/

Wildlife Conservation Society: www.wcs.org

Worldwide Fund for Nature (shark projects in several countries): www.wwf.org

Online resources

Australian Museum: www.australianmuseum.net.au/fishes

Chondrichthyan Tree of Life (Phylogeny, Atlas, Database and Anatomy): www.sharksrays.org

Convention on International Trade in Endangered Species of Wild Fauna and Flora (CITES): www.cites.org

Convention on the Conservation of Migratory Species, Memorandum of Understanding on the Conservation of Migratory Sharks: www.cms.int/sharks

CSIRO marine animal tracking, Australia: https://coastalresearch.csiro.au/?q=node/210

CSIRO shark research in Australia: www.csiro.au/en/Research/OandA/Areas/Marine-resources-and-industries/Marine-biodiversity/Shark-research

eOceans citizen science website and smartphone app: www.eoceans.co

FAO Fisheries Department: www.fao.org/ipoa-sharks/background/sharks/en/ (fact sheets, identification guides, catch data)

Fish Base: www.fishbase.org

Florida Museum of Natural History: www.floridamuseum.ufl.edu/discover-fish/sharks/

Global FinPrint reef shark and ray BRUVS survey: www.globalfinprint.org

Guidelines for shark and ray recreational fishing: www.rac-spa.org/sites/default/files/doc_fish/gl_shark_ray_en.pdf

Guy Harvey Research Institute, Nova Southeastern University, tracking portal: www.ghritracking.org

International Shark Attack file: www.floridamuseum.ufl.edu/shark-attacks/

IUCN Red List of Threatened Species: www.iucnredlist.org

IUCN Shark Specialist Group publications and conservation strategies: www.iucnssg.org/publications.html

Pew Charitable Trusts: www.pewtrusts.org/en/projects/archived-projects/global-shark-conservation

ReefQuest Center for Shark Research: www.elasmo-research.org

Responsible Shark Anglin: www.sharktrust.org/pages/faqs/category/angling-project

Save our Seas Foundation: www.saveourseas.com

Sea Angling Diary Project (citizen science, UK): www.seaangling.org

Shark fin identification, for commercially-traded CITES species: www.identifyingsharkfins.org

sharkPulse citizen science mobile phone and web application for monitoring wild shark populations: http://baseline3.stanford.edu/SharkPulse/

Shark References bibliographic online database: www.shark-references.com

Shark Share International/Otlet (coordinates sharing of biological samples): www.otlet.io

Shark Smart and Sea Sense, Western Australia: Sharksmart.com.au

Smithsonian National Museum of Natural History: www.ocean.si.edu/ocean-life-ecosystems/sharks-rays

Tagging of Pelagic Predators (Pacific, incl. Salmon, White and Shortfin Mako Sharks): www.topp.org

WISEscheme (training to minimise disturbance to marine wildlife, UK): www.wisescheme.org

Index

Page numbers in **bold** text refer to plate pages